Ulrich
6.4

NEUTRON DOSIMETRY

Proceedings of the Eighth Symposium,
Paris, November 13 – 17 1995

Proceedings Editors:

H.G. Menzel
J.L. Chartier
R. Jahr
A. Rannou

ISBN 1 870965 43 4
CONF 951119
EUR 16519 EN
RADIATION PROTECTION DOSIMETRY Vol. 70 Nos. 1–4 1997
Published by Nuclear Technology Publishing

NEUTRON DOSIMETRY

**Proceedings of the Eighth Symposium
Paris, November 13–17 1995**

All rights reserved. No part of this book may be reproduced, stored in a retrieval system or transmitted in any form or by any means, electronic, electrostatic, magnetic, mechanical, photocopying, recording or otherwise, without permission in writing from the publishers.

British Library Cataloguing in Publication Data

A catalogue record of this book is available at the British Library

ISBN 1 870965 43 4 EUR 16519 EN
COPYRIGHT © 1997 Nuclear Technology Publishing/European Commission

Radiation Protection Dosimetry

ISSN 0144-8420

Editor-in-Chief:
Dr J.A. Dennis, UK

Executive Editor:
Mr E.P. Goldfinch, UK

Staff Editor:
Mrs M.E. Calcraft, UK

Consultant Editors:
Prof. Dr G. Dietze, Germany
Prof. Y. Horowitz, Israel
Dr J. Stather, UK
Dr G.A. Swedjemark, Sweden

Editorial Board Members
Dr R.M. Alexakhin, Russia
Prof. Dr K. Becker, Germany
Dr M.A. Bender, USA
Dr A. Birchall, UK
Prof. Dr J. Böhm, Germany
Dr A.J.J. Bos, The Netherlands
Dr L. Bötter-Jensen, Denmark
Dr G. Busuoli, Italy
Mr M.W. Carter, Australia
Dr M.W. Charles, UK
Mr G. Cowper, Canada
Dr W.G. Cross, Canada
Dr A. Delgado, Spain
Dr Li Deping, People's Republic of China
Prof. Dr B. Dörschel, Germany
Prof. Dr K. Duftschmid, Austria
Miss F.A. Fry, UK
Mr J.A.B. Gibson, UK
Mr R.V. Griffith, USA
Dr K. Harrison, UK
Mr J.R. Harvey, UK
Dr H. Ing, Canada

Dr K. Irlweck, Austria
Prof. Dr W. Jacobi, Germany
Dr R.L. Kathren, USA
Dr E. Kunz, Czech Republic
Dr D.C. Lloyd, UK
Dr H.F. Macdonald, UK
Mr T.O. Marshall, UK
Dr J.C. McDonald, USA
Dr M. Moscovitch, USA
Prof. Y. Nishiwaki, Japan
Mr K. O'Brien, USA
Dr P. Pihet, France
Dr G. Portal, France
Dr A.S. Pradhan, India
Dr D.F. Regulla, Germany
Prof. Dr A.R. Scharmann, Germany
Mr J.M. Selby, USA
Dr F. Spurný, Czech Republic
Dr R.H. Thomas, USA
Mr I.M.G. Thompson, UK
Dr L. Tommasino, Italy
Mr J.W.N. Tuyn, Switzerland

Published by Nuclear Technology Publishing, P.O. Box 7, Ashford, Kent, TN23 1YW, England.

Advertising office: Mrs L. Richmond   Subscription office: Mrs M.L. Mears

Subscription rates, 1996: Vols. 63–68 inclusive, UK £590.00 p.a.; outside UK US$ 1270.00 p.a.
1997: Vols. 69–74 inclusive, UK £620.00 p.a.; outside UK US$ 1330.00 p.a.

Orders and remittance should be sent to:

Subscription Department, Nuclear Technology Publishing, P.O. Box No 7, Ashford, Kent, TN23 1YW, England
Telephone (01233) 641683   Facsimile (01233) 610021

COPYRIGHT ©1997 Nuclear Technology Publishing

Legal disclaimer: The publisher, the editors and the Editorial Board accept no responsibility for the content of the papers, the use which may be made of the information or the views expressed by the authors.

Typeset by Photographics, Vine Yard, Vine Passage, Honiton, Devon, England
Printed by Geerings of Ashford Ltd., Cobbs Wood House, Chart Road, Ashford, Kent, England

Radiation Protection Dosimetry

INSTRUCTIONS TO AUTHORS

SCOPE: Radiation Protection Dosimetry covers all aspects of personal and environmental dosimetry and monitoring, for both ionising and non-ionising radiations. This includes the biological aspects, physical concepts, biophysical dosimetry, external and internal personal dosimetry and monitoring, environmental and workplace monitoring, accident dosimetry, and dosimetry related to the protection of patients. Particular emphasis is placed on papers covering the fundamentals of dosimetry such as units, radiation quantities and conversion factors. Papers covering archeological dating are included only if the fundamental measurement method or technique, such as thermoluminescence, has direct application to personal dosimetry measurements. Papers covering the dosimetric aspects of radon or other naturally occurring radioactive materials and low level radiation are included. Animal experiments and ecological sample measurements are not included unless there is a significant relevant content related to dosimetry in man.

Scientific or Technical Papers should be full papers of a theoretical or practical nature with comprehensive descriptions of the work covered.

Scientific or Technical Notes should be brief, covering not more than 4 printed pages (one page contains about 800 words or equivalent in figures) and are likely to cover work in development or topics of lesser significance than full papers.

Letters to the Editor should be written as letters with the authors' names and addresses at the end and should be marked 'For Publication'.

LANGUAGE: All contributors should be in **English**. Spelling should be in accordance with the Concise Oxford Dictionary. However please use dosemeter rather than dosimeter, for consistency within the journal. Authors whose mother tongue is not English are requested to ask someone with a good command of English to review their contribution before submission

TITLES should be brief and as informative as possible. A short title of not more than 50 characters for a running head should be supplied.

AUTHORS' names and addresses (with full postal address) should appear immediately below the title

ABSTRACTS containing up to 150 words should be provided on a separate page, headed by the title and authors' names.

SCRIPTS must be typewritten with at least the original copy **double** spaced. **Three** additional copies (may be photocopies) should be provided for refereeing purposes to minimise the time required for refereeing. Headings should be given to main sections and sub-sections which should not be numbered. The title page should contain just the title, authors' names and addresses and a short running title. Manuscripts should be written in the third person and not the first. If your manuscript is prepared using a computer or word processor it would be helpful if you could also send a copy of the computer disc (please specify software).

FIGURES AND TABLES should not be inserted in the pages of manuscript but should be supplied on separate sheets. One high quality set of illustrations and figures, suitable for direct reproduction, e.g. black ink or good quality black and white prints of line drawings and graphs, should be provided with original typed manuscript. These should be approximately twice the final printed size (full page printed area = 19cm x 15cm). The lettering should be of such a size that the letters and symbols will remain legible after reduction to fit the printed area available. Tables should be typed. Tables should be lightly lined in pencil. All figures and tables should be numbered, using Arabic numerals, on the reverse side of each copy. Numbered captions or titles should be typed on a separate sheet. Figures and tables should be kept to the minimum consistent with clear presentation of the work reported. Half-tone photographs should only be included if absolutely necessary. Figures generated by computer graphics are generally NOT suitable for direct reproduction. Photocopies of all figures and tables should accompany each copy of the manuscript for refereeing purposes. Colour figures can be reproduced at cost.

UNITS, SYMBOLS AND EQUATIONS: SI units should be used throughout but other established units may be included in brackets (Note that cGy is **not** acceptable). Any Greek letters or special symbols used in the text should be identified in the margin on each occasion they are used. Isotope mass numbers should appear at the upper left of the element symbol e.g. ^{90}Sr. Equations should be fully typed. FOOTNOTES should only be included if absolutely necessary. They should be typed on a separate sheet and the author should give a clear indication in the text by inserting (see footnote) so that they may appear on the correct page.

ABBREVIATIONS which are not in common usage should be defined when they first appear in the text.

REFERENCES should be indicated in the text by superior numbers in parenthesis and the full reference should be given in a list at the end of the paper in the following form, in the order in which they appear in the text:-

1. Crase, K.W. and Gammage, R.B. *Improvements in the Use of Ceramic BeO in TLD*, Health Phys. **29**(5) 739-746 (1975).
2. Clarke, R.H. and Webb G.A.M. *Methods for Estimating Population Detriment and their Application in Setting Environmental Discharge Limits*. Proceedings of Symposium - Biological Implications of Radionuclides Released from Nuclear Industries, Vienna, March 1979. IAEA-SM-237/6, 149-154 (1980)
3. Aird, E.G.A.A. *An Introduction to Medical Physics*. William Heineman Medical Books Ltd (London). ISBN 0 433 003502. (1983)
4. Duftschmid, K.E. *TLD Personnel Monitoring Systems - The Present Situation*. Radiat. Prot. Dosim. **2**(1) 2-12 (1982).

All the authors' names and initials (unless there are more than 10 authors), the title of the paper, the abbreviated title of the journal, volume number, page numbers and year should be given. Abbreviated journal titles should be in accordance with the current World List of Scientific Periodicals. If all of this information is not available the reference should not be cited.

PROOFS will be sent to any nominated author for final proof reading and must be returned within 3 days of receipt using the addressed label which will be provided. Type-setting or printer's errors should be marked in red. Any other changes should be marked in green but if they are significant they may be charged to the authors. Authors' changes marked in red may not be accepted. The Editor reserves the right to make editorial corrections to manuscripts. An order form for additional reprints will accompany proofs.

SUBMISSION: All manuscripts (original and three copies) and correspondence should be addressed to Mr E.P. Goldfinch, Executive Editor, Nuclear Technology Publishing, P.O. Box No 7, Ashford, Kent TN23 1YW, England. It is *essential* that they are accompanied by six fully addressed adhesive labels addressed to the author nominated to receive proofs and correspondence. These will be used for acknowledgement of receipt of the manuscript, notification of acceptance, return of proofs to authors and supply of reprints. Papers will be considered only on the understanding that they are not currently being submitted to other journals. The Publishers, The Editor-in-Chief and the Editorial Board do not accept responsibility for the technical content, the use of that content or the views expressed by authors.

CORRESPONDENCE: Please ensure that you provide telephone, FAX and E-mail numbers if available. Please quote the manuscript number in any correspondence once receipt of your manuscript has been acknowledged.

COMPUTER MANUSCRIPTS: If your manuscript is prepared using a computer or word processor, publication **may** be quicker if you submit a copy of the disc with the manuscript copies. If you do this please specify the software used and the version of that software.

COPYRIGHT: Authors submitting manuscripts do so on the understanding that if accepted for publication, copyright of the article shall be assigned to Nuclear Technology Publishing unless other specific arrangements are made.

GENERAL: In order to ensure rapid publication it is most important that **all** of the above instructions are complied with in **full**. Failure to comply may result in considerable delay in publication or the **return** of manuscripts to the author. In case of difficulty with illustrations and figures please consult the photo-reprographic section of your establishment. If illustrations of a quality high enough for direct off-set photographic reproduction cannot be supplied they may be redrawn by the publishers at the request of authors if all relevant details are provided. A charge will be made if requirements are extensive.

Proceedings of the Eighth Symposium on Neutron Dosimetry, Paris, November 13–17 1995

PROGRAMME COMMITTEE

W.G. Alberts	PTB, Braunschweig, Germany	L. Lindborg	SRPI, Stockholm, Sweden
A. Alekseev	IHEP, Moscow, Russia	F. Lucci	ENEA, Frascati, Italy
J. Barthe	CEA, Saclay, France	R. Sabattier	CHR, Orléans, France
D. Bartlett	NRPB, Chilton, UK	H. Schraube	GSF, Neuherberg, Germany
D. Blanc	UPS, Toulouse, France	F. Spurný	NPI, Praha, Czech Republic
R. Dollo	EDF, Paris la Défense, France	D. Thomas	NPL, Teddington, UK
R.A. Gabbauer	OSUH, Colombus, USA	M.N. Varma	DOE, Washington, USA
N. Golnik	IAE, Otwock Swierk, Poland	A. Waker	AECL, Chalk River, Canada
P. Gourmelon	IPSN, Fontenay aux Roses, France	A. Wambersie	UCL, Brussels, Belgium

CHAIRPERSONS

F. Barbry	CEA, Saclay, France	H.G. Menzel	EC, Brussels, Belgium
J. Barthe	CEA/DTA/DAMRI, Saclay, France	P. Pihet	IPSN, Fontenay aux Roses, France
D. Blanc	University of Toulouse, France	A. Rannou	IPSN, Fontenay aux Roses, France
J.L. Chartier	IPSN, Fontenay aux Roses, France	Th. Schmitz	KFA, Jülich, Germany
R. Gahbauer	University of Columbus, USA	H. Schraube	GSF, Neuherberg, Germany
N. Golnik	IAE, Otwock Swierk, Poland	H. Schuhmacher	PTB, Braunschweig, Germany
P. Gourmelon	IPSN, Fontenay aux Roses, France	F. Spurný	NPI, Praha, Czech Republic
J.R. Harvey	UK	J.E. Tanner	PNL, Richland, USA
C.R. Hirning	Ontario Hydro, Whitby, Canada	D.J. Thomas	NPL, Teddington, UK
R. Jahr	PTB, Braunschweig, Germany	L. Tommasino	ANPA, Roma, Italy
S. Kharlampiev	Institute for High Energy Physics, Russia	J.C. Vareille	University of Limoges, France
L. Lindborg	SSI, Stockholm, Sweden	A.J. Waker	AECL, Chalk River, Canada
A. Mazal	Centre de Prontonthérapie d'Orsay, France	A. Wambersie	UCL, Brussels, Belgium
J.C. McDonald	PNL, Richland, USA	J. Zoetelief	TNO, Arnhem, Netherlands

SCIENTIFIC SECRETARIES

J.L. Chartier	IPSN, Fontenay aux Roses, France
R. Jahr	PTB, Braunschweig, Germany
H.G. Menzel	EC, Brussels, Belgium

LOCAL ORGANISATION

A. Rannou	IPSN, Fontenay aux Roses, France

Jointly organised by the European Commission (EC), the Institut de Protection et de Sûreté Nucléaire (ISPN) and the US Department of Energy, with sponsorship from the Physikalisch-Technische Bundesanstalt (PTB), Braunschweig and Berlin.

Previous Proceedings published in **Radiation Protection Dosimetry** on behalf of the Commission for the European Union

EXOELECTRON EMISSION AND ITS APPLICATIONS 166 pp, Softback, Proceedings of the VIIth International Symposium, Strasbourg, March 1983.
Price US$66.50

INDOOR EXPOSURE TO NATURAL RADIATION AND ASSOCIATED RISK ASSESSMENT 440 pp, Softback, Proceedings of an International Seminar, Anacapri, October 1983,
Price US$133.00

MICRODOSIMETRIC COUNTERS IN RADIATION PROTECTION 120 pp, Softback, Proceedings of a Workshop, Hamburg/Saar, May 1984,
Price US$47.50

RADIATION PROTECTION QUANTITIES FOR EXTERNAL EXPOSURE 166 pp, Softback, Proceedings of a Seminar, Braunschweig, March 1985,
Price US$66.50

MICRODOSIMETRY 400 pp, Softback, Proceedings of the Ninth Symposium on Microdosimetry, Toulouse, May 1985,
Price US$133.00

DOSIMETRY OF BETA PARTICLES AND LOW ENERGY X RAYS 134 pp, Softback, Proceedings of a Workshop, Saclay, October 1985,
Price US$57.00

ENVIRONMENTAL AND HUMAN RISKS OF TRITIUM 192 pp, Softback, Proceedings of a Workshop, Karlsruhe, February 1986,
Price US$76.00

ETCHED TRACK NEUTRON DOSIMETRY 130 pp, Softback, Proceedings of a Workshop, Harwell, May 1987,
Price US$47.50

ACCIDENTAL URBAN CONTAMINATION 192 pp, Softback, Proceedings of a Workshop, Roskilde, June 1987,
Price US$76.00

NEUTRON DOSIMETRY 498 pp, Softback, Proceedings of the Sixth Symposium on Neutron Dosimetry, Neuherberg, October 1987,
Price US$142.50

NATURAL RADIOACTIVITY 560 pp, Softback, Proceedings of the Fourth International Symposium on the Natural Radiation Environment, Lisbon, December 1987,
Price US$161.50

BIOLOGICAL ASSESSMENT OF OCCUPATIONAL EXPOSURE TO ACTINIDES 400 pp, Softback, Proceedings of a Workshop, Versailles, May 1988,
Price US$123.50

IMPLEMENTATION OF DOSE-EQUIVALENT QUANTITIES INTO RADIATION PROTECTION PRACTICE (ISBN 1 870965 03 5) 166 pp, Softback, Proceedings of a Seminar, Braunschweig, June 1988,
Price US$66.50

IMPLEMENTATION OF DOSE-EQUIVALENT METERS BASED ON MICRODOSIMETRIC TECHNIQUES (ISBN 1 870965 04 1) 156 pp, Softback, Proceedings of a Seminar, Schloss Elmau, October 1988,
Price US$57.00

MICRODOSIMETRY (ISBN 1 870965 05 X) 460 pp, Softback, Proceedings of the Tenth Symposium on Microdosimetry, Rome, May 1989,
Price US$152.00

STATISTICS OF HUMAN EXPOSURE TO IONISING RADIATION (ISBN 1 870965 08 6) 280 pp, Hardback, Proceedings of a Workshop, Oxford, April 1990,
Price US$114.00

RESPIRATORY TRACT DOSIMETRY (ISBN 1 870965 09 4) 268 pp, Hardback, Proceedings of a Workshop, Albuquerque, July 1990,
Price US$114.00

SKIN DOSIMETRY - RADIOLOGICAL PROTECTION ASPECTS OF SKIN IRRADIATION (ISBN 1 870965 12 4) 212 pp, Hardback, Proceedings of a Workshop, Dublin, May 1991,
Price US$114.00

DOSIMETRY IN DIAGNOSTIC RADIOLOGY (ISBN 1 870965 11 6) 316 pp, Hardback, Proceedings of a Seminar, Luxembourg, March 1991,
Price US$152.00

AGE DEPENDENT FACTORS IN THE BIOKINETICS AND DOSIMETRY OF RADIONUCLIDES (ISBN 1 870965 15 9) 254 pp, Hardback, Proceedings of a Workshop, Schloss Elmau, November 1991,
Price US$114.00

GUIDEBOOK FOR THE TREATMENT OF ACCIDENTAL INTERNAL CONTAMINATION OF WORKERS (ISBN 1 870965 22 1) 50 pp, Softback, A Joint Publication for the CEC and the USDOE,
Price US$38.00

NEUTRON DOSIMETRY (ISBN 1 870965 16 7) 486 pp, Hardback, Proceedings of the Seventh Symposium on Neutron Dosimetry, Berlin, October 1991,
Price US$152.00

THE NATURAL RADIATION ENVIRONMENT (ISBN 1 870965 14 0) 800 pp, Hardback, Proceedings of the Fifth International Symposium, Salzburg, September 1991,
Price US$228.00

RADIATION EXPOSURE OF CIVIL AIRCREW (ISBN 1 870965 13 2) 140 pp, Hardback, Proceedings of a Workshop, Luxembourg, June 1991,
Price US$57.00

TEST PHANTOMS AND OPTIMISATION IN DIAGNOSTIC RADIOLOGY AND NUCLEAR MEDICINE (ISBN 1 870965 26 4) 416 pp, Hardback, Proceedings of a Workshop, Wurzburg, June 1992,
Price US$152.00

MICRODOSIMETRY (ISBN 1 870965 21 3) 500 pp, Hardback, Proceedings of the Eleventh Symposium on Microdosimetry, Gatlinburg, September 1992,
Price US$171.00

DECISION MAKING SUPPORT FOR OFF-SITE EMERGENCY MANAGEMENT (ISBN 1 870965 25 6) 320 pp, Hardback, Proceedings of a Workshop, Schloss Elmau, October 1992,
Price US$114.00

INTAKES OF RADIONUCLIDES - DETECTION, ASSESSMENT AND LIMITATION OF OCCUPATIONAL EXPOSURE (ISBN 1 870965 28 0) 370 pp, Hardback, Proceedings of a Workshop, Bath, September 1993,
Price US$152.00

INDIVIDUAL MONITORING OF IONISING RADIATION - THE IMPACT OF RECENT ICRP AND ICRU PUBLICATIONS.
(ISBN 1 870965 29 9) 232 pp, Hardback, Proceedings of a Workshop, Villigen, May 1993,
Price US$85.50

INDOOR RADON REMEDIAL ACTIONS -THE SCIENTIFIC AND PRACTICAL IMPLICATIONS. (ISBN 1 870965 30 2) 400 pp, Hardback, Proceedings of a Workshop, Rimini, Italy, June 27 to July 2 1993,
Price US$171.00

QUALITY CONTROL AND RADIATION PROTECTION OF THE PATIENT IN DIAGNOSTIC RADIOLOGY AND NUCLEAR MEDICINE. (ISBN 1 870965 37 X) 512 pp, Hardback. Proceedings of a Workshop, Grado, Italy, September 29 – October 1 1993,
Price US$171.00

ADVANCES IN RADIATION MEASUREMENTS: APPLICATIONS AND RESEARCH NEEDS IN HEALTH PHYSICS AND DOSIMETRY. (ISBN 1 870965 33 7), 310 pp, Hardback. Proceedings of an International Workshop, Chalk River, Ontario, Canada, October 3-6, 1994
Price US$142.50

DESIGN, CONSTRUCTION AND USE OF TISSUE EQUIVALENT PROPORTIONAL COUNTERS.
(ISBN 1 870965 43 4), 116 pp, Hardback. EURADOS Report
Price US$70.30

RADIATION RISK, RISK PERCEPTION AND SOCIAL CONSTRUCTIONS (ISBN 1 870965 44 2), 150 pp, Hardback. Proceedings of a Workshop, Oslo, Norway, October 19 – 20 1995
Price US$93.10

FUTURE PUBLICATIONS:

REAL TIME COMPUTING OF THE ENVIRONMENTAL CONSEQUENCES OF AN ACCIDENTAL RELEASE FROM A NUCLEAR INSTALLATION.
(ISBN 1 870965 49 3), Proceedings of the Fourth International Workshop, Aronsborg, October 6 – 11 1996.
Price not yet known.

INTAKES OF RADIONUCLIDES -OCCUPATIONAL AND PUBLIC EXPOSURE
(ISBN 1 870965 52 3) Proceedings of a Workshop, Avignon, France, September 15 – 18 1997
Price not yet known

Contents

Contents vii

Editorial

    Eightth Symposium on Neutron Dosimetry
    H.G. Menzel, J.L. Chartier, R. Jahr and A. Rannou xv

PHYSICAL DATA FOR RADIATION PROTECTION AND THERAPY DOSIMETRY

    Nuclear Data Needs for Neutron Therapy and Radiation Protection
    M.B. Chadwick, P.M. DeLuca Jr. and R.C. Haight (INVITED PAPER) 1

    Kerma Measurements in Polyenergetic Neutron Fields
    W.D. Newhauser, U.J. Schrewe, H.J. Brede, M. Matzke and P.M. DeLuca Jr. 13

    Comparison of Various Dose Quantities in Tissue and Tissue Substitutes at Neutron Energies between 20 MeV and 100 MeV
    U.J. Schrewe, W.G. Alberts, W.D. Newhauser, H.J. Brede and P.M. DeLuca Jr. 17

    Double Differential Cross Section Measurements of the (n,px), (n,dx), (n,tx) and (n,αx) Reactions on ^{12}C:Kerma Determination at Energies between 30 and 75 MeV
    I. Slypen, S. Benck, V. Corcalciuc and J.-P. Meulders 21

    Relative Neutron Sensitivity of Tissue-Equivalent Ionisation Chambers in an Epithermal Neutron Beam for Boron Neutron Capture Therapy
    J.T.M. Jansen, C.P.J. Raaijmakers, B.J. Mijnheer and J. Zoetelief 27

    W Values in Propane-Based Tissue-Equivalent Gas
    I.K. Bronić 33

    W Values of Protons Slowed Down in Molecular Hydrogen
    B. Grosswendt, G. Willems and W.Y. Baek 37

    Stopping Powers of He and Ar Gases for 50-200 keV ^{3}He$^+$ Ions
    A. Fukuda 47

    Calculations of the Giant-Dipole-Resonance Photoneutrons using a Coupled EGS4-Morse Code
    J.C. Liu, W.R. Nelson, K.R. Kase and X.S. Mao 49

    Calculation of Radial Dose Distributions for Heavy Ions by a New Analytical Approach
    J. Chen and A.M. Kellerer 55

NEEDS AND PERFORMANCES REQUIRED FOR MONITORS IN NUCLEAR INDUSTRY INDIVIDUAL ELECTRONIC DOSEMETERS

    Electronic Neutron Dosemeters: History and State of the Art
    J. Barthe, J.M. Bordy and T. Lahaye (INVITED PAPER) 59

CONTENTS

Needs and Performance Requirements for Neutron Monitoring in the Nuclear Power Industry
C.R. Hirning and A.J. Waker (INVITED PAPER) 67

Single Diode Detector for Individual Neutron Dosimetry using a Pulse Shape Analysis
J.M. Bordy, T. Lahaye, F. Landre, C. Hoflack, S. Lequin and J. Barthe 73

Advanced Detectors for Active Neutron Dosemeters
J.C. Vareille, B. Barelaud, J. Barthe, J.M. Bordy, G. Curzio, F. d'Errico, J.L. Decossas, F. Fernandez-Moreno, T. Lahaye, E. Luguera, O. Saupsonidis, E. Savvidis and M. Zamani-Valassiadou . 79

Efficiencies of an SSNTD and an Electronic Neutron Dosemeter with the Same (n,α)(n,p) Converter
E. Savvidis and M. Zamani 83

Separation of the Neutron Signal from the Gamma Component in (n-γ) Fields using Differential Pulse Analysis Techniques with a Double Silicon Diode
F. Fernández, E. Luguera, C. Domingo and C. Baixeras 87

Study of a Moderator Type Electronic Neutron Dosemeter for Personal Dosimetry
T. Moiseev 93

Feasibility Study of an Individual Electronic Neutron Dosemeter
M. Luszik-Bhadra, W.G. Alberts, E. Dietz and B.R.L. Siebert 97

ADVANCES IN BUBBLE DETECTORS

Advances in Superheated Drop (Bubble) Detector Techniques
F. d'Errico, W.G. Alberts and M. Matzke (INVITED PAPER) 103

Superheated Emulsions: Neutronics and Thermodynamics
F. d'Errico, R.E. Apfel, G. Curzio, E. Dietz, E. Egger, G.F. Gualdrini, S. Guldbakke, R. Nath and B.R.L. Siebert 109

Fast Discrimination of Neutrons from (α,n) and Fission Sources
R.E. Apfel, F. d'Errico and J.D. Martin 113

ADVANCES IN ETCHED TRACK DETECTORS

Methodological Studies on the Optimisation of Multi-Element Dosemeters in Neutron Fields
W.G. Alberts, B. Dörschel and B.R.L. Siebert 117

Future Developments in the NRPB PADC Neutron Personal Monitoring Service
R.J. Tanner, D.T. Bartlett, L.G. Hager and J. Lavelle 121

The Present Status of Etched Track Neutron Dosimetry in Europe and the Contribution of CENDOS and EURADOS
J.R. Harvey, R.J. Tanner, W.G. Alberts, D.T. Bartlett, E.K.A. Piesch and H. Schraube 127

Etched Track Size Distributions Induced by Broad Neutron Spectra in PADC
J. Jakes, J. Voigt and H. Schraube 133

Analysis of Acceptance Testing Data for more than 800 Sheets of CR-39 Plastic Assessed for the DRPS Approved Neutron Dosimetry Service
M. Jackson and A.P. French 139

CONTENTS

Polycarbonate Track Detectors with a Flat Energy Response for the Measurement of
High Energy Neutrons at Accelerators and Airflight Altitudes
K. Józefowicz, B. Burgkhardt, M. Vilgis and E. Piesch 143

An Automated Neutron Dosimetry System Based on the Chemical Etch of CR-39
J.R. Harvey, A.P. French, M. Jackson, M.C. Renouf and A.R. Weeks 149

Angular Response to Neutrons of the Electrochemically Etched CR-39 with Different Radiators
J. Bednář, K. Turek and F. Spurný 153

A Personal Neutron Dosimetry System Based on Etched Track and Automatic Readout by Autoscan 60
A. Fiechtner, K. Gmür and C. Wernli 157

Ten Years on: the NRPB PADC Neutron Personal Monitoring Service
D.T. Bartlett, J.D. Steele, R.J. Tanner, P.J. Gilvin, P.V. Shaw and J. Lavelle 161

ADVANCES IN TLD DETECTORS

Dose Equivalent Response of Personal Neutron Dosemeters as a Function of Angle
J.E. Tanner, J.C. McDonald, R.D. Stewart and C. Wernli 165

Measurement of Neutron Dose on a Fusion Reactor Shield using TLD-300 Phosphors
M. Angelone, P. Batistoni, A. Esposito, M. Martone, M. Pelliccioni, M. Pillon and V. Rado 169

TL Dosimetry in High Fluxes of Thermal Neutrons using Variously Doped LiF and KM_gF_3
G. Gambarini, M. Martini, A. Scacco, C. Raffaglio and A.E. Sichirollo 175

INDIVIDUAL NEUTRON DOSIMETRY IN NUCLEAR INDUSTRY AND APPLICATIONS OF NEUTRON SOURCES

Evaluation of Individual Neutron Dosimetry by a Working Group in the French Nuclear Industry
A. Rannou, A. Clech, A. Devita, R. Dollo and G. Pescayre 181

Neutron Dosimetry at the Reactor Facility VENUS
P. Deboodt, F. Vermeersch, F. Vanhavere and G. Minsart 187

Some Examples of Industrial Uses of Neutron Sources
J.L. Szabo and J.L. Boutaine 193

TEPC AND IONISATION CAVITY

TEPC Performance in the CANDU Workplace
A.J. Waker, K. Szornel and J. Nunes 197

Microdosimetric Investigations in Realistic Fields
J.F. Bottollier-Depois, L. Plawinski, F. Spurný and A. Mazal 203

Neutron Energy Deposition Spectra at Simulated Diameters down to 50 nm
E. Anachkova, A.M. Kellerer and H. Roos 207

Review of Recent Achievements of Recombination Methods
N. Golnik 211

CONTENTS

Recombination Index of Radiation Quality of Medical High Energy Neutron Beams
N. Golnik, E.P. Cherevatenko, A.Y. Serov, S.V. Shvidkij, B.S. Sychev and M. Zielczyński 215

Concept of a Neutron Dosemeter Based on a Recoil Particle Track Chamber
U. Titt, A. Breskin, R. Chechik, V. Dangendorf, B. Großwendt and H. Schuhmacher 219

RADIATION FIELD ANALYSIS AND MEASUREMENTS AT WORKPLACES

Workplace Radiation Field Analysis
H. Klein (INVITED PAPER) 225

Neutron Dose Equivalent Rates Calculated from Measured Neutron Angular and Energy Distributions in Working Environments
P. Drake and D.T. Bartlett 235

A Database of Neutron Spectra, Instrument Response Functions and Dosimetric Conversion Factors for Radiation Protection Applications
O.F. Naismith and B.R.L. Siebert 241

Neutron Fluence Measurements with a Liquid Scintillator
P.J. Binns, J.H. Hough and B.R.S. Simpson 247

Neutron and Photon Spectra and Dose Rates around a Shielding Cask Placed in a Salt Mine to Simulate a Nuclear Waste Package
K. Knauf, A.V. Alevra, J. Wittstock, H.J. Engelmann, M. Khamis and N. Niehues 251

Response of Neutron Dosemeters in Radiation Protection Environments: An Investigation of Techniques to Improve Estimates of Dose Equivalent
O.F. Naismith, B.R.L. Siebert and D.J. Thomas 255

Directional Information on Neutron Fields
M. Matzke, H. Kluge and M. Luszik-Bhadra 261

Application of Neutron Spectrometry in the DRPS Neutron Dosimetry Service
K. Barlow, M. Jackson, A. French and J.R. Harvey 265

A New Approach to Low Level Monitoring in Mixed Radiation Fields
S. Pszona 269

SPECIFICATIONS OF NEUTRON SPECTROMETRIC INSTRUMENTS

ROSPEC − A Simple Reliable High Resolution Neutron Spectrometer
H. Ing, T. Clifford, T. McLean, W. Webb, T. Cousins and J. Dhermain 273

Specification of Bonner Sphere Systems for Neutron Spectrometry
M. Kralik, A. Aroua, M. Grecescu, V. Mares, T. Novotny, H. Schraube and B. Wiegel 279

Improved Neutron Spectrometer Based on Bonner Spheres
A. Aroua, M. Grecescu, S. Prêtre and J.-F. Valley 285

High Energy Response Functions of Bonner Spectrometers
A.V. Sannikov, V. Mares and H. Schraube 291

CONTENTS

Measurements with the PTB 'C' Bonner Sphere Spectrometer in the PSI Villigen
55 MeV Neutron Field for Spectrometry and Calibration Purposes
A.V. Alevra and U.J. Schrewe 295

Calibration of a Neutron Spectrometer in the Energy Range 144 keV to 14.8 MeV with ISO Energies
W. Rosenstock, T. Köble, G. Kruziniski and G. Jaunich 299

REALISTIC NEUTRON CALIBRATION FIELDS

Recent Developments in the Specification and Achievement of Realistic Neutron Calibration Fields
J.L. Chartier, B. Jansky, H. Kluge, H. Schraube and B. Wiegel 305

Results of a Large Scale Neutron Spectrometry and Dosimetry Comparison Exercise
at the Cadarache Moderator Assembly
D.J. Thomas, J.-L. Chartier, H. Klein, O.F. Naismith, F. Posny and G.C. Taylor 313

Current Status of an ISO Working Document on Reference Radiations: Characteristics
and Methods of Production of Simulated Practical Neutron Fields
*J.C. McDonald, W.G. Alberts, D.T. Bartlett, J.-L. Chartier, C.M. Eisenhauer, H. Schraube,
R.B. Schwartz and D.J. Thomas* 323

Scattered Neutron Reference Fields Produced by Radionuclide Sources
H. Kluge, A.V. Alevra, S. Jetzke, K. Knauf, M. Matzke, K. Weise and J. Wittstock 327

Advances in Realistic Neutron Spectra: Progress in Fluence Monitoring of the DD Reaction
D. Paul, G. Pelcot and C. Itié 331

GRENF - The GSF Realistic Neutron Field Facility
H. Schraube, B. Hietel, J. Jakes, V. Mares, G. Schraube and E. Weitzenegger 337

CHARACTERISTICS OF CALIBRATION FACILITIES

Dosimetric Characteristics of the IHEP Neutron Reference Fields
A.G. Alekseev and S.A. Kharlampiev 341

SILENE, A Tool for Neutron Dosimetry
B. Tournier, F. Barbry and B. Verrey 345

Determination of Neutron Room Scattering Corrections in RCL's Calibration Facility at KAERI
S.C. Yoon, S.Y. Chang, J.S. Kim and J.L. Kim 349

Tests of Instruments for Neutron and Gamma Ray Measurements at Several Radionuclide Neutron Sources
F. Spurný, I. Votočková, J.-F. Bottollier-Depois, J. Kurkdjian and D. Paul 353

A New Expression for Determination of Fluences from a Spherical Moderator
Neutron Source for the Calibration of Spherical Neutron Measuring Devices
M. Khoshnoodi and M. Sohrabi 357

DEVELOPMENT OF NEW AMBIENT MONITORING INSTRUMENTS

The Neutron Fluence and H*(10) Response of the New LB 6411 Rem Counter
B. Burgkhardt, G. Fieg, A. Klett, A. Plewnia and B.R.L. Siebert 361

A Spherical Neutron Counter with an Extended Energy Response for Dosimetry
H. Toyokawa, A. Uritani, C. Mori, N. Takeda and K. Kudo 365

DOSIMETRIC QUANTITIES AND COMPUTATIONAL DOSIMETRY

Computational Dosimetry
B.R.L. Siebert and R.H. Thomas (INVITED PAPER) 371

On the Conservativity of $H_p(10)$
G. Leuthold, V. Mares and H. Schraube 379

Ambient Dose Equivalent Conversion Factors for High Energy Neutrons Based on the ICRP 60 Recommendations
A.V. Sannikov and E.N. Savitskaya 383

Calculation of the Personal Dose Equivalent, $H_p(10)$, for Neutrons in the MIRD Phantom
R.A. Hollnagel 387

Organ Doses and Dose Equivalents for Neutrons above 20 MeV
V. Mares, G. Leuthold and H. Schraube 391

DOSIMETRY OF COSMIC RADIATIONS FIELDS

Dosimetry for Occupational Exposure to Cosmic Radiation
D.T. Bartlett, I.R. McAulay, U.J. Schrewe, K. Schnuer, H.-G. Menzel, J.-F. Bottollier-Depois, G. Dietze, K. Gmür, R.E. Grillmaier, W. Heinrich, T. Lim, L. Lindborg, G. Reitz, H. Schraube, F. Spurný and L. Tommasino (INVITED PAPER) 395

The Cosmic Ray Induced Neutron Spectrum at the Summit of the Zugspitze (2963m)
H. Schraube, J. Jakes, A. Sannikov, E. Weitzenegger, S. Roesler and W. Heinrich 405

Experimental Approach to the Exposure of Aircrew to Cosmic Radiation
F. Spurný 409

Results of Dosimetric Measurements in Space Missions
G. Reitz, R. Beaujean, J. Kopp, M. Leicher, K. Strauch and C. Heilmann 413

Dosimetry of High Energy Neutrons and Protons by ^{209}Bi Fission
W.G. Cross and L. Tommasino 419

DOSIMETRIC MEASUREMENTS AROUND HIGH ENERGY ACCELERATORS

Measurements of Neutron Spectra at the Stanford Linear Accelerator Center
V. Vylet, J.C. Liu, S.H. Rokni and L.-X. Thai 425

Shielding Measurements for a Proton Therapy Beam of 200 MeV: Preliminary Results
A. Mazal, K. Gall, J.F. Bottollier-Depois, S. Michaud, D. Delacroix, P. Fracas, F. Clapier, S. Delacroix, C. Nauraye, R. Ferrand, M. Louis and J.L. Habrand 429

Neutron Measurements around a High Energy Lead Ion Beam at CERN
A. Aroua, T. Buchillier, M. Grecescu and M. Höfert 437

CONTENTS

Secondary Dose Exposures during 200 MeV Proton Therapy
P.J. Binns and J.H. Hough 441

ACCIDENT DOSIMETRY

Criticality Accident Dosimetry: An International Intercomparison at the SILENE Reactor
R. Médioni and H.J. Delafield (INVITED PAPER) 445

Biological Dosimetry after a Criticality Accident
M. Fatôme, D. Agay, S. Martin, J.C. Mestries and E. Multon (INVITED PAPER) 455

Characterisation of a Clothing Material for Gamma Dosimetry in Mixed Neutron Gamma Fields
M. Benabdesselam, P. Iacconi, D. Lapraz, A. Serbat, J. Dhermain and J. Laugier 461

Chromosome Aberrations Scoring for Biological Dosimetry in a Criticality Accident
P. Voisin, D. Lloyd and A. Edwards 467

NUCLEAR PARTICLE DOSIMETRY FOR RADIATION THERAPY

Fast Neutron Radiotherapy: The University of Washington Experience and Potential Use of Concomitant Boost with Boron Neutron Capture
K.J. Stelzer, K.L. Lindsley, P.S. Cho, G.E. Laramore and T.W. Griffin (INVITED PAPER) 471

Energy Spectra in the NAC Proton Therapy Beam
F.D. Brooks, D.T.L. Jones, C.C. Bowley, J.E. Symons, A. Buffler and M.S. Allie 477

Calculated Fluence Spectra at Neutron Therapy Facilities
M.A. Ross, P.M. DeLuca Jr, D.T.L. Jones, A. Lennox and R.L. Maughan 481

Biophysical Investigations of Therapeutic Proton Beams
R. Becker, J. Bienen, U. Carl, P. Cloth, M. Dellert, V. Drüke, W. Eyrich, D. Filges, M. Fritsch, J. Hauffe, W. Hoffmann, H. Kobus, R. Maier, M. Moosburger, P. Olko, H. Paganetti, H.P. Peterson, Th. Schmitz, K. Schwenke and R. Sperl 485

Microdosimetric Studies on the Orsay Proton Synchrocyclotron at 73 and 200 MeV
V.P. Cosgrove, S. Delacroix, S. Green, A. Mazal and M.C. Scott 493

Comparison of HETC and PTRAN with Phantom Measurement Data
R. Becker, J. Bienen, P. Cloth, M. Dellert, V. Drüke, W. Eyrich, D. Filges, M. Fritsch, J. Hauffe, U. Heinrichs, W. Hoffmann, H. Kobus, R. Maier, M. Moosburger, H. Paganetti, N. Paul, H.P. Peterson, Th. Schmitz, K. Schwenke and R. Sperl 497

Investigation of Efficiency of Thermoluminescence Detectors for Particle Therapy Beams
P. Bilski, M. Budzanowski, W. Hoffmann, A. Molokanov, P. Olko and M.P.R. Waligórski 501

Measurement of the Heat Defect in Water and A-150 Plastic for High Energy Protons, Deuterons and α Particles
H.J. Brede, O. Hecker and R. Hollnagel 505

3D Treatment Planning for 14 MeV Neutrons
R. Schmidt, T. Frenzel, A. Krüll, L. Lüdemann and T. Matzen 509

CONTENTS

Evaluation of the Dose Algorithm Based on the Scatter Model Applied for Neutron Therapy Treatment Planning
M. Yudelev, R.L. Maughan, T. He and D.P. Ragan 513

RADIATION THERAPY AND BIOLOGICAL EFFECTIVENESS

Specification of Absorbed Dose and Radiation Quality in Heavy Particle Therapy (A Review)
A. Wambersie and H.G. Menzel (INVITED PAPER) 517

A Unified System of Radiation Bio-Effectiveness and its Consequences in Practical Applications
D.E. Watt 529

Risk Scaling Factors from Inactivation to Chromosome Aberrations, Mutations and Oncogenic Transformations in Mammalian Cells
A.S. Alkharam and D.E. Watt 537

Nanodosimetric Results and Radiotherapy Beams: A Clinical Application?
L. Lindborg and J.-E. Grindborg 541

BORON NEUTRON CAPTURE THERAPY

BNCT: Status and Dosimetry Requirements
R. Gahbauer, N. Gupta, T. Blue, J. Goodman, J. Grecula, A.H. Soloway and A. Wambersie (INVITED PAPER) 547

Microdosimetric Analysis of Absorbed Dose in Boron Neutron Capture Therapy
C. Kota and R.L. Maughan 555

Characterisation of an Accelerator-Based Neutron Source for BNCT of Explanted Livers
S. Agosteo, P. Colautti, M.G. Corrado, F. d'Errico, M. Matzke, S. Monti, M. Silari and R. Tinti . . . 559

Reactor Based Epithermal Neutron Beam Enhancement at Řež
M. Marek, J. Burian, J. Rataj, J. Polák and F. Spurný 567

Fricke-Infused Agarose Gel Phantoms for NMR Dosimetry in Boron Neutron Capture Therapy and Proton Therapy
G. Gambarini, C. Birattari, D. Monti, M.L. Fumagalli, A. Vai, P. Salvadori, L. Facchielli and A.E. Sichirollo 571

Author Index 577

List of Participants 579

Radiation Protection Dosimetry is rated 1st out of 650 UK journals and 30th out of 4992 journals published throughout the world, in the International Energy Agency's Energy Technology Data Exchange Journal Productivity Listings (1993-1995), published annually by the ETDE Operating Agent in Oak Ridge, TN, USA. It is 1st in the world out of all radiation protection related journals.

Radiation Protection Dosimetry is abstracted or indexed in APPLIED HEALTH PHYSICS ABSTRACTS AND NOTES, Chemical Abstracts, CURRENT CONTENTS, Energy Information Abstracts (Cambridge), EXCERPTA MEDICA (EMBASE), Health and Safety Science Abstracts (Cambridge), INIS ATOMINDEX (hard copy and CD ROM), INSPEC, Nuclear Energy (Czech Republic), QUEST and Referativmaja Zhurnal.

Editorial

Eighth Neutron Dosimetry Symposium.

The Symposium in Paris continued the series of seven successful Symposia on Neutron Dosimetry which had previously been held in Germany, the last one taking place in Berlin in 1991 (see Radiation Protection Dosimetry, Vol. 44 (1992)). It was jointly organised by the European Commission (EC, Brussels), the Institut de Protection et de Sûreté Nucléaire (IPSN, Fontenay-aux-Roses), the Physikalisch-Technische Bundesanstalt (PTB, Braunschweig) and the US Department of Energy (DOE, Washington D.C.). The Symposium was co-sponsored by Electricité de France and NUKEM. The excellent local organisation provided by IPSN, along with the Parisian atmosphere symbolised by the venue near the Eiffel tower, contributed substantially to the success of this Symposium.

Neutron dosimetry is still a fascinating subject for research in the general field of dosimetry mainly because of the complexity of the nuclear interactions of neutrons with matter; these interactions are nuclide-specific and the range of practical interest extends from thermal neutron energies of some tens of meV up to several hundred MeV. Neutron fields are always accompanied by other long-range radiations such as gamma rays and, at higher energies, protons. The biological effects are complex and manifest themselves in strongly energy-dependent quality and radiation weighting factors to be used for radiation protection purposes.

In times of decreasing funds in the area of dosimetry, however, research must be justified by the importance of applications and needs in practice. Therefore, an additional aim of the Symposium was to intensify the contacts and discussions between the researchers, legal authorities, users and practitioners.

In radiation protection, nuclear industries need neutron dosimetry technology during the production of fuel elements for nuclear power stations and for removal, transport and interim storage of burnt fuel elements. Neutrons of energies below about 15 MeV are produced not only by fission but also by (α,n) processes. On the other hand, neutron exposures during the operation of modern nuclear power stations are normally negligible, due to the shielding construction of the stations. Collective neutron doses are generally small. The total annual number of personal neutron dosemeters issued in the EU countries can be estimated to be of the order of 10,000 per year. In spite of these comparatively small numbers, the expertise in this field must be maintained, preferably by active research, and the credibility of neutron dosimetry methods must be guaranteed by suitable quality management. Practical applications still suffer, amongst other things, from the fact that satisfactory small real-time personal dosemeters are not available and also that assessed doses by neutron dosemeters of different design, exhibit large discrepancies for the same exposure. The latter problem can, by and large, be solved by means of computer programs which derive suitable correction factors for the instrument readings from the response characteristics of the instruments and a suitably large collection of neutron spectra found in various applications. An alternative and promising approach relies on the concept of realistic neutron calibration fields which, under laboratory conditions, makes well-characterised (specified) neutron fields available, replicating as closely as possible those encountered at workplaces, and consequently, enables the calibration of dosimetric devices to be satisfactorily carried out in conditions simulating those of their practical use. This should be included in a 'code of practice' which should be developed in the framework of a future international quality management programme.

The largest collective neutron dose is received by tens of thousands of airline employees during their flights at altitudes above approximately 10 km. These persons are not monitored at all, which can be justified by the fact that over-exposures can almost be excluded apart from the effects of solar flares. These are, however, extremely rare events. About half of the total dose is due to neutrons which may have energies up to some 100 MeV. In view of the elevated dose level, the dosimetry research programme in this energy range, which is also important for workers at high energy accelerators, should be continued.

In radiation therapy, modern neutron therapy centres increasingly use higher neutron energies up to about 70 MeV. Low energy neutrons are mainly used in connection with boron neutron capture therapy. Since other

EDITORIAL

particles like protons of 60 to 250 MeV or heavy ions of still higher energies are becoming more and more important, dosimetric characterisation of these beams, as well as investigations about neutron contamination, were included as topics of this Symposium. The general trend to apply higher energies in radiation therapy requires corresponding research in order to decrease the uncertainties of dose measurements and the treatment planning programme. In particular, improved basic data for stopping powers, ranges and nuclear cross-sections for various projectiles and various tissues as targets are needed. Common to all topics is the field of computational dosimetry. Its increasing involvement in many studies has been stressed in different fields as a result of the progress in computing power and the distribution of general codes providing help in designing, optimising or even replacing experiments. Nevertheless, such a useful tool requires appropriate experimental verification and estimation of uncertainties.

The Symposium addressed a wide range of specific topics and provided again a forum of multi-disciplinary exchanges of information. The large number of participants demonstrated the continued importance of this research field.

We would like to thank IPSN for the excellent local organisation which was a basic prerequisite for holding the Symposium. Special thanks are due to the sponsors, the EC, for the financial support granted to young scientists, the Members of the Programme Committee, and those scientists who cooperated in the preparation and holding of the Symposium and in the evaluation of the papers published in these proceedings. We also express our thanks to Mr E.P. Goldfinch for his particular fortitude in preparing this publication.

H.G. Menzel
J.L. Chartier
R. Jahr
A. Rannou

NUCLEAR DATA NEEDS FOR NEUTRON THERAPY AND RADIATION PROTECTION

M. B. Chadwick†*, P. M. DeLuca, Jr.‡ and R. C. Haight§
†University of California, Nuclear Data Group
Lawrence Livermore National Laboratory, Livermore, CA 94550, USA
‡Department of Medical Physics
University of Wisconsin, Madison, WI 53706-1532, USA
§University of California, Physics Division
Los Alamos National Laboratory
Los Alamos, NM 87545, USA

INVITED PAPER

Abstract — New nuclear data are required for improved neutron and proton radiotherapy treatment planning as well as future applications of high energy particle accelerators. Modern neutron radiotherapy employs energies extending to 70 MeV, while industrial applications such as transmutation and tritium breeding may generate neutrons exceeding energies of 100 MeV. Secondary neutrons produced by advanced proton therapy facilities can have energies as high as 250 MeV. Each use requires nuclear data for transport calculations and analysis of radiation effects (dosimetry). The nuclear data needs supportive of these applications are discussed, including the different information requirements. As data in this energy region are sparse and likely to remain so, advanced nuclear model calculations can provide some of the needed information. In this context new evaluated nuclear data are presented for C, N, and O. Additional experimental information, including integral and differential data, are required to confirm these results and to bound further calculations. The required new data to be measured and the difficulties in carrying out such experiments are indicated.

INTRODUCTION

Nuclear data are fundamental to our understanding of dosimetry in radiation therapy and radiological protection when energetic neutrons and protons are used. The type, accuracy, and specificity of the needed information vary with the application. For proton radiotherapy, neutron production and non-elastic cross section information are essential to predicting shielding requirements and designing beam delivery devices. Neutron therapy dosimetry and more importantly neutron transport calculations for dosimetry demand detailed microscopic charged particle production information for prediction and interpretation of absorbed dose to the patient. Albeit with less precision and accuracy, similar data are essential for radiation protection applications. Recently proposed applications in accelerator based transmutation of nuclear waste[1-3] and production of tritium[4] also require considerable nuclear data. In this report recent work[5,6] on the calculation and evaluation of nuclear data needed for fast neutron and proton radiotherapy is discussed.

Below 20 MeV and particularly below 15 MeV neutron energy, extensively evaluated nuclear data are readily available for most elements and reactions. Evaluated libraries are highly recommended as these data, while not always complete or perfect, are under continuous scrutiny by experimentalists and theorists and therefore are considered the *best available*. Covering the neutron energy range from 10^{-11} to 20 MeV, the libraries contain detailed information on total, elastic, inelastic, and partial reaction cross sections. Most evaluated libraries do not include detailed information on charged particle emission spectra and angular distributions, since they were originally designed for neutron transport applications. However, some evaluations included such data. The *US standard* data library is the **E**valuated **N**uclear **D**ata **F**ile Library version B-VI, ENDF/B-VI, available from the National Nuclear Data Center (NNDC) at the Brookhaven National Laboratory. Similar libraries are available through the Nuclear Energy Agency Data Bank in Paris, the International Atomic Energy Agency Nuclear Data Section in Vienna, the Nuclear Data Centre in Obninsk (Russia), and through centres in Tokaimura (Japan), Beijing (China), other national centres, and national laboratories such as at Los Alamos and Lawrence Livermore (the ENDL library). Data are available electronically in the ENDF/B format, and many of these institutions, and their data, are accessible through the World Wide Web.

It is generally difficult to obtain detailed evaluated information on the charged particles emitted in reactions (induced by neutrons or other energetic particles) or on integral quantities such as kerma factors. At present, such information exists in the form of reports and articles (for example Refs. 5, 6, 7, 8 and 9). Data for some of these evaluations are expected to be available electronically soon. These data bases are of particular interest because they treat higher neutron energies and they were created principally for medical applications.

---

*Permanent address: Group T2, MS 0243, Los Alamos National Laboratory, Los Alamos, NM 87545, USA.

Note that the present ENDF evaluations result from the abiding needs of the nuclear power industry and nuclear weapons applications, and this information evolved from forty years of intensive experimental and nuclear model research. Unfortunately, extension of the existing evaluated data libraries to higher energies with the same comprehensive experimental detail is unlikely. Hence, nuclear modelling and semi-empirical techniques must be used to expand and interpolate the existing charged particle information for purposes of estimating kerma factor values[9,10]. The weakness in basing such efforts on limited charged particle data or inadequate nuclear models was demonstrated by comparisons of integral kerma factor measurements for carbon and oxygen with values determined from the expanded nuclear data base[11,12].

Previously, the most accurate and well-benchmarked calculations of higher energy neutron reactions on biologically important elements were those of Brenner and Prael[13]. New experimental results, together with our recent improvements in nuclear model calculation capabilities, make this an opportune time to perform a new analysis. The charged particle emission spectra measured at UC-Davis[14,15], which were available to Brenner and Prael, are now supplemented by measurements from Louvain-la-Neuve[16] and Los Alamos[17]. Also, recent measurements of elastic scattering data[18-21], along with their optical model analyses and compilations of elastic and inelastic scattering[19], are particularly useful for this work. Nuclear modelling improvements were incorporated into the FKK-GNASH code and described in detail in Refs. 5, 22 and 23. By including nuclear structure and nuclear reaction mechanisms specific to light nuclei and making use of recent advances in the multistep theory of pre-equilibrium nuclear reactions, this code was used to model neutron reactions on biological elements.

In this report some new results are summarised of nuclear model based estimates of microscopic cross section information for carbon, nitrogen, and oxygen. Results are compared with existing microscopic and integral measurements. The elemental results are combined to provide integral kerma factor values for the critical dosimetric mixtures A-150 plastic and ICRU muscle. These calculations and evaluations were driven by the nuclear data needs of the rapidly developing all-particle stochastic treatment planning initiative — PEREGRINE[24]. Finally, additional calculations and measurements needed to extend the nuclear data base to 100 MeV for medical and industrial applications are suggested.

NUCLEAR REACTION MODELS

Even though the measured database for neutron reactions on C, N, and O has grown in recent years, the measurements are still relatively sparse. To determine the kerma factors and cross sections needed for radiation transport calculations, a comprehensive account of the emission spectra and angular distributions of secondary particles produced in the reaction is needed. Nuclear theory and models can interpolate between measurements and extrapolate to where measurements have not been made. This can be done in practice by optimising the models and the model parameters used to account for the existing data. Thus validated, the model calculations can then be used with some confidence in regions where measurements do not exist.

Elastic and inelastic scattering

Elastic scattering processes are important in two ways. Firstly, elastic scattering generally occurs in more than 50% of the scattering events, and the scattered neutron energy and angular distributions must be known to describe the neutron transport through matter. Secondly, the target nucleus' recoil kinetic energy contributes to the kerma (and absorbed dose). Kinematics determines that the kinetic energy transferred to the recoil nucleus varies inversely with the target nuclear mass, and so the elastic partial kerma contribution can be significant for reactions on light nuclei.

Optical model analyses are used to determine the elastic scattering cross sections and angular distributions. They are also used to determine particle transmission coefficients and the reaction cross section (which is equivalent to the total non-elastic cross section at these energies). Reference 5 shows how these theoretical analyses describe the elastic and reaction cross sections fairly well. However, since there are a number of measurements of the reaction and total cross sections, small renormalisations to the calculated reaction and elastic cross sections (and transmission coefficients) were made to better describe these measurements (except for carbon, where the calculations were not modified)[5,6].

Non-elastic reactions

Non-elastic reaction mechanisms describe the many different possible nuclear fragmentation processes and are the dominant contribution to the kerma factors above 20 MeV. To calculate these cross sections, the FKK-GNASH nuclear reaction modelling code is used, which implements pre-equilibrium and equilibrium decay theories.

The interaction of the projectile neutron with the target nucleus is described as taking place through a number of stages. Initially the neutron interacts with a nucleon within the nucleus, exciting a particle–hole pair. The excited nucleons may then undergo further interactions, until all the energy brought in by the projectile is shared among the target nucleons in an equilibrated state. However, it is also possible for particles to be emitted from the early stages of the reaction, giving the pre-equilibrium secondary particles which are typically of high energy and exhibit a forward-peaked angular distribution. After the pre-equilibrium phase of the reaction, the residual nuclei (which are usually left in excited states) then decay by equilibrium particle or gamma ray emission.

The pre-equilibrium emission of nucleons was calculated using the quantum-mechanical Feshbach-Kerman-Koonin (FKK) multistep direct and multistep compound theory[25]. Pre-equilibrium emission of deuterons and alpha particles was determined with the cluster model of Kalbach[26]. In these calculations, model parameters were adjusted (within their uncertainties) to optimise agreement with measured data.

Additionally, in some cases the neutron, proton, deuteron, and alpha particle angle-integrated 'primary' (or first particle) pre-equilibrium spectra were pre-evaluated, by considering measured data, as well as predictions from the FKK and semiclassical[22] theories. This procedure incorporated restrictions from energy and flux conservation (unitarity), and systematic observations suggesting that the ratio of neutron to proton differential (angle-integrated) pre-equilibrium cross sections is approximately 2:1[5,27]. The advantage of this new method for determining pre-equilibrium cross sections within an evaluation is that it does not rely solely on model calculations, but utilises the experimental information where it exists. These pre-determined pre-equilibrium spectra are then included as an input into the FKK-GNASH code, so that the final results have the usual advantages of nuclear data based on model calculations — they naturally conserve energy, unitarity, angular momentum, and parity. Following 'primary' pre-equilibrium emission, but before equilibrium emission, emission of a second nucleon is allowed using the multiple pre-equilibrium theory[22]. This mechanism becomes important above a few tens of MeV incident energy, and is necessary for determining accurately the emission spectra, as well as the magnitude of 'exclusive' reaction cross sections (such as (n,2n), and the (n,n'3α) for C).

Following pre-equilibrium emission, Hauser-Feshbach theory is used to describe the sequential decays of the residual nuclei by particle or gamma ray emission. All possible sequential reaction pathways are included. These calculations incorporate a full conservation of angular momentum and parity, and use the low-lying measured nuclear level schemes for the nuclei (both target, intermediate, and residual) involved. Above a certain excitation energy (usually of the order of 10 MeV for the light nuclei, but determined in each case from level density analyses[5,6]), measured data are no longer complete and instead a statistical level density description is used. This continuum level density was obtained from the model of Ignatyuk *et al*[28], and is matched continuously onto the known low-lying levels.

Comparison with previous calculations

Many previous efforts applied nuclear reaction model codes to estimate neutron cross sections and kerma factors on biologically important elements. The present work differs from most of these in two principal ways: (1) a greater use is made of measurements (some of which were made only recently) in generating these evaluated data; and (2) the underlying physics in the modelling code was optimised for describing reactions on light nuclei. In most previous calculations detailed comparisons with measured data (e.g. emission spectra) are not made, and it is difficult to assess the accuracy of these results. However, two previous works do provide detailed comparisons with experimental data: that of Brenner and Prael, and that of Dimbylow[29].

Using an early version of the modelling code used in this work Dimbylow[29] calculated neutron reactions up to 50 MeV on a range of biologically important elements. Comparisons of these results with the UC-Davis data showed that while the gross features of the data were described, high energy deuteron and proton emission were significantly underestimated. These deficiencies were due to the use of rather crude pre-equilibrium models in that early code, as well as the neglect of multiple pre-equilibrium processes.

Brenner and Prael used an intranuclear cascade (INC) model to describe the fast-particle emission, followed by a Fermi break-up model to describe equilibrium decay, for neutrons up to 80 MeV on C, N, and O. Importantly, they recognised that nuclear models specifically designed for light nuclei must be used to obtain satisfactory results, and included models for direct (high energy) deuteron and alpha emission. They also included nuclear structure information of the decaying nuclei in the Fermi break-up calculations (but not in the INC calculations). Through extensive comparisons with data, particularly the UC-Davis charged particle emission spectra, they demonstrated that their model calculations describe the data well. To some extent the agreement with measurements is surprising since the INC theory assumes that the mean free path of the bombarding particle is large compared to the nucleon–nucleon separation, which is not true for incident energies below 100 MeV. Unlike Brenner and Prael's work, the present work also considers elastic scattering and gamma ray production. As we shall show in the next section, our calculations account for the measured data somewhat more accurately than those of Brenner and Prael.

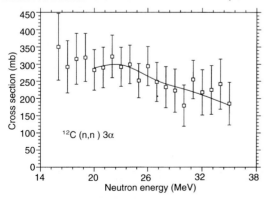

Figure 1. A comparison of calculated (solid line) and measured (□)[30] cross section values for the ^{12}C(n,n'3α) reaction.

RESULTS

Cross sections

In order to establish the accuracy of the evaluated cross section data, extensive comparisons with measured data must be made. Such data can be classified as 'exclusive' data, where a combination of ejectiles with a uniquely determined residual nucleus are specified (e.g. (n,2n), (n,pn), (n,n'3α) reactions, etc.) or as 'inclusive' data, where all possible exclusive reactions which contribute to the emission of a particle of interest are summed together (e.g. the inclusive proton emission cross section, also known as the proton production cross section and written as (n,xp), includes exclusive reactions such as (n,np), (n,2p), etc.). An example of comparison with exclusive data is shown in Figure 1 for the important $^{12}C(n,n'3\alpha)$ where our calculations are seen to describe the measurements well. At the incident energies included, the 3α break-up cross section accounts for a significant fraction of the non-elastic processes. Comparisons with other exclusive reaction channels are made in Refs 5 and 6. Above 20 MeV, however, most measurements are of inclusive emission spectra of light particles (A ≤ 4), and it is these data on which we concentrate since they are critically important for correctly determining the dosimetry and the transport.

Inclusive emission spectra of charged particles with A ≤ 4, for C, N, and O were measured by Subramanian et al[14,15] at neutron energies of 27, 40, and 61 MeV. Recently, Slypen et al[16] reported inclusive A ≤ 4 charged particle spectra for C at 43, 63, and 73 MeV, while Haight et al[17] measured inclusive alpha particle spectra for carbon up to 40 MeV with a pulsed neutron

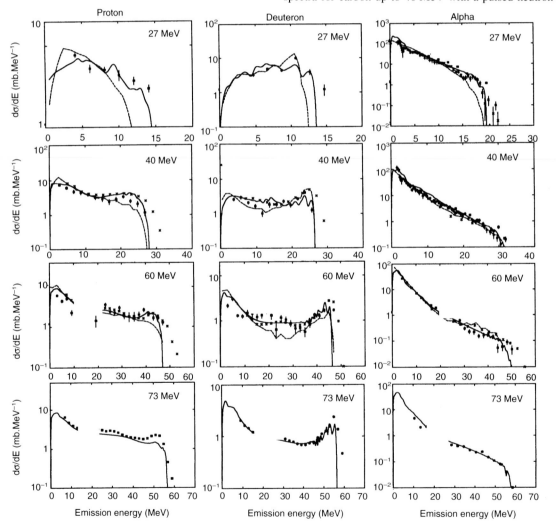

Figure 2. Calculated carbon angle-integrated inclusive spectra (solid lines) for proton, deuteron, and alpha particle emission compared with measured values (References 14, 16, 17, 31) and previous calculations (dashed line)[13].

source that is continuous in neutron energy. These authors measured the double-differential cross sections at a range of angles, and thence determined the angle-integrated emission spectra. The angular distributions are important for correctly accounting for the transport effects of the radiation. Since charged particle ranges are not large, the angle-integrated spectra are most important in determining the partial kerma factors. Therefore it is these data with which we compare our calculations most extensively.

Figures 2 and 3 show the calculated and measured angle-integrated emission spectra of protons, deuterons, and alphas for C and O. The figures show Brenner and Prael's calculations (dashed lines) for comparison. It is evident that our model calculations account for the measurements rather well, generally agreeing with the experimental data better than those of Brenner and Prael (particularly for the deuterons). The preponderance of alpha particles is related to the tight binding and the relative stability of the alpha particle. The structure that is seen in the calculated cross sections, particularly at the highest emission energies but also sometimes at the lowest incident energies, is due to our inclusion of the experimental discrete low-lying nuclear levels in the calculations.

In the case of oxygen, Figure 3 shows two sets of data for the angle-integrated 60 MeV proton spectrum: the original values obtained from UC-Davis[32] and newly derived results which were obtained[6] from angle-integration of the original UC-Davis double-differential data. These newly derived 'experimental' results are seen to be generally lower than the original UC-Davis values. This is because the UC-Davis measurements for oxygen only extended to angles as high as 65°. To integrate their differential data they assumed a constant value from 65° back to 180°, which overestimates the backward angle cross sections since angular distributions are generally very forward-peaked at these energies. For carbon the UC-Davis data did not suffer from this problem since a more comprehensive angular range was measured, though nitrogen, like oxygen, was measured only to 65°. For a more realistic angle-integration scheme, the shape of the angular distributions from our model calculations was used to

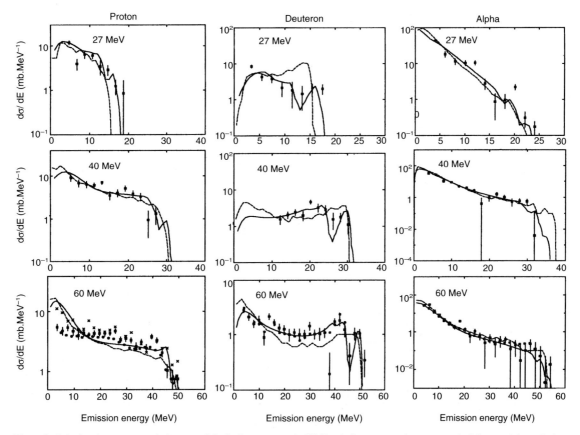

Figure 3. Calculated oxygen angle-integrated inclusive spectra (solid lines) for proton, deuteron, and alpha particle emission compared with measured values[14] and previous calculations (dashed lines)[13].

extrapolate the measured data from 65° to 180°. This approach is justifiable since our calculated angular distributions generally account for the measured distributions well[5,6]. It is evident that the model calculations are in better agreement with these newly derived angle-integrated data. In the future the plan is to perform new angle-integrations of all the N and O UC-Davis data.

Clearly a high priority for future experiments is to determine emission spectra over a wider angular range.

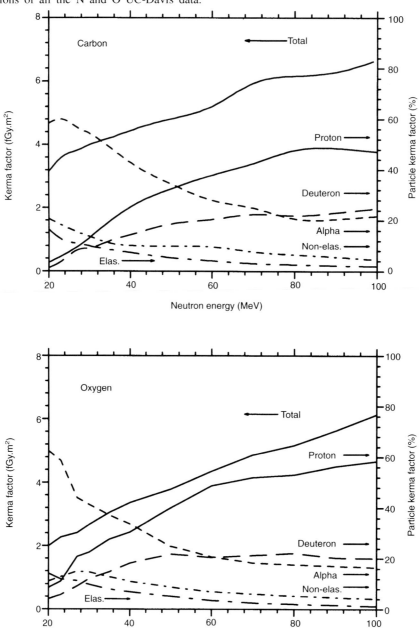

Figure 4. Calculated carbon and oxygen values for total kerma and for that due to different ejectiles plotted against neutron energy. Total kerma is given on the left ordinate while percentage kerma values by particle type are given on the right ordinate.

KERMA FACTORS

Elemental kerma factor

An important consequence of these microscopic cross section calculations is the determination of integral radiation dosimetry quantities — kerma factors. Figure 4 shows total and ejected particle partial kerma factors for C and O. Such detailed information is a direct result of the inclusive modelling technique. Note that the proton kerma increases rapidly with energy and exceeds that for alpha particles at 50 MeV (C) and 42 MeV (O). Another striking consequence is the contribution of deuterons which exceeds that for alpha particles at higher neutron energies, reaching 25%(C) and 20%(O) of the total kerma at 100 MeV. This rapid increase in proton and deuteron kerma at higher energies due to pre-equilibrium mechanisms was not recognised in the earlier extension to 32 MeV of ENDF/B-VI[10]. Table 1 compares the present cross sections and those of that evaluation.

Total kerma factor values above 20 MeV neutron energy combined with the Howerton evaluated[33] values below 20 MeV are shown for C, N, and O in the lower panel of Figure 5. Nitrogen values exceed those for O and in some cases that for C due to the more positive Q value for the $^{14}N(n,p)$ reaction. Table 2 provides evaluated elemental kerma factors derived from the present work for C, N, O, P, and Ca along with H values from Resler[34] based largely on the evaluation of Arndt.

Comparisons of integral calculations and measurements are a necessary but not sufficient guide to the reliability of modelling results. Although kerma determinations above 20 MeV are sparse, recent results greatly increased the available experimental data base. Figure 6 shows measured kerma factor values compared with the present calculations for C and O. While there is still some spread in the experimental values, the agreement is generally quite good, especially with the more recent work of Schrewe et al[44]. The importance of measuring complete back-angle particle emission data and the power of accurate modelling is clearly shown by the significant correction to the oxygen value of Brady and Romero. As mentioned, the original double differential proton, deuteron, and alpha particle

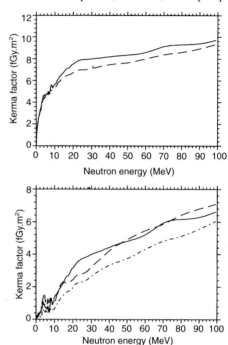

Figure 5. Kerma factor values for carbon (—); nitrogen (- - -), and oxygen (- · - · -), lower panel, and for A-150 plastic (—) and ICRU muscle (- - -), upper panel, plotted against neutron energy. Values below 20 MeV neutron energy are from Ref. 33, while others are from Refs 5 and 6.

Table 1. Exclusive reaction cross sections (in mb) for n + ^{12}C up to 30 MeV, for the most important reaction channels, compared with ENDF/B-VI[10].

Energy (MeV)	(n,n')	(n,2n)	(n,p)	(n,d)	(n,np)	(n,α)	(n,n'3α)
Present LLNL calculations							
20	122	0	29	13	7	20	289
23	97	1	23	28	24	12	298
27	69	4	21	52	45	5	252
30	58	8	23	46	55	3	228
Axton ENDF/B-VI evaluation							
20	90	0	13	53	26	42	299
23	80	4	5	25	24	15	305
27	60	17	1	9	8	7	254
30	53	24	0	4	4	4	228

Table 2. Total kerma factors for neutrons between 20 and 100 MeV in units of $fGy.m^2$.

Energy (MeV)	H	^{12}C	^{14}N	^{16}O	^{31}P	^{40}Ca
20	47.0	3.14	2.42	1.99	1.45	1.99
23	46.5	3.58	2.74	2.27	1.63	2.29
27	45.5	3.82	2.90	2.41	1.82	2.63
30	44.4	3.99	3.21	2.67	1.95	2.85
35	43.2	4.20	3.72	3.05	2.16	3.10
40	42.1	4.43	4.21	3.36	2.40	3.31
50	40.2	4.80	4.90	3.78	2.90	3.74
60	38.9	5.20	5.45	4.35	3.43	4.13
70	38.0	5.93	5.87	4.86	3.87	4.35
80	37.5	6.16	6.41	5.17	4.27	4.74
90	37.2	6.28	6.80	5.63	4.67	5.08
100	37.2	6.67	7.13	6.14	4.99	5.42

emission data were re-integrated using model calculations to estimate data beyond 65°.

Direct kerma determinations for C and O employ geometrically identical instrumentation frequently operated simultaneously[11,32,44,45]. Systematic uncertainties, particularly those associated with the neutron fluence are reduced by forming the ratio of kerma factors. Figure 7 shows ratios of all measured values as well as values derived from the present calculations and those from White *et al*[46]. Uncertainties were derived from the original values combined in quadrature.

Kerma factors of mixtures

Kerma factors for A-150 plastic[47] and ICRU muscle[48] are determined from the elemental kerma factors for C, N, and O combined with a recent evaluation of the hydrogen kerma factors[34]. Percentage kerma values by particle type for A-150 plastic and ICRU muscle are shown in Figure 8. Similar information organised by elemental contribution is given in Figure 9. For the latter figures, the present results were joined with the Livermore evaluation below 20 MeV

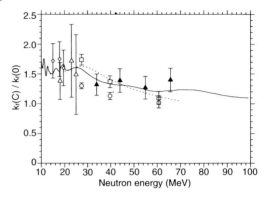

Figure 7. Calculated and measured carbon to oxygen kerma factor ratio values plotted against neutron energy. The present results are shown as a solid line, while the evaluation of White *et al*[46] is represented by a dashed line. Other symbols and References: (□) Brady with Dimbylow corrections[29,30], (○) Brady with Romero corrections (32), (△) DeLuca *et al*[11], (△) Hartmann *et al*[45], (▲) Schrewe *et al*[44].

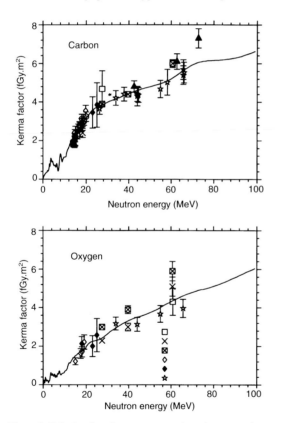

Figure 6. Calculated carbon, upper panel, and oxygen, lower panel, kerma factor values plotted against neutron energy. Values below 20 MeV neutron energy are taken from Howerton[33]. Other symbols indicate all known measured kerma factor values. References for symbols: Carbon: (□) 35, (×) 36, (⊠) 31, (+) 37, (◇) 11, (∗) 38, (■) 39, (◆) 45, (△) 40, (▲) 16, (○) 41, (●) 42, (☆) 44, (★) 43. Oxygen: (□) 6, (×) 29, 30, (⊠) 31, (◇) 11, (◆) 45, (☆) 44.

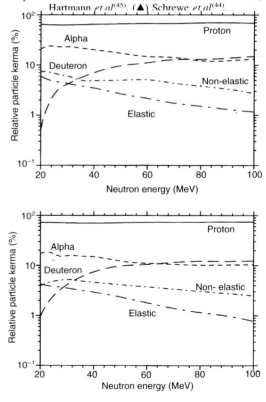

Figure 8. Calculated percentage kerma due to protons, deuterons, alpha particles, non-elastic recoils and elastic recoils in A-150 plastic, upper panel, and ICRU muscle, lower panel, plotted against neutron energy.

neutron energy. Kerma from elements other than C, N, and O was approximated by distributing their mass fractions in the mixture amongst those for C, N, and O according to the original relative mass fractions of C, N, and O. For A-150 plastic the mass fractions are H, 0.101327; C, 0.807676; N, 0.036511; and O, 0.054487; while for ICRU muscle they are H, 0.101997; C, 0.124525; N, 0.035425; and O, 0.738043. Again the contribution of deuteron kerma at higher energies is evident. At similar high energies, the kerma for A-150 plastic is dominated by carbon and that for ICRU muscle by oxygen. Total kerma factor values for A-150 plastic and ICRU muscle are plotted against neutron energy in Figure 5.

CONCLUSIONS AND FUTURE WORK

These results demonstrate that advances in nuclear modelling can greatly enhance our knowledge of microscopic cross section information essential to neutron radiotherapy and other applications. The critical need for experimental information to normalise and verify the calculations is also apparent. The present work concentrated upon elements essential for neutron transport and dosimetry for therapy applications. Much more information is needed. Table 3 shows the principal needs organised by dosimetry and transport. As for C, N, and O, there is a scarcity of microscopic cross section information for the stated elements. A needed high priority measurement is charged particle emission spectra from neutron reactions on oxygen, for a wide range of emission angles (since the UC-Davis data extended only to 65°, making an angle-integration somewhat ambiguous).

Besides C, N and O, Si and Ca require the most complete information for dosimetry. Detailed ejectile data are needed for Si to support applications in electronics. As bone is almost always present in the radiated field in neutron radiotherapy, Ca data are essential for accurate dosimetry. As P, K, and N are present at less than a few per cent in most tissues and dosemeters, less comprehensive data are required. For neutron transport calculations, Fe and W are the elements next in importance to C, N, and O. It is anticipated that the cross sections above 20 MeV for Cu and Pb are similar enough to Fe and W for values to be readily substituted for purposes of neutron transport. Information about Si, Ca, and N is required for transport calculations due their presence in significant quantities in concrete and air.

Acquisition of all this essential information by measurements is unlikely. Suitable facilities are increasingly scarce and funding is very limited. These facts emphasise the need for expanded modelling efforts. Measurements, and to an extent calculations, are also complicated by the neutron spectrum encountered at energies above 20 MeV. Monochromatic neutron sources are not possible. An example of this is shown by the most recent kerma determinations by Schrewe et al[44] who used a 71.8 MeV proton beam bombarding a non-stopping (2 mm) Be target to generate neutrons.

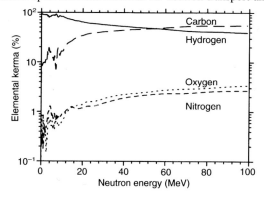

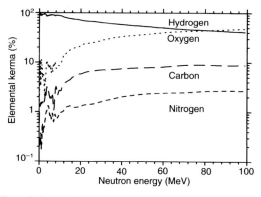

Figure 9. Calculated percentage kerma due to hydrogen, carbon, nitrogen, and oxygen in A-150 plastic, upper panel, and ICRU muscle, lower panel, plotted against neutron energy. Values below 20 MeV used the evaluations of Howerton[33] and Resler[34].

Table 3. Required neutron cross sections information for radiotherapy.

	Absorbed dose				Transport		Importance*
Element	p	α	d	recoil	Element	n,γ	
H	✓				H	✓	1
C	✓	✓	✓	✓	C	✓	1
O	✓	✓	✓	✓	O	✓	1
Si	✓	✓	✓	✓	Fe(Cu)	✓	2
Ca	✓	(✓)	(✓)	✓	W(Pb)	✓	2
P	✓	(✓)			Si	✓	3
K	✓	(✓)			Ca	✓	3
N	✓	(✓)			Pb	✓	3
					N	✓	3

*Importance:
(1) Need best accuracy, most complete data, small neutron energy intervals.
(2) Need good accuracy, wider spacing of neutron energies.
(3) High accuracy is not needed; approximate data are adequate.

The measured neutron spectrum[49–51] at the measurement point convolved with our kerma factor for carbon values is shown in Figure 10. Also plotted are values deduced from a LAHET[52] model calculation of the neutron spectrum. This calculation includes a complete description of the target and collimators used as well as transport through intervening air. No scatter from material behind the detectors was included. The resulting energy spectrum was convolved with the same carbon kerma factors. The agreement between measured and calculated values is quite good for the principal neutron peak. However, LAHET fails to predict the numerous pre-equilibrium neutrons observed. Use of an intranuclear cascade code at these energies is not appropriate, even one that includes modelling of pre-equilibrium neutron emission, and emphasises the need for extension of the present modelling efforts to neutron source characterisations. Figure 10 also shows the running sum of the percentage carbon kerma as a function of neutron energy. Without some form of neutron energy discrimination, off-energy neutrons contribute almost 50% of the total kerma that could be detected, requiring a large correction to the measured detector response. Assuming that low pressure proportional counters, which have slow timing characteristics, are employed to measure kerma, using neutron time-of-flight to isolate the detector response with energy would require very large flight paths and very intense neutron sources. For realistic distances and source intensities, corrections for the contribution of off-energy neutrons greatly increase the experimental uncertainties in the final result.

This paper concentrated on neutron nuclear data. We are currently working on proton data up to 250 MeV needed for proton therapy applications, and have completed preliminary evaluated data libraries for C, N, O, Ca, and P. Nuclear reactions are important here since they deplete protons from the incident beam, and the secondary particles that are produced must be characterised in order to assess radiation transport and dosimetry effects. These libraries, therefore, are optimised for an accurate representation of the non-elastic cross section, and for neutron and photon production. As well as comparing the calculated emission spectra with experimental data, they are also benchmarked against radionuclide production measurements. Results will be published shortly.

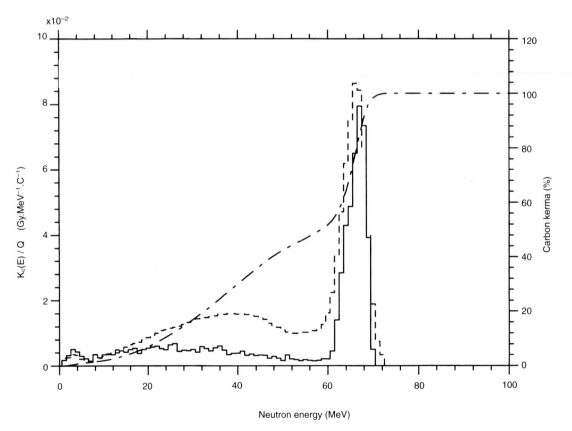

Figure 10. Neutron spectrum due to 71.8 MeV proton bombardment of a 2 mm Be target convolved with the kerma factor for carbon. Values measured by Schrewe et al[44] are shown as dashed line. Values calculated using the LAHET code system are given as a solid line[52]. A running sum expressed as a percentage of the Schrewe results is given as a dashed-dotted line.

In summary, advanced nuclear modelling of all microscopic cross sections for C, N, and O is shown to replicate accurately experimental information, differential and integral. As such, very complete nuclear data files are generated and are suitable for use in transport calculations. Extension of this work to other important elements is essential to complete the data base needed for medical and industrial applications. Finally, some new experimental data, particularly microscopic cross section information measured to back angles, are needed to bound and verify the modelling results.

ACKNOWLEDGEMENT

This work was performed in part under the auspices of the US Department of Energy by the Lawrence Livermore National Laboratory under contract number W-7405-ENG-48, by the Los Alamos National Laboratory under contract W-7405-ENG-36, and by the University of Wisconsin under contract DE-FG02-86ER60417.

REFERENCES

1. Arthur, E. et al. *The Los Alamos Accelerator Transmutation of Nuclear Waste (ATW) Concept*. Los Alamos National Laboratory Document LA-U-92-2020, April (1992).
2. Bezdecny, J. A., Vance, K. M. and Henderson, D. L. *Preliminary Analysis of the Induced Structural Radioactivity Inventory of the Base-Case Aqueous Accelerator Transmutation of Waste Reactor Concept*. Nucl. Technol. **110**, 369–395 (1995).
3. Lisowski, P. W., Bowman, C. D., Arthur, E. D. and Young, P. G. *Nuclear Physics Information Needed for Accelerator Driven Transmutation of Nuclear Waste*. In: Proc. Int. Conf. on Nuclear Data for Science and Technology, Jülich, Germany, 13–17 May 1991. Ed. S. M. Qaim (Heidelberg: Springer-Verlag) pp. 92–94 (1992).
4. Glanz, J. *Los Alamos Wins One in Tritium Race*. Science **270**(5234), 227–228 (1995).
5. Chadwick, M. B., Blann, M., Cox, L. J., Young, P. G. and Meigooni, A. S. *Evaluated Cross Section Libraries and Kerma Factors for Neutrons up to 100 MeV on Carbon*. Lawrence Livermore National Laboratory Report UCRL-ID-120829 (1995). Chadwick, M. B., Blann, M., Cox, L. J., Young, P. G and Meigooni, A. S. *Calculation and Evaluation of Cross Sections and Kerma Factors for Neutrons up to 100 MeV on Carbon*. Nucl. Sci. Eng. **123**, 17–37 (1996).
6. Chadwick, M. B. and Young, P. G. *Calculation and Evaluation of Cross Sections and Kerma Factors for Neutrons up to 100 MeV on ^{16}O and ^{14}N*. Nucl. Sci. Eng. **123**, 1–16 (1996).
7. Caswell, R. S., Coyne, J. J. and Randolph, M. L. *Kerma Factors for Neutron Energies Below 30 MeV*. Radiat. Res. **83**, 217–254 (1980).
8. International Commission on Radiation Units and Measurements. *Photon, Electron, Proton, and Neutron Interaction Data for Body Tissues*. ICRU Report 46 (Bethesda, Maryland: International Commission on Radiation Units and Measurements (1992).
9. Axton, E. J. *An Evaluation of Kerma in Carbon and the Carbon Cross Sections*. National Institute of Standards and Technology Report NISTIR 4838 (1992).
10. Fu, C. Y., Axton, A. J. and Perey, F. G. *Description of Evaluation for Natural Carbon Performed for ENBF/B-VI*. Oak Ridge National Laboratories: Evaluated Data for 0–32 MeV Neutrons on Elemental Carbon (Oak Ridge National Laboratories) (1989).
11. DeLuca, Jr, P. M., Barschall, H. H., Burhoe, M. and Haight, R. C. *Carbon Kerma Factor for 18- and 20-MeV Neutrons*. Nucl. Sci. Eng. **94**, 192–198 (1986).
12. DeLuca, Jr, P. M., Barschall, H. H., Sun, Y. and Haight, R. C. *Kerma Factors of Oxygen, Aluminium, and Silicon for 15 and 20 MeV Neutrons*. Radiat. Prot. Dosim. **23**, 27–30 (1988).
13. Brenner, D. G. and Prael, R. E. *Calculated Differential Secondary-Particle Production Cross Sections After Nonelastic Neutron Interactions With Carbon and Oxygen Between 15 and 60 MeV*. Atomic Data and Nucl. Data Tables **41**, 71–74 (1989).
14. Subramanian, T. S. et al. *Double Differential Inclusive Hydrogen and Helium Spectra From Neutron-Induced Reactions on Carbon at 27.4, 39.7, and 60.7 MeV*. Phys. Rev. C**28**, 521–528 (1983). Also Romero, J. L. private communication to M. B. Chadwick (1994).
15. Subramanian T. S. et al. *Double Differential Inclusive Hydrogen and Helium Spectra from Neutron-Induced Reactions at 27.4, 39.7, and 60.7 MeV: Oxygen and Nitrogen*. Phys. Rev. C**34**, 1580–1587 (1986).
16. Slypen, I., Corcalciuc, V., Ninane, A. and Meulders, J. P. *Kerma Values Deduced From Neutron Induced Charged-Particle Spectra of Carbon From 40–75 MeV*. Phys. Med. Biol. **40**, 73–82 (1995). Slypen, I., Corcalciuc, V. and Meulders, J. P. *Proton and Deuteron Production in Neutron-Induced Reactions on Carbon*. Phys. Rev. C **51**, 1303–1311 (1995). *Charged Particles Produced in Fast Neutron Induced Reactions on ^{12}C in the 45–80 MeV Energy Range*. Nucl. Instrum. Methods Phys. Res. A**337**, 431–440 (1994).
17. Haight, R. C. et al. *Alpha-Particle Emission From Carbon Bombarded With Neutrons Below 30 MeV*. In: Proc. Int. Conf. on Nuclear Data for Science and Technology, Gatlinburg, Tennessee, 9–13 May 1994. Ed. J. K. Dickens (American Nuclear Society) p. 311–313 (1994). Also private communication to M. B. Chadwick (1994).
18. Hjort, E. L., Brady, F. P., Romero, J. L., Drummond, J. R., Sorenson, D. S., Osborne, J. H., McEachern, B. and Hansen, L. F. *Measurement and Analysis of Neutron Elastic Scattering at 65 MeV*. Phys. Rev. C **50**, 275–281 (1994).
19. Islam, M. S. and Finlay, R. W. *Calculated Differential Elastic and Inelastic Neutron Scattering by ^{12}C and ^{16}O Between 30 and 60 MeV*. Nuclear Data and Atomic Data Tables **58**, 245–248 (1994).
20. Meigooni, A. S., Petler, J. S. and Finlay, R. W. *Scattering Cross-sections and Partial Kerma Factors for Neutron Interactions With Carbon at $20 < E_n < 65$ MeV*. Phys. Med. Biol. **29**, 643–659 (1984).
21. Olsson, N., Ramström, E. and Trostell, B. *Cross Sections and Partial Kerma Factors for Elastic and Inelastic Neutron Scattering from Nitrogen, Oxygen and Calcium at $E_n = 21.6$ MeV*. Phys. Med. Biol. **35**, 1255–1270 (1990).

22. Chadwick, M. B., Young, P. G., George, D.C. and Watanabe, Y. *Multiple Preequilibrium Emission in Feshbach-Kerman-Koonin Analyses.* Phys. Rev. C**50**, 996–1005 (1994).
23. Young, P. G., Arthur, E. D. and Chadwick, M. B. *Comprehensive Nuclear Model Calculations: Introduction to the Theory and Use of the GNASH Code.* Los Alamos National Laboratory Document LA-MS-12343 (1992). Chadwick, M. B. and Young, P. G. Phys. Rev. C**47**, 2255–2265 (1993).
24. White, R. M., Chadwick, M. B., Hartmann Siantar, C. L. and Westbrook, C. L. *Advances in Nuclear Data and All-Particle Transport for Radiation Oncology.* In: Proc. Int. Conf. on Nuclear Data for Science and Technology, Gatlinberg, Tennessee, 9–13 May 1994. Ed. J. K. Dickens (American Nuclear Society), p. 1023 (1994).
25. Feshbach, H., Kerman, A. and Koonin, S. *The Statistical Theory of Multi-Step Compound and Direct Reactions.* Ann. Phys. (N.Y.) **125**, 429–476 (1980).
26. Kalbach, C. *The Griffin Model, Complex Particles, and Direct Nuclear Reactions.* Z. Phys. A**283**, 401 (1977).
27. Kalend, A. M. *et al. Energy and Angular Distributions of Neutrons from 90 MeV Proton and 140 MeV Alpha-Particle Bombardment of Nuclei.* Phys. Rev. C**28**, 105–119 (1983).
28. Ignatyuk, A. V., Smirenkin, G. N. and Tishin, A. S. *Phenomenological Description of the Energy Dependence of the Level Density Parameter.* Sov. J. Nucl. Phys. **21**, 255–257 (1975).
29. Dimbylow, P. J. *Neutron Cross-section and Kerma Values for Carbon, Nitrogen, and Oxygen From 20 to 50 MeV.* Phys. Med. Biol. **25**(4), 637–649 (1980).
30. Antolković, B., Šlaus, I., Plenković, D., Macq, P. and Meulders, J. P. *Study of the Reaction $^{12}C(n,3\alpha)n$ from Threshold to $E_n = 35$ MeV.* Nucl. Phys. A**394**, 87–108 (1983).
31. Brady, F. P. and Romero, J. L. *Neutron Induced Reactions in Tissue Resident Elements.* Final Report to the National Cancer Institute, Grant No. 1R01 CA 16261 (University of California, Davis) (1979).
32. Romero, J. L., Brady, F. P. and Subramanian, T. S. *Neutron Induced Charged Particle Spectra and Kerma From 25 to 60 MeV.* In: Proc. Int. Conf. on Nuclear Data for Basic and Applied Sciences, Santa Fe, New Mexico, 13–17 May 1985. Ed. P. G. Young (Gordon and Breach) p. 687 (1986).
33. Howerton, R. J. *Calculated Neutron KERMA Factors Based on the LLNL ENDL Data File.* Lawrence Livermore National Laboratory Report UCRL-50400 **27**, revised (1986).
34. Resler, D. A. Lawrence Livermore National Laboratory, private communication to P.M. DeLuca, Jr. (September 1995).
35. Antolković, B., Šlaus, I. and Plenković, D. *Determination of the Kerma Factors for the Reaction $^{12}C(n,n'3\alpha)$ at $E_n = 10–35$ MeV.* Radiat. Res. **97**, 253 (1984).
36. Binns, P. J. *Microdosimetry for a Fast Neutron Therapy Beam.* Ph.D. Thesis, University of Cape Town, Republic of South Africa (1993).
37. Bühler, G., Menzel, H., Schuhmacher, H. and Guldbakke, S. *Dosimetric Studies with Non-hydrogenous Proportional Counters in Well Defined High Energy Neutron Fields.* In: Proc. Fifth Symp. on Neutron Dosimetry, Munich/Neuherberg, FRG, EUR-9762 (Luxembourg: Commission of the European Communities), pp. 309–329 (1985).
38. Goldberg, E., Slaughter, D. R. and Howell, R. H. *Experimental Determination of Kerma Factors at $E_n = 15$ MeV.* Lawrence Livermore Laboratory Report UCID-17789 (1978).
39. Haight, R. C., Grimes, S. M., Johnson, R. G. and Barschall, H. H. *The $^{12}C(n,\alpha)$ Reaction and the Kerma Factor of Carbon at $E_n = 14.1$ MeV.* Nucl. Sci. Eng. **87**, 41–47 (1984).
40. McDonald, J. C. *Calorimetric Measurements for the Carbon Kerma Factor for 14.6-MeV Neutrons.* Radiat. Res. **109**, 28–35 (1987).
41. Pihet, P., Guldbakke, S., Menzel, H. G. and Schuhmacher, H. *Measurement of Kerma Factors for Carbon and A-150 Plastic: Neutron Energies from 13.9 to 20.0 MeV.* Phys. Med. Biol. **37**, 1957–1976 (1992).
42. Schell, M. C., Pearson, D. W., DeLuca, Jr, P. M. and Haight, R. C. *Measurement of Dose Distributions of Linear Energy Transfer in Matter Irradiated by Fast Neutrons.* Med. Phys. **17**(1), 1–9 (1990).
43. Wuu, C. and Milavickas, L. *Determination of the Kerma Factors in Tissue-equivalent Plastic, C, Mg, and Fe for 14.7 MeV Neutrons.* Med. Phys. **14**(6), 1007–1014 (1987).
44. Schrewe, U. J., Newhauser, W. D., Brede, H. J., Dangendorf, V., DeLuca, Jr, P. M., Gerdung, S., Nolte, R., Schmelzbach, P., Schuhmacher, H. and Lim, T. *Measurement of Neutron Kerma Factors in C and O: Neutron Energy Range of 20 MeV to 70 MeV.* Radiat. Prot. Dosim. **61**, 275–280 (1995).
45. Hartmann, C. L., DeLuca Jr, P. M. and Pearson, D. W. *Measurement of Neutron Kerma Factors in C, O, and Si at 18, 23, and 25 MeV.* Radiat. Prot. Dosim. **44**, 25–30 (1992).
46. White, R. M., Broerse, J. J., DeLuca, Jr, P. M., Dietze, G., Haight, R. C., Kawashima, K., Menzel, H. G., Olsson, N. and Wambersie, A. *Status of Nuclear Data for Use in Neutron Therapy,* Radiat. Prot. Dosim. **44**, 11–20 (1992).
47. Smathers, J. B., Otte, V. A., Smith, A. R., Almond, P. R., Attix, F. H., Spokas, J. J., Quam, W. M. and Goodman, L. J. *Composition of A-150 Tissue-equivalent Plastic.* Med. Phys. **4**, 74–77 (1977).
48. International Commission on Radiation Units and Measurements. *Physical Aspects of Irradiation.* ICRU Report 10b (NBS Handbook 85) (Bethesda, Maryland: International Commission on Radiation Units and Measurements) (1964).
49. Alvera, A. V. and Schrewe, U. J. *Measurements With the PTB "C" Bonner Sphere Spectrometer in the PSI Villigen 55 MeV Neutron Field for Spectrometric and Calibration Purposes.* PTB Report (to be published).
50. Nolte, R., Schuhmacher, H., Brede, H. J. and Schrewe, U. J. *Measurements of High Energy Neutron Fluence with Scintillation Detectors and Proton Recoil Telescopes.* Radiat. Prot. Dosim. **44**, 101–104 (1992).
51. Schuhmacher, H., Siebert, B. R. L. and Brede, H. J. *Measurements of Neutron Fluence for Energies Between 20 MeV and 65 MeV Using a Proton Recoil Telescope.* In: Proc. NEANDC Specialists Meeting in Neutron Cross Section Standards for the Energy Region Above 20 MeV, Uppsala, NEANDC-305 (Paris: NEANDC) 123–134 (1991).
52. Prael, R. E. and Lichtenstein, H. *Users Guide to LCS: The LAHET Code System.* Los Alamos National Laboratory Technical Report LA-UR-89-3014 (1989).

KERMA MEASUREMENTS IN POLYENERGETIC NEUTRON FIELDS

W. D. Newhauser[†], U. J. Schrewe[†], H. J. Brede[†], M. Matzke[†] and P. M. DeLuca, Jr[‡]
[†]Physikalisch-Technische Bundesanstalt
Bundesallee 100
38116 Braunschweig, Germany
[‡]University of Wisconsin — Madison
1300 University Avenue, 1530 MSC
Madison, Wisconsin 53706, USA

Abstract — Absorbed dose was measured as a function of neutron energy with a small spherical proportional counter (PC) irradiated in the pulsed-beam broad energy spectrum neutron fields of the Physikalisch-Technische Bundesanstalt (PTB) that were produced with the p + Be reaction and extended to 20 MeV neutron energy. Time-of-flight (TOF) discrimination methods augmented the traditional microdosimetric pulse height (PH) analysis and yielded absorbed dose as a function of lineal energy y and neutron TOF. Below 0.7 keV.μm^{-1} lineal energy, unfolding procedures greatly improve the time resolution, e.g. from 225 ns full width at half maximum (FWHM) to 65 ns FWHM at 1.2 keV.μm^{-1}. The overall time resolution from unfolded TOF spectra is approximately 30 ns FWHM. The absorbed dose was normalised to neutron spectral fluence and, on the assumption that kerma is numerically equal to absorbed dose, yielded relative neutron fluence-to-kerma conversion coefficients as a function of energy that are in good agreement with values from previous work.

INTRODUCTION

The measurement of neutron fluence-to-kerma conversion coefficients, commonly called kerma factors, is sometimes complicated by the lack of monoenergetic neutron sources[1]. Many of the experimental integral kerma factor values are based on total kerma measurements made with proportional counters (PC) in pseudo monoenergetic neutron fields. Above about 20 MeV, where neutrons are produced from many reaction channels, corrections for kerma from off-energy neutrons are large and significantly increase the uncertainty in the desired nominal energy kerma factor. By measuring kerma as a function of neutron energy instead of just total kerma, the additional information makes possible a more accurate determination of kerma at the energy of interest.

In recent years several groups investigated time-resolved microdosimetry with PCs, i.e. the combined use of pulse height (PH) and time-of-flight (TOF) techniques. Randers-Pehrson et al[2] studied the time response of an A-150 plastic Rossi-type[3] PC (Model LET 1/2, Far West Technology, Goleta, California) and also that of a wall-less cylindrical PC at neutron energies of up to 25 MeV. They reported a 14 ns full width at half maximum (FWHM) time resolution for the cylindrical counter.

Schrewe et al[4–7] measured kerma as a function of neutron energy using Rossi-type PCs in various pulsed neutron fields with nominal energies of 11 keV to 60 MeV. In addition to conventional PH analysis, they applied TOF methods and improved the PC's time resolution by negatively biasing its helix in order to reduce the active detection volume from that of the entire cavity to that within the very much smaller cylindrical region bounded by the helix. They reported a 20 ns FWHM time resolution with the helix mode operation and an 80 ns FWHM for the normal mode. With a similar counter, also operated in helix mode, Binns et al achieved a 10 ns FWHM time resolution by taking into account amplitude-related timing defects[8,9].

Newhauser et al[10,11] measured kerma factors as a function of neutron energy using several cylindrical single-wire PCs and also Rossi-type PCs with combined PH and TOF methods. That work demonstrated the suitability of the PCs for TOF spectroscopy in the pulsed white neutron fields at the Los Alamos National Laboratory's Weapons Neutron Research Facility, which extend from less than 1 MeV to several hundred MeV. An empirical correction for the lineal energy dependent timing properties of the PCs (and associated timing electronics) resulted in a 60 ns FWHM time resolution.

In this work, we report on advances in the measurement of kerma as a function of neutron energy using a PC with TOF and unfolding methods.

MEASUREMENTS

Absorbed dose was measured with a custom built Rossi-type PC that is very similar to the design in Ref. 12. The counter was operated with 550 V anode bias and contained propane-based tissue-equivalent gas[13] at 9 kPa pressure. An internal alpha particle source provided the calibrations to lineal energy and absorbed dose. The cavity wall provides charged particle equilibrium (CPE) conditions up to 11 MeV neutron energy. On the assumption that kerma and absorbed dose are numerically equal under CPE conditions, the measured absorbed dose yields the corresponding kerma[14].

For the kerma measurements, 19.08 MeV cyclotron-produced protons bombarded a 5 mm thick Be target (Q = −1.85 MeV), generating neutron spectral fluences with nearly constant intensity from several MeV up to the maximum neutron energy, 17.2 MeV. The 1.71 μs pulse separation corresponds to a TOF frame overlap at 710 keV at 20 m flight path distance. The average proton beam current on the target was approximately 780 nA.

The time response information needed for unfolding of absorbed dose as a function of TOF was estimated from an additional measurement that was made close to the source, where the differences in neutron flight times are negligible. A Van de Graaff generator produced 1.248 MeV deuterons that bombarded a tritiated Ti target (Q = 17.6 MeV) and produced neutrons with 17.0 MeV nominal energy. The source-to-detector distance was 50 cm, the pulse time separation 1.6 μs, and the average deuteron beam current was approximately 1.8 μA on the target. Brede et al[15,16] and Klein et al[17] described the PTB pulsed neutron facility in detail. For this work, the 2 ns time duration of the proton and deuteron beam pulses is negligible.

Schrewe et al[5,7] previously described the data acquisition hardware and the signal processing electronics. A conventional leading-edge discriminator and a time-to-amplitude convertor provided the flight time measurement. PC spectra were acquired in three 512 by 512 matrices, each corresponding to a differing lineal energy range. These were combined into a single matrix with 256 TOF rows and 64 PH columns. The PH events were sorted into bins of increasing width, where the logarithms of the bin widths are equal, and the TOF events were binned linearly. These bin sizes afforded reasonable computation times and adequate counting statistics, both in the measurement to be unfolded and in the response functions.

UNFOLDING METHODS

Although the PC time response is a complex function of the bombarding particles' type and energy, it is assumed here that it depends solely on lineal energy. From the measurement made close to the neutron source, a response matrix was generated for each lineal energy interval. For the unfolding the iterative GRAVEL algorithm from Matzke[18] was chosen, which is based on the SAND-II alogrithm from Berg and McElroy[19]. For comparison, the matrix inversion method as implemented in the GAMMASUN code from Alt[20] was tested. The latter produced essentially identical solutions to those from the iterative method at intermediate lineal energies. Both methods produced oscillating solutions in the large and small lineal energy regions. However, by applying a small amount of Fourier filtration to the input TOF spectra, the iterative method produced solutions free from oscillation. The optimum amount of filtration, found empirically, was applied uniformly to all input spectra, except those from which the response matrices were produced, which were not smoothed.

RESULTS

The absorbed dose d(y,t) measured at 50 cm source-to-detector distance is shown in Figure 1 as a function of lineal energy y and time t, expressed in terms of the channel number. This measurement shows that the time response of the PC depends strongly on lineal energy, comprises several components, and does not lend itself to simple analytical models, e.g. a gaussian distribution. The time resolution is best at large lineal energies. Above 6 keV.μm^{-1}, the peak is narrow and slightly asymmetric. Below that lineal energy, the peak broadens and becomes more asymmetric with an increasingly large tail. Between 0.7 keV.μm^{-1} and 3.7 keV.μm^{-1}, the peak appears to comprise the superposition of two main components, the first with a short rise time and the latter with a short fall time. At 1.5 keV.μm^{-1} the components are approximately equal and lead to a nearly flat-topped peak of 300 ns duration. Below 0.7 keV.μm^{-1}, the time resolution worsens rapidly.

Figure 2 reveals the neutron and photon components in the kerma measurement at 20 m flight path distance. At small lineal energies the photon events appear as a broad and curved peak. After 100 unfolding iterations, the same measurement appears in Figure 3. The differences appear mainly below about 10 keV.μm^{-1}, where the PC's inherent time response is poor and the response matrices contain significant off-diagonal components. The photon peak in the unfolded spectrum provides a conservative estimate of the overall time resolution of the detector for neutrons. At 1.2 keV.μm^{-1}, the photon peak is 65 ns FWHM.

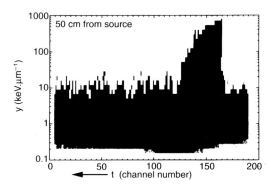

Figure 1. Measured microdosimetric spectrum in an A-150 plastic detector irradiated with a pulsed source of 17 MeV neutrons. The detector was located 50 cm from the neutron source. Grey level represents logarithm of yd(y,t), where the absorbed dose d is a function of lineal energy y and time t.

DISCUSSION

Examination of the response matrices at higher lineal energies suggests a 30 ns FWHM absorbed dose weighted average time resolution in the unfolded spectrum. This is considerably better than the 80 ns FWHM resolution prior to unfolding and to the 60 ns FWHM resolution obtained with the empirical correction in Ref. 10. The time resolution from this work is somewhat larger than was obtained with similar counters that were operated in the helix-mode, specifically the 20 ns FWHM value from Schrewe et al[7] and the 10 ns FWHM value from Binns et al[8]. However, the helix mode operation has several disadvantages compared to the normal mode: (1) the active volume of the gas is not defined by a well known boundary such as the detector wall, complicating the calibration to lineal energy and absorbed dose, (2) the chord-length distribution of particles crossing a cylinder (i.e. the region bounded by the helix) can make the spectra more difficult to interpret, and (3) the smaller collection volume reduces sensitivity, necessitating a larger neutron fluence.

Kerma factors k(E) from this work and those from Caswell et al[21] appear in Figure 4. Kerma values from the present work were relative and have been normalised to the corresponding 14 MeV value from Ref. 21 in order to facilitate comparison. The shape or energy dependence of the unfolded values is in reasonable agreement with values from Caswell et al. Uncertainties are estimated at 1 fGy.m^2 (2σ confidence interval) and are attributed to several sources, including the statistical fluctuations in the measurement.

CONCLUSION

The measurement of kerma as a function of energy can improve the accuracy of kerma factor determinations. Such methods are well suited for use in broad spectrum neutron fields and make possible the determination of kerma factors at many energies with a single irradiation. Time-resolved microdosimetry with PCs also provides response functions for unfolding neutron spectral fluence from a single-parameter (pulse height) PC spectrum. Taylor[22] recently discussed this application and presented a calculated response function below 20 MeV.

This work is the first result from a series of similar measurements. It is anticipated that further improvements in the analysis methods will reduce the uncertainties, and the authors expect of absolute exper-

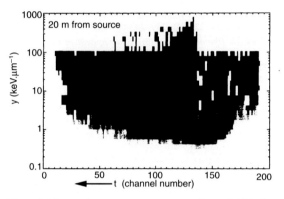

Figure 2. Measured two-parameter spectrum in an A-150 plastic detector located 20 m from a broad-spectrum pulsed neutron source. Grey level represents the logarithm of yd(y,t), where the absorbed dose d is a function of lineal energy y and time t. Photon events associated with the neutron source are observed below 10 keV.μm^{-1} near time channel 165.

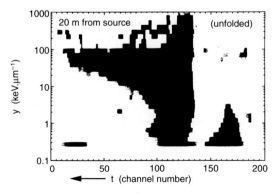

Figure 3. Iteratively unfolded two-parameter spectrum in an A-150 plastic detector located 20 m from a broad-spectrum pulsed neutron source. Grey level represents the logarithm of yd(y,t), where the absorbed dose d is a function of lineal energy y and time t. Photon events associated with the neutron source are observed below 10 keV.μm^{-1} near time channel 165.

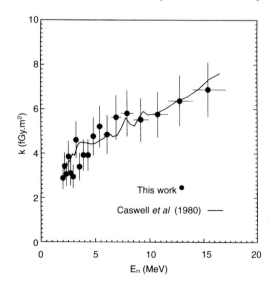

Figure 4. Neutron fluence-to-kerma coefficients k in A-150 plastic as a function of neutron energy E_n from a proportional counter measurement that was iteratively unfolded. Results are normalised to the value of k at 12 MeV from Caswell et al[21].

imental kerma factor values for A-150 plastic, C, Si, SiO$_2$, Al, AlN, and Al$_2$O$_3$ in the 5 MeV to 20 MeV energy range.

ACKNOWLEDGEMENTS

This work was partially funded by the European Commission through Grant No FI 3P-CT92-0045. The authors would like to thank R. Alt for assistance with the unfolding, F. Langner for technical support during the measurements, the many developers of the GNUPLOT plotting program, and P. Mikulik for the PM3D grey level plotting program.

REFERENCES

1. Brady, F. P. and Romero, J. L. *Summary of Monoenergetic Neutron Beam Sources for Energies > 14 MeV*. Nucl. Sci. Eng. **106**, 318–331 (1990).
2. Randers-Pehrson, G., Finlay, R. W., Dicello, J. F. and McDonald, J. C. *A Technique for Time-resolved Microdosimetric Spectroscopy*. In: Proc. 8th Symp. on Microdosimetry, Jülich, Germany, 27 Sept.– 1 Oct. 1982. Eds J. Booz and H. G. Eberts. EUR 8395, pp. 1169–1177 (Brussels: CEC) (1983).
3. Rossi, H. H. and Rosenzweig, W. *A Device for the Measurement of Dose as a Function of Specific Ionization*. Radiology **64**, 404–411 (1955).
4. Schrewe, H., Schuhmacher, U. J., Brede, H. J. and Dietze, G. *Determination of Photon and Neutron Dose Fractions with Tissue-equivalent Proportional Counters*. Nucl. Instrum. Methods. **31**(1/4), 143–147 (1990).
5. Schrewe, U. J., Brede, H. J. and Dietze, G. *Investigation of Tissue-equivalent Proportional Counters in Mixed Neutron-Photon Fields also Applying Time-of-flight Techniques*. Radiat. Prot. Dosim. **23**(1/4), 239–243 (1988).
6. Schrewe, U. J., Brede, H. J. and Dietze, G. *Dosimetry in Mixed Neutron-Photon Fields with Tissue-equivalent Proportional Counters*. Radiat. Prot. Dosim. **29**(1/2), 41–45 (1989).
7. Schrewe, U. J., Brede, H. J., Langner, F. and Schuhmacher, H. *The Use of Microdosimetric Detectors Combined with Time-of-flight Techniques*. Nucl. Instrum. Methods A**299**, 226–230 (1990).
8. Binns, P. J., Hough, J. H. and Simpson, B. R. S. *Time-resolved Microdosimetry in a Quasi-monoenergetic Neutron Beam*. Radiat. Prot. Dosim. **44**(1/4), 67–71 (1992).
9. Binns, P. J. and Hough, J. H. *Kerma Associated with High Energy Neutrons*. Radiat. Prot. Dosim. **52**(1/4), 105–109 (1994).
10. Newhauser, W. D. *Neutron Kerma Factor Measurements in the 25-MeV to 85-MeV Neutron Energy Range*. PhD Thesis, University of Wisconsin-Madison (Madison: Medical Physics Publishing) (1995).
11. Newhauser, W. D., DeLuca, Jr, P. M., Haight, R. C. and Lisowski, P. W. *Neutron Kerma Coefficient Measurements in Si and Fe in the 25-MeV to 85-MeV Neutron Energy Range*. (in preparation).
12. Oliver, Jr, D. O., Quam, W. M. and Wilde, W. O. *Emperical Dose Quality Distributions of Californium-252*. Health Phys. **22**, 341–349 (1972).
13. Srdoc, D. *Experimental Technique of Measurement of Microscopic Energy Depositions in Irradiated Matter using Rossi Counters*. Radiat. Res. **43**, 302–319 (1970).
14. Caswell, R. S. *Deposition of Energy by Neutrons in Spherical Cavities*. Radiat. Res. **27**, 92–107 (1966).
15. Brede, H. J., Cosack, M., Dietze, G., Gumpert, S., Guldbakke, S., Jahr, R., Schlegel-Bickmann, D. and Schölermann, H. *The Braunschweig Accelerator Facility for Fast Neutron Research, I: Building Design and Accelerators*. Nucl. Instrum. Methods **169**, 349–358 (1980).
16. Brede, H. J., Dietze, G., Kudo, K., Schrewe, U. J., Tancu, F. and Wen, C. *Neutron Yields from Thick Be Targets Bombarded with Deuterons or Protons*. Nucl. Instrum. Methods A**274**, 332–344 (1989).
17. Klein, H., Barrenscheen, G., Dietze, G. and Siebert, B. R. L. *The Braunschweig Accelerator Facility for Fast Neutron Research, II: Date Acquisition and Analysis*. Nucl. Instrum. Methods **169**, 359–367 (1980).
18. Matzke, M. *Unfolding Pulse Height Spectra: The HEPRO Program System*. PTB Report PTB-N-19 (Physikalisch-Technische Bundesanstalt, Braunschweig) (1994).
19. Berg, S. and McElroy, W. N. *A Computer Automated Iterative Method for Neutron Flux Determination by Foil Activation — SAND-II*. Technical Report (Air Force Weapons Laboratory, Research and Technology Division) (1967).
20. Alt, R. *Winkelauflösendes in situ γ-Spectrometer*. Diplomarbeit, Technischen Universität Carolo-Wilhelmina zu Braunschweig (1995).
21. Caswell, R. L. and Coyne, J. J. *Kerma Factors for Neutron Energies below 30 MeV*. Radiat. Res. **83**, 217–254 (1980).
22. Taylor, G. C. *An Analytical Correction for the TEPC Dose Equivalent Response Problem*. Radiat. Res. **61**, 67–70 (1995).

COMPARISON OF VARIOUS DOSE QUANTITIES IN TISSUE AND TISSUE SUBSTITUTES AT NEUTRON ENERGIES BETWEEN 20 MeV AND 100 MeV

U. J. Schrewe†, W. G. Alberts†, W. D. Newhauser†, H. J. Brede† and P. M. DeLuca, Jr‡
†Physikalisch-Technische Bundesanstalt
Bundesallee 100, D-38116 Braunschweig, Germany
‡ University of Wisconsin
1300 University Avenue, Madison, WI 53706, USA

Abstract — Using kerma factor data for H, C, N, and O in the neutron energy range from 20 MeV to 100 MeV provided by recent calculations and evaluations, the absorbed dose conversion factor $\alpha_{t,p}$ from A-150 tissue substitute plastic (p) to ICRU standard tissue (t) was calculated. The results show a constant value of $\alpha_{t,p} = 0.92$ with uncertainties of about 5% from 20 MeV to 50 MeV and of 7% above 50 MeV. Dose distributions from measurements with low pressure proportional counters are used to determine the mean quality factors in C and O. The dose distributions and the quality factors exhibit strong similarities for C and O. Combined with mean quality factors of A-150 plastic and with evaluated kerma factors, the ratio $\beta_{t,p}$ of dose equivalent in ICRU tissue to that in A-150 plastic was calculated. The result indicates that, due to the excess of C in A-150 plastic, the TEPC may overestimate the ambient dose equivalent by about 15% for neutron energies above 20 MeV. These findings confirm the suitability of the TEPC as dose equivalent meter for radiation protection and also suggest that for neutron therapy dosimetry the excess of C in A-150 plastic should be taken into account by using $\alpha_{t,p} = 0.92$ at neutron energies above 20 MeV.

INTRODUCTION

Dosimetric quantities for neutron therapy and radiation protection are defined for body tissues or tissue substitutes such as ICRU standard tissue[1]. Tissues and tissue substitutes consist mainly of O, C and H, and also contain smaller fractions of other elements. Since the neutron absorbed dose, D, and the dose equivalent, H, depend on the composition, V, the uncertainties can be reduced by measuring D and H in materials with tissue-like compositions. Thus, a conducting tissue substitute plastic (A-150) is common for dosemeters used in neutron dosimetry. A-150 plastic has the correct H content but the C and O mass fractions differ strongly from those in tissue.

In neutron therapy, conversion factors are applied to convert D measured in A-150 plastic to the corresponding value in body tissue. The conversion factors are usually calculated from the fluence-to-kerma conversion coefficients, briefly denoted as kerma factor k_i, combined with the relative mass fractions of each constituent element i. The biological effectiveness of neutrons in therapy, however, also depends on the composition of the secondary charged particle spectra in tissue. In order to estimate the radiation quality on the basis of microdosimetric distributions, low pressure proportional counters made of A-150 plastic (TEPC) are used. The interpretation may be influenced by the fact that the mass composition of A-150 plastic is different from that of body tissues.

For radiation protection dosimetry the ICRU has proposed operational quantities including the ambient dose equivalent, H*(d), defined as $H = \overline{Q} D$ at a depth d inside a 30 cm diameter sphere of ICRU standard tissue[1], where $\overline{Q}$ denotes the mean quality factor. Q is defined as a function of the linear energy transfer, L[2,3]. The TEPC is well suited for use as an ambient dose equivalent meter since it provides an estimate of D(L) distributions. However, there are various limitations, including the different mass composition of A-150 plastic and the ICRU sphere which are discussed in the present paper.

Kerma factors can be calculated from cross section data or alternatively, they may be measured with PCs since under charged particle equilibrium conditions the kerma, K, is numerically equal to D. Kerma factors of H are known with high accuracy up to 1 GeV. For C, N, and O, however, complete experimental cross section data in the energy range up to 100 MeV are not available. Chadwick et al[4,5] recently reported on calculations and evaluations of cross sections for ^{12}C, ^{14}N and ^{16}O, the most abundant isoptopes of the respective elements including kerma factors which agree well with measured data[6,7] and with kerma factor calculations of Savitskaya and Sannikov[8].

ABSORBED DOSE CONVERSION FACTORS

The absorbed dose in tissue, D_t, can be estimated from the absorbed dose in A-150 plastic, D_p, with $D_t = \alpha_{t,p} D_p$, where $\alpha_{t,p}$ denotes the A-150 plastic to tissue absorbed dose conversion factor. In monoenergetic neutron fields, and under charged particle equilibrium conditions, the conversion factor can be calculated from the kerma factors of the individual elements:

$$\alpha_{t,p} = \left(\sum_i t_i\, k_i\right) \Big/ \left(\sum_i p_i\, k_i\right) \qquad (1)$$

where t_i and p_i denote the elemental mass fractions in tissue and A-150 plastic, respectively. If the uncertainties in the mass fractions are negligible compared to the uncertainties in the kerma factors, and if the uncertainties are not correlated, the uncertainty in $\alpha_{t,p}$ is given by:

$$\Delta\alpha_{t,p} = \left[\sum_i \left(\frac{t_i k_i}{k_t} - \frac{p_i k_i}{k_p}\right)^2 \left(\frac{\Delta k_i}{k_i}\right)^2\right]^{\frac{1}{2}} \alpha_{t,p} \qquad (2)$$

In neutron therapy, values of $\alpha_{t,p}$ for each tissue irradiated are required. Kerma factor tables which allow the derivation of $\alpha_{t,p}$ for energies below 30 MeV for various tissues were published by the ICRU[9]. Hence we restrict the discussion here to the general trend of $\alpha_{t,p}$ at higher energies and present, as an example, $\alpha_{t,p}$ for ICRU standard tissue.

For A-150 plastic the uncertainties of the C and O kerma factors predominate over the uncertainty of $\alpha_{t,p}$. A-150 plastic also contains small mass fractions of F and Ca. Since kerma factor values in F were not available, values for P were used instead and large uncertainties were adopted. The mass fractions of the elements used in the calculation and the adopted kerma factor uncertainties are given in Table 1. The calculated values of $\alpha_{t,p}$ are shown in Figure 1 along with $\alpha_{t,p}$ values that are based on the experimental kerma factors in C and O[6] combined with evaluated kerma factors for H, N, F and Ca.

DOSE EQUIVALENT IN STANDARD TISSUE AND A-150 PLASTIC

Ambient dose equivalent H*(d) is defined as $H = \overline{Q}D$ at the reference point inside the ICRU sphere with $\overline{Q}$ given by [1]:

$$\overline{Q} = \frac{1}{D}\int_0^\infty D_L Q(L) dL \qquad (3)$$

where D_L denotes the dose distribution in L and $Q(L)$ the quality factor[2,3]. The TEPC can be used to estimate D_L. While the different compositions of the TEPC wall (A-150 plastic) and standard tissue influence D and may be converted with $\alpha_{t,p}$ their influence on D_L requires further consideration.

H in ICRU standard tissue (t) and in A-150 plastic (p) were calculated and the ratio of these values defined as $\beta_{t,p}$, the dose equivalent factor. It should be emphasised that H*(d) is defined in standard tissue and that the quantity H calculated here for A-150 plastic is only useful for estimating the limitations and uncertainties of ambient dose equivalent meters.

For neutron energies below 20 MeV $\beta_{t,p}$ was calculated by using:

$$\beta_{t,p} = \left(\sum_i t_i k_i \overline{Q}_i\right) \bigg/ \left(\sum_i p_i k_i \overline{Q}_i\right) \qquad (4)$$

where values of $k_i \overline{Q}_i$ of H, C, N, and O were taken from Siebert and Schuhmacher[10]. As comparable data

Table 1. Mass fractions and kerma factor uncertainties used in the calculation of $\alpha_{t,p}$.

Element	t_i	p_i	$t_i - p_i$	Energy range (MeV)	$\Delta k_i/k_i$ (%)	Ref.
H	0.101	0.101	0.000	0 – 19	1	9
				20 – 69	2	11
				70 – 100	4	8
C	0.111	0.776	−0.665	0 – 19	5	12
				20 – 49	8	5
				50 – 100	10	5
N	0.026	0.035	−0.009	0 – 19	10	4
				20 – 49	16	4
				50 – 100	20	4
O	0.762	0.052	+0.710	0 – 19	5	12
				20 – 49	8	4
				50 – 100	10	4
F	0.000	0.017	−0.017	0 – 29	20	9
				30 – 49	40	13
				50 – 59	60	13
				60 – 100	80	13
Ca	0.000	0.018	−0.018	0 – 19	10	9
				20 – 29	20	8
				30 – 49	30	8
				50 – 100	40	8

The symbol t_i denotes the mass fraction of element i in ICRU standard tissue, p_i the mass fraction of element i in A-150 plastic, and $\Delta k_i/k_i$ the relative uncertainty of the kerma factors used to estimate $\Delta\alpha_{t,p}$ in Equation 2 (see text).

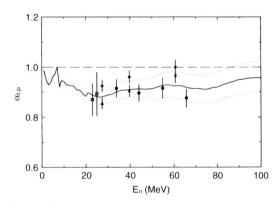

Figure 1. Absorbed dose conversion factor $\alpha_{t,p}$ for A-150 plastic to ICRU standard tissue. The solid line is based on evaluated kerma factor data. The dotted lines correspond to a 68% confidence level. The data points are derived from experimental kerma factors of C and O that have been published for $E_n >$ 20 MeV. The symbols refer to experimentals works cited elsewhere[5]: (□) Hartmann et al, (◇) Brady and Romero, (x) Romero et al, (●) Schrewe et al.

for F and Ca are not available in the literature and their contributions to $\beta_{t,p}$ are small, these elements were not considered.

Alternatively, Equation 4 can be expressed as:

$$\beta_{t,p} = 1 + \sum_i (t_i - p_i) \frac{k_i Q_i}{k_p Q_p} = 1 + \sum_i \zeta_i \quad (5)$$

This formula has advantages since the difference in the mass fractions of H, $(t_H - p_H)$, is negligible, as are the corresponding values for N, F and Ca. Thus, $\beta_{t,p}$ can be derived from evaluated kerma factors and the experimental quality factors of C, O, and A-150 plastic.

The absorbed dose probability density functions $d(y)$ in C and O in the quasi-monoenergetic neutron fields of energies of 34 MeV and 55 MeV were measured[6]. From these measurements the mean quality factors were calculated using Equation 3 on the assumption that $d(y)$ is equal to D_L, and with the quality factors definitions from ICRP 21[2] or from ICRP 60[3]. The results of $\overline{Q}_C$ and $\overline{Q}_O$, $\overline{Q}_C/\overline{Q}_O$ and $\overline{Q}_P$ are given in Table 2.

$\overline{Q}_C$ and $\overline{Q}_O$ decrease monotonically with increasing neutron energy while $\overline{Q}_p$ remains almost constant. In order to make possible an extrapolation of $\beta_{t,p}$ to lower or higher neutron energies, the relation $\overline{Q}_i = m_i E_n^{-1} + c_i$ was adopted for C and O and $\overline{Q}_p = m_p E_n + c_p$ for A-150 plastic. The parameters m_i, m_p, c_i and c_p were determined from a least squares fit to experimental data. The uncertainty $\Delta\beta_{t,p}$ was approximated by:

$$\Delta\beta_{t,p} = \left\{ (\beta_{t,p} - 1)^2 \left[\left(\frac{\Delta k_p}{k_p}\right)^2 + \left(\frac{\Delta \overline{Q}_p}{\overline{Q}_p}\right)^2 \right] \right. $$
$$\left. + \sum_i (\zeta_i)^2 \left[\left(\frac{\Delta k_i}{k_i}\right)^2 + \left(\frac{\Delta \overline{Q}_i}{\overline{Q}_i}\right)^2 \right] \right\}^{\frac{1}{2}} \quad (6)$$

DISCUSSION

The absorbed dose conversion factor $\alpha_{t,p}$ shown in Figure 1 is nearly constant for neutron energies greater than 20 MeV. In the energy range between 20 MeV and 100 MeV, the average value of $\overline{\alpha}_{t,p}$ is 0.92 ± 0.02 (68% confidence level). When including the uncertainties of the elemental kerma factor data, a constant value of $\overline{\alpha}_{t,p} = 0.92$ with a total uncertainty of 5% for 20 MeV $< E_n <$ 50 MeV and 7% above 50 MeV was found in fair agreement with the recommendation of the European neutron therapy protocol, $\alpha_{t,p} = 0.95$.

The influence of the excess of C in A-150 plastic on the determination of the mean radiation quality of

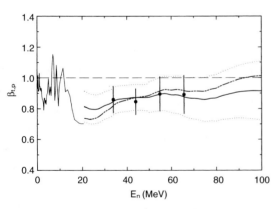

Figure 2. Dose equivalent correction factor $\beta_{t,p}$ corrects for differences in compositions of ICRU standard tissue and A-150 plastic. The thin line for neutron energies below 20 MeV was calculated from data of Siebert and Schuhmacher[10]. The data points (●) were derived from experimental quality factors in C and O[6] using the Q(L) relation from ICRP 60[3] combined with evaluated kerma factors of C and O (see text). The thick line, for 20 MeV $< E_n <$ 100 MeV, was calculated from interpolated mean quality factors derived using the Q(L) relation from ICRP 60[3] (the thin dotted line represents its uncertainties) while the broken thick line was obtained using the same procedure but the Q(L) relation from ICRP 21[2].

Table 2. Mean quality factors $\overline{Q}_C$ in C and $\overline{Q}_O$ in O derived from the Q(L) relations from ICRP 21[2] and from ICRP 60[3] and measured dose distributions.

Factor	Reference (ICRP no.)	E_n (MeV)			
		33.9	44.0	54.9	65.7
k_c		41.4 (3.3)	45.6 (3.7)	49.2 (4.9)	56.2 (5.6)
$\overline{Q}_c$	21	14.30	12.50	11.20	10.10
$\overline{Q}_c$	60	14.00	12.80	11.90	10.40
k_o		29.6 (4.4)	35.0 (5.3)	40.1 (6.0)	45.6 (6.8)
$\overline{Q}_c$	21	13.80	12.00	11.40	10.20
$\overline{Q}_c$	60	14.00	11.70	11.40	10.00
k_p		80.1 (2.8)	82.0 (3.0)	83.5 (4.0)	88.3 (4.6)
$\overline{Q}_p$	21	7.3	7.0	6.6	6.3
$\overline{Q}_p$	60	8.1	7.7	7.2	6.8
$\overline{Q}_C/\overline{Q}_O$	21	1.04	1.04	0.98	0.99
$\overline{Q}_C/\overline{Q}_O$	60	1.00	1.09	1.04	1.04
$\beta_{t,p}$	21	0.82 (0.08)	0.86 (0.08)	0.92 (0.10)	0.92 (0.10)
$\beta_{t,p}$	60	0.86 (0.09)	0.85 (0.08)	0.90 (0.11)	0.89 (0.12)

The mean quality factors in A-150 plastic, $\overline{Q}_p$, are linearly interpolated from measured data. The kerma factors are taken from evaluations or calculations as described in Table 1. The dose equivalent correction factors $\beta_{t,p}$ were calculated using Equation 5. k_c, k_o and k_p are given in pGy.cm^2.

therapy beams on the basis of microdosimetric distributions measured with TEPC can be estimated from the values of $\overline{Q}_C/\overline{Q}_O$, which are given in Table 2. The average value is 1.01 for the Q(L) relation from ICRP 21 and 1.04 for the Q(L) relation from ICRP 60, which implies that the relative dose fractions from different secondary charged particle types in C and O are nearly identical. It can therefore be concluded that the influence on the radiation quality determination in terms of RBE must be negligibly small.

The extrapolated values of the dose equivalent ratio $\beta_{t,p}$ show a minimum of 0.7 near 20 MeV while otherwise its value is about 0.85. Other limitations of the TEPC as an ambient dose equivalent meter appear to be of much greater significance than the differences in A-150 plastic and ICRU standard tissue. For example, neutron interactions with the ICRU sphere cannot be approximated by the TEPC wall of only a few mm thickness. Siebert and Schuhmacher[10] showed that the phantom's influence decreases with increasing neutron energy in the 10 MeV to 20 MeV region while relative dose contributions of both C and O increase. This is the reason for the minimum $\beta_{t,p}$ value around 20 MeV.

At greater neutron energies, however, the neutron interactions with C and O become increasingly similar and the trend in $\beta_{t,p}$ towards 100 MeV shows that the influence on $\beta_{t,p}$ caused by the different composition of the TEPC wall and the ICRU sphere becomes negligibly small. Hence, the use of A-150 plastic is a material well suited for constructing ambient dose equivalent meters for use in the high energy range, compared with the other limitations discussed by Siebert and Schuhmacher[10] and Schuhmacher et al[14].

REFERENCES

1. ICRU. *Determination of Dose Equivalents Resulting from External Radiation Sources*. Report 39 (Bethesda, MD, USA: International Commission on Radiation Units and Measurements) (1985).
2. ICRP. *Data for Protection against Ionizing Radiation from External Sources: Supplement to ICRP Publication 15*. Publication 21 (Oxford: Pergamon Press) (1973).
3. ICRP. *1990 Recommendations of the International Commission on Radiological Protection*. Publication 60 (Oxford: Pergamon Press) (1991).
4. Chadwick, M. B. and Young, P. G. *Calculation and Evaluation of Cross Sections and Kerma Factors for Neutrons up to 100 MeV on ^{16}O and ^{14}N*. Nucl. Sci. Eng. (submitted).
5. Chadwick, M. B., Cox, L. J., Young, P. G. and Meigooni, A. S. *Calculation and Evaluation of Cross Sections and Kerma Factors for Neutrons up to 100 MeV on Carbon*. Nucl. Sci. Eng. (submitted).
6. Schrewe, U. J., Newhauser, W. D., Brede, H. J., Dangendorf, V., DeLuca Jr, P. M., Gerdung, S., Nolte, R., Schmelzbach, P., Schuhmacher, H. and Lim, T. *Measurement of Neutron Kerma Factors in C and O: Neutron Energy Range of 20 MeV to 70 MeV*. Radiat. Prot. Dosim. **61**, 275–280 (1995).
7. Slypen, I., Corcalciuc, V. and Meulders, J. P. *Kerma Values Deduced from Neutron-induced Charged-particle Spectra of Carbon from 40 to 75 MeV*. Phys. Med. Biol. **40**, 73–82 (1995).
8. Savitskaya, E. N. and Sannikov, A. V. *High Energy Neutron and Proton Kerma Factors for Different Elements*. Radiat. Prot. Dosim. **60**, 135–146 (1995).
9. ICRU. *Photon, Electron, Proton and Neutron Interaction Data for Body Tissues*. Report 46 (Bethesda, MD. USA: International Commission on Radiation Units and Measurements) (1977).
10. Siebert, B. R. L. and Schuhmacher, H. *Quality Factors, Ambient and Personal Dose Equivalent for Neutron, Based on the New ICRU Stopping Power Data for Protons and Alpha Particles*. Radiat. Prot. Dosim. **58**, 177–183 (1995).
11. White, R. M., Broerse, J. J., DeLuca Jr, P. M., Dietze, G., Haight, R. C., Kawashima, K., Menzel, H. G., Olsson, N. and Wambersie, A. *A Status of Nuclear Data in Neutron Therapy*. Radiat. Prot. Dosim. **44**, 11–20 (1992).
12. Howerton, R. J. *Calculated Neutron KERMA Factors Based on the LLNL ENDL Data File*. UCRL-50400 27: revised (University of California, Lawrence Livermore National Laboratory, CA, USA) (1986).
13. Dimbylow, P. J. *Neutron Cross-section and Kerma Value Calculation for C, N, O, Mg, Al, P, S, Ar and Ca from 20 to 50 MeV*. Phys. Med. Biol. **27**, 989–1001 (1982).
14. Schuhmacher, H., Alberts, W. G., Alevra, A. V., Klein, H., Schrewe, U. J. and Siebert, B. R. L. *Characterisation of Photon-Neutron Radiation Fields for Radiation Protection Monitoring and Optimisation*. Radiat. Prot. Dosim. **61**, 81–88 (1995).

DOUBLE DIFFERENTIAL CROSS SECTION MEASUREMENTS OF THE (n,px), (n,dx), (n,tx) AND (n,αx) REACTIONS ON ^{12}C: KERMA DETERMINATION AT ENERGIES BETWEEN 30 AND 75 MeV

I. Slypen†, S. Benck†, V. Corcalciuc‡ and J.-P. Meulders†
†Institut de Physique Nucléaire, Université Catholique de Louvain
Chemin du Cyclotron 2, B-1348 Louvain-la-Neuve, Belgium
‡Institute of Atomic Physics, PO Box MG6, Heavy Ion Department
Bucharest, Romania

Abstract — The double-differential cross sections for the (n,px), (n,dx), (n,tx) and (n,αx) reactions on carbon have been measured at 42.5, 62.7 and 72.8 MeV incident neutron energies at laboratory angles from 20° to 160° in 10° steps. The energy differential cross sections are deduced and consequently the partial and total kerma values. Experimental results are compared with existing data and theoretical calculations. Preliminary results on carbon are also shown concerning the light charged particle spectra induced by neutrons from the low energy tail of the incident neutron spectrum.

INTRODUCTION

Due to evident experimental difficulties, double differential cross section measurements of the (n,px), (n,dx), (n,tx) and (n,αx) reactions are rather scarce in the neutron energy range between 30 and 80 MeV[1]. Nevertheless these data are valuable for studies of nuclear interaction mechanisms and also provide essential information for various applications[2,3]. In particular, kerma factors are used to relate the dose measured

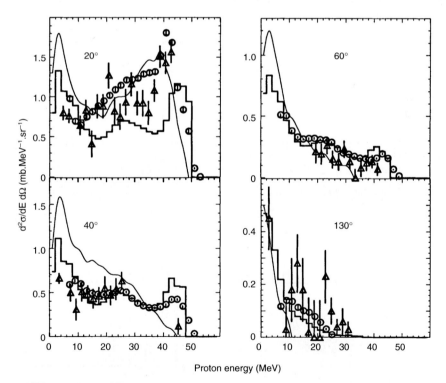

Figure 1. Measured ^{12}C(n,px) double differential cross sections at several laboratory angles for 62.7 MeV incident neutron energy (open dots with the respective statistical errors). Experimental results of UC-Davis[17,18] at 60 MeV incident neutron energy are shown as open triangles. Calculations for 60 MeV neutron energy from Brenner and Prael[15] are shown as continuous lines and from Chadwick[13] as histograms.

with dosemeters made of tissue substitute to the dose delivered to human tissue. As the main composition of human tissue differs from that of the tissue substitute mainly in the carbon and oxygen percentages, accurate kerma values for carbon and oxygen are essential for the determination of the neutron absorbed dose. Moreover, in the case of carbon, the predicted kerma factors vary considerably with the model used[4].

The double differential cross sections for the ^{12}C(n,px), ^{12}C(n,dx), ^{12}C(n,tx) and ^{12}C(n,αx) reactions at three incident neutron energies, 42.5, 62.7 and 72.8 MeV are here reported. The energy differential cross sections and consequently the partial and total kerma factors are deduced. In the next section, the experimental set-up is briefly described. Experimental results are then shown. Some preliminary results of work in progress concerning fast neutron induced light charged particle production on ^{12}C at other than the above mentioned incident neutron energies are also covered.

EXPERIMENTAL SET-UP

The fast neutron beam facility of Louvain-la-Neuve cyclotron, CYCLONE, has been previously described[5-7]. The ^{7}Li(p,n)^{7}Be$_{gs}$ (Q = −1.644 MeV) reaction produces, at 0° laboratory angle, a quasi-monoenergetic neutron beam consisting of a well-defined peak (full width at half maximum 2 MeV, containing about 50% of the produced neutrons) followed by a flat continuum of lower energy neutrons. The present measurements were performed at three different incident proton energies, 45, 65 and 75 MeV, producing on a 3 mm thick lithium target neutron spectra with the main peak centred at 42.5, 62.7 and 72.8 MeV respectively[7,8].

Four ΔE-E telescopes, each consisting of a 0.1 mm thick NE102 scintillator and a thick CsI(Tl) crystal, are used to detect the charged particles produced by a 1 mm thick carbon target. Angular distributions from 20° to 160° in 10° steps are measured. To select the charged particles produced by the neutrons from the monoenergetic peak, the time of flight information and the energy calibration of the detectors were used. Absolute cross sections were determined relative to the measured H(n,p) differential cross sections.

More details about the experimental set-up are given elsewhere[7-10].

DATA REDUCTION AND EXPERIMENTAL RESULTS

Double differential cross sections for the ^{12}C(n,px), ^{12}C(n,dx), ^{12}C(n,tx) and ^{12}C(n,αx) reactions and their angular distributions from 20° to 160° laboratory angles (15 angles at 42.5 and 62.7 MeV, and 10 at 72.8 MeV incident neutron energy) were obtained at the mentioned three incident neutron energies[11,12]. Low energy cutoffs in the energy spectra are 5, 9, 10 and 15 MeV for the protons, deuterons, tritons and alphas respectively.

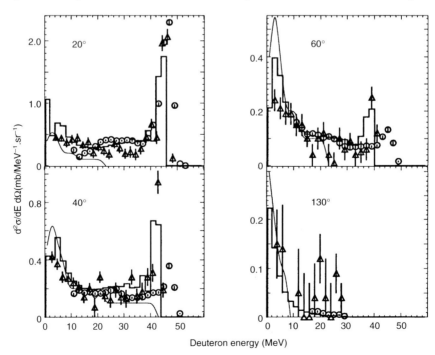

Figure 2. Same as in Figure 1 for deuterons.

They are mainly due to the ΔE detector thickness and the energy threshold of the E detector.

Figures 1 and 2 show some of the measured energy spectra for protons and deuterons compared with existing experimental data and theoretical calculations. The calculations done by Chadwick[13] are performed with Feshbach-Kerman-Koonin and GNASH codes[14] which include equilibrium, pre-equilibrium and direct reaction mechanisms. No supplementary normalisation was necessary between experimental data and theoretical calculations. The overall agreement is good. Generally the theory of Brenner and Prael[15] predicts less high energy charged particles than are experimentally observed[1,11,16,19]. The agreement between calculations

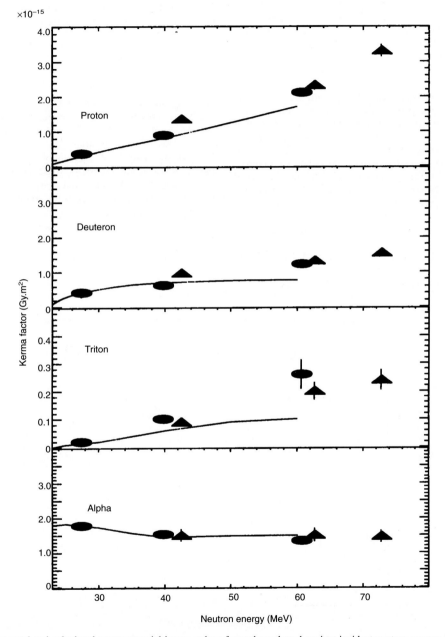

Figure 3. Measured and calculated neutron partial kerma values for carbon plotted against incident neutron energy. The triangles and dots represent respectively the present work and UC-Davis data[20]. Theoretical calculations from Brenner and Prael[15] are shown as continuous lines.

from Chadwick and our data is generally good except for protons at forward angles. The angle integration of the experimental angular distributions provides the energy differential cross sections. For completeness, the measured angular range was extended to very forward (2.5° and 10°) and backward angles (170° and 175°), with the theoretical cross sections of Brenner and Prael[15]. The theoretical calculations of 60 MeV neutron energy[15] have been extrapolated in order to complete the 72.8 MeV data at the above mentioned forward and backward angles. For the five missing angles at 72.8 MeV (80°, 90°, 100°, 130° and 150°), extrapolations were used for our experimental cross sections measured at 62.7 MeV.

Based on the obtained energy differential cross sections[7,11,12,16], partial and total kerma values were calculated. In order to do this, the energy differential cross sections were extrapolated below the detection energy threshold with theoretical cross sections[15]. The results of the present work are presented in Figure 3 together with existing experimental data and theoretical calculations. Note that for UC-Davis partial kerma factors, no extrapolation below the energy threshold of the detection system has been performed[20], therefore the comparison should be regarded with caution. The continuous lines represent the kerma values calculated from the energy differential cross sections of Brenner and Prael[15].

The total kerma values are shown in Figure 4. The UC-Davis data have been revised by the authors[21] in the sense that the cross sections are extrapolated below the energy threshold of the detection system and they also include an estimated non-elastic contribution. Unfortunately, the revised partial kerma factors have not been published. The total kerma values measured with low pressure proportional counters[22] are also shown. The theoretical total kerma values from Brenner[23] include the elastic-recoil but not the inelastic-recoil contributions. Nevertheless, these contributions are included in our experimental values and represent less than 9% at 42.5, 5% at 62.7 MeV and 4% at 72.8 MeV of the total kerma factors. This partly explains the observed differences between the experiment and the theory. In our case the contribution of the extrapolation in the low energy part with theoretical cross sections amounts to about 5, 8 and 75% of the total cross sections for, respectively, protons, deuterons and alpha particles. In the latter case, this important contribution is mainly due to the $^{12}C(n,3\alpha)$ reaction channel. Recommended total kerma values and their minimum uncertainties[24] are also shown for comparison.

Generally, our kerma values agree with previous data based on microscopic cross sections but are clearly higher than those obtained by integral measurements of Ref. 22.

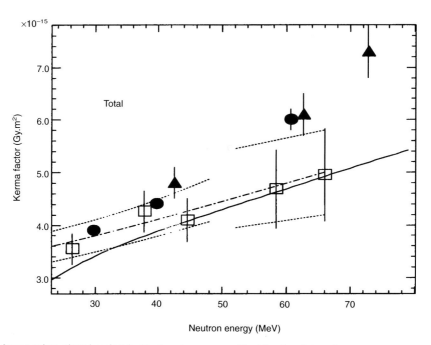

Figure 4. Total kerma values plotted against incident neutron energy. The triangles, dots and open squares represent the experimental data of the present work, UC-Davis[21] and Schrewe et al[22] respectively. The continuous line shows Brenner predictions[23]. The dashed and dotted lines present the recommended kerma values and their minimum uncertainties[24].

WORK IN PROGRESS

The event by event data taking, the good time resolution (about 0.8 ns) and the long acquisition time are arguments to extend our analysis to events induced by neutrons from the low energy tail of the incident neutron spectra. Double differential cross sections can be extracted at several incident neutron energies between 25 and 70 MeV. The set of cross sections obtained will be homogeneous in what concerns data reduction procedures and their subsequent treatment.

As an example, Figure 5 shows the energy spectra for proton and deuteron production induced by 50 ± 2.5 MeV incident neutrons on carbon. The calculations of Ref. 15 agree quite well with our data at the forward angles but as already said, they predict less than observed high energy particles at backward angles. On the other hand, Chadwick's predictions are generally in good agreement with the experiment except mainly for the protons at the forward angles.

ACKNOWLEDGEMENTS

The authors acknowledge many discussions with Dr M. Chadwick and the permission to use his theoretical calculations. This work was supported by the Institut des Sciences Nucléaires, Belgium and partly by the Commission of the European Communities, contract FI3P-93-0084-BE.

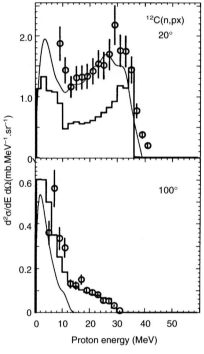
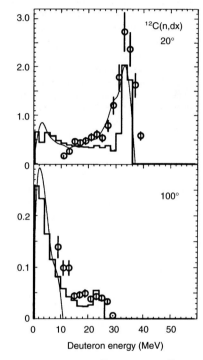

Figure 5. Measured double differential cross sections at two laboratory angles for ^{12}C(n,px) and ^{12}C(n,dx) reactions at 50 ± 2.5 MeV incident neutron energy. The statistical errors are 15%. Calculations at 50 MeV neutron energy from Chadwick[13] and from Brenner and Prael[15] are presented as respectively histograms and continuous lines.

REFERENCES

1. Subramanian, T. S., Romero, J. L. and Brady, F. P. *Experimental Facility for Measurement of Charged Particle Continua arising from Neutron Induced Reactions.* Nucl. Instrum. Methods **174**, 475–489 (1980).
2. Wells, A. H. *A Consistent Set of Kerma Values for H, C, N and O for Neutrons of Energies from 10 to 80 MeV.* Radiat. Res. **80**, 1–9 (1979).
3. Meigooni, A. S., Petler, J. S. and Finlay, R. W. *Scattering Cross-sections and Partial Kerma Factors for Neutron Interactions with Carbon at $20 < E_n < 65$ MeV.* Phys. Med. Biol. **29**, 643–659 (1984).
4. Schuhmacher, H., Brede, H. J., Henneck, R., Kunz, A., Meulders, J.-P., Pihet, P. and Schrewe, U. J. *Measurement of Neutron Kerma Factors for Carbon and A-150 Plastic at Neutron Energies of 26.3 and 37.8 MeV.* Phys. Med. Biol. **37**(6), 1265–1281 (1992).
5. Bol, A., Leleux, P., Macq, P. and Ninane A. *A Novel Design for a Fast Intense Neutron Beam.* Nucl. Instrum. Methods **214**, 169–173 (1983).

6. Dupont, C., Leleux, P., Lipnik, P., Macq, P. and Ninane, A. *Study of Collimated Fast-neutron Beam.* Nucl. Instrum. Methods A **256**, 197–206 (1983).
7. Slypen, I. *Mesures Expérimentales de Sections Efficaces Doublement Différentielles de Particules Chargées Légères Induites par des Neutrons Rapides sur Carbone aux Énergies 42.5, 62.7 et 72.8 MeV.* PhD Thesis, Université Catholique de Louvain (1995).
8. Slypen, I., Corcalciuc, V., Ninane, A. and Meulders, J.-P. *Charged Particles Produced in Fast Neutron Induced Reactions on ^{12}C in the 45-80 MeV Energy Range.* Nucl. Instrum. Methods A **337**, 431–440 (1994).
9. Slypen, I., Corcalciuc, V. and Meulders, J.-P. *Geometry and Energy Loss Features of Charged Particle Production in Fast-neutron Induced Reaction.* Nucl. Instrum Methods B**88**, 275–281 (1994).
10. Slypen, I., Corcalciuc, V., Ninane, A., Meulders, J.-P. and Scott, M. *Fast Neutrons Interaction with Light Nuclei. Nuclear Reactions Induced by Fast Neutrons on ^{12}C for $E_n = 45$-80 MeV.* NATO Adv. Study Inst. 'Topics in Atomic and Nuclear Collisions'. Rom. J. Phys. **38**, 419–435 (1993).
11. Slypen, I., Corcalciuc, V. and Meulders, J.-P. *Proton and Deuteron Production in Neutron-induced Reactions on Carbon at $E_n = 42.5, 62.7$ and 72.8 MeV.* Phys. Rev. C**51**(3), 1303–1311 (1995).
12. Slypen, I., Corcalciuc, V., Meulders, J.-P. and Chadwick, M. *Triton and Alpha Production in Neutron-induced Reactions on Carbon at $E_n = 42.5, 62.7$ and 72.8 MeV.* Phys. Rev. C**53**(3), 1309–1318 (1996).
13. Chadwick, M. B. private communication (1995).
14. Chadwick, M. B., Blann, M., Cox, L. J., Young, P. G. and Meigooni, A. *Evaluated Cross-section Libraries and Kerma Factors for Neutrons up to 100 MeV on ^{12}C.* UCRL-ID-120829 (1995).
15. Brenner, D. J. and Prael, R. E. *Calculated Differential Secondary-particle Production Cross-sections after Nonelastic Neutron Interactions with Carbon and Oxygen between 15 and 60 MeV.* At. Nucl. Data Tables **41**, 71–99 (1989).
16. Slypen, I., Corcalciuc, V. and Meulders, J.-P. *Kerma Values Deduced from Neutron-induced Charged-particle Spectra of Carbon from 40 to 75 MeV.* Phys. Med. Biol. **40**, 73–82 (1995).
17. Subramanian, T. S. *et al. Double Differential Inclusive Hydrogen and Helium Spectra from Neutron-induced Reactions on Carbon at 27.4, 39.7 and 60.7 MeV.* Phys. Rev. C**28**, 521–528 (1983).
18. Romero, J. L. private communication (1994).
19. Young, P. G., Arthur, E. D. and Chadwick, M. B. *Los Alamos National Laboratory Report* LA-MS-12343 (1992).
20. Brady, F. P. and Romero, J. L. *Final Report to the National Cancer Institute.* Grant no 1RO1 CA6261 (1979).
21. Romero, J. L., Brady, F. P. and Subramanian, T. S. *Neutron Induced Charged Particle Spectra and Kerma from 25 to 60 MeV.* Radiat. Effects **94**, 13–25 (1985).
22. Schrewe, U. J., Brede, H. J., Henneck, R., Gurdung, S., Kunz, A., Menzel, H. G., Meulders, J.-P., Pihet, P., Schumacher, H. and Slypen, I. *Determination of Kerma Factors for A-150 Plastic and Carbon for Neutron Energies above 20 MeV.* In: Proc. Int. Conf. on Nuclear Data for Science and Technology, Jülich 1991. Ed. S. M. Quaim (Berlin: Springer) pp. 586–588 (1992).
23. Brenner, D. J. *Neutron Kerma Values above 15 MeV Calculated with a Nuclear Model Applicable to Light Nuclei.* Phys. Med. Biol. **29**, 437–441 (1983).
24. White, R. M., Broerse, J. J., DeLuca, P. M. Jr, Dietze, G., Haight, R. C., Kawashima, K., Menzel, H. G., Olsson, N. and Wambersie, A. *Status of Nuclear Data for Use in Neutron Therapy.* Radiat. Prot. Dosim. **44**, 11–20 (1992).

RELATIVE NEUTRON SENSITIVITY OF TISSUE-EQUIVALENT IONISATION CHAMBERS IN AN EPITHERMAL NEUTRON BEAM FOR BORON NEUTRON CAPTURE THERAPY

J. T. M. Jansen†, C. P. J. Raaijmakers‡, B. J. Mijnheer‡ and J. Zoetelief†
†TNO Centre for Radiological Protection and Dosimetry
PO Box 9034, 6800 ES Arnhem, The Netherlands
‡The Netherlands Cancer Institute, Antoni van Leeuwenhoek Huis
Plesmanlaan 121, 1066 CX Amsterdam, The Netherlands

Abstract — In recent years, boron neutron capture therapy has received renewed interest as new promising boron compounds and epithermal neutron beams have become available. For establishment of the different dose components in epithermal neutron beams, various techniques are applied, including the use of tissue-equivalent (TE) ionisation chambers for the determination of the intermediate and fast neutron dose. This paper is directed towards the calculation of the relative neutron sensitivity of TE ionisation chambers, k_T, for the free-in-air spectrum at the HB11 epithermal beam facility of the High Flux Reactor in Petten. The k_T value obtained from the present calculations is 0.87. This value is about 10% lower than data currently applied for similar conditions.

INTRODUCTION

Boron neutron capture therapy (BNCT) has recently gained renewed interest due to availability of new promising boron compounds and epithermal neutron beams. Borocaptate sodium (BSH) in combination with a thermal neutron beam has been used in Japan for the treatment of brain tumours since 1967. To reduce the effects from the rapid attenuation of thermal neutrons in tissue the skull was opened during irradiation. The results as reported by, for example, Hatanaka and Nakagawa[1] for treatment of brain tumours with BNCT are promising. A better penetration can be achieved by increasing the number of intermediate neutrons in the beam[2], i.e. by using an 'epithermal' neutron beam. Intermediate neutrons are moderated in tissue resulting in a peak in the thermal neutron fluence rate at depths of approximately 2 to 3 cm. By using an epithermal neutron beam, opening of the skull for BNCT can be avoided. An epithermal neutron beam (HB11) with a kerma weighted average neutron energy of about 10 keV has been installed at the High Flux Reactor (HFR) in Petten, the Netherlands[3,4] with the aim of performing clinical trials on treatment of glioma employing BSH.

The total absorbed dose in tissue irradiated with an epithermal neutron beam can be subdivided into various components, i.e. dose due to thermal ($E_n < 0.5$ eV), intermediate (0.5 eV $\leq E_n \leq$ 10 keV) and fast ($E_n >$ 10 keV) neutrons as well as dose resulting from gamma rays. Since the contribution to the absorbed dose from intermediate neutrons is relatively small compared to that from fast neutrons, the separation between these dose components is considered not to be practical[5]. The combined dose components are referred to as fast neutron dose.

As the different dose components have a different relative biological effectiveness their separate determination is required. For this purpose various dosimetric techniques are applied, e.g. activation foils for the determination of the thermal neutron fluence[5,6], thermoluminescence dosemeters[5] and non-hydrogenous ionisation chambers[5,6] for the assessment of the photon absorbed dose, and tissue-equivalent (TE) ionisation chambers flushed with methane-based TE gas for the determination of the fast neutron dose[5,6]. Since the detectors used for mixed field dosimetry are usually sensitive to more than one dose component dual or triple detector techniques are commonly employed. Following the nomenclature recommended by the ICRU[7,8] for mixed field dosimetry and extending this with a third component the following relations result[5]:

$$R_U = h_U D_\gamma + k_U D_n + k'_U k_f \Phi \quad (1)$$

$$R_T = h_T D_\gamma + k_T D_n + k'_T k_f \Phi \quad (2)$$

where the subscript U refers to a detector with a low sensitivity to neutrons and subscript T to a detector with approximately equal sensitivity to neutrons and photons; R_U and R_T are the readings of the dosemeters in the mixed field relative to their responses for the photons used for calibration; h_U and h_T are the responses of the dosemeters to the photons in the mixed field relative to those for the photons used for calibration; k_U and k_T are the responses of the dosemeters to the fast neutrons in the mixed field relative to those for the photons used for calibration; D_γ and D_n are the photon and the fast neutron dose in tissue, respectively; k'_U and k'_T are the readings of the dosemeters due to the thermal neutrons in the mixed field relative to their responses to the photons used for calibration, k_f is the kerma factor for the thermal neutrons and Φ is the thermal neutron fluence. Since the thermal neutron sensitivity of the ionisation chambers generally applied in BNCT dosimetry deviates strongly from theoretical values[5,6], the thermal neutron sensitivity is not incorporated in the k_U and k_T values. Instead empirical values ($k'_U k_f$ and $k'_T k_f$) are used to correct for thermal neutron responses. For accurate determination of the different dose components the relative sensitivities of the detectors employed have to be determined. In addition, the thermal neutron fluence has to be measured independently.

The present contribution is directed towards the calculation of the relative neutron sensitivity k_T of TE/TE ionisation chambers in the neutron energy range from about 2.6×10^{-8} to 20 MeV since accurate data are lacking in the lower neutron energy region, i.e. below about 0.1 MeV. The h_T value of TE/TE chambers is commonly assumed to be equal to unity. In addition, average k_T values of TE/TE chambers are calculated for neutron energies above about 1 eV for the actual spectrum determined free-in-air in the epithermal HB11 neutron beam at the HFR in Petten.

MATERIALS AND METHODS

According to ICRU Report 45[7], the relative neutron sensitivity, k_T, of a TE ionisation chamber can be expressed by:

$$k_T = \frac{W_c}{W_n} \left(\frac{[(\bar{L}/\rho)_m/(\bar{L}/\rho)_g]_c}{(r_{m,g})_n} \right) \left(\frac{[(\mu_{en}/\rho)_t/(\mu_{en}/\rho)_m]_c}{(K_t/K_m)_n} \right) \quad (3)$$

where the subscripts c and n refer to the calibration and mixed field situation, respectively; the subscripts m, g and t indicate the chamber wall material, chamber gas and ICRU muscle tissue, respectively; W represents the average energy required to form an ion pair in the chamber gas; $(\bar{L}/\rho)_c$ is the mean restricted collision mass stopping power for the slowing down electrons produced by the photons used for calibration; $(r_{m,g})_n$ is the gas-to-wall absorbed dose conversion factor for an ionisation chamber in a mixed field; (μ_{en}/ρ) is the mass energy absorption coefficient; and K is the kerma.

The following simplifications and assumptions are made for the present calculation of k_T. The ratio of the gas-to-wall absorbed dose conversion factor and the ratio of mean restricted collision mass stopping powers, $(r_{m,g})_n/[(\bar{L}/\rho)_m/(\bar{L}/\rho)_g]_c$ is 1.00 ± 0.02 for neutron energies employed for fast neutron therapy[8]. This might not be directly applicable to the present situation. The calculations of Burger and Makarewicz[9] suggest that cavity size effects are not maximal for infinite size. At energies below the values for which they performed calculations, the cavity size will approach more and more infinite size. Therefore, it is assumed that the ICRU recommendation for fast neutrons also holds true for the wide range of neutrons considered in the present study. The ratio $[(\mu_{en}/\rho)_t/(\mu_{en}/\rho)_m]_c$ can be calculated using data from ICRU Report 44[10] and equals 1.000 for ^{60}Co γ rays and 1.003 for ^{137}Cs γ rays. For the present calculations, this ratio is taken to be equal to unity. A value of 29.2 eV has been taken for W_c for the methane-based TE gas which is based on the value for W_{air} of 33.97 eV from Boutillon and Perroche-Roux[11] and the estimate for W_{air}/W_g of 1.166 derived from measurements with ionisation chambers filled with TE gas and air[12]. Data on kerma factors are obtained from ICRU Report 44[10].

To arrive at values of W_n for various neutron energy bins, the Caswell–Coyne code for calculation of energy deposition and ion yield[13] is used after replacing the functions and numerical values for W of charged particles by those presented by Taylor et al[14]. W_n values for the free-in-air spectrum in the epithermal neutron beam were originally calculated using the Caswell–Coyne code. However, for the broad range of neutron energies (about 9 decades) and the limited number (600) of linear energy bins, erroneous results are obtained due to the fact that the important low energy parts of the spectrum were all in one large energy bin. Therefore, an in-house computer code was developed for the calculations of W_n and k_T for the broad spectrum of the epithermal neutron beam. An additional (small) advantage of the in-house code is that the energy bins can be adapted to bins of the actual neutron spectrum. The average W_n values are calculated from:

$$W_n = \frac{\sum_i \bar{N}_i W_{n,i}}{\sum_j \bar{N}_j} = \frac{\sum_i \left(\frac{N_{n,i} k_{f,i}}{W_{n,i}} \right) W_{n,i}}{\sum_j \left(\frac{N_{n,j} k_{f,j}}{W_{n,j}} \right)} \quad (4)$$

where i and j indicate the summation over the neutron energy bins, $\bar{N}_i$ and $\bar{N}_j$ are the numbers of ion pairs produced by the neutrons in a specific energy bin, $N_{n,i}$ and $N_{n,j}$ are the numbers of neutrons in a specific energy bin, $k_{f,i}$ and $k_{f,j}$ are the kerma factors for a specific energy bin and $W_{n,i}$ and $W_{n,j}$ are the W values for a specific neutron energy bin. The numerator contains the weighting by kerma of the W value and the denominator is required for normalisation. The k_T values are calculated using Equation 3. The results obtained with the in-house code have been compared with those obtained with the Caswell–Coyne code when the spectrum for the latter calculations is subdivided into four parts allowing sufficient numbers of energy bins in each region.

The neutron spectrum was calculated using Monte Carlo simulations for the free-in-air condition at the HB11 epithermal neutron beam facility of the High Flux Reactor in Petten[4]. The experimental configuration was modelled using the Monte Carlo radiation transport code MCNP version 3B[15]. The three-dimensional geometry modelled in MCNP represented the final shutter assembly of the beam, the beam aperture and a simplified description of the treatment room. The epithermal beam was defined via a source plane at the rear of the final beam shutter with spectral intensity and angular parameters representing the beam, as described elsewhere[4,16].

RESULTS AND DISCUSSION

W_n values calculated for neutrons in the energy range 2.6×10^{-8} to 20 MeV are shown in Figure 1. Also shown in this figure are data published previously by Makarewicz and Burger[17]. The W_n values show a distinct peak around about 0.3 keV for both sets of calculations. The largest difference in W_n value between the present calculations and those of Makarewicz and Burger amounts to about 40% and occurs in the energy bin 0.14 to 0.26 keV. This difference is most likely due to the different mathematical expressions used for W values of charged particles as a function of energy. For the present calculations the following function is used[14,18]:

$$W_R(E_R) = A[\ln(E_R + B)]^{-D} + C\, E_R E_b/(E_R^2 + E_b^2) + W_\infty \quad (5)$$

where E_R is the specific energy in keV per nucleon for the particle (R) under consideration, W_∞ the W value at infinite particle energy, the parameters A, B, and D are similar to those used in an older expression suggested by Coyne et al[19] and the constant C and co-factor are used to model a 'bump' at a specific energy near E_b. For protons and alpha particles, the function with a 'bump' and for C, N and O the function without 'bump' is used. Numerical values for the parameters A, B, C, D, E_b and W_∞ are presented elsewhere[14].

Makarewicz and Burger[17] used the following relation for W as a function of charged particle energy:

$$W \propto 1/(1 - U/\epsilon') \quad (6)$$

where U is a constant, representing an ionisation threshold and ϵ' the kinetic energy of the charged particle in question.

According to Siebert et al[18] the expression given in Equation 5 gives a better quality of fit and includes a bump that is clearly observed in some experiments with protons and is suggested by theory. They conclude that their analytical representation should replace older ones such as that given in Equation 6.

It is expected that the largest differences between different calculations will occur at the largest W_n values, since at these neutron energies the charged particles with the lowest energies are produced. Both sets of W_n values indicate that the linear extrapolation of W_n values below 0.1 MeV made by Rogus et al[6] is not justified.

Relative neutron sensitivities of TE ionisation chambers calculated as a function of neutron energy are shown in Figure 2. The present calculations show that the highest k_T value is 1.04 (at about 18.5 MeV) and the lowest k_T is 0.19 (at about 3.6×10^{-4} MeV). For thermal neutrons k_T is approximately 0.94. These results are quite different from the values published by Rogus et al[6] as shown in Figure 2. The differences are mainly due to the approximations used for W_n values below 0.1 MeV by Rogus et al. They either extrapolated W_n values of Goodman and Coyne[20] linearly to neutron energies below 0.1 MeV or applied for this neutron energy region the W_p data of Leonard and Bohring[21] that result in infinite W values below about 10^{-3} MeV. More recent information of Huber et al[22] showed that W_p does not increase as rapidly as found by Leonard and Bohring.

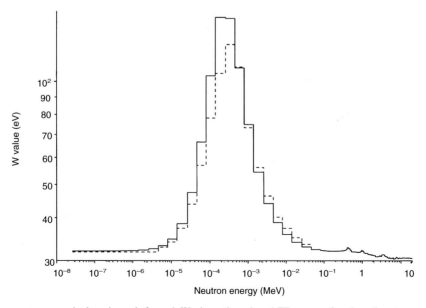

Figure 1. Average energy required per ion pair formed, W_n, in methane-based TE gas as a function of neutron energy. The figure shows the results from the present calculations (———) and data from Makarewicz and Burger[17] (– – – –).

Spectral data of the HB11 beam at the HFR in Petten have been used in combination with W_n and kerma factors for the neutron energy bins to calculate average W and k_T values with the code developed in-house. Since the thermal neutron response of the chamber is treated separately, the thermal part of the spectrum has not been included in the calculations. For comparison, the results obtained from the Caswell–Coyne code when the spectrum was subdivided in four regions are also shown in Table 1. It is concluded that the differences between the results from both codes are small, i.e. about 0.3% at maximum. The k_T value calculated using the W_n values of Makarewicz and Burger[17] for the total spectrum resulted in a value of 0.87 which is in good agreement with the k_T values calculated using the W_n values from the present study. The present k_T value is about 10% smaller than the value of 0.95 proposed by Rogus et al[6] and Raaijmakers et al[5] which is important in view of the required accuracy for dosimetry in radiotherapy. Work is in progress to determine k_T values for conditions that are more directly relevant for therapy, i.e. at various positions inside polymethylmethacrylate phantoms.

CONCLUSIONS

W_n values for methane-based TE gas have been calculated for neutrons in the energy range 2.6×10^{-8} to

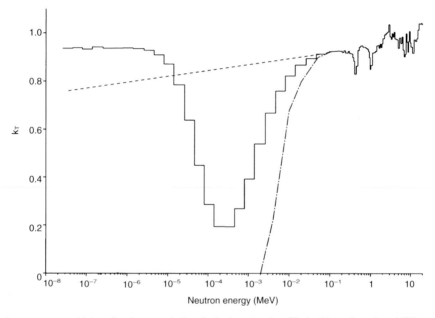

Figure 2. Relative neutron sensitivity of a tissue-equivalent ionisation chamber filled with methane-based TE gas as a function of neutron energy. The figure shows the results from the present calculations (———) and extrapolations made by Rogus et al[6] based on data on W_n of Goodman and Coyne[20] (– – – –) and on data on W_p of Leonard and Bohring[21] (–·–·–).

Table 1. W_n and relative neutron sensitivity, k_T, of a TE/TE ionisation chamber calculated for fast and intermediate neutrons free-in-air in the HB11 epithermal neutron beam facility at Petten. The results using the Caswell-Coyne code[13] for four sub-spectra and for the total spectrum by combining the data of the sub-spectra are also given.

Energy range (MeV)	W_n (eV)	k_T	W_n (eV) (Ref. 13)	Rel. number of ion pairs (Ref. 13)
2.4×10^{-6} to 1.2×10^{-4}	—	—	38.7	0.006
1.2×10^{-4} to 6.0×10^{-3}	—	—	52.6	0.052
6.0×10^{-3} to 3.0×10^{-1}	—	—	33.1	0.543
3.0×10^{-1} to $1.5 \times 10^{+1}$	—	—	31.9	0.399
Total spectrum	33.8	0.87	33.7	1.000

20 MeV. The results are in reasonably good agreement with data published by Makarewicz and Burger[17] but differ considerably from the results obtained with the extrapolations suggested by Rogus et al[6]. A value of 0.87 for the relative neutron sensitivity k_T of a TE/TE ionisation chamber free-in-air at the HB11 epithermal neutron beam in Petten has been calculated using a code developed in-house. A comparison of results obtained for the fast (and intermediate) neutron component of the epithermal beam between the in-house code and the Caswell-Coyne code after subdivision of the spectrum showed good agreement. The k_T value obtained using W_n data published by Makarewicz and Burger[17] is less than one per cent smaller than that using W_n values from the present study. The present k_T value is about 10% lower than those currently employed[5,6], which is important in view of the required accuracy for dosimetry in radiotherapy.

ACKNOWLEDGEMENT

This work has been supported in part by the Commission of the European Communities under contract number F13P-CT-920045 and by the Dutch Ministry of Housing, Physical Planning and the Environment.

REFERENCES

1. Hatanaka, H. and Nakagawa, Y. *Clinical Results of Long-Surviving Brain Tumor Patients who Underwent Boron Neutron Capture Therapy*. Int. J. Radiat. Oncol. Biol. Phys. **28**, 1061–1066 (1994).
2. Fairchild, R. G. *Development and Dosimetry of an "Epithermal" Neutron Beam for Possible Use in Neutron Capture Therapy*. Phys. Med. Biol. **10**, 491–504 (1965).
3. Moss, R. L. *Current Overview on the Approach of Clinical Trials at Petten*. In: Boron Neutron Capture Therapy: Toward Clinical Trials of Glioma Treatment. Eds D. Gabel and R. Moss (New York: Plenum Press) pp. 33–46 (1992).
4. Watkins, P., Konijnenberg, M. W., Constantine, G., Reif, H., Ricchena, R., de Haas, J. B. M. and Freudenreich, W. *Review of the Physics Calculations Performed for the BNCT Facility at the HFR Petten*. In: Boron Neutron Capture Therapy: Toward Clinical Trials of Glioma Treatment. Eds D. Gabel and R. Moss (New York: Plenum Press) pp. 47–58 (1992).
5. Raaijmakers, C. P. J., Konijnenberg, M. W., Verhagen, H. W. and Mijnheer, B. J. *Determination of Dose Components in Phantoms Irradiated with an Epithermal Neutron Beam for Boron Neutron Capture Therapy*. Med. Phys. **22**, 321–329 (1995).
6. Rogus, R. D., Harling, O. K. and Yanch, J. C. *Mixed Field Dosimetry of Epithermal Neutron Beams for Boron Neutron Capture Therapy at the MITR-II Research Reactor*. Med. Phys. **21**, 1611–1625 (1994).
7. ICRU. *Neutron Dosimetry for Biology and Medicine*. Report 26 (Bethesda, MD: ICRU Publications) (1977).
8. ICRU. *Clinical Neutron Dosimetry. Part 1: Determination of Absorbed Dose in a Patient Treated by External Beams of Fast Neutrons*. Report 45 (Bethesda, MD: ICRU Publications) (1989).
9. Burger, G. and Makarewicz, M. *Average Energy to Produce an Ion Pair in Gases (W-values) and Related Quantities of Relevance in Neutron Dosimetry*. In: Nuclear and Atomic Data for Radiotherapy and Related Radiobiology (Vienna: IAEA) STI/PUB/741, pp. 225–238 (1987).
10. ICRU. *Tissue Substitutes in Radiation Dosimetry and Measurement*. Report 44 (Bethesda, MD: ICRU Publications) (1989).
11. Boutillon, M. and Perroche-Roux, A. M. *Re-evaluation of the W Value for Electrons in Dry Air*. Phys. Med. Biol. **32**, 213–219 (1987).
12. Zoetelief, J. and Schraube H. *Experimental Procedures for the on-site Neutron Dosimetry Intercomparison ENDIP-2*. In: Proc. Fifth Symp. on Neutron Dosimetry. Eds H. Schraube, G. Burger and J. Booz. EUR 9762 (Brussels and Luxembourg: Commission of the European Communities) pp. 1179–1190 (1985).
13. Caswell, R. S. and Coyne, J. J. *Energy Deposition Spectra for Neutrons Based on Recent Cross Section Evaluations*. In: Proc. Sixth Symp. on Micodosimetry. Eds J. Booz and H. G. Ebert (Brussels: Commission of the European Communities) EUR-6064, Vol. II, pp. 1159–1171 (1978).
14. Taylor, G. C., Jansen, J. T. M., Zoetelief, J. and Schuhmacher, H. *Neutron W-values in Methane-Based Tissue-Equivalent Gas up to 60 MeV*. Radiat. Prot. Dosim. **61**, 285–290 (1995).
15. Briesmeister, J. *MCNP: A General Monte Carlo Code for Neutron and Photon Transport, Version 3A*. (LA-7396-M, Rev2) (Los Alamos National Laboratory, Los Alamos, NM, USA) (1986).
16. Konijnenberg, M. W., Dewit, L. G. H., Mijnheer, B. J., Raaijmakers, C. P. J. and Watkins, P. R. D. *Dose Homogeneity in Boron Neutron Capture Therapy Using an Epithermal Neutron Beam*. Radiat. Res. **142**, 327–339 (1995).
17. Makarewicz, M. and Burger, G. W_n *for Methane-Based TE Gas and Carbon Dioxide and* W_n *and Gas-to-Wall Dose Conversion Factors for TE/TE Cavities*. In: Proc. Fifth Symp. on Neutron Dosimetry, Eds H. Schraube, G. Burger and J. Booz (Luxembourg: Commission of the European Communities) EUR 9762, pp. 275–286 (1985).
18. Siebert, B. R. L., Grindborg, J. E. Grosswendt, B. and Schuhmacher, H. *New Analytical Representation of W Values for Protons in Methane-Based Tissue-equivalent Gas*. Radiat. Prot. Dosim. **52**, 123–127 (1994).
19. Coyne, J. J., Caswell, R. S., Zoetelief, J. and Siebert, B. R. L. *Calculations of Microdosimetric Spectra for Low Energy Neutrons*. Radiat. Prot. Dosim. **31**, 217–221 (1990).

20. Goodman, L. J. and Coyne, J. J. W_n *and Neutron Kerma for Methane-Based Tissue-equivalent Gas.* Radiat. Res. **82**, 13–26 (1980).
21. Leonard, B. E. and Bohring, J. W. *The Average Energy per Ion Pair, W, for Hydrogen and Oxygen Ions in a Tissue Equivalent Gas.* Radiat. Res. **55**, 1–9 (1973).
22. Huber, R., Combecher, D. and Burger, G. *Measurements of Average Energy Required to Produce an Ion Pair (W Value) for Low-Energy Ions in Several Gases.* Radiat. Res. **101**, 237–251 (1985).

W VALUES IN PROPANE-BASED TISSUE-EQUIVALENT GAS

I. K. Bronić
Rudjer Bošković Institute
PO Box 1016, 10001 Zagreb, Croatia

Abstract — W values (mean energy required to form an ion pair after complete dissipation of incident particle energy) for electrons, protons, alpha particles and heavy ions (^{12}C, ^{14}N, ^{16}O) in propane-based tissue-equivalent gas are summarised and analytically represented. Experimental data are compared with the W values calculated from W values for pure components by applying the mixing model with the Strickler's constants. The analysis of a rather limited number of W values shows that there is a need for new data for protons and alpha particles for all energies, and for new high energy W values for electrons and heavy ions.

INTRODUCTION

Interpretation of measured ionisation yield in terms of absorbed energy, which is required in microdosimetry, depends on a precise knowledge of the mean energy required to form an ion pair by an ionising particle after complete dissipation of its initial energy (W value). The W value thus relates the measured number of ion pairs and the energy deposited by an incident particle in a medium. The W value depends on the type of an ionising particle and its energy, and it is a characteristic of a medium. Therefore, it is important to know accurately the energy dependence of W values for various ionising particles in gases of interest in dosimetry. Propane-based tissue-equivalent (p-TE) gas, which consists of 54% propane, 40.5% CO_2, and 5.5% N_2, is one of the two TE gases mostly used in microdosimetry ionisation chambers and in low pressure proportional counters. Accurate W values for high energy electrons, for protons of about 100 keV, and for alpha particles of approximately 5 MeV and 700 keV are required for calibration of proportional counters.

Neutrons, as indirectly ionising particles, ionise medium through induced charged particles. In order to calculate a reliable W value for neutrons, it is necessary to know W values for all the contributing ions over wide energy range, especially for low energy ions. Two reports[1,2] summarise the experimental W values for various ionising particles in gases of interest for dosimetry and radiotherapy. For methane-based TE gas, W values are reasonably well known, although the knowledge is far from complete and satisfactory, as shown by Siebert et al[3] for protons and by Taylor et al[4] for ^{4}He, ^{12}C, ^{14}N and ^{16}O ions. Much less experimental data on W values exist for p-TE gas[2,5–7]. While the deviations among the results obtained by different authors require a very detailed comparative analysis of W values in methane and methane-based TE gas, the main problem for propane and p-TE gas is the lack of results by different groups.

METHOD

Analytical representation of energy dependence

One of the most important characteristics of the W value is its energy dependence: W values approach a constant value at high energies, and increase as the energy of ionising particles decreases. The energy dependence of the W value for electrons can be described[8] as

$$W(E) = \frac{W_\infty}{1 - (U/E)} \quad (1)$$

Here, W_∞ is the asymptotic high energy electron W value, E is electron energy, and U is the average kinetic energy of sub-ionisation electrons[8].

The experimental data for heavier particles can be fitted using the function

$$W(E) = \frac{A}{[\ln(E + B)]^C} + D \quad (2)$$

where E is energy of the particle in keV.amu^{-1}, and A, B, C, and D are fitting parameters. A function of the same form was used by Siebert et al[3] and Taylor et al[4]. The part of their equation describing a 'bump' in energy dependence of the W value is neglected here because of the lack of experimental data in the interesting energy region in p-TE gas.

Mixing model

The W value in a regular gas mixture, i.e. in a mixture without appreciable energy transfer among constituent molecules, can be calculated if the W values and concentration fractions for each component (W_i and C_i, respectively) are known[2,5,9]. An agreement between the experimental and calculated W values in methane-based TE gas was found regardless of the model concept[5]. In contrast, the smallest relative deviations for p-TE gas were obtained by applying the mixing model with empirical constants derived by Strickler[9]:

$$W = \frac{\sum f_i C_i W_i}{\sum f_i C_i} \quad (3)$$

RESULTS

W values for electrons

W values for electrons in propane-based TE (p-TE) gas have been measured in the energy range from about 20 eV to 1.5 keV by Combecher[6], and for 0.26, 1.2 and 5.6 keV electrons by Krajcar Bronić et al[10], with experimental uncertainties <2% and <5%, respectively. The experimental values were fitted using Equation 1 (Figure 1). By using only data by Combecher[6], the values of 26.9 ± 0.3 eV and 9.27 ± 0.16 eV for W_∞ and U, respectively, are obtained. By using all the data[6,10], the corresponding values were practically the same, 27.0 ± 0.3 eV and 9.24 ± 0.16 eV. Therefore, the high energy electron W value may be set to 27.0 ± 0.3 eV. However, new experimental data are needed for electrons having energy higher than several keV to confirm the constancy of W at high electron energies.

The experimental W value data in p-TE gas agree well with the W values calculated by Equation 3, except for very low electron energies[5]. Relative deviations between the experimental and calculated W values are less than 3%. The average ratio of W values for electrons in p-TE gas and those in propane is 1.07.

W values for protons

Only one set of W value data for protons in p-TE gas[7] is known over a limited energy range (25–375 keV) with uncertainties of ~3%. Data indicate a low minimum around 175 keV, in both propane and p-TE gas. An analysis of the W value data for protons in methane-based TE gas[3] showed a small 'bump' at ~400 keV following the shallow minimum at ~100 keV. A very narrow energy region of experimental data for p-TE gas does not enable a similar analytical analysis. A relatively good fit was obtained by a polynomial of the 4th order in this limited energy range, but any attempt at extrapolation to either lower or higher energies was meaningless.

The experimental W value data in p-TE gas agree well with the W values calculated by Equation 3 with relative deviations <2%[5]. The average ratio of W values in p-TE gas and those in propane is 1.07.

W value for alpha particles

For alpha particles in both propane and p-TE gas also only one set of data exists[7] at energies below 1 MeV, having uncertainties of ~3%. Only one experimental W value for 5 MeV alpha particles in propane is available[11]. The data for propane and for p-TE gas were fitted by Equation 2 (Table 1). The W value for He ions[7] shows somewhat stronger energy dependence in propane than in p-TE gas (Figure 2). The average ratio of the W values in p-TE gas and those in propane is 1.0, which is lower than the corresponding ratio for electrons and protons (1.07), as well as for heavy ions (see below).

W values in p-TE gas calculated by Equation 3 are also shown in Figure 2. Relative differences between the measured[7] and the calculated W values are larger than 3%, indicating that W values in either propane or p-TE gas include a systematic error[5]. The calculated W value for 5 MeV alpha particles in p-TE gas is 28.0 eV, giving the ratio of W values in p-TE gas and in propane[11] equal to 1.065. The ratio of calculated W values at lower energies and those measured in propane[7] is now 1.052. All these facts suggest that W values for alpha particles in p-TE gas[7] are systematically too low.

For methane-based TE gas Taylor et al[4] showed a statistical evidence of a 'bump' at ~200 keV.amu^{-1}. Unfortunately, in this energy region no experimental data exist for either propane or p-TE gas.

W values for heavier ions

As for protons and alpha particles, the only measurement of W values for heavy ions (^{12}C, ^{14}N, ^{16}O) was performed at low energies (25–375 keV) by Posny et al[7] with uncertainties of ~3%. Varma and Baum[12] reported a W value for 42 MeV oxygen ions in p-TE gas equal to 30.41 eV (uncertainty <0.1%). Unfortunately, they did not report a W value for propane. Thus, this high-energy W value is included in the fitting process only for O ions. The fitted curves for C and N ions, based on low energy data only, crossed the curve for O ions, thus giving high energy values higher than that for O ions (Table 1). Therefore, the 4-parameter function (Equation 2) was replaced by a three-parameter function, by fixing the parameter D at 20.5, i.e. at the value obtained for O ions. The new fits (Table 1) showed a

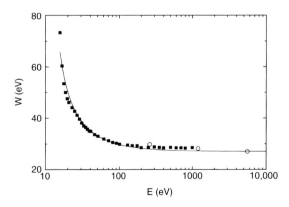

Figure 1. W value for electrons in propane-based TE gas. Experimental data (■) Combecher[6]; (○) Krajcar Bronić et al[10]. (—) Fitted line, Equation 1.

slight increase of W value with the increase of the ion mass (Figure 3), as was observed in other gases where more measurements exist[2,13].

For heavy ions, the relative differences between the measured and calculated (Equation 3) W values lie within ±2%, except for the lowest energies, <75 keV, where the measured values are up to 15% lower than the calculated W values. The ratios of W values in p-TE gas and those in propane are 1.016, 1.046 and 1.013 for ^{12}C, ^{14}N and ^{16}O ions, respectively, at energies above 75 keV.

DISCUSSION AND CONCLUSION

The present analysis that combined the energy dependence study of W values for various ionising particles in propane-based TE gas and the mixing relations, has led to some interesting conclusions, although the amount of experimental data is very limited.

W values for low energy electrons (up to several hundreds of eV) measured by Combecher[6] connect smoothly with the data by Krajcar Bronić et al[10] for 0.26, 1.2 and 5.6 keV electrons, leading eventually to the high energy W value recommended by ICRU[1] in all alkanes (methane through hexane) with an exception of propane[10]. In addition, the W values for photons having energy between 80 eV and 6 keV in propane[14], connects smoothly with Combecher's data. The ICRU[1] value (24.0 ± 0.5 eV) is lower than the W_∞ values obtained by fitting of the experimental W data in propane[6,10,14] (25.6 eV). The lower ICRU W value for propane leads, through Equation 3, to a W_∞ value for p-TE gas (25.9 eV) which is 1.1 eV lower than the experimental W value (27.0 eV), indicating again that the ICRU[1] W value for high energy electrons in propane is too low.

Taylor et al[4] fixed the value of parameter B of Equation 2 (for ^{4}He, ^{12}C, ^{14}N, and ^{16}O) to unity. The present results show that the value of parameter B in p-TE gas is, in any case, higher than that obtained by Taylor et al[4]. Values of parameter C are not significantly different from Taylor's values, especially taking into account the rather large standard deviations in the

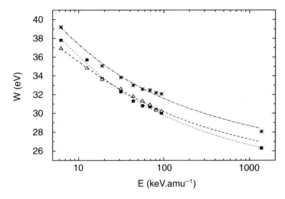

Figure 2. W value for alpha particles in propane and propane-based TE gas. Experimental data (■) propane[7]; (●) propane[11]; (△) propane-based TE gas[11]. W values in p-TE gas calculated by Equation 3 (∗). Fitted lines, Equation 2; (···) propane; (– –) p-TE, experimental data; (–··–) p-TE, calculated values.

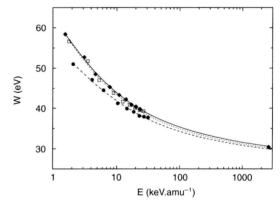

Figure 3. W value for heavy ions (^{12}C, ^{14}N, ^{16}O) in propane-based TE gas. Experimental data: (●) ^{12}C; (□) ^{14}N; (◆) ^{16}O. Fitted lines, Equation 2; (– –) ^{12}C, with fixed parameter D; (···) ^{14}N, with fixed parameter D; (—) ^{16}O.

Table 1. Coefficients A, B, C, and D, obtained by fitting the experimental data with Equation 2.

	A	B	C	D
Alpha particles				
propane[7,11]	37.44	3.95	0.72	17.45
p-TE gas[7]	34.45	5.17	0.68	18.13
p-TE calculated	34.46	3.16	0.63	18.38
^{12}C ions[7]	59.5	4.37	1.71	30.7
	42.1	2.60	0.73 ± 0.07	20.5 fixed
^{14}N ions[7]	53.8	3.18	1.51	30.5
	45.5	2.06	0.75 ± 0.09	20.5 fixed
^{16}O ions[7,12]	46.2	2.11	0.75 ± 0.15	20.5

present fits that are due to the small number of experimental data.

The values of parameter D, which should reflect the W value for infinitely energetic particles and should therefore be slightly different for different ionic species, for all ionic species is very low, even lower than the value of 26.2 eV, recommended by ICRU[1] for ~5.3 MeV alpha particles in propane. For all heavy ions (^{12}C, ^{14}N, ^{16}O) the present value is fixed at 20.5 eV, as obtained for ^{16}O ions. Taylor et al[4] fixed the value of D to 23 eV, which was also significantly lower than the value of 29.1 eV, assumed for high energy protons in methane-based TE gas[3]. Such low values of what is to be the high energy W value, are most probably due to the extrapolation of the fits obtained with the limited number of low energy experimental W value data to energies beyond the range of experimental data.

For the purpose of calibration of proportional counters, the following W value data in p-TE gas may be recommended:

high energy electrons (>6 keV):	27.0 ± 0.3 eV
protons of ~100 keV:	28.2 ± 0.3 eV
alpha particles, ~5 MeV:	28.0 ± 0.5 eV
~700 keV:	30.6 ± 0.6 eV

New W values for high energy electrons are needed in propane (to check the difference between the older data[1] and the new experimental value) and in p-TE gas (not existing for energies above ~6 keV). New W values for protons in propane and in p-TE gas are needed in the whole energy range. It would be interesting to look for the structure in energy dependence in the region 100–400 keV. New W values for alpha particles in p-TE gas are also needed, because the analysis of the presently available experimental data indicates that these values have some systemic bias. For heavy ions, new high energy data are needed.

ACKNOWLEDGEMENT

This work was performed under the Project grant 1-07-064 from the Ministry of Science, Republic of Croatia, and the Alexander-von-Humboldt Stiftung Research Fellowship.

REFERENCES

1. International Commission on Radiation Units and Measurements. *Average Energy Required to Produce an Ion Pair*. ICRU Report 31 (Washington, DC: ICRU Publications) (1979).
2. Srdoč, D., Inokuti, M. and Krajcar Bronić, I. *Yields of Ionisation and Excitation in Irradiated Matter*. In: Atomic and Molecular Data for Radiotherapy and Radiation Research. IAEA-TECDOC-799 (Vienna: IAEA) Ch. 8, pp. 547–631 (1995).
3. Siebert, B. R. L., Grindborg, J. E., Grosswendt, B. and Schuhmacher, H. *New Analytical Representation of W values for Protons in Methane-Based Tissue-Equivalent Gas*. Radiat. Prot. Dosim. **52**, 123–127 (1994).
4. Taylor, G. C., Jansen, J., Zoetelief, J. and Schuhmacher, H. *Neutron W Values in Methane-Based Tissue-Equivalent Gas up to 60 MeV*. Radiat. Prot. Dosim. **61**(1–3), 285–290 (1995).
5. Krajcar Bronić, I. and Srdoč, D. *A Comparison of Calculated and Measured W Values in Tissue-Equivalent Gas Mixtures*. Radiat. Res. **137**, 18–24 (1994).
6. Combecher, D. *Measurements of W values of Low-energy Electrons in Several Gases*. Radiat. Res. **84**, 189–218 (1980).
7. Posny, F., Chary, J. and Nguyen, V. D. *W Values for Heavy Particles in Propane and in TE Gas*. Phys. Med. Biol. **32**, 509–515 (1987).
8. Inokuti, M. *Ionisation Yields in Gases under Electron Irradiation*. Radiat. Res. **64**, 6–22 (1975).
9. Strickler, T. D. *Ionisation by Alpha Particles in Binary Gas Mixtures*. J. Phys. Chem. **67**, 825–830 (1963).
10. Krajcar Bronić, I., Srdoč, D. and Obelić, B. *The Mean Energy Required to Form an Ion Pair for Low-Energy Photons and Electrons in Polyatomic Gases*. Radiat. Res. **115**, 213–222 (1988).
11. Kemmochi, M. *Measurements of W Values for Alpha Particles in Tissue-Equivalent Gases*. Health Phys. **30**, 439–446 (1976).
12. Varma, M. N. and Baum. J. W. *Stopping Power and Average Energy to Form an Ion Pair for 42 MeV Oxygen Ions*. Phys. Med. Biol. **27**, 1449–1453 (1982).
13. Christophorou, L. G. *Atomic and Molecular Radiation Physics* (London: Wiley Interscience) p. 46 (1972).
14. Srdoč, D. *Dependence of the Energy per Ion Pair on the Photon Energy below 6 keV in Various Gases*. Nucl. Instrum. Methods **108**, 321–332 (1973).

W VALUES OF PROTONS SLOWED DOWN IN MOLECULAR HYDROGEN

B. Grosswendt, G. Willems and W. Y. Baek
Physikalisch-Technische Bundesanstalt
Bundesallee 100
D-38116 Braunschweig, Germany

Abstract — For protons completely slowed down in molecular hydrogen, the mean energy W required for forming an ion pair has been studied in the energy range between 1 keV and 10 MeV, partly by measurements and partly by calculations. The experimental W values were determined for collimated beams of monoenergetic protons with energies between 1 keV and 100 keV by measuring the positive ion yield they produce in a plane-parallel ionisation chamber filled with hydrogen gas. The calculated W values were determined (i) using the Monte Carlo method for energies between 1 keV and 100 keV and (ii) within the framework of an analytical model based on the continuous slowing down approximation in the energy range between 1 keV and 10 MeV, taking into account the proton stopping power, the effect of charge exchange, the different ionisation cross sections for protons and neutral hydrogen particles, and the contributions of secondary electrons to the total ionisation yield. The complete data set for W as a function of energy shows the typical decrease with increasing energy in the region of up to about 5 keV, followed by a smoothly oscillating shape for higher energies. It can be assumed that the oscillating shape of W as a function of energy is mainly caused by charge-exchange processes and by the ionisation yield contribution of secondary electrons.

INTRODUCTION

The metrology of the interaction of neutrons with matter is frequently based on the measurement of ionisation yields using gas-filled ionisation chambers or proportional counters. For the conversion of the measured ionisation yields into energy deposited it is necessary to know the effective value[1] of the mean energy W expended in a gas per ion pair formed[2] or the corresponding differential value ω which depends not only on the neutron field existing around the point of measurement but also on the chamber design (size of the measuring volume, gas filling, wall materials).

Basically, the effective values could be determined if the energy deposition spectrum of all kinds of secondary charged particles set in motion by neutrons inside the measuring volume or penetrating it, and the corresponding W or ω values were known with appropriate accuracy. In this context, the data of protons are of particular importance because these particles are set in motion by elastic neutron scattering in any kind of hydrogen-containing media.

Unfortunately, absolute values of W and ω as a function of proton energy are not well known for many gases of interest and are not yet completely understood, either for higher energies where constant values are commonly assumed, or for energies of less than 100 keV where W and ω strongly depend on energy.

The lack of accurate proton data for a great number of different gases, in contrast to the situation of electrons[2–4], also prevents generally applicable scaling rules from being defined to estimate W values for gases where no direct experimental data are available. This is particularly true for lower proton energies where charge-changing effects and secondary electrons lead to a structured energy dependence of W specific to the stopping gas[5,6]. At higher energies, as recommended by the ICRU[2], proton W values for a definite gas can be roughly estimated from those of α particles.

Molecular hydrogen is one example of a gas where only a little information can be gathered from the literature, despite the fact that it is frequently used as filling gas in proton recoil proportional counter spectrometry. Its values of W or ω for protons are almost unknown apart from the tentative value of W = 35.6 eV measured by Gerthsen[7] at energies between about 19.1 keV and 35.4 keV, and the value of ω = 35.3 eV published by Bakker and Segrè[8] at 340 MeV. The latter value was determined by measuring the ionisation yield produced by protons in molecular hydrogen relative to that in argon on the basis of a stopping power ratio between hydrogen and argon of 0.145 and ω = 25.5 eV for argon. When using instead ω = 26.6 eV for argon, as recommended by the ICRU[2], and a stopping power ratio of 0.143, as recently published in ICRU Report 49[9], the original ω value of Bakker and Segrè for hydrogen has to be replaced by ω = 36.3 eV. At energies greater than 100 keV, following the recommendations of the ICRU[2], the proton W value should be approximately that of 5.7 MeV α particles which is 36.43 eV.

Because of this lack of data, it is the aim of the present paper to provide absolute values of W and ω for protons in the energy range between 1 keV and about 10 MeV slowed down in hydrogen. This data set is based on measurements for energies of up to 100 keV and on calculations in the high energy region.

EXPERIMENTAL METHOD

Monoenergetic protons, produced in a low voltage discharge ion source and (after acceleration to the energy T) collimated to a narrow beam, entered a hydrogen gas filled parallel-plate ionisation chamber (56 cm in diameter; 10 cm in length at energies T < 10 keV and 20 cm at higher energies) in a direction perpendicular to the collecting electrodes.

The primary beam current was measured in a vacuum using a Faraday cup, and the accelerating potential by using a precise high voltage divider connected to a voltmeter, checked at low potentials by determining the primary proton energy by the retarding field method.

The positive charge carriers produced upon the slowing down of the primary protons within the chamber were collected on a grid located in front of the electrode opposite the beam entrance, and measured as a function of gas pressure while keeping the ratio F/p of the collecting field strength F and the pressure p constant. The final ionisation current was determined by extrapolation to zero pressure, in order to eliminate the influence of hydrogen gas streaming out of the ionisation chamber through the entrance aperture. The typical pressure range used for extrapolation was 4 to 8 hPa at 1 keV, 8 to 16 hPa at 10 keV, and 20 to 37 hPa at 100 keV. To reduce the influence of charge-carrier diffusion and recombination, an additional extrapolation to vanishing F/p was applied. For more details concerning the techniques of measuring W values, see the publications of Waibel and Grosswendt[10] and of Waibel and Willems[11,12], and for the correction with respect to proton backscattering, the paper of Grosswendt and Willems[13].

CALCULATION METHODS

The analytical model

The theory of W for protons is more complicated than that for electrons because the former must cover both the complete slowing down of the incident particles and that of secondary electrons and secondary heavier particles. Whereas the energy degradation of electrons is mainly governed by excitation, dissociation and direct ionisation processes, protons lose a part of their kinetic energies also by elastic scattering (in particular, in the very low energy region) and by electron capture processes which lead to the formation of neutral hydrogen or negatively charged hydrogen ions. Consequently, electron stripping and capture processes by neutral hydrogen projectiles and negatively charged ions must be taken into account. For protons slowed down in hydrogen, such processes, combined to form so-called charge-exchange cycles, lead to remarkable ionisation yield contributions up to primary energies of 1 MeV.

A rough estimate of the energy dependence of the mean number N(T) of ionisations produced upon complete slowing down of protons and, thus also of W, can be simply calculated within the framework of the Spencer-Fano theory in the continuous slowing down approximation[4–5] if the contribution of heavier particles is neglected:

$$N(T) = \mathbb{N} \int_{T_{min}}^{T} [\sigma_t(T') Q(T')/(dT/dx)_{T'}] \, dT' \quad (1)$$

Here $\mathbb{N}$ is the number density of target molecules, $\sigma_t(T) = \sigma_i(T) + \sigma_{ex}(T)$ is the sum of the total cross section $\sigma_{ex}(T)$ for possible charge-exchange cycles and of the weighted cross section $\sigma_i(T)$ for direct ionisation with respect to protons and neutral hydrogen projectiles, $(dT/dx)_{T'}$ is the linear total stopping power at energy T', and Q(T) is a factor to take into account the ionisation yield contribution by secondary electrons. The integration is performed in the limits of the initial proton energy T and a value T_{min} slightly smaller than the ionisation threshold energy, I, to include charge-changing effects.

The factor Q(T) representing the ionisation yield contribution by secondary electrons is given by the following equation:

$$Q(T) = 1 + [\sigma_{ex}(T)/\sigma_t(T)]E_{ex}/W_e(E_{ex})$$
$$+ [\sigma_i(T)/\sigma_t(T)] \int_{I}^{E_{max}} q(T,E) [E/W_e(E)] dE \quad (2)$$

where $E_{ex} = T/1836.15$ is the energy of electrons set free during the charge-exchange cycle, W_e the corresponding electron W value, q(T,E) the spectral distribution of secondary electrons due to direct ionising collisions, and $E_{max} = T/459.0$ the appropriate maximum electron energy within the classical limits.

The cross section $\sigma_{ex}(T)$ with respect to charge-exchange cycles can be determined if all possible cross sections $\sigma_{\mu\nu}(T)$ of changing the charge state μ of a hydrogen projectile into a state ν are known; μ or ν will be equal to +1 if the projectile is a proton, and 0 or −1 if it is a neutral hydrogen atom or a negatively charged ion. In the case of a particle beam in the charge-state equilibrium, which is assumed to be valid for our analytical model, $\sigma_{ex}(T)$ can be approximately calculated by the following equation[14]:

$$\sigma_{ex}(T) = f_0(T) \, \sigma_{0-1}(T)$$
$$+ f_1(T) [\sigma_{10}(T) + 2\sigma_{1-1}(T)] \quad (3)$$

Here, $f_0(T)$ and $f_1(T)$ are the equilibrium fractions of neutral hydrogen projectiles or protons of energy T given, according to Tawara and Russek[15], by Equations 4 and 5.

$$f_\mu(T) = w_\mu(T) \bigg/ \sum_{\nu=-1}^{1} w_\nu(T) \quad (4)$$

$$w_\mu(T) = \sigma_{j\mu}(T) [\sigma_{k\mu}(T) + \sigma_{kj}(T)] + \sigma_{jk}(T) \sigma_{k\mu}(T) \quad (5)$$

where μ, j and k can assume the values +1, 0 and −1, with the restrictions $j \neq k$ and j and $k \neq \mu$.

With $\sigma_\mu^+(T)$ being the direct ionisation cross section of a projectile of charge state μ, the total cross section $\sigma_i(T)$ for direct ionisation is given by Equation 6:

$$\sigma_i(T) = \sum_{\mu=-1}^{1} f_\mu(T) \, \sigma_\mu^+(T) \tag{6}$$

After this brief outline of the procedure for calculating the W values [for more details, see References 5 and 6], a few words must be said about the data used in the present work. The total linear stopping power dT/dx was determined almost completely on the basis of the mass collision and mass nuclear stopping powers recently recommended for protons and alpha particles by the ICRU[9]. Up to energies of 200 keV, the collision stopping power was calculated using the empirical fitting functions of Anderson and Ziegler[16] combined with the recommended parameters of the ICRU (Table 3.1 of Reference 9). For higher proton energies Anderson and Ziegler's parameters were used together with the appropriate fitting function. To obtain the nuclear stopping power, the scaled values of the ICRU (Table 4.1 of Reference 9) were used. Our data are therefore based on the classical collision theory with the assumption of a screened Coulomb scattering potential and Molière's screening function and length[17] for a bare positive charge.

The cross sections $\sigma_{\mu\nu}(T)$ of single or double electron capture by protons, electron stripping or electron capture by neutral hydrogen atoms, and single or double electron stripping by negatively charged hydrogen ions were calculated using the fitting formula and the parameters of Nakai et al[18]; the cross sections $\sigma_\mu^+(T)$ for direct ionisation by hydrogen atoms or protons using the semi-empirical function of Green and McNeal[19] or the fitting functions and parameters of Rudd et al[20].

The spectral distribution q(E, T) of secondary electrons of energy E set in motion by particles of energy T was obtained using the formulae and parameters of the proton single-differential ionisation cross section by Rudd et al[21] assuming that it is independent of the charge state of the hydrogen projectiles. As far as electron W values are concerned, we used Combecher's data[22] extrapolated to the high energy limit recommended by the ICRU[2].

The Monte Carlo model

Since our analytical model does not include potential contributions to the ionisation yield by heavier secondary particles and by highly excited molecules, which may dissociate into stable or metastable neutral fragments or react with hydrogen molecules, for instance, to form H_3^+ ions, it was supplemented by a Monte Carlo simulation (for 10^4 primary protons at each energy) organised in the following way.

Starting with the initial proton energy, the total proton track was subdivided into a series of straight line segments connecting successive interaction points. The distance l(T) between two interaction points of a projectile of energy T was determined from the mean free path length $\lambda(T)$ in the usual way:

$$l(T) = -\lambda(T) \ln(\eta) \tag{7}$$

Here, η is a random number uniformly distributed between 0 and 1 and $\lambda(T) = [N \times \sigma_\mu(T)]^{-1}$, where $\sigma_\mu(T)$ is the total interaction cross section of a projectile of charge state μ given by Equation 8:

$$\sigma_\mu(T) = \sigma_\mu^<(T) + \sigma_\mu^>(T)$$
$$+ \sigma_\mu^*(T) + \sigma_\mu^+(T) + \sum_{\nu=-1}^{+1} \sigma_{\mu\nu}(T) \tag{8}$$

where the value of the summation index is restricted to $\nu \neq \mu$. The cross sections $\sigma_\mu^>(T)$ and $\sigma_\mu^<(T)$ are those for single elastic scattering with a polar scattering angle greater than a minimum angle Θ_{min} (for its definition, see below) or with a scattering angle less than Θ_{min}, respectively, the latter of which is treated within the framework of the continuous slowing down approximation as described below; $\sigma_\mu^*(T)$ and $\sigma_\mu^+(T)$ are the cross sections for excitation or direct ionisation and $\sigma_{\mu\nu}(T)$ that of a charge-changing effect from a state μ to a state ν.

The type of process the projectile of state μ suffers after passing through the distance l(T) is sampled from the probabilities of the different interaction effects, which are equal to the ratios of the appropriate cross sections and the total one. The energy loss of a projectile of energy T taken into account after a definite process had occurred was the following.

Single elastic interactions with a scattering angle less than Θ_{min} were bundled together to lead to a total energy loss of $0.01 \times T$. In the case of an interaction with scattering angle greater than Θ_{min}, the energy loss was determined from the kinematics of elastic scattering within the classical limits after calculation of the actual polar angle of scattering in the centre-of-mass system on the basis of the appropriate differential cross section (for details, see Reference 13).

The energy loss in the case of impact excitation was calculated using the excitation cross sections and excitation energies summarised by Phelps[23] for protons and neutral hydrogen; because of the lack of appropriate data, the excitation by negatively charged hydrogen ions was treated in the same manner as that by protons. The error induced by this treatment is small because H⁻ ions make up only a minor fraction of projectiles contained in a hydrogen beam.

As far as direct ionisation processes were concerned, the energy loss was calculated by adding the ionisation threshold of hydrogen molecules of 15.37 eV and the kinetic energy of secondary electrons determined on the basis of the single differential cross sections by Rudd et al[21] for protons in hydrogen in the same way as in our analytical model.

In agreement with Nakai et al[18], the energy loss of protons with respect to charge-changing effects was assumed to be 1.84 eV and 36 eV for single or double electron capture, that of neutral hydrogen 14.7 eV and

13.6 eV for single electron capture or single electron stripping, and that of negatively charged hydrogen ions 1.47 eV and 15 eV for single or double electron stripping. The energy E_{ex} of secondary electrons set in motion by stripping processes was assumed to be the same as in the analytical model.

To investigate the contributions made by secondary hydrogen molecules to the ionisation yield, their initial energy was set equal to the energy transfer of elastic scattering with an angle greater than Θ_{min} and the histories were followed up in the same way as those of the primary particles assuming, however, a velocity half that of a primary at the same energy.

The total ionisation yield necessary to determine the W value was calculated by summing up the number of electrons or positively charged hydrogen molecules produced during complete particle degradation. The additional yield contributions made by secondary electrons were determined in the same manner as in the analytical model by using the appropriate electron W values, and those after impact excitation by assuming that a fixed fraction R_{exc}, of excited molecules with excitation energies $k_{exc} > 17.6$ eV in the case of proton interactions and $k_{exc} > 22.5$ eV in the case of neutral hydrogen interactions leads to additional ion pairs, a fact well known in the case of the ionisation yield formation by primary electrons as discussed by Gerhart[24].

The history of each particle was assumed to be at its end when the initial energy had been degraded to 10 eV.

As far as the cross sections are concerned, the same data sets were used with respect to direct ionisation and charge-changing interactions as in the analytical model. The elastic cross section $\sigma_\mu^>(T)$ was calculated by integration of the differential elastic cross section in the angular range $\Theta_{min} \leq \Theta \leq \pi$ assuming an energy-dependent value of Θ_{min} which was set either equal to the value of the scattering angle that leads to an energy transfer of less than 10 eV, or equal to 5° if the corresponding energy transfer was less than 10 eV. In the case of protons, the differential cross section was calculated in the same way as described in Reference 13 whereas in the case of neutral hydrogen atoms or negatively charged hydrogen ions, the screening of the attached electrons was taken into account. The cross section $\sigma_\mu^<(T)$ was calculated using the difference of the total elastic stopping cross section at energy T and the fraction that leads to scattering angles greater than Θ_{min} and division by $0.01 \times T$.

The excitation cross sections $\sigma_\mu^*(T)$ were determined using the electronic linear stopping power of the analytical model after subtraction of the contributions of direct ionisation and charge-changing effects (weighted by the projectiles' charge state equilibrium fractions of Equation 4) and division by the product of the number density of hydrogen molecules and the mean excitation energy derived from the appropriate data of Phelps[23] with respect to impact excitation by protons or neutral hydrogen atoms. In doing so, the electronic stopping power was interpreted to be that of the charge-state equilibrium of the projectiles as also proposed by Phelps.

RESULTS AND DISCUSSION

Figure 1 shows a few results of our model calculations with respect to the mean energy W per ion pair formed. At a glance, the typical behaviour of W for protons as a function of the initial energy T is obvious:

(i) W steeply decreases with increasing energy in the very low energy region,
(ii) it is approximately constant at high energies, and
(iii) it shows some structure at energies in between, caused by the ionisation yield contributions due to charge-changing effects and to the degradation of secondary electrons. The importance of these contributions is apparent from the figure at energies greater than about 20 keV as far as secondary elec-

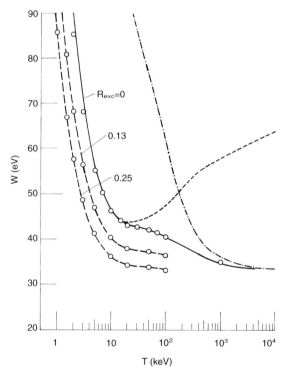

Figure 1. Calculated W values as a function of initial energy T: (—) results of the analytical model without restrictions; (------) results of the analytical model neglecting the ionisation yield contribution due to the degradation of secondary electrons; (-•-•-) results of the analytical model neglecting the ionisation yield contribution due to charge-changing processes; (O- - -O) results of the Monte Carlo simulation for different fractions R_{exc} of highly excited hydrogen molecules transformed into ion pairs.

trons are concerned, and at energies less than about 5 MeV with respect to charge-changing processes.

A further effect, also obvious from Figure 1, is a potential transformation of highly excited hydrogen molecules into ion pairs, which is of great importance in hydrogen gas as pointed out by Gerhart[24] when studying W values of low energy electrons. At 100 keV, for instance, the W value of protons will be reduced by about 4 eV if a fraction of $R_{exc} = 0.13$ of highly excited molecules is transformed into ion pairs and by about 7 eV in the case of $R_{exc} = 0.25$.

The reason for the strong influence of charge exchange and excitation processes on the W value becomes clear when considering the fraction of initial energy $\Delta E/T$ spent on these impact interactions upon the complete slow-down of hydrogen projectiles of energy T (see Figure 2). In the energy range 1 keV < T < 100 keV, about 50% of the initial energy is spent on excitation processes and up to about 20% on forming charge-exchange cycles, the most important of which is that of single electron capture by protons combined to single electron stripping by neutral hydrogen projectiles (cycle 1 ⇔ 0). The fraction of energy spent on direct ionising interactions becomes comparable to or greater than that of the former ones only at energies T > 15 keV as far as charge-changing processes are concerned, and at T > 100 keV with respect to impact excitation.

The last ionisation yield contribution, not yet mentioned but at least basically possible, is that by secondary hydrogen particles set in motion by the elastic scattering effect, which is of importance during the degradation of primary particles of energies less than about 10 keV. According to Figure 2, the energy fraction spent on elastic scattering is as high as 40% at T = 1 keV and still about 8% at 10 keV. Its contribution to the ionisation yield is, nevertheless, negligible since the energy of the greater part of secondary particles is less than a few hundred eV and therefore too low to lead to additional ion pairs. To demonstrate this fact, Figure 3 shows the results of the Monte Carlo simulation with regard to the spectral distribution $\Delta N/\Delta T'$ of the number N of secondary particles of energy T' set in motion upon the complete degradation of one proton of initial energy T. At a glance it can be seen from the figure that, independently of the primary energy, the overwhelming number of secondaries is produced in the energy range of $T' < 100$ eV and, because of this, does not contribute to the ionisation yield formation.

As a main result of our model calculations it can therefore be assumed that, because of additional competing interaction effects of the initial particles, the shape of the W value for hydrogen projectiles in molecular hydrogen as a function of energy T should show much more structure than the corresponding energy dependence of the W value for electrons (see, for instance, Reference 22). The real shape, however, must be taken from the experimental data summarised in Table 1, where W is the weighted mean of n single values W_ν (see Equation 9) measured at primary energy T using an average ratio F/p of the electric collection field strength and the gas pressure p.

$$W(T) = \sum_{\nu=1}^{n} s_\nu^{-2} W_\nu(T) \bigg/ \sum_{\nu=1}^{n} s_\nu^{-2} \quad (9)$$

Here, s_ν is the uncertainty of W_ν determined on the basis of the law of propagation of uncertainties taking into account those

(i) of the measurement of primary currents (≤1%),
(ii) of the energy of the primary particles (0.05%),
(iii) of the correction with respect to charge-changing effects in the beam guiding system of the accelerator (10%),
(iv) of the measurement of ionisation currents due to non-linearities of the current meter (0.3%), the extrapolation with respect to pressure (calculated for each measurement and in general less than 0.2%) and ion-collection efficiency (0.3%), and
(v) of the correction with respect to particle back-scattering inside the ionisation chamber (≤2%). In addition, an overall systematic uncertainty of 0.4% was taken into account.

The overall standard deviation, s_t, quoted, was

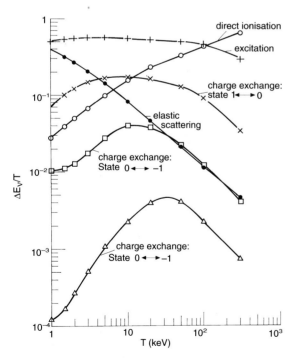

Figure 2. Fraction $\Delta E/T$ of the energy spent on possible interaction processes during the complete slowing down of primary particles of initial energy T calculated using the Monte Carlo method.

determined by quadratically adding a total systematic uncertainty of 0.4% and the standard deviation, s_w, of the weighted mean W, which is equal to $1/\Sigma s_\nu^{-2}$ if the summation is applied to ν from 1 to n.

The differential mean energy ω per ion pair formed which is also given in Table 1 can be defined by Equation 10 (see, for instance, Reference 1) and is obtained by numerical differentiation of the experimental W against T curve:

$$\omega(T) = W^2(T) \Big/ [W(T) - T(dW/dT)] \qquad (10)$$

The uncertainties of ω caused by the uncertainties of W and the numerical procedure of differentiation are estimated to be less than 1.5% at energies $1 \text{ keV} < T < 100 \text{ keV}$.

Before analysing the experimental results with respect to W, our results for the differential energy ω per ion pair formed should be considered, since

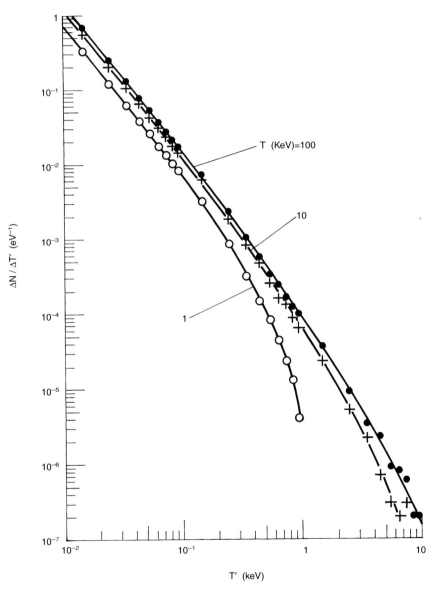

Figure 3. Spectral distribution $\Delta N/\Delta T'$ of the number N of secondary hydrogen projectiles of energy T' set in motion by elastic scattering upon the complete slowing down of one initial proton of energy T.

(i) some other experimental data sets for the relative energy dependence of ω are known from proportional counter measurements[25–27], and

(ii) ω is very sensitive to any structure of the W against T curve (because of Equation 10).

For this purpose, Figure 4 shows our experimental absolute values of Table 1 as a function of T in comparison with the relative values of Bennett[25] renormalised to a value of ω = 32.3 eV at energies T > 10 keV and the data of Breitung[26] renormalised to the same value of ω at T > 5 keV. A clear structure of ω over the whole energy range is apparent from the present data, quite similar to that of the other data sets for energies less than about 7 keV, confirming the structured behaviour of the mean energy per ion pair formed proposed by the theoretical calculations. The steep decrease of ω with increasing energy T in the very low energy range reflects the fact that the energy fraction spent on ion-producing interactions is rather small at low energies (see Figure 2) but increases with increasing energy; the structure obtained afterwards reflects the competition between the different possible interaction effects in part leading to ion pairs and in part representing wasted energy as far as the formation of ionisation yields is concerned.

To start with the discussion of our experimental W values (see Table 1), they should first be compared with the value of 35.6 eV measured by Gerthsen[7] in the energy region 19.1 keV ≤ T ≤ 35.4 keV. As is apparent from Table 1, in this energy range our measured data are smaller than that of Gerthsen by as much as 11.5%. The deviation is therefore much greater than the uncertainties to be ascribed to our experimental results (less than 1.5%), but not surprising if taking into account the uncertainty of Gerthsen's W value. This should be of the order of 10% to 15% (10% due to the measured ionisation yields as stated by Gerthsen and, in addition, about 5% with respect to the determination of the initial proton energy).

The comparison of our experimental W values and the results of the model calculations without taking ionisation yield contributions due to highly excited states into account (see Figure 1), also reveals that the measured data are appreciably smaller than the theoretical ones, by about 14 eV at 10 keV and 8 eV at 100 keV. These deviations are decreased to about 8 eV or 4 eV at the corresponding primary energies, assuming an additional ionisation yield contribution by transformation of the fraction R_{exc} = 0.13 of highly excited molecules into ion pairs.

Despite this improvement, the agreement between experimental and theoretical values is not satisfactory from the practical point of view. The deviations can, however, be understood in view of the rather large uncertainties of cross sections and even of the stopping power data used for performing the model calculations. As stated, for instance, by Rudd et al[20], the uncertainties of the total cross sections of direct impact ionisation of protons in molecular hydrogen are between 10% to 25% at energies around 70 keV and above, and even >25% at lower energies. A similar situation exists

Table 1. Experimental W values for protons slowed down in molecular hydrogen.

T (keV)	W (eV)	s_t (%)	n	$s_w \times n^{1/2}$ (%)	F/p (V.cm^{-1}.hPa^{-1})	ω (eV)
1.000	49.35	0.6	21	1.8	5.3	35.0
1.250	44.73	0.5	23	1.4	6.8	31.8
1.500	41.11	0.6	24	1.8	5.3	30.2
1.750	39.39	0.5	36	1.2	5.2	29.7
2.000	37.71	0.5	28	1.0	6.5	29.5
3.000	34.54	0.5	22	0.8	6.7	29.6
4.000	33.31	0.5	24	0.8	6.8	30.1
5.000	32.66	0.5	13	0.7	6.6	31.0
7.000	32.54	0.5	25	0.7	6.6	32.3
10.00	32.39	0.5	45	0.7	7.0	31.9
15.00	32.06	0.5	31	0.7	6.6	31.1
20.00	31.78	0.5	18	0.7	5.7	30.8
30.00	31.52	0.6	12	1.1	5.0	31.1
40.00	31.48	0.5	25	1.0	4.9	31.7
50.00	31.60	0.5	21	0.8	4.5	32.3
70.00	31.91	0.5	20	0.9	3.9	33.2
85.00	32.30	0.5	22	0.9	3.4	33.7
100.0	32.44	0.5	25	0.9	3.4	34.0

T, initial energy; W, weighted mean of n single measurements; s_t, overall standard deviation; s_w, standard deviation of the weighted mean; F/p, average ratio of the electric collection field strength and the gas pressure used during measurement; ω, differential value of the mean energy per ion pair formed.

as far as the cross sections for charge changing processes are concerned. According to Nakai *et al*[18], the uncertainties of the cross sections $\sigma_{10}(T)$, $\sigma_{-10}(T)$, and $\sigma_{-11}(T)$, for instance, are as high as 26%, 11% and 42%, respectively. The situation with regard to the stopping power data is not much better either. As quoted by the ICRU[9], the uncertainty of the nuclear stopping power is 10% to 20% at 1 keV and up to 10% at 100 keV; the uncertainty of the electronic stopping power is as high as 30% at 1 keV and up to 5% at 1 MeV. Moreover, its energy dependence, which is commonly assumed to be proportional to $T^{0.5}$ at energies $T < 10$ keV has been put into question by the measurements of Golser and Semrad[28] for helium projectiles and of Schiefermüller *et al*[29] for hydrogen particles.

Because of these rather large uncertainties of the cross sections it can be assumed that the greater part of the deviations between our measured and calculated W values is caused by the set of input data that had to be used in our model calculations. The original data were therefore somewhat modified in a tentative way to investigate the influence of cross section changes on the final results. To give an impression of these data, Figure 5 shows our experimental W values of Table 1 as a function of energy compared with the few experimental data known from literature[7,8], the recommendations by the ICRU[2] for energies $T > 100$ keV and theoretical values determined on the basis of our analytical model either with the original (Curve 1) or with modified input data (Curves 2 and 3). Curve 2 was obtained with a nuclear stopping power reduced by a factor of 2 and a total cross section with respect to the three charge-exchange cycles (see Equation 3) increased by a factor of 1.25, and Curve 3 with the original nuclear stopping power but the same modification of the charge-exchange cycles as in Curve 2 and, in addition, an electronic stopping power proportional to $T^{0.75}$ at energies $T < 10$ keV. As can be seen from the figure, the calculated W values presented by Curves 2 and 3 are clearly greater than our experimental results but agree in a very satisfactory way with the measurements by Gerthsen[7] between about 19 keV and 35 keV and the ω value of Bakker and Segrè[8] at 340 MeV and, in addition, they

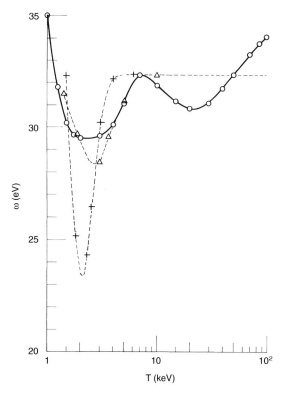

Figure 4. Differential value ω of the mean energy per ion pair formed as a function of energy T: (○) present experimental results; (△) data by Bennett[25] normalised to ω = 32.2 eV at $T > 10$ keV; (+) data by Breitung[26] normalised to the same ω value at $T > 5$ keV.

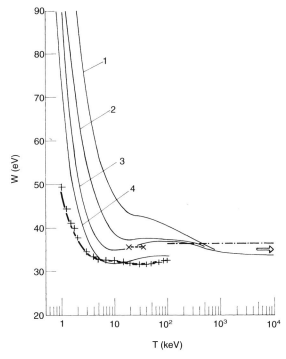

Figure 5. Mean energy W per ion pair formed as a function of energy T: (+) present experimental results; (×----×) experimental values by Gerthsen[7]; (⇒) ω value at 340 MeV by Bakker and Segrè[8]; (—●—●—) recommendation by the ICRU[2]; (Curves 1–3) results of the analytical model when using original (Curve 1) or modified stopping powers and cross sections with respect to charge-changing processes (for the details, see text); (Curve 4) results of the Monte Carlo calculation with the same modifications as in the analytical model (Curve 3) and, in addition, a transformation of highly excited molecules into ion pairs ($R_{exc} = 0.13$).

also confirm the recommendations by the ICRU at least in the energy range 100 keV ≤ T ≤ 1 MeV. Unfortunately, this agreement is not very decisive because of the large uncertainty of Gerthsen's W value and of that of our model calculations (see above), and the deviations of the theoretical values from our own experimental data are of much greater importance. These, however, decrease if we include in our model calculations also the transformation of highly excited states into ion pairs. As can be seen from Curves 3 and 4, a satisfactory agreement between our experimental and calculated W values (Curve 4) could be achieved by Monte Carlo simulation, at least at energies T > 3 keV, taking into account the same stopping power and cross section data as in Curve 3 and an additional yield contribution caused by the transformation of the fraction R_{exc} = 0.13 of highly excited states into ion pairs.

CONCLUSIONS

It was the aim of the present work to provide absolute values of the mean energy W required to form an ion pair and of the corresponding differential value ω for protons of initial energies T between 1 keV and 10 MeV, slowed down in molecular hydrogen. This aim could be safely achieved by measurements in the energy range 1 keV ≤ T ≤ 100 keV with resulting uncertainties of less than 1% as far as the W value is concerned (apart from a few exceptions at low energies) and of 1.5% with respect to ω.

The energy dependence of W at higher energies is, however, still uncertain for two reasons, the first being that no experimental data in the high energy region are known apart from the ω value at 340 MeV by Bakker and Segrè[8]. The second reason is that stopping power and cross section data necessary to perform reliable model calculations are either unknown or not known with sufficient accuracy. The energy dependence of our experimental W values at energies T > 50 keV and the results of our model calculations indicate, however, a further increase in W with increasing energy at T > 100 keV. From this point of view, a W value of (34.5 ± 1.5) eV at energies greater than about 500 keV seems to be more realistic than the value of 36.43 eV for α particles recommended by the ICRU[2] also for protons.

The final decision with respect to the real energy dependence of W at T > 100 keV must therefore be left to future work. This should apply

(i) to the measurement of W and ω up to high energies, at least up to about 5 MeV,
(ii) to the measurement of elastic and inelastic stopping powers at energies less than 100 keV and, in particular, at energies less than 10 keV,
(iii) to the accurate measurement of cross sections with respect to direct ionisation by protons and neutral hydrogen atoms and to those of all possible charge-changing processes, and
(iv) to a detailed investigation of excitation interactions and dissociative processes (with special emphasis on the formation of metastable hydrogen atoms), and of the formation of H_3^+ ions, for instance, by neutral hydrogen projectiles.

REFERENCES

1. Booz, J. *Precision and Accuracy of W-Values as Applied in Radiation Biophysics and Radiobiology*. In: Ionizing Radiation Metrology, Ed. E. Casnati (Editrice Compositori Bologna) pp. 419–435 (1977).
2. ICRU. *Average Energy Required to Produce an Ion Pair*. Report **31** (Washington, DC 20014: ICRU Publications) (1979).
3. Srdoč, D., Inokuti, M. and Krajcar-Bronić, I. *Yields of Ionization and Excitation in Irradiated Matter*. In: Atomic and Molecular Data for Radiotherapy and Radiation Research, IAEA-TECDOC-799 (Vienna: International Atomic Energy Agency) pp. 547–631 (1995).
4. Grosswendt, B. *Ionization Yields of Electrons and Protons in Gases*. In: Proc. 10th International Congress of Radiation Research, Würzburg (1995).
5. Baek, W. Y. and Grosswendt, B. *Energy Dependence of W Values for Protons in Gases*. Radiat. Prot. Dosim. **52**, 97–104 (1994).
6. Grosswendt, B. and Baek, W. Y. *The Energy Dependence of W for Protons of Energies up to 10 MeV*. Radiat. Prot. Dosim. **61**, 267–274 (1995).
7. Gerthsen, Chr. *Über Ionisation und Reichweite von H-Kanalstrahlen in Luft und Wasserstoff*. Ann. Phys. (5. Folge) **5**, 657–669 (1930).
8. Bakker, C. J. and Segrè, E. *Stopping Power and Energy Loss for Ion Pair Production for 340-MeV Protons*. Phys. Rev. **81**, 489–492 (1951).
9. ICRU. *Stopping Powers and Ranges for Protons and Alpha Particles*. Report **49** (Bethesda, Maryland 20814: ICRU Publications) (1993).
10. Waibel, E. and Grosswendt, B. *Zur Technik der Bestimmung von W-Werten in Gasen*. PTB-Mitt. **87**, 13–21 (1977).
11. Waibel, E. and Willems, G. *Ionisation Ranges and W Values for Low Energy Protons in Tissue-Equivalent Gas*. Radiat. Prot. Dosim. **13**, 79–81 (1985).
12. Waibel, E. and Willems, G. *W Values for Low-Energy Protons in Methane-Based Tissue-Equivalent Gas and its Constituents*. Phys. Med. Biol. **37**, 249–259 (1992).

13. Grosswendt, B. and Willems, G. *The Backscattering of Low Energy Protons in Gases of Importance for Ionisation Measurements*. Radiat. Prot. Dosim. **61**, 245–252 (1995).
14. Baek, W. Y. and Grosswendt, B. *W Values for Helium Particles in Nitrogen and Argon in the Energy Range from 10 keV to 10 MeV*. Phys. Med. Biol. **40**, 1015–1029 (1995).
15. Tawara, H. and Russek, A. *Charge Changing Processes in Hydrogen Beams*. Rev. Mod. Phys. **45**, 178–229 (1973).
16. Andersen, H. H. and Ziegler, J. F. *Hydrogen: Stopping Powers and Ranges in All Elements*, Vol. 3 of 'The Stopping and Ranges of Ions in Matter' (New York: Pergamon Press) (1977).
17. Molière, G. *Theorie der Streuung schneller geladener Teilchen I: Einzelstreuung am abgeschirmten Coulomb-Feld*. Z. Naturforsch. **2a**, 133–145 (1947).
18. Nakai, Y., Shirai, T., Tabata, T. and Ito, R. *Cross Sections for Charge Transfer of Hydrogen Atoms and Ions Colliding with Gaseous Atoms and Molecules*. At. Data Nucl. Data Tables **37**, 69–101 (1987).
19. Green, A. E. S. and McNeal, R. J. *Analytic Cross Sections for Inelastic Collisions of Protons and Hydrogen Atoms with Atomic and Molecular Gases*. J. Geophys. Res. **76**, 133–144 (1971).
20. Rudd, M. E., Kim, Y.-K., Madison, D. H. and Gallagher, J. W. *Electron Production in Proton Collisions: Total Cross Sections*. Rev. Mod. Phys. **57**, 965–994 (1985).
21. Rudd, M. E., Kim, Y.-K., Madison, D. H. and Gay, T. J. *Electron Production in Proton Collisions with Atoms and Molecules: Energy Distributions*. Rev. Mod. Phys. **64**, 441–490 (1992).
22. Combecher, D. *Measurement of W Values of Low-Energy Electrons in Several Gases*. Radiat. Res. **84**, 189–218 (1980).
23. Phelps, A. V. *Cross Sections and Swarm Coefficients for H^+, H_2^+, H_3^+, H, H_2, and H^- in H_2 for Energies from 0.1 eV to 10 keV*. J. Phys. Chem. Ref. Data **19**, 653–675 (1990).
24. Gerhart, D. E. *Comprehensive Optical and Collision Data for Radiation Action. I. H_2*. J. Chem. Phys. **62**, 821–832 (1975).
25. Bennett, E. F. *Low-Energy Limitations on Proton-Recoil Spectroscopy Through the Energy Dependence of w*. Trans. Am. Nucl. Soc. **13**, 269–274 (1970).
26. Breitung, W. *Untersuchung zur Energieabhängigkeit von w (Energieverlust/Ionenpaar) für Protonen in Wasserstoff, Methan und H_2-CH_4-Gemischen im Hinblick auf Neutronenspektrumsmessungen mit Rückstoßprotonen-Proportionalzählrohren unterhalb 25 keV*. Report KFK-1623 (Ges. für Kernforschung mbH. (KFK), Karlsruhe) (1972).
27. Werle, H., Fieg, G., Seufert, H. and Stegemann, D. *Investigation of the Specific Energy Loss of Protons in Hydrogen Above 1 keV with Regard to Neutron Spectrometry*. Nucl. Instrum. Methods **72**, 111–119 (1969).
28. Golser, R. and Semrad, D. *Observation of a Striking Departure from Velocity Proportionality in Low-energy Electronic Stopping*. Phys. Rev. Lett. **66**, 1831–1833 (1991).
29. Schiefermüller, A., Golser, R., Stohl, R. and Semrad, D. *Energy Loss of Hydrogen Projectiles in Gases*. Phys. Rev. A**48**, 4467–4475 (1993).

STOPPING POWERS OF He AND Ar GASES FOR 50–200 keV ^{3}He$^+$ IONS

A. Fukuda
Electrotechnical Laboratory
1-1-4 Umezono, Tsukuba, Ibaraki 305, Japan

Abstract — Absolute stopping powers of He and Ar gas were measured for ^{3}He$^+$ ions in the energy range between 50 and 200 keV. The data for Ar gas are of practical importance in the field of neutron metrology when using ^{3}He counters. The measurements were performed using a differentially pumped gas cell and an electrostatic energy loss spectrometer. The results are compared with the ^{4}He$^+$ stopping powers of He and Ar in the same energy range, previously measured by the present author. The results show that the measured values for ^{3}He$^+$ ions are almost equal to the values estimated theoretically from previous results for ^{4}He$^+$ ions in the lower energy region. There is a difference in the higher energy region.

INTRODUCTION

In designing and utilising a ^{3}He counter for neutron measurement, some Ar gas is added to control the range of ^{3}He$^+$ ions produced in the counter. For the determination of the partial pressure of the Ar gas to be added, stopping powers of Ar for ^{3}He$^+$ ions must be known. These were measured for energies from 50 keV to 200 keV. ^{3}He$^+$ stopping powers of He gas were also measured for comparison. A gas cell with a differential pumping system and a 127° electrostatic energy analyser were used for the measurement.

METHOD AND APPARATUS OF THE EXPERIMENT

The method and apparatus used for the absolute measurement of the stopping powers were the same as those of previous works[1–3]. A schematic view of the measuring device is shown in Figure 1. ^{3}He$^+$ ions are produced by a duoplasmatron type ion source and are accelerated by a Cockroft–Walton type accelerator. Analysed ions are led into a gas cell, the beam path length of which is 53.4 cm. The gas cell is set in a differentially pumped chamber and filled with Ar or He. The pressure of the gas is kept constant during an energy loss measurement. The pressure was varied from 0.1 Pa to 1.0 Pa in steps of 0.1 Pa. At each pressure, an energy loss spectrum was measured by an electrostatic cylindrical energy analyser placed inside the differentially pumped chamber. It has a beam path radius of 300 mm and is controlled by a computer that also controls an ion counting system. Data reduction and analysis are the same as in previous works[1–3].

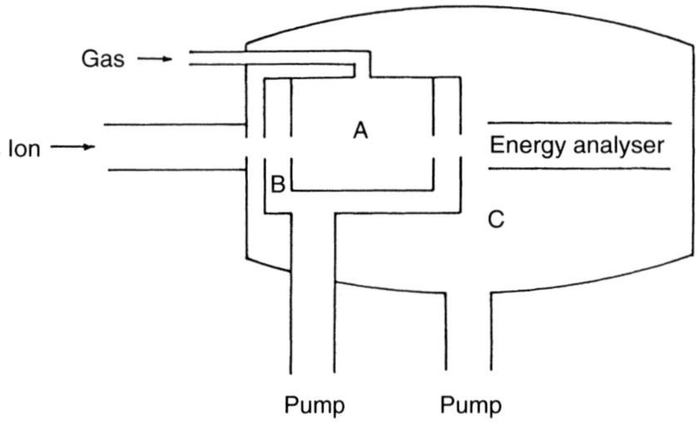

RESULTS AND DISCUSSION

The results are shown in Figures 2 and 3 along with results of earlier experiments for ^{4}He$^+$ ions. All the values are tabulated in Tables 1 and 2. The systematic and random errors are the same as in Reference 1. In the energy region studied in the present work, the theory of electronic stopping power is not established and dif-

Figure 1. Schematic side view of the chamber. ^{3}He$^+$ ions come from the accelerator, are analysed by a magnet and are led into the gas cell. The differentially pumped gas cell is supported in the large chamber which also contains the electrostatic energy loss spectrometer.

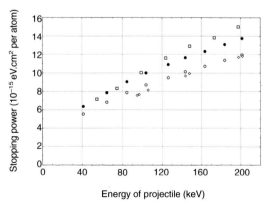

Figure 2. Electronic stopping powers of He for ^{3}He$^+$. (□) are the present results. (○) are electronic stopping powers of He for ^{4}He$^+$ from previous work[3], (◇) are by Baumgart et al[5] and (●) are stopping powers of He for ^{3}He$^+$ theoretically estimated using Equation 2 and the results of Reference 3.

ferent empirical fitted formulae are proposed by many researchers. One of the formulae is, for example, that of Lindhard and Scharff[4] given by Equation 1:

$$Se = \xi(8\pi e^2 a_0^2) \left(\frac{Z_1 Z_2}{(Z_1^{2/3} + Z_2^{2/3})^{3/2}} \right) \frac{v}{v_0} \quad (1)$$

where e is the elementary charge, a_0 and v_0 are the radius and velocity of an electron in the first Bohr orbit in hydrogen, Z_1 and Z_2 are the atomic numbers of the incident and target atoms, v is the velocity of the incident atom and ξ is a constant of about $Z_1^{1/6}$. According to Equation 1, the ratio of the stopping power of Ar for $^3He^+$ ($Se(^3He\text{-}Ar)$) and that of Ar for $^4He^+$ ($Se(^4He^+\text{-}Ar)$) is equal to the ratio of the velocities of the incident ions. Since the ratio of the velocities of the same projectile energy E equals the inverse ratio of the square root of the masses $m(^3He)$ and $m(^4He)$ of the ions, the stopping power ratio for $^3He^+$ and $^4He^+$ ions is given by Equation 2:

$$\frac{Se(^3He^+ - Ar)}{Se(^4He^+ - Ar)} = \frac{v(^3He^+)}{v(^4He^+)} =$$

$$\frac{\sqrt{\left(\frac{2E}{m(^3He^+)}\right)}}{\sqrt{\left(\frac{2E}{m(^4He^+)}\right)}} = \sqrt{\left(\frac{m(^4He^+)}{m(^3He^+)}\right)} = 1.15 \quad (2)$$

In the Figures 2 and 3 the measured $^3He^+$ stopping powers are compared with those for $^3He^+$ by Equation 2 using the previously measured data for ^4He ions[3]. A difference between the present experiments and calculated result in the high energy region is obvious from the figures. As the experimental data are larger than the calculated ones, the ratio of stopping powers given by Equation 2 must be larger than 1.15. An explanation of this fact could be that the stopping power of the energies studied in the present work is not proportional to the velocity of the projectile. Stopping power measurement of other rare gases for $^3He^+$ ions will be continued for a more detailed discussion.

ACKNOWLEDGMENTS

The author would like to thank Tatsuo Kakutani for his helpful assistance in measurements and in data reduction. He would also like to thank Dr Nobuhisa Takata, Dr Katsuhisa Kudo and Naoto Takeda for their helpful discussions.

Table 1. Measured stopping powers of He for $^3He^+$.

$^3He^+$ energy (keV)	Stopping power (10^{-15} keV cm^2 per atom of He)
54.70 ± 0.36	7.15 ± 0.12
74.43 ± 0.42	8.31 ± 0.14
99.07 ± 0.50	10.03 ± 0.17
123.67 ± 0.57	11.61 ± 0.18
148.29 ± 0.64	12.90 ± 0.21
173.05 ± 0.72	13.81 ± 0.23
197.72 ± 0.79	15.01 ± 0.26

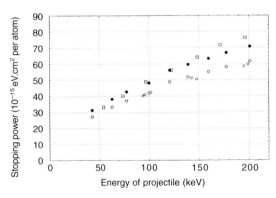

Figure 3. Electronic stopping powers of Ar for $^3He^+$. (□) are the present results. (○) are electronic stopping powers of Ar for $^4He^+$ from previous work[3], (◇) are by Baumgart et al[5] and (●) are stopping powers of Ar for $^3He^+$ theoretically estimated using Equation 2 and the results of Reference 3.

Table 2. Measured stopping powers of Ar for $^3He^+$.

$^3He^+$ energy (keV)	Stopping power (10^{-15} eV cm^2 per atom of Ar)
54.19 ± 0.36	33.06 ± 0.55
73.91 ± 0.42	40.12 ± 0.67
97.02 ± 0.49	48.72 ± 0.81
122.92 ± 0.57	56.03 ± 0.93
148.53 ± 0.64	64.1 ± 1.1
171.88 ± 0.72	71.7 ± 1.2
196.35 ± 0.79	76.2 ± 1.3

REFERENCES

1. Fukuda, A. *Stopping Powers of a Tissue-equivalent Gas for 40–200 keV Protons*. Phys. Med. Biol. **25**, 877–886 (1980).
2. Fukuda, A. *Stopping Powers of a Tissue-equivalent Gas for 40–200 keV He$^+$ and N$^+$*. Phys. Med. Biol. **26**, 623–632 (1981).
3. Fukuda, A. *Stopping Powers of Rare Gases for 40–200 keV Rare Gas Ions*. J. Phys. B: At. Mol. Phys. **14**, 4533–4544 (1981).
4. Lindhard, J. and Scharff, M. *Energy Dissipation by Ions in the keV Region*. Phys. Rev. **124**, 128–130 (1961).
5. Baumgart, H., Berg, H., Huttel, E., Pfaffe, E., Reiter, G. and Clausnitzer, G. *^{4}He Stopping Cross Sections in H$_2$, He, N$_2$, O$_2$, Ne, Ar, Kr, Xe, CH$_4$ and CO$_2$*. Nucl. Instrum. Methods **215**, 319–328 (1983).

CALCULATIONS OF THE GIANT-DIPOLE-RESONANCE PHOTONEUTRONS USING A COUPLED EGS4-MORSE CODE

J. C. Liu, W. R. Nelson, K. R. Kase and X. S. Mao
Stanford Linear Accelerator Center
PO Box 4349, Stanford, CA 94309, USA

Abstract — The production and transport of the photoneutrons from the giant-dipole-resonance reaction have been implemented in a coupled EGS4-MORSE code. The total neutron yield (including both the direct neutrons and evaporation neutrons) is calculated by folding the photoneutron yield cross sections with the photon track length distribution in the material. Empirical algorithms were developed to estimate the fraction and energy of the direct neutron. The EVAP4 code, incorporated as a MORSE sub-routine, was used to determine the energies of the evaporation neutrons. The emission of slow neutrons (<2.5 MeV) was assumed to be isotropic and the fast neutrons were emitted anisotropically in the form of $1+C\sin^2\theta$. Comparisons between the calculations and the measurements (spectra of the direct, evaporation and total neutrons; nuclear temperatures; direct neutron fractions) for lead, tungsten, tantalum and copper show satisfactory results over the photon energy range of interest.

INTRODUCTION

For electron accelerators (<30 MeV), neutrons from the photon-induced giant-dipole-resonance (GDR) reaction consist of a large portion of evaporation neutrons which dominate at low energies and a small fraction of direct neutrons which dominate at high energies. The energies of the evaporation neutrons can best be described by the Weisskopf's statistical model, on which the EVAP4 code[1] is based. The energies and yield of the direct neutrons can be calculated using the pre-equilibrium theories that a few codes[2–4] use. To our knowledge, there has been no comprehensive presentation of the photoneutron spectra and angular distributions from these theoretical calculations.

On the other hand, simplified calculations of the photoneutron spectra have been used for medical accelerators and other applications[5–8]. These calculations suffer from two major assumptions they used and which, therefore, render their applications unsuitable for other situations. The first assumption is the use of a Maxwellian fission spectrum with a constant nuclear temperature (T)[6,8], regardless of the photon energy, to approximate an evaporation neutron spectrum. The second assumption is about the fraction and the spectral and angular distributions of the direct neutron component. For example, the spectral and angular distributions were assumed to be the same as those of the evaporation neutrons[7,8], whereas the actual energy is higher and fast neutrons are emitted anisotropically[5,9]. The energy of the direct neutron was assumed to be the difference between the photon energy and the neutron separation energy[5,6], whereas the direct neutron energies are actually smaller than the difference and they are not monoenergetic either. The fraction of the direct neutrons was assumed to be constant[6], whereas the fraction should increase as the photon energy becomes higher, and should also vary for different targets[9,10].

In this study, the production and transport of the GDR photoneutrons, with regard to the spectral and angular distributions, in a coupled EGS4[11] and MORSE[12] code is the main subject. The approach is a compromise between the above-mentioned theoretical and simplified methods. Basically, empirical algorithms were developed based mainly on the Mutchler thesis[9] to estimate the fraction and energy of the direct neutron for each photon energy. The EVAP4 code, included as a MORSE sub-routine, is used to obtain the energies of the evaporation neu-

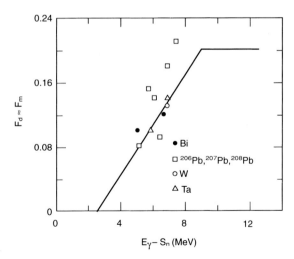

Figure 1. The direct neutron fraction plotted against the difference of the photon energy (E_γ) and the neutron separation energy (S_n) for a few high-Z isotopes[9]. The linear line shown was used as the function to estimate the direct neutron fraction.

trons. The emissions of slow neutrons (≤2.5 MeV) are assumed to be isotropic and an angular algorithm, also derived from Mutchler[9], is used for the anisotropic emissions of fast neutrons.

The main area of application at this stage is the radiotherapy accelerator and the materials of interest are lead, tungsten, tantalum, and copper. Therefore, the empirical algorithms were developed based on the measurements for these elements. There are some measurements of the spectral and angular distributions[9,10,13–16] of the photoneutrons for different beam energies and target materials. The Mutchler thesis[9] gave the most comprehensive information and, therefore, was used as the main basis to derive the empirical algorithms.

METHODS

Neutron production and yield

An EGS4 user code, using the Combinatorial Geometry routine, was used to generate a photon history file, which contains the parameters of position, direction, energy and track length for those photons capable of generating neutrons in each medium. The photon histories are then randomly sampled by MORSE with a user-written SOURCE sub-routine, which has incorporated the empirical algorithms and the EVAP4 code for the production and transport of photoneutrons. The neutron (both direct and evaporation) yield for the sampled photon was calculated by folding the photoneutron yield cross section[17] with the photon track length.

Direct neutron fractions

The direct neutron fraction, F_d, is defined to be the ratio of the neutrons from the direct process to the total neutrons. Mutchler[9] defined his measured direct neutron fraction, F_m, as the ratio of the direct neutrons >3 MeV to the total neutrons >0.4 MeV. Figure 1 plots the measured F_m values against the difference of the photon energy (E_γ) and the neutron separation energy (S_n) for a few high-Z isotopes[9]. It is known[9,10] that the higher the photon energy, the higher the direct neutron fraction. In the photon energy range of interest, the maximum of the measured fractions[5,9] was ~0.2.

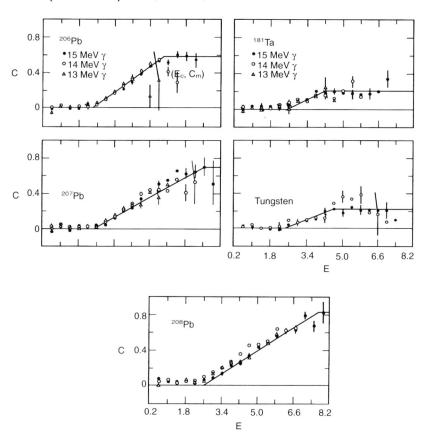

Figure 2. Angular distribution of the photoneutrons in the form $1+C\sin^2\theta$ where the measured relationships[9] between the constant C and the neutron energy for materials at different photon energies are shown.

Therefore, a linear relationship shown in Figure 1 was used as an empirical algorithm to estimate the direct neutron fraction F_d for all materials. It was also assumed that $F_d = F_m$ and only photons with energies greater than the S_n of the medium by 2.5 MeV can produce direct neutrons.

Direct neutron energies

It was also observed from Mutchler[9] that the direct neutron spectra from lead isotopes (undeformed nuclei) had peaks about 10–20% lower than the value of $(E_\gamma - S_n)$. The direct neutron spectra from tungsten and tantalum isotopes (deformed nuclei) were even softer. Therefore, for a photon energy E_γ and an isotope with a separation energy S_n, the resulting direct neutron spectrum was assumed to be constant between the neutron energy values of $(E_\gamma - S_n)$ and $D(E_\gamma - S_n)$, where D is a constant <1. The direct neutron intensity was assumed to be zero at other energies. It was found that $D = 0.7$ for undeformed nuclei and $D = 0.4$ for deformed nuclei could produce direct neutron spectra that are in agreement with the measurements. The energies of the evaporation neutrons were determined using EVAP4, after checking that the level density parameters in EVAP4 are consistent with the published values[18].

Neutron emission angle

The angular distribution of the direct neutrons was known to be in the form of $1+C\sin^2\theta$, where θ is the angle between the incoming photon and the emitted neutron and C is a constant dependent on neutron energy. Figure 2 reproduces the measured relationship[9] between the constant C and the neutron energy for materials at different photon energies. The value C_m and the neutron energy E_c, at which the value C reaches its maximum, vary for different elements/isotopes. The linear line in each sub-figure is used to estimate the value C at a neutron energy for an isotope. The function $1+C\sin^2\theta$ is then used to sample the neutron emission angle θ. In our angular algorithm, neutrons $\leqslant 2.5$ MeV are assumed to be emitted isotropically, i.e. $C = 0$. For isotopes without measured angular distributions, an isotropic emission is assumed.

SPECTRAL RESULTS AND COMPARISONS

Spectral calculations and comparisons were made for lead, tungsten, tantalum, and copper with different beam conditions. Detailed and satisfactory results have been described elsewhere[19] and only some examples are given below.

Lead (an undeformed nucleus)

Mutchler[9] measured the neutron spectra (and the direct neutron fraction F_m) from targets of ^{208}Pb, ^{207}Pb and ^{206}Pb isotopes (2 cm radius and 0.635 cm thick) hit by a quasi-monoenergetic photon beam peaked at 14 MeV. Good spectral agreements for lead between the calculations and the measurements are shown in Figure 3(a) for direct neutrons and in Figure 3(b) for total neutrons (the two curves were arbitrarily normalised at 1.5 MeV). The semilogarithmic plot in Figure 3(b) allows the determination of the nuclear temperature T from the slope of the evaporation neutron curve. The nuclear temperature of the evaporation neutrons between 0.9 and 3.5 MeV is 0.97 MeV; this agrees with the published values[9,10,15]. The calculated F_m is 0.13, ~20% lower than the measured value of 0.16. The direct neutron fraction F_d is 0.12. Note that if the direct neutron is not included, the average energy of the spectrum would be ~1.5 MeV, much smaller than the actual value of 2.28 MeV.

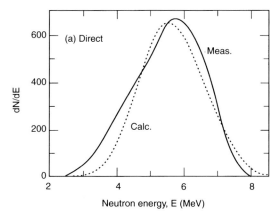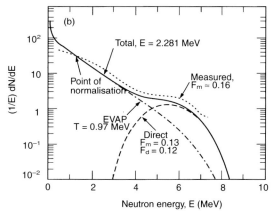

Figure 3. Comparisons between the calculations and the measurements[9] for direct neutrons (part a) and total neutrons (part b) from a lead target hit by quasi-monoenergetic 14 MeV photons. The nuclear temperature T for the evaporation neutrons is 0.97 MeV. The calculated F_m is 0.13 (20% lower than the measurement) and the direct neutron fraction F_d is 0.12.

To check that the algorithm can be applied to a much wider photon energy range, a comparison was made with one of the Kimura measurements[10], in which the neutron spectrum at 90° with respect to a 16.5 MeV electron beam hitting a 'thick' lead target (size $10 \times 10 \times 5$ cm^3) was measured. The calculated neutron spectrum is shown in Figure 4. The nuclear temperature is 0.88 MeV for total neutrons between 2 and 3.5 MeV and is 1.45 MeV for total neutrons between 4 and 6 MeV; within 3% of the measure-

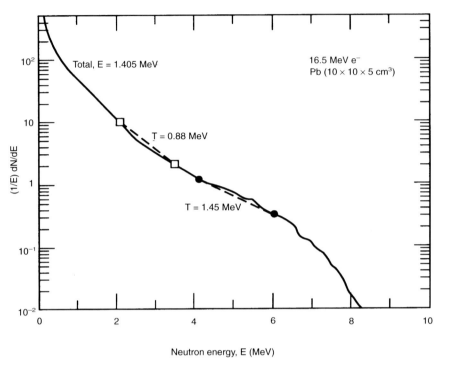

Figure 4. Calculated neutron spectrum at 90° relative to a 16.5 MeV electron beam hitting a thick lead target. The T values for two neutron energy regions are within 3% of the measurements[10].

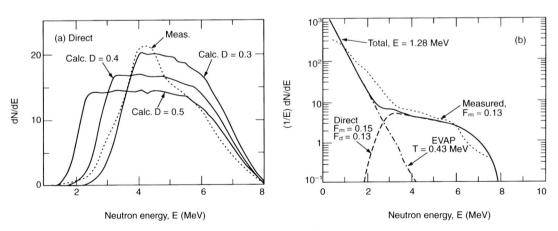

Figure 5. Comparisons between the calculations and the measurements[9] for direct neutrons (part a) and total neutrons (part b) from a tungsten target hit by quasi-monoenergetic 14 MeV photons. The algorithm with a D = 0.4 gave a reasonable direct neutron spectrum. The nuclear temperature T for the evaporation neutrons is 0.43 MeV. The calculated F_m is 0.15 (15% higher than the measured value) and the direct neutron fraction F_d is 0.13.

ments. The calculated F_m and F_d are only 0.07 and 0.06, respectively. These are lower than the Mutchler case, because the photon track length distribution inside a thick target is E_γ^{-2}.

Tungsten (a deformed nucleus)

The neutron spectrum from the Mutchler 14 MeV photons hitting a tungsten disc, 2.54 cm radius and 0.556 cm thick, was calculated. Spectral comparisons between the calculations and the measurements[9] are shown in Figure 5(a) for direct neutrons (with three D values) and in Figure 5(b) for total neutrons. Considering the large uncertainty for the measured direct neutron spectrum, the empirical algorithm with a D = 0.4 seems to produce a reasonable direct neutron spectrum. The nuclear temperature for the evaporation neutrons between 0.9 and 3.5 MeV is 0.43 MeV. The calculated total spectrum is in fair agreement with the measurements, except again at the high and low energy ends. The calculated F_m is 0.15, about 15% higher than the measured value. The direct neutron fraction F_d is 0.13.

Another comparison with the measured photoneutrons[18] from a tungsten target (1.8 cm radius and 2 cm thick) bombarded by a E_γ^{-1} bremsstrahlung beam with an endpoint of 24 MeV was also made, and the results are shown in Figure 6. The calculated nuclear temperature is 0.49 MeV and the direct neutron fraction is 0.14. These values agree well with the measurements, whose T value was 0.50 MeV by assuming a direct neutron fraction between 0.15–0.20.

SUMMARY AND CONCLUSIONS

The spectral and angular characteristics of the GDR photoneutrons are important for many applications. For example, photoneutrons seriously contaminate the radiation fields from medical accelerators operated above 8 MeV. In the beamline shielding design for synchrotron light facilities, the photoneutrons from gas bremsstrahlung hitting the beamline devices are one of the main radiation sources. In this study the production and transport of the GDR photoneutrons have been implemented using empirical spectral and angular algorithms in a coupled EGS4-MORSE code. General agreements between the calculations and the measurements (spectra, nuclear temperatures and direct neutron fractions) show that the simple algorithms can produce reasonable results. However, our approach may not be appropriate for elements with $Z < 20$, because of their more discrete features of the nuclear energy levels.

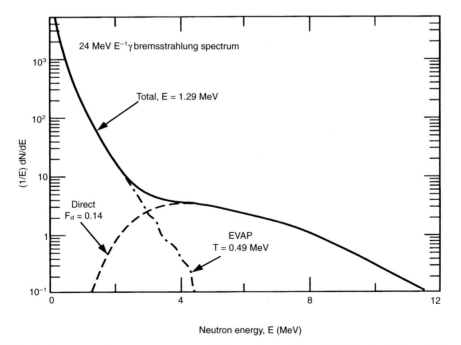

Figure 6. Calculated photoneutrons from a tungsten target (1.8 cm radius and 2 cm thick) bombarded by a E_γ^{-1} bremsstrahlung beam with an endpoint of 24 MeV. The calculated nuclear temperature T is 0.49 MeV and the direct neutron fraction is 0.14, which agree well with the measurements[18].

REFERENCES

1. ORNL. *EVAP Calculation of Particle Evaporation from Excited Compound Nuclei.* (Radiation Shielding Information Center, Oak Ridge National Laboratory, Oak Ridge, TN) PSR-10 (1974).
2. Blann, M. *Recent Progress and Current Status of Preequilibrium Reaction Theories and Computer Code ALICE* (Lawrence Livermore National Laboratory) UCRL-JC-109052 (1991).
3. Chadwick, M. B. and Young, P. G. *Photonuclear Reactions in the GNASH Code: Benchmarking Model Calculations on Lead up to 140 MeV.* (Lawrence Livermore National Laboratory) UCRL-ID-118721 (1994).
4. Fasso, A., Ferrari, A. and Sala, P. R. *Designing Electron Accelerator Shielding with FLUKA.* In: Accelerator Shielding Specialist Meeting, Arlington, TX, April 1994.
5. NCRP. *Neutron Contamination from Medical Electron Accelerators* (Bethesda, MD: National Council on Radiation Protection and Measurements) NCRP Report 79 (1984).
6. Agosteo, S., Foglio Para, A., Silari, A., Torresin, A. and Tosi, G. *Monte Carlo Simulations of Neutron Transport in a Linac Radiotherapy Room.* Nucl. Instrum. Methods Phys. Res. B**72**, 84–90 (1992).
7. Armstrong, T. W., Colborn, B. L. and Johnson, D. L. *Transport Calculations of Radiation Streaming through Shielding Penetrations for a Free-Electron-Laser Accelerator Facility.* In: Proc. ANS Topical Meeting on New Horizons in Radiation Protection and Shielding, Pasco, WA, 26 April–1 May 1992.
8. Sanford, T. W. L., Lorence, L. J., Halbleib, J. A., Kelly, J. G., Griffin, P. J., Poukey, J. W., McAtee, W. H. and Mock, R. C. *Photoneutron Production Using Bremsstrahlung from the 14-TW Pulsed-Power HERMES III Electron Accelerator* Nucl. Sci. Eng. **114**, 109–213 (1993).
9. Mutchler, G. S. *The Angular Distributions and Energy Spectra of Photoneutrons from Heavy Elements.* Ph.D. Thesis, Massachusetts Institute of Technology (1966).
10. Kimura, I., Hayashi, S. A., Kobayashi, K., Yamamoto, S. and Shibata, T. *Measurements of Angular Distributions and Energy Spectra of Photoneutrons from Lead Targets Bombarded by High Energy Electrons.* Annu. Rep. Res. Reactor Inst. Kyoto Univ. **3**, 75–83 (1970).
11. Nelson, W. R., Hirayama, H. and Rogers, D. W. O. *The EGS4 Code System* (Stanford Linear Accelerator Center, Stanford, CA) SLAC-265 (1985).
12. Emmett, M. B. *The MORSE Monte Carlo Radiation Transport Code System* (Radiation Shielding Information Center, Oak Ridge National Laboratory, Oak Ridge, TN) ORNL-4972 (1975).
13. Byerly, P. R. Jr. and Stephens, W. E. *Photodisintegration of Copper.* Phys. Rev. **83**(1), 54–62 (1951).
14. Toms, M. E. and Stephens, W. E. *Photoneutrons from Lead.* Phys. Rev. **108**(1), 77–81 (1951).
15. Gayther, D. B. and Goode, P. D. *Neutron Energy Spectra and Angular Distributions from Targets Bombarded by 45 MeV Electrons.* J. Nucl. Energ. **21**, 733–747 (1967).
16. Poss, H. L. *Note on Angular Asymmetries in (γ,n) Reactions.* Phys. Rev. **79**, 539–540 (1950).
17. Dietrich, S. S. and Berman, L. B. *Atlas of Photoneutron Cross Sections Obtained with Monoenergetic Photons.* At. Data Nucl. Data Tables **38**, 199 (1988).
18. Barrett, R. F., Birkelund, J. R., Thomas, B. J., Lam, K. S. and Thies, H. H. *Systematic of Nuclear Level Density Parameters of Nuclei Deficient in One Neutron.* Nucl. Phys. A**210**, 355–379 (1973).
19. Liu, J. C., Nelson, W. R., Ken, K. R. and Mao, X. S. *Calculations of the Giant-Dipole-Resonance Photoneutrons Using a Coupled EGS4-MORSE Code* (Stanford Linear Accelerator Center, Standford, CA) SLAC-PUB-95-6764 (1995).

CALCULATION OF RADIAL DOSE DISTRIBUTIONS FOR HEAVY IONS BY A NEW ANALYTICAL APPROACH

J. Chen[†] and A. M. Kellerer[†‡]
[†]GSF-Institute of Radiation Biology
Postfach 1129, 85758 Oberschleißheim, Germany
[‡]Radiobiological Institute, University of Munich
Schillerstr. 42, 80336, Munich, Germany

Abstract — With the new analytical method recently developed for calculation of radially restricted LET, radial dose distributions are calculated for protons and heavy ions of different energies. In the entire range of radial distance, the results agree well with the results of comprehensive Monte Carlo simulations and with experimental data. A comparison with other models of radial dose distribution demonstrates the advantages of the new analytical approach.

INTRODUCTION

Different ionising radiations produce biological effects by the same initial physical mechanisms, i.e. by electronic collisions and the resulting excitations and ionisations in the irradiated tissue. However, considerable differences in the biological effectiveness of different radiations result from different microscopic distributions of primary radiation products. The radial dose distribution is one useful parameter to characterise these distributions, especially when one deals with high energy heavy ions which are of growing importance for radiotherapy and for radiation protection in space. They are equally important for computations related to recoil particles of fast neutrons.

A new convenient analytical approach has recently been developed for the calculation of radially restricted LET[1]. The method provides a better fit than other models and is valid over the entire range of radial distance from track centre to the maximum radial distance travelled by the most energetic secondary electrons. Since the radially restricted LET is an integrated form of the radial dose distribution, it can be utilised to derive the radial dose distributions around particle trajectories. The new approach will here be compared with other models.

MONTE CARLO SIMULATION

The radial dose distribution, D(r), is defined as the average energy imparted per unit mass by a charged particle within a cylindrical shell of radii r and r + dr concentric to the particle trajectory. D(r) is an absorbed dose, conditional on the location of the particle. A Monte Carlo code[1,2] has been used to compute radial dose distributions. In the present calculations a code modification is employed to include four main Auger electron lines[3]. A comparison of the results with experimental data[4–6] in Figure 1 indicates good agreement.

The upper panel of Figure 2 gives D(r) plotted against r for protons from 1 MeV to 1 GeV. Since radial doses are commonly assumed to be inversely proportional to the square of the radial distance, $D(r)r^2$ is plotted in the lower panel. It is seen that the $1/r^2$ approximation holds over a wide range of r for higher particle energies. For small radial distances, however, the $1/r^2$ approximation is invalid. Measurements of the lateral distributions of absorbed dose by observations of the intensity of the emitted fluorescent light in single photon counting technique[7] support this finding and agree with the dependencies in the lower panel of Figure 2.

THE NEW ANALYTICAL APPROACH

While radial dose distributions can be obtained by Monte Carlo simulations, such computations are complicated and time consuming. Approximations are, therefore, usually employed. The radial dose distri-

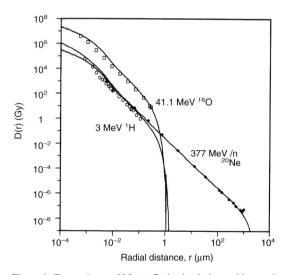

Figure 1. Comparisons of Monte Carlo simulations with experimental data for ions in tissue-equivalent material. (○) Wingate and Baum[4], (□) Varma et al[5], (◆) Varma and Baum[6].

bution is closely related to the radially restricted LET, L_r. Using the new approach[1], an approximation proposed by Rossi, one can calculate the radial dose distribution analytically:

$$D(r) = \frac{1}{2\pi r \rho} \frac{dL_r}{dr} \quad (1)$$

$$L_r = L_\Delta + \frac{\rho N_A Z}{M_A} \Delta(r) \sum_i \int_{\Delta(r)}^{T_{max}} \sigma_i(T,E) dT$$

$\Delta(r)$ is the energy of electrons with a csda range r. The energy restricted LET, L_Δ, is defined according to the new draft document of ICRU[8]. $\sigma_i(T,E)$ is the differential ionisation cross section, E being the kinetic energy of heavy ions and T the kinetic energy of secondary electrons. The constants, ρ, N_A, Z, and M_A, have their usual meaning.

For electrons in water an equation for $\Delta(r)$ is found by fitting data given by ICRU Report 16[9]:

$$\Delta(r) = \exp[1.8488 + 0.56717 \ln r - 0.019612(\ln r)^2 + 0.002836(\ln r)^3]$$

if the units keV and μm are used. $\sigma_i(T,E)$ can be taken from the analytical formulas given by Rudd[10].

The analytical approximation for protons in water is compared in Figure 3 with the simulation (broken lines). The agreement is generally good, and the Rossi approximation appears to be valid over the entire range of radial distances from track centre to the maximum range of secondary electrons.

COMPARISON WITH EARLIER APPROACHES

There are several analytical models of the radial dose distribution. The model of Chatterjee and Schaefer[11] is the most widely used; it employs the $1/r^2$ approximation:

$$D(r) = \frac{L_\infty}{2\pi\rho(1 + 2\ln(r_p/r_c))} \frac{1}{r^2}$$

$$r_p = 0.768E - 1.925\sqrt{E} + 1.257 \quad (2)$$

$$r_c = 0.0116\beta$$

β is the speed of the particle in fractions of the speed of light; E is the MeV.amu^{-1}; the core radius, r_c and the penumbra radius, r_p, are in μm. The model was developed for ions in water of unit density, and for radial distances in the penumbra region, $r_c < r < r_p$. The results of the Chatterjee–Schaefer model are given in Figure 4. Outside the core region they agree fairly well with the simulated results.

Katz and his colleagues have, much earlier, studied the radial dose distribution around heavy ions[12–14]. They utilised the classical free electron model and classical kinematics for the angular distribution of electrons ejected by impact ions. Their fairly simple formula for the radial dose is:

$$D(r) = \frac{NZ^{*2}e^4}{\rho m v^2} \frac{1}{r^2} \left(1 - \frac{r}{r_{max}}\right) \quad (3)$$

where v is the velocity of impact ions, m and e the electron mass and charge, N the electron density and r_{max} the maximum delta ray range. The effective ion charge Z^*e is calculated according to Barkas' formula[15]:

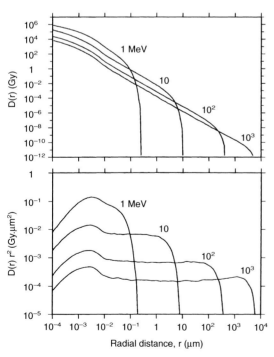

Figure 2. Results of Monte Carlo simulations for protons from 1 MeV to 1 GeV in water vapour (1 g.cm^{-3}).

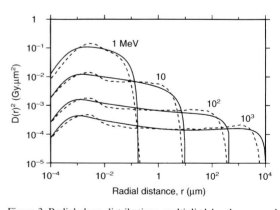

Figure 3. Radial dose distributions multiplied by the squared radial distance. Comparison of the Rossi approximation (solid lines) with Monte Carlo simulations (broken lines). Protons in water vapour.

$$Z^*e = Ze(1 - \exp(-125\beta Z^{-2/3})) \quad (4)$$

Equation 3 accounts to a certain degree for deviations from the $1/r^2$ approximation at large radial distances. Recently the model was modified[16,17] by a better range–energy relation. For the distribution of secondary electrons the free electron model was modified by the assumption of a mean binding energy, I(I = 10 eV was taken for water). The modified radial dose distribution was then given by the formula:

$$D(r) = \frac{NZ^{*2}e^4}{\alpha\rho mc^2\beta^2} \frac{\left(1 - \frac{r + kI^\alpha}{r_{max} + kI^\alpha}\right)^{1/\alpha}}{r(r + kI^\alpha)} \quad (5)$$

The values $k = 6 \times 10^{-6}$ cm keV^{-1} and $\alpha = 1.079$ or $\alpha = 1.667$ are used for ions whose secondary electrons have a maximal kinetic energy smaller or larger than 1 keV. Results (solid lines) are plotted in Figure 5. The model is fairly good outside the core region.

As the two earlier approaches that have here been considered, other earlier models are also applicable for a limited range of radial distances and particle energies. In contrast the Rossi approximation is valid over the entire range of radial distance from the track centre to the maximum range of secondary electrons. The approximation is somewhat more demanding because it requires the differential cross sections for the particle in the specified receptor material, but it is, otherwise, of simple and general form that makes it readily applicable in dosimetric and microdosimetric computations for heavy ion beams or neutron fields.

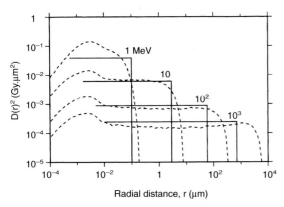

Figure 4. Radial dose distributions multiplied by the squared radial distance. Comparison of the Chatterjee–Schaefer approximation (solid lines) with Monte Carlo simulations (broken lines). The core radius is 0.535 nm for 1 MeV protons, 1.68 nm for 10 MeV protons, 4.97 nm for 100 MeV protons, and 10.2 nm for 1 GeV protons.

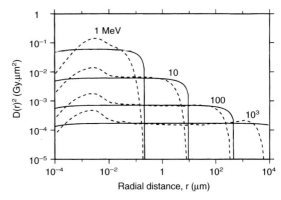

Figure 5. Radial dose distributions multiplied by the squared radial distance. Comparison of the Zhang approximation (solid lines) with Monte Carlo simulations (broken lines).

REFERENCES

1. Chan, J., Kellerer, A. M. and Rossi, H. H. *Radially Restricted Linear Energy Transfer for High-energy Protons: a New Analytical Approach.* Radiat. Environ. Biophys. **33**, 181–187 (1994).
2. Kellerer, A. M., Hahn, K. and Rossi, H. H. *Intermediate Dosimetric Quantities.* Radiat. Res. **130**, 15–25 (1992).
3. Moddeman, W. E., Carlson, T. A., Frause, M. O., Pullen, B. P., Bull, W. E. and Schweizer, G. K. *Determination of the K-LL Auger spectra of N_2, O_2, CO, NO, H_2O and CO_2.* J. Chem. Phys. **55**, 2317–2336 (1971).
4. Wingate, C. L. and Baum, J. W. *Measured Radial Distributions of Dose and LET for Alpha and Proton Beams in Hydrogen and Tissue-equivalent Gas.* Radiat. Res. **65**, 1–19 (1976).
5. Varma, M. N., Baum, J. W. and Kuehner, A. V. *Radial Dose, LET, and W for ^{16}O ions in N_2 and Tissue-equivalent Gases.* Radiat. Res. **70**, 511–518 (1977).
6. Varma, M. N. and Baum, J. W. *Energy Deposition in Nanometer Regions by 377 MeV/nucleon ^{20}Ne Ions.* Radiat. Res. **81**, 355–363 (1980).
7. Lassmann, M., Roos, H. and Kellerer, A. M. *Lateral Dose Distribution of Proton Beams at Energies from 4 to 11 MeV.* Radiat. Prot. Dosim. **52**, 51–55 (1994).
8. Allisy, A., Jennings, W. A., Kellerer, A. M., Muller, J. W. and Rossi, H. H. *Radiation Quantities and Units, Interaction Coefficients — Dosimetry*, ICRU News (December 1993).
9. ICRU. Linear Energy Transfer Report 16 (Washington, DC: ICRU Publications) (1970).
10. Rudd, M. E. *Cross Sections for Production of Secondary Electrons by Charged Particles.* Radiat. Prot. Dosim. **31**, 17–22 (1990).
11. Chatterjee, A. and Schaefer, H. J. *Microdosimetric Structure of Heavy Ion Tracks in Tissue.* Radiat. Environ. Biophys. **13**, 215–227 (1976).

12. Butts, J. J. and Katz, R. *Theory of RBE for Heavy Ion Bombardment of Dry Enzymes and Viruses*. Radiat. Res. **30**, 855–871 (1967).
13. Kobetich, E. J. and Katz, R. *Width of Heavy Ion Tracks in Emulsion*. Phys. Rev. **170**, 405–411 (1968).
14. Katz, R. and Kobetich, E. J. *Particle Tracks in Emulsion*. Phys. Rev. **186**, 344–351 (1969).
15. Barkas, H. *Nuclear Research Emulsions*. Vol. 1, Chapter 9, p. 371 New York: Academic Press (1963).
16. Zhang, C. X., Dunn, D. E. and Katz, R. *Radial Distribution of Dose and Cross Sections for the Inactivation of Dry Enzymes and Viruses*. Radiat. Prot. Dosim. **13**, 215–218 (1985).
17. Zhang, C. X., Liu, X. W., Li, M. F. and Luo, D. L. *Numerical Calculation of the Radial Distribution of dose abound the Path of a Heavy Ion*. Radiat. Prot. Dosim. **52**, 93–96 (1994).

ELECTRONIC NEUTRON DOSEMETERS: HISTORY AND STATE OF THE ART

J. Barthe†, J. M. Bordy‡ and T. Lahaye‡
†Département des Applications et de la Métrologie des Rayonnements Ionisants
Commissariat à l'Energie Atomique, Division des Technologies Avancées
Centre d'Etudes de Saclay, 91193 Gif-Sur-Yvette Cedex, France
‡Institut de Protection et de Sûreté Nucléaire
Département de Protection de l'Homme et Dosimétrie, Service de Dosimétrie
Centre d'Etudes de Fontenay, 92265 Fontenay-aux-Roses Cedex, France

INVITED PAPER

Abstract — A review of the development of electronic neutron dosemeters is given, with an assessment of their characteristics. Comment is made on necessary lines for future development.

INTRODUCTION

The neutron dose received by personnel in the nuclear industry is a few per cent of the gamma ray dose and only a very small number of staff are involved. In many cases, and mainly in nuclear power plants, the neutron dose is determined as a fraction of the gamma ray dose depending on the work station placement. In nuclear fuel reprocessing units, the situation is more complex, the neutron spectrum being strongly dependent on the worker's location.

Personal dosemeters and survey instruments now used are generally not very accurate and solid state or passive dosemeters are often unsatisfactory. The energy dependence of TLD albedo and NTA film dosemeters is well documented. Unless properly calibrated, personal dosemeters can be in error by more than one order of magnitude[1,2]. Bubble dosemeters, insensitive to gamma rays, are now ready to meet the requirements for accuracy and energy independence.

GENERAL INFORMATION ON NEUTRON SENSORS AND ASSOCIATED ELECTRONICS

Classification of phenomena and associated processes used for neutron dosimetry

(1) Dose measurement with passive or solid state dosemeters needs a set of dosemeters and readers.
(2) Active or electronic dosemeters can directly display dose, dose rate and give warning signals during routine work.
(3) Bifunctional active and passive dosemeters can be used simultaneously as passive and/or active dosemeters and do not need applied voltage during the dose integration period.

Dosemeter	Associated reading system
PIN diodes[3,4]	Sequential measurement of relative direct voltage
Tissue-equivalent proportional counters[5,6]	Multichannel analyser
Double or single diodes with converter[7–12]	Single or differential counting
Rise time single diode with converter[13]	Counting and statistical counting
CCD[14] or computer memory chips[15]	Rise time discriminator and counting
Bubble dosemeter[16,17]	Acoustic pulse analysis and counting

PIN diodes and bubble dosemeters can integrate the dose without any power supply.

Present and commercially available personal neutron sensors and dosemeters

Albedo double diode dosemeter	Aloka (Japan)
PIN diodes	Anguelardt (USA) and Czech Metrology Institute (Czech republic)
Bubble dosemeters	Apfel Industries (USA) and BTI (CANADA) and Kurchatov Institute (RUSSIA)

Personal dosemeters under study

Gas devices

Multichamber TE proportional counter	Columbia University, New York, USA
Multichannel TE proportional counter	SDOS/IPSN Fontenay, France

Streamer chamber	SDOS/IPSN Cadarache, France
Semiconductor devices	
SRAM and DRAM dosemeters	KFA Jülich, Germany
Opposite diodes on the same woofer	Autonomous University of Barcelona, Spain
Dual TED and diode system	University of Thessaloniki, Greece
Double diode dosimetry and spectrometry system	LEPOFI, University of Limoges, France
Rise time single diode with converter	SDOS/IPSN Fontenay, France

The first research on individual electronic neutron dosemeters appears to have started in 1962. From 1974 to 1995 (from INIS data base) there were found:

- 4309 papers on personnel dosimetry including all types of radiation and principles,
- 1127 papers on neutron personnel dosimetry including all principles,
- 128 papers on individual neutron dosemeters based on semiconductors and gas counters,
- 45 papers exclusively on neutron electronic semiconductor individual dosemeters.

To these can be added patents, internal reports and unpublished meeting proceedings.

DESCRIPTION OF SOME REPRESENTATIVE PERSONAL NEUTRON DOSEMETERS

Multi-element tissue-equivalent proportional counter

The principle of this dosemeter was proposed by P. Kliauga of Columbia University (New York) in 1992. It consists of 200 elements (right cylinders, diameter = height = 3.6 mm) in 25 groups. Each element corresponds to a microdosimetric proportional counter and each group has a common wire anode. The counter radiation cross section is 5.3 times that of a spherical counter with the same volume. The mechanical structure is very complex as is shown in the corresponding schematic diagram (Figure 1).

Multichannel tissue-equivalent proportional counter MC-TEPC[18,19]

This dosemeter, based on a microdosimetric tissue-equivalent proportional counter, was designed and developed by J. M. Bordy and J. Barthe at SDOS in 1992–93. As in the Kliauga counter, the sensitivity is enhanced by increasing the counter radiation cross section. The 8 mm thick cathode is drilled with some 250 4 mm diameter holes and electron swarms, created by interactions of secondary charged particles in the counting gas, are drifted along the holes axis to 10 high-potential wire anodes by an electrokinetic field. This electric field is obtained by a bias current flowing from the internal to the external face of the cathode. Electrons are multiplied around and collected by the wire anodes. Its sensitivity is between 5 and 7 times that of a multi-cellular and spherical counter of the same external size. A view of the counter interior is shown in Figure 2 and the chord length distributions for the MC-TEPC and two other common counters are given in Figure 3. It must be noted that, for MC-TEPC, the mean and most probable chord lengths are practically identical.

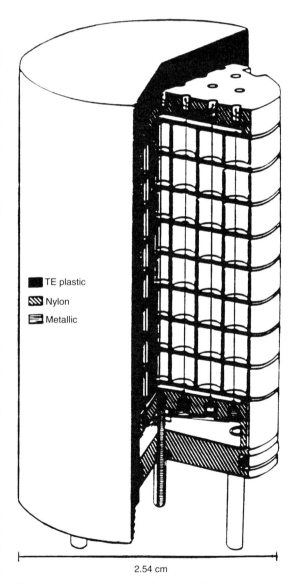

Figure 1. Schematic view of the multi-element tissue-equivalent proportional counter proposed by P. Kliauga (1992).

EG&G individual neutron dosemeter[20]

This dosemeter was first described by W. Quam (Battelle Institute and EG&G) in 1981. The radiation detector consists of three cylindrical TEPCs, diameter = 1.9 cm and length = 13.3 cm. The overall monitor is 4.6 cm × 7.2 cm × 20.3 cm, and weighs 0.630 kg. The associated electronics operates with a linear hybrid circuitry and a 256 channel ADC. A CMOS microprocessor is used for calculations and display of dose, dose equivalent and mean quality factor. Neutron dose determination is within ±20% for LETs above 10 keV.μm^{-1}.

Differential double diode neutron dosemeter[21,22]

This dosemeter (Figure 4) was designed by J. L. Decossas and J. C. Vareilles of the LEPOFI laboratory (Limoges University) in 1988. The experimental results presented here were obtained at the IPSN/SDOS labora-

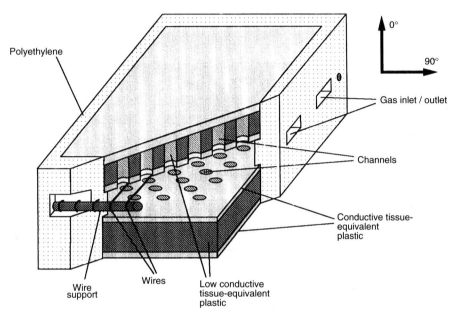

Figure 2. Schematic view of the multichannel tissue-equivalent proportional counter (MC-TEPC) developed at the SDOS (1992–1993).

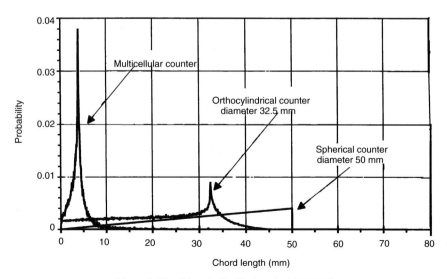

Figure 3. Chord length distributions in MC-TEPC.

tory. The dosemeter consists of two identical diodes, one covered with a thin film of ^{10}B-doped polyethylene. Neutrons interact with the converter by two reactions: (n,p) with hydrogen and (n,α) with ^{10}B. Secondary charged particles lose their energy in the depleted zone and are detected. A differential method allows eliminating the photon component and decreasing the neutron detection threshold. However, for low energies (below

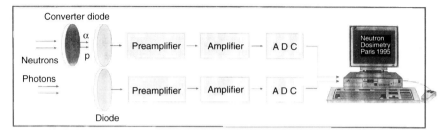

Figure 4. Principle of the differential double diode neutron dosemeter (LEPOFI).

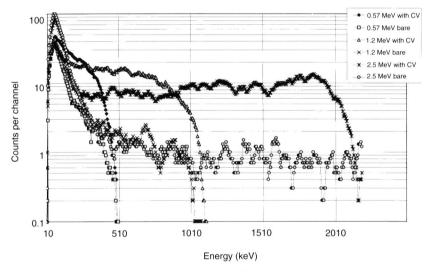

Figure 5. Pulse height spectra for different neutron energies.

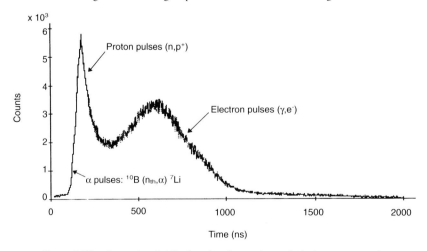

Figure 6. Rise time pulse distribution showing proton and electron components.

400 keV), non-identical diodes can lead to negative doses! The diode responses with and without a converter for three different neutron energies are given in Figure 5.

Rise time discrimination diode neutron dosemeter[13]

The principle of such a dosemeter, developed and studied at IPSN/SDOS since 1993, is based on a single diode covered with a boron-doped polyethylene converter. Discrimination of secondary charged particles is obtained from the difference in pulse rise time due to the difference in carrier mobility in the depleted zone and substrate respectively. The difference in path lengths is related to the LET of secondary charged particles: high LET and short path length for protons and α particles in depleted zone only; low LET and long path length for electrons in depleted zone and all the substrate (Figure 6). In other words, discrimination between electrons and other charged particles is obtained from rise time for low amplitude pulses and

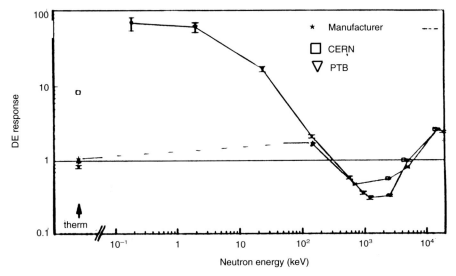

Figure 7. Neutron energy response of the Aloka dosemeter.

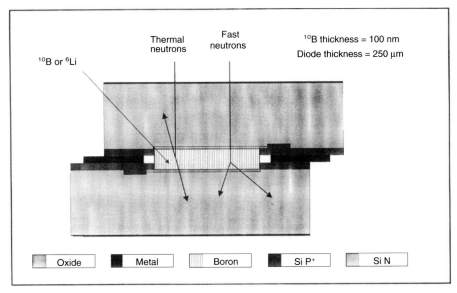

Figure 8. Principle of the double diode spectrometric monitor (LEPOFI, 1994).

peak height for high amplitude pulses. At present, the energy detection threshold is as low as 100 keV for incident neutrons. It is assumed that there are very few direct interactions of photons and neutrons with silicon.

The 'neutron blind zone' was decreased from 0.6 MeV to 0.2 MeV by using a specifically designed moderator and gamma ray shield. For a realistic mixed field (reactor), the 'K spectral shape factor', determined from the time resolved electron to proton spectra, allows reducing the neutron energy response by approximately a factor of 10 compared to the double diode system.

Aloka double diode monitor[23,24]

The PDM-303 dosemeter was designed and developed by the Aloka company in Japan. It is based on an albedo detector consisting of two spectrometric silicon diodes. The first diode is covered with a hydrogenous converter of a few tens of μm to detect recoil protons from fast neutron interactions in the converter. The second diode, doped with ^{10}B, is sensitive to thermal and slow neutrons. Its response to slow neutrons is increased by a 1 cm thick polyethylene moderator. Discrimination between gamma photons and neutrons is obtained by adjusting the detection threshold. Figure 7 gives the energy response of the monitor (measurements performed by Aloka, CERN and PTB). A lower neutron detection threshold leads to taking high amplitude pulses induced by photons into account.

Double diode spectrometric monitor[25]

The principle of such a monitor was developed at LEPOFI and LAAS in 1994 (Figure 8). The basic idea is to measure the exact incident neutron energy and derive the dose equivalent (or dose equivalent rate). Two face to face diodes are in contact with a 100 nm thick boron-doped hydrogenous converter. Secondary charged particles are produced by neutron interactions in the converter and detected by these diodes within a 4π angle. Charge collection is complete and independent of angle. Collected charges per interaction are proportional to the sum of energy lost by the incident particle and nuclear reaction energy.

This sensor has many advantages: it operates as a neutron spectrometer, a large amount of energy is produced per interaction, and the dose and dose equivalent can be precisely determined from the experimental spectrum. Its disadvantages are that the sensitivity decreases with increasing neutron energy, measured energy is obtained after subtraction of the nuclear reaction energy and negative doses or dose equivalents can be obtained.

SRAM dosemeter[26]

The principle of this dosemeter was first proposed by Davis in 1985. Experimental tests were made at KFA and reported by Schroeder in 1993. Incident particles, neutrons and photons, create secondary charged particles, protons and electrons, respectively, in the polyamide layer which protects the semiconductor (SCD) and in the SCD. These particles are detected in the depleted zone of each SCD and the voltage drop of a SRAM chip is directly proportional to the deposited charges. Dose information is obtained by determination of the conduction state. Each change of state corresponds to an energy of a few hundred keV deposited in a thin layer of a few μm. A refresh time of a few milliseconds can be used.

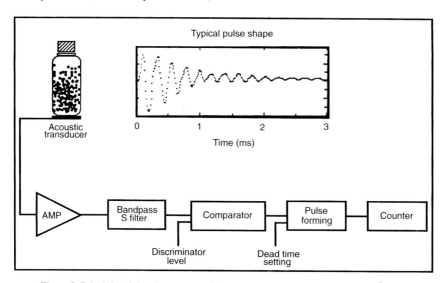

Figure 9. Principle of the electro-acoustic bubble dosemeter (Apfel Industries 1989).

ERADCOM electronic PIN diode dosemeter[27]

This dosemeter, designed for the US Army by Ramondetta and the RCA company in 1978, is probably the first manufactured and used in the world. It is based on the direct voltage increase of a PIN diode with the integrated dose. The dose rate is determined stepwise from the voltage difference at the end of each time period. To avoid effects due to temperature, corrections are made continuously. A direct high bias current (100 mA) is used, probably to reduce the dose annealing effect. The sensitivity is low and around 100 mV.Gy^{-1}, the detection threshold about 10 mGy and the estimated error below 25% from −40 to +50°C.

Electronic bubble dosemeter[16]

Bubble dosemeters are designed and manufactured by two companies and one institute.

The principle is the following: a great number of very small droplets of a liquid with a low heat of vaporisation are incorporated in an organic gel or polymer. For a specific temperature and pressure, the liquid droplets are superheated in a metastable state. This unstable equilibrium can be destroyed by an external energy supply.

At the interaction point between droplet and high LET particle, such as a neutron or heavy charged particle, the liquid droplet becomes vaporised as a visible gas bubble. The overall absorbed dose is correlated to the number of bubbles.

If the gel viscosity is low, bubbles can reach the surface and increase the upper gas volume. The corresponding dose can be derived by measuring the gas volume thus produced with a simple optical scaling system.

A sound is emitted for each bubble being created and the bubbles are counted in real time by an electroacoustic transducer (Figure 9). The dose rate corresponds to bubble frequency.

Bubble dosemeters generally have low sensitivity to photon radiation.

CONCLUSIONS

More than 50 researchers have been, or now are, working on individual electronic neutron dosemeters. At present, two types of individual electronic dosemeters are commercially available.

(1) Electronic diode dosemeter: Aloka (Japan).
(2) Bubble electronic sound counting dosemeters: Apfel industries (USA).

Situation of other systems

Beyond our present knowledge and technology.
Not meeting standards or requirements, mainly in sensitivity and energy response.
Too expensive: the worldwide estimated market is less than 20,000 devices.
In progress: two or three monitors should be operational in the next 5 years.

Future guidelines

Optimising present and studying new physical detection concepts.
Decreasing background noise of sensors and associated electronics.
Optimising use of miniaturised electronic devices and computers.

REFERENCES

1. Griffith, R. V. *Individual Neutron Monitoring — Needs for the Nineties*. Radiat. Prot. Dosim. **44**(1/4), 259–266 (1992).
2. Vallario, E. J. and Faust, L. *Advances in Personnel Neutron Dosimetry 2–3*. Health Phys. Soc. Newsletter **11**(8), 3–4, **11**(9), 4 (1983).
3. Dulieu, P. *Mesure de Flux de Neutrons Intermédiaires et Rapides au Moyen de Diodes Silicium*. Note CEA R-391 (1962).
4. Aoyama, T., Oka, Y. and Honda, K. *A Neutron Detector using PIN Photodiodes for Personal Neutron Dosimetry*. Nucl. Instrum. Methods A **314**(3), 500–594 (1992).
5. Brackenbush, L. W. and Endres, G. W. R. *Personnel Monitors using Tissue Equivalent Proportional Counters*. In: Proc. 5th Symp. on Neutron Dosimetry, Munich, Germany, 1984 (Luxembourg: CEC) pp. 359–368 (1985).
6. Boutruche, B., Bordy, J. M., Barthe, J. and Portal, G. *A New Concept of a High Sensitivity TEPC for Individual Neutron Dosimetry*. Radiat. Prot. Dosim. **52**(1–4), 335–338 (1994).
7. Tyree, W. H. and Falk, R. B. *Personal Neutron Dosimeter*. In: Proc. 9th DOE Workshop on Personnel Neutron Dosimetry. pp. 154–161 (1982).
8. Brackenbush, L. W. and Quam, W. *Hydrogenous Semiconductor Neutron Detectors*. Radiat. Prot. Dosim. **10**(1–4), 303–305 (1985).
9. Wall, B. F. *Fast Neutron Dosimetry using Wide Based n$^+$pp$^+$ Silicon Diodes*. Report EUR-4896, pp. 343–365 (1968).
10. Eisen, Y., Engler, G., Ovadia, E., Shamai, Y., Baum, Z. and Levi, Y. *A Small Size Neutron and Gamma Dosemeter with a Single Silicon Surface Barrier Detector*. Radiat. Prot. Dosim. **15**(1), 15–30 (1986).
11. Yoshida, Y., Suzuki, T. and Ishikura, T. *Silicon Radiation Detectors*. Fuji-Electric-Review **34**(4), 147–152 (1988).
12. Nakamura, T., Horiguchi, M. and Yamadera, A. *A New Type Active Personal Dosemeter with a Solid State Detector*. In: Proc. 7th Int. Cong. of IRPA, 1988, Sydney, Australia (Oxford: Pergamon Press) **3**, 286–289 (1988).

13. Bordy, J. M., Lahaye, T., Landre, F., Hoflack, C., Lequin, S. and Barthe, J. *Single Diode Detector for Individual Neutron Dosimetry using a Pulse Shape Analysis.* Radiat. Prot. Dosim. **70**(1–4), 73–78 (This issue) (1997).
14. Pierschel, M., Ehwald, K. E., Heinemann, B. and Januschewski, F. *A BCDD-based Dosimeter for Mixed Radiation Field.* Nucl. Instrum. Methods, A**326**(1/2), 304–309 (1993).
15. Davis, J. L. *Use of Computer Memory Chips as the Basis for a Digital Albedo Neutron Dosimeter.* Health Phys. **49**(2), 259–265 (1985).
16. Apfel, R. E. and Lo, Y. C. *Practical Neutron Dosimetry with Superheated Drops.* Health Phys. **56**(1), 79–83 (1989).
17. Ing, H., Cundari, K., Cousins, T. and Rushton, L. P. *Preliminary Measurements of Radiation from a Critical Assembly Using the Bubble Detector.* AECL Report — 9336 (1986).
18. Barthe, J., Bordy, J. M., Mourgues, M., Lahaye, T., Boutruche, B. and Segur, P. *New Devices for Individual Neutron Dosimetry.* Radiat. Prot. Dosim. **54**(3–4), 365–368 (1994).
19. Bordy, J. M., Barthe, J., Boutruche, B. and Segur, P. *Assessment of the Characteristics of a New Type of TEPC for Personal Neutron Dosimetry.* Radioprotection **29**(1), 11–28 (1994).
20. Quam, W. M., Del-Luca, T., Plake, W., Graves, G. and De Vore, T. *Pocket Neutron Rem Meter.* In: Proc. 8th DOE Workshop on Personnel Neutron Dosimetry, pp. 46–50 (1981).
21. Barelaud, B., Decossas, J. L., Makovicka, L. and Vareille, J. C. *Capteur Electronique pour la Dosimetrie des Neutrons.* Radioprotection **26**(2), 307–328 (1991).
22. Barthe, J., Lahaye, T., Moiseev, T. and Portal, G. *Personal Neutron Diode Dosemeter.* Radiat. Prot. Dosim. **47**(1–4), 397–399 (1993).
23. Nakamura, T. and Tsujimura, N. *Development, Characterisation of Real-time Wide-energy Range Personal Neutron Dosimeter.* Nippon-Genshiryoku-Gakkai-Shi. **36**(4), 337–345 (1994).
24. Alberts, W. G., Dietz, E., Guldbakke, S. and Kluge, H. *Response of an Electronic Personal Neutron Dosemeter.* Radiat. Prot. Dosim. **51**(3), 207–210 (1994).
25. Barelaud, B., Nexon-Mokhtari, F., Barrau, C., Decossas, J. L., Vareilles, J. C. and Sarrabayrousse, G. J. *Study of an Integrated Electronic Monitor for Neutron Fields.* Radiat. Prot. Dosim. **61**(1–3), 153–158 (1995).
26. Schröeder, O., Schmitz, T. and Pierschel, M. *Microdosimetric Dosemeters for Individual Monitoring Based on Semiconductor Detectors.* Radiat. Prot. Dosim. **52**(1–4), 431–434 (1994).
27. Ramondetta, P. W. and Groever, E. O. Jr *Digital Neutron Dosimeter.* Health Phys. **35**(6), 835–843 (1978).

NEEDS AND PERFORMANCE REQUIREMENTS FOR NEUTRON MONITORING IN THE NUCLEAR POWER INDUSTRY

C. R. Hirning† and A. J. Waker‡
†Health Physics Department, Ontario Hydro
1549 Victoria Street East, Whitby, Ontario L1N 9E3, Canada
‡Radiation Biology and Health Physics Branch
AECL, Chalk River Laboratories
Chalk River, Ontario K0J 1J0, Canada

INVITED PAPER

Abstract — Neutrons do not contribute greatly to the collective occupational dose at Ontario Hydro's nuclear generating stations. Nonetheless, there is a need to monitor workers' exposure to neutrons in some locations. At present, neutron area surveys and personal dosimetry are both performed using integrating rem meters (of the Andersson-Braun design), which workers take with them when they will be working in an area where neutrons are known to be present. This is not a satisfactory method for neutron monitoring or dosimetry, both because of the inconvenience of carrying around a large, heavy instrument, and because rem meters are known to be inaccurate in many situations. There is a further problem associated with the calibration of the rem meters, which is done in a fast neutron field quite unlike the neutron fields found in nuclear generating stations. However, rem meters continue to be used, owing to the lack of suitable alternatives and also to the low priority neutron dosimetry has traditionally received in the company. The properties of an ideal neutron dosemeter and of an ideal survey meter for use in nuclear generating stations are described. These properties are compared with those of currently available dosemeters and instruments, and it is shown where changes and improvements are needed. The need for a calibration field that is more representative of the energy spectra to be measured in practice is also discussed.

INTRODUCTION

In the CANDU (CANadian Deuterium Uranium) nuclear generating stations operated by Ontario Hydro, the need for personal neutron dosimetry and portable neutron area monitors (survey meters) is limited to the measurement of low doses received at infrequent intervals in specific locations. When the need does arise, however, the performance requirements on the instruments are demanding and difficult to satisfy. As a consequence of this combination of factors, there has been little progress over the past 25 years in improving the techniques used for neutron dose and dose rate measurement.

Although the discussion and examples given in this paper are related to the environment of a CANDU reactor, most of the needs identified are common to the nuclear power industry generally. The paper concludes with some recommendations on areas where performance improvement is most needed.

THE NEED FOR NEUTRON MONITORING

At Ontario Hydro, neutron dosimetry and monitoring are presently both done using the Andersson-Braun type of dose equivalent meter, or rem meter (manufactured by Nuclear Research Corporation, and commonly called a 'Snoopy'). The rem meters have been modified by the addition of a scaler to integrate the dose equivalent, in increments of 10 μSv. When a worker goes to perform a task in an area where neutron fields are known to be present (and such areas are few in number), he/she is expected to obtain a rem meter from stores and to verify its operation with a checker containing an ^{241}Am-Be neutron source. The scaler is then zeroed and the rem meter taken to the work site, where it is expected to be kept in a location where the neutron dose rate is representative of that being received by the worker. At the end of the job, the worker reads the dose indicated by the scalar and enters it onto a card or into a computer terminal, from where it is subsequently transferred to the official dose record of the worker.

There are several problems associated with the present approach; among them are the well-known radiological limitations of the Andersson-Braun rem meter. It has been recognised for a long time that this instrument over-responds by a factor of about 3 in the intermediate energy range below 100 keV, and that its angular response is non-isotropic[1]. The rem meters used at Ontario Hydro are calibrated relative to unmoderated ^{241}Am-Be neutrons incident at right angles to the moderator axis, so variations in energy and angular response might be expected to result in large errors in the measurement of the stongly moderated neutron fields typical of a nuclear power station. However, recent measurements in power reactor environments[2,3] have shown that the rem meter response is often (but not always) within a factor of two of the expected ambient dose equivalent (calculated with the 'old' quality factors[4]). A more appropriate calibration field, more representative of the fields to be measured, would help to reduce the inaccuracy further.

The most important problems, however, are related to human factors. First, there is the inconvenience to the worker of having to obtain the rem meter and take it to the work site. The process of signing the instrument out and performing a source check takes time and delays the start of work. Then, the instrument is large and heavy; it has been implicated in cases of back injury and is very difficult to use in areas where space is limited. There is little assurance that it will be positioned in a location where the dose rate is representative of the dose rate at the worker's position, or that it will be relocated each time the worker moves. In response to these problems, the health physics staff at one of Ontario Hydro's generating stations are proposing to permit an alternative technique based on survey measurements and stay times to estimate neutron doses. This technique may contribute even more uncertainty to the neutron doses than would result from the rigorous use of the rem meter, but the station staff believe that a less accurate technique that is used consistently is better than a more accurate technique that is used inconsistently.

It would seem, then, that there are good reasons for Ontario Hydro to adopt better devices for neutron dosimetry and monitoring. Why has there not been more progress towards improving the situation?

The answer lies largely in the fact that neutrons do not contribute greatly to worker dose. The total effective dose equivalent (from both external and internal sources of radiation) received at Ontario Hydro's nuclear generating stations (including up to 20 operating reactors) over the five-year period 1990–94 was 70.63 person.Sv. The total collective neutron dose assigned over this time period was 0.30 person.Sv, or 0.42% of the total. Not only is the average neutron dose small, but the neutron doses received by individual workers are also small. Figure 1 shows the distribution of recorded neutron doses, again for the period 1990–94. The vast majority of these doses were in the range 0.01–0.20 mSv, and only 12 out of the total of 2783 recorded doses exceeded 1.0 mSv. The highest single dose recorded was 2.55 mSv. Summing these doses to give annual neutron doses per worker does not greatly change the picture, since most of the workers who receive any neutron dose receive it only a few times per year.

These doses were measured using the rem meter technique described earlier, so they are subject to large uncertainties. However, many of the random uncertainties will cancel out in collective dose calculations, and the systematic uncertainties are expected to overestimate the true doses. Neutron doses may therefore be of even less importance than these numbers would suggest.

In summary, neutron monitoring at Ontario Hydro has not been of sufficient concern to make its improvement a matter of priority. The improvement of other aspects of the radiation protection programme has been considered more important, and has received attention and resources. Nonetheless, the issue of improvements to neutron monitoring keeps coming back, mainly because of widespread worker dissatisfaction with the operational problems associated with the present technique.

OPERATING ENVIRONMENT

Radiation fields

There has been a rapid increase recently in the amount and quality of information available on the neutron radiation fields that may be encountered in nuclear generating stations[5–8]. In Canada, a series of in-station measurements has been carried out at Ontario Hydro and New Brunswick Power by researchers from the Chalk River Laboratories of Atomic Energy of Canada Limited[3,9]. These measurements included neutron energy spectra with Bonner spheres, lineal energy spectra with a tissue-equivalent proportional counter, and dose equivalent with various other instruments that were available at the time, including rem meters, bubble dosemeters and etched track detectors.

At the measurement locations in CANDU stations, neutron dose equivalent rates ranged over about 4–1800 $\mu Sv.h^{-1}$. Neutron-to-gamma dose equivalent rate ratios ranged from 0.14–1.6, but most were near the bottom end of this range. The neutron energy spectra varied from location to location within a station and from station to station. Two examples of ambient dose equivalent spectra are shown in Figures 2(a) and 2(b) to illustrate the range of spectral shapes that was measured at Ontario Hydro. The spectrum shown in Figure 2(a) was measured outside the door of a reactor vault airlock at Darlington Nuclear Generating Station. Nearly all of the dose (93%) was delivered by neutrons with an energy of greater than 100 keV, but essentially none above 2 MeV. The spectrum shown in Figure 2(b) was measured in a moderator dump valve room at Pickering Nuclear Generating Station. In this case, 59% of the dose was delivered at energies greater than 100 keV, and 34% of

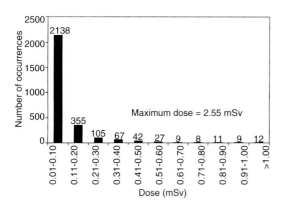

Figure 1. Frequency distribution of recorded neutron doses at all Ontario Hydro nuclear generating stations during 1990–94.

the dose at energies below 1 keV. Although these data support the expectation that most neutron dose is due to neutrons in the range 100 keV–1 MeV[10], they also show that there can be significant fractions at lower energies. Instruments that significantly under- or over-respond to these lower energies will introduce additional uncertainty into dose estimation.

These studies did not attempt to quantify the angular distribution of the neutron fluence. However, it appears from the observation that the rem meter readings were nearly independent of their orientation that the fields were not strongly directional in most of the measurement locations. There is evidence from operational surveys by station staff of highly directional beams in some locations, such as reactivity mechanism decks.

Owing to these variations in the neutron fields, a single calibration source and geometry would not be suitable for all locations if the response of the instrument depends strongly on energy or angle of incidence.

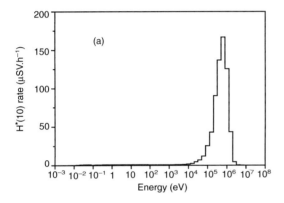

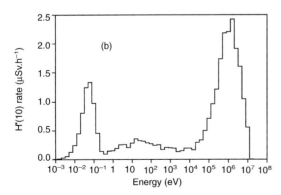

Figure 2. Measured neutron energy spectra in terms of ambient dose equivalent at two sites at Ontario Hydro. (a) Outside the door to a reactor vault airlock at Darlington Nuclear Generating Station. (b) In a moderator dump valve room at Pickering Nuclear Generating Station.

Physical conditions

The work environment of a nuclear generating station poses some severe challenges to the operation of sensitive and delicate instruments. Temperatures at accessible locations inside the station range from about 20°C to over 50°C. Relative humidities range from very dry in reactor containment to nearly 100% inside personal protective clothing. There are numerous sources of electromagnetic radiation, ranging from communications transmitters to the ignition systems on fork-lift trucks. There are also many sources of acoustical noise and mechanical vibration. These can cause serious interference with detectors that are microphonic, and high ambient noise levels can make it difficult to hear audible alarms. There is great potential for equipment to be dropped, struck or squeezed, so it must be sufficiently rugged to withstand typical impact and crushing forces. Finally, some tasks must be carried out in, or accessed through, confined spaces, so the volume of instruments becomes important.

OPERATIONAL NEEDS

The design of both personal neutron dosemeters and neutron survey meters for use in nuclear power stations must take into account the environmental factors described in the previous section. In addition, there are some operational needs that are specific to each category of instrument. These requirements are derived from Ontario Hydro's own operational experience, but, not surprisingly, they are similar to those specified in the IEC standards for portable neutron dose equivalent rate meters[11] and direct reading personal neutron dose equivalent rate meters[12].

Personal neutron dosemeter

A personal dosemeter for neutron radiation should be designed and calibrated to measure the operational quantity personal dose equivalent at 10 mm depth[13], although the suitability of the proposed radiation weighting factors used in the definition of this quantity has been the subject of some debate. Since the doses to be measured are low, high accuracy is not necessary and a factor of 2 in the overall uncertainty should be an acceptable design target. However, good reproducibility of a dose measurement under the same exposure conditions is important, if workers are to have confidence in the results. Since neutron doses are received infrequently, it is desirable to provide dosemeters that are suitable for short-term use. The detection limit should not be greater than 10 μSv. Ideally, the dosemeter should be direct-reading and capable of transferring its reading electronically to a data base. In addition, there are the usual desirable attributes of low system cost, ease of use and light weight (<200 g is the common standard for electronic personal dosemeters for photons and electrons).

Neutron survey meter

The operational dose quantity to be measured by a survey meter is the ambient dose equivalent at 10 mm depth, H*(10). There are concerns that this quantity, as presently defined, does not satisfy the objective for an operational quantity of being a conservative estimate of the effective dose under all expected conditions of use[14]. These concerns may lead to a re-definition of the proposed radiation weighting factors[15]. In order that surveys may be conducted quickly in areas of low neutron dose rate (but potentially high gamma dose rate), the sensitivity of a neutron survey meter should be sufficiently high to ensure a fast response. It is also important that power consumption be low enough to ensure that battery lifetime is equal to at least the length of a work shift (12 h in some cases), and that there be a clear indication when the battery is nearing the end of its useful life or is in need of recharging. Finally, there are similar requirements to those for a personal dosemeter: reasonable cost, ease of use and light weight (<2 kg would be a substantial improvement over existing rem meters).

SUITABILITY OF AVAILABLE TECHNIQUES

For personal neutron dosimetry

None of the commonly used passive neutron dosemeters, including nuclear track emulsions, etched track detectors and albedo thermoluminescence dosemeters (TLD), satisfy the need identified above for a short-term, direct-reading neutron dosemeter. They are also subject to a number of other well-known limitations related to their radiological properties, including high energy thresholds (nuclear track emulsions), poor energy responses (albedo TLD), poor angular responses (etched track detectors) and high detection thresholds (nuclear track emulsions, etched track detectors).

A more recent technique for neutron dosimetry based on superheated drops of liquid in a viscous or semiviscous medium has shown promise with respect to sensitivity and energy response. In the version of this type of dosemeter that is made by Bubble Technology Industries (Chalk River, Ontario, Canada), the bubbles produced by neutron interactions with the superheated drops stay in place and are counted by eye (if not too numerous) or by automated counting equipment. Recent experience with these dosemeters at Ontario Hydro has shown that they exhibit some serious limitations. Although they are temperature compensated to some degree, the compensation does not extend to the highest temperatures that may be experienced in the generating stations. A mechanical shock to the dosemeter gives rise to a cloud of small bubbles that can obscure neutron-produced bubbles. Reproducibility from dosemeter to dosemeter is sometimes poor under field conditions, and may reflect variations in the energy detection threshold when there is a significant neutron fluence at that energy. Lastly, the proper use of the bubble detector is dependent upon careful adherence to procedures, which increases the potential for human error.

Different applications of the same basic principle are used in dosemeters made by Apfel Enterprises (New Haven, Connecticutt, USA). In the survey meter/dosemeter design that counts bubbles as they are produced using an acoustical detector (Model 201 Neutron Monitor), the neutron dose is displayed digitally and has the capability of electronic transfer to a computer. However, in a high-noise or high-vibration environment, the circuit that discriminates against external signals in the acoustic detector causes the dosemeter to under-respond to neutron dose[16]. Its weight (0.41 kg) and dimensions ($133 \times 85 \times 46$ mm^3) are somewhat high for convenient use as a personal dosemeter, but would be very satisfactory for use as a survey meter. Apfel Enterprises also produces a passive, direct-reading personal neutron dosemeter called a Neutrometer™[17]. In this device, the bubbles leave the medium in which they are produced and exert a pressure on a small piston. The movement of the piston with respect to a scale provides a measure of the neutron dose received. Used in this way, the minimum measurable dose is about 50 μSv, which is higher than desirable. A lower detection limit is achievable by operating the dosemeter in the horizontal position, but this would not seem to be practical for routine use.

There has been considerable effort in recent years towards applying two established radiation detection techniques to personal neutron dosimetry. The first of these techniques is the use of silicon detectors, either singly or in pairs. An electronic neutron dosemeter using a single silicon detector, presumably in contact with a converter layer, has been commercially available from Aloka (Tokyo, Japan) for several years (Model PDM-303). The difficulty of achieving a suitable energy response with a single detector is clearly shown by test results for this dosemeter[18]; the response varies from a low of 0.3 at about 1.2 MeV to a high of over 50 at about 1 keV and below. The possibility of improving the energy response of a single diode detector using pulse shape analysis is the subject of one of the papers in this issue[19]. Another possibility that was raised a number of years ago[20], but apparently not pursued since, is the development of a hydrogenous semiconductor.

In the approach that uses two silicon detectors, one detector is sensitive only to photons and the other is made sensitive to neutrons as well by the addition of a suitable converter (e.g. Refs 21, 22). The difference in the number of pulses produced by the two detectors is a measure of the neutron dose. This technique is still in the developmental stage, and it remains to be seen whether a satisfactory energy response can be achieved, particularly in the intermediate neutron energy region and in the presence of high photon dose rates. The extension of this approach to multi-element dosemeters, each element with a different energy response, may result in a better combined energy response[23].

The second technique is based on the measurement of microdosimetric event spectra using tissue-equivalent proportional counters. The main challenge with this approach is to miniaturise the proportional counter without losing too much sensitivity. Progress has been made by developing a multicellular counter[24,25], but the technology has not yet approached the stage of commercial feasibility.

At an even earlier stage of development is the use of commercial memory chips and gallium arsenide diodes as the basis of a personal neutron dosemeter[26–28]. The aim of this research is to use solid state detectors to measure lineal energy spectra, in a way analogous to tissue-equivalent proportional counters. With a static random access memory chip, it is possible to compensate for the low sensitivity of a single active element by measuring the signals from a large number of elements in the same device. The response of these devices to intermediate energy neutrons has not yet been reported.

For neutron area surveys

It is evident from the discussion earlier in this paper that neutron survey meters that employ a large, heavy moderator around a thermal neutron detector do not meet the needs of nuclear generating stations very well. While a survey meter with a spherical moderator would be expected to offer some improvement in angular response over the cylindrical moderator presently used at Ontario Hydro, it would not offer a substantial reduction in weight or size. Lighter, more compact instruments are needed, while retaining the sensitivity and ruggedness of rem meters.

Among commercially available survey meters, there are two designs that result in a substantially lighter instrument than the Andersson-Braun rem meter. One, the Dineutron by Nardeux, uses two smaller moderating spheres of different diameters to obtain some basic spectral information that is used to correct the response. The result is not entirely successful, since the instrument shows a large variation in energy response[29] and its design precludes an isotropic angular response. The other, the REM500 by Health Physics Instruments (Goleta, California, USA), uses a tissue-equivalent proportional counter 57 mm in diameter as its detector. This instrument is currently undergoing evaluation at the Chalk River Laboratories for potential use in CANDU generating stations. Its performance in actual CANDU fields is described in another paper in this issue[30]. The main conclusions are that its energy response could be made acceptably good with an appropriate calibration, but that its sensitivity may not be high enough for routine monitoring.

OPPORTUNITIES FOR IMPROVEMENT

To meet the needs of the nuclear power industry for personal neutron dosemeters, emphasis must be placed on the development of active devices with better energy response characteristics in the intermediate energy region and with good dose detection limits. There are some basic problems with the devices that have been constructed to date, and these will have to be overcome to result in a commercially attractive dosemeter. There appear to be some promising avenues of investigation currently in progress, and it may not be long before at least one of these leads to a successful conclusion.

Improved neutron survey meters could result from techniques similar to those being investigated for use as personal dosemeters, but the need for high sensitivity is greater in order to minimise the time needed to conduct surveys and keep doses 'as low as reasonably achieveable'. There appears to be good potential for the use of tissue-equivalent proportional counters in this application, if the development of a more sensitive, multi-element counter works out well.

While dosemeters and survey meters with nearly ideal dose equivalent responses would be the most desirable outcome of further research and development, this may be too much to hope for. It seems more likely that there will continue to be significant differences in their response to standard calibration fields compared with realistic power station fields. With better and more detailed information on the energy spectra of the station fields and on the instrument response, it becomes possible to calculate calibration correction factors with some accuracy. The alternative, which may be simpler and less expensive in the long run, is to develop realistic calibration fields that are representative of the range of fields to be found in the power station environment (as described by Chartier[31]). Measurements in such fields would provide both more accurate calibration factors and a good estimate of the uncertainty in those factors.

ACKNOWLEDGEMENTS

The authors thank J. Chase, M. Haynes, K. Lemkay and R. Manley for their helpful comments on a draft of this paper.

REFERENCES

1. Rogers, D. W. O. *Why Not to Trust a Neutron Remmeter.* Health Phys. **37**, 735–742 (1979).
2. Aroua, A., Boschung, M., Cartier, F., Gmür, K., Grecescu, M., Prêtre, S., Valley, J.-F. and Wernli, Ch. *Study of the Response of Two Neutron Monitors in Different Neutron Fields.* Radiat. Prot. Dosim. **44**, 183–187 (1992).
3. Nunes, J. C., Waker, A. J. and Arneja, A. *Neutron Spectrometry and Dosimetry in Specific Locations at Two Candu Power Plants.* Radiat. Prot. Dosim. **63**, 87–104 (1995).

4. ICRP. *Recommendations of the International Commission on Radiological Protection.* ICRP Publication 26 (Oxford: Pergamon Press) (1977).
5. Delafield, H. J. and Perks, C. A. *Neutron Spectrometry and Dosimetry Measurements Made at Nuclear Power Stations with Derived Dosemeter Responses.* Radiat. Prot. Dosim. **44**, 227–232 (1992).
6. Bartlett, D. T., Britcher, A. R., Bardell, A. G., Thomas, D. J. and Hudson, I. F. *Neutron Spectra, Radiological Quantities and Instrument and Dosemeter Responses at a Magnox Reactor and a Fuel Reprocessing Installation.* Radiat. Prot. Dosim. **44**, 233–238 (1992).
7. Aroua, A., Boschung, M., Cartier, F., Grecescu, M., Prêtre, S., Valley, J.-F. and Wernli, Ch. *Characterisation of the Mixed Neutron-Gamma Fields Inside the Swiss Nuclear Power Plants by Different Active Systems.* Radiat. Prot. Dosim. **51**, 17–25 (1994).
8. Lindborg, L., Bartlett, D., Drake, P., Klein, H., Schmitz, Th. and Tichy, M. *Determination of Neutron and Photon Dose Equivalents at Work-Places in Nuclear Facilities in Sweden.* Radiat. Prot. Dosim. **61**, 89–100 (1995).
9. Nunes, J. C., Waker, A. J. and Lieskovský, M. *Neutron Fields Inside Containment of a CANDU 600-PHWR Power Plant.* Health Phys. (submitted).
10. Cross, W. G. *Suitability of CR-39 Dosimeters for Personal Dosimetry Around CANDU Reactors.* Report AECL-10678 (Atomic Energy of Canada Limited, Chalk River Laboratories, Canada) (1992).
11. IEC 1005. *Portable Neutron Dose Equivalent Rate Meters for Use in Radiation Protection.* (International Electrotechnical Committee (IEC), Geneva) (1990).
12. IEC 1323. *Radiation Protection Instrumentation — Neutron Radiation — Direct Reading Personal Dose Equivalent and/or Dose Equivalent Rate Meters* (International Electrotechnical Committee (IEC), Geneva) (1995).
13. Portal, G. and Dietze, G. *Implications of the New ICRP and ICRU Recommendations for Neutron Dosimetry.* Radiat. Prot. Dosim. **44**, 165–170 (1992).
14. Hollnagel, R. A. *Calculated Effective Doses in Anthropoid Phantoms for Broad Neutron Beams with Energies from Thermal to 10 MeV.* Radiat. Prot. Dosim. **44**, 155–158 (1992).
15. Leuthold, G. and Schraube, H. *Critical Analysis of the ICRP Proposals for Neutron Radiation and a Possible Solution.* Radiat. Prot. Dosim. **54**, 217–220 (1994).
16. Szornel, K. *Evaluation of Apfel 201 Neutron Monitor Based on a Superheated Drop Detector.* Report number COG-94-297 (AECL, Chalk River Laboratories, Chalk River, Ontario K0J 1J0, Canada) (1994).
17. Apfel, R. E. *Characterisation of New Passive Superheated Drop (Bubble) Dosemeters.* Radiat. Prot. Dosim. **44**, 343–346 (1992).
18. Alberts, W. G., Dietz, E., Guldbakke, S. and Kluge, H. *Response of an Electronic Personal Neutron Dosemeter.* Radiat. Prot. Dosim. **51**, 207–210 (1994).
19. Bordy, J. M., Lahae, T., Landre, F., Hoflack, C., Lequin, S. and Barthe, J. *Single Diode Detector for Individual Neutron Dosimetry using a Pulse Shape Analysis.* Radiat. Prot. Dosim. **70**(1–4), 73–78 (This issue) (1997).
20. Brackenbush, L. W. and Quam, W. *Hydrogenous Semiconductor Neutron Detectors.* Radiat. Prot. Dosim. **10**, 303–305 (1985).
21. Barelaud, B., Paul, D., Dubarry, B., Makovicka, L., Decossas, J. L. and Vareille, J. C. *Principles of an Electronic Neutron Dosemeter Using a PIPS Detector.* Radiat. Prot. Dosim. **44**, 363–366 (1992).
22. Barthe, J., Bordy, J. M., Mourgues, M., Lahaye, T., Boutruche, B. and Segur, P. *New Devices for Individual Neutron Dosimetry.* Radiat. Prot. Dosim. **54**, 365–368 (1994).
23. Alberts, W. G., Dörschel, B. and Siebert, B. R. L. *Methodical Studies on the Optimisation of Multi-Element Dosemeters in Neutron Fields.* Radiat. Prot. Dosim. **70**(1–4), 117–120 (This issue) (1997).
24. Bordy, J. M., Barthe, J., Boutruche, B. and Segur, P. *A New Proportional Counter for Individual Neutron Dosimetry.* Radiat. Prot. Dosim. **54**, 369–372 (1994).
25. Bordy, J. M., Barthe, J., Lahaye, T. and Boutruche, B. *Improving a Multi-Cellular Tissue Equivalent Proportional Counter for Personal Neutron Dosimetry.* Radiat. Prot. Dosim. **61**, 175–178 (1995).
26. Schröder, O. and Schmitz, T. *Can a Personal Dosemeter for Neutron Radiation Based on a Semiconductor Chip Match the New ICRP Recommendations?* Radiat. Prot. Dosim. **54**, 361–365 (1994).
27. Schröder, O. and Schmitz, T. *The Application of Commercial Semiconductor Chips for Personal Neutron Dosimetry.* Radiat. Prot. Dosim. **61**, 9–12 (1995).
28. Schröder, O. *Counting Protons — A Simple Approach to a Personal Neutron Dosemeter?* Radiat. Prot. Dosim. **61**, 179–182 (1995).
29. Rimpler, A. *Dose Equivalent Response of Neutron Survey Meters for Several Neutron Fields.* Radiat. Prot. Dosim. **44**, 189–192 (1992).
30. Waker, A. J., Szornel, K. and Nunes, J. *TEPC Performance in the CANDU Workplace.* Radiat. Prot. Dosim. **70**(1–4), 197–202 (This issue) (1997).
31. Chartier, J. L., Posny, F. and Buxerolle, M. *Experimental Assembly for the Simulation of Realistic Neutron Spectra.* Radiat. Prot. Dosim. **44**, 125–130 (1992).

SINGLE DIODE DETECTOR FOR INDIVIDUAL NEUTRON DOSIMETRY USING A PULSE SHAPE ANALYSIS

J. M. Bordy†, T. Lahaye†, F. Landre†, C. Hoflack†, S. Lequin† and J. Barthe‡
†Institut de Protection et de Sûreté Nucléaire
Département de Protection de la santé de l'Homme et de Dosimétrie-SDOS
ISPN-BP no 6, F92265 Fontenay-aux-Roses CEDEX, France
‡CEA DTA DAMRI, CE Saclay, BP no 56, F91193 Gif sur Yvette CEDEX, France

Abstract — A new method using one PN junction covered with a hydrogenous converter (^{10}B-loaded) is proposed for individual neutron dosimetry. This method is based on a pulse shape analysis to discriminate the photon signal from the neutron signal. This method allows drastic reduction of the photon sensitivity (by a factor of 1000). Additionally, when using a photon correction factor, the photon response becomes negligible. By applying a neutron correction factor to the low energy events, the gap in neutron sensitivity for intermediate energy can be partly filled. Lead shields used to surround the detector allow the remaining photon sensitivity to be decreased by a factor of two. An especially designed hydrogenous moderator placed at the top of the detector allows the neutron sensitivity to be increased by a factor of two for 250 keV neutron energy.

INTRODUCTION

Knowledge of effective dose due to neutrons for radiation protection purposes is of importance in controlled areas of nuclear power plants, research laboratories and facilities involved in the nuclear fuel cycle, since neutrons often account for a substantial fraction of the total effective dose[1]. Due to the high variation in neutron energy spectra in the laboratories, area monitoring is not sufficient and personal neutron dosimetry is definitively required[2].

Semiconductor detectors are widely used in different areas such as gamma ray and alpha spectrometry. For personal dosimetry, PN junctions are usually associated with (n, heavy charged particle) converters. They can measure personal dose equivalent in a phantom through recoil nuclei counting[3–5] or LET spectrometry[6,7]. It is assumed that personal dose equivalent in a phantom is a good estimate of the personal dose equivalent. Additionally, their small size and low power consumption remain a valuable advantage.

The two main problems of individual neutron dosemeters are (i) to provide a high sensitivity to neutron and a low response variation as a function of the neutron energy, and (ii) to separate the contribution of neutron and photon components from the radiation field to the total dose equivalent. Up to now, available individual dosemeters do not fulfil these requirements. However, semiconductor detectors are able to meet them at least partly.

The method proposed in this paper is based on a PN junction covered with a polyethylene ^{10}B-loaded converter. Neutron dose equivalents are calculated by recoil nuclei counting (mainly protons and alpha particles). The photon rejection is based on a pulse shape analysis.

PHOTON REJECTION PRINCIPLE[8]

It is a well-known fact that shapes of electronic pulses generated by nuclear radiation detectors depend significantly on the species of the detected particles. Various techniques, using this property, have been employed to achieve particle identification through pulse shape analysis (scintillators, gas detectors, etc)[9,10].

A PN junction can be roughly divided into two parts: the depleted layer and the remainder of the silicon detector, called the 'wafer' in this paper.

In our case, protons and alpha particles, generated in the converter, interact essentially inside the depleted layer and not significantly in the wafer, while electrons, generated by photons in the silicon, interact in both zones. Charge carrier velocities are different in these two regions, they are greater in the depleted layer because the electric field is higher in this region. For this reason, rise times of heavy charge particle pulses are shorter than those of photon pulses.

There are roughly three kinds of pulses (Figure 1):

(1) 'Fast pulses' (a) are due to interactions occurring exclusively in the depleted layer. For particles stopped inside the depleted layer, collection times are short and depend slightly on penetration depth.
(2) 'Slow pulses' (b) are due to interactions that occur exclusively in the wafer. The collection mechanism is the diffusion. Rise time depends on the diffusion time, i.e. the carrier mobilities and the distance between the point of interaction and the depleted layer. These pulses are mainly due to electrons, generated by photons, interacting exclusively in the wafer.
(3) 'Mixed pulses' (c) are also found. They present two slopes. These are due to electrons, generated by photons, interacting in both wafer and depleted layer. The highest slope is due to interactions that occur in the depleted layer.

EXPERIMENTAL SET-UP

The device (Figure 2) consists of 3 parts: (i) the detector and preamplifier (Intertechnique PSC 762), (ii) the amplifier step and the pulse shape analyser, all of them made of commercial NIM standard board, (iii) three multichannel analysers record the spectra corresponding to So1, So2 and So3.

The detector (Figure 3) consists of a Canberra PIPS diode (a) (300 μm thickness, bulk resistivity: 600 Ω.cm, sensitive area: 1 cm²) covered with a polyethylene converter ^{10}B-loaded (b) (5×10^{15} bore, 30 μm thick). Conductive rubber rings (c) and an aluminium plate (d) are used to apply a reverse bias voltage to the junction — for 12 V the depleted layer is about 30 μm thick. Two printed circuits (e) join the different pieces together.

A polyethylene moderator (f) (1 cm thick) drilled with holes covers the detector. The total sensitive area of the diode covered by polyethylene is 0.5 cm². Lead shields (g) cover both sides of the detector.

This detector has been irradiated on a phantom according to the requirements of the calibration procedure for individual dosemeters. The 'slab phantom' (30 cm × 30 cm × 15 cm) made of polymethyl methacrylate has been used here.

ELECTRONIC PRINCIPLE OF THE RISE TIME MEASUREMENT

The rise time measurements are carried out with a time amplitude converter — TAC. The rise time is measured between a 'start' and a 'stop' signal related to the event being measured. The amplitude of the logic pulses generated by the TAC is related to the interval between these two signals. The 'start' signal is generated when the analogue input pulse crosses a low-level threshold. This marks the arrival time of the event. The 'stop' signal is produced by a timing single channel analyser when the peak of the input pulses is detected; this technique uses the zero-crossing of a bipolar pulse related to the input pulse. Thus the rise time is virtually independent from the input signal amplitude.

METHOD AND EXPERIMENTAL RESULTS

The dosemeter sensitivity (S) in terms of personal dose equivalent in the phantom is as follows:

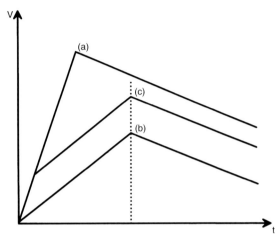

Figure 1. Schematic view of the pulse shapes at the output of the preamplifier. Figure 4 provides typical values of rise time.

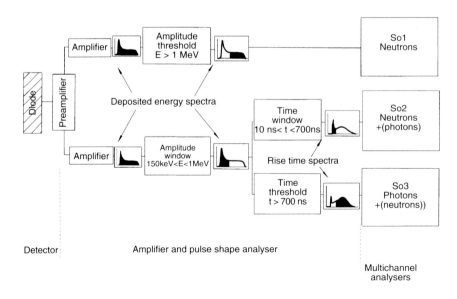

Figure 2. Block diagram of the experimental set-up.

$S = N/H_p(10,0°)_{ref}$

where $H_p(10,0°)_{ref}$ is the reference individual dose equivalent at a 10 mm depth in phantom and 0° of incidence, and N is the number of pulses due to heavy charged particles.

Let N(E,t) the number of pulses as a function of the deposited energy in the detector (E) and the rise time of the pulses (t). Two kinds of spectra are defined:

(1) The 'deposited energy spectrum', d N(E)/dE, with

$$N(E) = \int_{t=0}^{t=\infty} [\partial N(E,t)/\partial t] dt$$

(2) The 'rise time spectrum', d N(t)/dt,

with $N(t) = \int_{E=0}^{E=\infty} [\partial N(E,t)/\partial E] dE$

The deposited energy spectrum is calibrated by using alpha sources such as ^{244}Cm ($E_\alpha \approx 5.8$ MeV), a photoelectric peak of ^{241}Am gamma ray ($E_\gamma \approx 60$ keV) and monoenergetic neutrons between 250 keV and 2.5 MeV. In the latter case the maximum deposited energy corresponds to the neutron energy according to the neutron to proton elastic scattering theory. Thus, a wide range of energy is covered.

Rise time measurement requires time and electric consumption. The deposited energy spectrum is divided into three parts partly to avoid these problems.

(1) Due to the electronic background, superimposed on pulses, a good accuracy in rise time measurement is obtained when the pulse amplitude is greater than 5 times the electronic background magnitude (BG = 50 keV). Below 3 times (150 keV), it is no longer possible to separate neutron and photon pulses. Therefore pulses are not taken into account below 150 keV.

(2) Between 150 keV and 1 MeV, pulses are due to photons or to neutrons. Rise time is also measured for each pulse. The rise time spectrum is divided into two parts:

(i) Below 700 ns, pulses are considered as pulses due to neutrons:

$$So2 = \int_{E=0.15\,MeV}^{E=1\,MeV} \int_{t=0}^{t=700\,ns} \frac{d^2 N(E,t)}{dE\,dt} dE\,dt$$

(ii) Above 700 ns, pulses are considered as pulses due to photons:

$$So3 = \int_{E=0.15\,MeV}^{E=1\,MeV} \int_{t=700\,ns}^{t=\infty} \frac{d^2 N(E,t)}{dE\,dt} dE\,dt$$

(3) Above 1 MeV, the rise time is not measured because the higher amplitude of photon pulses is about 650 keV (without pulse shape analysis). Thus, using 1 MeV as threshold value, pulses are exclusively due to neutrons. The total number of pulses above 1 MeV is

$$So1 = \int_{E=1\,MeV}^{E=\infty} \frac{d\,N(E)}{dE} dE$$

Thus, the number of pulses due to heavy charge particles should be:

$$N = So1 + So2 \qquad (1)$$

The rise time threshold (700 ns) has been chosen experimentally. It is a compromise solution, one can see that a small number of pulses with rise time lower than 700 ns are due to photons (Figure 4). In mixed n,γ radiation fields, one must subtract this residual photon sensitivity from the estimated number of neutron pulses N. To estimate the number of photon pulses counted as neutron pulses in a mixed (n,γ) radiation field, the detector is exposed to pure photon radiation. In this particular case $So1_\gamma$, $So2_\gamma$ and $So3_\gamma$ are respectively the

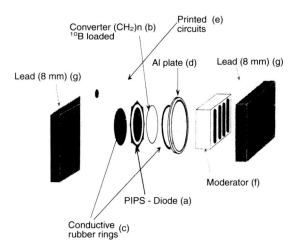

Figure 3. Three-dimensional view of the detector.

So1, So2 and So3 counts. As $So1_\gamma$ is equal to zero and N must be equal to zero (the neutron dosemeter must not measure photon dose equivalent), a 'gamma correction factor' $k_\gamma = So2_\gamma/So3_\gamma$ is used. Table 1 gives the k_γ values for different photon radiation fields. The highest values have been found for 250 keV and ^{226}Ra. The latter being a broad spectrum source. Its value ($k_{\gamma_{Ra}}$) is used in this paper, on the assumption that it is a realistic photon spectrum. Therefore it is assumed that whatever the case the residual photon component is never underestimated.

Equation 1 is therefore modified as follows:

$$N = So1 + So2 - k_{\gamma_{Ra}} So3 \quad (2)$$

As mentioned previously lead shields have been added on both sides of the detector to decrease the photon sensitivity. An 8 mm thick shield reduces the photon sensitivity by a factor of two.

Figure 5 shows deposited energy spectra plotted from the So1 outputs for 0.570 MeV monoenergetic neutrons. It is compared to a spectrum achieved without photon rejection. It can be seen that the photon component is drastically reduced. Only a few pulses, recorded between 0.15 and 0.25 MeV, are not due to neutrons. Some of these pulses could be due to recoil silicon[11] and photons.

A dip in neutron sensitivity is noticed around 250 keV (Figure 6). An attempt has been made to obviate this problem by applying a weighting factor k_w and by adding an hydrogenous moderator.

The moderator allows the sensitivity to be increased around 0.25 MeV (Figure 6) but this sensitivity remains low. Additionally, the standard deviation for this energy is very large (50%) instead of 5% for the upper energies and 10% for 0.144 MeV. A better knowledge of the sensitivity for 0.25 MeV is also necessary and requires extra measurements.

Using k_w, Equation 2 is put in the form:

$$N = So1 + k_w (So2 - k_{\gamma_{Ra}} So3) \quad (3)$$

The gap in neutron sensitivity can be partly filled by increasing the value of the weighting factor (Figure 6). For k_w values up to 5, the increase in sensitivity below 1 MeV is faster than above 1 MeV; thus the variation in neutron sensitivity is reduced. With $k_w > 5$, the ratio between the minimum and the maximum sensitivities is not significantly reduced because the sensitivity around 1.2 MeV becomes greater than that at 2.5 MeV.

DISCUSSION AND CONCLUSION

The pulse shape analysis allows drastical reduction in the photon response of semiconductor detectors. Using the photon correction coefficient, k_γ, this photon response can be reduced to a very low level. The detection threshold in terms of neutron energy usually met within semiconductor detectors around 700 keV[1] does not exist any more. At the very most, a low response persists around 0.25 MeV.

A weighting function, decreasing as the deposited energy increases, could be used up to 1 MeV instead of the single weighting factor, k_w. This requires a better knowledge of the deposited energy spectrum under 0.25 MeV. This could be obtained thanks to a decrease in the electronic background magnitude. In this case, the rise time measurement would be more accurate, the photon pulse rejection would be better and therefore the number of recorded photon pulses under 0.25 MeV would decrease and finally the k_γ value will decrease. A new device with a lower magnitude for the background signal is being manufactured.

More than statistical uncertainties, given in the previous section, systematic errors due to the k_γ variation

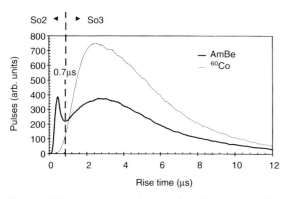

Figure 4. Rise time spectra obtained for cobalt and AmBe sources.

Table 1. Gamma correction factor for various photon sources.

Sources	250 keV	^{137}Cs	^{60}Co	^{226}Ra
k_γ (±10%)	1.2×10^{-2}	7.8×10^{-3}	6.0×10^{-3}	1.2×10^{-2}

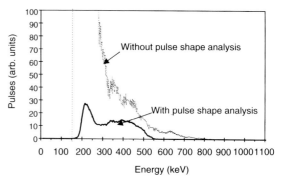

Figure 5. Deposited energy spectrum obtained for 0.570 keV monoenergetic neutrons with and without pulse shape analysis.

as a function of photon energy are the major problem. Indeed, the k_γ is arbitrarily chosen. Also, if the k_γ value underestimates the photon contribution, some photon pulses can be counted as neutron pulses, therefore the neutron component is overestimated and vice versa (see Equation 3). This problem is very important for intermediate neutron energies due to the low neutron sensitivity in this case. Again, a decrease of the k_γ can solve this problem, because this reduces the magnitude of this effect on neutron energy dependence.

Using the pulse shape analysis, the semiconductor detectors would be able to meet, at least partly, the requirements of individual neutron dosimetry. They can separate neutron and photon signals. Using a specially designed moderator and a weighting function the dip in neutron sensitivity could be partly filled. In any case, one has to keep in mind that this 'weighting method' cannot be used if photon pulses are not previously eliminated and does not improve the accuracy of the measurement for intermediate neutrons. Shape and thickness of the moderator have to be improved to increase the sensitivity and therefore the precision for neutron energies around 0.25 MeV. Calculation using Monte Carlo codes are in progress to provide helpful analysis of the neutron interaction in moderators.

ACKNOWLEDGEMENTS

The authors are grateful to Dr P. Burger at Canberra Semiconductor (N.V. Olen, Belgium), Dr F. Schultz and M. Richter at MGP instrument (Lamanon, France), Dr J. Guillon at Fimel (Velizy, France) and Dr J. Dhermain at the DGA/CEB (Arcueil, France).

This study was partly supported by the Commission of European Communities (contract FI3P CT93 0072).

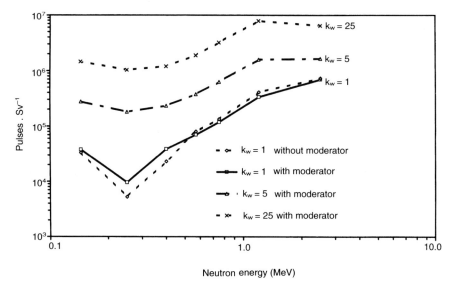

Figure 6. Dose equivalent response obtained with and without moderator and by using the weighting factor k_w. Uncertainties are given in the text.

REFERENCES

1. Alberts, W. G., Bordy, J. M., Chartier, J. L., Jahr, R., Klien, H., Luszik-bhadra, M., Posny, F., Schuhmacher, H. and Siebert, B. R. L. *Neutron Dosimetry*. Radioprotection **31**(1), 37–65 (1996).
2. Medioni, R., Bermann, F., Bordy, J. M. and Portal, G. *S.S.N.T.D. Calibration in a Neutron Source Fabrication Hot Laboratory*. In: Proc Tenth DEO Workshop on Personnel Neutron Dosimetry, September 1983. PNL-SA-12352, Battelle (1983).
3. Barthe, J., Lahaye, T., Moiseev, T. and Portal, G. *Personal Neutron Diode Dosemeter*. Radiat. Prot. Dosim. **47**(1–4), 397–399 (1993).
4. Zamani, M. and Savvidis, E. *A Real-time Personal Neutron Dosemeter Based on PIPS and a Double Layer Neutron Converter*. Radiat. Prot. Dosim. **63**(2), 299–303 (1996).
5. Fernandez, F., Lugera, E., Domingo, C. and Baixeras, C. *Separation of the Neutron Signal from the Gamma Component in (n,γ) Fields using Differential Pulse Analysis Techniques with a Double Silicon Diode*. Radiat. Prot. Dosim. **70**(1–4), 87–92 (This issue) (1997).
6. Orlic, M., Lazarevic, V. and Boreli, F. *Microdosimetric Counter based on Semiconductor Detectors*. Radiat. Prot. Dosim. **29**(1–2), 21–22 (1989).

7. Schroder, O. and Schmitz, T. *The Application of Commercial Semiconductor Chips for Personal Neutron Dosimetry*. Radiat. Prot. Dosim. **61**(1–3), 9–12 (1995).
8. Barthe, J., Bordy, J. M., Lahaye, T. and Mourgues, M. *New Principle of Single Diode Neutron Dosimeter Based on Time Resolution*. In: Proc. 17th Congress, IRPA Regional Congress on Radiological Protection, Portsmouth, June 1994. (Ashford, UK: Nuclear Technology Publishing) pp. 97–100 (1994).
9. Knoll, G. F. *Radiation Detection and Measurement*. 2nd edn (New York: John Wiley) (1989).
10. Blanc, D. *Les Rayonnements Ionisants* (Paris: Masson) (1990).
11. Tanner, J. E., Witt, R., Tanner, R. J., Bartlett, D. T., Burgess, P. H., Edwards, A. A., More, B. R., Taylor, G. C. and Thomas, D. J. *An Assessment of the Feasibility of using Monte Carlo Calculations to Model a Combined Neutron/Gamma Electronic Personal Dosemeter*. Radiat. Prot. Dosim. **61**(1–3), 183–186 (1995).

ADVANCED DETECTORS FOR ACTIVE NEUTRON DOSEMETERS

J. C. Vareille[a], B. Barelaud[a], J. Barthe[b], J. M. Bordy[c], G. Curzio[d], F. d'Errico[d], J. L. Decossas[a], F. Fernandez-Moreno[e], T. Lahaye[c], E. Luguera[e], O. Saupsonidis[f], E. Savvidis[f] and M. Zamani-Valassiadou[f]
[a]Faculté des Sciences
123 Avenue Albert Thomas, 87060 Limoges, France
[b]CEA-DTA DAMRI
BP 52, 91193 Gif sur Yvette, France
[c]Institut de Protection et de Sureté Nuclaire-DPHD SDOS
IPSN B.P. 6, 92265 Fontenay aux Roses CEDEX, France
[d]Dipartimento di Costruzioni Meccaniche e Nucleari
Università degli Studi di Pisa, Via Diotisalvi 2, 56126 Pisa, Italy
[e]Departament de Fisica, Universitat Autonoma de Barcelona
08193 Cerdanyola (Barcelona), Spain
[f]Physics Department, NPD, Aristotle University of Thessaloniki
54006 Thessaloniki, Greece

Abstract — Different options were investigated in the development of active neutron dosemeters for personnel and area monitoring. In particular, two different classes of detectors, superheated emulsions and silicon diodes, were studied as radiation sensors for such devices. The detectors were analysed with respect to their neutron sensitivity and their overall suitability for practical dosimetry. Superheated emulsions proved adequate in terms of their dose equivalent response and their photon discrimination: future work should now focus on the development of a rugged device. Significant improvements were also achieved with the diode sensors, although more research will be necessary for the design of a simple, single structure readily usable as a neutron dosemeter.

INTRODUCTION

Advances in the instrumentation for neutron dosimetry have led to new moderator-type survey meters as well as passive personnel dosemeters with improved and extended range energy response[1]. Nevertheless, the need is still felt for both light-weight area monitors and active personnel dosemeters. Our research focused on two different neutron detection methods that hold promise for the development of such systems: superheated emulsions and silicon diodes. The intrinsic nature of these detectors is quite different, nevertheless the two systems appear suitable for an optimised approach to the different neutron dosimetry needs. In fact, a variety of workplaces exists differing in radiation field quality and strength, and also in environmental conditions such as acoustical or electromagnetic noise, vibration, humidity and temperature. Therefore, various versions of these detectors were studied and characterised in terms of their response to monoenergetic neutrons, and to broad-spectrum benchmark fields. Based on these investigations, improvements were adopted and planned as described here and, in greater detail, in separate publications of the laboratories participating in this CEC-supported research[2-6].

SUPERHEATED EMULSIONS

These detectors, commercialised as superheated drop detectors (SDD) or bubble detectors, are uniform emulsions of microscopic, over-expanded halocarbon droplets dispersed in a compliant matrix. Upon neutron irradiation, nuclear interactions inside or next to the droplets nucleate the phase transition of the superheated liquid therein, generating macroscopic bubbles which may be registered by various methods[2].

When the SDD droplets are sufficiently superheated and contain chlorine, they can be nucleated by the products of the $^{35}Cl(n,p)^{35}S$ neutron capture reaction. Through this mechanism, fast neutron sensitive halocarbon-12 emulsions also respond to thermal and intermediate neutrons, as required of a neutron dosemeter. A prototype neutron area monitor was developed based on these emulsions and on the acoustical bubble counting method (Figure 1). The SDD is thermally controlled: this allows for the twofold advantage of avoiding external temperature effects while ensuring an improved dose equivalent response (Figure 2). The energy dependence was determined and optimised by means of experimental investigations with mono-

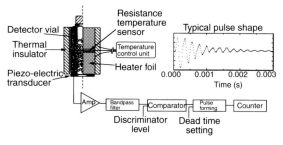

Figure 1. Superheated drop (bubble) detector vial and related acoustical pulse counting electronics.

energetic neutrons and by Monte Carlo detector simulations[7]. A neutron sensitivity of about 5 bubbles per μSv was achieved combined with a virtually complete gamma discrimination. The device was extensively tried by means of numerical and in-field tests of the response to broad neutron spectra: benchmark fields, nuclear power plants, spent fuel storage sites, and relativistic stray radiation areas. All these indicated a remarkably constant dose equivalent response regardless of neutron energy distributions. Results were generally within 10% of the 'true' value, while conventional meters over-read reference dose values by 50% on average (with much higher deviations). The prototype is a fairly delicate system, though, and can be operated reliably when environmental conditions are not extreme.

Through the active control of their superheat, SDD emulsions were also tried for neutron spectrometry. A system was developed allowing for temperature regulation within 0.2°C by means of thin heating strips operated by a time proportioning controller and monitored by a platinum resistance sensor. Two different emulsions were employed which are insensitive to thermal neutrons and present different response thresholds. By varying their operating temperature, accurately defined detection thresholds are correspondingly generated: virtually any desired value in the 0.01 to 10 MeV range. Tests with radionuclide neutron sources proved that these response curves can be used to scan neutron spectra, and indicated the potential of this original approach. Measured data were unfolded by means of various codes and results were always consistent with the source spectra.

SILICON DIODES

Various types of sensors, shielding, radiation converters and also pulse shape analysis and data processing, were tried for the development of neutron individual dosemeters based on silicon diodes (also referred to as Planar Implanted Positive Silicon, or PIPS).

Differential method

The Classical Differential Method (CDM)[8] was first studied, employing two identical p-n junctions, one bare and one covered by a ^{10}B loaded polyethylene converter (typically 35 μm thick). In the CDM (Figure 3), the difference of the two diode counts is taken, as it should theoretically be free of electronic background and γ contribution. Unfortunately, the two diodes are never truly identical and to avoid the γ sensitivity it was necessary to set a threshold at 600 keV, despite the removal of most metal parts and the reduction of the depleted layer to 10 μm. In the 1–4 MeV neutron energy range, the sensitivity in terms of individual dose equivalent[1] falls between 0.5 and 1 pulse.μSv^{-1} (Figure 4). Thermal neutrons are also detected thanks to the ^{10}B converter. However, the device fails as a personal dosemeter as it does not cover the important 10–600 keV energy range. Two

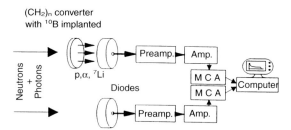

Figure 3. Block diagram of the Classical Differential Method.

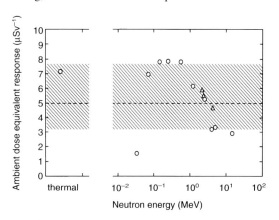

Figure 2. Ambient dose equivalent response of a halocarbon-12 emulsion stabilised at 31.5°C. The shaded area indicates the region falling within a factor 1.5 of the reference value of 5 bubbles per μSv (dashed line). (○) monoenergetic beams, (△) radionuclide sources.

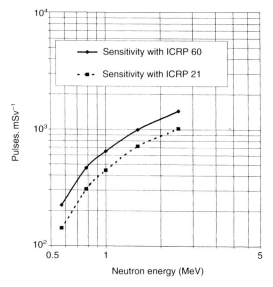

Figure 4. Fluence response of the CDM coupled diodes as a function of neutron energy (normalised to 1 cm^2).

possibilities were investigated to improve its performance: a double diode detector to minimise the differences between the diodes, and some moderating shields with thick ^{6}LiF converters to yield a higher sensitivity in the intermediate range (from 10 keV to 1 MeV).

Two p-n junctions, expected to present the same background, were applied to the 'front' and 'back' sides of a silicon substrate (bulk resistivity 500 Ω.cm)[4]. The device was custom-made by Canberra Industries, Inc., with diodes of 2 cm^2 sensitive area. The two junctions were partly depleted (35 μm with a polarisation of 10 V) and a 40 μm thick polyethylene converter was placed over the front one. Again, computation of dose equivalent quantities was based on the difference between the two diode counts, and again a deposited energy threshold was necessary, but the new value was only 250 keV. This threshold was necessary since diodes have different photon sensitivity. In fact, the back one records the additional pulses due to secondary electrons generated in the wafer. This device is not sensitive to thermal neutrons since the converter does not contain ^{10}B. Its main characteristics are: neutron dose equivalent response constant within ±50% in the range 570 keV–15 MeV, and detection threshold of 10 μSv at 1.2 MeV. A Monte Carlo simulation confirmed these experimental results and the value of the detection threshold (Figure 5). Thus, an improvement of the CDM was achieved, but problems still arise from the different photon sensitivity of the two diodes.

The previous variation on the differential method did not yield an adequate sensitivity to intermediate neutrons, therefore new solutions were adopted. A 3 μm thick ^{6}LiF converter evaporated on a 500 μm polyethylene foil was developed[5,9] and applied to two detectors: a 30 μm depleted PIPS diode in a special PMMA casing (meant to reduce the γ contribution by minimising the metal parts)[10] and a commercial PIPS diode with 100 μm depleted zone in a standard metallic mount. Results were promising: the neutron component of the signal was well identified since γ ray, α particle and ^{3}H regions are clearly separated in the energy spectra. The sensitivity varies between 100 and 10 pulses.μSv^{-1} from 73 keV to 14 MeV (Figure 6). The dose equivalent response was improved by controlling, via the converter, the albedo contribution from the phantom. The γ ray sensitivity of the two detectors is similar and does not affect the neutron dose measurements since it can be discriminated by means of a threshold on the low energy channels.

Pulse shape analysis

A further method was tried using a single p-n junction covered with a hydrogenous ^{10}B loaded converter[11]. This relies on pulse shape analysis to discriminate photon and neutron signals[3]: magnitude and rise time of the pulses depend on geometrical and electrical characteristics of the junction. Heavy charged particles generated in the converter interact essentially in the depleted layer of the detector, whereas photons interact in the depleted layer or in the rest of the wafer. Since charge carrier velocities are different in these two regions, due to different electric fields, heavy charged particle pulses have a faster rise time. Standard NIM electronics can be used in the rise time analysis. This method allows for a strong reduction of the photon sensitivity (by a factor of 1000). The maximum deposited energy of the photon spectra is about 225 keV, while it was 650 keV in the CDM. Lead shields surrounding the detector decrease the residual photon sensitivity by a factor of two, while a hydrogenous moderator placed over the detector increases the neutron sensitivity by a factor of two for 250 keV neutrons. Moreover, algorithms are under study which suppress the photon signal and compensate for the dip in neutron sensitivity at intermediate energies (Figure 7).

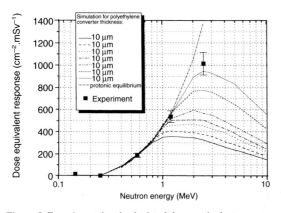

Figure 5. Experimental and calculated dose equivalent response to normally incident neutrons as a function of neutron energy (normalised to 1 cm^2).

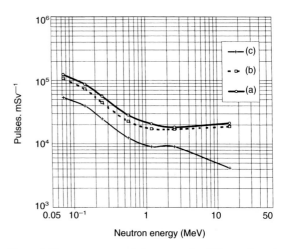

Figure 6. Energy response of the system for different channel thresholds (normalised to 1 cm^2): (a) integration over the whole spectrum, (b) threshold at the lower part of alpha particle peak, (c) threshold between alpha and ^{3}H peaks.

Coincidence method

Finally, a coincidence method was devised based on a 'Diode/Reactive-Layer/Diode' sandwich achieved by microelectronics techniques[6]. The objective is a multi-area dosemeter which identifies different neutron energy ranges. The principle is analysing the total energy deposited in silicon detectors working in coincidence by $^{10}B(n,\alpha)^{7}Li$ or $^{6}Li(n,\alpha)^{3}H$ reactions in the reactive layer. ^{10}B has a high cross section for neutron capture, but the excited state from this reaction yields complex spectra of difficult analysis. ^{6}Li generates simpler spectra for two reasons: it does not lead to excited states, and emitted particles are lighter and more energetic. The response of the system was first simulated numerically to optimise the design of the reactive layer (nature and thickness from 10 nm to 500 nm) and the dead layer thickness of the diode (from 10 nm to 500 nm). The main goal was to obtain a correct sensitivity to intermediate neutrons. Preliminary experiments were then carried out on a device assembled with two separate diodes and a boron layer. A paraffin-shielded Am-Be source was employed and measured spectra showed the feasibility of the method while confirming its complexity.

CONCLUSIONS

This work analysed the potential of two classes of detectors for neutron dosimetry. Superheated emulsions were studied in terms of the optimal combination between thermodynamic and nuclear properties of different sensitive liquids. They allowed for the development of high sensitivity neutron detectors with an almost constant dose equivalent response and a complete photon rejection, but they currently constitute delicate systems which cannot be operated in extreme environments. Overcoming this limitation will be the focus of future work. Silicon diodes appear suitable for the development of rugged devices capable of withstanding rough working conditions. However, at present, no simple, single electronic structure provides a satisfactory dose equivalent response, especially in the 10–600 keV neutron energy range. Nevertheless, encouraging preliminary data and useful indications were drawn from this work for the development of improved diode sensors. Further research will be devoted to optimisation of detector encasing materials, choice of radiation converters and moderators, reduction of depleted zone thickness, and pulse shape analysis.

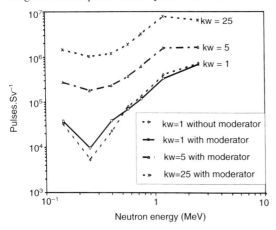

Figure 7. Dose equivalent response of the NIM standard device (normalised to 1 cm^2). k is an arbitrary weighting factor applied to the low energy deposition events to increase the response to intermediate neutrons.

REFERENCES

1. Alberts, W. G. *Neutron Protection Dosimetry for Area and Individual Monitoring*. In: Proc. Int. Conf. on Neutrons in Research and Industry, Crete 1996. SPIE Proc. Series (in press).
2. d'Errico, F., Alberts, W. G. and Matzke, M. *Advances in Superheated Drop (Bubble) Detector Techniques*. Radiat. Prot. Dosim. **70**(1–4), 103–108 (This issue) (1997).
3. Bordy, J. M., Lahaye, T., Landre, L., Hoflack, C., Lequin, S. and Barthe, J. *Single Diode Detector for Individual Neutron Dosimetry using a Pulse Shape Analysis*. Radiat. Prot. Dosim. **70**(1–4), 73–78 (This issue) (1997).
4. Fernandez, F., Luguera, E., Domingo, C. and Baixeras, C. *Separation of the Neutron Signal from the Gamma Component in (n-γ) Fields using Differential Pulse Analysis Techniques with a Double Silicon Diode*. Radiat. Prot. Dosim. **70**(1–4), 87–92 (This issue) (1997).
5. Savvidis, E. and Zamani Valasiadou, M. *Efficiencies of an SSNTD and Electronic Neutron Dosemeter with the Same (n,α) (n, p) Converter*. Radiat. Prot. Dosim. **70**(1–4), 83–86 (This issue) (1997).
6. Barelaud, B., Nexon, F., Decossas, J. L., Vareille, J. C. and Sarrabayrousse, G. *Study of an Integrated Monitor for Neutron Fields*. Radiat. Prot. Dosim. **61**(1–3), 153–158 (1995).
7. d'Errico, F., Alberts, W. G., Dietz, E., Gualdrini, G. F., Kurkdjian, J., Noccioni, P. and Siebert, B. R. L. *Neutron Ambient Dosimetry with Superheated Drop Detectors*. Radiat. Prot. Dosim. **65**(1–4), 397–400 (1996).
8. Barelaud, B., Decossas, J. L., Makovicka, L. and Vareille, J. C. *Capteur Électronique pour la Dosimétrie des Neutrons*. Radioprotection **26**(2), 307–328 (1991).
9. Zamani, M. and Savvidis, E. *A Real Time Personal Neutron Dosemeter based on PIPS Diodes and a Double Layer Neutron Converter*. Radiat. Prot. Dosim. **63**(4), 299–303 (1996).
10. Paul, D., Barelaud, B., Dubarry, B., Makovicka, L., Decossas, J. L. and Vareille, J. C. *Gamma Interference on an Electronic Dosemeter Response in a Neutron Field*. Radiat. Prot. Dosim. **44**(1–4), 371–374 (1992).
11. Barthe, J., Bordy, J. M., Lahayet, T. and Mourgues, M. *New Principle of Single Diode Neutron Dosemeter Based on Time Resolution*. In: Proc. IRPA Regional Congress, Portsmouth, 1994, pp. 97–100 (Ashford, UK: Nuclear Technology Publishing) (1994).

EFFICIENCIES OF AN SSNTD AND AN ELECTRONIC NEUTRON DOSEMETER WITH THE SAME (n,α) (n,p) CONVERTER

E. Savvidis and M. Zamani
University of Thessaloniki
Physics Dept, Nuclear Physics Division
Thessaloniki 540 06, Greece

Abstract — Efficiencies of an SSNTD (solid state nuclear track detector) and an electronic neutron dosemeter were studied under the same field of neutrons when an appropriate (n,α) (n,p) converter was used. Characteristics of each system related to neutron dosimetry were studied. CR-39 was used as the SSNTD. The same characteristics were examined when PIPS (passivated implanted planar silicon) diodes were used as detectors. Both dosimetric systems work with a high and continuous response from thermal up to fast neutron energy range (14 MeV). The angular response approaches isodirectional standards. The two systems show good linearity up to the 10 mSv doses which are used for calibration in the frame of Eurados joint irradiations. The lowest detectable dose is discussed for both neutron dosemeters. This work presents results from on-phantom irradiations. Comparing characteristics of the two detection methods, the electronic system presents higher sensitivity and response from the SSNTD although the whole behaviour of the two systems is very similar.

INTRODUCTION

For many years research has continued on the use of solid state nuclear track detectors (SSNTD) for individual neutron dosimetry[1,2]. In parallel, electronic devices have been developed[3]. New requirements[4] for more sensitive and precise instruments for measuring the radiation exposure of occupationally exposed workers has been demanded. The work was oriented on suitable modifications of the response functions of the instruments.

The characteristics of two different neutron detection systems for use in individual neutron dosimetry have been studied. The one based on SSNTDs[5] is known to be insensitive to photons and highly efficient for charged particles. The other, based on Si diodes of the PIPS type, is sensitive to charged particles and also photons. For the purpose of neutron dosimetry diodes with a thin depletion zone[3] (about 30 μm) have been used, as well as commercially available diodes. A threshold technique was applied in order to restrict the photon contribution.

In both experimental techniques a new converter was used which gives high and continuous response from thermal neutrons up to several decades of MeV (energies for which other systems have a very low response). For fast neutrons a polyethylene (PE) moderator was used in addition to the converter system. Both systems were tested under monoenergetic neutrons and also real reference fields. Experiments under free-in-air conditions as well as on-phantom were performed[2,5]. The behaviour of the dose equivalent response was investigated as a function of various parameters concerning dosimetry. The dose equivalent response is defined as: R = M/H* (10) (or H_p (10)), i.e. the quotient of the reading of the detector and the relevant dose equivalent. In the case of SSNTDs the track density was taken as a reading of the detector. The results given for diodes were taken by a spectroscopy system using pulse height analysis: thus the detector reading was expressed as a number of pulses. From the definition of the dose equivalent response it was concluded that it must be independent of neutron energy.

EXPERIMENTAL

The most frequently used systems for individual dosimetry consist of the detector and the converter. For the experiment a new converter was developed which in its final form is based on ^{6}LiF evaporated on a hydrogen rich material as PE. Computer simulation[6] and experimental results[1,5] have proved that 500 μm of PE and 3–5 μm of ^{6}LiF give the best response for thermal as well as for fast neutrons. For low neutron energies, alpha particles and tritons originating from the ^{6}Li(n,α)t reaction are counted. In the range of fast neutrons the same reaction works by using a PE moderator of 0.5–1.0 cm, and proton recoils from (n,p) elastic scattering are also detected.

As SSNTD, CR-39 with a high efficiency for alpha particles is used. When diodes are used for particle detection a spectrum containing alpha and triton peaks is obtained[7]. However, diodes have high efficiencies to alpha particles and also to protons (higher than SSNTDs[3]). They also detect photons which are well distinguished from the rest of the particle spectrum and so a threshold technique can be applied[7].

RESULTS AND DISCUSSION

The ambient dose equivalent response of both dosemeters studied is presented in Figure 1 as a function of the neutron energy. The behaviour of the two systems is very similar, corresponding to the combination of the two cross sections, one of the ^{6}Li(n,α)t reaction and the other of the (n,p) elastic scattering. The response of the diodes (Figure 1(a)) is 2–3 times higher than that of the SSNTD (Figure 1(b)). Both curves show a continuous response to low energy neutrons (in fact from thermal neutrons) up to 14 MeV. This behaviour is due to the converter characteristics, in contrast to the commonly used proton recoil based converters, which respond to protons above 200 keV[1]. The good response above 2.5 MeV must also be mentioned. The other advantage of the converters is their high response to the albedo component when irradiations take place on-phantom. This contribution was especially studied[8] in order to know their contribution to the dose equivalent, but also for the control of the flatness of the ambient dose response curve. For the results presented in Figure 1(c) 1 mm Cd was used in the

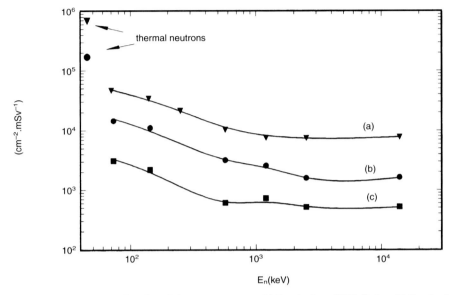

Figure 1. Dose equivalent response as a function of the neutron energy. (a) Results from PIPS diodes. (b) Results from SSNTDs (CR-39). (c) Results from SSNTDs with 1 mm Cd in the rear of the detector (phantom side). For both detectors the same converter is used.

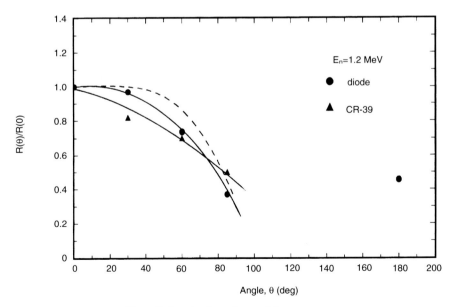

Figure 2. Angular dependence of the response.

rear of the detector (phantom side). From Figure 1(c) a flat response in the range of 5×10^2 to 10^4 keV is found. Its value is higher for diodes and considerably lower for SSNTDs. Below 5×10^2 keV the response remains 2–3 times higher than in the flat region.

Concerning other characteristics of importance in dosimetry, the angular dependence of the response as well as the linearity of the systems were studied. The angular dependence of the response is presented in Figure 2 for both systems at $E_n = 1.2$ MeV. The dashed curve represents calculated function $R(\Theta)/R(0)$ as a function of the angle of incidence[9]. The linearity of both systems was tested for dose equivalents from 0.2 mSv up to 10 mSv. In Figure 3 we present results at $E_n = 1.2$ MeV. The results for CR-39 for this energy were taken from free-in-air irradiations (Eurados-Cendos intercomparison) while those of diodes from in-phantom irradiations. For low ambient dose equivalents CR-39 shows large deviations due to the detector background. Nevertheless, the system shows good linearity above 1 mSv. Concerning diodes they show full linearity from low doses.

The lowest detectable limit for SSNTDs is found to be 5–50 µSv in the best case[5]. For diodes this limit depends on the part of the spectrum taken into consideration. The best value is obtained when results are taken by integration of the triton peak and it is 0.5 µSv[7].

CONCLUSIONS

Two different neutron detection techniques with SSNTDs and Si diodes were studied with a new converter for their application in individual neutron dosimetry. The results show a similar behaviour of both systems comparing their energy response, angular response and linearity. Diodes present a higher response than SSNTDs. The photon response can be overcome by using an electronic threshold without considerable influence on the response to neutrons. Diodes have a lower detectable limit than SSNTDs. SSNTDs can be used as passive dosemeters with high efficiency. Diodes can be developed to a real-time neutron dosemeter with equally high efficiency.

ACKNOWLEDGEMENTS

This work was supported by the CEC through contracts B170020C and F13P-CT930072. Some of the irradiations were performed in the frame of a EURADOS-CENDOS intercomparison.

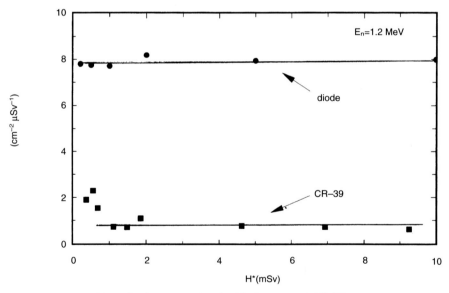

Figure 3. Linearity results of PIPS diodes and of SSNTDs.

REFERENCES

1. Piesch, E., Al-Najjar, A. and Ninomiya, K. *Neutron Dosimetry with CR-39 Track Detectors using Electrochemical Etching: Recent Improvements, Dosimetric Characteristics and Aspects of Routine Application.* Radiat. Prot. Dosim. **27**, 215–230 (1989).
2. Alberts, W. G. *Investigation of Individual Neutron Monitors on the Basis of Etched-track Detectors.* In: The 1990 Eurados-Cendos Exercise, PTB (May 1992).
3. Barelaud, B., Paul, D., Dubbary, B., Makovicka, L., Decossas, J.-L. and Vareille, J.-C. *Principles of an Electronic Neutron Dosemeter using a PIPS Diode.* Radiat. Prot. Dosim. **44**, 363–366 (1992).

4. ICRP. *The 1990 Recommendations of the International Commission on Radiological Protection*, Publication 60 (Oxford: Pergamon) (1991).
5. Zamani, M., Sampsonidis, D. and Savvidis, E. *An Individual Neutron Dosemeter with (n,α) and (n,p) Converter*. Radiat. Meas. **26**, 87–92 (1996).
6. Savvidis, E. *PhD Thesis*, Thessaloniki (1990).
7. Zamani, M. and Savvidis, E. *A Real-Time Personal Neutron Dosemeter Based on PIPS Diodes and a Double Layer Neutron Converter*. Radiat. Prot. Dosim. **63**, 299–303 (1996).
8. Savvidis, E., Sampsonidis, D. and Zamani, M. *Separation of Albedo Neutron Component of a CR-39 Neutron Dosemeter during on Phantom Irradiation*. Presented at 11th Int. Symp. on Solid State Dosimetry, Budapest. 1995.
9. Sieberg, B. R. L. and Morhart, A. *A Proposed Procedure for Standardising the Relationship between the Directional Dose Equivalent and Neutron Fluence*. Radiat. Prot. Dosim. **44**, 47–51 (1989).

SEPARATION OF THE NEUTRON SIGNAL FROM THE GAMMA COMPONENT IN (n–γ) FIELDS USING DIFFERENTIAL PULSE ANALYSIS TECHNIQUES WITH A DOUBLE SILICON DIODE

F. Fernández, E. Luguera, C. Domingo and C. Baixeras
Grup de Física de les Radiacions, Departament de Física
Universitat Autònoma de Barcelona, E-08193 Cerdanyola del Vallés, Spain

Abstract — An individual electronic neutron dosemeter based on a combination of a polyethylene converter with a double diode (Canberra CD-NEUT-200-DBL) has been studied. The dosemeter has been irradiated by monoenergetic neutron beams of energies from 73 keV to 2.5 MeV with dose equivalents between 0.3 and 5.7 mSv. The differential method has been applied to separate the neutron response from the gamma contribution. The energy response of the dosemeter has been studied for these neutron energies. Linearity has been studied for 1.2 MeV neutrons. Simulation of the response of this dosemeter using a Monte Carlo code agrees with the experimental results obtained and confirms an energy threshold of the order of 250 keV for neutron detection. It is inferred from simulation that the neutron dose equivalent response varies about ±50% in the energy range from 570 keV to 15 MeV, if the actual device with a 20 μm thick polyethylene converter is used.

INTRODUCTION

Many devices based on silicon diodes have been extensively developed in recent years in order to perform photon and neutron dose measurements for individual as well as for area monitoring[1–4]. Specially designed dosemeters have been applied to individual neutron dosimetry[5–7] with the aim of fulfilling the ICRP 60 requirements on their implementation. The main problem of such dosemeters is to separate the neutron component from the gamma one, as in practice they coexist in real fields. The differential method[8], based on the use of two diodes of identical characteristics, has often been employed in order to effect this separation. The first of these diodes, placed behind and in contact with a converter, detects the charged particles produced in the converter as a result of interaction of neutrons, as well as primary electrons and photons which should be present in the beam. In addition, electrons and photons originated by interactions in the depleted zone and in the sensor's surroundings may also be recorded. The second diode is not covered by any converter, and therefore it detects only primary electrons and photons if present, as well as particles (electrons, photons) originated by interactions in the diode itself and in the sensor's surroundings. The difference between the response of the two diodes is, thus, mainly due to particles originated by interaction of neutrons in the converter.

Although it should be theoretically possible to detect thermal and fast neutrons using two identical diodes, the existence in practice of differences in their electronic background and in their photon sensitivity leads to an energy cut-off, which increases the dosemeter detection threshold and, consequently, constrains its field of application to fast neutrons. This problem might be overcome if a double diode built on the two opposite sides of a single silicon block is used. Several advantages should appear if such a device is used:

(a) The electronic characteristics of the diodes should be identical

(b) A double diode has a global size smaller than two single diodes of the same characteristics, so that the case containing them would be smaller and, therefore, a reduction of the number of secondary particles originating in the case could be achieved.

(c) The number of electrical connections is reduced from four (in two single diodes) to three (in a double diode), so that the electronic noise might decrease and the reference voltage applied to both diodes would be identical.

In this work, a Canberra CD-NEUT-200-DBL double diode, consisting of two silicon diodes on the same woofer in reverse position, has been used as detector for a real time neutron dosemeter. The detector characteristics ensure, *a priori*, that no differences between the electrical properties of the diodes should appear. A polyethylene converter about 40 μm thick, which originates a fluence rate of emerging protons roughly proportional to the neutron dose rate in the energy range 100 keV to 5 MeV[9] when irradiated in a neutron field, is used in this dosemeter. The aim of this work is to calibrate the dosemeter following the ICRU[10] recommendations, to achieve the separation of the gamma component from the neutron one, to study the energy response of the dosemeter to normally incident neutrons and to compare this response with that obtained from simulation, and to study the dosemeter response as a function of the neutron dose equivalent for a given incident neutron energy.

EXPERIMENTAL PROCEDURE

Dosemeter arrangement

A dosemeter arrangement composed of a 40 μm thick layer of polyethylene converter followed by a double diode detector has been used. The double diode detector (CD-NEUT-200-DBL) consists of two Canberra diodes, with an effective area of 2 cm^2 and a bulk resistivity of about 500 Ω.cm^{-1}, located on the opposite sides of a single silicon block. The depleted zone of each of the two diodes is, in the actual configuration, 30 μm deep for a polarisation tension of 10 V. These depleted zones are separated one from the other by 222 μm of Si. Ortec preamplifiers and a Canberra multichannel analyser with amplifier are used and connected to a Toshiba T-180 portable computer for data acquisition, as indicated in Figure 1. Data analysis is performed using the Ortec Maestro II software.

Irradiations

Two types of irradiation were used to perform the energy calibration of the personal electronic dosemeter: photons of about 1 MeV from a ^{60}Co source for low energy loss rates, and alpha particles of about 5 MeV from a ^{231}Am, ^{239}Pu, ^{244}Cm combined source for high energy loss rates. It has been found that the energy in keV is given by the channel number divided by 1.6.

The dosemeter was irradiated, once calibrated, during a joint irradiation experiment in March 1994 at the Bruyère-le-Chatel Nuclear Research Centre, to monoenergetic neutron beams of 73 keV, 144 keV, 250 keV, 570 keV, 1.2 MeV and 2.5 MeV with dose equivalents (H*(10), according to the ICRP 60) ranging from 0.3 to 5.7 mSv. The dosemeter was placed for irradiation on the centre of the front face of a 30 × 30 × 15 cm^3 PMMA slab phantom, as recommended by the ICRU[10] for calibration purposes, and irradiated by normally incident neutrons. The net spectrum, that corresponds mainly to protons originated in the converter, is calculated by subtracting the integrated spectrum obtained in the back diode from that obtained in the front one.

NEUTRON RESPONSE CALCULATION

A code based on a Monte Carlo method, written in Fortran77 language, has been used in a personal computer to simulate the passage of a normally incident neutron beam through a polyethylene converter plus a 30 + 222 + 30 μm Si detector. The Monte Carlo procedure employed has been developed by this group and is based on general principles outlined elsewhere[11,12]. The code assumes that, in the energy range involved, the only relevant physical process that originates secondary charged particles in the polyethylene converter is elastic scattering between neutrons and protons. Neutron–proton collisions are simulated from a given incident neutron fluence, using the appropriate cross sections and conservation laws, and the emerging protons are followed in order to discover whether they reach any (or both) of the diodes. The simulated net spectrum is obtained by subtracting the proton spectrum in the back diode from that in the front one, following the same procedure as in the differential method applied in the experiment. The dosemeter equivalent response in terms of H'(10) (cm^{-2}.mSv^{-1}), which equals the ambient dose equivalent H*(10) for normally incident neutrons, is obtained for each incidence energy from the net spectrum, taking into account the area of the diode surface and using the ICRP 60 conversion factors[13]. In addition, it has been necessary to consider the proton registration energy threshold (239 keV) in order to take into account the characteristics of the device. The transport of the gamma and the electron components originated in the sensor's surroundings, mainly due to neutron interactions in the diode aluminium capsule and in the phantom, has not been taken into account as the amount of energy deposited in the diodes by these components is very small and falls below cut-off. On the other hand, the transport of the gamma

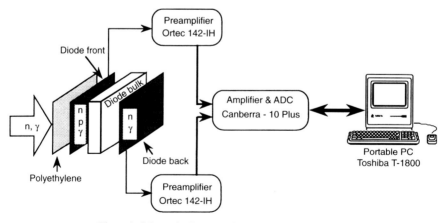

Figure 1. Schematic diagram of the experimental set-up.

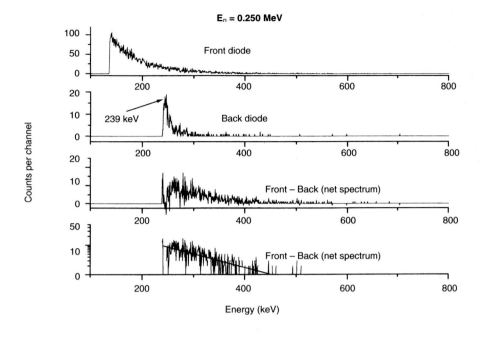

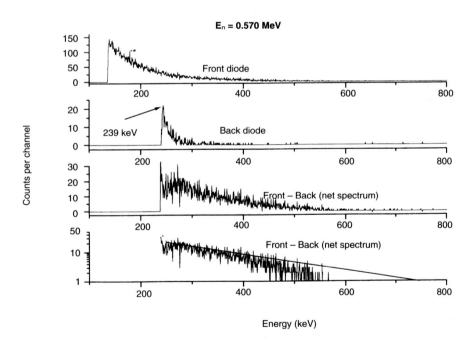

Figure 2. Spectra obtained for each diode, together with the net spectra, for neutron energies of 0.250 and 0.570 MeV.

component present in the main beam is not considered as it does not contribute to the response of our dosemeter because of the use of the differential method.

RESULTS AND DISCUSSION

Analysed spectra

The spectra for each diode (front and back) of the detector are presented in Figure 2, together with the net spectra, for neutron incidence energies of 0.250 and 0.570 MeV. The spectra for neutron incidence energies of 0.073 and 0.144 MeV are not displayed as they have a very similar behaviour to that at 0.250 MeV. The figure shows the existence of a shift between the spectra obtained in both diodes, due probably to differences between their capacities. This fact prevents us from applying the differential method to proton energies below 239 keV and, consequently, neutrons of energies below 500 keV have a small registration efficiency. This fact is accounted for, in practice, by considering an energy cut-off for neutron detection that should be of the order of 250 keV. Analysis is thus reduced to neutron energies of 0.570, 1.2 and 2.5 MeV.

The exponential behaviour of the net spectrum originated by 0.250 MeV neutrons observed in the figure, which is also found in the spectra of 0.073 and 0.144 MeV, confirms that these spectra originate from the contribution of electromagnetic radiation. A small deviation from this exponential behaviour appears in the 0.570 MeV neutron net spectrum. This deviation suggests the presence of a small signal in the detector due to recoil protons with energy greater than 239 keV. This result agrees with the fact that 570 keV neutrons should have a small registration efficiency.

Figure 3 displays the net spectra obtained for neutrons of 570 keV, 1.2 MeV and 2.5 MeV. In the case of 1.2 MeV neutrons, the non-exponential behaviour of the net spectrum is evidence of the presence of recoil protons with a maximum energy that equals the incidence neutron energy. For 2.5 MeV neutrons the presence of a maximum in the number of recorded protons around 1.8 MeV is observed, although in an inconclusive way due to lack of statistics. This value agrees with that obtained by Monte Carlo simulation for neutrons of this energy. The presence of a peak in the number of recorded protons that corresponds to the most probable proton energy, originating from a given neutron energy, is hidden in some cases by statistical fluctuations due to the small number of recorded events. In fact, this experiment is part of an intercomparison between different groups and the exposure dose equivalent values were fixed beforehand and were low because a higher response of the dosemeter was expected on the basis of former works.

Linearity

A fit of the integrated counts of the net spectrum at 1.2 MeV plotted against reference directional dose equivalent ($H'(10)$) is shown in Figure 4. Errors in the integrated counts are calculated as the square root of the number of counts and errors in the dose equivalent values are 10% of the nominal value, according to the exposure characteristics. The intercept value obtained of 0.010 mSv may be considered as the dosemeter detec-

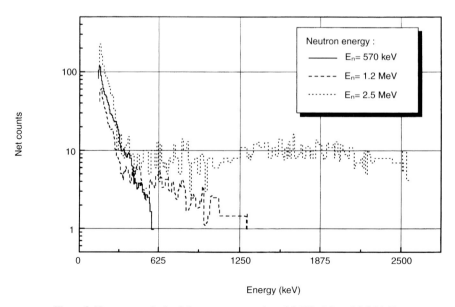

Figure 3. Net spectra obtained for neutron energies of 0.570, 1.2 and 2.5 MeV.

tion limit for this neutron energy, accounting for our proton registration energy cut-off. On the other hand, taking into account the slope (1105 counts per mSv) and the area of the diode surface, a value of 553 $cm^{-2}.mSv^{-1}$ is obtained for the mean equivalent dose response value for this energy, accounting for our proton registration energy cut-off. This value is a factor of eight greater than that obtained with a SSNTD dosemeter[14] for the same energy, and agrees with that obtained by Barthe et al[6], using a similar dosemeter configuration.

Energy dependence

Figure 5 shows the simulated dose equivalent response $R_{H'}$ for normally incident neutrons as a function of neutron energy for this dosemeter configuration and several polyethylene converter thicknesses, together with the experimental values obtained. The excellent agreement between the simulated and the experimental results is to be emphasised. The dosemeter response may become flatter if the polyethylene converter thickness is reduced, as deduced from the figure. Once this thickness is selected, it is obvious from the results obtained that it is necessary to reduce as much as possible the energy cut-off introduced in order to increase the response to 144, 250 and 570 keV neutrons. It does not seem probable that a net signal for neutrons below 250 keV can be obtained using the actual technology. This fact would imply that it is not possible to improve much the response flatness in the energy range studied.

CONCLUSIONS

The characteristics of our real time personal neutron dosemeter gathered from the results of the present study can be summarised as follows:

(1) The dosemeter displays, on its actual configuration, a low response value for low energy neutrons and a high response value for photons. These effects make it necessary in practice to introduce an energy threshold of the order of 250 keV for neutron detection.
(2) It is inferred from simulation that the neutron dose equivalent response varies about ±50% in the energy range from 570 keV to 15 MeV, if the actual device with a 20 μm thick polyethylene converter is used.
(3) The use of ^{10}B in the polyethylene converter would allow thermal neutrons to be recorded, as the energy of the alpha particle originated in the n + ^{10}B reaction is over the energy cut-off that has been introduced.

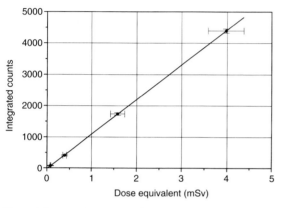

Figure 4. Integrated counts as a function of the directional dose equivalent (H'(10)) for 1.2 MeV neutrons.

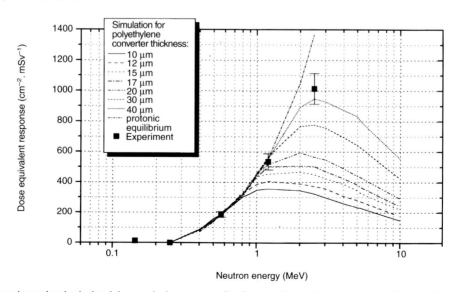

Figure 5. Experimental and calculated dose equivalent response $R_{H'}$ for normally incident neutrons as a function of neutron energy.

ACKNOWLEDGEMENTS

This work is partially funded by a contract with the European Union (FI3P-CI93–0072). The authors wish to thank Mr Jordi López, from Grup de Física de les Radiacions, for his technical support on setting up all the electronic devices. The cooperation of the staff at the neutron source facilities at the Bruyère-le-Chatel Nuclear Research Centre is gratefully recognised.

REFERENCES

1. Eisen, Y., Engler, G., Ovadia, E., Shamai, Y., Baum, Z. and Levi, Y. *Single Silicon Surface Barrier Detector.* In: Workshop on Personal Neutron Dosimetry held jointly with CENDOS, Acapulco, Mexico (1983).
2. Eisen, Y., Engler, G., Ovadia, E. and Shamai, Y. *Combined Real Time Wide Energy Range Neutron Dosemeter and Survey Meter for High Neutron Dose Rates with Si Surface Barrier Detectors.* Nucl. Instrum. Methods **211**, 171–178 (1983).
3. Fasasi, M., Jung, M. and Siffert, P. *Thermal Neutron Dosimetry with Cadmium Telluride Detectors.* Radiat. Prot. Dosim. **23**(4), 429–431 (1988).
4. Nakamura, T., Horiguchiond, M. and Yamano, T. *A Real Time Wide Energy Range Personal Neutron Dosemeter with Two Silicon Detectors.* Radiat. Prot. Dosim. **20**(3), 149–156 (1989).
5. Matsumoto, T. *Pin Diode for Real Time Dosimetry in a Mixed Field of Neutron and Gamma rays.* Radiat. Prot. Dosim. **35**(3), 193–197 (1991).
6. Barthe, J., Lahaye, T,, Moisseev, T. and Portal, G. *Personal Neutron Diode Dosemeter.* Radiat. Prot. Dosim. **47**(1–4), 397–399 (1993).
7. Jung, M., Teissier, C., Siffert, P. and Raffasoe, C. *Fast Neutron Monitoring with Silicon Detectors.* Strahlenschutz: Physik und Messlechnik, **1**, 151–158 (Karlsruhe) (1994).
8. Paul, D., Barelaud, B., Dubarry, B., Makovicka, L., Vareille, J. C. and Decossas, J. L. *Gamma Interference on an Electronic Dosemeter Response in a Neutron Field.* Radiat. Prot. Dosim. **44**(1–4), 371–374 (1992).
9. Makovicka, L. *Contribution à la Dosimétrie Neutron-Gamma: Étude d'un Ensemble Radiateur-Detecteur Type CR-39.* Thèse d'Etat. Université de Limoges, France (1987).
10. ICRU. *Measurements of Dose Equivalents from Extermal Photon and Electronic Radiations.* Report 47 (Bethesda: ICRU Publications) (1992).
11. Fernández, F., Domingo, C., Luguera, E. and Baixeras, C. *Experimental and Theoretical Determination of the Fast Neutron Response using CR-39 Plastic Detectors and Polyethylene Radiators.* Radiat. Prot. Dosim. **44**(1–4), 337–340 (1992).
12. Fernández, F., Bouassoule, T., Domingo, C., Luguera, E. and Baixeras, C. *Response of a CR-39 Fast Neutron Dosemeter with a Polyethylene Radiator Improved with Makrofol.* Radiat. Prot. Dosim. **66**(1–4), 343–347 (1996).
13. ICRP. *1990 Recommendations of the International Commission on Radiological Protection.* Report 60 (Oxford: Pergamon Press) (1991).
14. Fernández, F., Baixeras, C., Domingo, C. and Luguera, E. *EURADOS-CENDOS Joint Neutron Irradiation, 1990. Results from Grup de Física de les Radiacions.* EURADOS-CENDOS Report 1992–02 (1992).

STUDY OF A MODERATOR TYPE ELECTRONIC NEUTRON DOSEMETER FOR PERSONAL DOSIMETRY

T. Moiseev
Institute for Physics and Nuclear Engineering, Laboratory 7
Magurele, PO Box Mg-6, Sect. 5, Bucharest, Romania

Abstract — A moderator type neutron dosemeter is described and its response to 'realistic neutron fields' and fission sources is investigated, for irradiations with and without a poly-methyl methacrylate (PMMA) phantom. Changes in response induced by the use of the moderator are observed throughout the experiments, with respect to the energy distribution of the neutron fluence and the dose equivalent. The experimental dosemeter shows an over-response to thermal neutrons and for on phantom irradiations, and a low sensitivity for fast neutrons. Compared to a classical neutron dosemeter based on silicon diodes (without moderator), the actual design shows, for the same active area of the silicon detector, a sensitivity (counts per mSv) from 10 to 40 times higher for 'realistic neutron fields'.

INTRODUCTION

As the development of electronic neutron dosemeters has arrived at a point where the classical design[1-3] does not allow an accurate response in terms of the dose equivalent for a wide neutron energy range (from thermal neutrons up to 14 MeV) a new approach to the problem has been made by surrounding the silicon detector with a small size moderator.

Moderator type neutron dose equivalent monitors using passive detectors (working on the Bonner sphere principle) have already been described[4] and it was shown that their response per fluence unit as a function of neutron energy is in good agreement with the fluence to dose equivalent curve (calculated from ICRP 21[5] data) up to 5 MeV.

While the use of passive detectors (TLD type) requires full moderation of neutrons over the whole neutron energy range from thermal up to 14 MeV, the use of silicon diodes as charged particle detectors led to the concept of a 'selected moderation', which consists of moderating only neutrons up to a certain energy (a few hundred keV) and detecting higher energy neutrons directly as fast neutrons, using the same detector.

This is possible by using a radiator made by a thin polyethylene foil with a deposited ^{10}B layer, positioned on the active surface of the silicon diode. This radiator allows the detection of thermal neutrons through the ^{10}B(n,α)^{7}Li reaction and the detection of fast neutrons through the protons produced by fast neutron elastic scattering on the hydrogen nuclei in polyethylene.

The 'selected moderation' of neutrons requires moderators smaller in size than those used for 'full moderation', allowing their use for personal dosimetry.

THE MODERATOR TYPE ELECTRONIC DOSEMETER

The detection assembly (moderator + diodes) shown in Figure 1 consists of two implanted silicon diodes, each having an active area of 19.6 mm^2, one of the diodes being provided with a 45 μm thick polyethylene radiator deposed with a 2 μm layer of B_2O_3 (96% ^{10}B enriched). The diode with radiator detects neutrons and the background signal (usually, generated by gamma rays), while the diode without radiator detects only the background signal. The pulses from the diodes are fed through two acquisition chains and then treated by a differential method, in order to subtract the background and gamma ray generated pulses.

Each diode is covered with an aluminium shielding (0.3 mm thick), and both diodes are included in a hemispherical moderator provided with a cylindrical opening (Figure 1).

As known from the response of the Bonner spheres, a 2 inch diameter sphere will moderate neutrons in the range from 10^{-2} eV up to about 1 keV, while higher energy neutrons will have fewer interactions, as their mean free path in polyethylene is of the order of 1.8 cm at 500 keV; and it increases with neutron energy (Figure 2).

The cylindrical opening in the moderator was

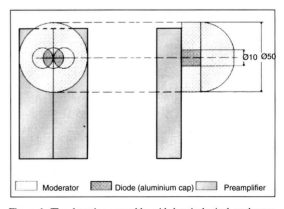

Figure 1. The detection assembly with hemispherical moderator.

intended to provide direct detection for fast neutrons at normal and near normal incidence.

The low amplitude threshold of the system is at 100 keV and an informational cut was imposed at 290 keV in order to discriminate gamma ray generated pulses down to a sensitivity of 10 counts per mSv for a ^{137}Cs gamma ray source (for measurements with or without phantom).

THE IRRADIATION TESTS

The irradiation tests have been performed using the 'Realistic Neutron Fields' from CEA/CEN Cadarache-France and the fast neutron beam from the U120 Cyclotron facility (I.P.N.E., Bucharest); and also neutron fission sources such as ^{241}Am-Be and ^{252}Cf.

The realistic neutron fields ('Canel+', 'Canel+ with water shield' and 'Sigma') have been designed by Monte Carlo computation and measured by the Bonner spheres method, in order to obtain a neutron fluence distribution and a neutron dose equivalent distribution close to those encountered in the proximity of accelerators and shielded neutron sources[7].

The 'Canel+ field is generated by a ^{238}U converter (a 12 cm thick spherical shell) irradiated by the 14.7 MeV neutron emission of an accelerator target placed at its centre. The ^{238}U shell is covered by a 34 cm thick iron shell and this whole assembly is placed inside a 10 cm thick cylindrical polyethylene duct.

The Sigma field is generated by six Am-Be sources (16 Ci) placed in a graphite cube ($1.5 \times 1.5 \times 1.5$ m^3).

The main parameters characterising these fields are described below and indicated in Table 1[8]. They are:

(i) spectral distribution of the neutron fluence, Φ; and ambient neutron dose equivalent $H^*(10)$ for three neutron energy ranges, thermal (0.025 eV–1 eV), intermediate (1 eV–10 keV) and fast neutrons (10 keV–14 MeV);

(ii) mean neutron energy, E (keV);
(iii) effective energy, E_H(keV), which is the dose equivalent average neutron energy.

The following irradiation tests have been performed for each neutron field at normal incidence:

(a) free-in-air irradiation of the detection system with and without moderator;
(b) on-phantom irradiations using a poly-methyl methacrylate (PMMA) phantom ($30 \times 30 \times 15$ cm^3) for the detection system with and without moderator. The detection system has been placed at the centre front face of the PMMA phantom.

RESULTS

For each measurement, the values of ambient dose equivalent $H^*(10)$ and respectively personal dose equivalent $H_p(10)$ were provided by CEA/IPSN/DPHD/S.DOS/GDN Cadarache (France) for the realistic neutron fields and by I.P.N.E. Bucharest (Romania) for the fast neutron beam. These were

Table 1. Neutron fields characteristics for the realistic neutron beams[8].

Neutron field characteristics	Canel+	Canel+ with H$_2$O shield	Sigma
ϕ Thermal	36%	43%	97%
ϕ Intermediate	20%	26%	1%
ϕ Fast	44%	31%	2%
Mean energy	150 keV	80 keV	72 keV
H* thermal	6%	10%	46%
H* intermediate	3%	7%	4%
H* fast neutrons	92%	83%	50%
Effective energy	610 keV	68 keV	1323 keV

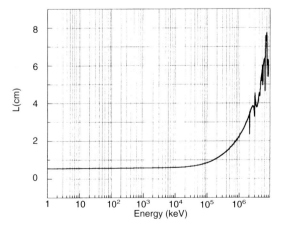

Figure 2. Neutron mean free path L(cm) in polyethyelene[6].

Table 2. Sensitivity to realistic neutron fields, for irradiations without phantom. Values for the detection assembly with and without moderator are shown and also the sensitivity of a classical electronic dosemeter[1,2].

Neutron field (free-in-air irradiations)	Sensitivity (counts per mSv)		
	Moderator type dosemeter		Classical dosemeter
	No moderator	With moderator	
Canel+	337	397	48
Canel+ with H$_2$O shield	773	720	18
Sigma	1353	2766	170

obtained from neutron fluence monitoring and the corresponding fluence to dose equivalent conversion factors (calculated from ICRP 21 and ISO 8529/1989[9] data).

The typical charged particle spectra obtained are shown in Figures 3 and 4.

The dosemeter's response for the above irradiation tests is shown in Table 2, and is compared with the response of a differential detection system with two passivated implanted Si detectors (Canberra PIPS type), provided only with a radiator (and no moderator),

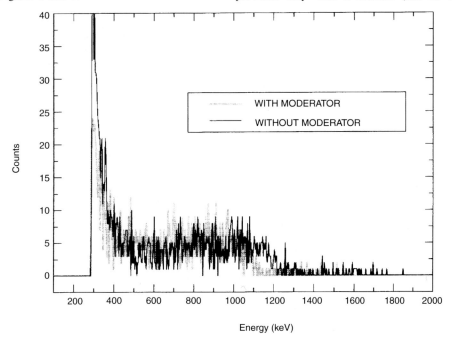

Figure 3. Typical charged particles spectra from free-in-air irradiations at Canel+ with H_2O shield, for the detection assembly with and without moderator. The ambient dose equivalent value has been normalised at 1.63 mSv.

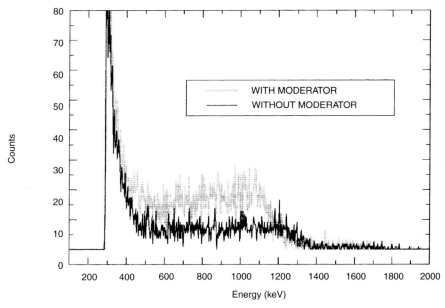

Figure 4. Typical charged particle spectra from on-phantom irradiation on Canel+ with H_2O shield, for the detection assembly with and without moderator. The ambient dose equivalent value has been normalised at 1.72 mSv.

developed at CEA/IPSN/DPHD/SDOS Fontenay-aux-Roses[2].

To enable comparison, the silicon diode's active area has been normalised to 19.6 m².

DISCUSSION AND CONCLUSIONS

The proposed moderator type neutron dosemeter investigated shows a low sensitivity for fast neutrons (Table 3) which is mainly due to the small active area of the diode.

For thermal neutrons (results for the Sigma field, Table 2) an over-response was obtained when using the moderator, which is an expected result, even for measurements free-in-air. A cadmium shielding of a certain extent on the moderator's surface could overcome this inconvenience.

For the Cadarache realistic neutron fields (Tables 2 and 4) the sensitivity of the moderator system is higher than that obtained for the classical system using only the radiator. This is mainly due to the higher ^{10}B concentration in the radiator, but also to the presence of the moderator.

In conclusion, the moderator type neutron dosemeter investigated shows a high sensitivity for realistic neutron fields where other detection systems using silicon detectors proved to have a gap in response.

The small active area of the silicon detectors, as well as the differential system used allows a very low gamma ray sensitivity of the detection system (10 counts per mSv), but it has also the disadvantage of providing a low sensitivity to fast neutrons.

The overall response obtained for irradiations on-phantom could also be used, after proper calibration of the dosemeter, allowing an increase in sensitivity.

Table 3. Sensitivity to fast neutrons.

Neutron source	Sensitivity (counts per mSv) Moderator type dosemeter	
	No moderator	With moderator
^{241}Am-Be	135	196
^{252}Cf	137	199
Ul20 Cyclotron (E = 5.4 MeV)	92	116

Table 4. Sensitivity to realistic neutron fields for irradiations on phantom. Values for the detection system with and without moderator are shown.

Neutron field (On-phantom irradiations)	Sensitivity (counts per mSv) Moderator type dosemeter	
	No moderator	With moderator
Canel+	631	763
Canel+ with H$_2$O shield	1489	2785

REFERENCES

1. Barelaud, B., Paul, D., Dubarry, B., Makovika, L., Decossas, J. and Vareille, J. C. *Principles of an Electronic Neutron Dosemeter Using a PIPS Detector*. Radiat. Prot. Dosim. **44**, 363–366 (1992).
2. Barthe, J., Lahaye, T., Moiseev, T. and Portal, G. *Personal Neutron Diode Dosemeter*. Radiat. Prot. Dosim. **47**(1–4), 397–399 (1993).
3. Alberts, W. G., Dietz, E., Guldbakkle, S. and Kluge, H. *Response of an Electronic Personal Neutron Dosemeter*. Radiat. Prot. Dosim. **51**(3), 207–210 (1994).
4. Esposito, A., Manfredotti, C., Pellicioni, M., Ongaro, C. and Zanini, A. *A Moderator Type Dose Equivalent Monitor for Environmental Neutron Dosimetry*. Radiat. Prot. Dosim. **44**(1/4), 207–210 (1994).
5. ICRP. *Data for Protection against Ionizing Radiation from External Sources*. Publication 21 (Oxford: Pergamon Press) (1971).
6. Langner, I., Schmidt, J. and Woll, D. *Tables of Evaluated Neutron Cross-Sections for Fast Reactor Materials* (Gesellschaft fur Kernforschung M.B.H., Karlsruhe) (1968).
7. Buxerolle, M., Chartier, J. L., Kurkdjian, J., Medioni, R., Massoutie, M., Posny, F., De Matos, E. and Sueur, M. *Experimental Simulation and Characterisation of Neutron Spectra for Calibrating Radiation Protection Devices*. Radiat. Prot. Dosim. **23**(1/4), 285–288 (1988).
8. Chariter, J. L., Posny, F. and Buxerolle, M. *Experimental Assembly for the Simulation of Realistic Neutron Spectra*. Private communication (1991).
9. ISO 8529:1989. *Neutron Reference Radiations for Calibrating Neutron-measuring Devices used for Radiation Protection Purposes and for Determining their Response as a Function of Neutron Energy*.

FEASIBILITY STUDY OF AN INDIVIDUAL ELECTRONIC NEUTRON DOSEMETER

M. Luszik-Bhadra, W. G. Alberts, E. Dietz and B. R. L. Siebert
Physikalisch-Technische Bundesanstalt, Bundesallee 100
38116 Braunschweig, Germany

Abstract — A combined active/passive personal neutron dosemeter is proposed. It consists of a position-sensitive silicon diode, which is covered on four different positions by different thermal and fast neutron absorbers, moderators and converters — including a passive CR-39 detector. MCNP calculations were used to optimise the set-up. First measurements with a simplified electronic device were performed with neutrons in the energy range from thermal up to 14.8 MeV and with photons up to 7 MeV. It is shown that by the use of the readout in different positions of the diode and by setting different electronic thresholds, several response functions can be achieved, whose neutron energy dependence differs in such a way that rough information on the neutron spectrum can be obtained.

INTRODUCTION

The increase in the quality factor for neutrons, the overall reduction of dose limits for radiation workers, and the public's increased awareness of problems related to radiation calls for new individual neutron dosemeters with reduced lower detection limits, which offer more information on the radiation field and higher reliability.

The CR-39 track dosemeter[1], recently developed in the PTB, already provides intrinsic spectrometric information which, as a consequence, improves the dosimetric reliability in workplaces with changing neutron spectra. However, these dosemeters have not yet been adopted for routine use for mainly two reasons. As a result of the material-dependent background, the dose limit cannot be reduced to values significantly smaller than 0.2 mSv so that a monthly readout is unreasonable. Also, the high costs of the evaluation of these dosemeters which does not take place fully automatically are an obstacle to routine use.

An electronic device giving an alarm when dose limits are exceeded would further improve the dosimetric reliability. Its alarm and active dose indication would help to avoid unwanted irradiation according to the ALARA (As Low As Reasonably Achievable) principle. However, lasting problems due to the failure of electronic devices under special conditions (for example, electromagnetic fields) and their poor dose-equivalent response as a function of neutron energy have resulted in their exclusive use as personal neutron dosemeters appearing questionable.

A combined dosemeter is therefore proposed consisting of a capsule which contains a silicon diode and, in addition, CR-39 track detectors. An active instrument would allow the passive track detectors to be evaluated after prolonged wearing periods. Since the CR-39 track detectors show low fading and their intrinsic background does not increase very much during one year, their minimum detectable dose of 0.2 mSv could serve as a threshold also for the annual dose.

In recent years, several studies have been undertaken with a view to designing a neutron dosemeter using one or more silicon diodes. It has been found that the photon sensitivity of silicon diodes poses a serious problem for their use as neutron dosemeters. Suppression or subtraction of the photon yield is possible. Usual techniques are pulse shape discrimination[2] or the use of two diodes with different photon-to-neutron response ratios[3]. Since these techniques require, however, advanced electronics and sensors, a different approach was chosen.

The dosemeter envisaged consists of a position-sensitive electronic sensor whose front and rear sides are covered with different neutron absorbers and converters in defined areas. The absorbers are used in a way similar to that proposed by Eisen et al[4] for etched track detectors, but an additional area without thermal neutron absorber is provided as a window for thermal neutrons. Furthermore, a thermal neutron converter of ^{6}Li instead of ^{10}B is chosen for the higher energy of the charged particles produced in the ^{6}Li(n,α)^{3}H reaction (2.74 MeV ^{3}H and 2.0 MeV alphas). This allows a high electronic threshold of 1.5 MeV to be set and thus the photon sensitivity to be reduced. The response up to 1.5 MeV is exclusively induced by neutron reactions with ^{6}Li. The response above 1.5 MeV results from the reading of a zone which is covered by a hydrogen-containing converter only. The design is aimed at obtaining several response functions which are as linearly independent as possible and thus allow rough information on the neutron spectrum in the field to be obtained. In the following, MCNP calculations are described which were performed in the neutron energy range from thermal up to 1.2 MeV to optimise the dosemeter set-up. To check the calculations, measurements were first carried out with a simplified device with thermal and 144 keV neutrons. Additional measurements were conducted with photons with energies up to 7 MeV and with high energy neutrons of 2.5 MeV, 5 MeV and 14.8 MeV.

DOSEMETER SET-UP AND OPTIMISATION BY MCNP CALCULATIONS

A sketch of the dosemeter configuration under test is shown in Figure 1. For a first feasibility study, it is assumed that a single position-sensitive silicon diode can be used. In front of the silicon diode (315 μm thick, 3 cm² sensitive area) a batch of thermal and fast neutron converters was placed at 1 mm distance. The thermal neutron converters (section (c) in Figure 1) consist of three LiF layers behind different shieldings. The positions are denoted by 'Albedo', 'Closed' and 'Front'. The fast neutrons are detected separately in position 4 (see Figure 1). Fast neutron converters and moderators are different sheets of hydrogen-containing plastics. They consist of a polyethylene sheet 0.5 mm thick (with the evaporated LiF layers), followed by two CR-39 sheets each 0.7 mm thick. The one is used as a passive CR-39 detector and the other one serves as a CR-39 converter designed in a way similar to that proposed in Reference 1 for track detectors.

For the optimisation of the dosemeter design (material of the thermal neutron absorbers, size of the windows and thickness of the LiF layers), the response functions in the three positions ('Albedo', 'Closed' and 'Front') were computed using the neutron–photon transport code MCNP (version 4A, Reference 5). For these calculations, the dosemeter was positioned on the front face of a 30 cm × 30 cm × 15 cm ISO water slab phantom whose centre was irradiated perpendicularly by a parallel monoenergetic neutron beam 14 cm in radius. Cross sections were taken from ENDF/B-IV for ^{7}Li, ^{10}B and ^{14}N, from ENDF/B-III for ^{11}B, from ENDL-85 for ^{1}H, ^{12}C, ^{16}O, ^{nat}Fe and ^{nat}Si and from the LASL sub-library for ^{6}Li. For thermal neutrons the MCNP hydrogen-bound data for polyethylene and water at 300 K were used. In the three positions with LiF layers, the number of ^{6}Li(n,α)^{3}H reactions was calculated by MCNP for neutron energies between thermal and 1.2 MeV. On the assumption that all charged particles in the front half space can be detected by the diode, this determines the fluence response of the dosemeter. This fluence response was converted into dose equivalent response using the fluence-to-dose conversion factors for $H_p(10,0°)$ as given in Reference 6. The results of the calculations for four designs (denoted by A, B, D and E) are shown in Figure 2.

At present, the set-up case E shown in Figure 1 is the best design. In cases A and B, smaller circular front and albedo windows 4 mm in diameter and ^{nat}LiF layers 1 μm thick (also 4 mm in diameter) were used. In case A, the thermal neutron shielding consisted of 1 mm Cd and in case B, of 2 mm polystyrol with 50% boron carbide. The dose equivalent response calculated behind the albedo window was flatter for the boron plastic shielding, but not much better in comparison with that measured for the ALOKA dosemeter[7] which is already commercially available. In the case of the cadmium shielding, more pronounced differences are observed for the readings in the different positions in the thermal neutron region. For case D (case C not shown here), a shielding of both 1 mm cadmium and 2 mm boron plastics was used and the windows were increased to 5 mm in diameter. More pronounced differences for the different positions are now observed in the higher neutron energy region. In case E, the albedo window was opened wider (see Figure 1) and ^{6}LiF layers 3 μm thick (5 mm in diameter) were used.

Important for the practical use are spectrometric capabilities in the neutron energy regions where neutrons chiefly contribute to the dose in fields of routine surveillance, e.g. in the thermal energy region and in the energy region from 50 keV up to 1.5 MeV. For case E, the ratio of the readings behind the albedo window to that behind the front window is an increasing function of neutron energy between 50 keV and 1.2 MeV. This

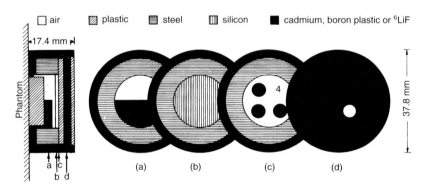

Figure 1. Sketch of the dosemeter set-up case E (see text). The four sections (a) to (d) show: (a) the albedo window, (b) the silicon diode, (c) the front side of the plastic converter with three LiF layers and, (d) the cadmium sheet with front window. The four positions on section (c) are, counterclockwise: (1) 'Albedo', (2) 'Closed', (3) 'Front' and (4) 'Fast' without ^{6}Li converter.

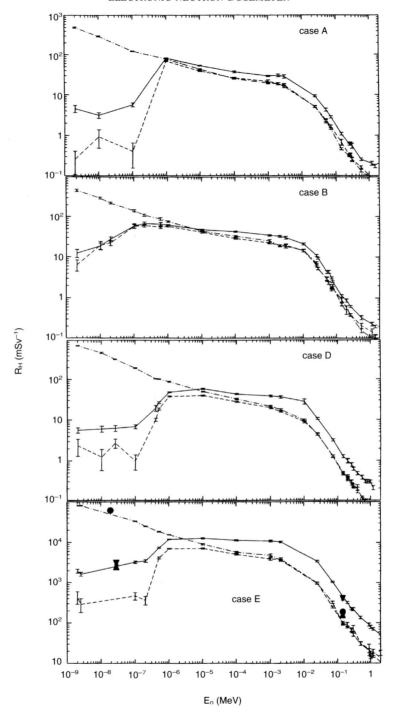

Figure 2. Dose equivalent response, calculated for different dosemeter configurations (see text) and measured for case E in the positions 'Albedo' (▼), 'Closed' (▲) and 'Front' (●). All cases: (———) Albedo, (– – –) Closed, (—·—·—) Front.

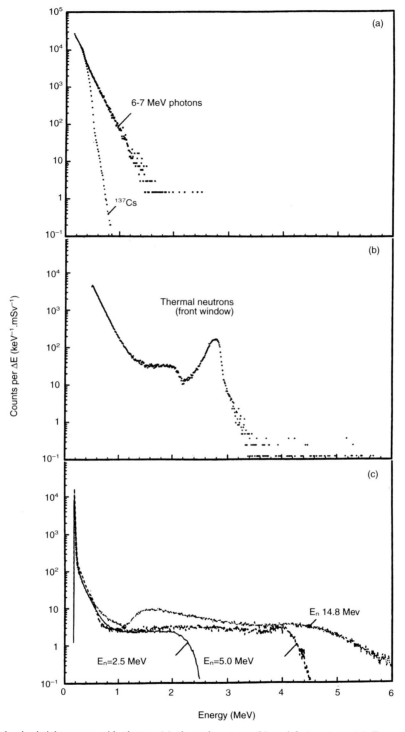

Figure 3. Measured pulse-height spectra with photons (a), thermal neutrons (b), and fast neutrons (c). For ease of comparison, all spectra shown are normalised to 1 mSv of photon or neutron dose equivalent.

may be sufficient for a rough characterisation of the neutron spectrum in this energy range.

MEASUREMENTS WITH NEUTRONS AND PHOTONS

First measurements with conventional nuclear laboratory electronics and a dosemeter set-up resembling the one shown in Figure 1 (case E) were performed. A partially depleted PIPS diode with a minimum depleted layer of 100 μm supplied by Canberra was used.

In order to simulate a position-sensitive diode, for first test measurements, just one ^{6}LiF layer was used, which was rotated into the three different positions. For initial measurements with photons and high energy neutrons (neutron energy above 1.5 MeV), the dosemeter was used without LiF layers.

In all cases, a bias voltage of 9 V was used and an energy calibration was performed with 5.48 MeV α particles from an ^{241}Am source in vacuum. Figure 3 shows measured pulse-height spectra (channels converted to energy after zero-point correction) for irradiations with photons (a), thermal neutrons (b) and high energy neutrons (c). For ease of comparison, all spectra shown were normalised to 1 mSv of photon or neutron dose equivalent.

A threshold of 1.5 MeV has to be set for this neutron dosemeter to be sufficiently insensitive to photons. Although the dose equivalent response of ^{137}Cs photons (660 keV) above a threshold of 1.5 MeV is negligible, there is still a strong response due to photons with energies from 6 MeV to 7 MeV (see Figure 3(a)). Integration of the counts for these high energy photons above an electronic threshold of 1.5 MeV yields a response of 730 mSv^{-1} which is almost comparable to the fast neutron response at 2.5 MeV neutron energy (see below). Photons with energies from 6 MeV to 7 MeV from the ^{16}O (n,γ) reaction contribute to the dose especially at reactors and have to be taken into account when neutron dosemeters are developed. In this measurement, photons with energies from 6 MeV to 7 MeV were produced by the reaction ^{19}F(p,αγ)^{16}O at the PTB's van de Graaff accelerator using a LiF target and a proton energy E_p = 1.95 MeV.

In order to check the MCNP calculations, measurements were performed in the low energy range with thermal and 144 keV neutrons at the research and measurement reactor of the PTB. The spectrum measured with thermal neutrons (see Figure 3(b)) shows a peak indicative of 2.74 MeV ^{3}H particles whose energy is reduced only slightly due to energy losses in the 3 μm ^{6}LiF layer, the 1 mm air layer in front of the diode and the 50 nm entrance window of the diode. Higher energy losses are expected for the 2.05 MeV α particles which are only in part detected when an electronic threshold at 1.5 MeV is used. The values measured for the different windows using an integration threshold at 1.5 MeV are shown in Figure 2 (case E). Since not all α particles are detected, the experimental values should be lower than the values calculated by MCNP. Within the experimental and theoretical uncertainties (only statistical uncertainties shown) and small differences between the set-up used in the calculations and the set-up verified, sufficient agreement is, however, reached. The higher measured value for thermal neutrons in the closed position can be explained by inexact positioning of the shieldings during these first experiments.

The measurements with fast neutrons of energies E_n > 1.5 MeV were performed at the PTB's van de Graaff accelerator. In the irradiation with 5 MeV neutrons, a spectrum of particles, chiefly protons with energies up to 5 MeV is produced, but the maximum energy observed is only 4.5 MeV (see Figure 3(c)). Since in silicon 4.5 MeV protons have a range of 178 μm, the

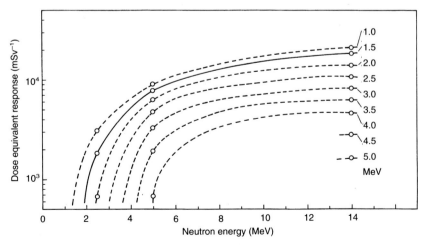

Figure 4. Fast neutron response functions using different electronic thresholds as indicated in the figure on the right side. The lines drawn are fits by eye only.

diode (full nominal thickness: 315 μm) is only approximately half depleted. The particles detected at higher energies (14.8 MeV neutron irradiation) are probably more heavier charged particles (d, α, etc). Silicon recoil nuclei are also detected. In the case of 14.8 MeV and 5.0 MeV neutron energy, the maximum energy of silicon recoil nuclei (central collision) is 2.0 MeV and 0.67 MeV. The shoulders observed a little below these energies indicate the contribution of silicon recoil nuclei in both measured spectra.

The dose equivalent response using different integration thresholds is determined from the measured spectra (Figure 3(c)) and shown in Figure 4 as a function of neutron energy. It increases by a factor of ten between 2.5 MeV and 14.8 MeV (dose equivalent responses are 1800 mSv^{-1} and 18300 mSv^{-1} for a threshold at 1.5 MeV). If it is possible to determine a set of response functions by setting different thresholds as shown in Figure 4, it will, however, be possible to obtain spectral information.

CONCLUDING REMARKS

It has been shown that the readout of different positions of a diode and the use of different electronic thresholds allow a neutron dosemeter with spectrometric properties to be made available. Further experiments and calculations are necessary to determine the dose limits at which a combination of the readings leads to reliable results. Particularly in a field which contains neutrons both above and below 1.5 MeV, the high energy neutron response (above 1.5 MeV) has to be determined for all windows (also with ^{6}LiF layers) and high energy contributions have to be subtracted before the ratio of the thermal neutron induced readings in the different windows is looked at. The uncertainties due to such subtraction procedures can be minimised by selecting thicker ^{6}LiF layers.

The studies performed serve as a starting point. The detector set-up, the shieldings and the detector specifications (thinner sensitive layer to reduce the photon sensitivity), for example, should be further optimised.

ACKNOWLEDGEMENTS

The authors wish to thank Dr S. Guldbakke, W. Sosaat and G. Urbach for performing the neutron and photon irradiations and for valuable discussions.

REFERENCES

1. Luszik-Bhadra, M., Alberts, W. G., d'Errico, F., Dietz, E., Guldbakke, S. and Matzke,M. *A CR-39 Track Dosemeter for Routine Individual Neutron Monitoring.* Radiat. Prot. Dosim. **55**, 285–293 (1994).
2. Barthe, J., Bordy, J. M., Lahaye, T. and Mourgues, M. *New Principle of Single Diode Neutron Dosemeter Based on Time Resolution.* In: Proc. 17th IRPA Regional Congress, Portsmouth, pp. 97–100 (1994).
3. Dubarry, B., Barelaud, B., Decossas, J. L., Mackovicka, L., Paul, D. and Vareille, J. C. *Electronic Sensor Response in Neutron Beams.* Radiat. Prot. Dosim. **44**, 367–370 (1992).
4. Eisen, Y., Shamai, Y., Ovadia, E., Karpinovitch, Z., Faermann, S. and Schlesinger, T. *A Rem Equivalent Personnel Neutron Dosimeter for Neutron Energies of 1 eV to 14 MeV.* Health Phys. **41**, 349–362 (1981).
5. Briesmeister, J. F. (Ed.) *MCNP^{TM} — A General Monte Carlo N-Particle Transport Code, Version 4A.* Report LA-12625-M (Los Alamos National Laboratory, Los Alamos) (1993).
6. Siebert, B. R. L. and Schuhmacher, H. *Quality Factors, Ambient and Personal Dose Equivalent for Neutrons, Based on the New ICRU Stopping Power Data for Protons and Alpha Particles.* Radiat. Prot. Dosim. **58**, 177–183 (1995).
7. Alberts, W. G., Dietz, E., Guldbakke, S. and Kluge, H. *Response of an Electronic Personal Neutron Dosemeter.* Radiat. Prot. Dosim. **51**, 207–210 (1994).

ADVANCES IN SUPERHEATED DROP (BUBBLE) DETECTOR TECHNIQUES

F. d'Errico†, W. G. Alberts‡ and M. Matzke‡
†Dipartimento di Costruzioni Meccaniche e Nucleari (DCMN)
Università degli Studi di Pisa, Via Diotisalvi 2, I-56126 Pisa, Italy
and Yale University School of Medicine, Division of Radiological Physics
333 Cedar Street, New Haven, CT 06510, USA
‡Physikalisch-Technische Bundesanstalt (PTB)
Bundesallee 100, D-38116 Braunschweig, Germany

INVITED PAPER

Abstract — State-of-the-art neutron dosemeters based on superheated drop (bubble) detectors are described. These are either active systems for area monitoring, which rely on the acoustical recording of drop vaporisations, or passive pen size ones for personal dosimetry, based on optical bubble counting. The technological solutions developed for the construction of robust devices for health physics applications are described with special emphasis on methods adopted to reduce mechanical shock and temperature sensitivity of the detectors. Finally, a review is given of some current research activities. In particular, a new approach to neutron spectrometry is presented which relies on the thermal effects for the definition of the response matrix of the system.

INTRODUCTION

Although almost twenty years have passed since their invention, superheated drop detectors (SDD)[1], or bubble damage detectors[2], are regarded as relatively new in the field of neutron dosimetry. Indeed, they still constitute a challenging topic for research and development (R&D), but they have also proven an effective answer to some operational radiation protection dosimetry needs. In fact, several devices suitable for practical applications have appeared on the market offering high neutron sensitivity, dose equivalent response and photon discrimination.

Early studies focused on the delicate chemistry, or rather alchemy, of the detectors and aimed at the reproducible realisation of homogeneous and stable suspensions of halocarbon droplets. The first difficulty to overcome was the emulsification of a superheated liquid, a metastable phase, in another fluid. The latter, in turn, had to be an immiscible and inert host, so that droplets would neither dissolve nor lose their properties through chemical reactions. Two laboratories, at Yale University[1] and at AECL Chalk River[2], succeeded in the effort, which led to the establishment of the present manufacturing companies[3,4]. The two approaches were slightly different, as host media were either an aqueous gel, formulated to provide the desired visco-elasticity, or a stiffer polyacrylamide compound.

At present, a variety of emulsions can be produced with different superheated halocarbons and has been characterised in terms of response to neutrons and to other types of radiation. Current R&D concentrates on the achievement of simple and robust systems able to operate reliably in the typical working environment. Although some analogies are obviously present, the two manufacturers have followed different paths developing either active, real-time counters, or passive, integrating pen detectors. These are described along with the most recent results of research on SDD-based neutron spectrometry.

ACTIVE DEVICES

A peculiar method to count bubbles in superheated drop detectors is by acoustically recording the pulses radiated during drop vaporisations[5]. The phase transition of a superheated droplet is an 'explosive' event and the associated pressure pulse can easily be picked up, e.g. by means of piezo-electric transducers in contact with the detector. However, a problem arises immediately: the discrimination of bubble pulses against spurious noise.

Early counters employed two transducers — one in contact with the detector vial and the other attached to the external casing of the meter, and performed anti-coincidence counting of the signals produced by the bubbles only[6]. Devices based on this approach were commercialised for a short period. In fact, it was soon clear that external vibrations could paralyse the counters, preventing them from detecting true pulses in a noisy environment.

Current designs still rely on the double transducer scheme, but adopt a comparative pulse-shape analysis of the signals[7]. Pulses from the two piezo-electric transducers are amplified and fed to rectifier detectors producing their envelopes, which are then processed by analogue/digital converters (Figure 1). When the signal from the SDD vial exceeds an adjustable threshold and presents the typical decay pattern of a bubble pulse, then the two digitised tracks are compared. The detector pulse is only accepted when its shape clearly differs from the signal in the noise channel. Vibration dampers de-couple the vial from the meter case, thus increasing the effectiveness of the scheme.

Advantages of this somewhat elaborate method are the possibilities to acquire and display counts and count rates in real time and to apply correction factors to the recorded data. First of all, measurements are adjusted for the dependence of the detector response on its operating temperature. This is a well documented phenomenon[8,9], intrinsic to the nature of superheated emulsions, and can be corrected or compensated for[10]. In commercial active counters, a correction is applied by means of a multiplying factor of about $0.05°C^{-1}$. This is based on the measured variation of response to Am-Be neutrons and is fairly accurate over the 15–35°C range. Results are also corrected for the non-linearity of the dose response, i.e. a gradual loss of sensitivity due to the progressive drop depletion occurring during the detector's use. The factor to be applied can be assessed from the exponential asymptotic trend in the cumulative response[11]. The correction is of the order of a few points per cent when the detector has accumulated some thousand counts, the exact value depending on the initial number of drops (usually, tens of thousands).

These counters employing dichlorofluoromethane emulsions have given excellent dosimetric results in a variety of radiation environments[7], but problems arise when ambient levels of noise and temperature exceed, respectively, 100 dB and 40°C (above ca. 45°C homogeneous nucleation of dichlorofluoromethane occurs, i.e. the spontaneous vaporisation of the droplets). These environmental conditions are not unusual in workplaces such as the containment of nuclear power plants. This limits the applicability in circumstances, e.g. during in-service inspections, when it would be advantageous to employ such a light device (below 1 kg) instead of moderator type 'rem counters' (above 10 kg). To make this possible, R&D now aims at improved noise insulation of the detector and at its active temperature control, which can also ensure an optimal energy dependence of the dose equivalent response[8,12].

PASSIVE DEVICES

Of the various types of passive SDD systems, the polymer-based, pen size version for personal dosimetry[2] greatly contributed to the increasing diffusion of this technology. This was tested extensively and successfully[13,14], and appears to be a valid candidate for the replacement of traditional TLD or film-based neutron dosemeters.

Relatively simple compared to the active devices, these detectors present some ingenious technological solutions. They rely on the optical counting by eye or automated cameras of the bubbles which remain immobile in the polymer matrix after their formation. Two important features have been recently introduced: the possibility of re-condensing the bubbles at the end of a measurement, and a passive temperature compensation system.

Early models could no longer be used when about a hundred bubbles had formed. These would crowd the dosemeters, overlap each other and be very difficult to count. Even in the presence of a few bubbles, after some time the detectors could not be used optimally since bubbles would keep on expanding, inflated by gas and vapour released by the polymer matrix. The situation improved substantially with the adoption of a system allowing for the mechanical pressurisation of the detector at the end of each use. This is a screw-in piston that applies a pressure higher than the vapour tension of the emulsified halocarbon, thus re-condensing the bubbles to the liquid phase. The procedure virtually resets the device, and when the cap is unscrewed the pressure is released and the detector 'sensitised' as the drops are again in a metastable liquid state.

In order to minimise the temperature dependence of the response, an effective and elegant solution was proposed[10] and is now implemented in passive devices. In a sealed vial, a low boiling point liquid is introduced over the free surface of the sensitive emulsion (Figure 2). The vapour tension of this liquid varies with ambient temperature, and the pressure thus applied to the detector compensates the temperature effects. In fact,

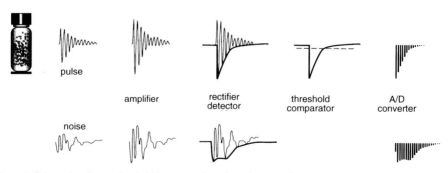

Figure 1. Schematic of an active bubble counter based on the acoustical recording of bubble vaporisations.

the neutron detection efficiency depends on the overall degree of superheat of the emulsion, i.e. on the combination of pressure and temperature values. The dependence on these quantities is illustrated in Figure 3 showing that the increase in response of a dichlorofluoromethane detector warmed up from 20 to 25°C is almost exactly counterbalanced by raising the operating pressure from 1 to 2 bar[15]. In dosemeters with the pressure compensation, the variation of sensitivity per degree Celsius is reduced from 5 to about 1%.

CURRENT RESEARCH AND DEVELOPMENT

Potential of SDD techniques

The previously described active and passive superheated drop (bubble) detector systems are designed for operational health physics. The stringent requirements of ruggedness and ease-of-use that apply put heavy constraints on the possibilities offered by this technology. In particular, the temperature dependence of the response is treated as a negative factor and different methods are implemented to remove it, as was previously reported. However, a different approach was successfully followed in recent years. A systematic investigation was carried out on the correlation between nuclear and thermodynamic properties of the detectors[16]. The response to different radiation types and energies was studied for a variety of sensitive emulsions as a function of their degree of superheat. These data provided further insight into the nature of the detectors and constitute the basis for their optimal use. For example, it was found that monochloropentafluoroethane detectors at room temperature are sensitive to electrons, and therefore detect both photons and protons through their secondaries[17]. For this reason, these emulsions are currently optimised for the development of three-dimensional dosimetry systems for medical applications[18]. As regards neutron dosimetry, it was found that dichlorofluoromethane emulsions, sensitive from thermal to fast neutrons, present an almost constant ambient dose equivalent response when stabilised at about 31°C[8]. A prototype dosemeter based on this principle was successfully tested[12,19] and implementation of the approach in advanced active counters is envisaged.

Neutron spectrometry

The results most relevant to the field of neutron detection came from the development of a new spectrometer concept, called BINS (Bubble Interactive Neutron Spectrometer). For such a specialised instrument it appeared possible to design a delicate device and take advantage of all the characteristics of the superheated emulsions, particularly their ill-famed temperature dependence. The latter is exploited to control the response of the detectors as it was shown that virtually any desired detection threshold can be generated in the 0.01–10 MeV neutron energy range and used for differential fluence measurements[9].

A BINS prototype was made which employs two detectors of different types, the acoustical bubble counting method, and a time proportioning controller to stabilise the detectors at various working temperatures. With this apparatus, the selected emulsions of octafluorocyclobutane and dichlorotetrafluoroethane are successively set at 25, 30, 35 and 40°C, and thus provide two series of four response functions covering, respectively, the 0.01–1 MeV and the 1–10 MeV ranges. These threshold curves are nested and spaced in quasi-isolethargic energy bins. As a consequence, the response matrix of the spectrometer is a virtually orthogonal system, i.e. the response functions are linearly independent, which allows for highly effective few channel unfolding procedures. The experimentally determined response curves are reported in Figure 4, plotted on a linear-logarithmic scale along with the smooth interpolating functions employed in the spectrometry trials described hereafter.

Tests of BINS were conducted in three neutron fields

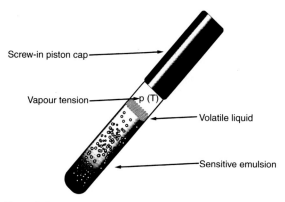

Figure 2. Passive bubble detector with re-pressurisation cap and temperature compensation system.

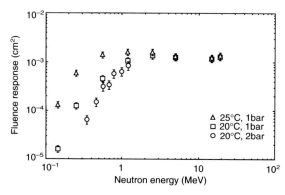

Figure 3. Fluence response of a dichlorofluoromethane emulsion as a function of pressure, temperature and neutron energy.

of different energy distributions. At the PTB, the two well characterised reference fields produced by ^{252}Cf and Am-Be sources[20] were utilised to provide indications in the high energy region. At CEA Cadarache, Canel+, a benchmark realistic neutron field with a prevailing low energy component, was employed during an intercomparison of spectrometry systems organised by EURADOS Working Group 7[21].

Two unfolding codes, MSITER and UNFANA, were used to obtain neutron spectra in the 0.2-12 MeV range. MSITER relies on the least-squares method and the covariance representation of uncertainties[22], its algorithm is derived from the well known STAY'SL code[23]. MSITER converges towards the most likely neutron spectrum and its related uncertainty matrix, or covariance matrix, by minimising a χ^2 expression. The latter consists of two biquadratic terms containing, respectively, the differences between measured and calculated count rates, and the differences between known *a priori* and adjusted neutron fluence. Each term is weighted by the inverse of the corresponding uncertainty matrix. The adjusted neutron spectrum strongly depends on the information provided *a priori* and on its uncertainty matrix. The other unfolding code, UNFANA, is based on the principle of maximum entropy[24]. It is defined 'analytical' Monte Carlo since it proceeds by sampling possible spectra drawn from an exponential distribution. The latter replaces the probability distribution used by classic Monte Carlo methods. UNFANA is a very efficient method which minimises computing time. It does not require any *a priori* information as it extracts from measured data the maximum information on differential fluence, with the associated uncertainties, imposing the condition that χ^2 be equal to the number of degrees of freedom, i.e. the number of detectors employed.

For these unfolding trials, an uncorrelated uncertainty of 10% on the scaling factor of each response function was assumed and added quadratically to the 10% uncertainties on each measured data point. In the fluence adjustment runs with MSITER, the PTB and CEA reference spectra were taken as *a priori* information with an uncorrelated relative uncertainty of 2% in each energy group. Unfolded results are plotted in Figures 5-7, showing that input and output spectra virtually

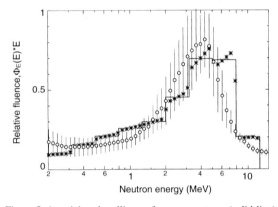

Figure 5. Americium-beryllium reference spectrum (solid line) versus spectra measured with BINS and unfolded with MSITER (*) and UNFANA (○).

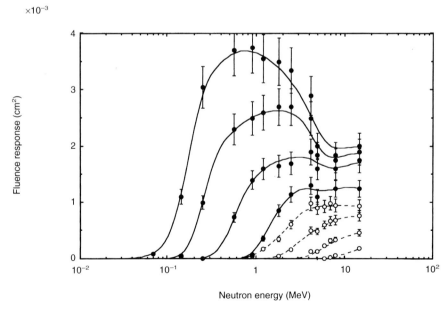

Figure 4. Measured response matrix and relative interpolating functions for neutron spectrometer BINS (solid lines and black circles, octafluorocyclobutane; dashed lines and white circles, dichlorotetrafluoroethane).

coincided in all cases. The value of χ^2 per degree of freedom was less than 1, indicating complete consistency between known *a priori* and BINS-measured data. Also reported in Figures 5-7 are the results from the unfolding with UNFANA. In the absence of any binding pre-information, this code generated smoother spectral distributions which better reflect the intrinsic resolution of the current BINS prototype.

Results were very encouraging: despite the pronounced differences between the three spectra chosen for these tests, curves unfolded with the two methods differed from each other by less than the associated uncertainties. These trial runs indicated that a higher number of thresholds is necessary in order to improve the resolution, especially in the high energy region. However, even with the few channels we employed, the fluence integrals assessed with either code over broader energy intervals were affected by much smaller uncertainties thanks to the negative correlations between adjacent groups.

CONCLUSIONS

Superheated drop (bubble) detectors offer a viable approach to neutron dosimetry in mixed radiation fields. Several devices have been developed for health physics applications, either for ambient dosimetry or for personal monitoring. These appear suitable for practical use as two fundamental problems posed by this technology have been practically overcome, i.e. the temperature sensitivity and, in the case of acoustical bubble detection, the discrimination of external noise. However, in the attempt to develop rugged instruments, some of the unique features offered by these detectors, such as the possibility of modifying their response through their thermal control, have been neglected. This was the focus of our research, which demonstrated the possibility to exploit the correlation between nuclear and thermodynamic properties and develop instrumentation of new conception, for neutron dosemetry and spectrometry but also for the detection of other radiation types.

ACKNOWLEDGEMENTS

The authors wish to express their gratitude to Prof. Giorgio Curzio, Università degli Studi di Pisa and Prof. Robert Apfel, Yale University, for their advice and support. Their thanks are also extended to Mr Jesus Martin, AE New Haven, and Dr Harry Ing, BTI Chalk River, for valuable discussions and to Mr Hermann Kluge, PTB Braunschweig, for his help with the neutron source irradiations. This work was partially supported by the Commission of the European Communities and by the National Institutes of Health.

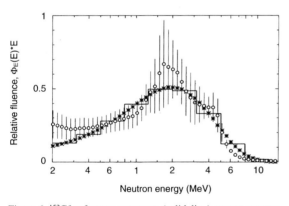

Figure 6. ^{152}Cf reference spectrum (solid line) versus spectra measured with BINS and unfolded with MSITER ($*$) and UNFANA ($\bigcirc$).

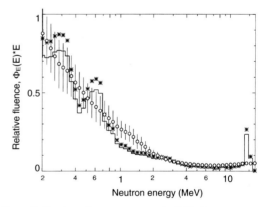

Figure 7. Canel+ reference spectrum (solid line) versus spectra measured with BINS and unfolded with MSITER ($*$) and UNFANA ($\bigcirc$).

REFERENCES

1. Apfel, R. E. *The Superheated Drop Detector*. Nucl. Instrum. Methods **162**, 603–608 (1979).
2. Ing, H. and Birnboim, H. C. *A Bubble-Damage Polymer Detector for Neutrons*. Nucl. Tracks Radiat. Meas. **8**(1–4), 285–288 (1984).
3. Apfel Enterprises Inc. (AE), New Haven, Connecticut, USA.
4. Bubble Technology Industries (BTI), Chalk River, Ontario, Canada.
5. Apfel, R. E. and Roy, S. C. *Instrument to Detect Vapor Nucleation of Superheated Drops*. Rev. Sci. Instrum. **54**(10), 1397–1400 (1983).
6. Ipe, N. E., Donahue, R. J. and Busick, D. D. *The Active Personnel Dosemeter — Apfel Enterprises Superheated Drop Detector*. Radiat. Prot. Dosim. **34**(1/4), 157–160 (1990).

7. Apfel, R. E., Martin, J. D. and d'Errico, F. *Characteristics of an Electronic Neutron Dosemeter Based on Superheated Drops.* Health Phys. Suppl. **64**(6), S49 (1993).

8. d'Errico, F. and Alberts, W. G. *Superheated Drop (Bubble) Detectors and Their Compliance with ICRP 60.* Radiat. Prot. Dosim. **54**(3/4), 357–360 (1994).

9. d'Errico, F., Alberts, W. G., Curzio, G., Guldbakke, S., Kluge, H. and Matzke, M. *Active Neutron Spectrometry with Superheated Drop Detectors.* Radiat. Prot. Dosim. **61**(1/3), 159–162 (1995).

10. Apfel, R. E. *Characterisation of New Passive Superheated Drop (Bubble) Dosemeters.* Radiat. Prot. Dosim. **44**(1–4), 343–346 (1992).

11. d'Errico, F. and Apfel, R. E. *A New Method for Neutron Depth Dosimetry with the Superheated Drop Detector.* Radiat. Prot. Dosim. **30**, 101–106 (1990).

12. d'Errico, F., Alberts, W. G., Dietz, E., Gualdrini, G. F., Kurkdjian, J., Noccioni, P. and Siebert, B. R. L. *Neutron Ambient Dosimetry with Superheated Drop Detectors.* Radiat. Prot. Dosim. **65**(1–4), 397–400 (1996).

13. Spurny, F. and Votockova, I. *The Response of Bubble Damage Neutron Detectors in Reference Neutron Fields.* Radiat. Prot. Dosim. **65**(1–4), 393–396 (1996).

14. Rannou, A., Clech, A., Devita, A., Dollo, R. and Pescayre, G. *Evaluation of Individual Neutron Dosimetry by a Working Group in the French Nuclear Industry.* Radiat. Prot. Dosim. **70**(1–4), 181–186 (This issue) (1997).

15. d'Errico, F., Alberts, W. G., Apfel, R. E., Curzio, G. and Guldbakke, S. *Applicability of Superheated Drop (Bubble) Detectors to Reactor Dosimetry.* In: Reactor Dosimetry. Eds H. Farrar, E. P. Lippincott, J. G. Williams and D. W. Vehar. ASTM STP 1228, pp. 225–232 (Philadelphia: ASTM) (1994).

16. d'Errico, F., Apfel, R. E., Curzio, G., Dietz, E., Egger, E., Gualdrini, G. F., Guldbakke, S., Nath, R. and Siebert, B. R. L. *Superheated Emulsions: Neutronics and Thermodynamics.* Radiat. Prot. Dosim. **70**(1–4), 109–112 (This issue) (1997).

17. d'Errico, F. and Egger, E. *Proton Beam Dosimetry with Superheated Drop (Bubble) Detectors.* In: Hadrontherapy in Oncology. Eds U. Amaldi and B. Larsson. Excerpta Medica, International Congress Series 1077, pp. 488–494 (Amsterdam: Elsevier Science) (1994).

18. d'Errico, F., Nath, R. and Apfel, R. E. *Superheated Drop Detectors for the Three Dimensional Dosimetry of Brachytherapy Sources.* In: Advanced Diagnostic Modalities and New Irradiating Techniques in Radiotherapy. Eds G. Gobbi and P. Latini, pp. 401–405 (Napoli: L'Antologia) (1994).

19. Lindborg, L., Bartlett, D., Drake, P., Klein, H., Schmitz, T. and Tichy, M. *Determination of Neutron and Photon Dose Equivalent at Work-Places in Nuclear Facilities in Sweden.* Radiat. Prot. Dosim. **61**(1–3), 89–100 (1995).

20. Kluge, H., Alevra, A. V., Jetzke, S., Knauf, K., Matzke, M., Weise, K. and Wittsock, J. *Scattered Neutron Reference Fields Produced by Radionuclide Sources.* Radiat. Prot. Dosim. **70**(1–4), 327–330 (This issue) (1997).

21. Thomas, D. J., Chartier, J. L., Klein, H., Naismith, O. F., Posny, F. and Taylor, G. C. *Results of a Large Scale Neutron Spectrometry and Dosimetry Comparison Exercise at the Cadarache Moderator Assembly.* Radiat. Prot. Dosim. **70**(1–4), 313–322 (This issue) (1997).

22. Matzke, M. *Unfolding of Pulse Height Spectra: The HEPRO Program System.* Report PTB-N-19 (Physikalisch-Technische Bundesanstalt: Braunschweig) (1994).

23. Perey, F. G. *Least-Squares Dosimetry Unfolding: The Program STAY'SL.* Report ORNL/TM-6062, (Oak Ridge, Tenn) (1977).

24. Weise, K. *Mathematical Foundation of an Analytical Approach to Bayesian-Statistical Monte Carlo Spectrum Unfolding.* Physikalisch-Technische Bundesanstalt, Braunschweig, Report PTB-N-24 (1995).

SUPERHEATED EMULSIONS: NEUTRONICS AND THERMODYNAMICS

F. d'Errico[1][2], R. E. Apfel[3], G. Curzio[1], E. Dietz[4], E. Egger[5], G. F. Gualdrini[6], S. Guldbakke[4], R. Nath[2] and B. R. L. Siebert[4]
[1]Dipartimento di Costruzioni Meccaniche e Nucleari (DCMN)
Università degli Studi di Pisa, Via Diotisalvi 2, I-56126 Pisa, Italy
[2]Yale University School of Medicine, Division of Radiological Physics
333 Cedar Street, New Haven CT 06510, USA
[3]Yale University, Department of Mechanical Engineering
PO Box 2159 Yale Station, New Haven CT 06520, USA
[4]Physikalisch-Technische Bundesanstalt (PTB)
Bundesallee 100, D-38116 Braunschweig, Germany
[5]Paul Scherrer Institut (PSI), Abteilung für Strahlenmedizin
CH-5232 Villigen PSI, Switzerland
[6]Ente per le Nuove Tecnologie, l'Energia e l'Ambiente (ENEA)
Via del Colle 16, I-40136 Bologna, Italy

Abstract — The results of some recent theoretical and experimental investigations on the physics of superheated emulsions are presented. Computational fluid thermodynamics allowed for a detailed description of the temporal and spatial history of the energy deposition process by a charged particle in a superheated liquid. Despite the assumptions it is based upon, this model gives information in agreement with experimental data on bubble nucleation. The experimental findings concern the role of interfacial reactions between drops and emulsifer, the existence of inhibition temperatures for the detector's response, and the progressive sensitisation to protons.

INTRODUCTION

Neutron sensitive emulsions of over-expanded halocarbon droplets, known as superheated drop detectors[1] or bubble detectors[2], are receiving increasing acceptance in the field of ambient and personal neutron dosimetry. Thanks to their neutron sensitivity and photon discrimination, they provide a viable solution to various mixed field dosimetry needs. On the other hand, their physics re-introduced the need to confront some of the unsolved problems associated with the classic bubble chamber. Some theoretical developments of recent years are presented here, also new experimental evidence which poses new challenges to the current theory.

THEORY

When Hahn and Peacock first observed neutron induced cavitation in superheated liquids[3], Glaser's electrostatic theory[4] of the bubble chamber was already being replaced by Seitz's 'temperature spike' model[5]. However, the complete dynamic model for the bubble nucleation process by ionising radiation in a metastable superheated liquid was only solved 30 years later by Sun et al[6].

Seitz's theory suggests that when a heavy charged particle slows down moving through a liquid, its kinetic energy is transferred as thermal energy to extremely small regions (temperature spikes) through the intermediary of δ rays. The intense heating induces localised boiling, creating trails of microscopic vapour cavities which develop into macroscopic bubbles when the density of energy deposition is high enough. Although the process is extremely complex, involving aspects of atomic and nuclear physics as well as fluid thermodynamics, in Seitz's approach these are avoided under the approximation of a static, equilibrium phase transition of the superheated liquid. It is assumed that a spherical vapour cavity of radius r embedded in a liquid of surface tension σ and vapour tension p_v expands indefinitely when $(p_v - p_e) > 2\sigma/r$, where p_e is external pressure. Therefore, a critical radius $r_c = 2\sigma/\Delta p$ defines the discriminant between growing bubbles and those collapsing under the action of external forces. Various expressions of the formation energy for such critical bubble have been proposed, such as[7]:

$$W = (4\pi r_c^2 \sigma - \frac{4}{3}\pi r_c^3 \Delta p) + \frac{4}{3}\pi r_c^3 \rho_v h_{fg} + 2\pi \rho_l r_c^3 \dot{r}^2 + F$$

$$= W_{Gibbs} + H + E_{wall} + F$$

where ρ_v is vapour density, ρ_l liquid density, h_{fg} latent heat of vaporisation, $\dot{r}$ vapour wall velocity. W_{Gibbs} is the minimum reversible work required for bubble formation or Gibbs free energy[8], H the vaporisation energy, E_{wall} the kinetic energy imparted to the liquid by the motion of the vapour wall, F the energy imparted to the liquid during the growth of the bubble by the viscous forces. Neglecting the last two terms[7,9], and substituting for $r_c = 2\sigma/\Delta p$, we find:

$$W = \frac{16 \pi \sigma^3}{3 \Delta p^2} \left[1 + \frac{2p_v}{\Delta p}\left(1 + \frac{h_{fg}}{R^*T}\right)\right]$$

where R^* is the gas constant. It may be immediately observed that when the degree of superheat Δp increases, both the critical radius and the minimum formation energy decrease. Several authors have tried to correlate this formation energy with that deposited by charged particles within a distance ($k\, r_c$) along their track through the superheated liquid[7]. Values of k between 1 and 13 have been proposed for the expression $E_{min} = k\, r_c\, (dE/dx)$. Although this semi-empirical approach allows for an estimate of the measured bubble formation energy in superheated liquids, it is far from reflecting the physics of the bubble nucleation.

The actual phenomenon involves a first phase with the generation of a strong shock wave resulting from the heating of a small region to temperatures and pressures far beyond their critical values. When the hot, high pressure region has expanded sufficiently, the critical parameters are achieved at a certain radius. Then an interface separating liquid and vapour can be defined and demarked by a temperature-dependent surface tension. The vapour bubble continues to expand and reaches a radius of critical size if the initial neutron–nucleus interaction had been sufficiently energetic for the given degree of superheat.

The complete temporal and spatial history of the energy deposition process by a charged particle in a superheated liquid was determined through computational fluid dynamics techniques[6]. The problem was solved assuming that the behaviour of the medium can be described by the usual macroscopic fluid equations (continuum mechanics approximation) and that the energy is deposited instantly and uniformly along an infinite line (in the immediate vicinity of a heavy charged particle). The flow fields, as a function of time t and radial distance r, of a viscous, heat-conducting and compressible fluid subject to the singular initial condition of a sudden energy deposition are then governed by five fluid dynamic equations: three conservation equations (mass, momentum and energy), one equation of state, and its associated specific internal energy equation (treating the medium as a Horvath–Lin fluid[10]). These are sufficient to solve for the five unknowns: temperature, T; pressure, p; velocity, u; specific volume, v (density $\rho = 1/v$); and specific internal energy, e. The conservation equations in a cylindrical co-ordinate system are:

$$\frac{\partial \rho}{\partial t} + \frac{1}{r}\frac{\partial(\rho r u)}{\partial r} = 0$$

$$\rho\left(\frac{\partial u}{\partial t} + u\frac{\partial u}{\partial r}\right) = \rho f_r - \frac{\partial p}{\partial r} + \frac{4}{3}\mu\frac{\partial}{\partial r}\left(\frac{1}{r}\frac{\partial(ur)}{\partial r}\right)$$
$$+ \frac{4}{3}\frac{\partial \mu}{\partial r}\left(\frac{\partial u}{\partial r} - \frac{1}{2}\frac{u}{r}\right)$$

$$\rho c_v\left(\frac{\partial T}{\partial t} + u\frac{\partial T}{\partial r}\right) + T\frac{\partial p}{\partial T}\left(\frac{\partial u}{\partial r} + \frac{u}{r}\right) = \frac{4}{3}\mu\left(\frac{1}{r}\frac{\partial(ur)}{\partial r}\right)^2$$
$$- 4\mu\frac{u}{r}\frac{\partial u}{\partial r} + \frac{1}{r}\frac{\partial}{\partial r}\left(kr\frac{\partial T}{\partial r}\right) + \rho q$$

where μ is viscosity, k thermal conductivity, c_v specific heat capacity at constant volume, f_r body force. The thermal equation of state and its associated equation are:

$$p = \frac{R^*T}{v-b} - \frac{a}{Tv(v+c)}$$

$$e = e_0 + \int_{T_0}^{T} c_v(T)\, dT + \frac{2a}{cT}\ln\left(\frac{v}{v+c}\right) - \frac{2a}{cT_0}\ln\left(\frac{v_0}{v_0+c}\right)$$

where a, b, c are three parameters in Horvath–Lin's equation, R^* gas constant, e_0, T_0 and v_0, the undisturbed values of e, T and v, respectively.

The above system of non-linear partial differential equations with its boundary and initial conditions was solved by means of a hybrid computational method which automatically moves the nodes into regions of steep gradients. The critical radius was assessed as $r_c^c = \sigma/\Delta p$ for the cylindrical bubble assumed in the model. Once converted to the spherical case ($r_c^s = 2\sigma/\Delta p$), computed values correspond to about 2/5 of Seitz's critical radius. It is obvious that simulations do not mirror exactly the real physical phenomena, for example the energy is not deposited uniformly along an infinite line but within a limited region. However, it is noteworthy that a critical radius equal to 2/3 of Seitz's had already been shown to improve consistency with the experimental findings[11]. Other information of great interest provided by the numerical simulations are the estimates of the time and length scale for the collapse of vapour cavities under the action of surface tension and viscous forces. These quantities are virtually impossible to determine experimentally, their values being respectively of the order of 0.1 ns and 10 nm.

EXPERIMENTS

As reported above, many of the nucleation phenomena occur at the nanometre and nanosecond levels and cannot be detected directly. Nevertheless, some experiments provide valuable insight into the operation of superheated emulsions.

In this work, the correlation existing between density of energy transfer and efficiency of radiation induced vaporisation was examined by determining the response to thermal neutrons of dichlorofluoromethane (CCl_2F_2) emulsions as a function of temperature. These emulsions contain ^{35}Cl that undergoes the exoergic capture reaction $^{35}Cl(n,p)^{35}S$ with low energy neutrons. Of the reaction products, the proton receives 598 keV and deposits it over 13 µm, a relatively long range compared to the critical radius, whereas the sulphur ion receives 17 keV and deposits it within 43 nm. Experiments with a pure thermal neutron beam at the PTB revealed the temperature (i.e. degree of superheat) at which this energy deposition pattern becomes adequate to nucleate the bubble formation. The pure thermal neutron field was generated by extracting neutrons from the FMRB research reactor[12] and guiding them through a bent beam line made of glass tubes with nickel-plated inner walls. Fast neutrons escape tangentially, while thermal ones are reflected and travel along the vacuum line until they emerge through a thin aluminium window. With these neutrons, the capture process is the only reaction which can trigger bubble formation and it introduces a well defined amount of energy. Measurements carried out at carefully controlled temperatures between 10°C and 40°C showed that the sensitivity to thermal neutrons is thermodynamically inhibited below 15°C, while it arises sharply above ~22°C (Figure 1).

These observations are consistent with the more general correlation between nuclear and thermodynamic properties of superheated emulsions, previously found investigating the sensitivity of several detectors as a function of neutron energy and degree of superheat[13]. Among these findings are the responses of octafluorocyclobutane (C_4F_8) and dichlorotetrafluoroethane ($C_2Cl_2F_4$) shown in Figure 2: two sets of four curves are reported, each one identified by the corresponding degree of superheat ΔT (expressed in Kelvin above the boiling temperature). It is evident that the higher the superheat, the lower the energy that neutron charged secondaries must impart to the droplets in order to nucleate their evaporation. An apparent discrepancy is that octafluorocyclobutane at 32 ΔT presents higher response and lower threshold than dichlorotetrafluoroethane at 36 ΔT. This suggests that chemical reactions at the interface between suspended droplets and emulsifier gel alter the effective degree of superheat by affecting the surface tension of the emulsions. The phenomenon is not considered in current numerical models. It was observed with chlorocarbons and results were particularly pronounced with hydrocarbons (e.g. isobutane, C_4H_{10}, or monochlorodifluoroethane, $C_2H_3ClF_2$), which dissolve progressively in hydrogenated gels.

Figure 2 also documents the progressive appearance of a maximum in the response of highly superheated emulsions (cf. curves for C_4F_8). This is correlated with the arising detection efficiency for neutron-recoiled protons entering the drops from the surrounding hydrogenated gel. In order to verify this hypothesis, experiments were carried out with 62 MeV proton beams at

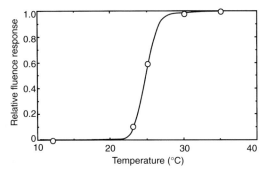

Figure 1. Relative fluence response of a dichlorofluoromethane emulsion to thermal neutrons as a function of operating temperature.

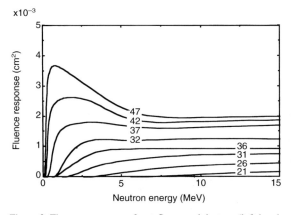

Figure 2. Fluence response of octafluorocyclobutane (left hand side) and dichlorotetrafluoroethane (right hand side) emulsions as a function of degree of superheat (reported on each curve) and neutron energy.

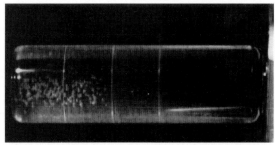

Figure 3. Photograph of a dichlorotetrafluoroethane emulsion irradiated with 62 MeV protons entering from the left (marks on the vial are spaced in 1 cm intervals).

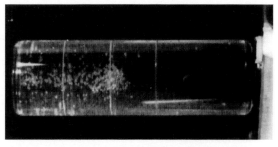

Figure 4. Photograph of a monochloropentafluoroethane emulsion irradiated with 62 MeV protons entering from the left (marks on the vial are spaced in 1 cm intervals).

the OPTIS facility of PSI Villigen[14]. Dichlorotetrafluoroethane and monochloropentafluoroethane ($CClF_2CF_3$) emulsions were irradiated at 22 °C (their ΔT being 18 and 61K, respectively) with the vial axis parallel to the beam. Protons had a range of about 2.5 cm in the sensitive emulsions and were stopped entirely inside the detectors. In the lower superheat ones, bubble formation occurred only within the first 2 cm of penetration (Figure 3). In that region, protons had energies higher than 20 MeV and triggered nuclear reactions creating heavy charged particles[15]. The detection of part of the proton range is consistent with the neutron response of these emulsions, only sensitive to high energy neutron recoils (cf. Figure 2). Conversely, the high superheat $CClF_2CF_3$ emulsions detected protons along their full range with a higher density of bubbles near the end, corresponding to the Bragg peak (Figure 4). Thus, it is shown that the high detection efficiency for low energy protons contributes to the neutron response of the most superheated emulsions in Figure 2, whose maxima correspond to Bragg peak structures for hydrogen recoils.

CONCLUDING REMARKS

The advent of superheated emulsions reopened a chapter of nuclear physics that had been abandoned along with the bubble chambers. Although the emulsions cannot provide the track structure information available from the chambers, the two systems share the same operating principles. These were investigated both theoretically and experimentally, and the physics of the interactions between charged particles and superheated liquids is now better understood. Computational fluid dynamics describe the very early stages of bubble formation, when phenomena occur at the nanometre and nanosecond levels. New findings also came from experiments with well characterised neutron and proton beams of different energies. These showed the role of interfacial reactions between drops and emulsifer, the existence of inhibition temperatures for the detector's response, and the progressive sensitisation to protons. These findings allow for an optimal use of the superheated emulsions, and may lead to the development of instrumentation of new conception for the characterisation of radiation fields.

ACKNOWLEDGEMENTS

The authors wish to express their gratitude to Dr H. Friedrich, PTB Braunschweig, Dr H. Schraube, GSF München, and Dr R. Cherubini, INFN Legnaro, who contributed to this research with their irradiation facilities and valuable discussions. This work was partially supported by the Istituto Nazionale di Fisica Nucleare, and the National Institutes of Health.

REFERENCES

1. Apfel, R. E. *The Superheated Drop Detector.* Nucl. Instrum. Methods **162**, 603–608 (1979).
2. Ing, H. and Birnboim, H. C. *A Bubble-Damage Polymer Detector for Neutrons.* Nucl. Tracks Radiat. Meas. **8**(1–4), 285–288 (1984).
3. Hahn, B. and Peacock, R. N. *Ultrasonic Cavitation Induced by Neutrons.* Nuovo Cimento **28**(2), 335–340 (1963).
4. Glaser, D. A. *Progress Report on the Development of Bubble Chambers.* Nuovo Cimento Suppl. **9**(11), 361–368 (1954).
5. Seitz, F. *On the Theory of the Bubble Chamber.* Phys. Fluids **1**(1), 2–13 (1958).
6. Sun, Y. Y., Chu, B. T. and Apfel, R. E. *Radiation-Induced Cavitation Process in a Metastable Superheated Liquid.* J. Comput. Phys. **103**(1), 116–140 (1992).
7. Bell, C. R., Oberle, N. P., Rohsenow, W., Todreas, N. and Tso, C. *Radiation-Induced Boiling in Superheated Water and Organic Liquids.* Nucl. Sci. Eng. **53**, 458–465 (1974).
8. The Collected Works of J. W. Gibbs. Vol. 1, p. 254 (New Haven, CT: Yale University) (1957).
9. Norman, A. and Spiegler, P. *Radiation Nucleation of Bubbles in Water.* Nucl. Sci. Eng. **16**, 213–217 (1963).
10. Horvath, C. and Lin, H. J. *A Simple Three-parameter Equation of State with Critical Compressibility-factor Correlation.* Can. J. Chem. Eng. **55**, 450–456 (1977).
11. West, C. *Cavitation Nucleation by Energetic Particles.* Report AERE-R 5486 (Harwell, UK: AEA) (1967).
12. Alberts, W. G. and Dietz, E. *Filtered Neutron Beams at the FMRB — Review and Current Status.* Report PTB-FMRB-112 (PTB, Braunschweig) (1987).
13. d'Errico, F., Alberts, W. G., Curzio, G., Guldbakke, S., Kluge, H. and Matzke, M. *Active Neutron Spectrometry with Superheated Drop Detectors.* Radiat. Prot. Dosim. **61**(1/3), 159–162 (1995).
14. d'Errico, F. and Egger, E. *Proton Beam Dosimetry with Superheated Drop (Bubble) Detectors.* In: Hadrontherapy in Oncology. Eds U. Amaldi and B. Larsson, Excerpta Medica, International Congress Series 1077, pp. 488–494 (Amsterdam: Elsevier Science) (1994).
15. Pearlstein, S. *Nuclear Data for Neutron and Proton Interaction with ^{12}C in the Energy Range 0–10 GeV.* Health Phys. **65**(2), 185–189 (1993).

FAST DISCRIMINATION OF NEUTRONS FROM (α,n) AND FISSION SOURCES

R. E. Apfel†‡, F. d'Errico§ and J. D. Martin‡
†Yale University, New Haven, CT 06520-8286, USA
‡Apfel Enterprises, Inc., 25 Science Park, New Haven, CT 06511, USA
§Dipartimento di Costruzioni Meccaniche e Nucleari (DCMN)
Università degli Studi di Pisa, Via Diotisalvi 2, I-56126 Pisa, Italy

Abstract — Numerical and experimental investigations were carried out in order to test the possibility of rapidly distinguishing (α,n) from fission neutron sources by means of superheated drop detectors (SDDs). This was achieved by measuring the ratio between the responses of two detectors operating at 30°C: SDD-1000, which has a threshold at about 0.5 MeV, and SDD-6000, which has a threshold near 4 MeV. The approach holds promise as it appears suitable for the development of compact instrumentation for safeguards verification purposes.

PLUTONIUM ASSAY

Assessing the plutonium content of Pu contaminated materials (PCM) is generally implemented by a non-electronic assay of the radiation emitted from the material or by an electronic approach in which the material is 'interrogated' with external neutron or gamma sources[1]. In the passive approach, the neutron leakage due to spontaneous fission or (α,n) reactions is measured. The relevant fissioning Pu isotopes are 238, 240, and 242, whereas (α,n) neutrons derive from the interaction of alpha particles from plutonium-238, -239, -240, -242 and americium-241, with light elements such as Be, B, F, Al, C, and O. Data for these two cases are shown in Tables 1 and 2. Normally, one attempts to measure spontaneous fission events resulting from ^{240}Pu and ^{242}Pu, and to discriminate them from the (α,n) neutrons by coincidence counting of emitted neutron pulses. Counting of neutron coincidences involves elaborate and bulky instrumentation that is not as readily portable as the one described in what follows.

Neutron energy spectra are another characteristic property of the neutron emission processes. The method proposed herein relies on the fact that despite the interaction with the materials of the PCM, the energy distributions of (α,n) neutrons retain a high energy component which differentiates them from spontaneous fission spectra. Discriminating these degraded fission and (α,n) neutrons on the basis of neutron energy has always been considered not practicable or at least quite inefficient. Among the main reasons for this is the fact that no system was available to provide the necessary sensitivity and response thresholds. Activation neutron detectors undergoing fast neutron threshold reactions do exist, but their sensitivity is too low for practical use.

Superheated drop detectors (SDD)[2] and bubble damage detectors[3] are known to offer a number of favourable features for neutron detection, including (a) high sensitivity, (b) isotropic response, and (c) photon discrimination. Almost two decades after their invention[4], they have reached an advanced level of development. SDDs have been shown to respond to fast neutrons, up to 66 MeV[5], and to present distinct energy thresholds depending on their composition. Moreover, some chlorine-bearing materials offer a thermal neutron sensitivity[6]. Among the latter, SDD-100 has a nearly dose equivalent response[7]. SDDs are intrinsically temperature sensitive. Therefore, general purpose instrumentation requires either passive compensation, or real-time monitoring of the temperature and the application of an appropriate correction factor. For optimal performance, temperature control is recommended[8].

The SDD materials can be employed in both non-

Table 1. Mean neutron energy $\bar{E}$ of spontaneous fission neutron sources.

Fissioning isotopes	Average neutron energy (MeV)
^{238}Pu	1.9
^{240}Pu	1.88
^{242}Pu	1.88
^{241}Am	1.95
^{242}Cm	1.96
^{244}Cm	1.98
^{252}Cf	2.1

Table 2. Maximum neutron energy from (α,n) reactions of 5.3 MeV alphas with various light target materials.

Isotope	Maximum neutron energy (MeV)
^{9}Be	10.88
^{10}B	6.14
^{13}C	7.30
^{18}O	5.93
^{19}F	4.66
^{27}Al	2.73

electronic and electronic neutron detectors. In the former, the response is measured after the irradiation by counting bubbles or measuring bubble volume displacement[3,9]. In the latter, the events are counted in real time by the acoustic signature accompanying the neutron-induced nucleation of the superheated drop[4,10]. The latter method forms the basis for the measurements described in this work.

NUMERICAL WORK

In PCMs, spectra of fission and (α,n) neutrons are more or less degraded by the surrounding materials. Nevertheless, it appears possible to use SDD materials with appropriate thresholds to distinguish between the two sources. In order to verify this hypothesis, transmission spectra of two typical fission and (α,n) sources (Cf and Am-Be, respectively) were computed by MCNP Monte Carlo simulations for various shielding configurations[11], as tabulated below:

Source	Shielding
californium or americium-beryllium	none
californium or americium-beryllium	iron annulus with 25 cm wall
californium	15 cm D_2O sphere

Using Fabry and Eisenhauer's post processing code DETAN, these spectra were folded with the measured energy response functions of SDD-1000 and SDD-6000 at 30°C[12]. This temperature corresponds to the thresholds allowing for a maximum discrimination between fission and (α,n) spectra, as shown in the two cases of Figures 1 and 2. The DETAN computations yielded the expected readings from the two detectors to be compared with data from the measurements of this work.

EXPERIMENTAL WORK

A series of tests were performed at the National Institutes of Standards and Technology (NIST) in order to verify our predictions for the relative responses of SDD-1000 and SDD-6000 to fission and (α,n) sources in the various shielding configurations.

Thermal stabilisation of the detectors was achieved through the use of foil heating elements wrapped around the SDD cartridges, operated by a simple ON/OFF solid state controller (Minco, Inc., MI). Operating temperature was $30.5 \pm 0.5°C$. Each detector was surrounded by a styrofoam insulator, which introduced a negligible neutron attenuation. Counts were acquired on a simple, multichannel instrument that records the acoustic pulses accompanying the neutron-induced vaporisations in the SDDs.

Satisfactory measurements yielding adequate statistics were made with the bare sources and with the heavy water moderated sources. The Am-Be source, on the other hand, was relatively weak and the iron shielding rather thick; therefore, the accumulated statistics for

Table 3. Response ratio: SDD-1000/SDD-6000.

Source-Shielding	Predictions	Measurements
Cf-Bare	54	41 ± 4
Cf-D_2O	41	26 ± 3
Cf-Fe	572	306 ± 75
AmBe-Fe	142	243 ± 65
AmBe-Bare	15	15 ± 2
AmBe-D_2O	16	*

*Not available due to the impossibility of inserting the large Am-Be source in the D_2O sphere.

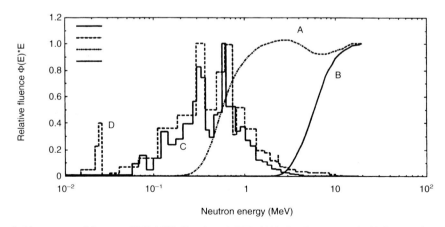

Figure 1. Threshold responses of detectors SDD-1000 (line A) and SDD-6000 (line B) compared with iron-moderated californium (line C) and americium-beryllium (line D) spectra.

these measurements were not as good, as reflected in the associated uncertainties.

RESULTS AND CONCLUSIONS

A comparison between the predictions, based on the SDD responses folded with the various spectra, and the experimental results is given in Table 3.

The NIST tests confirmed the potential of this approach, demonstrating the feasibility of discriminating between fission and (α,n) sources by means of superheated drop compositions with appropriate thresholds. Even with the strong spectral degradation introduced by the iron shield, the predictions in Table 1 indicate a factor of four difference between the moderated californium results and the americium-beryllium ones. This derives from a high energy portion of the moderated Am-Be spectrum which appears small in Figure 1, but still constitutes a substantial relative difference from the moderated californium case. The experimental results from the iron shielded sources suffered from poor statistics and prompted the development of SDD vials with significantly higher sensitivity which will be employed in a new series of measurements.

ACKNOWLEDGEMENTS

The authors would like to thank Drs Robert Schwartz and Charles Eisenhauer, NIST, Gaithersburg, MD, for their assistance with the experiments and the numerical simulations. Their thanks are also expressed to Prof. V. Sangiust, Politecnico di Milano, and Dr P. Schillebeeckx, Euratom Ispra, for valuable discussions. This work is supported by grant no DE-FG05-94ER81707-A003 of the US Department of Energy.

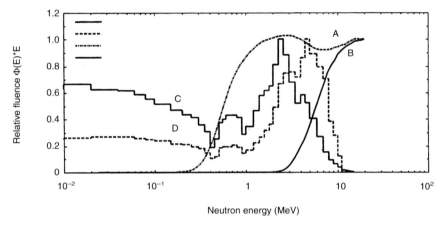

Figure 2. Threshold responses of detectors SDD-1000 (line A) and SDD-6000 (line B) compared with heavy-water-moderated californium (line C) and americium-beryllium (line D) spectra.

REFERENCES

1. Birkhoff, G. *Monitoring of Pu Contaminated Solid Waste Streams. A Technical Guide to Design and Analysis of Monitoring Systems.* EUR 10026 (Luxembourg: Office for Official Publications of the European Communities) (1985).
2. Apfel, R. E. *The Superheated Drop Detector.* Nucl. Instrum. Methods **162**, 603–608 (1979).
3. Ing, H. and Birnboim, H. C. *A Bubble-Damage Polymer Detector for Neutrons.* Nucl. Tracks Radiat. Meas. **8**(1–4), 285–288 (1984).
4. US patent, 4, 143, 274 (1979).
5. d'Errico, F. and Alberts, W. G. *Superheated Drop (Bubble) Detectors and Their Compliance with ICRP 60.* Radiat. Prot. Dosim. **54**(3/4), 357–360 (1994).
6. Apfel, R. E. *Photon-Insensitive, Thermal to Fast Neutron Detector.* Nucl. Instrum. Methods **179**, 615 (1981).
7. Apfel, R. E. and Lo, Y.-C. *Practical Neutron Dosimetry with Superheated Drops.* Health Phys. **56**, 79 (1989).
8. d'Errico, F., Alberts, W. G., Dietz, E., Gualdrini, G. F., Kurkdjian, J., Noccioni, P. and Siebert, B. R. L. *Neutron Ambient Dosimetry with Superheated Drop Detectors.* Radiat. Prot. Dosim. **65**(1–4), 397–400 (1996).
9. Nath, R., Meigooni, A., King, C., Smolen, S. and d'Errico, F. *Superheated Drop Detctor for Determination of Neutron Dose Equivalent to Patients Undergoing High-energy X-ray and Electron Radiotherapy.* Med. Phys. **20**, 781 (1993).
10. Apfel, R. E., Martin, J. D. and d'Errico, F. *Characteristics of an Electronic Neutron Dosemeter Based on Superheated Drops.* Health Phys. Suppl. **64**(6), S49 (1993). Also US Patent 4,143,274.
11. d'Errico, F., Apfel, R. E. and Eisenhauer, C. M. *Passive Neutron Assay of Plutonium Contaminated Materials by means of*

Neutron Energy Discrimination with Superheated Drop Detectors. Università degli Studi di Pisa Report DCMN 006(94) (1994).

12. d'Errico, F., Alberts, W. G., Curzio, G., Guldbakke, S., Kluge, H. and Matzke, M. *Active Neutron Spectrometry with Superheated Drop (Bubble) Detectors.* Radiat. Prot. Dosim. **61**(1–3), 159–162 (1995).

METHODOLOGICAL STUDIES ON THE OPTIMISATION OF MULTI-ELEMENT DOSEMETERS IN NEUTRON FIELDS

W. G. Alberts[†], B. Dörschel[‡] and B. R. L. Siebert[†]
[†]Physikalisch-Technische Bundesanstalt
POB 3345, D-38023 Braunschweig, Germany
[‡]Technische Universität Dresden, Inst.f. Strahlenschutzphysik
D-01062 Dresden, Germany

Abstract — A mathematical formalism for evaluating a multi-sensor dosemeter is presented. The formalism is tested for a personal dosemeter in a series of neutron spectra taken from a catalogue and an example for estimating the uncertainties is given.

INTRODUCTION

The fluence response of a single-sensor personal dosemeter for neutrons generally cannot match the shape of the fluence-to-dose equivalent conversion factors as a function of neutron energy. The safe use of such a dosemeter requires at least some knowledge of the radiation environment. In view of this situation many attempts have been made to use dosemeters based on more than one sensor. The intrinsic spectrometric capabilities of the multi-element dosemeter provided by the different response characteristics of the various sensors are, for instance, used to characterise the spectrum by ratios of sensor readings which depend on incident neutron energy. Derived correction factors are then used to interpret the dosemeter reading in order to yield a value for one selected dose equivalent quantity (e.g. Eisen et al[1], Piesch and Burgkhardt[2]). This approach, however, is hampered by being non-additive. In another approach, the readings of all sensors are linearly combined using weighting factors to be determined in such a way that the dose equivalent response is as independent as possible of the exposure conditions, which are usually unknown.

Over a decade ago, two different approaches were considered for the determination of the appropriate weighting factors. Siebert et al[3] described a concept for designing a personal dosemeter consisting of a limited set of linearly independent sub-detectors and weighting factors ('calibration constants') obtained by a least squares fit or linear programming. Schuricht and Dörschel[4] investigated in a similar approach the optimisation of combination dosemeters by subdividing one reference spectrum into as many energy intervals as there were sensors in the dosemeter and thus determining the weighting factors for this spectrum by solving a set of linear equations. They evaluated the uncertainties due to systematic errors of response functions and weighting factors, also taking into account the uncertainties due to random measurement errors.

In this paper an example of a three-element personal dosemeter is analysed on the basis of such considerations, showing the expedience of calculational dosemeter performance testing in a series of neutron spectra taken from a catalogue and stating the uncertainties.

FORMALISM

The discussion in this paper is restricted to the consideration of the energy dependence of the dosemeter responses; the formalism can in principle be extended to include the angle dependence as well as responses to different types of radiations, e.g. in mixed neutron–photon fields[3,4].

An 'ideal' dosemeter would have a fluence response $R_\Phi(E)$ (that is M/Φ at energy E, M being the detector reading and Φ the neutron fluence) proportional to the fluence-to-personal dose equivalent conversion factor: $h_{p\Phi}(E) = NR_\Phi(E)$ (N calibration factor). It would then measure personal dose equivalent H_p correctly in any spectrum:

$$H_p = \int h_{p\Phi}(E)\, \Phi_E(E)\, dE = N \int R_\Phi(E)\, \Phi_E(E)\, dE = NM$$

For an ideal multisensor dosemeter with J elements the same equation would lead to

$$h_{p\Phi}(E) = \Sigma_j w_j R_j(E) \qquad j = 1, \ldots, J \qquad (1)$$

where the weighting factors or 'calibration constants', w_j, could be found by simply solving a set of J Equations 1 at arbitrary energies E_j.

Present dosemeters are more or less far from ideal and Equation 1 is usually expanded into an overdetermined system of I equations ($I > J$), e.g. by subdividing the energy range into energy bins (discretisation):

$$h_i = \Sigma_j w_j R_{ij} \text{ or } \mathbf{h} = \mathbf{Rw} \qquad (2)$$

where h_i and R_{ij} denote mean values in the ith energy bin and the bold letters denote vectors and matrices. The solution of Equation 2 can be characterised by $\mathbf{w}^T = (w_1, \ldots, w_J)$ and an uncertainty covariance matrix $\mathbf{C_w}$. A measured value of dose equivalent H would then be determined from measured values $\mathbf{m}^T = (m_1, \ldots, m_J)$ by $H = \mathbf{w}^T \mathbf{m}$ and its variance is given by

$$s^2(H) = \mathbf{m}^T \mathbf{C_w} \mathbf{m} + \mathbf{w}^T \mathbf{C_m} \mathbf{w} \qquad (3)$$

where $\mathbf{C_m}$ is the given uncertainty covariance matrix associated with the measured values.

Equation 3 reflects the full covariance analysis for the method. Aiming at practical dosimetry applications in this paper, a simple procedure is suggested for finding the solution for such a problem and evaluating the uncertainties. For solving Equation 2 a simple least-squares procedure is proposed in conjunction with a selective weighting of the I equations. The weighting has the purpose of optimising the dosemeter for spectra found in workplaces. Such a weighting can be done interactively with a computer. The overall quality of the solution found by this procedure is then examined by testing whether the ratios of $H = \mathbf{w}^T\mathbf{m}$ and $H = \int h_\Phi(E) \Phi_E(E) \, dE$ are around unity within acceptable limits for a variety of spectra $\Phi_E(E)$ (ICRP 35 states such limits as a factor of 1.5 in both directions). With this approach the first term on the right-hand side of Equation 3 is studied, which characterises fit model errors; for the second term describing the uncertainties of the measurement a diagonal matrix $\mathbf{C}_m$ is assumed (no correlations), with diagonal elements $s^2(m_j)$, the standard uncertainties associated with m_j. Then

$$s^2(H) = \Sigma_j w_j^2 s^2 (m_j) \qquad (4)$$

An alternative method for finding the weighting factors w_j is to replace Equation 2 by

$$h_k = \Sigma_j w_j R_{kj} \qquad (5)$$

where k denotes a neutron spectrum out of a whole set of K spectra expected to describe the range of spectra at the workplaces considered, and R_{kj} is the mean fluence response and h_k the mean conversion factor in that spectrum. The selection of spectra, depending on the expected application for the dosemeter, will influence the resulting w_j. The variance analysis for Equation 5 is analogous to that for Equation 2 and is not further described here.

There is an equivalence between Equations 2 and 5. If every spectrum k is subdivided in energy bins i with group fluences Φ_{ik}, then the use of the spectra means a different weight for each equation of Type 2 determined by the K spectra used in Equation 5. The design aim for a dosemeter would then be to achieve a well conditioned and invertible 'structure matrix' $A_{ij} = \Sigma_i R_{ij} R_{ij}$ (details can be found in Reference 5).

APPLICATION TO A PERSONAL DOSEMETER

The example considered here is a personal neutron dosemeter with three elements based on CR-39 etched track detectors[6]. Figure 1 shows the fluence response of the three elements as a function of neutron energy together with the fluence-to-personal dose equivalent conversion factor $h_{p\Phi} = H_p(10)/\Phi$ for normal incidence on an ICRU tissue slab phantom[7]. The response curves are normalised in a way that $R_\Phi = h_\Phi$ for an Am-Be neutron spectrum. All three elements show a similar response to fast neutrons above about 80 keV. Whereas the element called 'fast' has no response below that energy, the element 'int' also responds to intermediate and thermal neutrons and the element called 'th' particularly to thermal neutrons. This example shows that the requirement of linearly independent response functions is not met.

Figure 2 shows the fluence response of combinations of two (int and fast) and all three components fitted according to Equation 2, subdividing the energy range from 2 MeV to 20 MeV into 104 logarithmically equidistant energy bins. The responses are shown as uncertainty bands calculated according to Equation 4 assuming an uncertainty associated with a measured dose equivalent of 0.2 mSv at each energy. The two-sensor fit yielded relative weightings of 27% int and 73% fast and the three-sensor fit 1% th, 33% int and 66% fast. These numbers and the response curves show the small additional effect of the thermal 'th' sensor. The ratio of the resulting response and the conversion function varies from 0.4 near 60 keV to 1.6 near 200 keV.

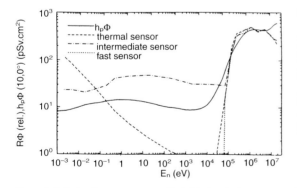

Figure 1. Relative fluence response, R_Φ(rel.), for single sensors and fluence-to-personal dose equivalent conversion factor for normally incident neutrons, $h_{p\Phi}(10,0°)$, as a function of neutron energy, E_n.

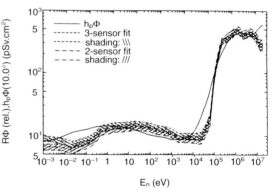

Figure 2. Relative fluence response, R_Φ(rel.), for sensor combinations and fluence-to-personal dose equivalent conversion factor for normally incident neutrons, $h_{p\Phi}(10,0°)$, as a function of neutron energy, E_n.

Whereas the comparison of response and conversion factor energy by energy is the hardest test for the energy dependence of the response of a dosemeter, it does not allow immediate conclusions as to the practical feasibility. This can be done by calculating the spectrum-averaged dose equivalent response $R_H = M/H_p(10) = R_\Phi/h_{p\Phi}$ for a series of spectra taken from a catalogue connected to the programme SPKTBIB[8]. From this catalogue two sets of spectra are considered: a set of 155 'reactor spectra' (related to reactors, fuel cycle, calibration sources) and a set of over 250 other spectra (e.g. from accelerators). R_H (which ideally should be unity in any spectrum) was calculated and normalised to R_H for the spectrum of an Am-Be source: $R_H(rel.) = R_H/R_H(Am-Be)$. For graphic representation, each spectrum has been characterised by its conversion factor $h_{p\Phi}$ serving as an indicator for spectrum hardness. $h_{p\Phi}$ varies from 14 to 450 pSv.cm² for the reactor fields and extends down to below 10 pSv.cm² for the other fields.

Figure 3 shows the relative responses for both fits described above in the over 400 spectra considered. The relative dose equivalent responses range from 0.6 to 1.25 in the 'reactor fields' where, as expected from the response functions of Figure 2, the three-element fit is only slightly better than that with two elements. If all considered spectra are included in the test, $R_H(rel.)$ varies from 0.55 to 1.3 with the observation that at the very low $h_{p\Phi}$ fields there is a large scatter of data and the results for the three-element fit are significantly larger than those for two elements (as opposed to the results in spectra with $h_{p\Phi} > 30$ pSv.cm²). The aforementioned ICRP 35 criterion for dosemeter performance is just slightly exceeded if one takes all spectra into consideration.

The response functions of Figure 1 were also used to fit weighting factors using Equation 5 and the same spectra in which the aforementioned feasibility studies were performed. Table 1 shows the results for various types of fits. A common result is the small contribution of the thermal sensor in this dosemeter. The fitted weighting factors show a relatively large spread depending on the basis of the equation system to be solved by the least squares fit.

Figure 4 shows the test for the weighting factors fitted with all spectra. Although the basis for the least squares fit for finding the appropriate w_j are the same spectra, at low $h_{p\Phi}$ values relative responses down to 0.3 are

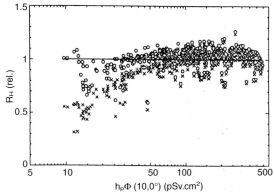

Figure 4. Relative dose equivalent response, $R_H(rel.)$, for sensor combinations (fit according to Equation 5) in all spectra considered. (○) 3-sensor fit with weighting, (×) 2-sensor fit without waiting.

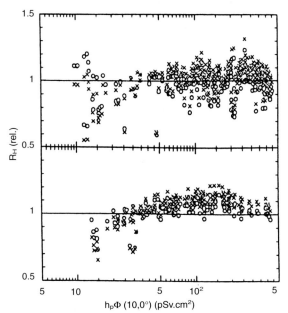

Figure 3. Relative dose equivalent response, $R_H(rel.)$, for sensor combinations (see Figure 2) in 'reactor spectra' (bottom) and in 'other' spectra (top), characterised by the conversion factor $h_{p\Phi}(10,0°)$. (○) 3-sensor fit, (×) 2-sensor fit.

Table 1. Fitted weighting factors w_j for two or three elements of the example dosemeter. Fits are based on energy bins (E), reactor spectra (R) or all spectra (A).

Fit basis	No of elements	Weighting factors for the single sensors		
		th	int	fast
E #1	2	—	0.280	0.756
E #2	3	0.011	0.321	0.636
A #1	2	—	0.159	0.821
A #2	2	—	0.290	0.703
A #3	3	−0.027	0.331	0.674
A #4	3	0.027	0.304	0.671
R #1	2	—	0.262	0.679
R #2	2	—	0.359	0.597
R #3	3	0.010	0.396	0.470

found for the fit of two sensors (A #1, see Table 1). A selective weighting of the equations and addition of the third sensor (A #4) resulted in the lowest responses at just below 0.6

UNCERTAINTIES

A full covariance analysis formalism for the fitting method described is not available at this time. It has been replaced by estimating separately the quality of the fitting method by looking at the R_H(rel.) in the set of spectra of interest. For improving the fit, weights of equations are modified in a way that does not provide defined input uncertainties. The variance due to the fit model is estimated from the spread of the data shown in Figures 3 and 4. For information about the total variance, the experimental uncertainty shown in Figure 2 would have to be added to these data.

CONCLUSION

An approach to generalising a multi-element dosemeter response has been described. The selected example of a personal dosemeter only perpendicularly irradiated is not representative for the general surveillance situation. As mentioned before, the method studied here can also be applied to the angle dependence of a personal dosemeter's response which will be a future object of consideration.

REFERENCES

1. Eisen, Y., Shamai, Y., Ovadia, E., Karpinovitch, Z., Faermann, S. and Schlesinger, T. *A REM Equivalent Personal Neutron Dosimeter for Neutron Energies of 1 eV-14 MeV*. Health Phys. **41**, 349–362 (1981).
2. Piesch, E. and Burgkhardt, B. *Albedo Dosimetry System for Routine Personnel Monitoring*. Radiat. Prot. Dosim. **23**, 117–120 (1988).
3. Siebert, B. R. L., Hollnagel, R. and Jahr, R. *A Theoretical Concept for Measuring Doses from External Radiation Sources in Radiation Protection*. Phys. Med. Biol. **28**, 521–533 (1983).
4. Schuricht, V. and Dörschel, B. *Optimization of Combination Dosemeters*. Kernenergie **27**, 200–204 (1984).
5. Alevra, A. V., Matzke, M. and Siebert, B. R. L. *Experiences from an International Unfolding Intercomparison with Bonner Speres*. In: Proc. 7th ASTM-EURATOM Symp. on Reactor Dosimetry, EUR 14356 EN (Dordrecht: Kluwer Academic Publishers) pp. 215–222 (1992).
6. Luszik-Bhadra, M., Alberts, W. G., Dietz, E., Guldbakke, S. and Kluge, H. *A Simple Personal Dosemeter for Thermal, Intermediate and Fast Neutrons Based on CR-39 Etched Track Detectors*. Radiat. Prot. Dosim. **44**, 313–316 (1992).
7. Sieber, B. R. L. and Schuhmacher, H. *Quality Factors, Ambient and Personal Dose Equivalent for Neutrons, Based on the New ICRU Stopping Power Data for Protons and Alpha Particles*. Radiat. Prot. Dosim. **58**, 177–183 (1995).
8. Naismith, O. F. and Siebert, B. R. L. *A Database of Neutron Spectra, Instrument Response Functions, and Dosimetric Conversion Factors for Radiation Protection Applications*. Radiat. Prot. Dosim. **70**(1–4), 241–245 (This issue) (1997).

FUTURE DEVELOPMENTS IN THE NRPB PADC NEUTRON PERSONAL MONITORING SERVICE

R. J. Tanner, D. T. Bartlett, L. G. Hager and J. Lavelle
National Radiological Protection Board
Chilton, Didcot, Oxon, OX11 0RQ, UK

Abstract — NRPB has been operating an etched track neutron personal monitoring service for ten years. This has, since its inception, used an etch regime which is a combination of a chemical etch and a 2 kHz electrochemical etch. Processing at most other laboratories has settled on either a chemical etch or a two frequency electrochemical etch. These other types of processing each have significant advantages, but attempts to introduce them for routine processing at NRPB have also shown them to have significant disadvantages. Recent improvements in the manufacture of PADC have, however, changed the situation markedly, leading to the prospect of a viable two frequency electrochemical etch for routine personal monitoring.

INTRODUCTION

NRPB has been operating a routine neutron personal monitoring service, based on a 2 kHz electrochemical etch of PADC (poly allyl diglycol carbonate, commonly referred to by the trade name CR-39) elements, since 1986[1,2]. The etch which has been employed differs significantly from most etches used for routine processing or in research laboratories, the majority of which are either chemical etches or etches which utilise a low frequency (50–100 Hz) followed by a high frequency (2–3 kHz) for the electrochemical stage. Both these other styles of processing, however, have tended to produce poor signal to noise ratios and the low energy threshold for chemical etches has been considered to be too high for the measurement of some workplace spectra. Consequently, the etch process currently used for routine monitoring by NRPB differs only slightly from that which was implemented at the start of the service[3].

Two frequency etching[4] has been the system favoured by many research laboratories, but it has not yet been introduced by any operational service using PADC in Europe. Its merits in terms of dosimetric performance are clear: the dose equivalent response is relatively high and flat over the range from 100 keV to 20 MeV, with the low energy threshold in particular being good relative to alternative etches. Background variability has, however, proved to be an intractable problem: higher mean backgrounds are compounded by an increased tendency to produce outliers, which has made background sampling unreliable. Consequently, the probable frequency of false positives in such a system has prevented NRPB from replacing the existing 2 kHz etch with a two frequency process.

Research into new materials has yielded a form of PADC[5] which performs very much better with the current NRPB etch regime in terms of signal to noise ratio[6]. The resultant improvement in the low dose detection threshold which has resulted from the introduction of this material for routine use in the NRPB service, has both enhanced the reliability of low dose measurement and reduced the frequency of false positive doses. Additionally the rejection rate for sheets of plastic has been considerably reduced.

The new material has, however, also shown very impressive results with two frequency etching, such that it now offers the possibility of improved signal to noise ratios for the majority of sheets of PADC, thereby both improving the dosimetry and reducing costs.

ETCH PARAMETERS

The etch parameters of the routine etch and the two frequency etch have both been optimised to give good signal to noise ratios with NRPB's routine plastic supply. These parameters are summarised in Table 1. The routine etch differs from that used originally in the NRPB Neutron Personal Monitoring Service primarily in the field strength which is applied (23.5 kV.cm^{-1} as opposed to 21 kV.cm^{-1}). The other main difference is that the chemical and electrochemical phases are operated contiguously at the same temperature (40°C) rather than the chemical phase being performed at 70°C and the electrochemical phase at 30°C.

Table 1. Etch parameters for the two etches developed at NRPB.

Etch type	Chemical + 2 kHz	Chemical + 50 Hz + 2 kHz
Etchant	5.0 N NaOH	5.26 N KOH
Temperature	40	50
Chemical etch (h)	11.5	4.0
Field strength (kV.cm^{-1})	23.5	19.6
50 Hz (h)	—	12.0
2 kHz (h)	8.0	2.0

LOW DOSE DETECTION

Theory

Perhaps the commonest parameter quoted when describing the smallest dose which can be measured, is the 'minimum detectable dose equivalent' (MDDE) or 'detection limit' (L_D)[7]. This is defined according to Equation 1, where σ_{n-1} is the standard deviation on the background distribution and R the response. It corresponds to the level at which implied doses are significant at the 97.5% confidence level and relies upon the assumption that the standard deviation can be measured with precision and that the response is easily defined:

$$L_D = \frac{2\sigma_{n-1}}{R} \quad (1)$$

A more sophisticated approach yields a more complex relation[8] which in the limit above reduces to Equation 2[9], where A is the read area on the dosemeter and L_Q is the dose level at which 2.5% of a sample of dosemeters would generate false negatives. This parameter is sometimes referred to as the 'determination limit'[7] or the minimum measurable dose equivalent (MMDE). In this relation, the background standard deviation and dose equivalent response are quoted in tracks per unit area rather than tracks per dosemeter:

$$L_Q = \frac{4\left(\sigma_{n-1} + \frac{1}{A}\right)}{R} \quad (2)$$

Background levels

Sheets of PADC used for the NRPB Neutron Personal Dosimetry Service are cut into 90 individually encoded elements, of which 10 are removed to sample the background. This measurement is then used to assess the pass/fail status of the sheet on the grounds that the sheet must have an MDDE of 150 μSv or less, which given that the lowest neutron $H_p(10)$ values quoted by the NRPB correspond to 200 μSv or greater, allows a significant margin for error. If the sheet is deemed to have passed this assessment, then this value for the background is used in the Neutron Personal Monitoring Service.

When the MDDE values for 42 sheets of plastic processed using both the 2 kHz etch and the two frequency etch are compared, the results are as shown in Figure 1. It can be seen that the results for the two frequency etch are on average lower, and do not extend to the high values encountered for the 2 kHz etch. Overall the average MDDE for the 2 kHz etch process is 105 μSv, whereas that for the two frequency etch is only 35 μSv. The value for the 2 kHz etch is strongly influenced by the high values obtained for some sheets, but it is clear from these results that the signal to noise ratio with PADC cured using the new cure cycle ceases to be a problem for the NRPB two frequency etch. This is in sharp contradiction to results published previously for PADC produced using an earlier cure cycle[10] but from the same manufacturer[11].

These results may be in part influenced by the poor quality of the sample for the 2 kHz etch: the last 13 monthly batches of PADC used in the NRPB service (286 sheets) have had a mean MDDE of 54 μSv, and a pass rate of 91.5%. These sheets, however, are not routinely quality accepted using the two frequency etch, and hence there are no comparative data. The pass rate for the 2 kHz etch of the sheets used in this work (76%) is clearly unrepresentative, so the 93% pass rate for the two frequency etch must also be questioned. When Spearman's rank correlation coefficient is computed for the two sets of MDDE data, however, the result is a value of 0.43, which with a Student's t of 3.0 must be regarded as being significant with greater than 99% confidence. Consequently, it may be inferred that if the PADC has performed as well for the 2 kHz etch as would have been expected, then the two frequency etch would probably have performed better itself.

FALSE POSITIVE INCIDENCE

The only true way of assessing the incidence of false positives in a service is to sample the issued dosemeters. If this is to be performed reliably, then dosemeters must be taken from a large number of sheets over a long period, which cannot be done for the two frequency etch since it is not in routine operation at NRPB. An easier method is to study the distribution of implied doses on the quality acceptance backgrounds. This is shown in Figure 2 which shows the implied doses for all the background dosemeters which were used for the MDDE measurements in Figure 1. Included in these data are results from sheets which would have failed the quality acceptance tests, so it is to be expected that a system

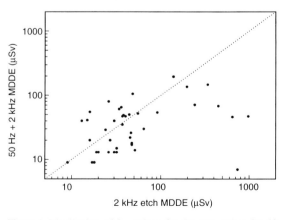

Figure 1. Distribution of L_D values for the two etches for 42 different sheets of PADC.

which was operating with either etch would inevitably perform better than represented in this figure. Given this condition, however, it is clearly seen that the distribution of backgrounds for the two frequency etch is much better than for the 2 kHz etch, with 3 (0.7%) implied false positives (implied dose ≥0.2 mSv) out of 420 dosemeters, compared to 9 (2.1%) for the 2 kHz etch. In practice it should be anticipated that these are the upper limits for false positive incidence since all the false positives shown in Figure 2 would have been eliminated at the system quality acceptance stage.

LOW DOSE MEASUREMENT

Systems with a high sensitivity have a greater potential for low dose measurement owing to their reduced coefficients of variation on dosed distributions. This simply results from their generating more tracks per unit dose, which if a system exhibits good linearity, aids performance in terms of actual dose measurement, as opposed to background discrimination: two systems may have the same value for the MDDE according to Equation 1, but that which has a higher sensitivity will have a lower L_Q according to Equation 2.

The mean values for the MDDE and L_Q for each etch are quoted in Table 2. These values are quoted both before and after quality acceptance of the plastic: the result for 'before' is hence representative of the etch, whereas that for 'after' is typical of the values which would be achieved in a service operating using such an etch. It is evident that the 2 kHz etch performs markedly worse for this sample of 42 sheets, but that after the pass/fail assessment at quality acceptance, the systems would be very similar. The difference is hence primarily caused by the sheets which have poor low dose characteristics with the 2 kHz etch, since both etches are clearly capable of producing good performance with most sheets.

RESPONSE FUNCTION

Measurements of the energy dependence of the response of PADC manufactured using the new cure cycle have shown there to be a significant increase for ^{252}Cf spontaneous fission neutrons, and also for monoenergetic neutrons with energies of 144 keV and 14.7 MeV[6]. The finer detail of the shape of the response function had not been measured, however, so additional irradiations were required to determine the actual shape of the response function. Such measurements are necessary before attempts can be made to investigate the expected response to known neutron spectra by folding the response function with the measured spectrum.

The responses to ^{252}Cf of the two etches with both types of PADC are given in Table 3. It can be seen that

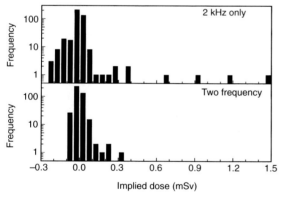

Figure 2. Distribution of implied doses for all background detectors used for quality acceptance testing using the two different etches.

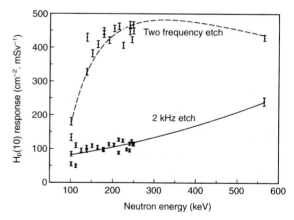

Figure 3. $H_p(10)$ response for the two etches over the energy range from 100 to 565 keV.

Table 2. Mean MDDE and L_Q values for both etches taken before and after quality acceptance.

Etch	Before quality acceptance		After quality acceptance	
	Mean MDDE (μSv)	Mean L_Q (μSv)	Mean MDDE (μSv)	Mean L_Q (μSv)
2 kHz	110	232	34	80
Two frequency	45	93	37	78

there is an increase in the response for both etches, with the response of the 2 kHz etch increasing by 23% and that for the two frequency etch by 35%. These results are repeatable over many sheets and batches of plastic, and indicate that there is a significant increase in the response in the region of 2 MeV.

Measurements have been made using a semicircular jig at the National Physical Laboratory which allows simultaneous irradiation of dosemeters with neutrons in the energy range from 100 to 250 keV. This utilises the variation in energy with angle of emission of the neutron from a ^{7}Li(p, n)^{7}Be reaction, for which the neutron emitted in the beam direction has an energy of 250 keV. In this field, which has been calibrated for relative fluence rate for angles of up to ±75° to the beam direction, the neutron energy can be calculated from kinematics to show that the energy at an angle of 75° corresponds to 100 keV. It is thus possible simultaneously to irradiate dosemeters with a range of monoenergetic neutrons of energies which approximately span the low energy threshold of most PADC neutron personal dosemeters.

The results from measuring the $H_p(10)$ response for normal incidence over the range from 100 to 565 keV are plotted in Figure 3. It can be seen that the response is higher for the two frequency etch over this whole range, but that it is dropping faster at 100 keV than is the case for the 2 kHz etch. This is probably related to the lower field strength which is employed. It is, however, indicative of the usefulness of this measurement, since the previous measurements with a monoenergetic 144 keV neutron source failed to detect this sudden drop because the response is still high at 144 keV.

When the results for the two etches are normalised and plotted over the range from 100 keV to 14.7 MeV, as shown in Figure 4, the response functions are seen to be quite similar in shape. Hence, it appears that the $H_p(10)$ response function for the two frequency etch is not flatter than that for the 2 kHz etch when measured with the new plastic. This is a very different result from those obtained previously, which showed the variation in response over the 144 keV to 15 MeV range to be less marked for the two frequency etch[12].

RESPONSE TO WORKPLACE SPECTRA

Folding these A-P response functions with measured spectra from the BNFL Sellafield site[13], PWR spectra from Ringhals and fuel storage spectra from CLAB in Sweden[14], and source spectra from bare ^{252}Cf and Am-Be sources, yields the results shown in Table 4. These data include spectra from locations in the reprocessing cycle, a Magnox reactor, pressurised water reactors and a fuel storage site. They cover a great range of spectrum hardnesses, with the $H_p(10)$ to fluence conversion factors ranging from 10 pSv.cm^2 to 375 pSv.cm^2, yet the calculated response values are constrained to the ranges 0.68–1.28 for the 2 kHz etch and 0.67–1.32 for the two frequency etch. The result for the ^{252}Cf spectrum with the two frequency etch is less satisfactory, since for the chosen calibration response, this spectrum would give a 40% overestimate.

Figure 5 shows the data from Table 4 plotted against the fluence to dose conversion factor. It is evident that the 2 kHz etch performs slightly better than the two frequency etch, but the difference is marginal. The degree of overestimate or underestimate is seen to be greater for the two frequency etch from its standard deviation of 0.24 on the normalised reading, compared to a value of 0.16 for the 2 kHz etch.

CONCLUSIONS

Two frequency etches have been strongly advocated as providing the most promising processing method for PADC neutron personal dosemeters. This promise, however, has not yet led to a European system which oper-

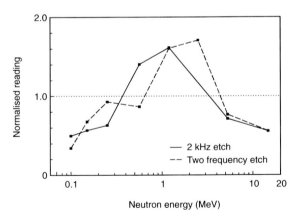

Figure 4. Normalised $H_p(10)$ response for the two etches over the range from 100 keV to 15 MeV.

Table 3. Sensitivity to ^{252}Cf neutrons at normal incidence (±SEM) for both etches and cure cycles.

	2 kHz (cm^{-2}.mSv^{-1})	50 Hz + 2 kHz (cm^{-2}.mSv^{-1})	2 kHz (mSv^{-1})	50 Hz + 2 kHz (mSv^{-1})
Old cure cycle	168 ± 4	506 ± 15	282 ± 7	850 ± 25
New cure cycle	207 ± 13	685 ± 8	347 ± 22	1151 ± 13

ates with real personal dosemeters: in practice, two frequency etches must first be shown to give better performance for most sheets of plastic, so that high reject rates which can be accepted in experimental systems are not experienced in a personal monitoring service.

The introduction of PADC produced using the new cure cycle has changed this picture considerably, since the plastic is now able to produce better signal to noise ratios and fewer rejected sheets than the current NRPB routine etch process. This could make such a service both more accurate at low dose levels and cheaper to run.

Measurements of the energy dependence of the response function have shown that the new material is no longer superior in terms of response to workplace spectra. This has resulted from the poorer shape of the response function with the new material, with the dose equivalent response now peaking in the region of 1–2 MeV. The difference in terms of expected results for workplace fields is not, however, very marked.

It would hence appear that the two frequency etch does now have the potential to replace the current system and improve dosimetric performance. The enhancements will be mainly in terms of low dose detection, since the precision of measurements will not be helped by the energy dependence of response characteristics. Higher sensitivity will, however, generate smaller fractional standard deviations at dose levels for which linearity considerations are not a problem.

The routine two frequency etches which operate in North America do so with plastic obtained from a North American plastics supplier, and most good results with

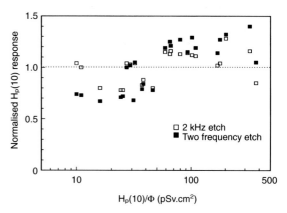

Figure 5. Responses for workplace spectra obtained by folding the measured spectrum with the response function, plotted against spectrum hardness.

Table 4. Result of folding known spectra with the response functions of the two etches.

Spectrum/Site	$H_p(10)/\phi$ (pSv.cm^{-2})	2 kHz only		Two frequency	
		Response (cm^{-2}.mSv^{-1})	Normalised	Response (cm^{-2}.mSv^{-1})	Normalised
Am(Be)	375	154	0.85	507	1.05
^{252}Cf	331	209	1.15	677	1.40
BNFL	10	188	1.04	357	0.74
	11	181	1.00	354	0.73
	27	187	1.04	484	1.00
	29	187	1.03	493	1.02
	32	188	1.04	508	1.05
	59	208	1.15	577	1.19
	65	205	1.13	607	1.25
	66	210	1.16	587	1.21
	80	205	1.13	618	1.27
	93	207	1.14	559	1.15
	102	203	1.12	628	1.29
	110	201	1.11	579	1.19
	171	185	1.02	554	1.14
	180	188	1.04	614	1.27
	203	232	1.28	640	1.32
Ringhals	16	145	0.80	325	0.67
	24	142	0.78	344	0.71
	25	142	0.78	347	0.72
	31	124	0.68	329	0.68
	37	150	0.83	381	0.79
	38	160	0.88	409	0.84
	46	145	0.80	379	0.78

such etches which have been produced by European laboratories have used the same supplier. Most work in the field of etched track, however, seeks to show that a particular style of processing is 'best', when in reality the results in this work show that it may be preferable to tailor the etch to the plastic: when considering signal to noise ratios as opposed to response functions it is not etches or types of PADC which are necessarily 'bad', but rather combinations of etches with plastics which work well.

A true understanding of the relationship between background levels, plastic type and etch process requires a detailed knowledge of the sources of background tracks. Atmospheric radon and its daughters are undoubtedly responsible for a component of the background tracks, but the complex relationship between cure cycle and the depth of the plastic which is studied, indicate that other mechanisms must be responsible for most of the tracks. When this problem is solved it may become possible truly to understand how to obtain optimum signal to noise ratios with a particular plastic.

ACKNOWLEDGEMENTS

The authors would like to thank Dr D. J. Thomas and staff at NPL for providing the 100–250 keV, 565 keV and 14.7 MeV neutron fluences.

REFERENCES

1. Bartlett, D. T., Steele, J. D. and Stubberfield, D. R. *Development of a Single Element Neutron Personal Dosemeter for Thermal, Epithermal and Fast Neutrons.* Nucl. Tracks **12**, 645–648 (1986).
2. Gilvin, P. J., Bartlett, D. T. and Steele, J. D. *The NRPB Neutron Personal Dosimetry Service.* Radiat. Prot. Dosim. **20**, 99–102 (1987).
3. Tanner, R. J., Gilvin, P. J., Steele, J. D., Bartlett, D. T. and Williams, S. M. *The NRPB PADC Neutron Personal Dosemeter: Recent Developments.* Radiat. Prot. Dosim. **34**(1/4), 17–20 (1990).
4. Piesch, E. and Urban, M. *Dosimetric Properties of Different CR-39 Plastics used as a Neutron Recoil Track Detector.* Nucl. Tracks **12**(1–6), 661–664 (1986).
5. Ahmad, S. and Stejny, J. *Polymerisation, Structure and Track Recording Properties of CR-39.* Nucl. Tracks Radiat. Meas. **19**, 11–16 (1991).
6. Tanner, R. J., Bartlett, D. T. and Hager, L. G. *The Energy and Angle Dependence of Neutron Response for PADC Manufactured using Two Different Cure Cycles.* Radiat. Meas. **25**(1–4), 467–468 (1995).
7. Currie, L. A. *Limits for Qualitative Detection and Quantitative Determination: Application to Radiochemistry.* Anal. Chem. **40**, 586–593 (1968).
8. Burgkhardt, B., Piesch, E. K. A. and Vilgis, M. *Uncertainty of Measurement and Lower Detection Limit of Track Etched Detector Systems: Experimental Verification and Consequences for Intercomparison Experiments.* Radiat. Prot. Dosim. **47**(1/4), 617–622 (1993).
9. Tanner, R. J. and Bartlett, D. T. *A New Criterion for Comparing Track Etch Dosimetry Systems.* Radiat. Meas. **25**(1–4), 445–448 (1995).
10. Adams, Jr, J. H. *A Curing Cycle for Detector Quality CR-39.* In: Solid State Nuclear Track Detectors. Eds P. Fowler and V. M. Clapham (Oxford: Pergamon Press) pp. 145–148 (1981).
11. Tanner, R. J., Bartlett, D. T., Williams, S. M., Steele, J. D. and Dean, S. F. *A Two Frequency Electrochemical Etch for Routine Processing of Neutron Dosemeters at the NRPB?* Nucl. Tracks Radiat. Meas. **19**(1–4), 217–220 (1991).
12. Bartlett, D. T., Tanner, R. J. and Steele, J. D. *NRPB Personal Neutron Dosemeter: Comparison of Etch Cycles in the 1988/89 EURADOS/CENDOS Irradiations.* In: Response of Proton-Sensitive Etched Track Detectors to Fast Neutrons: Results of a Joint Multilaboratory Experiment. Ed. H. Schraube. EURADOS/CENDOS Report 1990-01, GSF-Bericht 22/90 (1990).
13. Bartlett, D. T., Britcher, A. R., Bardell, A. G., Thomas, D. J. and Hudson, I. F. *Neutron Spectra, Radiological Quantities and Instrument and Dosemeter Responses at a Magnox Reactor and a Fuel Reprocessing Installation.* Radiat. Prot. Dosim. **44**(1/4), 233–238 (1992).
14. Lindborg, L., Bartlett, D. T., Drake, P., Klein, H., Schmitz, Th. and Tichy, M. *Determination of Neutron and Photon Dose Equivalent at Workplaces in Nuclear Facilities in Sweden.* Radiat. Prot. Dosim. **61**(1–3), 89–100 (1995).

THE PRESENT STATUS OF ETCHED TRACK NEUTRON DOSIMETRY IN EUROPE AND THE CONTRIBUTION OF CENDOS AND EURADOS

J. R. Harvey†, R. J. Tanner‡, W. G. Alberts§, D. T. Bartlett‡, E. K. A. Piesch|| and H. Schraube*
†Dosimetry Consultant, 9 Torchacre Rise, Dursley, Gloucestershire, GL11 4LW, UK
‡National Radiological Protection Board
Chilton, Didcot, Oxon OX11 0RQ, UK
§Physikalisch-Technische Bundesanstalt
Bundesallee 100, D-38116 Braunschweig, Germany
||Forschungszentrum Karlsruhe
Postfach 3640, D-76021 Karlsruhe, Germany
*GSF-Neuherberg, Institut für Strahlenschutz
D-85758, Oberschleissheim, Germany

Abstract — A working group, set up in 1981 under the auspices of CENDOS, to coordinate European activities in etched track neutron dosimetry remained active until 1994, in its later years under the auspices of EURADOS. The group facilitated the interchange of technical ideas amongst European groups, promoted the production of a summary report in 1984, organised a workshop in 1987, and arranged a joint irradiation with protons and five joint irradiations with neutrons over the period 1984–1992. Some key aspects of the work of the group are discussed as is the current status of etched track neutron dosimetry in Europe. The available systems and their dosimetric characteristics are considered. The systems are based on both chemical and electrochemical etching and the characteristics discussed include linearity, energy and angular dependence of response, lowest detectable doses and reproducibility. Future trends and research needs are identified.

INTRODUCTION

It was not until the late 1960s and early 1970s that the potential application of etched track techniques to neutron dosimetry was fully recognised. The interest was further increased when it was reported in 1978[1] that a commonly used plastic, CR-39 (a trade name for the polymer poly allyl diglycol carbonate, PADC) could be used to detect protons. This led to research activity at a number of laboratories in the early 1980s and ultimately to international cooperative activities. These were sponsored principally by two European organisations, CENDOS (the Cooperative European Research Project on Collection and Evaluation of Neutron Dosimetry Data) and EURADOS (the European Radiation Dosimetry Group).

HISTORICAL DEVELOPMENT OF CENDOS AND EURADOS ACTIVITIES

During 1981, CENDOS set up a working group on etched track detectors. This group, known as the 'Ion Recoil Neutron Dosimetry Group', promoted the development of solid state nuclear track detectors for neutron dosimetry within Europe, through meetings and joint irradiations. In 1985 CENDOS was combined with EURADOS and the 'Ion Recoil Neutron Dosimetry Group' was enlarged and became Committee Five on 'The Application of Track Detectors to Neutron Dosimetry'. The work of the group continued under the aegis of Committee 5 of EURADOS being associated for a period with work on ion chambers when the committee was entitled 'Basic Physical Data and the Characteristics of Radiation Protection Instrumentation'.

In 1991 the group became Working Group 8 of EURADOS and its membership and scope were increased to cover 'The development and improvement of techniques for the individual dosimetry of ionising radiation'. In 1994 the group was terminated in line with the new EURADOS policies. The authors of this report were asked to report on the work of the committee in its various guises and in addition to summarise the present state of etched track neutron personal dosimetry in Europe. The present report is a condensed version of a fuller report which is being produced.

ACTIVITIES OF THE GROUP

A very productive activity has always been the informal exchange of information between the various organisations represented on the working group. The first formal joint activity was to study the etch characteristics of tracks produced by monoenergetic protons generated in a Van der Graaff facility, the low fluxes required being produced by backscattering from a thin gold foil.

In 1984[2,3] two members of the group produced two comprehensive reports summarising the current status of etched track neutron dosimetry. The two documents which contain more than 200 references were recognised at the time to be an invaluable summary and are still frequently referred to. In 1987 the group organised an international workshop on etched track neutron dosimetry at Harwell[4].

A total of five joint neutron irradiations were organised by the working group, some being extended to scientists outside the European Union. The results from each joint irradiation were collated and published together with analyses of important aspects[5–9]. These joint irradiations proved invaluable to the participating laboratories and have been a key element lying behind much of the progress in this field.

The group was always conscious of the importance of spurious readings (backgrounds) which are often seen in unexposed dosemeters. The group accordingly organised two surveys of the background characteristics of participants' dosemeters[10,11]

THE PRESENT SITUATION IN EUROPE

Research into etched track based neutron dosimetry is being pursued in many countries in Europe including Denmark, France, Germany, Greece, Italy, Spain, Switzerland, the Czech Republic and the United Kingdom. Most groups are studying systems based on the detection of protons from hydrogenous radiators although at least two groups have the capability of detecting fission fragments from fissile radiators. There are at least four operational services which use etched track neutron dosimetry, two of which are in the United Kingdom, one in Italy and one in Switzerland. Trial services are in operation in Greece, Germany and Switzerland. The first commercial automatic etched track reading system is now available in the United Kingdom.

MATERIALS AND SYSTEMS

Many dosimetry centres have continued to use and improve electrochemical etch techniques. Typically, electrochemically etched pit sizes lie in the range 25 to 500 μm. These are, in the main, visible unaided and can be counted automatically by a number of simple, rapid, commercially available, comparatively cheap, image analysis systems. Several laboratories (amongst them ENEA Bologna, NRPB, FZK (formerly KfK)) and commercial dosimetry companies have used such systems reliably during the past ten years. Better discrimination against detritus or surface scratches, and the possibility of partial discrimination against background pits, can be obtained with the latest generation of high resolution CCD cameras, with single frame grab at high data rate, and the use of proven, modular, image analysis programs running on a PC. Such systems are available at relatively low cost and have been installed at, for example, GSF, PTB, and NRPB.

Several laboratories began their research in this field using the electrochemical etch parameters reported by Griffiths et al[12], namely AERE, GSF, ENEA Bologna, ENEA Rome, KfK (now FZK) and PSI. In general, the advantages of high sensitivity and very good proportionality of neutron energy response characteristics extending to below 100 keV, were to be set against the variable, and sometimes very large, background pit density. To overcome this, ENEA Rome and KfK sought to discriminate against non-radiation induced electrochemically etched pits on size, but with only limited success. Following the ENEA Rome development of 50 Hz etching, KfK devised a two frequency (two step) electrochemical etch procedure, with very good neutron energy range and improved signal to noise ratio. This approach gives excellent results providing the plastic is of suitably high quality. PTB developed a dosimetry system using this process.

The NRPB chose to develop etch and read procedures suited to commercially available plastic. A less stressful set of electrochemical etch parameters was established which led to the introduction in 1986 of an operational system with a minimum detectable dose of less than 200 μSv, relatively few outliers, and proportionality of reading of dose equivalent down to 150 keV[13]. The basis of the etch process was adopted by other Working Group laboratories: AERE, ENEA Bologna, and recently in part by GSF. In general the etch procedures in operational use have been modified and improved and research continues on electrochemical etching techniques to obtain better dosimetric characteristics.

Other groups, motivated by the desire to make the etch process simpler and cheaper, have developed systems which use chemical etch alone. One chemical etch system which uses accurate microscopic analysis of etch pit dimensions can be used to give spectrometric information in addition to personal dose[14]. Another, which uses a simple optical technique to increase the pit image size, obviating the need for electrochemical etch[15], is the basis of a commercial automatic reading system which is now being used or investigated by several dosimetry services.

Research on improving PADC manufacture did not lead to the commercial production of plastic of significantly better quality until the new cure cycles proposed by Ahmad and Stejny[16]. Use by the NRPB of PADC manufactured with one of the new cure cycles had led to consistently improved signal to noise ratios, with few outliers, for both single and two frequency electrochemical etching[17]. It should now be practical to use the latter technique operationally. Cellulose nitrate (e.g. LR-115) and the polycarbonate Makrofol have been studied and used principally in France and Germany.

PERFORMANCE OF SYSTEMS

Background levels and lowest detectable doses

The 800 PADC sheets accepted for use in the UK DRPS service had a mean background track density corresponding to 0.13 mSv with a standard deviation of 0.03 mSv. This is very similar to the standard deviation in background of the average sheet which passes the UK NRPB quality acceptance system (0.025 mSv).

The main problem is that the distribution of backgrounds about the mean is skewed to high doses. A mathematical model of a typical distribution of background readings has been developed by the NRPB. If the mathematical model is applied to the average and worst sheets to pass the NRPB quality acceptance system it will predict the expected rates of spurious doses greater than any given value. The predictions are shown in Figure 1. It can be seen that the frequency of significant spurious doses is not trivial if applied to a service issuing 10,000 dosemeters per year.

Linearity

All neutron dosimetry systems based on track counting are inherently linear to the extent permitted by the track counting statistics and the absence of edge and other distorting effects. An analysis of the results of the 1990 and 1992 joint irradiations would indicate that there are anomalies in some of the systems tested, although most of the systems were roughly linear over the important operational range from 0.5 to 5 mSv. For practical operational systems, lack of linearity is rarely a problem as long as the system response is accurately known over the entire dose range between low operational doses and what is often a relatively high dose used for routine calibration.

Sensitivity and reproducibility

A study of the results of the 1990 joint irradiation indicates that a typical system has a response of about 500 $cm^{-2}.mSv^{-1}$ in the neutron energy band of principal sensitivity. The fractional standard deviation of the disturbance of the readings of a number of dosemeters given the same dose cannot, on average, be less than $1/\sqrt{N}$, where N is the mean number of tracks in the counting area. For a counting area of 1 cm^2, therefore, the average percentage standard deviation of a series of measurements of 1 mSv cannot be less than 4.5%. The corresponding figure is 10% at 0.2 mSv. If the counting area is 4 cm^2 the reproducibility will be improved by a factor of two. Early data from operational systems suggest that the spread of readings experienced in routine calibrations have standard deviations which are a factor of two or three greater than these minimum figures.

Energy dependence

Examples of typical responses as a function of neutron energy are shown in Figure 2. The low energy threshold of a typical chemical etch system is around 200 keV[15] and of an electrochemical etch system is below 144 keV[8]. The low energy threshold displayed by an experimental system may not translate to a routine service and there is usually a trade-off between low energy threshold and average background level. A significant reduction in background can sometimes be attained by raising the low energy threshold. It is noted that in the environments around operating nuclear reactors, 10% or less of the dose equivalent may be associated with neutrons in the energy band between 100 keV and 300 keV[18], although proportions as high as 28% have been indicated[19].

Etched track detectors can be made sensitive to thermal neutrons by the addition of radiators containing boron, lithium, or nitrogen[13]. Nitrogen is already present in cellulose nitrate (and nuclear emulsion). Such detectors have advantages for use in albedo dosimetry since they have no photon sensitivity and can easily be combined with a fast neutron sensitive etched track dosemeter. One such device has sensitivity over a wide energy range[20].

Angular response

When the element or elements of a dosemeter lie in a single plane there is always a significant fall-off in sensitivity as the angle of incidence is increased. Typical angular responses, normalised to unity at 0°, are shown in Figure 3 where it can be seen that the response of planar dosemeters falls off more rapidly than required to measure $H_p(10)$. The errors stemming from poor angular response can be corrected for by the use of location specific correction factors based upon the angular and spectral characteristics of the neutron fields in which the dosemeters are worn. Derivation of the correction factors is not a trivial task and in some situations it can be more appropriate to use a dosemeter in which the angular response has been improved by the use of curved or multiple elements. Dosemeters have been developed with three elements[21] or more[14] which have improved angular response (see Figure 3). In general, dosemeters which have a lower fast neutron energy threshold have a better fast neutron angular response.

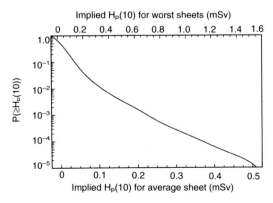

Figure 1. Probability of spurious doses greater than a given value of $H_p(10)$ occurring on undosed detectors.

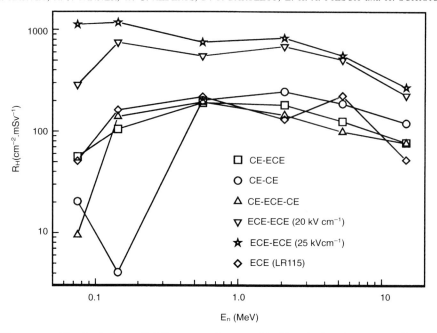

Figure 2. H*(10) response (closely similar to $H_p(10)$ response for normal incidence) of various systems for normally incident neutrons.

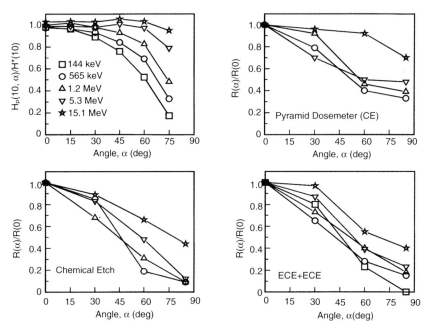

Figure 3. Quotient of fluence response at a given angle of incidence, α, to that at 0° (r): the ideal response is given by $H_p(10)/H^*(10)$, as shown in the top left figure. The two lower figures refer to planar dosemeters.

SUGGESTED AREAS OF RESEARCH

Although there have been significant improvements in etched track plastic production techniques, notably the recently devised cure cycles of Ahmad and Stejny, the polymerisation process in general and the origin of background tracks in particular are still imperfectly understood. Further research would be widely welcomed in this area since background levels are becoming increasingly significant as quality factors rise and dose levels fall.

Although improvements in energy response and sensitivity will always be welcome, greater improvements in overall accuracy are likely to stem from improvements in angular response and reduction in backgrounds. Attempts to improve angular response would be valuable since in many practical environments a significant fraction of the personal dose equivalent is associated with neutrons incident at oblique angles of incidence to the body surface.

The occasional high spurious reading produces problems in operational systems. This could be reduced by the use of multi-element dosemeters. An alternative approach is the use of a thermal neutron 'telltale' which is an etched track element (or part of an element) given very high thermal neutron sensitivity by means of an adjacent boron or lithium loaded radiator. Genuine fast neutron exposures can then be identified since they will always be associated with a measurable thermal neutron dose at the body surface, which will give a reading in the 'telltale' field. Both approaches may have operational disadvantages.

There are clearly general advantages in developing readers which read sub-areas within the overall etched track field. They can be used for example to identify those high background elements in which spurious tracks are restricted to part of the element. They can also be used for the detection of neutrons within defined energy regions such as thermal or very high energies.

It cannot be overemphasised that those involved in developing experimental etched track neutron dosimetry systems need to be aware of the situation which will arise if their system is eventually utilised in a routine service. In the experimental system each element can be carefully examined and assessed. The cost of the dosemeter and the reading system need not be considered. The implications of the occasional high background element can be ignored as can the effects of harsh environmental conditions. A routine service on the other hand needs to be able to assess large numbers of dosemeters quickly and economically, and the occurrence of occasional high background elements can have serious consequences. Many elements of reliable quality which are able to withstand the rigours of operational use, are hence required to make feasible an operational service. One routine regulatory monitoring service has been in successful operation for ten years. A continued requirement for a reliable, operationally robust, passive neutron dosemeter is envisaged.

ACKNOWLEDGEMENTS

The authors are indebted to the European Commission for funding cooperative activities, to EURADOS and CENDOS, to the irradiating laboratories, particularly PTB, Braunschweig, and GSF, Neuherberg, and to those members of the working group who have organised activities and published results. Chairmen, G. Portal and K. G. Harrison, played key rôles in the early years of the working group.

REFERENCES

1. Cartwright, B. G., Shirk, E. K. and Price, P. B. *A Nuclear-track Recording Polymer of Unique Sensitivity and Resolution.* Nucl. Instrum. Methods **153**, 457–460 (1978).
2. Tommasino, L. and Harrison, K. G. *Damage Track Detectors for Neutron Dosimetry: I. Registration and Counting Methods.* Radiat. Prot. Dosim. **10**, 207–218 (1984).
3. Harrison, K. G. and Tommasino, L. *Damage Track Detectors for Neutron Dosimetry: II. Characteristics of Different Detection Systems.* Radiat. Prot. Dosim. **10**, 219–235 (1984).
4. Bartlett, D. T., Booz, J. and Harrison, K. G. (Eds) *Etched Track Neutron Dosimetry.* Proc. Workshop on Etched Track Neutron Dosimetry, Harwell (UK) 12–14 May 1987. Radiat. Prot. Dosim. **20**(1–2) (1987).
5. Harrison, K. G. (Ed.) *Neutron Irradiations of Proton-sensitive Track Detectors: Results of a Joint Irradiation Organised by CENDOS.* AERE Harwell Report AERE R 11926. CENDOS Report 1985-02 (1985).
6. Piesch, E. K. A. (Ed.) *Neutron Irradiations of Proton-sensitive Track Etch Detectors: Results of the Joint European/USA/Canadian Irradiations.* Kernforschungszentrum Karlsruhe Report KfK 4305, EURADOS-CENDOS Report 1987-01 (1987).
7. Scharube, H. (Ed.) *Response of Proton-sensitive Etched Track Detectors to Fast Neutrons: Results of a Joint Multilaboratory Experiment Organised by EURADOS-CENDOS (1988/1989).* GSF-Bericht 22/90, EURADOS-CENDOS Report 1990-01 (1990).
8. Alberts, W. G. (Ed.) *Investigation of Individual Neutron Monitors on the Basis of Etched-track Detectors: The 1990 EURADOS-CENDOS Exercise.* PTB-Bericht PTB-N-10, EURADOS-CENDOS Report 1992-02 (1992).
9. Schraube, H., Alberts, W. G. and Weeks, A. (Eds) *Fast and High-Energy Neutron Detection with Nuclear Track Detectors: Results of the European Joint Experiments 1992/3.* GSF-Bericht 15/95, EURADOS Report 1995-1 (1995).

10. Lembo, L. (Ed.) *Results of a Survey of Backgrounds of Etched Track Neutron Dosemeters Organised by EURADOS-CENDOS in 1988*. ENEA Report PAS-FIBI-DOSI(89) 1 (1989).
11. Bartlett, D. T. (Ed.) *Results of a Survey of Backgrounds of Etched Track Neutron Dosemeters for PADC (CR39) from Different Manufacturers Organised by EURADOS-CENDOS in 1989/90*. NRPB Report NRPB-M342 (1992).
12. Griffiths, R. V., Thorngate, J. H., Davidson, K. J., Rueppel, D. and Fisher, J. C. *Monoenergetic Neutron Response of Selected Etch Plastics for Personnel Neutron Dosimetry*. Radiat. Prot. Dosim. **1**(1), 61–71 (1981).
13. Gilvin, P. J., Bartlett, D. T. and Steele, J. D. *The NRPB PADC Neutron Personal Dosimetry Service*. Radiat. Prot. Dosim. **20**, 99–102 (1987).
14. Worley, A., Fews, A. P. and Henshaw, D. L. *The Bristol University Neutron Dosimetry System*. Radiat. Prot. Dosim. **20**(1/2), 95–98 (1987).
15. Harvey, J. R., French, A. P., Jackson, M., Renouf, M. C. and Weeks, A. R. *An Automated Neutron Dosimetry System Based on the Chemical Etch of CR39*. Radiat. Prot. Dosim. **70**(1–4), 149–152 (This issue) (1997).
16. Ahmad, S. and Stejny, J. *Polymerisation, Structure and Track Recording Properties of CR39*. Nucl. Tracks Radiat. Meas. **19**(1–4), 11–16 (1991).
17. Tanner, R. J., Bartlett, D. T., Hager, L. G. and Lavelle, J. *Future Developments in the NRPB PADC Neutron Personal Monitoring Services*. Radiat. Prot. Dosim. **70**(1–4), 121–126 (This issue) (1997).
18. Delafield, H. J. and Perks, C. A. *Neutron Spectrometry and Dosimetry Measurements Made at Nuclear Power Stations with Derived Dosemeter Response*. Radiat. Prot. Dosim. **44**(1/4), 227–232 (1992).
19. Lindborg, L., Bartlett, D. T., Drake, P., Klein, H., Schmitz, Th. and Tichy, M. *Determination of Neutron and Photon Dose Equivalents at Workplaces in Nuclear Facilities in Sweden*. Radiat. Prot. Dosim. **61**, 89–100 (1995).
20. Luszik-Bhadra, M., Alberts, W. G., Dietz, E., Guldbakke, S. and Matzke, M. *A Wide-Range Neutron Dosemeter Based on a CR-39 Track Detector*. Nucl. Tracks Radiat. Meas. **22**, 671–674 (1993).
21. Weeks, A. R., Ford, T. D. and Harvey, J. R. *An Assessment of Two Types of Personal Neutron Dosemeter which Utilise the AUTOSCAN 60 Automatic Etch-Track Reading System*. UK Nuclear Electric, Technology Division Report TEPZ/REP/0084/93 (1993).

ETCHED TRACK SIZE DISTRIBUTIONS INDUCED BY BROAD NEUTRON SPECTRA IN PADC

J. Jakes, J. Voigt and H. Schraube
GSF-Forschungszentrum Neuherberg – Institut für Strahlenschutz
D-85758 Oberschleissheim, Germany

Abstract — Detectors based on the PADC plastic, made by Pershore and TASL, were irradiated with protons, alpha particles, and monoenergetic neutrons from an accelerator, and with broad-spectra neutrons from Am-Be and ^{252}Cf neutron sources. Latent tracks were etched in a two-step electrochemical procedure and measured using an image analysis system. Track size distributions were determined showing systematic changes with the quality of charged particle spectra.

INTRODUCTION

Because of the unique registration features of PADC plastic (this material is more commonly referred to as CR-39) for detecting neutrons in mixed neutron–gamma radiation fields, dosemeters based on this material became attractive for individual neutron dosimetry. For this purpose, however, predictable dosimetric characteristics are needed, meeting specific requirements regarding the lower limit of detection, and the dosemeter's response to the energy and angle of incident neutron radiation. Methods were developed permitting rough spectrometric measurements by evaluating separately thermal, intermediate and fast neutron components using suitable radiators, a multi-element technique, or a combination of both[1,2]. Extensive attention has been paid to the dosemeter's background and methods of its reduction[3]. In order to obtain satisfactory values of minimum detectable dose well below 0.1 mSv, the total background induced by alpha particles from radon and its progeny, and caused by mechanical defects, has to be reproducible within given statistics and kept as low as possible. There are different ways of minimising the background induced by alpha particles. One of them, as proposed by Hankins[4], is based on track size analysis. Experimental results published by Azimi-Garakani et al[5] and Bernardi et al[6] indicate a correlation between the track size and neutron energy. It has further been shown[7] that the evaluation of the PADC dosemeter's response in terms of etched track size distributions represents a promising way for subtracting the background consisting of spurious etch pits, caused by mechanical defects, and alpha radiation, without using a chemical pre-etch. Moreover, a qualitative estimate of the presence of neutrons with energies over 10 MeV is possible.

The present work was intended to study the track size distributions induced in PADC plastics by monoenergetic protons and alpha particles, and by quasi-monoenergetic neutrons and neutrons of broad energy spectra. Latent tracks were etched under strictly constant etching conditions. Track size distributions in plastics from different manufacturers were measured and evaluated.

EXPERIMENTAL PROCEDURES

Detectors used in this study were made of Stejny cured plastic[7], manufactured by Pershore Mouldings (PM500), and Tastrak plastic, manufactured by TASL. Both materials are characterised by a reasonably low background. All samples were laser cut and coded. For each energy, four samples in a stack were irradiated in a single run under the same experimental conditions. Bare samples were irradiated at normal incidence with 0.45 MeV protons and 1.26 MeV alpha particles produced in an accelerator. During irradiation, samples were exposed to a vacuum of 0.13 mPa (10^{-6} torr) for about 3 min. Another set of samples was irradiated with quasi-monoenergetic neutrons of energies of 0.57, 1.2, 5.3, and 15.1 MeV, respectively, with neutrons from an Am-Be source, and neutrons from bare and heavy water moderated ^{252}Cf. Samples exposed to neutrons were covered with a polyethylene protective foil, approximately 50 μm thick, mounted between two radiators made of 1.5 mm thick PMMA, and sealed in a dry nitrogen atmosphere inside an aluminised polyethylene sachet with a wall thickness of around 100 μm.

Two-stage electrochemical etching was used with 6N potassium hydroxide as an etchant at a temperature of 65°C. The first stage took 5 h at 100 Hz and was followed by the 2 kHz stage lasting 1 h. The applied electric field strength was 20 kV.cm^{-1}.

The area or diameter of a look-through projection of the dielectric treeing of an etched track onto the surface of a sample is most commonly referred to as the track size. Throughout this work, the diameter is used as the parameter for the track size. The track sizes were measured using an image analysis system (Elas 16, MueTec) that provides a complete set of image and data analysing functions, and is capable of fully automated scanning of the etched area and an unattended evaluation of 20 samples at a high processing speed. The track size distributions of single samples in each stack were then summed up to get a representative distribution for a particular particle and energy, and the frequency distributions of track sizes were calculated.

DISCUSSION

The results obtained show differences in the track sizes among particles of different quality. As can be seen from Figure 1, relatively narrow peaks represent the distributions for both cases, and the mean track size of alpha particles is systematically larger compared with that of protons. The number of tracks over a selected threshold can thus be used as a parameter indicating the presence of particles of a given quality. As given in Table 1, the number of tracks over 100 μm in diameter is higher in samples irradiated with alpha particles for both the materials studied in this work. This is consistent with results published on samples of American Acrylics exposed to alphas from radon and its progeny[8], and can be accounted for by differences in duration of the dielectric treeing. Although the track formation in the case of accelerator irradiated samples is a surface phenomenon, a similar effect is observed in samples exposed to neutrons, where charged particles and latent tracks arise within the whole irradiated volume.

A complete set of experiments was performed on PM500 plastic. Before irradiating with the broad spectra neutrons, samples were exposed to the quasi-monoener-

Table 1. Probabilities of occurrence of tracks with diameters over 100 μm.

Charged particles	Probability	
	PM500	TASTRAK
Protons 0.45 MeV	2.22×10^{-1}	6.14×10^{-2}
Alphas 1.26 MeV	7.13×10^{-1}	5.02×10^{-1}

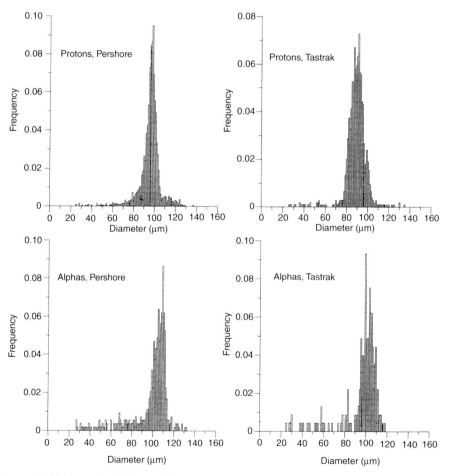

Figure 1. Frequency distributions of track sizes of electrochemically etched tracks induced by protons and alphas, respectively, in PM500 and Tastrak plastics.

ETCHED TRACK SIZE DISTRIBUTIONS INDUCED BY BROAD NEUTRON SPECTRA IN PADC

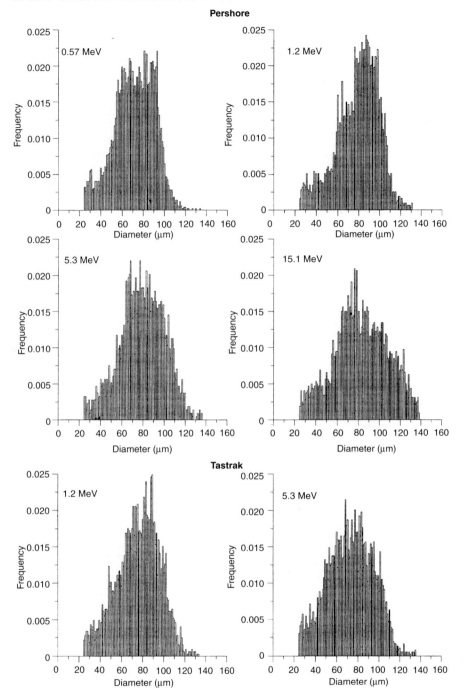

Figure 2. Frequency distributions of track sizes of electrochemically etched tracks induced by 0.57, 5.3, and 15.1 MeV neutrons, respectively, in PM500 and Tastrak plastics.

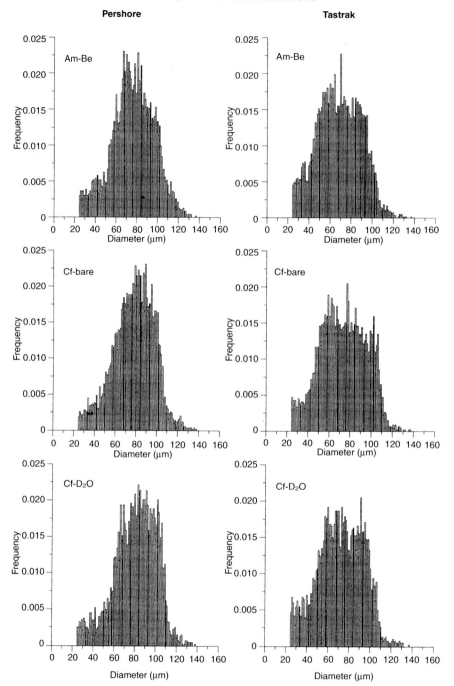

Figure 3. Frequency distributions of track sizes of electrochemically etched tracks induced by neutrons of broad spectra in PM500 and Tastrak plastics.

getic neutrons. Systematic changes in the shape of the distributions with increasing neutron energy from 0.57 MeV to 15.1 MeV were observed, which is consistent with the results published on American Acrylics[8] and is accounted for by a higher number of heavy recoils and the presence of alpha particles in charged particles spectra of 15.1 MeV neutrons. Changes in the distributions induced by 1.2 MeV and 5.3 MeV neutrons are only due to a higher contribution of heavy recoils and were observed to be insignificant. For this reason, supplementary experiments on Tastrak plastics were performed confirming results gained on PM500 material. Particularly in this energy region, etching conditions have to be carefully selected and kept constant.

The track size distributions induced by the broad spectra neutrons indicate the presence of protons, alphas, and heavy recoils. The number of tracks over selected thresholds lie on average between the same parameters for 1.2 MeV and 5.3 MeV neutrons. Distributions induced in Tastrak seem to be broader as compared with that induced in Stejny cure material.

CONCLUSIONS

The results obtained on PADC materials produced by different manufacturers showed a dependence of the track size on the radiation quality and proved to provide a method that can be used to distinguish between low and high energy neutron fields. This is true, provided there are low track densities, both for the neutron irradiated samples, where the latent tracks are homogeneously distributed throughout the irradiated volume, and for the plastics irradiated by the charged particles, where the occurrence of the latent tracks is a surface phenomenon. The latter was confirmed by the systematic shift in the peak position in proton and alpha particle irradiated samples. The evaluation of the track size distributions of electrochemically etched tracks without applying additional radiators could hardly be used in neutron spectrometry; however, this can be used in detecting the presence of heavy recoils and alpha particles exceeding the natural background. Careful selection of etching conditions and their consistent reproducibility is required.

ACKNOWLEDGEMENTS

The authors are grateful to Dr H. Hietel of the Institut für Strahlenschutz, GSF Neuherberg, for his support and fruitful discussions during irradiation on the accelerator.

Table 2. Probabilities of occurrence of tracks with diameters over selected threshold. A = 100 μm; B = 120 μm.

Neutron energy and spectrum	Probability			
	PM500		TASTRAK	
	A	B	A	B
0.57 MeV	3.79×10^{-2}	1.01×10^{-3}	—	—
1.2 MeV	1.25×10^{-1}	8.60×10^{-3}	1.03×10^{-1}	5.53×10^{-3}
5.3 MeV	1.61×10^{-1}	1.41×10^{-2}	1.12×10^{-1}	7.15×10^{-3}
15.1 MeV	2.62×10^{-1}	7.69×10^{-2}	—	—
Cf-bare	1.10×10^{-1}	1.05×10^{-2}	1.22×10^{-1}	5.40×10^{-3}
Cf-D$_2$O	1.68×10^{-1}	9.94×10^{-3}	7.39×10^{-2}	5.96×10^{-3}
Am-Be	1.12×10^{-1}	8.84×10^{-3}	5.04×10^{-2}	3.22×10^{-3}

REFERENCES

1. Bartlett, D. T., Steele, J. D. and Stubberfield, D. R. *Development of a Single Element Neutron Personal Dosemeter for Thermal, Epithermal and Fast Neutrons.* Nucl. Tracks **12**, 645–648 (1968).
2. Luszik-Bhadra, M., Alberts, W. G., Dietze, G. and Guldbakke, S. *Aspects of Combining Albedo and Etched Track Techniques for Use in Individual Neutron Monitoring.* Radiat. Prot. Dosim. **46**, 31–36 (1993).
3. *Results of a Survey of Backgrounds of Etched Track Neutron Dosemeters for PADC (CR-39) from Different Manufacturers.* Ed. D. T. Bartlett. Organised by EURADOS-CENDOS in 1989/90. NRPB M342 (1990).
4. Hankins, D. E. *Lawrence Livermore National Laboratory Results for the CENDOS-EURADOS Joint Irradiation Programme 1988/89.* In: Results of a Joint Multi-laboratory Experiment. Ed. H. Schraube. GSF-Bericht 22/90, 4/15-4/19 JSSN 0721-1694 (1990).
5. Azimi-Garakani, D., Boschung, M. and Wernli, G. *Track-Size Distribution of Electrochemically Etched CR-39.* Nucl. Tracks Radiat. Meas. **19**(1–4), 443–444 (1991).
6. Bernardi, L., Tommasino, L., Torri, G. and Azimi-Garakani, D. *Results of the Third EURADOS-CENDOS Joint Neutron Irradiation: ENEA/DIRECTORATE of Nuclear Safety and Radioprotection.* In: Results of a Joint Multilaboratory Experiment. Ed. H. Schraube. GSF-Bericht 22/90, 4/9-4/14 JSSN 0721-1694 (1990).

7. Ahmad, S. and Stejny, J. *Polymerisation, Structure and Track Recording Properties of CR-39.* Nucl. Tracks Radiat. Meas. **19**(1–4), 11–16 (1991).
8. Jakes, J., Voigt, J. and Schraube, H. *Evaluation of the CR-39 Response on the Basis of Track-size distributions.* Nucl. Tracks Radiat. Meas. **25**(1–4), 437–440 (1995).

ANALYSIS OF ACCEPTANCE TESTING DATA FOR MORE THAN 800 SHEETS OF CR-39 PLASTIC ASSESSED FOR THE DRPS APPROVED NEUTRON DOSIMETRY SERVICE

M. Jackson and A. P. French
Defence Radiological Protection Service (DRPS), Institute of Naval Medicine
Alverstoke, Hampshire, PO12 2DL, UK

Abstract — The Defence Radiological Protection Service (DRPS) Approved Dosimetry Service issues over 1200 planar CR-39 neutron dosemeters to the Ministry of Defence (Navy) each month. Results of sheet acceptance and calibration data suggest that average sheet background track densities have not varied significantly over the two-year period of the investigation. Average sheet sensitivity values have decreased significantly over the same period due to possible variations in the ageing characteristics of the CR-39 plastic. The data also suggest that sheets with higher initial sensitivities do not age as significantly as sheets with lower initial sensitivity values. The consequence of lowering the neutron dosemeter's limit of detection was investigated. Results indicate there would be an increase in the number of sheets rejected for use in the neutron dosimetry service by a factor of 2.5.

INTRODUCTION

The CR-39 plastic used in the DRPS Approved Neutron Dosimetry Service is supplied by Track Analysis Systems Limited (TASL, Bristol, England) and is known commercially as Tastrak PADC. The plastic is chemically etched and analysed on the Autoscan 60 reader which has been described previously[1-3]. It is essential that only high quality CR-39 plastic is used in a neutron dosimetry service as spuriously high background track density readings could give rise to neutron dose results significantly above the planar neutron dosemeter's limited of detection (0.2 mSv).

The DRPS undertakes acceptance testing on each sheet of Tastrak PADC plastic which is based on the assessment of sheet sensitivity and background track densities. This paper reviews the acceptance testing results for over 800 sheets of CR-39 plastic. The variation in the average background track density for 56 batches of Tastrak plastic have been reviewed together with the sheet sensitivities for 45 of the 56 batches. The data have also been used to review the impact of reducing the dosemeter's limit of detection (LOD) from 0.2 mSv to 0.1 mSv.

METHOD

Sheet acceptance tests

The DRPS undertakes routine acceptance testing of each sheet of the CR-39 plastic. The testing comprises randomly selecting eleven of the one hundred and eight etched track detectors that are available from each sheet. Nine detectors are used to determine the average sheet background track density together with a value for the associated standard deviation. The other two detectors are used to determine an approximate sheet sensitivity. A more accurate assessment of the sheet sensitivity is determined at a later date if the sheet has 'passed' the DRPS acceptance test criteria. For sheets that have passed acceptance testing, a further six detectors are selected at random and irradiated midway through the three-month issue period to a personal dose equivalent, $H_p(10)$ of 7.5 mSv using a 111 GBq ^{241}Am-Be calibration source. The sensitivity is assessed at the midpoint of the issue period in order to minimise any errors associated with sheet ageing effects, for example, loss of sheet sensitivity. The ageing characteristics have been investigated previously by the DRPS[4] and have been found to be dependent on the sheet of CR-39 plastic.

DRPS sheet acceptance test criteria

Sheet acceptance parameters are used to reject sheets with:

(i) spuriously high background elements,
(ii) high average background track densities,
(iii) low sensitivities,
(iv) two initial sensitivity elements irradiated to 7.5 mSv with track densities which differ by more than 380 tracks.cm^{-2}.

The first criterion is met by ensuring that the sheet acceptance parameter, A in mSv is less than 0.2 mSv for each sheet of plastic where A = 2B/S. B represents the standard deviation associated with the average sheet background and S represents the initial sheet sensitivity estimated from the two elements irradiated to 7.5 mSv. The sheet acceptance parameter ensures that sheets with spuriously high background elements are rejected as they have high values for the standard deviation of the background track density. Additionally, sheets are rejected if the average background track density exceeds 150 tracks.cm^{-2}, the sheet sensitivity is less than 300 tracks.cm^{-2}.mSv^{-1} or the difference between initial calibration elements (irradiated to a personal dose of 7.5 mSv) from any given sheet exceeds 380 tracks.cm^{-2}. The value for the maximum background track density has been obtained from operational experience that sheets with high average track densities have a high probability of containing elements with spuriously high background counts. The allowed sheet sensitivity variation (380 tracks.cm^{-2}) is based on a 95% confidence level and has been derived from the analysis of detectors from ten sheets of CR-39 plastic irradiated to a personal dose equivalent, $H_p(10)$ of 7.5 mSv. A difference in track density which exceeds 380 tracks.cm^{-2} is considered to be due to one or both of the elements having spuriously high background track density readings.

RESULTS AND DISCUSSION

Variation of batch background track density

For accepted sheets, the variation in the average background track density over the 56 batches of Tastrak PADC plastic is detailed at Figure 1. The average background track density of the 697 sheets is 54 tracks.cm^{-2} with a standard deviation of 14 tracks.cm^{-2}. Individual sheet background track densities varied from 14.8–150 tracks.cm^{-2}. There is a very strong correlation between the average sheet background track density and the background standard deviation for sheets that have failed the DRPS acceptance testing criteria which is shown at Figure 2, the correlation coefficient being 0.9. The data support the acceptance test criteria that sheets with elements that have high average background track densities (greater than 150 tracks.cm^{-2}) will have a high probability of containing rogue background elements. There is a much weaker correlation between the average sheet background track density and standard deviation for sheets that have passed the acceptance test criteria. This indicates that the acceptance test criteria adopted at the DRPS is successful in screening out poor quality sheets. It is essential that poor quality sheets are rejected, as elements with high background track densities could give rise to significant anomalous doses. For example, the maximum rogue background track density observed from the 858 sheets analysed was 2012 tracks.cm^{-2} which equates to a personal dose, $H_p(10)$ of 6.8 mSv.

Number of sheets failing acceptance testing criteria

The total number of sheets failing the acceptance testing criteria was 161 out of 858 (19%). The percentage of sheets in a batch of plastic which failed acceptance testing criteria varied from 0% to 62%. The DRPS has undertaken a review of the number of rejected sheets from a batch and the average batch background track density. There was no correlation between these two factors suggesting that the problems relating to poor quality CR-39 plastic are dependent on the manufacturing process for each sheet rather than the whole batch.

TASL quality assurance data from certificate of conformity

The CR-39 plastic from TASL is supplied with a quality assurance Certificates of Conformance. The information detailed on the certificates includes the number of tracks of a certain size (6–17 μm or 17–40 μm) counted per unit area on small amounts of Tastrak PADC plastic after exposure to an alpha emitting source. The plastic is chemically processed for either one hour at 98°C or six hours at 75°C and analysed on the TASL reading system[5]. Whilst these data could be useful for radon dosimetry services, the DRPS have not found any correlation between the quality assurance data provided by TASL and the DRPS sheet acceptance testing results. This is not entirely surprising because of the different calibration sources and reading systems employed by TASL and DRPS[3,5].

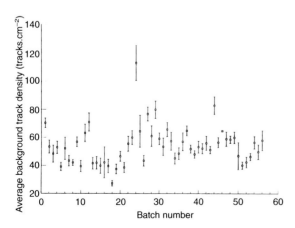

Figure 1. Variation of average background track density with batch of Tastrak PADC for sheets which have *passed* acceptance testing criteria.

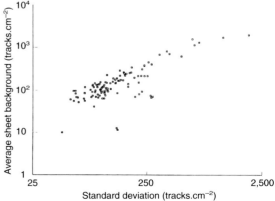

Figure 2. Correlation between the average sheet background and the background standard deviation for sheets which have *failed* acceptance testing criteria.

Variation of sheet sensitivity with batch of CR-39 plastic

The variation of the average sheet sensitivity between different batches of Tastrak PADC for sheets which have passed acceptance testing criteria are detailed in Figure 3. The sheet sensitivity values have been determined at the mid-point of the neutron dosemeter issue period. The average batch sheet sensitivity is 429 tracks.cm^{-2}.mSv^{-1} with a standard deviation of 90 tracks.cm^{-2}.mSv^{-1}. The results indicate that there is a considerable amount of variation in the average sheet sensitivity readings for the 45 batches of plastic analysed. There has been an apparent decrease in the average sheet sensitivity, determined at the mid-point of the issue period, over a period of two years. This could be due to a number of reasons:

(i) there has been a general decrease in the initial sensitivity of the CR-39 plastic,
(ii) there may have been a gradual change in the manufacturing process which has resulted in an increase in the ageing characteristics of the plastic resulting in reduced sensitivity determined at midpoint of the issue period.

Figure 4 suggests that there has been some apparent decrease in the initial sensitivity of the CR-39 plastic from batch 38 onwards. These data are approximate values for the sheet sensitivity and should be treated with caution as they have been determined from a small number of elements. The mid-issue sheet sensitivity also suggests that there has been an apparent decrease in sensitivity from batch 38 onwards. The data suggest that it is a combination of a reduced initial sensitivity and increased ageing which are contributing to this effect. The DRPS are currently investigating these observations with the manufacturer. The decrease in sheet sensitivity observed is not due to variations in the Autoscan 60 reader sensitivity. Quality assurance procedures used at the DRPS ensure the stability of the reader, standard elements are used to establish the variation (if any) of the reader on a daily basis.

Ageing effects of Tastrak PADC plastic

The average percentage decrease in the sheet sensitivity of the 45 batches of plastic, analysed after the etched track detectors had been sealed inside a radon proof pouch for $1\frac{1}{2}$ months was 33% with a standard deviation of ±14%. The amount of ageing varied from 8% to 58% over the batches of plastic investigated. All sheets of Tastrak PADC were of a similar age when acceptance tests were undertaken. The relatively large percentage standard deviation is possibly due to the fact that the percentage ageing is determined by comparing the sheet sensitivity value calculated from a very small number of etched track detectors used during acceptance testing criteria to that obtained when the detectors are calibrated midway through a three-month issue period. The data indicate that the later batches of CR-39 plastic have demonstrated greater ageing effects. Figures 3 and 5 suggest that sheets with higher initial sensitivities have better ageing characteristics. There is a correlation of 0.7 between these two parameters. The DRPS will be undertaking further work to investigate whether the sheet acceptance testing criteria could be refined: rejecting sheets that have a higher probability of ageing significantly over the issue period by increasing the minimum allowed sheet sensitivity parameter.

Limit of detection of the DRPS planar neutron dosemeter

The DRPS has also undertaken a review of the implications of reducing the limit of detection (LOD) of the neutron dosemeter to anticipate future monitoring

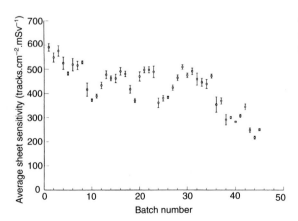

Figure 3. Variation in sheet sensitivity based on calibration 6 weeks after issue with batch Tastrak PADC for sheets which have *passed* acceptance testing criteria.

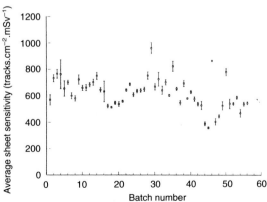

Figure 4. Variation of initial sheet sensitivity based on calibration before issue with batch of Tastrak PADC for sheetes which have *passed* acceptance testing criteria.

requirements as a result of the ICRP 60 recommendations for dose limits[6]. If the LOD for the neutron dosemeter was reduced to 0.1 mSv, the sheet failure rate would increase from approximately 20% to 50%. The requirement for a lower limit of detection would therefore have significant cost implications as it would increase the number of sheets failing acceptance tests by a factor of approximately 2.5.

CONCLUSIONS

The main difficulty with providing a reliable neutron dosimetry service based on CR-39 detectors is the quality of the plastic. The DRPS have developed a combination of parameters to ensure that the quality of the CR-39 plastic meets the strict requirements for a personal neutron dosimetry service. A review of the sheet acceptance data for CR-39 sheets failing the acceptance tests suggests that a maximum limit on the average background track density must be included as this parameter has a high correlation with the standard deviation for the sheet background track density. A high value for the standard deviation is undesirable because it indicates that there is a high probability that the sheet will contain rogue elements. Reassuringly, for passed sheets there is very little correlation between the value for the average sheet background track density and its associated standard deviation.

There was no correlation between the average batch background track density for accepted sheets and the number rejected from the batch, which suggests that the problems relating to the quality of the plastic are sheet rather than batch dependent.

Due to the nature of the tests undertaken by the manufacturer, the quality assurance Certificates of Conformance would probably be useful for radon services, however, the DRPS have found the data to be of little use for the neutron dosimetry service.

The initial and mid-issue sensitivity assessments suggest that latterly there has been an overall reduction in both. The data suggests that the reduction in the midpoint sensitivity is due to a combination of effects of a lower initial sensitivity and an increased ageing of the plastic. Further work will be undertaken with the supplier to investigate whether the data can be correlated to storage, manufacturing conditions or mixing of the plastic constituents.

The results confirm other neutron dosimetry service findings[7] that the physical characteristics of the CR-39 plastic, in particular background track density, sensitivity and ageing is variable and must be constantly monitored. Further development work into the manufacturing process of CR-39 plastic is required to address these issues so that the CR-39 neutron detection system can meet future monitoring requirements, which may also include a requirement for a lower LOD. Currently, the only practical method for reducing the LOD would result in an increase in the number of rejected sheets by a factor of 2.5.

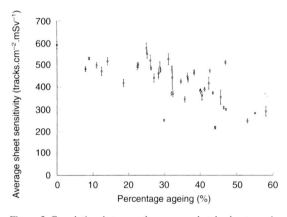

Figure 5. Correlation between the average batch sheet sensitivity and the percentage ageing effect.

REFERENCES

1. Weeks, A. R., Ford, T. D. and Harvey, J. R. *An Assessment of Two Types of Personal Neutron Dosemeter which Utilize the AUTOSCAN 60 Automatic Etch-Track Reading System.* Nuclear Electric Report TEPZ/REP/0084/93.
2. Wilson, M., French, A., Harvey, J. R. and Wernli, C. *An Assessment of Thermal Neutron Sensitivity Dosemeters Based on a Commercial Etched Track Reading System.* In: Strahlenschutz: Physik und Messtechnik (W. Koelzer and R. Maushart) **2**, 529–534 (1994).
3. Harvey, J. R., French, A. P., Jackson, M., Renouf, M. C. and Weeks, A. R. *An Automated Neutron Dosimetry System Based on the Chemical Etch of CR39.* Radiat. Prot. Dosim. **70**(1–4), 149–152 (This issue) (1997).
4. Jackson, M., French, A., Harvey, J. R. and Weeks, A. R. *Some Environmental Trial Results for a Planar and Isotropic Neutron Dosemeter.* Radiat. Meas. **25**(1–4), 461–462 (1995).
5. Worley, A., Fews, A. P. and Henshaw, D. L. *The Bristol University Neutron Dosimetry System.* Radiat. Prot. Dosim. **20**(1–2), 95–98 (1987).
6. ICRP. *1990 Recommendations of the International Commission on Radiological Protection.* Publication 60. Ann ICRP **21**(1–3) (Oxford: Pergamon) (1990).
7. Gilvin, P. J., Bartlett, D. T. and Steele, J. D. *The NRPB PADC Neutron Personal Dosimetry Service.* Radiat. Prot. Dosim. **20**(1–2), 99–102 (1987).

POLYCARBONATE TRACK DETECTORS WITH A FLAT ENERGY RESPONSE FOR THE MEASUREMENT OF HIGH ENERGY NEUTRONS AT ACCELERATORS AND AIRFLIGHT ALTITUDES

K. Józefowicz†, B. Burgkhardt‡, M. Vilgis‡ and E. Piesch‡
†Institute of Atomic Energy (IEA)
05-400 Otwock-Świerk, Poland
‡Karlsruhe Research Center (FZK)
POB 3640 D-76021 Karlsruhe, Germany

Abstract — For the polycarbonate detector routinely used at FZK, the two-step electrochemical etching was finally optimised after changing the detector thickness from 300 μm to 500 μm and applying a field strength of 42 kV.cm^{-1}. On the basis of calibrations with monoenergetic neutrons in the neutron energy range from 2 to about 70 MeV the response of Makrofol is now 130 tracks.cm^{-2}.mSv^{-1} within ± 30%. The linearity of detector response has been checked. The increase of the track density due to the neutron component of the natural background radiation at ground level has been measured during 1.5 y. Additional calibrations were performed in broad neutron beams at the PSI and CERN. The detector is suitable for monitoring the long-term occupational exposure at high energy particle accelerators and the exposure of the aircrews with a lower limit of detection 0.2 mSv per year. Polycarbonate is the only neutron detector which is insensitive to the high energy proton component of cosmic radiation.

INTRODUCTION

Polycarbonate detectors were the first etched track detectors able to record the neutron induced recoils and alpha particles. Among them Makrofol polycarbonate, manufactured by Bayer AG, is a commercial material, produced in large batches of good quality and reproducibility.

The application of chemical etching made possible the detection of neutrons with an energy threshold of about 1.5 MeV[1]. Later on a combined chemical and electrochemical etching (ECE) was introduced in 1978[2]. This technique was adopted by the personnel monitoring service at the FZK. For a large scale routine application an electrochemical etching system was developed, consisting of a high voltage/high frequency generator and a system of etching cells. This first commercially available system is now used worldwide by different laboratories applying ECE for polycarbonates and CR-39.

In the 1980s, this ECE technique was improved by adopting the two-step ECE technique[3]. For Makrofol DE of 300 μm thickness it was found that a field strength higher than 27 kV.cm^{-1} not only improved the neutron response and energy dependence, but also increased the background significantly. With the change of the detector thickness from 300 μm to 500 μm it was possible to apply a higher field strength. In comparison to the first ECE regime applied at FZK, the recent optimisation increased the neutron sensitivity of polycarbonate by one order of magnitude and significantly extended the energy range.

OPTIMISED EXPERIMENTAL PROCEDURE

Several kinds of Makrofol have been examined; Makrofol DE 1-4 475 has been chosen as the best for further experiments. This material, of thickness of about 490 μm has been used for the final optimisation of the two-step ECE technique. The background and the response of Makrofol to ^{252}Cf and ^{241}Am-Be neutrons were examined for field strength ranging from 35 to 51 kV.cm^{-1} (Figure 1).

With respect to a low background, the optimum was found at 42 kV.cm^{-1}. For this field strength the optimisation of the first ECE step, at 100 Hz, has been done. The removed layer, the track diameter, the background as well as the response to ^{252}Cf and ^{241}Am-Be neutrons have been investigated as a function of etching time, resulting in an optimum at 20 μm of the removed layer, after 5 h, when the saturation effect can be observed (Figure 2). The second step of ECE, used for enlarging the tracks up to diameters of 300 μm maximum, was 1 h at 2 kHz. The new optimised conditions of the two-step ECE for 500 μm thick Makrofol are given in Table 1.

NEUTRON ENERGY DEPENDENCE

In order to examine the energy dependence of the neutron response the Makrofol detectors were irradiated with monoenergetic neutrons in the range from 0.3 MeV to 19 MeV at the PTB Braunschweig, and with ^{252}Cf and ^{241}Am-Be neutrons at the FZK and IEA. Additional calibration irradiations were performed at the Eurados Intercomparison in 1992 at PSI Villigen with 44 MeV and 66 MeV neutron beams[4]; the results of measurements were corrected for the low energy neutrons contribution. On the basis of the optimised ECE conditions (Table 1) the resulting neutron fluence response (in tracks per 10^6 neutrons) and response to ambient neutron dose equivalent, $H^*(10)$, according to ICRP 60[5], are presented in Figures 3 and 4, respectively. In the neutron energy range between 2 and about 70 MeV the dose equivalent response of Makrofol DE, 490 μm thick, has been found to be 130 tracks.cm^{-2}.mSv^{-1} and independent of energy within ± 30% (see Figure 4).

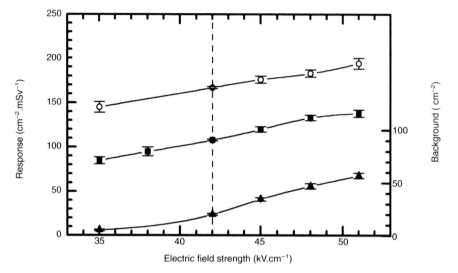

Figure 1. Background and neutron response of Makrofol DE as a function of electric field strength. (■) Cf, (○) Am-Be, (▲) background.

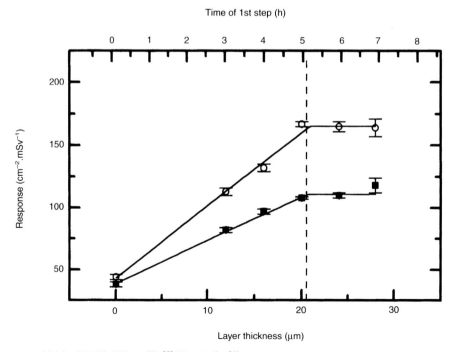

Figure 2. Response of Makrofol DE 475 to (■) ^{252}Cf and (○) ^{241}Am-Be neutrons for various thickness of the material layer removed in the first step of ECE.

Polycarbonate foils of about 500 μm thickness can also be bent easily into a cylindrical shape in order to reduce the angular dependence of the response.

HIGH ENERGY NEUTRONS

The dosimetric properties of Makrofol DE have also been confirmed by intercomparison irradiations using broad high energy neutron spectra at the PSI and CERN. Table 2 shows the ratio of ambient neutron dose equivalent measured with Makrofol and CR-39 track detectors, H_M, and the reference value. The reference dose, H_{ref}, means the reading of the TEPC; track detectors were calibrated with ^{252}Cf neutrons.

The results for the CERN irradiations in the hadron field behind the concrete shield are of special interest. With respect to the LET distribution this spectrum has been found to be similar and representative for the neutron spectrum of cosmic rays in typical airflight altitudes of 10–14 km[6]. The H_M/H_{ref} ratio for Makrofol DE has been found to be 0.83[7], compared to the value of 0.35 for CR-39, using an optimised two-step ECE.

Behind the iron shield about 90% of the neutron fluence is below 20 MeV. Nevertheless, CR-39 underestimates H_{ref} with 0.53, and the Makrofol result is about 55% of CR-39 result.

MONITORING OF COSMIC RADIATION

Due to the low background track density of about 15 cm^{-2} and the sufficient sensitivity for the cosmic neutron spectrum, Makrofol DE detectors offer the possibility of measuring the neutron component of the natural background radiation at ground level as well as at airflight altitudes.

The increase of the background has been examined during a period of 1.5 y for one batch of Makrofol, stored in the laboratory in vacuum sealed aluminium/plastic covers (to protect them from α particles from radon and its decay products). The results in Figure 5

Table 1. Optimised two-step ECE conditions for Makrofol DE 475.

Temperature:	40°C		
Etching solution:	50% 6N KOH + 50% ethanol		
1st step:	42 kV.cm^{-1},	100 Hz,	5 h
2nd step:	42 kV.cm^{-1},	2000 Hz,	1 h
post-etching (PE)			0.5 h

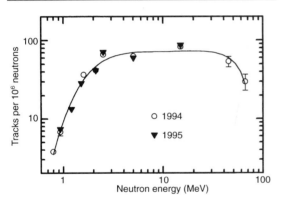

Figure 3. Energy dependence of the fluence response of Makrofol DE 475. (The line is an eye guide only.)

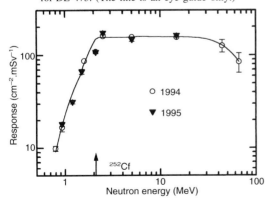

Figure 4. Response of Makrofol DE 475 to the ambient neutron dose equivalent H*(10), according to ICRP 60[5].

Table 2. Relative reading H_M/H_{ref} of etched track detectors in high energy neutron radiation fields with broad neutron energy spectrum at the PSI and CERN.

	Neutron field	H_M/H_{ref}	
Energy	Fluence fraction	Makrofol	CR-39
Accelerator PSI			
44 MeV	50% < 24 MeV	1.08	0.41
66 MeV	50% < 36 MeV	0.93	0.39
Hadron field CERN			
behind concrete	50% < 20 MeV	0.83	0.35
behind iron	90% < 20 MeV	0.29	0.53

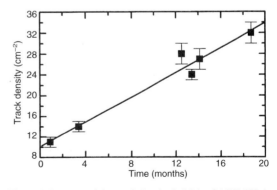

Figure 5. Increase of the track density in Makrofol DE 475 due to neutrons from cosmic rays at ground level.

show the increase in the background of 14 ± 2 tracks.cm^{-2} in one year, which is equal to about 0.1 mSv. Measurements performed with passive rem counters of a 30 cm diameter PE sphere[8] resulted in values of about 0.06 mSv.y^{-1}. This value is lower than the Makrofol value because of an underestimation of the high energy neutron component above 10 MeV. Regulla and David[9], comparing in-flight measurements during intercontinental flights, found the ratio of readings of Anderson-Braun rem counters to TEPC between 0.42 and 0.7.

The dose rate of the cosmic neutron component increases with the altitude above ground and the geomagnetic latitude to values higher than $2~\mu$Sv.h^{-1}. Experiments during intercontinental flights have shown that neutron doses between 1 and 5 mSv per year are expected for the aircrews, based on a total flight time of 1000 hours per year.

With respect to an application at airflight altitudes, Makrofol DE is insensitive to protons and thus to the high energy proton component of the cosmic ray spectrum.

The expected neutron response of our Makrofol detector in airflight altitudes has been found to be 86% of the ^{252}Cf response[10] by folding the response in Figure 3 with the measured cosmic neutron spectrum at 12.3 km altitude, by Hewitt *et al*[11], tabulated in the IAEA Compendium 318[12]. The data about anisotropy of cosmic neutrons spectra are in preparation[10] but not yet available. The angular response of Makrofol has been investigated in the lower energy range[13] (Figure 6). The use of a combination of two or three detectors arranged perpendicular to each other may provide an isotropic response.

Makrofol detectors are cheap and promising passive detectors for long-term monitoring of the air crews. Within an EC in-flight measurement programme the Makrofol detector has been applied for long term experiments at high airflight altitudes[14].

LOWER DETECTION LIMIT: LINEARITY

The consistent batch quality of the material, with respect to response and low background of about 15 ± 3 cm^{-2} results in a lower detection limit of about 0.15 mSv which is comparable with that of good CR-39 systems[15]. The change of the BG track density from batch to batch has been found to be between 10 and 35 tracks.cm^{-2} depending also on the storage period between production and application (see Figure 5).

The lowest level of detection, H_{LLD}, defined here as 3 times the background standard deviation, is shown in Figure 7 as a function of neutron energy for CR-39 (two values of electric field strength) and Makrofol DE (optimised ECE). The comparison shows the advantage of CR-39 in the low neutron energy range, but confirms the excellent performance of Makrofol for energies above 2 MeV.

In order to check the linearity of track counting, Makrofol detectors have been irradiated with ^{252}Cf neutrons, the doses ranging from 0.1 to 20 mSv. Up to about 600 tracks.cm^{-2} the observed track densities were close to the expected ones. For higher track densities the correction for non-linearity is about 20% at 20 mSv for visual

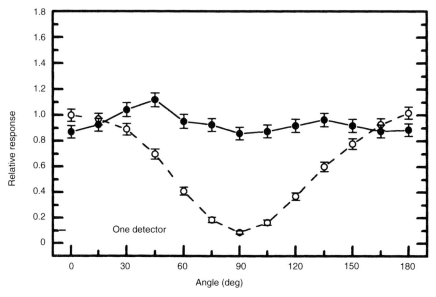

Figure 6. Angular response of Makrofol DE 475 to ^{252}Cf neutrons for one detector (○) and two detectors perpendicular to each other (●).

track counting and 50% at 10 mSv for automatic scanning.

CONCLUSIONS

In contrast to other etched track detectors Makrofol polycarbonate is a dosemeter with consistent batch quality of the material, high reproducibility, low background, flat energy dependence in the high energy region and sufficient sensitivity for the cosmic neutron spectrum. Polycarbonate detectors, with a low uncertainty of measurement, offer the best stability against ambient influences up to temperatures of more than 50°C. At high energy particle accelerators, for instance at Brookhaven National Laboratory, polycarbonates replaced the formerly used Kodak NTA films[16]. In addition, this new type of detector appears to be a suitable device to monitor the long-term neutron exposures at high energy particle accelerators as well as at typical airflight altitudes with a lower detection limit of 0.2 mSv per year. Up to now, TEPC devices above all have been used to monitor the dose rate for airflight routes[6,17]. In comparison to active detector systems, passive polycarbonate detectors may alternatively be used to estimate the neutron dose. The advantages of passive detectors are:

(a) the insensitivity to the high energy proton component of the cosmic rays;
(b) the easy long-term dose accumulation instead of computer assisted evaluation of count rates, the recording of flight time, altitudes and routes;
(c) the easy portable and exchangeable passive detector of about 10 g instead of installed equipment of 10–15 kg; and
(d) the alternative use as an individual monitor for representative individuals and routes, respectively.

ACKNOWLEDGEMENTS

This work was done at the Forschungszentrum Karlsruhe, Hauptabteilung Sicherheit/Dosimetrie, within the programme of Polish–German co-operation in the years 1993–1995. The financial support of the Polish Committee for Scientific Research is also acknowledged.

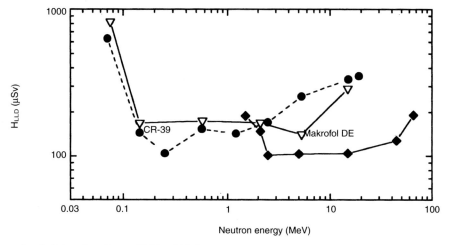

Figure 7. Lowest level of detection, H_{LLD}, of Makrofol DE and AA CR-39 as a function of neutron energy using optimised two-step ECE, for CR-39 at field strength of 20 kV.cm^{-1} ($\triangledown$) and 25 kV.cm^{-1} ($\bullet$).

REFERENCES

1. Józefowicz, K. *Energy Dependence of the Efficiency of Neutron Detection in Polycarbonate by Recording Atom Recoil Tracks.* In: Neutron Monitoring for Radiation Protection Purposes (Vienna: IAEA) STI/PUB/318, Vol. II, p. 183 (1973).
2. Hassib, G. M., Kasim, S. A. and Piesch, E. *Neutron Energy Dependence of Different Track Etch Detectors.* Radiat. Effects **45**, 57 (1979).
3. Józefowicz, K. and Piesch, E. *Electrochemically Etched Makrofol DE as a Detector for Neutron Induced Recoils and Alpha Particles.* Radiat. Prot. Dosim. **34**(1/4), 25–28 (1990).
4. EURADOS Intercomparison 1992. Report to be published.
5. Siebert, B. R. L. and Schuhmacher, H. *Quality Factors, Ambient and Personal Dose Equivalent for Neutrons, Based on the New ICRU Stopping Power Data for Protons and Alpha Particles.* Radiat. Prot. Dosim. **58**, 177–183 (1995).
6. Hoefert, M., Stevenson, G. R. and Alberts, W. G. *Messungen von Strahlenfeldern und Dosimetrie in grossen Hoehen.* Report FS-94-71-T, Vol. II, pp. 541–548 (TUV Rheinland GmbH) (1994).

7. Hoefert, M. Actual information on the Karlsruhe results for the CERN calibration irradiations (1995).
8. Burgkhardt, B. Piesch, E. and Urban, M. *Measurement of the Neutron Dose Equivalent Component of the Natural Background using Electrochemically Etched Polycarbonate Detector and Boron-10 Radiator*. In Proc. 13th Int. Conf. on Solid State Nuclear Track Detectors, Rome, 1985. Nucl. Tracks **12**, 573–576 (1986).
9. Regulla, D. and David, J. *Measurements of Cosmic Radiation on Board Lufthansa Aircraft on the Major Intercontinental Flight Routes*. Radiat. Prot. Dosim. **48**(1), 65–72 (1993).
10. Siebert, B. PTB-Braunschweig, private communication.
11. Hewitt, J. E., *et al. Ames Collaborative Study of Cosmic Ray Neutrons: Mid Latitude Flights*, Health Phys. **34** (1976).
12. Griffith, R. V., Palfalvi, J. and Madhvanath, U. *Compendium of Neutron Spectra and Detector Responses for Radiation Protection Purposes*. IAEA Technical Report Series 318 (Vienna: IAEA) (1990).
13. Józefowicz, K., Burgkhardt, B., Piesch, E. and Vilgis, M. *Dalsze badania odpowiedzi neutronowej poliwęglanowego detektora śladów cząstek Makrofol DE 1–4 475*. Raport B IEA Nr: B. 61/94 (in Polish).
14. O'Sullivan, D. (to be published).
15. Burgkhardt, B., Piesch, E. and Vilgis, M. *Uncertainty of Measurement and Lower Detection Limit of Track Etched Detector Systems: Experimental Verification and Consequences for Intercomparison Experiments*. Radiat. Prot. Dosim. **47**, 617–622 (1993).
16. Musolino, S., Rohrig, N. and Baker, T. *Intercomparison of Dosimetry Measurements at the Alternating Gradient Synchrotron from 1986–1988*. Health Phys. **65**, 96–102 (1993).
17. EURADOS Report. *Exposure of Aircrew to Cosmic Radiation* (to be published).

AN AUTOMATED NEUTRON DOSIMETRY SYSTEM BASED ON THE CHEMICAL ETCH OF CR-39

J. R. Harvey†, A. P. French‡, M. Jackson‡, M. C. Renouf§ and A. R. Weeks∥
†Dosimetry Consultant, 9 Torchacre Rise
Dursley, Gloucestershire GL11 4LW, UK
‡Defence Radiological Protection Service
Institute of Naval Medicine
Alverstoke, Hants PO12 2DL, UK
§BICRON◆NE
Bath Road, Beenham, Reading, Berkshire RG7 5PR, UK
∥Magnox Electric plc
Berkeley Centre
Berkeley, Gloucestershire GL13 9PB, UK

Abstract — The dosimetric characteristics of two types of personal neutron dosemeter have recently been extensively assessed. The effects of exposure to various extreme environments have also been studied. Both types of dosemeter utilise chemically etched elements which are read in an automated reader, the Autoscan 60, which uses an edge illumination system to increase the pit image size. One type of dosemeter contains three elements in a pyramid structure. The other uses one or two elements in a planar structure. The results indicate that both types of dosemeter can be used to assess accurately the personal dose from neutrons in a range of harsh environments. Formal approval for the operational use of the system has been received.

THE ETCH AND READ SYSTEM

The approach discussed here obviates the need for electrochemical etch[1] by using a simple optical system to increase the apparent size of chemically etched pits[2]. The system has been incorporated into an automated etch and read system available from BICRON◆NE, Beenham, Reading, Berkshire, RG7 5PR, UK. A composite photograph of elements of the system is shown in Figure 1.

The etched track elements and etch procedures

The individual poly allyl diglycol carbonate (CR-39) etched track elements, 20 mm × 25 mm × 1.5 mm, are pre-etched for 1 h in a mixture of 60% (by volume) methanol and 40% 6.25 N sodium hydroxide at 70°C. This pre-etch, which removes about 50 μm from each face, polishes the surface and removes superficial alpha particle tracks and scratches. The elements are then etched for 6 h in 6.25 N sodium hydroxide at 70°C. Up to sixty elements can be pre-etched and etched in a stainless steel carousel which is then transferred directly to the reader after washing and drying. Many carousels can be pre-etched and etched simultaneously.

The automatic reading system

When a carousel is loaded into the reader each etched track element in turn is elevated to the read position where bright white light is channelled into one edge of the element. The light is totally internally reflected from the faces of the element except where a defect or pit exists where it is reflected and refracted. Pits and defects therefore appear as bright spots of light when the element is viewed from a position normal to its surface by a television camera/monitor system through a low power zoom microscope. The intensity of the light is so great that the system registers images much larger than the true pit size which can easily be counted by a simple image analysis system. The reader can, if required,

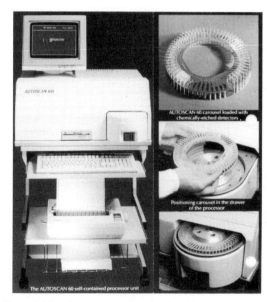

Figure 1. The Autoscan reader showing details of carousel handling.

count the number of tracks in each of 16 sub-areas and in addition give a coarse track size spectrum. If there is significant overlap of track images the magnification of the viewing lens is automatically increased by a factor of two. This process can be repeated up to five times giving an overall increase in magnification of up to 32 times. Each element, which is normally read in about 30 s, has a seven digit decimal number engraved on it together with a binary dot code which is read by the reader.

THE DOSEMETERS

The planar dosemeter

The planar dosemeter was developed by the UK Defence Radiological Protection Service and consists of two etched track elements located side by side within a rectangular Perspex insert of approximately the same thickness as the etched track elements. This is contained in a pouch of Mayblam PE within a modified NRPB neutron dosemeter holder designed for nuclear track films. The use of two elements in a planar dosemeter leads to a significant reduction in the likelihood of a serious spurious high reading due to high background plastic.

The pyramid dosemeter

Each pyramid dosemeter contains three elements lying on the sides of a pyramid plastic foam structure with faces at an angle of about 40° to the base. The use of three elements in a single dosemeter makes it possible to correct the poor angular response exhibited by all planar etched track and electrochemical etch neutron dosemeters. There is space for up to two more elements in its base which can be used for ancillary functions such as thermal neutron detection if required. This pyramid structure is contained in a moulded plastic dosemeter originally designed as the UK NRPB radon dosemeter. The overall diameter is 58 mm and the overall height, 20 mm. The basis of the design of the dosemeter and in particular the choice of a pyramid face angle of order 35° to 40° is described elsewhere[3].

PERFORMANCE OF THE DOSEMETERS

The principal dosimetric characteristics of both types of dosemeter have recently been established[4-6]. The quantity measured in the investigations was the Directional Dose Equivalent, $H'(10)$, a commonly used standard analogue of the quantity recommended for personal dosimetry of penetrating radiation, the Personal Dose Equivalent, $H_p(10)$. The characteristics of both types of dosemeter are summarised below.

Sensitivity and background

A full analysis[7] of 800 sheets accepted for use in a routine service indicates that the mean sensitivity of the planar dosemeter for normally incident ^{241}Am-Be neutrons is 429 $cm^{-2}.mSv^{-1}$ with a standard deviation of 90 $cm^{-2}.mSv^{-1}$. The mean background track density is 54 cm^{-2} with a standard deviation of 14 cm^{-2}. Dividing the background track density by the mean sensitivity of 429 $cm^{-2}.mSv^{-1}$ gives 0.126 ± 0.033 mSv for the planar dosemeter for normally incident radiation. The corresponding figure for the pyramid dosemeter is about 0.21 ± 0.06 mSv for radiation incident at any angle.

Linearity

The system is linear to about 20 mSv for ^{241}Am-Be neutrons and has been shown to be linear up to 10 mSv for neutrons of energy 1.2 MeV, 5.3 MeV and 15.1 MeV[4]. Between 20 mSv and 200 mSv correction is needed as track overlap starts to become significant. Low and high dose linearity relationships for ^{241}Am-Be neutrons are shown in Figures 2 and 3. Between 200 mSv and 20 Sv the track density is such that tracks can easily be counted by eye using an optical microscope or automatically with a sight-system analyser.

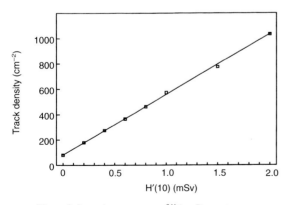

Figure 2. Low dose response, ^{241}Am-Be neutrons.

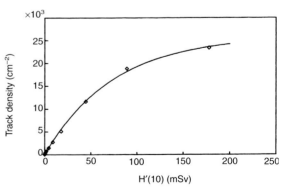

Figure 3. High dose response, ^{241}Am-Be neutrons.

Reproducibility

In a single carefully controlled experiment where 30 elements were given identical exposures of 4.61 mSv, the standard deviation of the distribution of readings about the mean was 4.0%[4]. This is close to the minimum statistically attainable for the mean number of tracks per elements, 580, within the counting area of 1/4 cm^2.

Energy response

Initial measurements[4] over a wide energy range gave the data shown in Figure 4. The results of subsequent measurements of the detailed structure of the low energy threshold of the planar dosemeter[5] are shown in Figure 5.

Thermal neutron response

Dosemeters can be made sensitive to thermal and near thermal neutrons by locating lithium-loaded radiators in contact with the etched track element. This sensitivity is due to tritons produced in the ^{6}Li(n,α)^{3}H reaction, the only charged particle from the reaction which has a range (50 μm) comparable with the thickness of etch plastic removed by the pre-etch process. Following early studies with lithium fluoride[8], experiments were performed with plastic radiators loaded with given percentages of lithium. These gave thermal neutron sensitivities between 3000 and 5000 tracks.cm^{-2}.mSv^{-1} per 1% loading of normal abundance lithium by weight[6].

Angular dependence

Angular response data[4] for both dosemeters are shown in Figures 6 and 7. It is clear that the planar dosemeter shows a fall-off in sensitivity with increasing angle of incidence similar to that shown by all planar etched track and electrochemical etch dosemeters. The pyramid dosemeter, however, shows a relatively angle-independent response as predicted by design studies[3].

Location correction factors

In order to improve the accuracy of either dosemeter it is possible to modify the readings on the basis of environmental surveys of the spectral and directional characteristics of the neutron fields in locations where the dosemeters are worn. Such correction factors have been established by the use of a transportable neutron spectrometry system and are used routinely within the UK Defence Radiological Protection Service[9].

Environmental tests

Both types of dosemeter have been tested for fading, ageing and background changes in three environments: room temperature; 6°C and 10% humidity; and 40°C and 90% humidity[10]. The results are complex and appear to be batch dependent. Generally changes are not great at room temperature and at 6°C, but significant changes were observed in the 40°C environment. In general any changes which might occur in practical operating environments over normal issue periods can easily be corrected for.

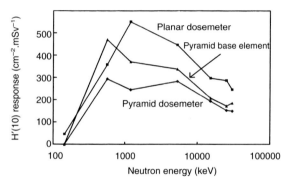

Figure 4. Response of pyramid dosemeter, pyramid base element and planar dosemeter as a function of neutron energy.

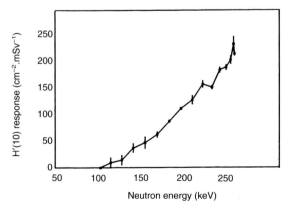

Figure 5. Fine detail of low energy response of planar dosemeter.

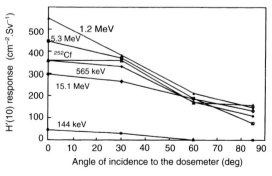

Figure 6. Angular response of planar dosemeter.

Dosemeters have also been tested for exposure to environmental contaminants[10]. These included exposure in a vacuum cleaner bag for several days of use, being dropped from 5 m, being exposed for 20 min to 23 mW.cm^{-2} of UV radiation, both before and after irradiation, and being exposed to a normal washing machine procedure. None of these environments gave rise to statistically significant changes in sensitivity or background.

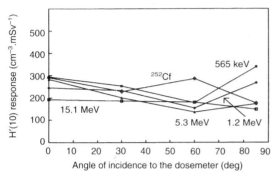

Figure 7. Angular response of pyramid dosemeter.

CONCLUSIONS

Dosemeters which use the Autoscan system have been shown to have dosimetric and environmental characteristics which are appropriate for routine personal dosimetry. The tests summarised in this note have led to UK Health and Safety Executive approval of the personal neutron dosimetry system developed by the UK Defence Radiological Protection Service which is based on the Autoscan 60. The service is being applied to the UK Ministry of Defence (Navy) submarine flotilla.

ACKNOWLEDGEMENTS

The authors are indebted to Working Group 8 of EURADOS and to EURADOS for sponsoring the provision of a range of well-calibrated neutron fields principally at the Physikalisch Technische Bundesanstalt, Braunschweig, Germany and the GSF-Munchen, Institut fuer Strahlenschutz, Neuherberg, Germany[11]. Irradiation facilities were also provided by the National Physical Laboratory and Defence Radiological Standards Centre (DRaSTaC), AWE, Aldermaston, UK and Berkeley Centre, Berkeley, UK.

REFERENCES

1. Tommasino, L. and Armellini, C. *New Etching Technique for Damage Track Detectors*. Radiat. Eff. **20**(4), 253–255 (1973).
2. Harvey, J. R. and Weeks, A. R. *Recent Developments in a Neutron Dosimetry System Based on the Chemical Etch of CR-39*. Radiat. Prot. Dosim. **20**(1/2), 89-93 (1987).
3. Harvey, J. R. *A Three Element Etched Track Neutron Dosemeter with Good Angular and Energy Response Characteristics*. Radiat. Prot. Dosim. **44**(1/4), 325–328 (1992).
4. Weeks, A. R., Ford, T. D. and Harvey, J. R. *An Assessment of Two Types of Personal Neutron Dosemeter which Utilise the AUTOSCAN 60 Automatic Etch-Track Reading System*. UK Nuclear Electric, Technology Division Report TEPZ/REP/0084/93 (1993).
5. Jackson, M., French, A., Harvey, J. R. and Weeks, A. R. *Low Energy Neutron Response — Planar and Isotropic Neutron Dosemeters, (March 1994)*. In: CR-39 Neutron Dosimetry Report 9/94 (Defence Radiological Protection Service, Institute of Naval Medicine, Alverstoke, Gosport, Hants, PO12 2DL, UK) (1994).
6. Wilson, M., French, A., Harvey, J. R. and Wernli, C. *An Assessment of Thermal Neutron Sensitive Dosemeters Based on a Commercial Etched-Track Reading System*. Presented at Conf. on 'Radiation Protection Physics and Measuring Technique', 24–26 May 1994, Karlsruhe (1994).
7. Jackson, M. and French, A. *Analysis of Acceptance Testing Data for more than 800 Sheets of CR-39 Plastic Assessed for the DRPS Approved Dosimetry Service*. Radiat. Prot. Dosim. **70**(1–4), 139–142 (This issue) (1997).
8. Wernli, C. and Langen, K. *The New Personal Dosimetry System at the Paul Scherrer Institute*. In: Strahlenschutz-Physik und Messtechnik, Eds. W. K. Oelzer, R. Maushart, Vol II, pp. 529–534 (Koln: Verlag Tüv Rheinland GmbH).
9. Barlow, K., Jackson, M., French, A. and Harvey, J. R. *The Application of Neutron Spectrometry in the DRPS Neutron Dosimetry Service*. Radiat. Prot. Dosim. **70**(1–4), 265–268 (This issue) (1997).
10. Jackson, M., French, A., Harvey, J. R. and Weeks, A. R. *Some Environmental Trial Results for a Planar and Isotropic CR39 Neutron Dosemeter*. Radiat. Meas. **25**(1–4), 461–462 (1995).
11. Harvey, J. R., Tanner, R. J., Alberts, W. G., Bartlett, D. T., Piesch, E. and Schraube, H. *The Present Status of Etched Track Neutron Dosimetry and the Contribution of CENDOS and EURADOS*. Radiat. Prot. Dosim. **70**(1–4), 127–132 (This issue) (1997).

ANGULAR RESPONSE TO NEUTRONS OF THE ELECTROCHEMICALLY ETCHED CR-39 WITH DIFFERENT RADIATORS

J. Bednář, K. Turek and F. Spurný
Nuclear Physics Institute, Department of Radiation Dosimetry
Na Truhlářce 39/64, 180 86 Praha 8, Czech Republic

Abstract — The method of a single bent plastic detector has been used for the detailed angular response study of the electrochemically etched CR-39 track detector. For several monoenergetic and radionuclide neutron sources the angular responses have been determined experimentally for three different situations with respect to the radiator covering the detector surface; polyethylene (2 mm), CR-39 (0.5 mm) and for a bare detector irradiated free-in-air. The set of experimental data obtained for each situation provides a good base for the application of the least squares fit method. This way the angular response has been expressed as a function of angle α of neutron incidence by a simple empirical formula $\rho(\alpha) = a + b\cos(2\alpha + \pi)$, where parameters a,b depend on neutron energy and experimental conditions. For all neutron sources used in this study the normalised angular response functions $R(\alpha) = A + B\cos(2\alpha + \pi)$, $R(\pi/2) = 1$ have been compared with respect to different kinds of radiator.

PRINCIPLE OF MEASUREMENT

The method of a single bent detector previously described[1] was used to obtain the detector response over the full range of incident angles. The track densities ρ were counted consecutively in narrow strips perpendicular to the length of a large rectangular detector. The transformation of the coordinate l of each strip (Figure 1) into the angle of incidence α given by the equation

$$\alpha = \mathrm{arctg}\left[(\cos(l/r) - r/d)/\sin(l/r)\right] \quad (1)$$

enables the detector response to be determined for any angle of incidence using only one exposure. The set of experimental data (several tens as a rule for a single detector) can easily be fitted using the least squares method. Taking into account the shape of the distribution of experimentally obtained track density values, the empirical relation

$$\rho(\alpha) = a + b\cos(2\alpha + \pi) \quad (2)$$

was chosen for fitting. The treatment of data (including correction for the variable distance between the detector and source) was carried out using a PC program.

EXPERIMENTAL

On the basis of the principle described above, the angular responses were determined for different neutron sources using CR-39 material, (Pershore Mouldings, 500 μm, standard grade, 32 h cure). Detectors of size 30 × 120 mm² were covered with proton radiators of 2 mm polyethylene and bent on a cylinder of 60 mm diameter (Figure 1). The neutron sources and irradiation facilities are given in Table 1. The distance between the polyethylene surface and the neutron source was at least 300 mm in order to consider the neutron source as a point source. Detectors were etched electrochemically using 30.2% KOH at 60°C in two steps; 5 h at 50 Hz, 22 $kV_{peak}.cm^{-1}$ and then 30 min at 10 kHz, 25 $kV_{peak}.cm^{-1}$. The etched surface of size 20 × 110 mm² was scanned in steps of 1 mm. In such a way, about 100 strips have been counted on a single detector,

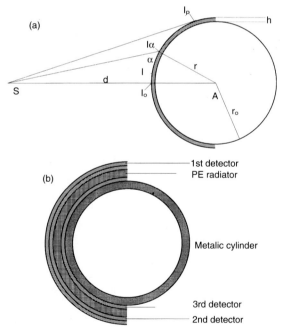

Figure 1. Scheme of irradiation geometry (a) and the arrangement of detectors (b). S, source; A, axis of cylinder (made of aluminium, wall 3 mm thick); I_o, perpendicular impact point; I_p, parallel impact point; I_α, impact point under angle of incidence α; h, detector thickness (500 μm); r_o, cylinder radius (30 mm); $r = r_o + h$, effective radius; d, distance SA; l, distance $I_o I_\alpha$.

153

each corresponding to ~20 mm². Tracks were counted visually, at a magnification of 50× to 100×. The maximum track diameter was about 80–90 μm.

RESULTS AND DISCUSSION

The experimental track densities obtained for monoenergetic neutrons are presented in Figure 2(a). Despite their empirical origin, the fitting curves expressed by Equation 2 provide a quite acceptable approximation of the experimental data. In the next step all fitting curves were normalised to the response at perpendicular incidence using equation

$$R(\alpha) = A + B \cos(2\alpha + \pi) \quad (2')$$

with $R(\alpha) = \rho(\alpha)/\rho(\pi/2)$ (see Figure 2(b)).

The less significant angular dependence has been observed in the case of air radiator (for $E_n \geq 1.2$ MeV). It is probably due to the remarkably higher relative contribution of heavy recoils (C, N and O) and a different angular distribution of scattered particles compared to recoil protons. For all neutron energies and neutron sources the table contains the response to perpendicularly incident neutrons $\rho(\pi/2)$ calculated from Equation 2, the energy response R_Φ of the detectors related to neutron fluence, in tracks per neutron, and the dose equivalent response R_H, in tracks·cm^{-2}·mSv^{-1}. The last three columns present the coefficients A, B determining the normalised angular response (Equation 2'), and the response R(0) for lateral irradiation relative to response at perpendicular incidence. For neutron energies higher than 2.5 MeV the value R(0) seems to increase with energy even for CR and PE radiators, i.e. the angular response starts slightly to improve. The

Table 1. Results of experiments: energy response and normalised angular response. $R(\alpha) = A + B \cos(2\alpha + \pi)$, $R(\pi/2) = A + B = 1$.

E_n (MeV)	Irradiat. facility	Radiator	$\rho(\pi/2)$ (cm^{-2})	Energy response		Normalised ang. response		
				R_Φ (10^{-4} tr.n^{-1})	R_H (tr.cm^{-2}·mSv^{-1})	A	B	R(0)
0.565	PTB	Air	2048	0.47	240	0.51	0.49	0.02
		CR-39	2400	0.55	280	0.5	0.5	0
		PE	2533	0.59	300	0.5	0.5	0
1.2	PTB	Air	898	0.56	170	0.63	0.37	0.25
		CR-39	1752	1.1	330	0.53	0.47	0.06
		PE	2301	1.5	440	0.51	0.49	0.02
2.5	PTB	Air	—	—	—	—	—	—
		CR-39	4000	1.4	350	0.53	0.46	0.07
		PE	7996	2.7	690	0.5	0.5	0
5	PTB	Air	813	0.18	45	0.97	0.02	0.95
		CR-39	3801	0.86	210	0.59	0.41	0.18
		PE	4948	1.1	280	0.58	0.41	0.16
14.8	PTB	Air	1313	0.63	150	0.81	0.19	0.62
		CR-39	1107	0.53	130	0.82	0.17	0.65
		PE	2670	1.3	310	0.76	0.24	0.52
Am-Be	NPI	Air	291	0.22	58	0.76	0.24	0.52
		CR-39	1251	0.94	250	0.58	0.42	0.16
		PE	1441	1.1	290	0.56	0.45	0.11
^{252}Cf	NPI	Air	236	0.28	81	0.75	0.25	0.5
		CR-39	469	0.55	160	0.59	0.4	0.19
		PE	1383	1.64	480	0.53	0.48	0.05
Pu-Be	JINR Russia	Air	2197	0.19	54	0.72	0.27	0.45
		CR-39	5923	0.51	150	0.62	0.39	0.25
		PE	14164	1.2	350	0.59	0.4	0.19
^{252}Cf Moderated 305mm PE	JINR Russia	Air	705	—	190	0.78	0.22	0.55
		CR-39	884	—	240	0.55	0.45	0.1
		PE	1537	—	420	0.54	0.46	0.08

values of R(0) for radionuclide sources confirm this assumption.

CONCLUSIONS

The technique described in this work provides a method of measurement of angular response using a single detector and offers a sufficient number of experimental data for good fitting. The simple empirical equation used here is not the only possible fit and some others, more precise, can be found. Nevertheless, for some calculations considering angular dependence even this expression can be useful. As was found, the relative (normalised) angular response does not differ substantially for radiators of recoil protons. More detailed discussion considering quantitatively the actual angular distribution of all charged particles generated by neutrons in experiment should be necessary for general conclusions.

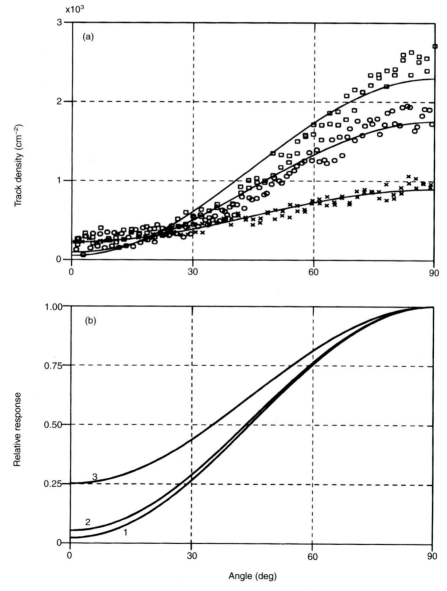

Figure 2. (a) Experimental data obtained for 1.2 MeV neutrons and their fit. (□) 2 mm PE, (○) 0.5 mm CR-39, (×) bare detector. (b) Corresponding angular dependence of response normalised to perpendicular incidence. Curve 1, 2 mm PE; curve 2, 0.5 mm CR-39; curve 3, bare detector.

ACKNOWLEDGEMENT

The present work was partially supported by the Project X 242.4 (Scientific and Technological Co-operation between Germany and Czech Republic). The authors would like to thank Dr S. Guldbakke (PTB Braunschweig), Dr B. Burgkhardt (FZ Karlsruhe) and Dr V.P. Bamblevskij (JINR Dubna) for irradiations.

REFERENCE

1. Turek, K., Bednář, J. and Piesch, E. *Determination of the Neutron Angular Response using a Single Etched Track Detector.* Radiat. Prot. Dosim. **59**(3), 205–211 (1995).

A PERSONAL NEUTRON DOSIMETRY SYSTEM BASED ON ETCHED TRACK AND AUTOMATIC READOUT BY AUTOSCAN 60

A. Fiechtner, K. Gmür and C. Wernli
Division for Radiation Hygiene
Paul Scherrer Institut
CH-5232 Villigen, Switzerland

Abstract — An instrument for reading chemically etched CR-39 elements, the Autoscan 60, has become commercially available. Based on this readout system a new personal neutron dosemeter badge has been developed. PN3 detectors (NE Technology) are used with a two step etching process. The radiators are polyethylene (PE), polyethylene with ^{6}Li and aluminium. In a nuclear reaction of ^{6}Li with thermal neutrons alpha particles and tritons (^{3}H) are produced. While the range of alpha particles (9 μm) is too short to be detected with a two step etching procedure, tritons (range = 52 μm in CR-39) easily reach the sensitive layer of the detector and increase the response for thermal neutrons to a value determined by the ^{6}Li concentration in the PE radiator. Aluminium inhibits the production of recoil protons and hence, causes a lower response to fast neutrons as compared to PE. From the ratio of the readings behind PE and Al, spectral information can be derived, especially in accident situations when the track density is high and a radiation field analysis is required. For extremely high track densities a microscope is used for evaluation, since automatic counting by Autoscan 60 saturates at doses above 50 mSv. The system has been tested for reproducibility, linearity, energy dependence up to 66 MeV neutron energy, angular dependence, radon background response and multiple use of CR-39 detectors.

INTRODUCTION

For routine application, etching and evaluation of detectors should be easy and not time consuming. Electrochemical etching simplifies counting of tracks but the etching process is cumbersome. Chemical etching is easy on a large scale and since there is a fully automatic reader available, the Autoscan 60[1], this system allows for a routine application with restricted man power.

To develop an appropriate badge one has to consider the prevalent neutron fields. At Paul Scherrer Institut (PSI) the accelerator neutron fields have an energy range up to 600 MeV with local maxima at energies of 1–5 and 50–100 MeV behind the concrete shieldings and in the thermal energy region at the exit of labyrinths[2]. A possible badge configuration is one CR-39 detector covered with different radiators for a multi-field evaluation. Further experiments were done to investigate the effect of different radiators (i.e. beryllium, carbon, and lead) on the response for high energy neutrons.

IRRADIATIONS

For irradiations with thermal neutrons the thermal column of the PSI zero power reactor PROTEUS was used. In this configuration, not enough space was available to use any phantom. Irradiations with fast neutrons were performed in the calibration laboratory of PSI with the neutron sources ^{252}Cf, bare and moderated with a D$_2$O sphere, and Am-Be. Irradiations with neutron energies of 144 keV and 565 keV were done by PTB Braunschweig. Detectors were mounted on a 30 × 30 × 15 cm^3 PMMA slab phantom. High energy irradiations were performed at PSI with two 'quasi-monoenergetic' neutron beams with dominant contributions at 35 and 66 MeV, produced by 40 and 71 MeV protons, respectively, from the Philips variable energy cyclotron hitting a Be target of 2 mm thickness[3]. Here a small (10 × 10 × 15 cm^3) PMMA phantom was used.

DESCRIPTION OF THE ETCHING PROCESS AND THE DETECTOR SYSTEM

Detector material

The CR-39 used were PN3 detectors from NE Technology, England. The detectors were 1.5 mm thick and 20 × 25 mm^2 in size. All detectors were cut by the supplier and have a decimal and binary code for automatic identification by the reading system.

Etching procedure

A two-step etching process was used for chemical etching of the PN3 detectors. In a first step, the 'pre-etch', the detectors were etched for 30 min[1] in a mixture of 60% methanol and 40% 6.25N sodium hydroxide at 70°C to polish the surface and to remove alpha particle tracks and scratches. On each side of the CR-39 detector 15 μm were removed during the pre-etch. Afterwards the detectors were etched for six hours at 70°C in 6.25N sodium hydroxide. Finally the detectors were neutralised in a weak hydrochloric acid solution and washed with hot and cold distilled water.

Reading system Autoscan 60

After drying, the detectors were read in the Autoscan 60. The detectors can be read in the same carousel as used for etching. One carousel can hold up to sixty detectors which can be read within 30 min. For the readout, the detectors were elevated automatically into the path of strong light channelled into one edge of the detector. The light is internally reflected from the faces of the detector except at defects or tracks where it is refracted. When the detector is viewed from a position in front of its surface, the tracks appear as bright spots of light. The intensity of the light is so high that the camera–monitor system registers images much larger than the true pit size so that pit images can easily be counted by an image analysis system without the necessity for an electrochemical etch.

Radiators

Different radiators were used for different energy regions:

(i) For fast neutrons the recoil protons produced in PE yielded a higher response.
(ii) For the measurements of thermal neutrons a PE radiator containing natural lithium was used. In a nuclear reaction of ^{6}Li with thermal neutrons alpha particles and tritons (^{3}H) are produced. The range of the alpha particles is too short (9 μm) to be detected but the tritons with a range of 52 μm in PN3 detectors can easily reach the sensitive layer of the detector. The increase of the response for thermal neutrons (Table 1) depends on the lithium concentration in PE.
(iii) Aluminium was used to get spectral information in the fast neutron energy region because it inhibits the production of recoil protons outside CR-39 and therefore, causes a lower response to fast neutrons (factor 2–3) as compared to PE[4].

MEASUREMENTS

Linearity of Autoscan 60

The Autoscan 60 can measure linear doses up to 15 mSv by counting the spots (Figure 1). For doses higher than 15 mSv the illuminated spots overlap and the Autoscan 60 measures the total brightness of the spots and gives an estimate for the number of tracks up to 50 mSv. For higher doses a microscope is used for evaluation.

System and radon background

The system background was tested over several detector production sheets and found to be 50–60 cm^{-2} and nearly constant for different sheets. Since the radon background is very important for the storage of detectors the radon background response was determined. Uncovered detectors were irradiated in a defined radon atmosphere and the radon background response was found to be about 2 cm^{-2}.kBq^{-1}.h^{-1}.m^3. When exposed in a badge no effect of the radon exposure was found.

Energy dependence

The energy dependence $R_H(E)$ for the personal dose equivalent response, shown in Figure 2 was measured

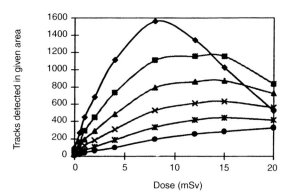

Figure 1. Linearity of Autoscan 60. –♦– 1.6 cm^2; –■– 0.8 cm^2; –▲– 0.4 cm^2; –+– 0.2 cm^2; –✶– 0.1 cm^2 and –●– 0.05 cm^2.

Table 1. Test of different radiators for thermal neutrons.

Radiator	PE (0.1% Li$_{nat}$)	PE (1% Li$_{nat}$)	PE (7.5% Li$_{nat}$)	Al (2.5% Li$_{nat}$)
Response [mSv^{-1}.cm^{-2}]	40±30%	530±10%	15300±30%	26700±30%

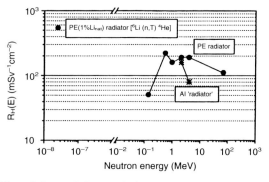

Figure 2. Personal dose equivalent response as a function of neutron energy with different radiators.

up to 66 MeV neutron energy. The response for thermal neutrons was determined with PE containing 1% natural lithium.

Angular dependence

The dependence of the response on the angle α of neutron incidence is an important factor for personal neutron dosimetry. The normalised angular dependence $R_H(\alpha)$ (Figure 3) for personal dose equivalent was determined with a ^{252}Cf neutron source and PE radiators. The fluence to dose conversion factors were taken from the ISONEUT draft version[5]. The irradiations were performed on an ISO water slab phantom (30 × 30 × 15 cm³). For obliquely incident neutrons the response decreases rapidly to almost 20% of $H_p(10)$. Earlier experiments have shown that angular dependence is much less pronounced for high energy neutrons[6,7].

Multiple use of detectors

PN3 detectors are quite expensive and in routine application the doses of the workers are often not above background. Therefore, the possibility of a second use of the non-irradiated detectors was investigated. Irradiation and etching were carried out as follows:

Case 1: Background-Etching-Readout-Background-Etching-Readout
Case 2: Background-Etching-Readout-Irradiation-Etching-Readout
Case 3: Irradiation-Etching-Readout-Irradiation-Etching-Readout

If detectors were not irradiated by etched twice (case 1) the background increased by 100% but was yet low enough to enable further dosimetry. If a detector was only irradiated in the second period of use, the response was increased by about a factor of 1.5 (case 2). If irradiated twice (case 3) it seems that the detector integrates the dose, but this effect needs further investigation.

Measurements of thermal neutrons

To detect thermal neutrons the nuclear reaction of ^{6}Li with thermal neutrons is applied. Li concentrations (the actual Li contents may deviate significantly from the nominal values) from 0.1% Li_{nat} up to 7.5% Li_{nat} in Al and PE were tested in thermal neutron irradiations at the zero power reactor PROTEUS. The results are shown in Table 1. The different method of production of the radiators containing Li may be one reason for the fact that the response is not proportional to the nominal Li concentration.

EXPERIMENTS WITHOUT PRE-ETCH OF THE DETECTORS

The response of the dosemeter to high energy neutrons was tested using various radiators. The 35 and 66 MeV spectra at PSI were used and a continuous mixed spectrum behind a concrete shielding with energies up to 1 GeV was available at CERN[8]. The reference dosimetry at PSI was carried out by PTB Braunschweig employing absolute beam monitoring techniques, while at CERN the reference dose was evaluated using calculational methods and a tissue-equivalent proportional counter.

In view of the short ranges of the reaction particles from the radiators, the detectors used in this experiment were not pre-etched. This has to be taken into account if practical applications are planned.

Table 2 shows the results of these investigations. The figures given are track densities per mSv relative to the response of three detectors with PE radiators in the radiation field of the Am-Be neutron source. In the neutron beams at PSI, relatively high nominal doses had been envisaged, resulting in track densities in the range of 1000 to 10,000 cm^{-2}. For consistency reasons, all the detectors of the high energy experiment were assayed manually on a Leitz microscope. The uncertainty of the results in Table 2 amounts to ±10% with the exception

Table 2. Responses for high energy neutrons with different radiators (normalised at PE radiator and Am-Be neutron source).

Radiator	Neutron spectra			
		High energy peak		
	Am-Be	35 MeV	66 MeV	CERN
PE	1.00	0.45	0.18	0.1
Be	0.53	0.59	0.29	—
C	0.53	0.41	0.18	0.2
^{7}LiF	0.47	0.41	0.23	0.3
Al	0.29	0.23	0.12	0.2
Pb	0.47	0.23	0.06	0.2

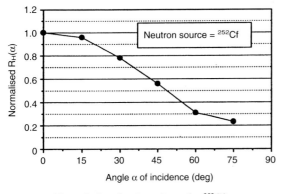

Figure 3. Angular dependence for ^{252}Cf.

of the CERN results, where the uncertainty is estimated at ±50% due to very low track densities of the order of 150 to 250 cm^{-2}.

First of all, with PE used as radiator, the decrease of the response at high energies known from earlier investigations[5] is confirmed. With a Be radiator the response seems less energy dependent, at the cost of a 50% reduction of response for Am-Be. C and ^{7}LiF perform well between PE and Be, while Al and Pb yield minimum responses.

Due to its toxicity, Be is ruled out from application in a personal dosemeter badge. It may be feasible, however, to employ track density ratios of different radiators as a spectral index which could form the base for calculational correction in dose assessment. Alternatively, spectral information can be gained from analysing the track area distribution. Although this approach may be not feasible in routine dosimetry, it is possible to improve dose evaluation in special cases, applying image analysis methods.

POSSIBLE BADGE CONFIGURATION AND ROUTINE APPLICATION

One possibility for a badge configuration consists of one detector with three different radiators in a planar array (Figure 4), contained in a polyamide holder. The three radiators may be PE for fast and high energy neutron detection, Al for spectral information in the fast neutron energy range, and one special radiator for the detection of thermal neutrons. There are different possibilities for the measurement of thermal neutrons. The response of CR-39 with a PE(1% Li$_{nat}$) radiator for thermal neutrons is approximately twice the response of CR-39 with PE for fast neutrons. Since the effective cross section of Li for fast neutrons is very low, Li has no impact on the response of PE(1% Li$_{nat}$) for fast neutrons. Therefore, the response to thermal neutrons is given by the difference of the track densities behind PE and PE(1% Li$_{nat}$). This may lead to big uncertainties of the readings for thermal neutrons if the fraction of fast neutrons is high. Since the Al(2.5% Li$_{nat}$) radiator has a higher Li concentration and Al reduces the response to fast neutrons, the response for thermal neutrons would be high for this radiator, even in a radiation field with a wide neutron energy spectrum.

CONCLUSIONS

The new personal neutron dosimetry system is still in the test phase. There are some residual problems such as the pronounced angular dependence of a planar badge configuration, the low response for intermediate and high energy neutrons, the uncertain availability and cost of the PN3 detectors in the future, and the reproducibility of their response. A first routine use at PSI is planned for 1996, in parallel with the old fission track system.

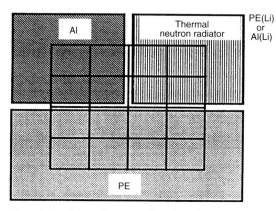

Figure 4. Possible radiator configuration for a detector in routine use.

REFERENCES

1. Wilson, M., French, A., Harvey, J. R. and Wernli, C. *An Assessment of Thermal Neutron Sensitive Dosemeters based on a Commercial Etched-track Reading System.* Strahlenschutz: Physik und Messtechnik, Band 2, **26**. Jahrestagung Karlsruhe, 24–26 Mai 1994. pp. 529–534.
2. Wernli, C., Duvoisin, J., Hauswirth, G. and Janett, A. *Experience with a Track Etch Detector in Personnel and Environmental Dosimetry.* Radiat. Prot. Dosim. **6**(1–4), 225–228 (1984).
3. Schuhmacher, H. and Alberts, W. G. *Reference Neutron Fields with Energies up to 70 MeV for the Calibration of Radiation Protection Instruments.* Radiat. Prot. Dosim. **42**(4), 287–290 (1992).
4. Wernli, C. and Langen, K. *The New Personnel Neutron Dosimetry System at the Paul Scherrer Institut.* In: Proc. Eighth Int. Congr. of the International Radiation Protection Association, 17–22 May 1992, Montréal, Canada.
5. ISO/TC85/SC2/WG2/SG3 WD 8529-3 ISONEUT draft version of 26 July 1995. *Reference Neutron Radiations: Calibration of Area and Personal Dosimeters and the Determination of Their Response as a Function of Neutron Energy and Angle of Incidence.*
6. Azimi-Garakani, D., Langen, K. and Wernli, C. *The Response of Various Neutron Dosemeters at High Energy Neutrons.* Nucl. Instrum. Methods Phys. Res. **A320**, 368–374 (1992).
7. Azimi-Garakani, D. and Wernli, C. *Study of the Angle Response of Various PADC Materials at Different Neutron Energies.* Nucl. Instrum. Methods Phys. Res. **B72**, 79–83 (1992).
8. Höfert, M. and Stevenson, G. R. *Individual Monitoring in High Energy Radiation Fields.* Radiat. Prot. Dosim. **54**(3/4), 303–306 (1994).

TEN YEARS ON: THE NRPB PADC NEUTRON PERSONAL MONITORING SERVICE

D. T. Bartlett, J. D. Steele, R. J. Tanner, P. J. Gilvin, P. V. Shaw and J. Lavelle
National Radiological Protection Board
Chilton, Oxon, OX11 0RQ, UK

Abstract — Development of the NRPB PADC neutron dosimetry system was carried out during the years 1982 to 1985. In 1985 the monitoring service was set up, and following 9 months of operational trials, was given approval by the UK Health and Safety Executive in April 1986. This paper traces the developments, and outlines the present situation. At an early stage it was judged that the most critical component of the system was likely to be the reliable supply of plastic of suitable quality. However, the 'quality' of plastic is very dependent on the processing and read procedures chosen. A commercial manufacturer of PADC was selected, and a two stage etch (chemical, electrochemical) plus a chemical post-etch, was developed which gave adequate dosemeter response characteristics, combined with a low background track density. Rigorous quality control procedures needed to be applied to the sheets of PADC in order to satisfy an operational criterion of a minimum detectable dose of 200 µSv. The choice of reader was made largely on the basis of known reliability. Procedures were kept as simple as possible with the further aim of maximising commonality with other NRPB monitoring systems. The intention was to optimise the system from an operational point of view. This meant, for example, the choice of etch parameters which did not necessarily give the greatest values of sensitivity, but gave values which were less dependent on small changes in plastic properties or etch parameters.

INTRODUCTION

The discovery by Cartwright, Shirk and Price[1] in 1978 of the sensitivity to recoil protons of the plastic termed Columbia Resin 39 (CR-39), the trade mark of the manufacturing company (now also known generally by the initials PADC — poly allyl diglycol carbonate) was hailed by many in neutron personal dosimetry as a breakthrough. The use of the technique of electrochemical etching[2] which amplified the size of tracks greatly, so that they were visible to the naked eye and easily counted by simple automatic systems, seemed to have completed the picture. The detection and dosimetry of fast neutrons down to about 100 keV, and, using converters, of neutrons, from thermal to 10 keV, was achievable by simple passive means. At NRPB in 1982, a project was set up to develop and introduce a personal monitoring system using these techniques.

SYSTEM DEVELOPMENT AND CURRENT POSITION

At an early stage, many laboratories discovered that in using the proposed etch regimes and plastic cured by standard procedures, for example the so-called Adams' cure[3], good energy response characteristics and high sensitivity were only obtained, in general, at the expense of high and variable background track densities. Improved plastic quality was desirable. At NRPB, it was decided to develop etch conditions which would make the best of the then commercially available plastic, and to optimise process parameters for routine operational conditions, whilst continuing the search to obtain improved plastic[4]. An etch regime was developed which comprised a chemical etch and a long electrochemical etch at a relatively low field strength. The etch regime currently used is as follows: etchant temperature 40°C; chemical etch $11\frac{1}{2}$ h; electrochemical etch 8 h; field strength 23 kV.cm^{-1}rms, 2 kHz.

The overnight etch regime has proved to be operationally beneficial. The optimisation of the etch process parameters to minimise the number of high or variable backgrounds, needed to be combined with rigorous quality control of the plastic sheets. The chosen manufacturer was asked to control closely the thickness of the sheet, but all attempts to reduce the background and its variability had limited effect. Each sheet had therefore to be sampled, and rejection criteria (based on average background track density and its variability) and sampling procedures were developed heuristically. These were aimed to obtain a minimum detectable dose of less than 200 µSv. The present system achieves a minimum measurable dose equivalent [4 (σ + 1/a)/S] (where σ is the standard deviation on background pit density, a is the area read, and S is the detector calibration factor as dose equivalent per unit pit density) of less than 100 µSv[5]. A particular problem with the operation of a PADC personal monitoring service is the occurrence of false positives. It is not possible to identify all sheets which may contain a higher than expected background. The observed distribution of background tracks is not Poisson, and prediction from sampling results is inexact. Until recently the frequency of occurrence of false positives, that is false reported doses of 200 µSv or greater on unexposed dosemeters was about 1 in 300. This caused concern to users, and required the re-estimation of doses[6]. The advent of improved plastic based on cure cycles proposed by Ahmad and Stejny[7,8] has reduced the number of rejected sheets to practically zero and the frequency of occurrences of false positives to less than 1 in 1000.

The detector is now encapsulated in a holder made

of nylon. Only the back surface of the PADC is etched. For energies in the range 100 keV to 7 MeV the neutrons are detected by tracks formed by recoil protons (mainly) produced in the PADC in front of this surface, and for neutron energies above about 7 MeV additionally from the front of the holder. For thermal (incident and albedo) neutrons and epithermal neutrons, protons are produced by capture reactions on the nitrogen in the dosemeter holder[9], giving rise to the lower energy component of the response characteristics (shown in Figure 1).

Dosemeters are read automatically by a reader supplied to NRPB specifications by Analytical Measuring Systems. This reads a central 1.68 cm² of the 2.5 cm² area which has been electrochemically etched. It uses a magnification of ×40, enabling identification of features down to 25 μm in diameter, a moving stage being employed to read 14, 4 × 3 mm sub-areas: data from these sub-areas are used to analyse for non-uniformity. High pit densities cause increasing numbers of closely adjacent pits which the reader cannot resolve individually. Thus, a correction factor is applied which counteracts this tendency to under-read at large doses.

Early consideration was given to a design of dosemeter with multiple elements at angles in order to improve the directional dependence of dosemeter response (as proposed for example by Schraube[10]). However, for operational reasons, it was decided to use a single planar element, and apply a modifying factor to the calibration factor in order to average the anterior-posterior and rotational response characteristics of the dosemeter. (This, in effect, assumes all wearers, on average, obtain 50% of their dose equivalent ($H_p(10)$) facing a unidirectional beam and 50% moving at random in a horizontal unidirectional beam, or working in a rotationally symmetric field.) The normalised readings of dosemeters for measured practical fields are given in Table 1. In dosemeter field trials, two PADC elements were inserted in each dosemeter holder. Close agreement was found between the readings of the two elements. This, together with the development of reassessment procedures[6] allowed the use of one element only.

SUMMARY OF SYSTEM PERFORMANCE AND CONCLUSIONS

Dosimetric characteristics of the NRPB PADC neutron personal monitoring system have been given above. The actual performance of the dosemeter in routine use is assessed by taking part in independent intercomparisons when these are available, and by a regular series of independent tests by a fictitious (dummy) customer. The dummy customer is issued dosemeters on a regular 13 week issue period, and irradiations are carried out, with or without tests of other influencing factors (effects of temperature for example) before returning the dosemeters for routine assessment. For the most recent dummy customer issue, the results were as follows: all of the unirradiated dosemeters (16 off) were reported as having zero dose; for 12 dosemeters irradiated at a dose equivalent (^{252}Cf) of conventional true value of 0.25 mSv, the mean reported dose equivalent showed a bias of 3%, and a coefficient of variation of 18%; for 12 irradiated at a conventional true value of 2.1 mSv, a bias of 2% and a coefficient of variation of 8%. It is considered that this performance is of a high standard. Nevertheless improvements to the performance are being actively pursued[11] and it is expected that the system will evolve further.

Table 1. Dosemeter readings in occupational fields.

Radiation field		Dosemeter reading		
		Per unit $H_p(10)^{(a)}$	Per unit $H_E^{(b)}$	Per unit $E^{(c)}$
Reprocessing plant	AP	1.6	2.2	1.9
	ROT	1.1	1.2	0.9
Magnox Reactor	ROT	1.8	1.5	1.1
PWR	AP	1.4	2.4	1.6
	ROT	1.25	1.3	0.8
Fuel storage	AP	1.5	2.5	1.8
	ROT	1.05	1.4	0.8
Cadarache (Simulated field)	AP	1.6	2.4	1.8

Notes: (Dosemeters reading normalised to average reading, AP and ROT, for ^{252}Cf, $^{(a)}H_p(10)$ average, $^{(b)}H_E$ average, $^{(c)}E$ average.

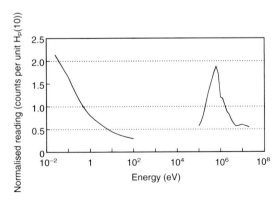

Figure 1. NRPB PADC dosemeter reading per $H_p(10)$ (old $Q(L)$) normalised to ^{252}Cf, normal incidence, slab calibration.

REFERENCES

1. Cartwright, B. G., Shirk, E. K. and Price, P. B. *A Nuclear Track Recording Polymer of Unique Sensitivity and Resolution.* Nucl. Instrum. Methods **153**, 457–460 (1978).
2. Tommasino, L. and Armellins, C. *New Etching Technique for Damage Track Detectors.* Radiat. Effects **20**, 253–255 (1973).
3. Adams, Jr, J. H. *A Curing Cycle for Detector Quality CR-39.* In: Solid State Nuclear Track Detectors. Eds P. Fowler and V. M. Clapham. pp. 145–148 (Oxford: Pergamon) (1981).
4. Bartlett, D. T. and Steele, J. D. *The NRPB CR39 Fast Neutron Personal Dosemeter.* In: Proc. 6th Int. Congr. IRPA, pp. 1111–1114 (Berlin: Fachverband für Strahlenschutz eV) (1984).
5. Tanner, R. J. and Bartlett, D. T. *A New Criterion for Comparing Track Etch Dosimetry Systems.* Radiat. Meas. **25**(1–4), 445–448 (1995).
6. Gilvin, P. J., Steele, J. D., Tanner, R. J. and Bartlett, D. T. *Re-assessment of PADC Neutron Dosemeter Results.* Nucl. Tracks. Radiat. Meas. **19**(1–4), 471–474 (1991).
7. Ahmad, S. and Stejny, J. *Polymerisation, Structure and Track Recording Properties of CR-39.* Nucl. Tracks Radiat. Meas. **19**, 11–16 (1991).
8. Tanner, R. J., Bartlett, D. T. and Hager, L. G. *The Energy and Angle Dependence of Neutron Response for PADC Manufactured using Two Different Cure Cycles.* Radiat. Meas. **25**(1–4), 467–468 (1995).
9. Bartlett, D. T., Steele, J. D. and Stubberfield, D. R. *Development of a Single Element Neutron Personal Dosemeter for Thermal, Epithermal and Fast Neutrons.* Nucl. Tracks **21**(1–6), 645–648 (1986).
10. Schraube, H. O. E. *Is It Possible to use CR-39 Detectors in Personal Dosimetry.* In: Strahlenschutzmesstechnik, Proc. 16th Annual Meeting of the Fachverband für Strahlenschutz, pp. 197–200, Report FS-83-30T (1983).
11. Tanner, R. J., Bartlett, D. T. and Lavelle, J. *Future Developments in the NRPB Neutron Personal Monitoring Service.* Radiat. Prot. Dosim. **70**(1–4), 121–126 (This issue) (1997).

DOSE EQUIVALENT RESPONSE OF PERSONAL NEUTRON DOSEMETERS AS A FUNCTION OF ANGLE

J. E. Tanner†, J. C. McDonald†, R. D. Stewart† and C. Wernli‡
†Pacific Northwest Laboratory
PO Box 999, Richland, WA 99352, USA
‡Paul Scherrer Institute, Villigen, CH-5232 Switzerland

Abstract — The measured and calculated dose equivalent response as a function of angle has been examined for an albedo-type thermoluminescence dosemeter (TLD) that was exposed to unmoderated and D_2O-moderated ^{252}Cf neutron sources while mounted on a $40 \times 40 \times 15$ cm^3 polymethylmethacrylate phantom. The dosemeter used in this study is similar to many neutron personal dosemeters currently in use. The detailed construction of the dosemeter was modelled, and the dose equivalent response was calculated, using the MCNP code. Good agreement was found between the measured and calculated values of the relative dose equivalent angular response for the TLD albedo dosemeter. The relative dose equivalent angular response was also compared with the values of directional and personal dose equivalent as a function of angle published by Siebert and Schuhmacher.

INTRODUCTION

At present, there is relatively little guidance available on the expected dependence of the dose equivalent response of personal dosemeters on the angle of radiation incidence for neutron exposures. International Commission on Radiation Units and Measurements (ICRU) Report 47[1] deals only with the determination of dose equivalent for photon and beta particle irradiations. This report contains tables of values for the photon air kerma to directional dose equivalent, but, until recently, no similar information was available for neutron irradiations.

Siebert and Schuhmacher[2] have published data for monoenergetic neutrons, that include fluence to directional and personal dose equivalent conversion factors. These data are also given as a function of the angle of incidence for the neutrons. The directional dose equivalent values are given for the ICRU sphere[3] and the personal dose equivalent values are given for the ICRU slab[1]. The ICRU slab phantom has a composition of 4-element ICRU tissue and has dimensions of $30 \times 30 \times 15$ cm$^{3\,[1]}$. More recently, Siebert and Schuhmacher have recalculated the personal dose equivalent for neutrons based on the new ICRU stopping power data for protons and alpha particles[4].

The present work made use of the calculated conversion factors to compute appropriate conversion coefficients as a function of angle for two commonly used reference neutron sources, namely the unmoderated and D_2O-moderated ^{252}Cf sources[5]. These data were then used to compare the measured angular dose equivalent responses of a TLD albedo neutron dosemeter. The dose equivalent response of this dosemeter was normalised to the value obtained at an irradiation angle of 0°.

The MCNP code[6] was used to model the detailed construction of the TLD albedo dosemeter and to calculate the dose equivalent response as a function of angle for the above mentioned neutron sources. The general features of the calculational model consisted of a point neutron source at a distance of 50 cm from the face of a polymethylmethacrylate (PMMA) phantom, on which the dosemeters were mounted for irradiation.

MATERIALS AND METHODS

Irradiations

The TLD albedo dosemeters were mounted on the surface of a $40 \times 40 \times 15$ cm^3 PMMA phantom. Irradiations were performed in a manner described previously by Jones et al[7]. Each set of dosemeters was exposed to approximately 2 mSv using either the unmoderated or D_2O-moderated ^{252}Cf source. Initial irradiations were performed using an angle between a line perpendicular to the phantom face and the source direction of 0°. Subsequent irradiations were performed by rotating the phantom clockwise to the various angles shown in the results below. Only clockwise rotational

Table 1. Calculated dosemeter responses for the Hanford Multipurpose Dosemeter, normalized to 0°.

Angle (deg)	Bare ^{252}Cf	Moderated ^{252}Cf
0	1.00	1.00
40	1.01	0.95
60	0.83	0.70
75	0.57	0.40
90	0.31	0.076
105	0.30	0.046
120	0.36	0.054
140	0.63	0.11
180	0.88	0.18

data were taken because the earlier results of Jones et al[7] indicated no significant angular asymmetries for a number of dosemeter types. The dosemeters were read out using an algorithm developed for evaluating personnel neutron dose equivalents at the Pacific Northwest Laboratory.

Calculations

The MCNP code, version 3B, was used to model the detailed geometry of the TLD albedo dosemeter. The dosemeter is comprised of five ^{6}LiF and ^{7}LiF chips with different filters. There are two ^{6}LiF chips, 'bare' and 'cadmium-covered'. The 'bare' chip has a tin filter between it and the phantom (or the person, when worn) while the 'Cd-covered' chip has the cadmium filter between it and the neutron source. The three ^{7}LiF chips are filtered with cadmium and tin in order to allow the measurement of photon dose equivalent at 0.07 and 10 mm depths and to allow the subtraction of the gamma ray response from the ^{6}LiF chips.

The neutron source was modelled as an isotropic point source with directional biasing toward the phantom to improve the calculational efficiency. The values calculated by MCNP were the energy deposited in the TLD material. Assuming the signal from the light output during processing is proportional to the energy deposited, these values give the relative dosemeter response. For frontal irradiations, MCNP transported 2 million neutrons to achieve relative errors of the order of 1–3%. For angles when the phantom was shielding the dosemeter, it was necessary to transport 10 million histories to achieve the same degree of precision. The overall uncertainty of the MCNP calculations is difficult to estimate since it includes uncertainties in cross sections and in the accuracy of the simulation of the source and detector. The total uncertainty in the calculations may amount to 5%.

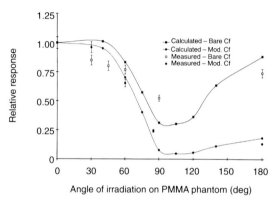

Figure 1. Calculated and measured response of the Hanford Multipurpose Dosemeter as a function of irradiation angle.

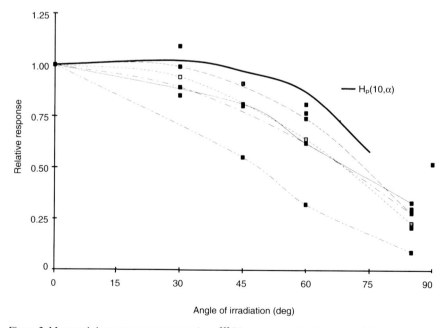

Figure 2. Measured dosemeter responses to bare ^{252}Cf compared with the personal dose equivalent.

RESULTS

The relative dosemeter responses calculated by MCNP for bare and moderated ^{252}Cf neutron sources are given in Table 1 for various irradiation angles. These values are compared against the measurements performed in PNL's Low Scatter Facility with the Hanford Multipurpose Dosemeter (Figure 1). The shape of the measured dosemeter response curve is predicted quite well by the calculations.

The most recent fluence to personal dose equivalent conversion factors for bare and moderated ^{252}Cf, normalised to an irradiation angle of 0°, are plotted in Figures 2 and 3, along with the measured responses of the Hanford Multipurpose Dosemeter and other DOE facility dosemeters[7]. The DOE dosemeters included TLD albedo, film and etched track detectors. The personal dose equivalent appears to form an upper limit of the dosemeter responses which would result in the dosemeter underestimating the desired operational quantity for neutron monitoring.

The personal dose equivalent conversion factor has only been recalculated for irradiation angles less than 90°. However, the previously reported values, which extend out to 180°, are used to provide information on the relative response. The personal dose equivalent, the directional dose equivalent and the effective dose equivalent computed for a male phantom[8] are compared with the calculated dosemeter response in Figure 4. The shapes of the curves are very similar to the dosemeter response curve except at the angles approaching 180° where the dosemeter response increases more rapidly than the personal dose equivalent and the directional dose equivalent. However, this increase may be less pronounced on the human body which, on average, has a greater depth than the slab phantom.

CONCLUSIONS

The angular response characteristics of current neutron personal dosemeters seem to be adequate to estimate the personal dose equivalent and the effective dose equivalent, or effective dose, for most angles of irradiation when the dosemeter is worn on the front of the torso. Unlike radiation survey instruments, personal dosemeters should not be designed to have an isotropic response. Instead, the ideal angular response would follow the operational quantity or the protection quantity.

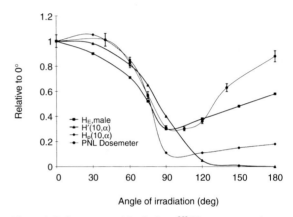

Figure 4. Reference quantities for bare ^{252}Cf neutrons as a function of irradiation angle.

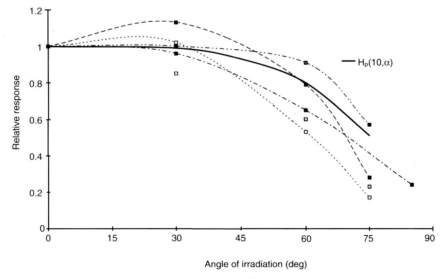

Figure 3. Measured dosemeter responses to D_2O-moderated ^{252}Cf compared with the personal dose equivalent.

REFERENCES

1. International Commission on Radiation Units and Measurements (ICRU). *Measurement of Dose Equivalents from External Photon and Electron Radiations*. ICRU Report 47 (Bethesda, MD: ICRU Publications) (1992).
2. Siebert, B. R. L. and Schuhmacher, H. *Calculated Fluence-to-Directional and Personal Dose Equivalent Conversion Coefficients for Neutrons*. Radiat. Prot. Dosim. **54**(3/4), 231–238 (1994).
3. International Commission on Radiation Units and Measurements (ICRU). *Radiation Quantities and Units*. ICRU Report 33 (Bethesda, MD: ICRU Publications) (1980).
4. Siebert, B. R. L. and Schuhmacher, H. *Quality Factors, Ambient and Personal Dose Equivalent for Neutrons, Based on the New ICRU Stopping Power Data for Protons and Alpha Particles*. Radiat. Prot. Dosim. **58**(3), 177–183 (1995).
5. International Organization for Standardization (ISO). *Neutron Reference Radiations for Calibrating Neutron-measuring Devices used for Radiation Protection Purposes and for Determining their Response as a Function of Neutron Energy*. ISO 8529 (Geneva: ISO) (1989).
6. Briesmeister, J. (Ed.). *MCNP — A General Monte Carlo Code for Neutron and Photon Transport*. Report LA 7396-M, Rev. 2 (Los Alamos, USA) (1986, Revised 1991).
7. Jones, K. L., Roberson, P. L., Fox, R. A., Cummings, F. M. and McDonald, J. C. *Performance Criteria for Dosimeter Angular Response*. Pacific Northwest Laboratory Report PNL-6452. (Pacific Northwest Laboratory, Richland, Washington) (1988).
8. Stewart, R. D., Tanner, J. E. and Leonowich, J. A. *An Extended Tabulation of Effective Dose Equivalent from Neutrons Incident on a Male Anthropomorphic Phantom*. Health Phys. **64**(4), 405–413 (1993).

MEASUREMENT OF NEUTRON DOSE ON A FUSION REACTOR SHIELD USING TLD-300 PHOSPHORS

M. Angelone†, P. Batistoni†, A. Esposito‡, M. Martone†, M. Pelliccioni‡, M. Pillon† and V. Rado†
†Associazione EURATOM-ENEA sulla Fusione
Centro Ricerche Frascati, C.P. 65-00044 Frascati (Rome), Italy
‡Istituto Nazionale Fisica Nucleare (I.N.F.N.)
Laboratori Nazionali Frascati, via E. Fermi 40, 00044 Frascati (Rome), Italy

Abstract — A benchmark experiment was carried out at the 14 MeV Frascati Neutron Generator (FNG) to measure the nuclear heating in an experimental assembly simulating a fusion reactor shield and its superconducting coils. TLD-300 detectors were used allowing the neutron and gamma contribution to the total nuclear heating to be separated by using the two-peak method. The experiment aimed at validating the EFF and FENDL neutron transport cross section files presently available for nuclear heating calculations of fusion reactor shields. The calculations were performed using the Monte Carlo code MCNP. The measured and calculated gamma absorbed dose agree well while differences up to 30% were found for the neutron absorbed dose when the EFF file is used. The discrepancy increases up to 50% when the FENDL file is used to calculate the neutron absorbed dose.

INTRODUCTION

Next step fusion devices like the International Thermonuclear Experimental Reactor (ITER)[1] will produce 1.5 GW of fusion power, most of which is released via 14 MeV neutrons produced by the D-T fusion reaction. The energy taken away by the neutrons is deposited into the reactor structure (nuclear heating) by means of the neutron (n,x) type reactions (where x stands for charged particles p,α etc.) and of the interactions of (prompt) gamma rays. Prompt gamma rays are produced by the neutron interactions. Since the nuclear heating can seriously compromise the properties of the superconducting coils (which produce the strong magnetic field necessary to operate the tokamak) the neutron fluence rate must be attenuated by means of a large shield. In ITER the shield ($\approx$1 m thick) is made essentially of water and stainless steel (SS). The optimisation of the shielding performances is a fundamental task of the Engineering Design Activities (EDA) of the ITER project and demands reliable information on the absorbed dose distribution throughout the shield. Calculation techniques and tools are available for nuclear heating studies, but up to now they have not been thoroughly tested against experimental data. Measured neutron and gamma absorbed doses are thus required to provide direct insight into the quality of the available calculation tools.

The absorbed dose measured in a shielding block simulating the shield of a fusion reactor like ITER by using TLD-300 ($CaF_2:Tm$) dosemeters is here reported. Since the experiment is carried out in a mixed gamma–neutron field, using the two main peaks of TLD-300 (the so called peak 3 and peak 5 which show different sensitivities to gamma and neutrons[2]), the phosphors can be used to separate the gamma and the neutron absorbed doses. TLD-300 was chosen because of its sensitivity to low gamma ray dose and of its high Z_{eff} (Z_{eff} = 16.2), which is not too far from that of stainless steel (Z_{eff} = 26.3). TLD-300 is widely used in health physics and radiotherapy, overall in gamma–neutron mixed fields, but as far as is known to the authors, this is the first attempt to use the TLD-300 dosemeters in this kind of experiment and to get experimental information on the neutron absorbed dose.

EXPERIMENTAL

The experiment was carried out at the 14 MeV Frascati Neutron Generator (FNG)[3] of ENEA Frascati which produces up to 1.0×10^{11} s^{-1} 14 MeV neutrons by using the D-T fusion reaction. The experimental assembly (Figure 1) was composed of a block (61 cm thick) made of 9 SS-316 plates, each 5 cm thick, alternated with 8 water equivalent material plates (Perspex), each 2 cm thick. This shielding block was followed by a smaller block, 30 cm thick, composed of SS-316 and copper plates, 2 cm thick, simulating the superconducting coils. The experimental assembly was positioned 5.8 cm from the FNG target. The nuclear heating was measured as a function of penetration depth inside the two blocks using TLD-300. The TLD-300 dosemeters were chips of 3.2×3.2 mm^2 and 0.9 mm thick, from the Harshaw company. Nine TLD-300 chips were located in each experimental position in a SS-316 holder and irradiated for 2.5 h. The average neutron yield was 6.5×10^{10} s^{-1} ±2%. For each experimental position the measured TL signal was the averaged value among the nine phosphors and the uncertainty was the associated standard deviation. The total measured doses ranged from 4.9×10^{-19} Gy per neutron up to 1.9×10^{-14} Gy.n^{-1}, which cover the whole linear response range for TLD-300. The experimental as well as the calculated total absorbed doses, the latter obtained using the Monte Carlo code MCNP[4], the EFF[5] and the FENDL[6] cross section files, are reported elsewhere[7]. In this paper the work of separating the neutron and gamma absorbed dose is stressed.

The TLD-300 dosemeters were calibrated in a secondary standard gamma radiation field, using a ^{60}Co source calibrated at 0.5% up to 3.5% depending on the dose level. The calibration ranged from 10 μGy up to 35 Gy. The peak area was used to measure the TL signal as is suggested by Gibson[8]. This allowed the two peaks to be used as independent dosemeters. The calibration was performed both for the total glow curve signal and for the signal due to peak 3 and peak 5 separately. The background signal was measured by means of 24 phosphors which followed the same history as the dosemeters used for the measurements.

The phosphors were read using a Vinten-Rialto reader with a heating rate of 10°C.s^{-1} and maximum reading temperature of 350°C. A pre-heating of 20 min at 80°C was used to eliminate peak 2 which has a fading time of 15 h. The annealing procedure was 1 h at 400°C.

NEUTRON–GAMMA DOSE SEPARATION AND EXPERIMENTAL RESULTS

Measuring the response of the two main peaks of TLD-300, it is possible to separate the neutron and gamma absorbed doses. The corresponding equations, accounting for the fact that the phosphors are embedded in stainless steel are:

$$R_\nu = h_\nu D_\gamma^* + k_\nu C_n D_n \quad (1)$$

where R_ν is the measured TLD response, normalised to a ^{60}Co gamma ray source, for peak ν (ν = 3 or 5), k_ν and D_n are the neutron sensitivity (for peak ν) referring to the ^{60}Co gamma ray and the neutron absorbed dose (in SS) respectively. C_n is the neutron kerma ratio between the TLD-300 and the surrounding material (SS), D_γ^* is the gamma ray absorbed dose (in SS) and h_ν is the photon sensitivity of peak ν in the mixed n-γ field, relative to the γ rays used for calibrating the detectors. In the following, using the results discussed in Ref.

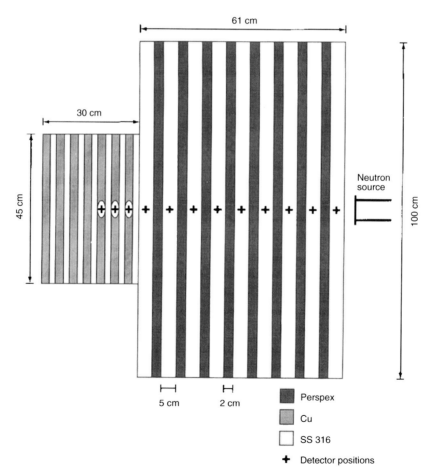

Figure 1. Lay-out of the experimental assembly.

7, h_ν is made = 1. Calculations have shown[7] that close to the neutron target the phosphor response (for peak 5) could be in the supralinear zone. The supralinear zone for peak 5 arises above 2 Gy[9]. To account for the expected supralinearity in Equation 1 one can put $D_\gamma^* = f_\nu D_\gamma$, where f_ν is the supralinear factor for peak ν as defined by Horowitz[10]. A study of the supralinear behaviour of our TLD-300 detectors was also carried out and the results are summarised in Figure 2. In our study of TLD-300 supralinearity it was found that $f_5 > 1$ for $D_\gamma > 2$ Gy and $f_3 = 1$ up to 15 Gy. These results substantially confirmed those reported elsewhere[9]. From the literature[9] we know that the supralinear response for the neutron dose arises for $D_n > 100$ Gy. Since in our experiment such a high neutron dose is never reached, in the following this effect is not considered for D_n. Solving Equation 1 we get for D_γ and D_n:

$$D_\gamma = \frac{\left(R_5 - \dfrac{k_5}{k_3} R_3\right)}{\left(f_5 - f_3 \dfrac{k_5}{k_3}\right)} \quad (2)$$

$$D_n = \frac{(f_5 R_3 - f_3 R_5)}{C_n(f_5 k_3 - f_3 k_5)} \quad (3)$$

Equations 2 and 3 depend on the C_n and k_ν parameters and some discussion is necessary about the data to be used. As already stated C_n is the kerma ratio between CaF_2 and the surrounding material. Its use allows the neutron absorbed dose inside the stainless steel to be obtained, which indeed is the goal of the experiment. This ratio can be calculated from data available in the literature or, as in the present work, by performing a neutron and gamma Monte Carlo transport calculation[7]. The calculated C_n values vary slightly with the experimental position (<5%) and the result of its averaged value is $C_n = 3.51 \pm 0.06$. As far as the k_ν values are concerned, from the literature it is known[2] that the neutron sensitivity of the two main peaks of TLD-300 varies with the neutron energy. In our experiment there is a strong variation of the neutron field, going from a position close to the target up to the deepest position into the coils. To account for this variation, the data available[11] for $k_\nu(E)$ were fitted, collapsed into group structure using (g is the group index):

$$k_\nu^g = \frac{\displaystyle\int_{E_g}^{E_{g+1}} k_\nu(E)dE}{\displaystyle\int_{E_g}^{E_{g+1}} dE} \quad (4)$$

and thus weighted for each experimental position j over the neutron flux spectra ϕ_g^j obtained from Monte Carlo calculation[7] (NG energy groups):

$$\langle k_\nu^j \rangle = \frac{\displaystyle\sum_1^{NG} k_\nu^g \phi_g^j}{\displaystyle\sum_1^{NG} \phi_g^j} \quad (5)$$

The $\langle k_\nu^j \rangle$ values are gathered in Table 1 together with other data of interest. The experimental results for D_γ and D_n are gathered in Table 2, where they are also compared with the calculated ones[7]. The measured D_γ

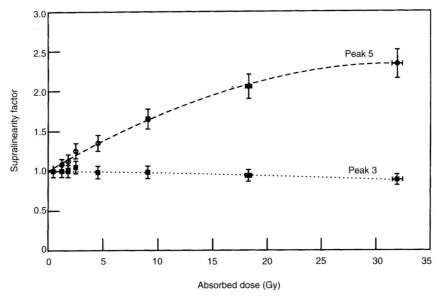

Figure 2. Measured supralinearity factor (f_ν) for peak 3 and peak 5 of TLD-300 plotted against absorbed dose (Gy).

and D_n are affected by an uncertainty less than ±15% and approximately equal to ±30% respectively. The total error was calculated using a quadratic propagation law. The overall contribution to the uncertainty for D_n comes from k_3 and k_5 which are known to ±15%, the uncertainty for f_5 is ±7%.

Since k_3 and k_5 are evaluated using the same set of data[11] and the same procedure when, as in Equation 2, the k_3 and k_5 values are used in the ratio, their contribution to the uncertainty cancel out. For this reason, the uncertainty for D_γ is smaller than for D_n.

DISCUSSION

As far as it is known to the authors, up to now the two-peak method with TLD-300 has been mainly employed in water phantoms. Water has a different performance (as a neutron and gamma shield) with respect to a stainless steel + water shield. The neutron energy released in water is mainly due to recoil of protons, as a consequence of the elastic scattering with hydrogen. In stainless steel the neutron energy is mainly released via nuclear reactions of the type (n,x). Since the energy released in collisions with hydrogen in water is greater than the energy released via (n,x) reactions in stainless steel, a greater energy is deposited around the collision point in water rather than in stainless steel. As far as the gamma ray energy deposition is concerned, water is rather 'transparent' to photons while, due to its higher Z value, stainless steel is much more effective for gamma ray attenuation, thus the D_γ/D_T ratio (D_T is the total absorbed dose) is higher in SS than in water. In our case D_γ/D_T is in the range 0.7–0.9. The neutron absorbed dose is about 30% of the total in the first position and drops down to less than 7% at distances, z, more than 25 cm, thus resulting in comparable or even less than the experimental uncertainty associated with each experimental response R_ν, which is less than ±12%. At distances more than z > 25 cm the neutron absorbed dose is thus masked by the quoted uncertainty on R_ν. This is an intrinsic limit of the two-peak method and to overcome it requires a great improvement in the measurement accuracy.

As far as the calculation is concerned, the present work shows a large difference (up to 25%) between the gamma and neutron absorbed dose obtained using EFF and FENDL cross section files (Table 2). The gamma dose calculated using the EFF cross section file agrees rather well with the measurements. For D_n the agreement is rather good in the first two positions but at deeper positions the measurement overestimates the calculated neutron dose by up to 30%. The results obtained using the FENDL file show larger differences which are up to 50% in the case of D_n and up to 22% for the gamma dose. A probable cause of the differences could be due to the kerma values used in FENDL since a com-

Table 1. Data and parameters used to solve Equations 2 and 3. The quoted uncertainty is from 6% up to 8% for R_3 and from 7% up to 12% for R_5.

Depth (cm)	$\bar{E}_n$ (MeV)	f_5	$\langle k_3 \rangle$	$\langle k_5 \rangle$	R_3 (Gy)	R_5 (Gy)
2.50 SS	7.40	1.65	1.09×10^{-1}	3.73×10^{-1}	9.38	17.75
9.51 SS	3.14	1.21	7.33×10^{-2}	2.58×10^{-1}	2.78	3.73
16.52 SS	1.93	1.08	6.24×10^{-2}	2.22×10^{-1}	1.09	1.21
23.40 SS	1.43	1.0	5.76×10^{-2}	2.06×10^{-1}	4.56×10^{-1}	4.85×10^{-1}

Table 2. Comparison between measured (after supralinearity correction) and calculated neutron and gamma absorbed doses (Gy per neutron). The total neutron yield was 6.5×10^{14}. The experimental uncertainty for D_γ is <±12% while for D_n it is ≈±30%. Calculation uncertainty was <±3%.

Exp. pos. (cm)	Measured	Calculated EFF	C/E	Calculated FENDL	C/E	FENDL/EFF
Neutron dose						
2.50	5.66×10^{-15}	5.49×10^{-15}	0.97	4.38×10^{-15}	0.77	0.80
9.51	1.01×10^{-15}	9.21×10^{-16}	0.91	6.97×10^{-16}	0.69	0.76
16.52	3.31×10^{-16}	2.30×10^{-16}	0.70	1.70×10^{-16}	0.51	0.78
23.40	8.74×10^{-17}	6.60×10^{-17}	0.76	5.17×10^{-17}	0.59	0.84
Gamma dose						
2.50	1.24×10^{-14}	1.30×10^{-14}	1.05	9.71×10^{-15}	0.78	0.75
9.51	4.04×10^{-15}	4.61×10^{-15}	1.14	3.56×10^{-15}	0.88	0.77
16.52	1.61×10^{-15}	1.78×10^{-15}	1.11	1.46×10^{-16}	0.90	0.82
23.40	6.89×10^{-16}	7.35×10^{-16}	1.07	6.42×10^{-16}	0.93	0.87

parison between calculated neutron and gamma energy spectra did not show any relevant difference between EFF and FENDL results within the quoted total errors.

New and accurate data on the neutron sensitivity are also requested to reduce the experimental uncertainty on the measured doses. It would be interesting to measure k_ν in the experimental assembly used and compare them with the data averaged over the neutron spectrum and used in the present work. The feasibility of such an experiment is presently under investigation. Since this was the first attempt to measure the neutron dose in this kind of experiment, more experimental data are necessary before drawing conclusions about the capability of the calculation to predict the absorbed doses in the stainless steel.

REFERENCES

1. Rebut, P. H. *ITER, the First Experimental Fusion Reactor*. Fus. Eng. Des. **30**, 85–105 (1995).
2. Hoffmann, W. and Prediger, B. *Heavy Particle Dosimetry with High Temperature Peaks of CaF_2:Tm and ^7LiF Phosphors*. Radiat. Prot. Dosim. **6**, 149–152 (1983).
3. Martone, M., Angelone, M. and Pillon, M. *The 14 MeV Frascati Neutron Generator*. J. Nucl. Mater. **212–215**, 1661–1665 (1994).
4. Briesmeister, L. (Ed.) *A General Monte Carlo N-particle Transport Code, Version 4-A*. (Los Alamos National Laboratory) LA 12625-M (Nov. 1993).
5. Gruppelaar, H. *Status of the European Fusion File (EFF)*. EFF-DOC-17 (July 1988).
6. Mann, F. M. *et al. Processing of FENDL-Pa/1.1*. Westinghouse Handford Company Report WHC-EP-0727 (February 1994).
7. Batistoni, P., Angelone, M., Pillon, M. and Rado, V. *Nuclear Heating Experiments for the Validation of the Fusion Reactor Shielding Performance*. Fus. Eng. Des. (submitted).
8. Gibson, J. A. B. *The Relative Tissue-kerma Sensitivity of Thermoluminescent Materials to Neutrons*. Report EUR 10105 EN (1985).
9. Rassow, J., Klein, C. and Meissner, M. *Supralinearity Behaviour of TLD-300 and TLD-700*. Radiat. Prot. Dosim. **23**(1/4), 409–412 (1988).
10. Horowitz, Y. *Thermoluminescence and Thermoluminescent Dosimetry*, Vol. II, pp. 1–41 (Boca Raton, Florida: CRC Press) (1984).
11. Dielhof, J. B., Bos, A. J. J., Zoetelief, J. and Broerse, J. J. *Sensitivity of CaF_2 Thermoluminescent Materials to Fast Neutrons*. Radiat. Prot. Dosim. **23**(1/4), 405–408 (1988).

TL DOSIMETRY IN HIGH FLUXES OF THERMAL NEUTRONS USING VARIOUSLY DOPED LiF AND KMgF$_3$

G. Gambarini†, M. Martini†, A. Scacco‡, C. Raffaglio† and A. E. Sichirollo§
†Dipartimento di Fisica dell'Università di Milano and INFN, Milano, Italy
‡Dipartimento di Fisica dell'Università 'La Sapienza', Roma, Italy
§Istituto Nazionale per lo Studio e la Cura dei Tumori, Milano, Italy

Abstract — ^{6}LiF enhanced dosemeters exposed to high fluences of thermal neutrons undergo irreversible radiation damage, preventing their subsequent utilisation. ^{6}LiF depleted dosemeters do not show such an effect, but in mixed fields of high fluxes with thermal neutrons they do not provide a simple way of discriminating among the contributions of the field components. The responses to thermal neutrons and to γ rays of some TL single crystals have been investigated. LiF:Mg,Cu,P and LiF:Cu^{2+} single crystals have shown a much higher sensitivity to thermal neutrons than to γ rays. KMgF$_3$, although promising for other dosimetry purposes, shows no useful features in such radiation fields.

INTRODUCTION

Thermoluminescence dosemeters (TLDs) play a fundamental role in the determination of the absorbed dose in tissue exposed to mixed fields of thermal neutrons and γ rays. In radiotherapy treatments utilising thermal neutrons, such as boron neutron capture therapy (BNCT), the spatial distribution of the absorbed dose is not uniform and three-dimensional determinations are necessary for good therapy planning. TLDs, for their small dimensions and their tissue equivalence for most radiations, are widely utilised because they provide the possibility of mapping the absorbed dose distribution without significantly perturbing the radiation field. The combination of ^{6}LiF-^{7}LiF dosemeters in (n$_{th}$,γ) mixed fields with a very low thermal neutron component allows discrimination between the contributions of the two field components. This choice has not proved itself useful for dose determinations in high fluxes of thermal neutrons, such as those required for radiotherapy, because ^{6}Li enhanced dosemeters undergo irreversible radiation damage, unaffected by conventional annealing. The problem of separating the contributions of thermal neutrons and γ rays in radiation fields with a high thermal neutron component is a very important topic, because γ rays are unavoidably present in neutron fields, and the dose in tissue from γ rays and from thermal neutrons, even at fluences of 10^{12}–10^{13} n.cm^{-2}, may be of the same order.

Some new phosphors have also been investigated, in order to test their sensitivity to thermal neutrons and with the purpose of examining whether the glow curve shapes after thermal neutron or γ irradiation are similar or if they show useful differences that may make discrimination easier. Some interesting results have been obtained with LiF:Mg,Cu,P and LiF:Cu^{2+} single crystals.

TLD IRRADIATION AND READING FACILITIES

All γ irradiations were performed at a rate of about 0.14 Gy.s^{-1} in a ^{137}Cs irradiator.

The exposures to thermal neutrons were performed in the swimming-pool-type facility of a TRIGA MARK II nuclear reactor, near the channel door where the thermal neutron fluence rate, with the reactor operating at 250 kW, was 1.44×10^8 n.cm^{-2}.s^{-1}, uniform in a region of 8×8 cm^2, and the cadmium ratio is about 91.

For TLD analysis a Harshaw/Filtrol system was used, composed of a Model 2000A detector, interfaced to a Model 2080 Picoprocessor, and a Model 3500 TLD reader from Harshaw/Bicron; these systems provide glow curve registration and graphic representation.

Some dosemeters from each of the investigated groups, with particular interest in crystals and new materials, were also analysed using a home-made high sensitivity spectrometer for wavelength resolved thermoluminescence measurements.

RESPONSE OF LiF:Mg,Ti AND LiF:Mg,Cu,P CHIPS

Owing to the high cross section of ^{6}Li for the reaction with thermal neutrons, ^{6}Li(n,α)^{3}H, LiF dosemeters of different isotopic composition, exposed in thermal neutron fields, present different sensitivities and different shapes of glow curve.

Commercial TLDs from Harshaw Chemical Co., in the form of chips ($3.1 \times 3.1 \times 0.9$ mm^3) have been investigated: TLD-600, with 96.5% ^{6}Li, TLD-700, with 99.99% ^{7}Li, and TLD-100 with the natural isotopic composition, i.e. 7.5% ^{6}Li.

Before each irradiation, all chips were annealed at 400°C for 1 h, 100°C for 2 h. No post-irradiation low temperature annealing was used, but all the readouts were made 48 h after irradiation to allow the decay of low temperature peaks. A constant heating rate of 4°C.s^{-1} up to 400°C was employed.

LiF:Mg,Cu,P phosphors have aroused great interest, in recent years, for many characteristics. In particular, as regards thermal neutrons, ^{6}LiF:Mg,Cu,P chips have shown[1] a response about 9 times higher than that of ^{6}LiF:Mg,Ti, linear in the low dose range. ^{6}LiF:Mg,Cu,P

(GR-206A) and ^{7}LiF:Mg,Cu,P (GR-207A), produced in the People's Republic of China (Beijing Radiation Detector Works) in the form of circular chips 4.5 mm diameter and 0.8 mm thick, were investigated. The recommended annealing procedure was used, i.e. 240°C for 10 min, followed by a rapid cooling to room temperature. The readout was performed, 3 days after exposure, using a constant heating rate of 4°C.s^{-1}, up to 240°C.

The dosemeters have been variously irradiated in the swimming-pool-type facility of the TRIGA MARK II reactor, in the same flux but for different times; the resulting fluences are in the range up to about 10^{12} n.cm^{-2}. In order to make comparisons, some chips were exposed in the same position and for the same time, but surrounded by a ^{6}LiF screen 0.7 g.cm^{-2} thick. This shield gives a γ ray attenuation of less than 1%, and a thermal neutron attenuation of a factor 10^3. The shapes of the flow curves of TLD-600 detectors irradiated up to different fluences are appreciably different, showing that the population of the various peaks, which is dependent on LET, is also dependent on dose. This fact is probably a peculiarity of LiF; already in 1978[2] such a dependence on LET and dose was found in the height ratio (peak 6 to peak 5) in thin LiF layers.

From the analysis of glow curves of shielded and unshielded dosemeters, it is deduced that for TLD-700 exposed in the above mentioned mixed (n$_{th}$,γ) field, the contributions to the total area from n$_{th}$ and γ rays are comparable (53% and 47% respectively).

In the investigated fluence interval, up to 1.6×10^{12} n.cm^{-2}, the TLD-600 response is nearly linear while GR-206A shows a noticeable lack of linearity, as is evi γ rays, and this is not an easily resolvable problem. the response, which is unaffected by annealing, was found. A similar behaviour to that in TLD-100 chips was observed. In Figure 2, the reduction factor of the response to γ rays of TLDs exposed to n$_{th}$ and afterwards annealed is shown as a function of the total n$_{th}$ fluence under previous exposure.

The above results show that ^{6}LiF chips after exposure to high fluences of thermal neutrons departs from linearity, moderately for the Mg,Ti doped phosphors and markedly for the Mg,Cu,P doped ones; moreover they reveal irreversible radiation damage which invalidates their new utilisation after such exposures. In contrast,

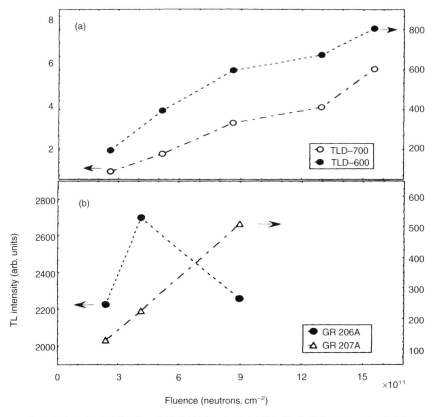

Figure 1. Response of (a) LiF:Mg,Ti (TLD-600 and TLD-700) and (b) LiF:Mg,Cu,P (GR-206A and GR-207A) exposed to high fluxes of thermal neutrons.

^{7}LiF chips, doped both with Mg,Ti or with Mg,Cu,P, do not show such radiation damage in the fluence investigated interval.

RESPONSE OF SOME TL SINGLE CRYSTALS TO THERMAL NEUTRONS

The observation of the poor reliability of the results, obtained in mixed fields of thermal neutrons by measuring the absorbed dose utilising the usual commercial TLDs, has caused our research to be directed to the investigation of other TL materials, not damaged by thermal neutrons and whose emission characteristics allow reliable dose determinations. The investigated materials are LiF single crystals, doped with Mg, Cu, P and doped only with Cu; and KMgF$_3$ doped with Eu, or Tl, or Ag.

Sample preparation and experimental procedure

Crystalline samples of both KMgF$_3$ and LiF have been obtained from the melt by using the Czochralski technique. The molten mass was formed by heating the starting powder in a platinum crucible and under nitrogen gas atmosphere; doped crystals are obtained by adding a proper amount of impurity to the melt. In general, the dopant concentration in the crystals does not coincide with that in the melt and is not constant along the ingot. Such effects are caused by known segregation phenomena during growth. For KMgF$_3$ doped samples, the starting powder consisted of a stoichiometric mixture of KF and MgF$_2$ to which was added TlCl 1 mol% or AgI 12 mol% or EuBr$_3$ 5 mol%. Single crystals of LiF:Mg,Cu,P were grown from molten LiF powder containing 0.2 mol% of MgF$_2$, 0.004 mol% of CuF$_2$ and 1.95 mol% of NH$_4$H$_2$PO$_4$. Impurity concentration, as well as the air environment for the growth, were chosen in agreement with those previously reported for the highest thermoluminescence sensitivity. LiF:Cu samples were obtained by adding to the melt 0.05 mol% or 0.5 mol% of CuF$_2$. All the samples were slabs cleaved from ingots and polished to obtain good thermal contact. Crystals were annealed at 500°C for 30 min, quenched to RT and irradiated with different fluences of thermal neutrons. The same procedure was followed for γ irradiations, which were performed before and after each neutron irradiation, to control sample characteristics and possible damage. TL measurements were performed in the range between RT and 400°C, with a heating rate of 4°C.s^{-1}.

Experimental results

LiF:Mg,Cu,P

The analysis of samples after subsequent γ and n$_{th}$ irradiations have shown a modification in the structure of the glow curve. The glow curves of LiF:Mg,Cu,P single crystals are shown in Figure 3.

After γ irradiation, the glow curve displays two pronounced peaks at 230°C and 350°C and two shoulders on both sides of the 230°C peak and above 350°C. These peaks are also present after β irradiation[3]. Thermal neutron irradiation mostly increases peaks at 260°C and 280°C and seems to quench the 350°C peak in favour of two peaks at 330°C and 370°C. This anomalous behaviour is observed to be persistent after a subsequent γ irradiation. The response to γ and neutrons decreases

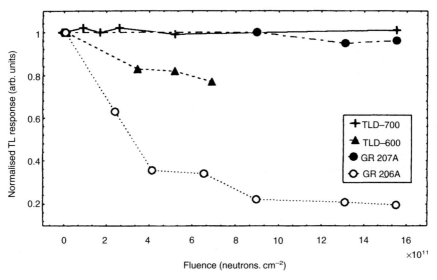

Figure 2. Comparison between commercial ^{6}Li enriched and depleted LiF, for γ ray irradiation after thermal neutron exposure at different fluences. TL intensities are normalised to the response after a 1 Gy γ irradiation.

with use. The absolute value of the thermoluminescence signal is much lower (by at least three orders of magnitude) than that obtained from commercial chips. The thermoluminescent emission is at 410 nm.

LiF:Cu

The glow curve of LiF:Cu displays a dominant peak at 150°C and much weaker secondary peaks at 220°C, 270°C and 330°C, whose relative height is doping-dependent. The area of the main peak is strongly dependent on the irradiation dose. The glow curve after thermal neutron exposure displays a strong population of the 270°C and 330°C peaks, proportional to the fluence. These samples display a good linearity for γ irradiation in the range 1–5 Gy in tissue. Light sensitisation was noticed after a 10^3 Gy dose and limited damage (about 10%) after strong neutron irradiation.

$KMgF_3$

Some fluoroperovskite crystals have been investigated because these phosphors, recently proposed as dosimetric materials[3], have shown some interesting features. In particular, Tl doped crystals have revealed[4] a linear response even up to doses of 10^3 Gy. $KMgF_3$ single crystals doped with Tl, or Eu, or Ag have been investigated. 3-D analysis of luminescence as a function of temperature and wavelength has been made, both after γ and after n_{th}-exposure. Tl and Ag doped crystals have shown their main emission in the spectral range 240–260 nm; that is, outside the wavelength range (350–700 nm) of the commonly used commercial readers. $KMgF_3$:Eu samples show a high peak centred at 360 nm, and the readout obtained from commercial readers is higher than that from LiF chips. In Figure 4 the isometric plot of a $KMgF_3$:Eu crystal after γ irradiation is shown.

The responses after a dose of 1 Gy in tissue from γ rays or from n_{th} are of the same order. With reference

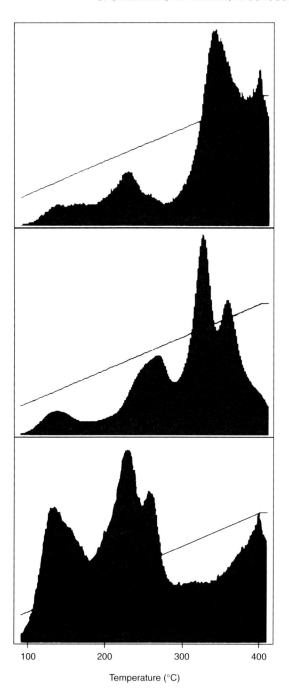

Figure 3. Glow curves for LiF:Mg,Cu,P crystals after consequential exposure to (a) 1 Gy γ irradiation, (b) thermal neutrons (fluence 3.66×10^{11} n.cm^{-2}) and (c) 5 Gy γ irradiation. Heating rate 4°C.s^{-1}, annealing at 500°C for 30 min before each irradiation.

Table 1. Comparison between different phosphors. TL response (R) after gamma or thermal neutron exposure.

Phosphor	R_γ/R_γ (TLD-700) $\times$ 10^{-2}	R_n/R_n (TLD-700) $\times$ 10^{-2}	R_n/R_n (TLD-600) $\times$ 10^{-3}
LiF:Mg,Cu,P	0.2	160	3.7
LiF:Cu 0.5 mol%	1.1	150	3.3
LiF:Cu 0.05 mol%	2.6	340	7.7
$KMgF_3$:Ag	5.8	3.3	0.07
$KMgF_3$:Eu	320	630	14.3
$KMgF_3$:Tl	4.4	5.3	0.12

to the possibility of discriminating between contributions in such mixed fields, no different emission characteristics after n_{th} or γ irradiation were discovered.

Final remarks

In Table 1 the responses (evaluated from total area) obtained with the investigated phosphors are reported, normalised to the mass and to the response of TLD-700 or TLD-600 chips. The responses to neutrons are determined on the basis of the first exposure to the same fluence of 4×10^{11} n.cm^{-2}.

Observing the values reported in the table, one can verify that the investigated LiF crystals show a response to thermal neutrons a little higher than TLD-700; in contrast, their response to γ rays is much lower. In the table the high response of KMgF$_3$:Eu crystals is also evident both for γ rays and thermal neutrons.

CONCLUSIONS

For high thermal neutron fluences, TLDs with high sensitivity to thermal neutrons are irreversibly damaged and can only be utilised once, and this makes reliable calibration difficult.

Dosemeters with low sensitivity for thermal neutrons are not damaged, but they require discrimination between contributions from thermal neutrons and from γ rays, and this is not an easily resolvable problem.

Single LiF crystals doped with Mg,Cu,P or with Cu present the promising characteristics of a considerably higher sensitivity to thermal neutrons than to gamma rays, they are therefore worthy of an exhaustive study to test if and how reproducibility and linearity features may be obtainable.

This work does not show conclusive results on TLD behaviour in n_{th} fields; for better defined deductions a broad systematic study is needed. It can only be concluded that the determination of thermal neutron and γ ray components with such dosemeters in mixed (n_{th},γ) fields with a high n_{th} component is especially difficult, and it is worthy of further study.

ACKNOWLEDGEMENTS

The authors are grateful to Dr G. Catolla (Tecnologie Avanzate, Torino, Italy) for allowing them to make measurements with Harshaw readers and to Mr C. Sanipoli (Università La Sapienza, Roma, Italy) for growing all crystals used in this work.

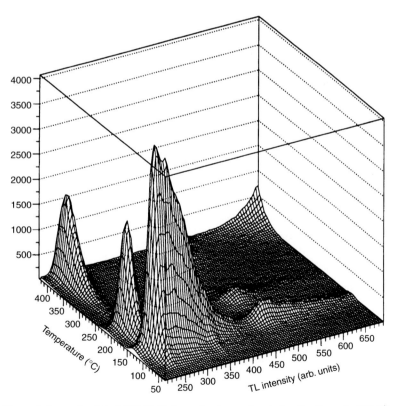

Figure 4. Isometric plot of KMgF$_3$:Tl thermoluminescent emission (heating rate 1°C.s^{-1}).

REFERENCES

1. Wang, S. S., Cai, G. G., Zhou, K. Q. and Zhou, R. X. *Thermoluminescent Response of $^6LiF(Mg,Cu,P)$ and $^7LiF(Mg,Cu,P)$ TL Chips in Neutron and Gamma Ray Mixed Fields.* Radiat. Prot. Dosim. **33**, 247–250 (1990).
2. Driscoll, M. H. *Studies of the Effect of LET on the Thermoluminescent Properties of Thin Lithium Fluoride Layers.* Phys. Med. Biol. **23**, 777–781 (1978).
3. Furetta, C., Bacci, C., Mendozzi, V., Sanipoli, C., Scacco, A., Cremona, M. and Montereali, R. M. *Modifications Induced in the Luminescence from Single Crystals of LiF:Mg,Cu,P.* Nucl. Instrum. Methods **B91**, 215–218 (1994).
4. Scacco, A., Furetta, C., Bacci, C., Ramogida, G. and Sanipoli, C. *Defects in γ-irradiated $KMgF_3:Tl^+$ Crystals.* Nucl. Instrum. Methods **B91**, 223–226 (1994).

EVALUATION OF INDIVIDUAL NEUTRON DOSIMETRY BY A WORKING GROUP IN THE FRENCH NUCLEAR INDUSTRY

A. Rannou[1], A. Clech[2], A. Devita[3], R. Dollo[4] and G. Pescayre[5]
[1]Institut de Protection et de Sûreté Nucléaire, IPSN F-92265 Fontenay aux Roses, France
[2]COGEMA, Centre de Marcoule, F-30206 Bagnols sur Cèze, France
[3]MELOX, F-30203 Bagnols sur Cèze, France
[4]EDF, Comité Radioprotection, F-92260 La Défense, France
[5]Commissariat à l'Energie Atomique, CEA Valduc, F-21120 Is sur Tille, France

Abstract — The difficulty of properly measuring neutron doses received by the workers in the nuclear industry is becoming more acute in the context of the revised recommendations of the ICRP Publication 60. A working group composed of representatives of the French nuclear industry was set up at the end of 1992 with the aim of assessing the extent of this problem and looking for appropriate solutions. The aims were (i) to define the needs as far as neutron dosimetry is concerned, (ii) to examine the different techniques used or currently available, and (iii) to draw up the specifications for the individual dosemeter which would ideally satisfy all the technical and practical requirements. This paper is intended as a first review of the analysis made.

INTRODUCTION

Exposure to neutron radiation in the nuclear industry is normally limited to a small number of workers operating in specific areas. Operational collective dose due to neutron exposure is almost negligible compared to other sources of external dose. However, individual neutron dosimetry is currently unsatisfactory and improvements are becoming more and more urgent in the context where both regulation and technological processes in the nuclear industry are evolving. Awareness of the neutron issue arose in France when the first 900 MWe Pressurised Water Reactors entered into operation (1977–1978). At that time, entry to the containment building was normally forbidden. Access was finally accepted under restricted conditions, for necessary and very brief interventions at reduced power. Afterwards, due to limited availability of the reactor, it has been admitted that access to the containment building could occur during nominal power. Occupational exposure to neutrons progressively increased in the mid-1980s, when the first discharges of irradiated fuel elements from the PWR began. In the meantime, reprocessing of irradiated fuel from French reactors has increased. More recently, industrial processing of mixed-oxide (MOX) fuel has started and increased still further the potential exposure to neutrons during fuel handling operations.

In 1991 the International Commission on Radiological Protection published new recommendations in its Publication 60[1], with great impact in particular for individual neutron monitoring: reduction of the annual dose limit averaged over 5 years from 50 to 20 mSv and assignment of increased values to the quality factor Q and radiation weighting factor W_R, particularly for neutrons in the range 100 keV to 2 MeV.

At the request of a Technical Committee from Institut de Protection et de Sûreté Nucléaire (IPSN) and Electricité de France (EDF) it was decided in the middle of 1992 that a Working Group (WG) would assess the extent of the neutron issue in the French nuclear industry and would look for appropriate solutions.

OBJECTIVES OF THE WORKING GROUP

The Working Group was initially formed with representatives from the public corporation that produces nuclear electricity in France (EDF), the private companies involved in processing and reprocessing nuclear fuel elements (MELOX and COGEMA respectively), the Commissariat à l'Energie Atomique (CEA) as a nuclear operator and researchers from the IPSN.

The objectives of the WG were to review the needs for neutron dosimetry in the nuclear industry, to consider the different techniques in routine use or under development, and to suggest specifications for individual dosemeters that would ideally meet both technical and practical requirements.

PRESENT SITUATION OF NEUTRON MONITORING IN THE FRENCH NUCLEAR INDUSTRY

In the nuclear power plant (EDF)

Exposure of the workers to neutrons is mainly related to entries into the containment building[2]. Neutron energy spectra in the reactor environment are well known, and can be divided in two components:

(1) High energy neutrons in the range 50 keV–2 MeV with a maximum intensity between 200 keV and 600 keV. The dose equivalent rates due to this component lie typically from 0.05 to 500 mSv.h^{-1}. High energy neutron beams mainly occur locally due to less effective shielding and thus are subject to strong variations in space.

(2) Neutrons from thermal to 10 eV with a maximum of intensity between 0.05 and 0.3 eV. The dose equivalent rates vary slightly around 0.05 mSv.h^{-1}. Due to multiple scattering, thermal neutrons constitute almost isotropic fields.

Entries into the containment building are limited both in frequency (less than one per month in each plant) and duration (about one hour). Gamma dose rates in the containment building are of the same order of magnitude as the neutron dose rates mentioned above. Collective dose received per entry is about 0.6 mSv (neutron: 0.26 mSv and gamma 0.34 mSv) distributed among three persons on average.

Discharge of irradiated fuel elements from the reactor and receipt of MOX fuel also contribute to exposure of the employees. Neutron energy spectra present a maximum in the region 200 keV–2 MeV. Dose rates typically measured are of the order of 0.1 mSv.h^{-1}. Individual doses received during each operation vary between 0.1 and 1 mSv. On average five entries are made each year on a reactor.

Exposure to neutrons at EDF is characterised by doses received during operations of short duration but delivered at relatively high dose rates. There are almost no fixed workplaces in nuclear generating stations. Workers generally move from place to place where radiation fields can change rapidly in intensity as well as in energy spectrum and angular distribution. This explains why individual dosimetry is particularly critical for this category of workers.

None of the passive techniques existing for individual monitoring was considered to be convenient owing to their limitations, i.e. high energy response dependence, poor lower limit of detection for neutrons or high sensitivity to photons. Thus, until now individual dose assessments have been extrapolated from ambient monitoring. While the reactor is functioning at reduced power, ambient measurements with rem counters are carried out in the different locations where workers will operate. In some cases, spectrometry systems are used to measure fluence spectra in working areas. Neutron dose rates are extrapolated to the nominal power of the reactor.

Individual doses are calculated taking into account mean residence duration for each worker in the different locations. Further measurements with rem counters and electronic gamma devices are made by the personnel during their operations. This allows calculation of individual dose assessments based on the ambient monitoring, and also prevents any risk of over exposure.

In the (re)processing utilities (COGEMA and MELOX)

Neutron dosimetry is relevant to all activities involving fissionable material, from fuel processing to conditioning and storage of radioactive waste. There has been increased potential for exposure in recent years owing to increased burnup of nuclear fuel and reprocessing of irradiated elements. Recently, the MELOX plant was built for processing MOX fuel. Risks associated with MOX fuel elements, in particular those due to neutrons, were considered early in the design of the plant.

On the one hand, because of their high radiotoxicity, fissionable compounds are handled in containment buildings and glove-boxes. Thus, neutron dose rates are relatively low (a few μSv.h^{-1} on average in the working areas). On the other hand, contrary to the situation in power plants where interventions are of short duration, workers in processing facilities are usually exposed all day long.

Individual dosimetry is based on the 'Cogebadge'[3], a dosemeter designed and developed by COGEMA. It consists of a photographic film and a card containing 4 thermoluminescence detectors (2 ^{6}LiF chips and 2 ^{7}LiF chips). Readout of the TLDs is performed on Harshaw readers. The calibration factor of this albedo dosemeter strongly depends on the energy spectra. An average value expected to be representative of most of the situations has been adopted and used universally for all workplaces.

The 'Cogebadge' may also be used in case of a criticality accident, when associated with activation of sulphur (ebonite) and gold (bare and under cadmium) contained in the dosemeter itself or in a criticality belt.

In the nuclear research centres (CEA)

Risks of exposure to neutrons at CEA are similar to those described above. Individual neutron dosimetry is currently based on two techniques: NTA film for fast neutrons >1.5 MeV and the CEA albedo dosemeter called PGPDIN[4]. Performance limitations of these two techniques have already been discussed. Readout of the albedo dosemeters is done by IPSN. CEA operators are responsible for providing the appropriate calibration factor for this dosemeter. Spectrometric measurements with multi-sphere or proton recoil systems are performed in some workplaces to determine it. However this factor is not systematically determined.

NEEDS FOR INDIVIDUAL ACTIVE DOSIMETRY SYSTEM

Priorities

The existing techniques for ambient monitoring in nuclear facilities were considered relatively satisfactory by the Working Group. On the contrary, routine individual neutron dosimetry, which is currently based either on personal albedo thermoluminescence detectors or on ambient measurements at workplaces, should be improved with high priority. In particular, owing to their characteristics (they are passive techniques and their calibration at each workplace is needed) the use of these dosemeters is not convenient in locations where the radiation field is not yet well known.

The dosemeter needed by operators as described below is direct reading (electronic) dosemeter. The needs in terms of number of dosemeters were estimated by each representative of the WG. This evaluation is given in Table 1.

General requirements

(a) The relevant dosimetric quantities to be measured and displayed are the personal dose equivalent and its rate due to neutrons ($H_p(10)$ and $\dot{H}_p(10)$). If the dosemeter is also capable of providing results for gamma radiation, the readout should be distinct.
(b) The basic quantity to be displayed must be the integrated personal dose equivalent. Readout of instantaneous dose equivalent rate could possibly be obtained for example by pressing a button. But in this case, basic readout should be restored in less than 5 s.
(c) Measurement must be performed directly, i.e. without any handling by the user.
(d) Exceeding of pre-set thresholds for the dose equivalent and the dose equivalent rate must be indicated by audible and visible alarms.
(e) Batteries must allow autonomous functioning of the device for more than one year (assuming 100 h of functioning).
(f) Life expectancy of the device must be better than 5 years in routine use.
(g) Weight and dimensions must be compatible with wearing by user in controlled areas, possibly in a pocket ($120 \times 90 \times 25$ mm^3 maximum).
(h) Automatic transfer of data to a computing system must be possible for dose processing and storage.

Dosemeter performance

The performance required of the active device is summarised in Table 2. The energy range of neutrons to be covered efficiently in radiation protection constitutes a major problem that is not yet solved with individual dosemeters. The WG considered that the response should be good at least in the regions of thermal and fast neutrons (see Table 2). However, the requirements regarding the dose sensitivity are expected to be more easily achievable. If the ideal solution of 5 decades of dose measurement is not possible, readout according 2 ranges of dose, suited to routine conditions and accident conditions respectively, seems to be a good compromise.

Switching from one range to another is done either by the user himself (when initialising the dosemeter) or automatically when the dose rate is higher than 10 mSv.h^{-1}. Readout of range should be unambiguous.

In normal conditions, alarms on dose rate and integrated dose must be setable by steps of 0.1 mSv.h^{-1} and 0.1 mSv respectively. The lower limit of detection is 10 µSv.

Accuracy on the measurements in the ranges mentioned in Table 2 should be the following:

(i) $H_p(10)$ must be determined with a maximum error of ±30% at 10 mSv.h^{-1}, whatever energy spectrum is considered.
(ii) $H_p(10)$ must be determined with a maximum error of ±30%, for monoenergetic neutrons of 0.1 eV and 1 MeV.

Moreover, the dosemeter should replicate the angular response of $H_p(10,\alpha)$ with less than ±20% of error for angles between −60° and +60°.

Environmental criteria

The dosemeter must be capable of efficient rejection of the photon component in mixed fields: where gamma dose rate and neutron dose rate are equal, the maximum error in the determination of neutron dose rate must be 10%.

Effect of the temperature in the range 10°C–40°C must be less than 10% on neutron dose determination.

The dosemeter must be waterproof; effect of moisture in the range 40–100% RH must be less than 10%.

The dosemeter must be not subject to electromagnetic interference, e.g. due to electric motors or communications transmitters.

Table 2. Specifications for an active individual neutron dosemeter — ranges to be covered.

Parameter	Range
Neutron energy	0.01 eV to 10 eV and 50 keV to 15 MeV
Dose equivalent rate:	
in routine conditions	10 µSv.h^{-1} to 10 mSv.h^{-1}*
in case of accident	1 mSv.h^{-1} to 1 Sv.h^{-1}*
Dose equivalent:	
in routine conditions	10 µSv to 10 mSv**
in case of accident	1 mSv to 1 Sv**

*Dynamic range of measurement.
**Range of readout.

Table 1. Rough estimate of the needs for individual neutron dosemeters in the French nuclear industry.

Corporation	Number of dosemeters
EDF	2000
COGEMA	40 to 100
MELOX	100
CEA	500
Total	about 3000

SHORT-TERM AND LONG-TERM STRATEGIES

It was quite clear to the members of the Working Group that no individual dosemeter currently available is capable of fulfilling the above requirements. Known limitations of systems, whatever technological principle is adopted, are mainly:

(1) A relatively high energy response dependence and a poor detection threshold in the intermediate energy range (generally the energy threshold appears in the region of several hundreds keV).
(2) An insufficient rejection of the photon component.
(3) The difficulty of reconciling sufficient neutron sensitivity and characteristics compatible with an individual dosemeter (weight, dimensions, autonomy...).

Considering the present situation the WG came to the following strategy:

(a) *Long-term objectives.* The specifications defined by the WG for an active, direct-reading, individual neutron dosemeter had to be submitted to manufacturers. The opinion of the WG was that the manufacturers potentially interested in new developments would be motivated by the bid since the French nuclear industry could guarantee to buy about 3000 devices every 5 years. However, it was foreseen that commercial availability of a suitable dosemeter would still be some years ahead. Thus a short-term solution was required.
(b) *Short-term objectives.* The bubble detector technique, which had already been introduced on a small scale in a few utilities, seemed to meet some of the requirements and could be a good compromise before finding the long-term solution. However it needed to be better characterised before being more largely recommended. The WG decided firstly to prepare a critical review of the bubble detector technique and secondly to test the systems available at that time, namely the following: BD 100R–PND ('rapid') and BDT ('thermal') from the BTI company* and those from APFEL company**. The test programme was designed and undertaken by the 'Centre Technique d'Homologation d'Instrumentation de Radioprotection' (CTHIR) at IPSN in collaboration with the Dosimetry Department at 'Etablissement Technique Central de l'Armement' (ETCA). A comprehensive report on this test programme and the results obtained have been given by Chemtob *et al*[5].

On the whole, bubble detector test results obtained in the laboratory were found to be rather good. It was then decided to carry on the tests in real field conditions in order to assess more realistically their applicability and reliability in practice. Several questions still existed, namely: To what extent are temperature, humidity, shocks..., critical parameters in real situations? How long is the life expectancy of the dosemeters used in routine? Do we need to test them periodically to ensure their good response? Finally, are the results consistent with other techniques like ambient monitors and passive dosemeters?

PRESENT STATUS

Bubble detectors

Bubble detectors have now been under test at EDF, CEA, MELOX and COGEMA for a few months. It is hoped that the operators will gain much feedback through this pilot programme and that a decision to recommend officially this type of dosemeter will arrive in the very near future.

Active direct reading dosemeter

Five private enterprises (or their representatives) and two national research laboratories finally answered the bid to develop a new active, direct-reading, individual neutron dosemeter. In fact, only three of them gave 'serious' proposals, among which one was finally chosen. Feasibility studies on a technique are in progress and should be concluded at the end of 1995. It is forecast that the performance of this technique will not meet entirely the requirements defined by the WG. However, it is expected that it will represent a better compromise than the bubble detector as long as a more 'intelligent' dosemeter does not exist.

DISCUSSION AND CONCLUSION

The need for active dosemeters in the nuclear industry has been increasing for many years[6]. Smaller and more versatile dosemeters are expected which could, in particular, store a large amount of information and facilitate data treatment within a centralised computing system. Such individual dosemeters already exist for gamma radiation and their implementation is under way in French nuclear utilities. The situation is not so good for neutron dose assessments and must be improved even more urgently since the passive techniques do show severe limitations.

French operators clearly expressed their willingness to support research which could lead in a reasonable period of time to a new active dosemeter for individual monitoring. Of course, the cost of a new dosimetry system has to be taken into consideration, but it cannot be simply compared to the present cost of passive dosimetry with nuclear emulsions and TLDs. Their potential benefit in terms of accuracy and direct dose control may

---

*Bubble Technology Industries (BTI), Highway 17, Box 100, Chalk River, Ontario, KO 1JO Canada.
**Apfel Entreprises, 25 Science Park, New Haven, CT 06511, USA.

lead to acceptance of a more substantial cost. Also relevant to cost is the particularity of the neutron issue: exposures to neutrons are limited to specific areas or operations. On the one hand, the number of workers exposed to neutrons is limited. On the other hand, measurements are difficult and need more sophisticated devices. It is thus acceptable to spend more money for this monitoring.

However, even if significant advances in instrumentation are made and if new progress is still expected in the future, it would be unreasonable to place our hope in the rapid development of an 'ideal' dosemeter with perfect energy response, good sensitivity to neutrons and efficient rejection of gamma component.

In the meantime, it is worthwhile directing our interest to less perfect detectors and, in parallel, stimulating research on new instrumentation. Since the dosemeter will be imperfect, strenuous efforts will have to be made regarding its calibration in a reference radiation field according to well established procedures. In particular, it is now recognised that such dosemeters can easily be calibrated in fields with broad energy spectra intended to approximate the conditions of their practical use[7].

Meanwhile we have also to encourage the definition of international standards for electronic dosemeters. Otherwise we risk defining specifications which could vary widely from one country to another.

ACKNOWLEDGEMENTS

The authors gratefully thank their colleagues from CEA, COGEMA, EDF, IPSN, and MELOX who collaborated in discussions and in the supply of information, and also those from ETCA for their valuable assistance of the Working Group.

REFERENCES

1. ICRP. *1990 Recommendations of the International Commission on Radiological Protection.* Publication 60 (Oxford: Pergamon) (1991).
2. Wolber, G., Guibbaud, Y. and Dollo, R. *La dosimétrie des Neutrons en Centrale Nucléaire.* Communication au Congrès EORTC'95, Orléans (France) 4–6 Juin 1995.
3. Espagnan, M., Truffert, H., Douillard, P. L., Chatenet, P., Pottier, A., Payan, G. and Marcellin, G. *The Dosimetry Laboratory of COGEMA Marcoule Nuclear Centre — Part 2, Dosimeters Used.* Harshaw Bicron TLD Users Symposium, Las Vegas, Nevada (USA), 13–17 March 1995.
4. Buxerolle, M. *Etude du D.I.N. Dosimètre Individuel de Neutrons.* Rapport CEA-R-5397 (1987).
5. Chemtob, M., Dollo, R., Coquema, C., Chary, J. and Ginisty, C. *Essais de Dosimètres Neutrons à Bulle, Modèle BD 100 R-PND et Modèle BDT.* Radioprotection **30**(1), 61–78 (1995).
6. Hirning, C. R. *The Increasing Need for Active Dosemeters in the Nuclear Industry.* Radiat. Prot. Dosim. **61**(1/3), 1–7 (1995).
7. Chartier, J. L., Kukdjian, J., Paul, D., Itié, C., Audoin, G., Pelcot, G. and Posny, F. *Progress on Calibration Procedures with Realistic Neutron Spectra.* Radiat. Prot. Dosim. **61**(1/3), 57–61 (1995).

NEUTRON DOSIMETRY AT THE REACTOR FACILITY VENUS

P. Deboodt, F. Vermeersch, F. Vanhavere and G. Minsart
SCK·CEN, The Belgian Nuclear Research Centre
Boeretang 200, B-2400 Mol, Belgium

Abstract — The reactor VENUS is a zero-power research reactor mainly devoted to studies on light water fuels. The need for undertaking a neutron spectrometric and dosimetric study became apparent when locally high neutron dose rates were measured. The spectrometric study is based on two approaches. The first is an experimental one in which the neutron spectrum was measured at three positions around the facility. The second is a theoretical one in which a numerical modelling of the neutron transport at the reactor site was performed in order to determine neutron spectra and fluence rates at different positions around the site. The measured and calculated spectra are interpreted in terms of the response of different individual and environmental dosemeters. These responses are confronted with the *in situ* measurements. The impact of the ICRP 60 recommendations on the determined dose rates is also studied.

INTRODUCTION

The need for a more reliable determination of neutron dose rates around nuclear facilities is gaining importance in the radiation protection field. The renewed interest is driven by the implications of the ICRP 60 recommendations that include the decrease of the annual dose limit combined with the increase of the quality factors for neutrons[1].

For our specific situation the interest in performing an in-depth study of neutron dosimetry around the VENUS critical facility is spurred on by the development of new research programmes involving MOX fuel with high Pu content. Our attention was also sharpened by the measurement of locally high neutron dose rates and the fact that the radiation field around the facility is a mixed neutron–gamma field characterised by a neutron-to-gamma dose rate ratio with values as great as 3.

The problems encountered in environmental and especially in individual neutron dosimetry are mainly due to the large spread in energy of the neutron spectrum and the inability to find a dosemeter with a reliable response covering the whole spectrum. A method to compensate for the non-ideal response curves of the instruments is the introduction of local correction factors to convert dose readings into effective dose received by the workers. To define such a correction factor, information is needed about the neutron spectrum at the specific workplaces. Therefore, an in-depth study was started of the dosimetric and spectrometric characteristics of the neutron field around the VENUS reactor.

The spectrometric study involved two parts; an experimental part and a theoretical part. The experimental part of the study involved a strong collaboration with the team of Dr J. L. Chartier (IPSN, Fontenay-aux-Roses) to determine the neutron spectrum around the facility at three specific locations. The theoretical part consisted in the simulation of the neutron field near the reactor and the working areas around it. This numerical simulation was performed at the SCK·CEN with the use of the TRIPOLI-3 code[2], probably its first use in such a context. The dosimetric study involved individual and environmental dosimetry measurements with different types of detectors.

SPECTROMETRIC MEASUREMENTS

The schematic view of the VENUS site is given in Figure 1. The reactor is shielded by a concrete wall

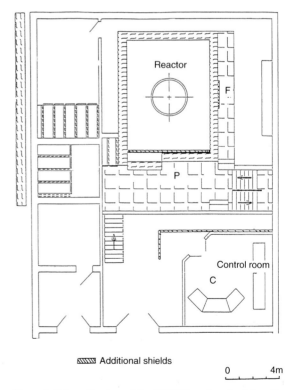

Figure 1. Schematic view of the VENUS reactor site, letters C, P and F refer to the locations where the neutron spectrum has been measured.

approximately 1 m thick. The core itself is placed in a cylindrical tank positioned just below a concrete access platform. This platform is surrounded by a concrete wall approximately 0.5 m thick. The entrance to the access platform is shielded by a rolling concrete door. The whole assembly is placed in a hall that contains areas for monitoring devices, workplaces and the control room. The reactor core loading consists of a combination of UO_2 and MOX fuel rods. The MOX fuel rods are longer than the UO_2 rods. As a consequence a substantial part of the MOX fuel rods are positioned above the water level of the reactor which gives rise to a high neutron leakage. Supplementary neutron shields were placed at several locations around the facility to reduce the neutron dose rates. The position of these shields is also shown in Figure 1.

Measurements of the neutron spectrum were performed by the team of Dr J. L. Chartier (IPSN, Fontenay-aux-Roses) at three different positions around the reactor designated by the letters P, F and C in Figure 1. The choice of positions P and F was determined by the need for high neutron fluxes for a reliable measurement of the spectrum. The choice of the position C is justified from the radiation protection point of view because it lies in the area where the operators are present during a power run of the reactor. Two different measuring techniques were used to cover the spectrum from 0.01 to 20 MeV. Proton recoil techniques were used for energies above 100 keV, while the multisphere technique was used for the lower energies down to the thermal part of the spectrum[3]. The spectrum at position C was determined only with the multisphere technique due to the low neutron flux in the control room, which made the use of the recoil detectors inpracticable.

The results of the measurement are presented in Figures 2–4. The spectra are characterised by a high thermal contribution, a low contribution in the intermediate energy range and a large contribution in the

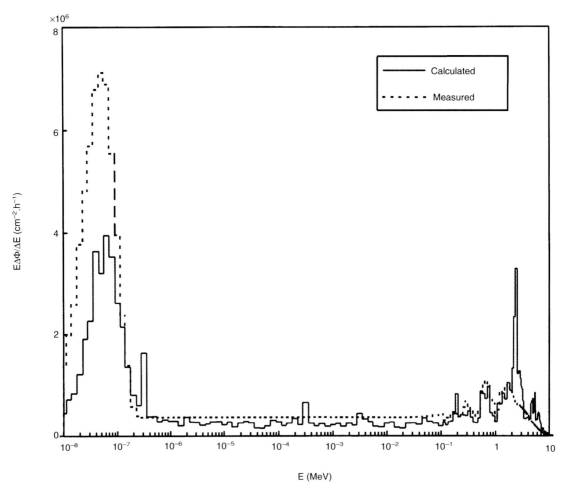

Figure 2. Measured and calculated spectrum at position P. Note the presence of the minima at 0.4 and 1 MeV in both spectra, and the presence of the 2.4 MeV peak only in the calculated one.

region of high neutron energies. Due to an electronics malfunction no proton recoil data were available above 2 MeV. As a consequence the spectrum above 2 MeV was determined from the multisphere data. This accounts for the smooth form of the spectrum above 2 MeV. Also, no detailed structure can be found in the measurement at position C due to the low spectral resolution of the multisphere technique, used at this position.

NEUTRON TRANSPORT MODELLING

The modelling of the neutron transport near the VENUS reactor was performed with the TRIPOLI-3 code. The code calculates the neutron transport via Monte Carlo simulation in a three-dimensional geometry. The geometry of the installation is entered in the program by defining a set of volumes containing the different materials that constitute the site. The present model contains 330 volumes defined by 220 surfaces. Materials used at the site are modelled as realistically as possible. A neutron source strength is determined as 1.5×10^{13} n.s^{-1}, through comparison of the measured and simulated spectra. The calculated spectra are compared with the measured ones in Figures 2–4. The general behaviour of the measured spectra is reproduced by the calculated ones. Detailed examination of the high energy part of the spectrum reveals the same structure in the measured and calculated spectrum at the positions P and F, i.e. the minima at 0.4 and 1 MeV. The only striking difference is the existence of a peak at 2.4 MeV in the calculated spectra that has not been detected in the measured spectra, probably due to the low spectral resolution of the multisphere technique at energies above 2 MeV. There are reasons to believe that the peak at 2.4 MeV is a real feature in the spectrum. Examination of the composition of the concrete used for the shielding shows the presence of large amounts of oxygen. The study of the total cross section of oxygen shows resonances at 0.4 and 1.0 MeV, and a low total cross section around 2.4 MeV. These correspond respectively with the minima detected at 0.4 and 1.0 MeV and the maximum at 2.4 MeV in the calculated spectra.

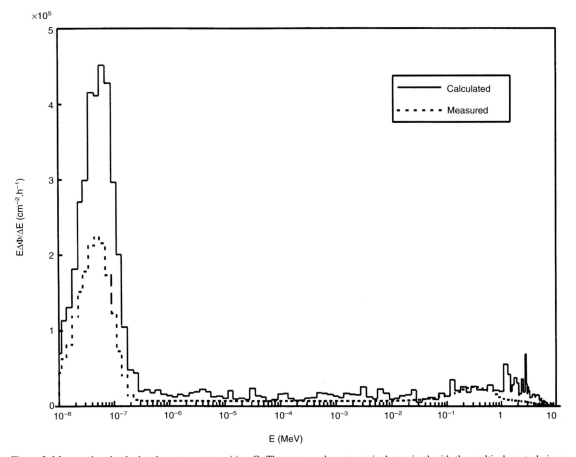

Figure 3. Measured and calculated spectrum at position C. The measured spectrum is determined with the multisphere technique over the whole energy range.

COMPARISON OF MEASURED AND DERIVED QUANTITIES

To facilitate the intercomparison of the different spectra and the comparison of the spectral with the dosimetric results, quantities such as total flux, dose rates and mean energy were derived from the calculated and measured spectra.

The equivalent dose rate $\dot{H}$ is calculated for different sets of flux to dose conversion factors. The first set is based on values published in ICRP 21[4] and determines $\dot{H}_{MADE}$ or the maximum ambient dose equivalent rate. The second set determines $\dot{H}*(10)$, the ambient dose equivalent rate at depth 10 mm, based on ICRP 51[5]. The third set is based on the values published recently[6], which take into account the radiation weighting factors of ICRP 60[7] and also determines $\dot{H}*(10)$. To determine the 'hardness' of the spectra the mean energy in terms of the fluence was calculated.

The results for the measured and calculated spectra are collected in Table 1.

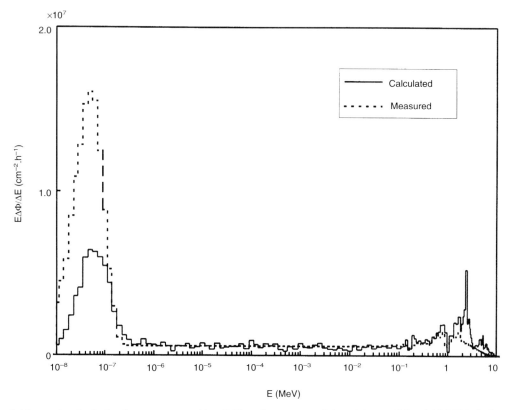

Figure 4. Measured and calculated spectrum at position F. Note the presence of the minima at 0.4 and 1 MeV in both spectra, and the presence of the 2.4 MeV peak only in the calculated one.

Table 1. Quantities determined from the measured and calculated spectra.

Quantities/Position	Quantities determined from the measured spectra			Quantities determined from the calculated spectra		
	P	C	F	P	C	F
ϕ (n/(cm^2.s))	5460	150	10800	4191	297	6795
H_{MADE} (mSv.h^{-1})	0.83	0.016	1.25	1.09	0.036	1.76
$H*(10)$ (mSv.h^{-1})*	0.84	0.017	1.22	1.07	0.037	1.80
$H*(10)$ (mSv.h^{-1})**	1.00	0.022	1.50	1.27	0.043	2.08
$\bar{E}$(MeV)	0.17	0.082	0.11	0.39	0.13	0.37

*Based on the ICRP 51 publication.
**Based on the recommendations of the ICRP 60[6].

Examining the results in Table 1 shows that the mean energy of the calculated spectra are higher than the ones determined from the measured spectra. The dose rate in the control room determined from the calculated spectrum is twice that determined from the measured spectrum. This difference is not caused by a difference in repartition of the spectrum but can be wholly attributed to the total neutron flux which is twice as high in the calculated spectrum. The fact that the 2.4 MeV peak is not detected during the measurements can partly account for this. From Table 1 it is also clear that the recommendations of ICRP 60 will lead to an augmentation of the dose rates by 20 to 30% for the typical spectra encountered around the reactor.

EVALUATION OF INDIVIDUAL AND ENVIRONMENTAL DOSEMETERS

The response of different individual and environmental neutron dose detectors was examined around the VENUS reactor site. The environmental monitoring was performed with a Harwell N91[8] and a Studsvik 2202 D[9]. For the individual dosimetry two types of detectors were used; the PGP-DIN albedo dosemeter[10], operated by the CEA and the BD-PND bubble detector[11] operated at the SCK·CEN.

Using the calculated and the measured spectra and the response functions of the above detectors the reading of each detector at the positions P, C and F was calculated. The response curves of the four detectors can be found in the published references. These theoretical dose readings are then confronted with the *in situ* dose rate measurements performed using the chosen detectors. Note that the PGP-DIN and the bubble detector were used here as environmental dosemeters at fixed positions (exposed in air). The results are presented in Table 2.

From Table 2 it is clear that the dose rates determined from the calculated spectra (column II, uncertainty ±15%) are higher than those determined from the measured ones (column I) for all detectors, except the PGP-DIN. The presence of the peak at 2.5 MeV and the greater mean energy of the calculated spectrum can account for this. The lower values of the PGP-DIN can be attributed to the fact that the response curve of the detector drops off rapidly for energies greater than 10 keV, so no contribution of the fast neutrons, which make up a significant part of these spectra, are detected.

Fair agreement is found between the *in situ* measurements and the quantities derived from the measured and calculated spectra. Only the *in situ* measurement of the PGP-DIN at the position P and F are inconsistent with the results derived from the spectral information. No explanation for this is found at the moment.

The bubble dosemeter reading (uncertainty ±10%) gives a good estimate of the equivalent dose rate for all measuring positions. If the results of Tables 1 and 2 are compared it can be seen that the Studsvik 2202 D is well suited to give a measure of $\dot{H}_{MADE}$ or $\dot{H}*(10)$ based on ICRP 51 for the typical spectra at the VENUS site. The Harwell N91 on the other hand seems to be adequate, with its current calibration, to determine $\dot{H}*(10)$ based on the ICRP 60 recommendations.

Clearly the low response of the PGP-DIN at high energies requires a large local correction factor to derive reliable dose estimates. Based on the results determined with the calculated spectrometric results we find local correction factors for the control room of 2.7 and 3.3 respectively for the conversion to $\dot{H}_{MADE}$ or $\dot{H}*(10)$ based on the ICRP 60. Based on the measured spectroscopic results we find 2.8 and 3.7 respectively. These large values agree with the type of spectrum that can be classified, on the basis of the mean energy, as a transport container spectrum[12].

The PGP-DIN and bubble detector were also used as a personal dosemeter by the operators in the control room during a VENUS reactor run. Such a run takes place about once a year for a period of twice 10 h. The bubble detectors gave an average dose rate of 0.019 mSv.h^{-1} in the control room, which is consistent with the results of the measured spectrum in Table 2. The PGP-DIN readings were higher namely 0.04 mSv.h^{-1} on average. It has to be pointed out that the PGP-DIN detector is read out only every three months while the bubble detector is reset after each 10 h of the VENUS reactor run.

Table 2. Calculated and measured dose equivalent response of neutron dosemeters and environmental monitors.

Dosemeters	(I) Derived dose equivalent measured spectrum (mSv.h^{-1})			(II) Derived dose-equivalent calculated spectrum (mSv.h^{-1})			(III) *in situ* detector readings (mSv.h^{-1})		
	P	C	F	P	C	F	P	C	F
Harwell N91	0.99	0.022	1.47	1.16	0.044	1.99	1.20	0.020	2.16
Studsvik 2202 D	0.84	0.017	1.20	1.07	0.036	1.80	0.73	0.011	1.24
PGP-DIN	0.22	0.006	0.46	0.16	0.013	0.24	1.36*	0.020*	16*
Bubble	0.77	0.015	1.06	1.03	0.031	1.72	1.09	0.019	1.33

*Calibration factor for PGP-DIN is 1 for all positions.

CONCLUSION

A study of the neutron dosimetry around the VENUS reactor was performed. It involved the spectrometric and dosimetric characterisation of the neutron field, and the investigation of the response of environmental and individual dosemeters. The specrometric characterisation was performed through measurements and through simulation. These two approaches supported and complemented each other well.

The analysis of the results showed that a large correction factor would be needed to convert the dose reading with the PGP-DIN into the dose received by the workers for the typical spectra at the site. The bubble detector seems, due to its spectral response, well suited to monitor the received dose. Another advantage of this type of detector is that its direct reading capability enables the ALARA principle to be applied.

The application of the ICRP 60 recommendations will introduce an augmentation of the equivalent dose in the order of 20 to 30% for the typical spectra at the VENUS reactor site.

REFERENCES

1. Portal, G. and Dietze, G. *Implications of New ICRP and ICRU Recommendations for Neutron Dosimetry*. Radiat. Prot. Dosim. **44**(1/4), 165–170 (1992).
2. *TRIPOLI 3.0, Code de Monte-Carlo tridimensionnel polycinetique*. Private communication (January 1993).
3. Posny, F., Chartier, J. L. and Buxerolle, M. *Neutron Spectrometry System for Radiation Protection: Measurements at Work Places and in Calibration Fields*. Radiat. Prot. Dosim. **44**(1/4), 239–242 (1992).
4. ICRP. *Data for Protection against ionizing Radiation from External Sources: Supplement to ICRP Publication 15*, Publication 21 (Oxford: Pergamon Press) (1971).
5. ICRP. *Data for use in Protection Against External Radiation*. Publication 15, (Oxford: Pergamon Press) Ann. ICRP **17**(2/3) (1987).
6. Siebert, B. R. L. and Schuhmacher, H. *Quality Factors, Ambient and Personal Dose Equivalent for Neutrons, based on the new ICRU Stopping Power Data for Protons and Alpha Particles*. Radiat. Prot. Dosim. **58**(3) 177–183 (1995).
7. ICRP. *1990 Recommendations of the International Commission on Radiological Protection*. Publication 60. Ann. ICRP **21**(1–3) (Oxford: Pergamon Press) (1990).
8. Leake, J. W. *Spherical Dose Equivalent Neutron Detector Type 0075. A Recommendation for Change in Sensitivity*. Nucl. Instrum. Methods **178**, 287–288 (1980).
9. Manual. *Neutron Dose Rate Meter 2202 D Studsvik* (AB Atomenergi, Sweden) (1973).
10. Buxerolle, M. *Etude du D.I.N. — Dosimètre Individuel des Neutrons*. Rapport CEA-R-5397 (1987).
11. Chemtob M., Dollo, R., Coquema, C., Chary, J. and Ginisty, C. *Essais de Dosimètres Neutrons à Bulle Modèle BD 100 R-PND et Modèle BDT.*, Radioprotection **30**(1), 61–78 (1995).
12. Aroua, A., Grecescu, M., Prêtre S. and Valey J.-F. *On the Use of some Spectrum Hardness Quantifier for Operational Neutron Fields*. Radiat. Prot. Dosim. **54**(2), 99–108 (1994).

SOME EXAMPLES OF INDUSTRIAL USES OF NEUTRON SOURCES

J. L. Szabo and J. L. Boutaine
CEA/DAMRI, BP 52,
91193 Gif-sur-Yvette Cedex, France

Abstract — Neutron techniques offer great potential for the solution of several problems in industry and not only in the nuclear industry. The use of neutron sources (sealed radionuclide sources, accelerators, neutron beams delivered by research reactors. etc.) is widespread. The type of neutron interaction with the matter investigated depends on the application. A short review of these applications is outlined, pointing out their main features. Some examples of the main industries which call on such techniques are mentioned. Finally, the main needs concerning neutron dosimetry linked to the French regulations on industrial devices containing neutron sources are described.

INTRODUCTION

Neutron sources (radionuclide sealed sources, accelerators or neutron beams delivered by research reactors, etc.) are used in different scientific and industrial areas for various purposes.

The choice of a suitable neutron source will depend on the type of the application in the following fields:

(i) nuclear oil well logging and mineral resources exploration,
(ii) moisture content measurement equipment for agronomic and industrial process control,
(iii) neutron shielding quality control in the nuclear fuel cycle industry,
(iv) analysis of materials involved in the nuclear industry,
(v) neutron radiography,
(vi) non-destructive detection of explosives and/or narcotics,

and other applications. This paper reviews briefly these applications, pointing out their main features. It concludes by highlighting the needs in neutron dosimetry in order to be in agreement with the safety regulations concerning the design and use of industrial gauges containing neutron sources.

NEUTRON SOURCES

The most important types[5] of neutron source used in industry are as follows.

Radionuclide sealed sources

These are mainly

(i) ^{252}Cf, a spontaneous fission source, emitting a neutron spectrum with an average energy of 2.2 MeV; and
(ii) ^{241}Am–Be, an (α,n) source, with a higher average energy of about 4.5 MeV.

^{252}Cf is preferred when emphasis is assigned to the neutron fluence rate because of its higher specific activity. However, the half-life of ^{252}Cf (2.64 years) is a disadvantage compared with that of ^{241}Am–Be (433 years).

Accelerators

In this category, neutrons are generally produced by four main reactions[1]:

Reaction	Q value (MeV)	Neutron energy range (MeV)
^{2}H(d,n)^{3}He	+3.27	1.65–7.75
^{3}H(d,n)^{4}He	+17.59	11.75–20.5
^{3}H(p,n)^{3}He	−0.763	0.3–7.6
^{7}Li(p,n)^{7}Be	−1.644	0.12–0.6

The D–D and D–T reactions have several advantages. The large positive Q values and the low atomic numbers make it possible to produce high yields of fast neutrons even at low incident energies. Moreover, the use of pulsed neutrons allows an increase in the areas of application since the neutron flux can reach 10^{10} n.s^{-1}.sr^{-1}.

Reactor neutron beams

These beams are generally used to obtain thermal neutrons for solid state physics research or radiography purposes, as the main advantage of this type of beam consists in a very high flux of usable neutrons.

For such applications the reactor Orphée, at the Saclay research centre, delivers a thermal beam having a thermal fluence rate (or flux density) of 10^9 n.cm^{-2}.s^{-1}. The beam cross section is 25×150 mm^2.

INDUSTRIAL APPLICATIONS

Nuclear oil well logging and mineral resources exploration

The aim of using neutron techniques consists in determining porosity and hydrocarbon content. Generally, fluid flows in porous media (oil, water and gas) contain hydrogen and successive slowing down processes with incident neutrons lead to a proportional signal measured by thermal neutron detectors.

The fairly widespread source of neutrons is the ^{241}Am–Be with a typical activity of 1 TBq (corresponding to about 6×10^7 n.s^{-1}.sr^{-1}), although some societies use D–T neutron generator tubes.

For metallic ores or some raw mineral materials exploration, there is the possibility of using a prompt neutron-induced γ ray technique to detect elements such as F, S, Al, W or U.

Moisture content gauges

This is by far the most common application of neutron gauges. Specific areas include: hydrology (snow), civil engineering (soil, roads, dams, airport runways) and agriculture (soil). In 1992 there were about 500 end users of this type of technique (sometimes associated with density measurements) registered by the French permanent Joint Ministerial Commission on Artificial Radioisotopes (C.I.R.E.A.)[2]. The source is usually ^{241}Am–Be with an average activity of 10 GBq (about 10^6 n.s.$^{-1}$.sr^{-1}).

Advantages of neutron systems for moisture measurements are simplicity, ruggedness and the ability to cover a large volume of material.

The CEA/DAMRI has developed a device specifically designed to measure the humidity of sands used in the industrial production of concrete. A gauge designed to determine the quantity of gas remaining in a gas bottle just before refilling has also been developed by our laboratory. Two sources of ^{241}Am–Be are used with a total activity of 20 GBq. The interaction of neutrons with hydrogen (from butane or propane) gives a signal proportional to the quantity of gas. Thus it allows an industrial on-line control.

Lastly, some needs for industrial process control can be mentioned, essentially to monitor electric furnaces (for example, by measuring the residual water content on electric batteries manufacturing lines).

Quality control of neutron absorber materials involved in the nuclear fuel cycle

These materials play a leading role in the nuclear fuel reprocessing industry to obviate the risk of criticality in the storage, the transport and the reprocessing of irradiated nuclear fuels.

These materials are selected on the basis of hydrogen content to moderate the incident neutrons, and neutron absorber content to capture thermal neutrons.

The CEA/DAMRI has designed several gauges to provide an objective proof through non-destructive inspection that the final products fully satisfy their intended objectives. Three main geometries have been considered:

(1) Backscattering geometry, when the criticality shield must reduce the reflection of neutrons.
(2) Transmission geometry, when the shield must reduce the interaction of exchanging neutrons.
(3) Monitoring geometry termed flux depression which is particularly well adapted for thin and non-hydrogeneous materials that also contain small amounts of neutron absorbers.

All these gauges contain a ^{252}Cf radioisotopic sealed source whose activity depends on the final control and the site constraints. ^{252}Cf has been chosen for its fission

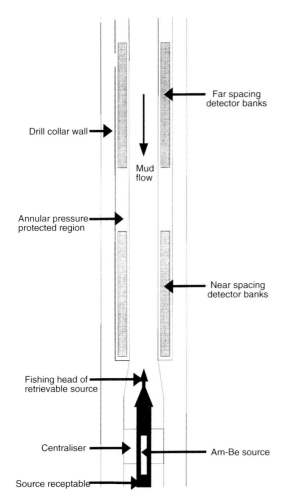

Figure 1. Schematic drawing of the measurement while drilling neutron porosity gauge (Schlumberger document).

spectrum which corresponds to the primary neutron spectrum encountered in a reprocessing plant.

Typical applications are as follows.

Fuel storage pools

Here, the criticality shield is borated steel. The aim is to maximise storage density of fuel containers.

Plutonium oxide powder storage cells

The criticality shields consist of colemanite mortar. The aim is to prevent interaction between the neutron emission of PuO_2 powders from reaching neighbouring containers stored in the cells.

Pulsed columns and annular tanks

The criticality shields consist of colemanite mortar. Circular, straight, ring-shaped and even conical shields have been installed in different reprocessing workshops.

Spent fuel assembly shipping containers

Different types of criticality shields are used for this application. They include colemanite mortar, hydrogen/boron compounds, mixed plastic/mineral powder granulates, polyethylene/boron plaster, boron aluminium, copper + B_4C sandwich materials, etc.

Analysis of nuclear materials

Irradiation with neutrons from a radionuclide source, followed by observation of the delayed gammas from the induced fission is an important technique. Nearly all fuel rods are assayed in this way. This method enables the detection of individual pellets of incorrect enrichment.

Another example concerns accelerators or radionuclide sources for neutron interrogation. With this pulsing method, the delayed fission neutrons and/or gammas can be used as a signature. In this case, the signal after the interrogating pulse of neutrons is recorded and ^{235}U, ^{239}Pu and ^{241}Pu (the odd fissionable nuclei) can thus be detected and quantified. This method was part of the subject of a recent thesis[3].

Neutron radiography

Generally, this method uses neutron beams delivered by research reactors. However in some cases a ^{252}Cf source or a sealed neutron tube can be used to install portable neutron radiography systems.

The following applications can be mentioned.

Explosive devices

Initiators, detonators, thrusters, etc. require precise alignment to ensure that the position of the bridgewire to the adjacent explosive is accurately defined.

Space applications

In the USA, the inspection of turbine blades represents most of neutron radiography. Mechanical components such as the titanium-niobium welds on the vernier control engines of the Space Shuttle or the rubber diaphragm seal found on liquid propellant tanks of satellites are also examined. In France, the same type of application concerns the launcher Ariane.

One can also mention the measurement of the water-permeability of concrete, the inspection of metal-tritium-helium systems, everything connected with aircraft inspection and so on in great variety. All these applications are extensively described elsewhere[4].

Non-destructive detection of explosives and drug contraband

Explosives contain a high level of O and N and a low level of C and H, whereas narcotic materials contain a high level of C, H and Cl and a low level of O and N. Thus, by detecting and measuring these elements, with a high accuracy, as well as the ratios C/O and (O+N)/C, one can detect such hidden materials.

The method used, commonly named Pulsed Fast Neutron Analysis (PFNA), is a non-intrusive technique based on neutron time-of-flight measurements. PFNA measurements determine elemental distributions in samples subjected to inspection by analysing the time-correlated gamma ray spectra induced by fast neutron interaction. This technique requires neutron generator tubes or accelerators. There is an extensive programme

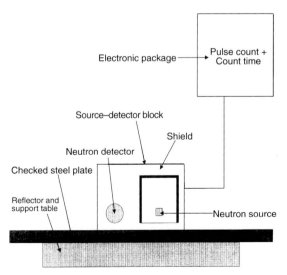

Figure 2. Gauge for checking the boron content of borated steel plates.

on this subject in the USA and many papers have been published[5].

The possibility must also be mentioned of using a neutron radionuclide source to detect hydrogenous matter in special environments. To that effect, the CEA/DAMRI has designed a gauge containing a ^{252}Cf source to detect contraband drugs. In this case, characterisation of elementary elements is impossible.

FRENCH REGULATIONS FOR GAUGES CONTAINING A NEUTRON SOURCE

In France, the AFNOR standard NF M 60-552[6] lays down the regulations for radioelement gauges and, *a fortiori*, for those which contain a radionuclide neutron source.

In addition to the usual characteristics such as mechanical resistance, fire resistance, etc., there is a large chapter relative to the radiological protection around such devices.

Without going into details, some particular points can be mentioned:

(i) dose equivalent measurements around the emitter unit with the source in its safety position;
(ii) measurements while the device is operating;
(iii) research of maximum value of dose equivalent rate;
(iv) measurements of dose rates at 5 cm or at 1 m from the surface of the emitter unit in 18 points with a specified distribution.

An average dose equivalent rate can be calculated from all these points, which affects the conformity or otherwise of the device. Because of the difficulty in doing precise neutron dosimetry, this last specific point is often difficult to obtain.

Finally, all end users in France must apply to the Joint Ministerial Commission on Artificial Radioisotopes (C.I.R.E.A) to become authorised to possess and manipulate gauges containing a radionuclide neutron source.

CONCLUSION

Through this brief review of industrial uses of neutron sources, the authors wish to make all experts in neutron dosimetry aware of the importance of their work in this area. This is particularly true concerning the security of all personnel working around neutron beams, end users of gauges and, more generally, the general public.

REFERENCES

1. Csikai, J. *Handbook of Fast Neutron Generators*, Vol. 1 (Florida: CRC Press) (1987).
2. Vidal, H. *Panorama des Utilisations*. In: Conf. on Sécurité des Sources Radioactives Scellées et des Générateurs Électriques de Rayonnement. 9–10 June, 1993 (Société Française de Radioprotection, Saclay) (1993).
3. Passard, C. *Application des Méthodes d'Interrogation Neutronique Active à l'Analyse en Ligne dans les Usines de Retraitement*. Thesis, Univ. Grenoble 1 (1993).
4. *Neutron Radiography*. Proc. 2nd World Conf., Paris, France, 16–20 June 1986 (D. Reidel Publishing Company) (1986).
5. *Radiation Measurements and Applications*. Proc. 8th Symp. on Radiation Measurements and Applications, Ann Arbor, MI, USA, 16–19 May 1994 (North-Holland) (1994).
6. *Radioelement Gauges — Appliances Intended for Permanent Installation*. AFNOR NF M 60-552 J. O, August 13 (1975).

TEPC PERFORMANCE IN THE CANDU WORKPLACE

A. J. Waker, K. Szornel and J. Nunes
AECL, Chalk River Laboratories
Chalk River, Ontario
K0J 1J0 Canada

Abstract — Tissue-equivalent proportional counters (TEPCs) have a number of features that make them an attractive option for neutron monitoring around power reactors. These features are principally the ability of the TEPC to operate in a mixed field environment, the direct determination of the dose equivalent from first principles and a reasonably well-understood response to the wide range of neutron energy encountered. A spherical TEPC from Far West Technology (5″ in diameter) and a commercial, TEPC-based, neutron monitor (REM 500) have been used to map the neutron fields at different locations in a CANDU 6 (Canadian Deuterium Uranium) power reactor operated by New Brunswick Power. Neutron ambient dose equivalent rates ranged between 8 μSv.h^{-1} and 500 μSv.h^{-1} as measured with a multisphere spectrometer and photon dose equivalent rates from 40 μSv.h^{-1} to 1.2 mSv.h^{-1} as determined with a TEPC. It is shown that, for the CANDU workplace, the energy response of TEPCs is still not known accurately enough to enable field correction factors to be derived from neutron fluence measurements or to avoid the need for more appropriate calibration procedures based on simulated 'workplace' neutron fields. The sensitivity of TEPCs is sufficient for workplace monitoring, but needs consideration when counters of 2″ diameter or smaller are proposed for use. For mixed-field work, the use of a TEPC for photon dosimetry as opposed to photon discrimination requires attention to count rates and instrument dead time.

INTRODUCTION

During the past decade, a focussed effort has been made to utilise tissue-equivalent proportional counters (TEPCs) and microdosimetric methods in radiation protection[1,2]. This effort has included intercomparison exercises carried out in standards laboratories[3] and workplace environments[4]. Results to date have generally confirmed that the photon discrimination, dose equivalent neutron energy response and 'active' nature of the TEPC make it a device worth investigating for its utility in workplace monitoring and commercial instruments are now beginning to emerge in the marketplace.

This paper deals with the assessment of the properties of two instruments based on tissue-equivalent proportional counters from the point-of-view of their use in mixed field monitoring in CANDU (Canadian Deuterium Uranium) reactors. The measurements used for this assessment were mainly carried out in the CANDU 6 power reactor, Pt. Lepreau Generating Station, operated by New Brunswick Power.

TEPC PERFORMANCE IN WORKPLACE FIELDS

Instruments and radiation fields

A laboratory system based on a 5″ diameter TEPC (Far West Technology, Goleta, CA) operating with a 2 μm simulated diameter and connected to a data acquisition and analysis system described by Waker was used in this assessment[5]. For the remainder of this document, it will be referred to as TEPC5; a commercial neutron monitor based on a 2.35″ diameter tissue-equivalent proportional counter (Health Physics Instruments, Goleta, CA) referred to by its commercial name REM 500; and a multisphere neutron spectrometer based on a ^{3}He proportional counter and described previously by Nunes and Waker[6]. Measurements were carried out at six different locations within the plant in areas where neutron fields were known to exist.

The multisphere spectrometer acted as a reference instrument. Data from the multisphere spectrometer were unfolded using the code STAY'SL[7]. This spectrometer has been used in an intercomparison of neutron spectrometers in realistic neutron fields organised by EURADOS at the Cadarache facility[8] and, in a preliminary analysis, found to yield values of the ambient dose equivalent within one standard deviation of the mean value of all spectrometry systems used in the intercomparison[9]. Neutron fluence spectra derived from the multisphere measurements were used to calculate ambient dose equivalent rates to compare with the ICRP 26-based dose equivalent rates determined by the TEPC5 and REM 500. The location of each measurement, the dose equivalent rates recorded by the instruments and their response compared to the multisphere spectrometer are given in Table 1. The fluence spectra recorded at the six locations in the power plant were all different with fluence averaged energies ranging from 75 keV to 174 keV. The fluence spectra and the locations have been described and discussed in detail by Nunes et al[10]. The different locations are identified in the following manner: Am-Be, 1.5 m from an americium beryllium source set up in an open, low scatter area of the turbine hall; HTAR, heat transport auxiliary room; Basement, against the containment wall in the reactor building basement; FMML, fuelling machine maintenance lock; FLA, new fuel loading area; HTP, at the base of a primary heat transport pump; PVG, heat transport pressuriser valve gallery.

TEPC and REM 500 neutron dose equivalent response

A TEPC does not measure ambient dose equivalent. The quantity measured by a tissue-equivalent proportional counter is the kerma in a tissue-equivalent gas cavity surrounded by a tissue-equivalent wall. This, in turn, is converted to a dose equivalent by multiplication with a mean quality factor derived from the pulse height distribution of signals from the TEPC. It is this dose equivalent that is listed in Table 1 and is compared with the ambient dose equivalent obtained from the multisphere spectrometer measurements. The ratio TEPC5/MS in Table 1 is then, in effect, the ambient dose equivalent response of a 5″ diameter thin walled TEPC in the various fields. These results are not corrected by a calibration factor determined by measuring the ambient dose equivalent response in a ^{252}Cf field or D_2O moderated ^{252}Cf field. The results in Table 1 indicate that the TEPC reading is, excluding the Am-Be measurement, on average, 41% of the ambient dose equivalent recorded by the multisphere spectrometer with values ranging from 27% to 49%. These values may appear low; however, the neutron spectra measured at the Pt. Lepreau plant are some of the softest that have been encountered in CANDU stations. The average response of the TEPC5 for 13 locations in three different plants is 0.49, which is close to the average response of 0.47, which was found for the same system (without ^{252}Cf ambient dose equivalent calibration) used in the EURADOS intercomparison at the Vattenfall nuclear generating complex at Ringhals, Sweden[4].

The REM 500 operates on the same principle as a conventional TEPC; however, the REM 500 is filled with propane rather than tissue-equivalent gas. This modification will tend to improve the dose equivalent response of the counter in the critical neutron energy region of 0.1 to 200 keV[11]. The REM 500 is shipped from the factory with a calibration that is based on measurements made in a ^{252}Cf field. To compare the performance of the TEPC5 and the REM 500 directly the TEPC5 results have been adjusted by a post-calibration factor, which would give it an ambient dose equivalent response of unity in a bare ^{252}Cf field. This factor has been measured to be 1.27 at 25 cm from a source of known activity in a large, but not scatter-free room. Comparing the last two columns of Table 1, the response of the TEPC5 in the reactor fields is on average 0.52 with a standard error of 21% and that of the REM 500 0.56 with a standard error of 12%. Field measurements are carried out under very difficult experimental conditions and it is not possible with these results to say unequivocally that the REM 500 is superior to the TEPC5 in terms of dose equivalent response. However the results are consistent with the use of a hydrogen rich counting gas improving the response of a TEPC in low energy workplace fields. The results for the Am-Be source are in agreement with the findings of Aroua et al[12] where they compared the REM 500 with another TEPC system in high energy neutron fields. The results listed in Table 1 also indicate that calibration of TEPCs in fission spectra fields is insufficient to offset the poor response of TEPCs to neutrons below a few hundred keV in energy. Calibration in a D_2O moderated ^{252}Cf would further improve the situation, but not substantially. Based on an earlier intercomparison[3] with a different 5″ TEPC the post-calibration factor for the TEPC5 would change from 1.27 to 1.33.

A combination of experimental measurements and calculations has led over time to a reasonably well understood energy response function for TEPCs. Nunes and Waker[10] have used this data to construct an ambient dose equivalent energy response function for the TEPC5. Using this response function and the spectral fluence distributions measured with the multisphere spectrometer it is possible to calculate the ambient dose equivalent response that should be obtained using the TEPC5. In Table 2, the ratio of the ambient dose equivalent response obtained by measurement with the

Table 1. Dose equivalent rates recorded by the TEPC5 and REM 500 at different locations inside the containment of a CANDU 6 power reactor. In columns 4, 5 and 6 are listed the ratio of the instrument readings to the ambient dose equivalent derived from measurements with a multisphere spectrometer (MS). TEPC5* represents the results of the TEPC5 corrected by calibration in a bare ^{252}Cf field.

Location	TEPC (μSv.h^{-1})	REM 500 (μSv.h^{-1})	TEPC5/MS	TEPC5*/MS	REM 500/MS
Am-Be	22	25	0.88	1.11	1.0
HTAR	22	36	0.4	0.5	0.65
Basement	33	39	0.49	0.62	0.57
FMML	10	22	0.27	0.34	0.59
FLA	3.1	4	0.36	0.46	0.46
HTP	208	292	0.42	0.53	0.59
PVG	115	117	0.51	0.64	0.50

TEPC5 to the value calculated using the energy response function as described above is shown for the different locations in the plant and for the Am-Be source. It is evident from Table 2 that agreement between the measured and calculated ambient dose equivalent response is very good for the high average energy Am-Be source and considerably poorer for the real workplace fields. The disagreement between measured and calculated response found for this series of measurements is greater than the approximate 10% difference observed for the EURADOS-Ringhals intercomparison[4]. It therefore seems that some uncertainty still exists with regard to the exact shape of the energy response function for TEPCs in the critical energy region between a few keV and 100 keV, precisely the energy region where obtaining good experimental data is difficult.

Analytical correction to the TEPC dose equivalent response

Taylor[13] has suggested a method to improve the dose equivalent response of a tissue-equivalent proportional counter by using the neutron-energy information contained in a microdosimetric event-size spectrum as an alternative to the use of the event-size spectra for the derivation of quality factors. Figure 1 shows the neutron component of the event-size spectra recorded with the TEPC5 in the Basement and HTP locations. For the Basement, approximately 70% of the fluence is from neutron energies less than 1 keV and only 15% of the fluence is above 100 keV, whereas for the HTP location, 30% of the fluence is below 1 keV and 40% above 100 keV[10]. Figure 1 illustrates quite well that, for the CANDU workplace, significant changes in the neutron energy spectrum do not lead to discernible changes in the shape of the microdosimetric event-size spectrum. This carries the implication that, with regard to nuclear power plant workplace fields, the diagnostic properties of a TEPC are of limited value. Taylor's correction has been applied to both workplace spectra shown in Figure 1 and leads to an enhancement factor of 1.49 for the Basement field and 1.48 for the HTP field. If this correction is applied to the measured TEPC

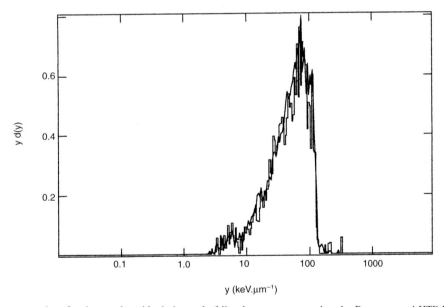

Figure 1. The neutron dose fraction per logarithmic interval of lineal energy measured at the Basement and HTP locations in a CANDU 6 power reactor with a 5″ diameter TEPC. The weak dependence of the shape of the microdosimetric spectra on changes in the primary neutron fluence spectrum is reflected in the fact that the two event-size spectra are almost indistinguishable. Line graph is HPT; step graph is Basement.

Table 2. The ratio, R, of the measured to calculated ambient dose equivalent for the TEPC5. In this case, 'calculated' refers to the folding of the ambient dose equivalent response function for the TEPC into the dose equivalent neutron spectra measured at each location with a multisphere spectrometer.

Location	Am-Be	HTAR	Basement	FMML	FLA	HTP	PVG
R	0.97	0.58	0.64	0.37	0.49	0.53	0.66

data, the response of the TEPC relative to the multi-sphere measurement becomes 73% for the Basement field and 62% for the HTP field. This is an improvement over what can be achieved with a calibration in a bare or moderated fission spectrum.

TEPC sensitivity

The need for real-time monitoring instruments in nuclear-power plants has been discussed previously by Hirning[14] and one of the attractions of a tissue-equivalent proportional counter for monitoring is that it is an active device. The sensitivity of a TEPC depends primarily on the physical size of the counter and increases as the square of the physical diameter[15]. Size is also of great importance from an ergonomic point of view, but here the requirement is for counters of small physical size and weight. Table 3 lists the sensitivity of the TEPC5 and REM 500 for the locations measured. The sensitivity of each instrument has been quantified in terms of pulses per second per μSv.h^{-1} measured with the multisphere spectrometer. For comparison, the sensitivity of an Eberline rem meter (PNR IV) used at the same locations is also listed. From the physical dimensions of the detectors, one would expect the sensitivity of the REM 500 to be a factor of 4.8 lower than that of the TEPC5. The fact that it is 3.7 times lower, averaged over all the fields, is plausibly a result of using 100% propane, with its higher hydrogen content, as the counting gas. This would result in a higher kerma and dose equivalent response of the REM 500 in these fields[11,12]. The proportional counter devices are 1 to 2 orders of magnitude below the sensitivity of the rem meter.

The question can be asked as to what is an appropriate sensitivity for an active area monitor. The answer is subjective in as much as it depends on how long a surveyor is prepared to wait for an accurate measurement of ambient dose equivalent, and this will depend on how high the total dose rate is at the location of the survey, which of course may not be known. It seems reasonable therefore that the sensitivity should be such that an accurate reading can be obtained in a few seconds in high dose rate locations. Furthermore, the uncertainty for a measurement time of a few seconds at low dose rate locations should still be, say, at the 100% level. For the measurements reported here, the location with the highest neutron dose equivalent rate was around the base of the heat transport pump (HTP) at 500 μSv.h^{-1}

and the lowest was in the new fuel loading area (FLA) at 8 μSv.h^{-1} as measured by the multisphere spectrometer. The uncertainty in the dose equivalent recorded by the TEPC5 in 1860 s at the HTP location was 2.3%, based on poissonian counting statistics[16]. An uncertainty of 20% would therefore be obtainable at this dose equivalent rate in measurements of 25 s duration. With the TEPC5 at the FLA location and for a measurement time of 3005 s, the uncertainty in the dose equivalent was 13.5%. An uncertainty of 100% would be obtainable in measurement times of around 55 s, which seems just acceptable. For the REM 500, similar uncertainties would be achievable in intervals about four times as long.

Mixed-field response

Another property of tissue-equivalent proportional counters of value in workplace monitoring is the use of pulse height analysis to discriminate between neutron and photon events. The neutron dose distributions shown in Figure 1 were obtained by the fitting of a pure ^{60}Co photon event-size spectrum to the low LET end of the measured mixed-field spectra and then subtracting the photon spectrum bin by bin of lineal energy. Figure 1 shows examples of where this fitting and subtracting procedure happend to have worked well, but this is not always the case. Figure 2 shows the measured event-size spectrum after photon event subtraction for the very low dose rate location at the new fuel loading area, FLA. Of course, a threshold could also be used to discriminate between photons and neutrons as indicated by the dashed line at 10 keV.μm^{-1} and in the example shown, this seems a more reasonable approach. However, in low energy neutron fields with low dose rates and poor counting statistics, photon discrimination is not precise. The difference between using the threshold technique and the fitting and subtraction method for the FLA field leads to a 24% difference in the recorded neutron dose equivalent which is similar in magnitude to the uncertainty resulting from the counting statistics.

Before leaving the topic of mixed-field measurement, it is important to note that in some locations encountered in CANDU stations the photon dose rate has been of the order of a few mSv.h^{-1} and high enough that, with the usual settings of the lower level discriminator of the analogue-to-digital converters (ADC) for the high-gain amplifier (around 150 mV, equivalent to

Table 3. Sensitivity, S, of the TEPC5, REM 500 and Eberline PNR-IV rem meter expressed in terms of pulses (counts) per second per μSv.h^{-1}.

Location	Am-Be	HTAR	Basement	FMML	FLA	HTP	PVG
S_{TEPC5}	0.04	0.014	0.019	0.017	0.018	0.015	0.014
$S_{REM\ 500}$	0.011	0.005	0.004	0.005	0.004	0.005	0.004
S_{PNR}	0.10	0.23	0.19	0.20	0.19	0.16	0.16

about 0.33 keV.μm^{-1}), the dead time has been greater than 25%. Any compact device for both photon and neutron dose equivalent determination based on a proportional counter should have some means of allowing adjustment of the lower pulse-height level discriminator and of indicating and correcting for instrument dead time. The REM 500 does not have dead-time correction, however, it is not operated as a mixed-field dosemeter. The ADC lower level discriminator on this device is set such that only events greater than 5 keV.μm^{-1}, and therefore very few photon-induced events, are recorded. For the PVG location with the highest gamma dose of 1.2 mSv.h^{-1} encountered in this series of measurements, the dead time of the REM500 was a fraction of 1%.

CONCLUSIONS

Considering the different attributes of tissue-equivalent proportional counters that have prompted the study of their use in workplace neutron monitoring, the measurements reported in this paper for the CANDU workplace indicate that:

(1) The TEPC5 dose equivalent energy response is still insufficiently known to be able to derive an accurate correction factor for a particular field from knowledge of the neutron fluence spectrum alone. Methods of calibration and correcting for the TEPC low energy response have been developed and improve the agreement between the dose equivalent measured with a TEPC and the ambient dose equivalent derived from multisphere measurements. Corrected TEPC5 measurements still underestimate the dose equivalent by up to 66%. The most promising strategy would seem to be to couple some of the design features of the REM 500, such as the use of a hydrogen rich counting gas with calibration in a so-called 'realistic' fields.

(2) The sensitivity of the TEPC5 is sufficient for CANDU workplace monitoring, although it is several orders of magnitude less than that of conventional rem meters. The sensitivity of the REM 500 (and, by implication, tissue-equivalent proportional counters of around 2″ diameter or less) is low enough that this is definitely a factor that needs some consideration when this device is used for surveying.

(3) Use of a TEPC as a mixed-field device to monitor both photon and neutron components necessitates a real-time indication of dead time and control of the ADC lower level discrimination. This constraint is relaxed if the TEPC is used in the manner of the REM 500, where photon interactions in a mixed field are discriminated against and not recorded.

In summary, these measurements carried out with a laboratory tissue-equivalent proportional counter system and the REM 500 show that low pressure tissue-equivalent proportional counters can form the basis of neutron monitors for use in the power-reactor environment. Comparison of the two instruments indicates how further improvements to the energy response might be

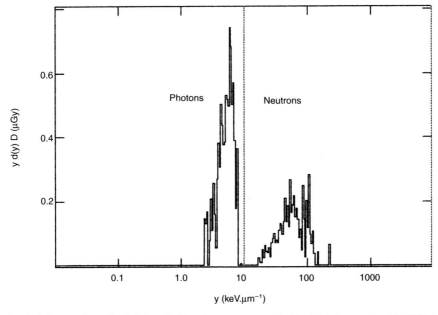

Figure 2. The absorbed dose per logarithmic interval of lineal energy measured at the FLA location in a CANDU 6 power reactor after fitting and subtraction of a ^{60}Co photon event-size spectrum.

achieved and the REM 500 demonstrates clearly that light, hand-held neutron monitors are feasible.

ACKNOWLEDGEMENTS

Funding for this work was provided by the Dosimetry and Dose Control Working Party of the CANDU Owners Group, COG. The authors wish to express their thanks to C. Nason, M. Lieskovský and R. Galbraith, health physics staff at Pt. Lepreau nuclear generating station for their enthusiastic co-operation and assistance during the measurement campaign. We also wish to acknowledge the help of Dr G. Taylor, National Physical Laboratory, UK for calculating the TEPC response correction factors.

REFERENCES

1. Booz, J., Edwards, A. A. and Harrison, K. G. (Eds) *Microdosimetric Counters in Radiation Protection*. Proceedings of a Workshop. Radiat. Prot. Dosim. **9**(3) (1984).
2. Menzel, H. G., Paretzke, H. G. and Booz, J. (Eds) *Implementation of Dose-Equivalent Meters Based on Microdosimetric Techniques in Radiation Protection*. Proceedings of a Workshop. Radiat. Prot. Dosim. **29**(1–2) (1989).
3. Alberts, W. G., Dietze, E., Guldbakke, S., Kluge, H. and Schuhmacher, H. *International Intercomparison of TEPC Systems Used for Radiation Protection*. Radiat. Prot. Dosim. **29**(1–2), 47–53 (1989).
4. Lindborg, L., Bartlett, D., Drake, P., Klien, H., Schmitz, Th. and Tichy, M. *Determination of Neutron and Photon Dose Equivalent at Workplaces in Nuclear Facilities in Sweden*. Radiat. Prot. Dosim. **61**(1–3), 89–199 (1995).
5. Waker, A. J. *Microdosimetric Radiation Field Characterisation and Dosimetry in a Heavy Water Moderated Reactor Environment*. Radiat. Prot. Dosim. **52**(1–4), 415–418 (1994).
6. Nunes, J. C. and Waker, A. J. *Multisphere Spectrometry and Analysis of TEPC and REM Meter Results Around a Heavy Water Moderated Reactor*. Radiat. Prot. Dosim. **59**(4), 279–284 (1995).
7. Perey, F. G. *Least-Squares Unfolding*: The Program STAY'SL. Oak Ridge National Laboratory Report ORNL/TM-60602, ENDF-254 (Radiation Shielding Information Centre, Oak Ridge National Labs, PO Box 2008, Oak Ridge, TN 37831-6362, USA) (1977).
8. Chartier, J. L., Kurkdjian, J., Paul, D., Itie, C., Audoin, G., Pelcot, G. and Posny, F. *Progress on Calibration Procedures with Realistic Neutron Spectra*. Radiat. Prot. Dosim. **61**(1–3), 57–62 (1995).
9. Thomas, D. J. National Physical Laboratory, UK. Private communication (1995).
10. Nunes, J. C., Waker, A. J. and Lieskovský, M. *Neutron Fields Inside Containment of a CANDU 600-PHWR Power Plant*. Health Phys. **71**(2), 235–247 (1996).
11. Khaloo, R. and Waker, A. J. *An Evaluation of Hydrogen as a TEPC Counting Gas in Radiation Protection Microdosimetry*. Radiat. Prot. Dosim. **58**(3), 185–191 (1995).
12. Aroua, A., Hoefert, M. and Sannikov, A. V. *On the Use of Tissue-Equivalent Proportional Counters in High Energy Stray Radiation Fields*. Radiat. Prot. Dosim. **59**(1), 49–53 (1995).
13. Taylor, G. C. *An Analytical Correction for the TEPC Dose Equivalent Response Problem*. Radiat. Prot. Dosim. **61**(1–3), 67–70 (1995).
14. Hirning, C. R. *The Increasing Need for Active Dosemeters in the Nuclear Industry*. Radiat. Prot. Dosim. **61**(1–3), 1–8 (1995).
15. Kliauga, P., Waker, A. J. and Barthe, J. *Design of Tissue-Equivalent Proportional Counters*. Radiat. Prot. Dosim. **61**(4), 309–322 (1995).
16. Menzel, H. G., Lindborg, L., Schmitz, T., Schuhmacher, H. and Waker, A. J. *Intercomparison of Dose Equivalent Meters Based on Microdosimetric Techniques: Detailed Analysis and Conclusions*. Radiat. Prot. Dosim. **29**(1–2), 55–68 (1989).

MICRODOSIMETRIC INVESTIGATIONS IN REALISTIC FIELDS

J. F. Bottollier-Depois†, L. Plawinski†, F. Spurný‡ and A. Mazal§
†Institute for Protection and Nuclear Safety, Human Health Protection and Dosimetry Division
Dosimetry Service, IPSN, BP n° 6, 92265 Fontenay-aux-Roses Cedex, France
‡Nuclear Physics Institute, Czech Academy of Sciences, Prague, Czech Republic
§Orsay Proton Therapy Centre, CPO, BP n° 65, 91402 Orsay Cedex, France

Abstract — Microdosimetric measurements have been obtained in different complex fields: behind concrete with a primary beam of 200 MeV protons to estimate the efficiency of the shielding, the neutron yield and the response of different kinds of detectors; and reference fields behind shielding with a high energy primary beam to simulate the radiation at aircraft altitude. In general, the LET spectra are composed of a low LET part which represents the photons and the muons and a high LET part corresponding to neutrons, the latter being the most important one. The value of the mean quality factor (deduced from the ICRP 21) has been found to be between 3 and 5 with concrete shielding and higher than 6 with iron shielding. The application of ICRP 60 has an important effect on the estimations of the ambient dose equivalent and the quality factor increases up to 60% depending on the field.

INTRODUCTION

Microdosimetric measurements have been made in different complex fields:

(i) realistic reference neutron fields (Realistic Neutron Spectra Facility at Cadarache) used for the calibration of radiation protection instruments;

(ii) behind concrete, with a primary beam of 200 MeV protons at the Proton Therapy Centre in Orsay to estimate the efficiency of the shielding, the neutron yield and the response of different kinds of detectors;

(iii) reference fields behind shielding, with a high energy primary beam (protons and pions with a momentum of 205 GeV.c^{-1} at CERN) to simulate the radiation at aircraft altitude (neutrons, charged particles, gamma) in the frame of a EURADOS/EC project.

THE SYSTEM

The equipment[1], named Nausicaa, is a proportional counter (TEPC) filled with low pressure tissue equivalent gas simulating a 3 μm biological target (Figure 1).

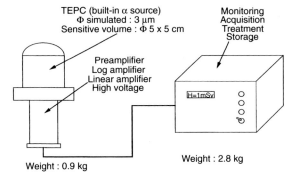

Figure 1. The Nausicaa system.

The sensitive volume is a right cylinder (5 cm diameter, 5 cm length). The walls are made of 1 cm thick tissue-equivalent plastic (A150). A logarithmic amplifier is used. Data are compacted in 200 intervals from a 4096 channel ADC. The measured parameter is the lineal energy (y) spectrum from which the following parameters are derived: the absorbed dose, the ambient dose equivalent (H*(10)), the mean quality factor calculated using ICRP 21 or ICRP 60 quality factor functions. An internal alpha source (^{244}Cm) is used for the lineal energy calibration. The calibration for the ambient dose equivalent is performed with a gamma ray source (^{60}Co) for the low y part and with a neutron source (AmBe) for the high part.

RESULTS

Proton Therapy Centre at Orsay

In order to characterise different shields, measurements were performed at the Centre de Protonthérapie d'Orsay[2] to evaluate the neutron yield when a 200 MeV proton beam interacts with different targets, the attenuation of varying thicknesses of concrete at different angles and the response of different detectors (Figure 2). The beam is produced by a synchrocyclotron with a frequency of 448 Hz, with an internal microstructure of 19.3 MHz. It is transported to an experimental area where a bunker of concrete has been assembled around different targets. The shielding around the target has been changed between 0 and 3 m of concrete (2.2 g.cm^{-3}), while it has been kept constant on the top (1 m concrete). Two targets have been used: copper, simulating materials of relatively high atomic number as present in proton beam delivery systems, and water, to represent the patient and some beam modifier devices like modulators and absorbers. In both cases the target was thick enough to stop the beam. The copper target was cylindrical with a diameter of 140 mm and 150 mm length (the range of the 200 MeV proton beam in copper is around 43 mm). The dimensions of the water target were 200 mm × 200 mm and 320 mm depth (range in water around 260 mm). The beam intensity during these measurements was set at 5 to 10 nA, depending on the configuration of the shielding.

Measurements show the efficiency of various shielding thicknesses for different directions between 0° and 90°. For instance, with the water target, at 0°, the absorbed dose attenuation factor is 210 with 3 m thick concrete shield and 7.6 with 0.5 m thick concrete at 90° (Table 1). The quality factor (ICRP 21)[3] depends on the target used and increases behind shielding (from 3.1 without shielding to 4.6 with 3 m thick concrete shield at 0°) because of the absorption of low LET particles like gamma rays.

Similar measurements have been performed using targets of Al, Fe and Pb at different depths of concrete[4]. For Siebert's measurements, the detector was under equilibrium inside the concrete. In both cases, the spectra have a general similar shape, either for no shielding or behind concrete. The low lineal energy component presents a peak between 1 and 10 keV.μm^{-1}, mainly with no shielding, while for thick shielding, the contribution of the neutron dose is around 30% of the total absorbed dose.

For a ^{60}Co source, the yd(y) spectrum shows a peak around 0.3 keV.μm^{-1}, and this peak moves towards higher lineal energies for lower photon energies (5 keV.μm^{-1} for 100 keV)[5]. In our case, a large photon spectrum corresponds to de-excitation of nuclei after evaporation of massive particles as well as inelastic interactions and neutron capture in concrete. Some

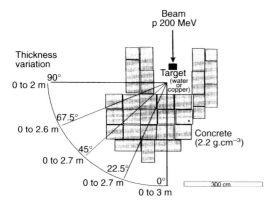

Figure 2. The Proton Therapy Centre (CPO) shielding at Orsay.

Table 1. Ambient dose equivalent (H*(10) in μSv.h^{-1} for 1 nA proton beam), absorbed dose (D in μGy.h^{-1} for 1 nA proton beam) and quality factor (Q) — always ICRP 21 — obtained at CPO with a water target.

		H*(10) (ICRP 21) (μSv.h^{-1} nA^{-1})	D (μGy.h^{-1} nA^{-1})	Q	High LET part of H*(10)
0°	no shielding	1043	334	3.1	78.2%
0°	3 m concrete	7.3	1.6	4.6	87.3%
90°	no shielding	532	144	3.7	83.2%
90°	0.5 m concrete	100	19	5.2	90.2%

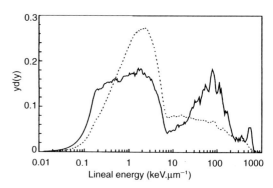

Figure 3. Relative absorbed dose distribution obtained at CPO with a water target at 0° with different shieldings: (- - -) no shielding, (———) 3 m concrete.

Table 2. Ambient dose equivalent (H*(10) in nSv for one PIC monitor count), absorbed dose (D in nGy for one PIC monitor count) and quality factor (Q) obtained at CERN on top iron and top concrete shields (detector position 2) with a 205 MeV.c^{-1} (protons and pions$^+$) primary beam.

	H*(10) (nSv.PIC unit^{-1})	D (nGy.PIC unit^{-1})	Q	High LET part of H*(10)
Top concrete				
ICRP 21	0.50	0.14	3.4	79.4%
ICRP 60	0.63	0.14	4.3	83.6%
Top iron				
ICRP 21	1.3	0.20	6.5	93.5%
ICRP 60	2.0	0.20	10.0	95.7%

spallation products contribute to the spectra for lineal energies higher than 100 keV.μm^{-1}[5], and they can be more easily identified for the measurements with no shielding.

Figures 3, 4 and 5 show the yd(y) spectra for different shielding, targets and angles. Without shielding, the smooth distribution around and above the proton edge shows the presence of high energy neutrons. In the shielded situation, the maximum values of the 'photon' and the 'neutron' components are similar, in agreement with Siebert's results for an aluminium target.

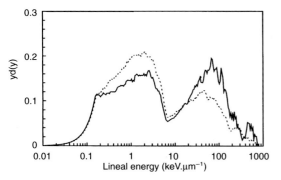

Figure 4. Relative absorbed dose distribution obtained at CPO with a water target at 90° with different shieldings: (– – –) no shielding, (———) 50 cm concrete.

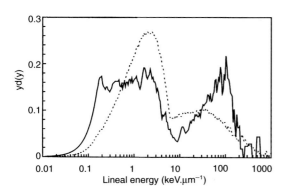

Figure 5. Relative absorbed dose distribution obtained at CPO with a copper target at 0° with different shieldings: (– – –) no shielding, (———) 3 m concrete.

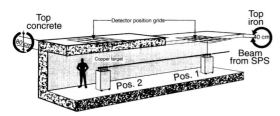

Figure 6. The mixed high energy field facility at CERN.

Mixed high energy fields at CERN

Mixed high energy particle fields have been produced at the super proton synchrotron at CERN in order to simulate fields encountered at flight altitudes of aircraft where there is a mixture of primary cosmic radiation and their secondary products generated in the atmosphere[6]. The primary beams are high energy protons and pions (momentum of 120 and 205 GeV/c) impinging on a copper target. Measurement positions were located at large angles to the beam, outside two different shields: 0.4 m of iron or 0.8 m of concrete (Figure 6).

The relative importance of the low y component is higher in the case of concrete shielding (Figures 7 and 8). The application of ICRP 60[7] has again an important effect, the quality factor increases from 3.4 (ICRP 21) to 4.3 with the concrete shielding and from 6.5 (ICRP 21) to 10 with the iron shielding (Table 2). The field encountered behind concrete is the most representative of a real field around 10,000 m altitude.

These results can be compared with those obtained with the HANDY system[8]. There is a good agreement

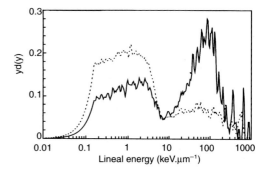

Figure 7. Relative absorbed dose distribution obtained at CERN on top iron (———) and top concrete (– – –) shields with a 205 GeV.c^{-1} (protons and pions$^+$) primary beam. The beam intensity is 956 units.s^{-1} for the 'concrete' and 1142 PIC units.s^{-1} for the 'iron' position.

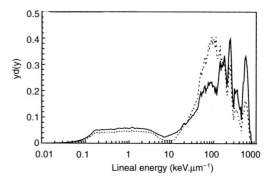

Figure 8. Relative dose equivalent distributions using ICRP 21 (———) and ICRP 60 (– – –) obtained at CERN on top concrete shield with a 205 GeV.c^{-1} (protons and pions$^+$) primary beam.

with the concrete shielding. With the iron shielding, spectra are different in the 'photon' part distribution and for the 'neutron–photon' ratio. The difference in the shape of the photon part can be explained by the effect of the beam intensity relative to the background[9]. The beam intensity is expressed in PIC units.s^{-1} using the PIC count normalised by the effective time of extraction during the measurement (2.5/14.4 × the time of measurement). The contribution of low y events is increasing when the beam intensity is decreasing because of muon background.

CONCLUSION

In spite of the fact that the TEPC response is not always ideal, for instance with low energy neutrons, TEPCs allow a direct estimate of the ambient dose equivalent in unknown complex particle fields (charged particles, neutrons, gamma rays) as encountered with cosmic radiation or around specific facilities which have not been sufficiently characterised.

The quality factor increases with the use of ICRP 60 factors in comparison with ICRP 21 in radiation fields with an important contribution of neutrons with energies below a few MeV.

ACKNOWLEDGEMENTS

The intercomparisons at CERN were partly supported by the Commission of European Communities and at the CPO by the ministry of health, CRAMIF and LNCC.

REFERENCES

1. Nguyen, V. D., Luccioni, C. and Parmentier, N. *Average Quality Factor and Dose Equivalent Meter Based on Microdosimetry Technique.* Radiat. Prot. Dosim. **10**(1–4), 277–282 (1985).
2. Mazal, A., Gall, K., Bottollier-Depois, J. F., Delacroix, D., Fracas, P., Delacroix, S., Nauraye, C. and Louis, M. *Shielding Measurements for a Proton Therapy Beam of 200 MeV: Preliminary Results.* Radiat. Prot. Dosim. **70**(1–4), 429–436 (This issue) (1997).
3. ICRP. *Recommendations of the International Commission on Radiological Protection.* Publication 21 Supplement to ICRP Publication 15 (Oxford: Pergamon) (1973).
4. Siebers, J. V., DeLuca, P. M., Pearson, D. W. and Coutrakon, G. *Measurement of Neutron Dose Equivalent and Penetration in Concrete for 230 MeV Proton Bombardment of Al, Fe and Pb Targets.* Radiat. Prot. Dosim. **44**(1–4), 247–251 (1992).
5. ICRU. *Microdosimetry*, Report 36 (Bethesda, MD: ICRU Publications) (1983).
6. EURADOS WG 11. *Exposure of Air Crew to Cosmic Radiation.* Report to be published.
7. ICRP. *1990 Recommendations of the International Commission on Radiological Protection.* Publication 60 (Oxford: Pergamon) (1990).
8. Schuhmacher, H., Alberts, W. G., Alevra, A. V., Klein, H., Shrewe, U. J. and Siebert, B. R. L. *Characterisation of Photon-Neutron Radiation Fields for Radiation Protection Monitoring and Optimisation.* Radiat. Prot. Dosim. **61**(1–3), 81–88 (1995).
9. Aroua, A., Hoefert, M. and Sannikov, A. *Effects of High Intensity and Pulsed Radiation on the Response of the HANDI-TEPC.* Radiat. Prot. Dosim. **61**(1–3), 113–118 (1995).

NEUTRON ENERGY DEPOSITION SPECTRA AT SIMULATED DIAMETERS DOWN TO 50 nm

E. Anachkova[†], A. M. Kellerer[†‡] and H. Roos[†]
[†]Strahlenbiologisches Institut der Universität München
Schillerstrasse 42, D-80336 München, Germany
[‡]Institut für Strahlenbiologie, GSF
Forschungszentrum für Umwelt und Gesundheit
Postfach 1129, D-85758 Oberschleißheim, Germany

Abstract — A special cylindrical tissue-equivalent proportional counter has been designed and constructed. Using a recently developed calibration technique with ^{37}Ar, its performance has been analysed for simulated sites with diameters down to 100 nm. Neutron energy deposition spectra have been obtained for simulated diameters 50–2000 nm with neutron beams from the ^{2}H(d,n)^{3}He reaction. The peak at high energies imparted due to heavy particle recoils and the proton edge are clearly seen in the spectra even at the smallest simulated diameters.

INTRODUCTION

In recent years there has been increased interest in the spectra of energy deposition by neutrons in small sites. Knowledge of the energy deposition in small subcellular dimensions is important for an improved understanding of the biological effects of densely ionising radiations. Numerous theoretical studies have utilised a variety of approaches and mathematical codes[1], but there are only a few experimental studies. The dose mean lineal energy, $\bar{y}_d$, in a 5.7 MeV neutron beam was determined by Lindborg et al[2] who used the variance–covariance method for simulated diameters down to 20 nm, employing as detectors tissue-equivalent spherical ionisation chambers. Kliauga[3] measured single-event distributions at simulated sizes below 300 nm with an ultra-miniature microdosimetric counter. Olko et al[4] reported measurements obtained with the same counter and compared them to computed results. It has been difficult, however, to judge the validity of these measurements in photon and neutron fields, as there was no reliable method to calibrate the ultra-miniature counter at simulated site sizes of the order of some nanometres and to determine its resolution. A reliable method to assess the detector response is therefore of crucial importance for measurements at small simulated sites. This investigation presents a solution of this problem and shows that reliable microdosimetric spectra can be determined at simulated sites smaller than usually assumed.

EXPERIMENT

Detector performance at small simulated sites

A cylindrical proportional counter with 60 mm length and 3 mm diameter has been designed and constructed using A-150 plastic as wall material. The central electrode, a gold plated tungsten wire of 20 μm diameter is held in place by supports made from Makrolon. No field tubes were assumed to be needed, because any effects of the insulators were likely to be negligible due to the large length-to-diameter ratio of 20:1 of the sensitive volume. However, the performance of a detector of such a simple construction requires careful testing. Measurements in the usual photon or neutron radiation fields are not appropriate for such a test, because one obtains broad energy spectra of charged secondaries in TEPCs, and these secondaries, in turn, deposit largely different energies in the counter. The relative variance of typical microdosimetric spectra under these conditions is usually too large — due to these large variations in the energy imparted — to permit meaningful conclusions on the characteristics of the detector response. It is effectively impossible, in this case, to recognise in the observed spectra the increase of the relative variance caused by the multiplication statistics, by the fluctuation of the number of ions, and by the non-uniformity of the gas gain.

The recently reported method[5] for calibrating and testing proportional counters with ^{37}Ar permits not only a reliable calibration, but more importantly it allows the determination of the detector resolution and detector performance with high accuracy. In our original investigations it has been employed to study detector function at simulated diameters from 0.1 to 2 μm. The K peak of ^{37}Ar was used to determine the calibration factor and the detector resolution for simulated diameters ranging from 0.3–2 μm. For smaller simulated sites the calibration in terms of the K peak is complicated, because one needs to account for the incomplete energy deposition by K Auger electrons. The L peak at 0.2 keV is essential for calibration at smaller simulated diameters, but it is close to the noise and its observation requires high gas gain and low detector noise. With the detector used in this study, calibration in terms of the L peak was possible for simulated diameters down to 0.1 μm. It was found that the distribution of the pulse heights in the observed L peak is well represented by the gamma distribution $f(x) = c\, e^{-x} x^{\lambda-1}$, where λ is the mean number of ionisations, x is the pulse height, and c is a constant. This corresponds to an exponential distribution of pulse height for single ions. The best fit was obtained with $\lambda = 6 \pm 0.5$, which agrees well with the W value of 32.32 eV for 200 eV electrons given by Waibel and Grosswendt[6]. The test with ^{37}Ar has thus shown that this counter with very simple construction possesses a resolution nearly equal to that of a counter with a uniform response and with optimal distribution[7] of pulse heights for single ions. At smaller simulated diameters the gas gain was limited by the maximum voltage that can be applied without electrical breakdown.

Operating conditions of the measurement of single event distributions

The operating parameters employed in the measurements of the energy deposition spectra are specified in Table 1. For completeness the reduced electric field strength on the wire surface and the gas gain, calculated with the Townsend formalism, are included. The problem usually encountered at small simulated diameters is the enlargement of the avalanche region around the multiplication wire and the resultant change of the gas multiplication for ionisations produced at different locations in the counter. The gradual change of the electron avalanche formation in its dependence on the radial distance is given in Figure 1 for the operating conditions in Table 1. If, however, the same data are presented in terms of the dependence of gas gain on the position of ionisation in the detector volume (Figure 2), it is realised that these changes are insignificant. The region with different gas gain around the central electrode amounts only to an insignificant fraction of the detector volume; the requirement for the uniformity of the gas multiplication is, therefore, met.

RESULTS AND DISCUSSION

Pulse-height spectra were measured at the Van de Graaf accelerator at GSF for neutrons produced by the ^{2}H(d,n)^{3}He reaction. The spectra were measured in three sections with different gas gain and electronic gain, and were then combined before normalisation. At simulated diameters of 0.5 μm or larger, calibration factors were obtained from the measurement of the position of the K peak of ^{37}Ar. A minor correction has been applied, to take into account the difference between the W value for the K Auger electrons (29.9 eV[6]) and the effective value, W_n, for 5 MeV neutrons (31 eV[8]). The spectra obtained at simulated diameters 0.5–2 μm are in good agreement with those published by other authors (e.g.

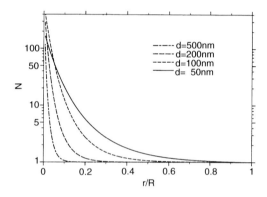

Figure 1. Part of the gas gain, N, that occurs outside the specified distance, r, from the detector axis. R is the counter radius. The data are obtained at the lower voltage specified in Table 1.

Table 1. Operating parameters of the detector: pressure, p, and voltage, U, at different simulated diameters, d, and the calculated reduced electric field strength, S, and the gas multiplication, M.

d (nm)	p (kPa)	U (V)	S (V.Pa^{-1}.m^{-1})	M	$\bar{y}_d$ (keV.μm^{-1})
500	15.8	600	755	108	55 ± 3
		700	881	301	
200	6.34	600	1888	280	60 ± 5
		700	2200	301	
100	3.17	600	3777	400	66 ± 6
		650	4092	680	
50	1.58	500	6355	167	70 ± 10
		550	6991	284	

N.B. The obtained values for $\bar{y}_d$ depend on the accuracy of the measurement in the low energy part of spectrum. The errors quoted with the values of $\bar{y}_d$ are rough indications of the uncertainty caused by the choice of the extrapolation to low energies.

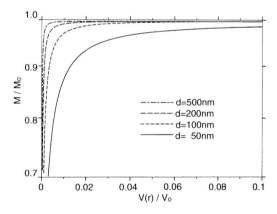

Figure 2. Dependence of the gas gain, M, on the distance r, from the detector axis. M_0 is the maximum gas gain, V_0 is the entire volume of the counter, V(r) is the part of the volume inside a cylinder with radius r.

Ref. 9). At simulated diameters from 0.2 to 0.1 μm, the calibration factor was determined by the L peak of ^{37}Ar and corrected for the different W values, as already described. At simulated diameters smaller than 0.1 μm, the calibration in terms of the L peak became uncertain because of the limited gas gain. Even at these small sites pulse-height spectra in neutron fields could still be determined, and in the high energy part their spectral features are similar to those that were found at a simulated diameter of 0.1 μm. The high energy peak due to recoils and the proton edge are clearly seen, although the latter becomes less steep. One can use these two features for the calibration of the microdosimetric distributions.

The dose distributions of lineal energy are given in Figure 3. The general features of the distributions, i.e. the observed broadening of the main peak with decreasing simulated sites, and its shift toward higher values of y are in good accordance with the calculations of Coyne and Caswell[1] which account for straggling. The calculated values of the dose mean lineal energy, $\bar{y}_d$, vary between 44 keV.μm^{-1} for a simulated diameter of 2 μm, and 70 keV.μm^{-1} for a simulated diameter of 50 nm. At simulated diameters 2–0.5 μm, the values are close to those obtained by Lindborg et al[2] with the pulse-height method, at smaller simulated sites they exceed the values he obtained with the variance–covariance method (Figure 4). Conversely, the values agree with those calculated by Coyne and Caswell[1] at small site sizes while they are smaller at the large sites. The computations[1], other than the variance–covariance measurements[2], do not include a γ-component. The pulse-height measurements eliminate part of the γ-component by the low energy discriminator.

The initial part of the spectra at small site sizes is hidden by the electronic noise. The magnitude of the resulting error is estimated in Table 1; for the site size 50 nm it corresponds to the two limiting cases of extrapolation indicated by the dotted lines in Figure 3. The reason for uncertainty was the limitation to the high voltage that causes the incompleteness of the spectra at low energies. The extrapolation of the measured distributions toward low lineal energies influences the values of $\bar{y}_d$.

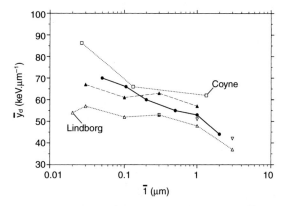

Figure 4. Dependence of the dose mean lineal energy, $\bar{y}_d$, for 5.7 MeV neutrons on the mean chord length, $\bar{l}$. The dots represent the values obtained in this study, the squares the calculations of Coyne and Caswell[1] for 4.9 MeV neutrons. The open triangles represent the measurements of Lindborg[2] with the pulse height method (▽) and with the variance–covariance method (△). Subtracting the gamma component from the latter measurements, one obtains the solid triangles (Lindborg, private communication).

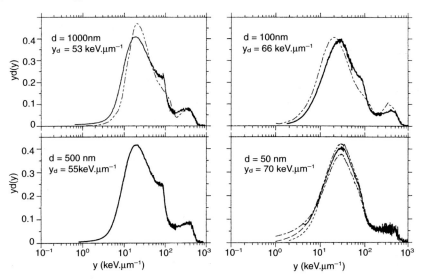

Figure 3. Dose distribution of lineal energy, yd(y), for neturons produced by the ^{2}H(d,n)^{3}He reaction, at simulated diameters 1000, 500, 100, and 50 nm. The dashed lines in the upper panel represent the calculations of Coyne and Caswell[1]. The dashed and the dotted-dashed lines in the distribution at 50 nm illustrate the uncertainty caused by the extrapolation toward low energies.

It is difficult to compare our results with the spectra obtained earlier with the ultra-miniature counter[3,4] for 15 MeV neutrons. However, it is of interest to note that the spectrum obtained for 50 nm with the ultra-miniature counter is much broader than our spectrum for 5.7 MeV neutrons, and that — other than in our case — the measured spectrum differs substantially from the computed distribution (see Figure 5).

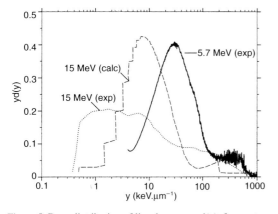

Figure 5. Dose distribution of lineal energy, yd(y), for neutrons at simulated diameter 50 nm. The full line represents our experimentally obtained distribution for 5.7 MeV neutrons. The dotted line indicates the distribution obtained with the ultra-miniature counter[3,4] for 15 MeV neutrons, and the dashed line the calculation of Olko et al[4].

CONCLUSIONS

This investigation demonstrates that one can perform reliable microdosimetric measurements for smaller simulated sites than is usually assumed. Reliable calibration and the quantitative determination of the detector response are a pre-condition for verifying the results. The calibration and testing of the resolution with ^{37}Ar is a suitable tool for this purpose.

ACKNOWLEDGEMENTS

The authors are grateful to the staff of the Van de Graaf accelerator of the GSF, Research Centre for Environment and Health, for their highly effective assistance in the measurements at their facility. This investigation was supported by EURATOM contract F13P-CT g20039; it is part of the cooperation within EURADOS Working Group 10. We are indebted to Dr. Lennart Lindborg for providing us with some unpublished results and for his interest in this study.

REFERENCES

1. Coyne, J. J. and Caswell, R. S. *Neutron Energy Deposition on the Nanometer Scale*. Radiat. Prot. Dosim. **44**(1–4), 49–52 (1992).
2. Lindborg, L., Marino, S., Kliauga, P. and Rossi, H. H. *Microdosimetric Measurements and the Variance–Covariance Method*. Radiat. Environ. Biophys. **28**(4), 251–263 (1989).
3. Kliauga, P. *Microdosimetry at Middle Age: Some Old Experimental Problems and New Aspirations*. Radiat. Res. **124**(S), S5-S15 (1990).
4. Olko, P., Morstin, K. and Schmitz, T. *Simulation of the Response of an Ultraminiature Microdosimetric Counter for Fast Neutrons*. Radiat. Prot. Dosim. **44**(1–4), 73–76 (1992).
5. Anachkova, E., Kellerer, A. M. and Roos, H. *Calibrating and Testing TEPC with ^{37}Ar*. Radiat. Environ. Biophys. **33**(4), 353–364 (1994).
6. Waibel, E. and Grosswendt, B. *W Values and Other Transport Data on Low Energy Electrons in Tissue Equivalent Gas*. Phys. Med. Biol. **37**(5), 1127–1145 (1992).
7. Kellerer, A. M. *Local Energy Spectra and Counter Resolution*. Ann. Res. Report NYO-2750-5 P-40:94-103 (1968).
8. ICRU. *Average Energy Required to Produce An Ion Pair* Report 31 (Bethesda, MD: International Commission of Radiation Units and Measurements) p. 30 (1979).
9. Pihet, P. Gerdung, S., Grillmaier, R. E., Kunz, A. and Menzel, H. G. *Critical Assessment of Calibration Techniques for Low Pressure Proportional Counters Used in Radiation Dosimetry*. Radiat. Prot. Dosim. **44**(1–4), 115–120 (1992).

REVIEW OF RECENT ACHIEVEMENTS OF RECOMBINATION METHODS

N. Golnik
Institute of Atomic Energy,
Dosimetry of Mixed Radiation,
PL-05-400 Otwock-Swierk,
Poland

Abstract — The recombination methods developed in recent years are reviewed. The discussion is focused on the methods which allow determination of the radiation protection quantities like quality factor, low LET dose component, H*(10) and crude microdosimetric spectra in terms of dose distributions in LET.

INTRODUCTION

Although recombination chambers have been used in radiation protection for a long time, remarkable progress has been made during the past few years[1]. These more advanced recombination methods are summarised, with special emphasis on their possible applications for monitoring of work places in the vicinity of nuclear facilities.

RECOMBINATION CHAMBER

Recombination chambers are high pressure tissue-equivalent ionisation chambers, designed and operated in such a way that ion collection efficiency in the chambers is governed by initial recombination of ions. The results illustrating this review were obtained with the large chamber of REM-2 type[1,2] (Figure 1), filled to a pressure of about 1 MPa with a gas mixture consisting of methane and 5% (by weight) of nitrogen.

RECOMBINATION METHODS

The output signal of the chamber is the ionisation current (or collected charge) as a function of collecting voltage (Figure 2). All the recombination methods require the measurement of the current (or charge) at least at two values of the collecting voltage. The highest voltage should provide the conditions close to saturation, i.e. almost all the ions generated in the chamber cavity should be collected. In practice, the maximum voltage is used that can be applied to the chamber without causing a considerable dark current or other unwanted effects[3].

Depending on the number of measured points of the saturation curve (see Figure 2) one can obtain different information:

(1) The ionisation current measured at maximum applied voltage is proportional to the absorbed dose, D (some small corrections for lack of saturation can be introduced when needed).
(2) Measuring at two voltages enables determination of the recombination index of radiation quality[4], Q_4,

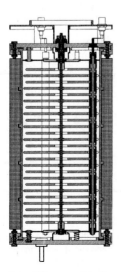

Figure 1. Cross section of the recombination chamber REM-2.

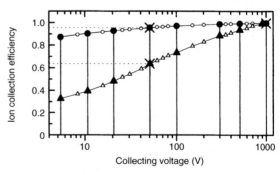

Figure 2. Example of saturation curves measured with REM-2 recombination chamber. Circles, ^{137}Cs reference gamma radiation. Triangles, 2.5 MeV neutrons. Crosses show two points needed for determination of ICRP 21 quality factor, solid points are needed for measurements of H*(10) with ICRP 60 quality factor and all the points are used for microdosimetric distributions (see text).

that approximates the quality factor, as defined in ICRP Report 21[5]. Then the ambient dose equivalent, $H^*(10)$, can be determined as $H^*(10) = D\,Q_4$.

(3) Since the relationship between Q_4 and the quality factor, Q_{ICRP60}, based on Report 60 of ICRP[6] is not linear, Q_4 should not be used as an assessment of Q_{ICRP60}. The measurement of ionisation current at six collecting voltages and mathematical analysis of the saturation curve enables separation of low LET and high LET dose fractions and determination of quality factor and $H^*(10)$ according to the ICRP60 Q(L) function.

(4) Measuring at 15 or more collecting voltages enables determination of crude microdosimetric spectra in terms of dose distributions in LET. The method is based on similar but more detailed mathematical analysis than those used when measurements are performed at 6 points. Somewhat better accuracy can be achieved in determination of $H^*(10)$; however, the time needed for measurements is longer and the mathematical analysis requires some experience.

RECOMBINATION INDEX OF RADIATION QUALITY

The recombination index of radiation quality (RIQ) denoted as Q_4, is a measurable quantity that directly approximates the radiation quality factor, as defined in ICRP Report 21[5], since the dependence of Q_4 on LET simulates well the ICRP 21 Q(L) function[1,4].

Determination of Q_4 involves two steps:

(1) Calibration of the chamber in a reference gamma radiation field from a ^{137}Cs source. During the calibration the recombination voltage, U_R, is determined in such a way that the ion collection efficiency $f_\gamma(U_R) \approx 0.96$.

(2) Determination of ion collection efficiency, $f_{mix}(U_R)$, investigated radiation field (this requires measurements at two voltages, as indicated by crosses in Figure 2).

The Q_4 is then calculated according to its definition as:

$$Q_4 = \frac{1-f_{mix}(U_R)}{1-f_\gamma(U_R)} \quad (1)$$

The large set of experimental results, collected during many years in different radiation fields, showed[1,4,7], that the values of Q_4 practically did not differ from the quality factor defined in ICRP Report 21. However, the earlier measurements were performed mostly in fields of isotopic sources and in the vicinity of accelerators where the reference values of quality factor were sometimes not well known. Recently, the measurements of Q_4 were performed in reference fields of monoenergetic neutrons[8] and the approximate equality $Q_4 \approx Q$ was well confirmed by the experimental results (see Figure 3). Since Q_4 is an additive quantity[1,4], this equality is valid for any composition of mixed radiation fields.

The above method became the basis for the design of a microcomputer controlled experimental set-up, intended for routine measurements of $H^*(10)$, described elsewhere[2].

MICRODOSIMETRIC APPROACH

When the whole saturation curve is measured (at least at 15 voltages), it might be used to derive[1,10,11] a crude (6–8 intervals) dose distribution versus LET. The value of Q_{ICRP60} can then be calculated as a dose weighted average of Q(L). However, more strictly, the distribution should be expressed in terms of restricted LET, L_Δ. The energy cut-off Δ might be estimated[1] from the theory of initial recombination to be of the order of 500 eV, which corresponds to a site size which is equivalent to about 70 nm of tissue.

Our microdosimetric approach[1,10] is based on the theory of initial recombination of ions and on the approximation that, in a certain range of operational conditions, the dependence of ion collection efficiency on gas pressure and on collecting voltage can be described by the same function for photons and for high LET particles. Theoretical and experimental investigations presented elsewhere[1] show that the above assumption is valid for our chamber, within few per cent of accuracy, in the investigated gas pressure range from 0.9 to 1.1 MPa.

In our method, the saturation curve for a mixed radiation is compared with the curve measured in a reference gamma radiation field. The ion collection efficiencies for mixed radiation, f_{mix}, and for reference gamma radiation, f_γ, have to be measured at the same collecting voltages. The data obtained are then expressed in the form of a dependence of ion collection efficiency f_{mix} on f_γ, and fitted by Equation 2:

$$f_{mix} = D_{low}\,f_\gamma + \sum_j D_j\,s_j \quad (2)$$

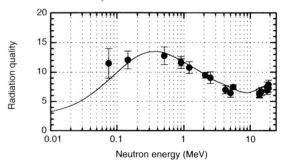

Figure 3. Values of recombination index of radiation quality Q_4 for monoenergetic neutrons (points) in comparison with the effective quality factor in ICRU sphere calculated[9] with ICRP21 Q(L) function (solid line).

where D_j are the dose fractions associated with the jth L_Δ interval, $L_0 = 3.5$ keV·μm^{-1} and s_j are the analytical functions of f_γ[10]:

$$s_j = \frac{1}{L_{\Delta,j+1} - L_{\Delta,j}} \int_{L_{\Delta,j}}^{L_{\Delta,j+1}} \frac{1}{1 + [L_\Delta (1 - f_\gamma)/(L_0 f_\gamma)]} dL_\Delta \quad (3)$$

Examples of such microdosimetric distributions are presented in Figure 4 for radiation fields from a ^{252}Cf source exposed free-in-air or in 10 cm spherical filters. Comparison of Figure 4(a) with 4(b) and 4(c) shows that the distributions obtained well reflect the influence of filters on the composition and energy spectrum of the ^{252}Cf radiation field. The paraffin filter increases the flux of thermal neutrons and this causes an increase of the low LET component. The iron filter shifts the energy of fast neutrons towards lower values and seriously decreases the gamma dose, so 80% of the energy is deposited in L_Δ range between 50 and 100 keV.μm^{-1}.

H*(10) WITH ICRP 60 QUALITY FACTOR

Taking into account the needs of radiation protection dosimetry we have proposed[8] a simplified method for the routine determination of H*(10) with the ICRP 60 quality factor. The ion collection efficiency is approximated now by a sum of two components related to low LET and high LET fractions of absorbed dose:

$$f_{mix} = D_{low} f_\gamma + D_{high} \frac{1}{1+[(1-f_\gamma)/f_\gamma] (L_{high}/L_0)} \quad (4)$$

where D_{low} and $D_{high} = 1 - D_{low}$ are the low LET and high LET dose fractions to the total absorbed dose and L_{high} is the effective value of L for the high LET component. As before, the f_{mix} and f_γ have to be measured at the same collecting voltages. The collected data are fitted by Equation 4 with parameters D_{low} and L_{high}.

Ambient dose equivalent is then calculated as:

$$H^*(10) = D[D_{low} + (1 - D_{low}) Q(L_{high})] \quad (5)$$

The experimental results, obtained in radiation fields of monoenergetic fast neutrons in the energy range from 75 keV to 19 MeV are presented in Figure 5, relative to the reference values. The energy dependence of the chamber response does not significantly differ from those when the ICRP 21 quality factors were used[2]; however, somewhat larger overestimation is observed for neutron energies around 1 MeV.

Preliminary investigations described elsewhere[1] indicate that the relative response of the chamber to H*(10) is close to 1 for thermal neutrons and overestimates the H*(10) within a factor of 1.5 for high energy neutrons, when compared with calibration with a ^{241}Am-Be source. Taking into account a possible experimental uncertainty up to 15%, one can expect that the accuracy of the measured H*(10) value will be between −25% and +70%, relative to a reference (conventionally true) value, for any (unknown) neutron spectrum. However, much better accuracy can be achieved in the majority of practical cases, when the neutron spectrum is broad or some information on neutron energy is available.

CONCLUSIONS

Experimental investigations performed in recent years clearly showed that a properly designed recombination

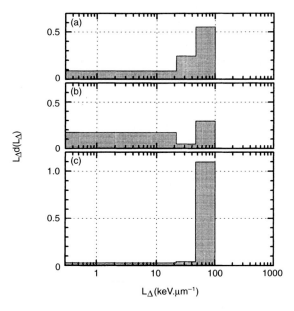

Figure 4. Dose distribution plotted against restricted LET for a ^{252}Cf radiation field: (a) the source free-in-air; (b) in a paraffin filter; (c) in an iron filter.

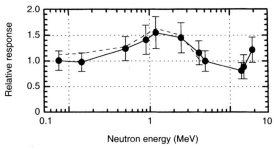

Figure 5. The ratio of H*(10) determined by the REM-2 chamber to the reference values calculated from the neutron fluence. Two different sets of fluence to H*(10) conversion factors[9,12] were used: solid line, based on old values of stopping powers; dashed line, based on the new data given in ICRU Report 49[14].

chamber well fulfils the requirements of the detector for an area monitor with a dose equivalent response weakly dependent on neutron energy. Since recombination chambers are the detectors of the total absorbed dose in mixed radiation fields, they can be considered as alternatives to TEPC instruments. It can be expected that in a majority of realistic radiation fields the recombination chamber should give a comparable response to the best TEPC systems.

ACKNOWLEDGEMENTS

The work was partially supported by the State Committee for Scientific Research under KBN grant No 4 S404 010 05.

REFERENCES

1. Golnik, N. *Recombination Methods in Dosimetry of Mixed Radiation.* IAE-20/A (Swierk, Poland) (1996).
2. Zielczynski, M., Golnik, N. and Rusinowski, Z. *A Computer Controlled Ambient Dose Equivalent Meter Based on a Recombination Chamber.* Nucl. Instrum. Methods A **370**, 563–567 (1996).
3. Zielczynski, M. and Golnik, N. *Recombination Index of Radiation Quality — Measuring and Applications.* Radiat. Prot. Dosim. **52**(1–4), 419–422 (1994).
4. Zielczynski, M., Golnik, N., Makarewicz, M. and Sullivan, A. H. *Definition of Radiation Quality by Initial Recombination of Ions.* In: 7th Symp. on Microdosimetry, EUR 7147, Eds J. Booz, H. G. Ebert and H. D. Hartfield (Luxembourg: CEC) pp. 853–862 (1981).
5. ICRP, International Commission on Radiological Protection. *Data for Protection against Ionising Radiation from External Sources.* Supplement to ICRP 15, Publication 21 (Oxford: Pergamon) (1973).
6. ICRP. *Recommendations of the International Commission on Radiological Protection.* Publication 60. Ann. ICRP **21**(1–3) (New York: Pergamon Press) (1991).
7. Golnik, N., Wilczynska, T. and Zielczynski, M. *Determination of the Recombination Index of Quality Employing High Pressure Ionisation Chambers.* Radiat. Prot. Doim. **23**(1–4), 273–276 (1988).
8. Golnik, N., Brede, H. J. and Guldbakke, S. *Response of REM-2 Recombination Chamber to H*(10) of Monoenergetic Neutrons* (in preparation).
9. Wagner, S. R., Grosswendt, B., Harvey, J. R., Mill, A. J., Selbach, H. J. and Siebert, B. R. L. *Unified Conversion Functions for the New ICRU Operational Radiation Protection Quantities.* Radiat. Prot. Dosim. **12**(2), 231–235 (1985).
10. Golnik, N. and Zielczynski, M. *Determination of Restricted LET Distribution for Mixed (n, γ) Radiation Fields by High Pressure Ionisation Chamber.* Radiat. Prot. Dosim. **52**(1–4), 35–38 (1994).
11. Golnik, N. *Microdosimetry Using A Recombination Chamber: Method and Applications.* Radiat. Prot. Dosim. **61**(1–3), 125–128 (1995).
12. Schuhmacher, H. and Siebert, B.R.L. *Quality Factors and Ambient Dose Equivalent for Neutrons Based on the New ICRP Recommendations.* Radiat. Prot. Dosim. **40**(2), 85–89 (1992).
13. Siebert, B. R. L. and Schuhmacher, H. *Quality Factors, Ambient and Personal Dose Equivalent for Neutrons, Based on the New ICRU Stopping Power Data for Protons and Alpha Particles.* Radiat. Prot. Dosim. **58**(3), 177–183 (1995).
14. ICRU. *Stopping Powers and Ranges for Protons and Alpha Particles.* Report 49 (Bethesda, MD: ICRU Publications) (1993).

RECOMBINATION INDEX OF RADIATION QUALITY OF MEDICAL HIGH ENERGY NEUTRON BEAMS

N. Golnik†, E. P. Cherevatenko‡, A. Y. Serov§, S. V. Shvidkij‡, B. S. Sychev§ and M. Zielczyński†
†Institute of Atomic Energy, 05-400 Otwock-Swierk, Poland
‡Joint Institute for Nuclear Research, Dubna, Russia
§Moscow Radiotechnical Institute, Moscow, Russia

Abstract — Recombination index of radiation quality (RIQ) was determined in high energy neutron beams from the 660 MeV phasotron of the Joint Institute for Nuclear Research, Dubna (JINR). Measurements were performed with a high pressure ion chamber at two collecting voltages and compared with calculations considering spectra of all secondary and tertiary charged particles. The neutron beams produced on beryllium, copper and lead targets were investigated. Only minor differences between RIQ values for these beams were observed. A high decrease in RIQ, from 6 to 3.5, has been observed in the first 5 cm depth in a water phantom, followed by a rather small progressive decrease towards 3.3 ± 0.3 for depths larger than 10 cm. Laterally small changes have been observed at a given depth between values on the beam axis and outside the border of the beam.

INTRODUCTION

The absorbed dose is a sufficiently good quantity to predict biological effects in radiation therapy with low LET radiations. However, with the development of neutron and high LET charged particle therapy, the need for determination of radiation quality became evident.

In this work the recombination index of radiation quality[1] (RIQ), measured by a recombination chamber is considered as a LET dependent parameter related to radiation quality. Originally, the RIQ (denoted as Q_R) has been introduced for purposes of radiation protection, hence its values are close to the radiation quality factor defined in ICRP Report 21[2]. It is considered that RIQ can also be used as an easily measurable indicator of radiation quality for purposes of radiation therapy, e.g. when different beams and irradiation conditions are preliminarily compared. Determination of the RIQ values can provide some qualitative information on RBE variations; however, the quantitative dependence between RIQ (or other physical indexes) and clinical RBE is not at present well established and the knowledge of RIQ does not change the need for additional information on RBE for final treatment planning. Additionally, the RIQ can be used for estimation of some correction factors needed for precise determination of absorbed dose[3,4].

Measurements in high energy neutron beams, presented in this work, were performed at the medical facility of the Joint Institute for Nuclear Research, Dubna (JINR). The experimental values of RIQ were compared with calculated ones, for different depths in a water phantom and different distances from the beam axis.

MEASUREMENTS

The neutron beams of the medical high energy particles facility at the JINR phasotron[5] are intended for radiation therapy of large, hypoxic tumours — both independently and in combination with protons. The beams are generated by steering the beam of 660 MeV protons onto a target. Targets made of beryllium, copper or lead were used for this work. The calculated neutron energy spectra for different targets (in the forward direction) are shown in Figure 1[5]. The influence of the collimator (10 cm in diameter) was estimated to be negligible at the distance of 9 m, where a water phantom (50 × 60 × 70 cm³) was placed. The beam intensity was monitored by a transmission chamber.

Measurements of Q_R were performed using a 3.8 cm³ in-phantom recombination chamber of F-1 type (see Figure 2) filled with TE gas up to 500 kPa. The disc-shaped chamber has three parallel-plate tissue-equivalent electrodes, 34 mm in diameter (total diameter of the chamber is 62 mm). The distance between electrodes is 1.75 mm. The sensitivity of the chamber is about 350 nC.Gy^{-1}. Its operational dose rate range is from

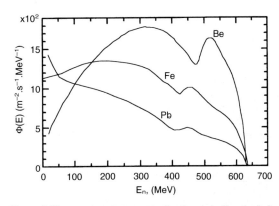

Figure 1. Neutron energy spectra (in forward direction) for three different targets used at the 660 MeV medical facility at JINR Dubna.

10^{-3} up to 1.5 Gy.s^{-1}. Two collecting voltages, $U_R = 30$ V and $U_S = 800$ V, were sequentially applied to the chamber electrodes.

RIQ was determined as[1]:

$$Q_R = \frac{1}{R}\left(1 - \frac{i(30\ V)}{i(800\ V)}\right) \quad (1)$$

where i is the ionisation current of the recombination chamber related to the monitor reading, $R = 1 - f_\gamma(U_R)$ and $f_\gamma(U_R)$ is the ion collection efficiency at collecting voltage U_R, determined in the gamma radiation field of the reference ^{137}Cs source. In experiments described here $R \approx 0.04$.

The proton beam current on the target was about 1 μA. The RIQ should be measured under conditions where initial recombination of ions largely predominates over the volume recombination. Hence, some measurements were repeated at 10 times lower intensity in order to check if the volume recombination could be neglected. For the same reason the measurements were performed at two pulse durations of 70 μs and 3500 μs. In the latter case the radiation pulse covered almost all the pulse cycle (4 ms). About two per cent of difference in ion collection efficiency was observed when measurements at maximum and minimum intensities were compared. Appropriate corrections to account for volume recombination[6] were introduced to the values of RIQ measured at higher intensity.

CALCULATIONS OF RIQ

The mean values of Q_R were calculated from the dependence of RIQ on LET[4,6] as:

$$\overline{Q}_R = \frac{1}{D}\int Q_R(L)d(L)dL \quad (2)$$

where

$$Q_R(L) = \frac{L/L_0}{1 + R(L/L_0 - 1)}$$

for $L \geq 3.5$ keV.μm^{-1} (3)

$Q_R(L) = 0.85 + 0.15\ L/L_0$

for $L < 3.5$ keV.μm^{-1} (4)

L is the unrestricted linear energy transfer (LET); $L_0 = 3.5$ keV.μm^{-1}, D is the absorbed dose and d(L) is the differential dose distribution versus LET.

The calculations of the spatial distribution of the Q_R value in the water phantom were performed by a consequent collisions method[7] taking into account the neutron energy spectra of the beams and spectra of all secondary and tertiary charged particles created in the phantom. The values of Q_R for secondary protons, charged π mesons and for heavy charged particles (d, ^{3}H, ^{3}He, ^{4}He) were calculated according to Equations 3 and 4, using the L values in water. For residual nuclei and for oxygen recoil nuclei the values of Q_R were made equal to 20. It was estimated that this simplification introduced only minor uncertainty to calculated Q_R values as the dose fraction from these particles is small and the values of Q_R for very high LET particles are close to 20 (with maximum value of 25 at infinite LET)[1,6].

Calculations were performed using two modes. First, the absorbed dose was made equal to a kerma value that was considered to give a lower limit of Q_R. Secondly, the contribution to ionisation according to ranges of charged particles was taken into account what was considered as the upper limit of Q_R. The difference between these two limits was interpreted as the uncertainty of the calculations and the mean value was judged to be the most probable value of Q_R.

RESULTS

The measurements of Q_R were performed on three separate occasions — in 1967[8], in 1990 after reconstruction of the 660 MeV phasotron[9] and in 1996.

The results obtained for three different targets are shown in Figure 3 in dependence on depth in the water phantom. There are two significant features of this figure. The shape of the $Q_R(z)$ curves is quite similar for all the targets. Identical values were observed in a steep gradient region for depths less than 3 cm (initial slope apparently about 1 cm^{-1}), followed by a slow progressive decrease with depth; between 10 and 30 cm depth in water (around 10^{-2} cm^{-1}). The curves representing $Q_R(z)$ for copper and beryllium targets differ less than 5%. The real difference (up to about 15%) is observed between Pb and Be targets at depths above 3 cm.

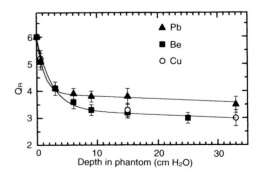

Figure 2. Cross section of the in-phantom chamber of F1 type (for historical reasons sometimes called the KR-13 chamber).

Figure 3. Depth distribution of recombination index of radiation quality in a water phantom for neutron beams generated by 660 MeV protons incident on three different targets.

The dependence of both Q_R and absorbed dose rate values on depth in the phantom for p(660 MeV) + Be neutrons only are presented in Figure 4. The experimental values of Q_R are compared with those resulting from calculation. It can be seen that in the region of the broad maximum of the dose rate, i.e. between 10 and 25 cm, the values of Q_R are practically independent of the depth in the phantom and $Q_R = 3.3 \pm 0.3$. The experimental results agree with the calculated ones within experimental and calculation uncertainties.

Radial distributions of Q_R and of absorbed dose rate at the depth of 15 cm in phantom are shown in Figure 5. Points in the figure represent the experimental values of Q_R determined by Zielczyński and co-workers[9]. The solid line with the shadowed area show the results of calculations of RIQ and the range of accuracy of the calculations[9]. Small variations of RIQ at the beam edge and outside the border of the beam probably reflect uncertainties of numerical integration and most likely have no physical meaning. The dose rate profile has been taken from the experiments of Abazov et al[5].

DISCUSSION

Some years ago it was recognised that microdosimetry could play an important role in providing a method for specifying the radiation quality for high LET radiotherapy. Series of microdosimetric measurements were performed for different beams of fast neutrons (with energy up to 65 MeV) using a tissue-equivalent proportional counter (TEPC). The results were compared in overview papers by Wambersie et al[10,11] and of Pihet et al[12]. It was shown[12] that small variations of the LET component between 50 and 150 keV.μm^{-1} considerably influence the radiation quality of the beams. Since the value of Q_R changes in this LET region from 9.3 at 50 keV.μm^{-1} to 16 at 150 keV.μm^{-1}, so the recombination chamber is a quite sensitive detector of such variations of LET. The differences in Q_R values measured at different conditions can also indicate an influence of scattered radiation or contamination of the beam.

Generally, even small variations in RIQ values, exceeding the experimental uncertainty, reflect some real differences in secondary charged particle spectra and should not be neglected. This also concerns the results presented in this work. The observed increase of RIQ for heavier targets is due to the lower energy of neutrons incident on the phantom, hence higher LET of secondary charged particles and lower contribution of low LET components generated by uncharged pion decay. Generally, beams with higher LET are preferred for radiation therapy of hypoxic tumours. In the case of the high energy neutron beams considered above, the differences in radiation quality are small in spite of considerably different energy spectra. Therefore, the main criterion for the choice of the target became the beam intensity and the beryllium target is the best from this point of view.

Higher values of Q_R close to the front wall of the phantom merely reflects the absence of proton equilibrium at small depths and relatively higher heavy particle contributions per unit dose, compared with larger depths.

CONCLUSIONS

The recombination index of radiation quality in high energy neutron beams decreases from 6 to 3.5 in the first 5 cm depth in a water phantom. The shape of the RIQ dependence on depth does not depend on target material and at depths above 10 cm in water RIQ equals to 3.3 ± 0.3 for the targets used for generation of the neutron beam in this study (beryllium, copper and lead).

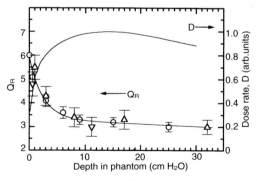

Figure 4. Depth distribution of RIQ and absorbed dose rate in a water phantom for neutrons from the reaction p(660 MeV) + Be. (△) data from 1967[8], (○) 1990[9], (▽) this work. Solid line, calculations.

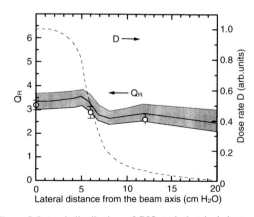

Figure 5. Lateral distribution of RIQ and absorbed dose rate at 15 cm in a water phantom for neutrons from the reaction p(660 MeV) + Be. Points, values of RIQ determined by Zielczyński et al[9]; solid line and the shadowed area, Monte Carlo calculations of RIQ with the range of accuracy[9].

REFERENCES

1. Zielczynski, M., Golnik, N., Makarewicz, M. and Sullivan, A. H. *Definition of Radiation Quality by Initial Recombination of Ions.* In: Proc. 7th Symp. on Microdosimetry, Oxford, September 1980 EUR 7147 (Luxembourg; CEC) Vol. 2, pp. 853–862 (1980).
2. International Commission on Radiological Protection. *Data for Protection against Ionizing Radiation from External Sources.* Supplement to ICRP 15, Publication 21 (Oxford: Pergamon) (1973).
3. Zielczynski, M. and Golnik, N. *Recombination Index of Radiation Quality — Measuring and Applications.* Radiat. Prot. Dosim. **52**(1/4), 419–422 (1994).
4. Zielczynski, M. and Golnik, N. *Energy Expended to Create an Ion Pair as a Factor Dependent on Radiation Quality.* In: Measurement Assurance in Dosimetry, STI/PUB/930 (Vienna: IAEA) pp. 383–391 (1994).
5. Abazov, V. M. et al. *Forming and Study of a High Energy Neutron Beam from JINR Phasotron.* JINR 18-88-392 (Dubna: Joint Institute for Nuclear Research) (in Russian) (1988).
6. Golnik, N. *Recombination Methods in the Dosimetry of Mixed Radiation.* IAE 20/A (Swierk: Institute of Atomic Energy) (1996).
7. Serov, A. Y. et al. *Spatial Distribution of the Dose Fields in Water Absorber Irradiated by High Energy Nucleons.* At. Energ. **56**(1), 36–40 (in Russian) (1984).
8. Zielczynski, M., Komochkov, M. M., Sychev, B. S. and Cherevatenko, A. P. *Measurement of the Quality Factor for High Energy Neutrons in the Tissue Equivalent Phantom.* JINR 16-3587 (Dubna: Joint Institute for Nuclear Research) (in Russian) (1967).
9. Zielczynski, M., Pliszczynski, T., Serov, A., Sychev, B. and Cherevatenko, E. *Quality Factor of the Therapeutic Neutron Beam of JINR Laboratory of Nuclear Problems.* JINR 16-90-265 (Dubna: Joint Institute for Nuclear Research) (in Russian) (1990).
10. Wambersie, A., Pihet, P. and Menzel, H. G. *The Role of Microdosimetry in Radiotherapy.* Radiat. Prot. Dosim. **31**(1/4), 421–432 (1990).
11. Wambersie, A. *Contribution of Microdosimetry to the Specification of Neutron Beam Quality for the Choice of the 'Clinical RBE' in Fast Neutron Therapy.* Radiat. Prot. Dosim. **52**(1/4), 453–460 (1994).
12. Pihet, P., Menzel, H. G., Schmidt, R., Beauduin, M. and Wambersie, A. *Biological Weighting Function for RBE Specification of Neutron Therapy Beams. Intercomparison of 9 European Centres.* Radiat. Prot. Dosim. **31**(1/4), 437–442 (1990).

CONCEPT OF A NEUTRON DOSEMETER BASED ON A RECOIL PARTICLE TRACK CHAMBER

U. Titt[†][§], A. Breskin[‡], R. Chechik[‡], V. Dangendorf[†], B. Großwendt[†] and H. Schuhmacher[†]
[†]Physikalisch-Technische Bundesanstalt, 38116 Braunschweig, Germany
[‡]Weizmann Institute of Science, 76100 Rehovot, Israel
[§]Universität Frankfurt, 60486 Frankfurt, Germany

Abstract — A new experimental approach for a neutron dosemeter and spectrometer is presented. It is based on an optically read out time projection chamber (TPC). Charged particles entering a low-pressure sensitive volume of about one litre create ionisation electrons along their track, which are detected and localised using an optoelectronic detection system. This allows type, energy, and direction of neutron-induced secondaries to be analysed. It will therefore be suited as an advanced active neutron monitor for the measurement of dosimetric quantities and will also supply spectrometric information. The method as well as the planned structure and its expected performance are presented. Results are reported that were obtained with a first experimental device which was built to study particle induced light and charge production in electron avalanches in various triethylamine-based gas mixtures.

INTRODUCTION

Despite past progress in radiation protection dosimetry, there is still demand for more reliable methods for assessing the exposure due to external irradiation in mixed neutron–photon fields.

One of the basic problems in the dosimetry of mixed radiation fields is the lack of adequate instruments, covering the whole range of radiation types and energies. Instead, depending on the radiation field, different methods have to be employed and combined. This makes the characterisation of radiation fields and accurate dosimetry complicated, time-consuming, and expensive. Here, a method is proposed which aims to overcome some of the present limitations on area monitoring of mixed radiation fields by directly measuring dosimetrically relevant quantities regardless of the type and origin of the primary radiation.

The proposed method is based on recording ionisation tracks of particles in gaseous detectors. It combines the optical readout method of proportional chambers[1,2] and the technique of cluster counting in drift chambers[3–5]. The result is a detection system allowing three-dimensional localisation of each ionisation electron in a low pressure gas.

This system has some similarities with low-pressure tissue-equivalent proportional counters (TEPC) which have shown important potential in dosimetry research, radiation protection dosimetry and characterisation of mixed radiation fields. The new particle track chamber discussed in this work would overcome limitations of the TEPC by measuring differential energy deposition in the gas cavity.

Work in a similar direction was reported by a group from Oak Ridge[6–8]. An ionisation chamber with an optical readout was developed there, in which, after the crossing of an ionising particle, a high RF voltage causes secondary light emission from the spots where free charges were deposited. The light is photographed by two cameras from orthogonal sides. The three-dimensional track structure is reconstructed from these two images. The inherent problem of this method is the large optical depth of field, required for viewing the whole volume of the chamber. This imposes compromises between the need for a sufficiently large solid angle for efficient light detection, and the focussing requirements for an optimum optical image. Satisfying these contradicting requirements seriously limits the obtainable resolution.

Another experimental approach was reported by Marshall *et al*[9]. They developed a cloud chamber technique, well known from particle physics, to visualise particle tracks in water vapour. They obtained an excellent spatial resolution of less than 5 nm (expressed in units for a density of 1 g.cm^{-3}). The drawback of this technique is the extremely low duty cycle due to the long recovery time of the chamber, so that it is not suitable for practical neutron spectrometry or dosimetry.

EXPERIMENTAL SET-UP

A schematic view of the experimental set-up is shown in Figure 1. The radiation sensor is a time projection chamber (TPC) 10 cm long and 10 cm in diameter, housed in a stainless steel vessel. The chamber is filled with triethylamine (TEA)-based gas mixtures at pressures p of 5 to 20 hPa. Charged particles which enter the gas volume produce ionisation electrons along their track. These electrons drift along an homogeneous electrical field with a reduced field strength E/p of about 10 V.cm^{-1}.hPa^{-1}, towards a charge and light amplification stage.

The amplification stage consists of a parallel gap, 3.2 mm wide, where a high electric field of 120 to 150 V.cm.hPa causes charge multiplication by a factor of up to 10^7. The excitation of TEA molecules as a result of inelastic collisions with the avalanche electrons leads to the emission of UV photons of about 280 nm in wavelength.

The electron drift and the subsequent amplification process provides a two-dimensional image of the electron distribution along the particle track, projected onto the plane of the anode of the amplification element. This image is viewed through a quartz window by an optical imaging system consisting of a UV lens and a gated image intensifier, fibre-optically coupled to a CCD camera. This readout method, electronically decoupled from the detector, provides the two-dimensional image of the recorded track with considerably reduced noise as compared with standard readout techniques.

The third coordinate, namely the charge distribution parallel to the drift field lines (z direction), is obtained by recording the time dependence of the light emission during the full drift time period with a photomultiplier coupled to a digitiser. The light dl produced within a time interval dt is related to the charge dq produced within the z interval dz by

$$\frac{dq}{dz} = \frac{dl}{dt} \frac{1}{v_z\, g}$$

where v_z is the drift velocity and g the light yield per initial electron.

A dedicated PC-based data acquisition and analysis system is in development to reconstruct the three-

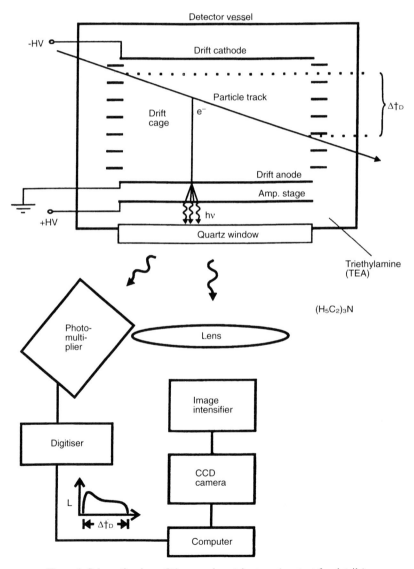

Figure 1. Schematic view of the experimental set-up (see text for details).

dimensional particle track from the CCD image and the dI/dt-recording.

SYSTEM PERFORMANCE

Estimation of the spatial resolution

Aiming for a spatial resolution of 10 nm in tissue (density of 1 g.cm^{-3}) about 1 mm resolution is required at a gas pressure of 10 hPa. The main parameter, which determines the localisation resolution is the electron diffusion during the drift process and within the avalanche. The influence of the lateral electron diffusion was estimated, following the elementary theory of electron drift and diffusion[10]. At a reduced electric field of $E/p = 10$ V.cm^{-1}.hPa^{-1}, the standard deviation of the electron diffusion is of the order of 0.5 mm for a drift path length of 10 cm.

The resolution of the optical system is limited by the UV lens (NYE Lyman α), which is of the order of one line pair per mm at its maximum aperture. The contribution of the image intensifier (Proxitonic BV2562QG) and of the CCD (EG&G RA0512J) to the effective resolution are negligible.

In the z direction the spatial resolution is determined by longitudinal diffusion (about 0.5 mm) and by the length of the drift path in the amplification element where the major part of the light is produced (a few tenths of a mm at a gain of 10^6).

Therefore, it is expected that the required resolution of 1 mm in all three dimensions will be obtained.

Sensitivity of the system

The ultimate goal is the detection of single electrons created along the track in the drift volume. For this purpose the light yield of the chamber and the sensitivity of the optical system have to be optimised in such a way that:

(i) a sufficient number of photons reach the photocathode of the image intensifier to ensure close to 100% detection efficiency by the optical amplification system, and
(ii) the optical amplification system illuminates a certain number of CCD pixels to a safe level above noise.

Based on the specifications of the optical system components as listed in Table 1 and at maximum amplification in the chamber, an average number of 750 photoelectrons emitted from the image intensifier photocathode for each electron entering the detector multiplication stage is estimated. Considering the gain and the sensitivity of the optical amplification elements, it is calculated that a 1 mm^2 light spot irradiates about 32 pixels and produces about 3×10^4 charge carriers in each one. These yields exceed by far the requirements of single electron detection efficiency and produce signals far above CCD noise and should allow the operation of the chamber in more moderate and comfortable amplification conditions.

The dynamic range of a single CCD pixel in the imaging system is about 25. For example, a carbon ion in its Bragg peak produces about 30,000 electrons along its path through the sensitive volume. The core of the particle track covers an area of about 80×5 mm^2 in the

Table 1. Typical system parameters of the TPC and its readout components, relevant for the detection sensitivity of single electrons.

Detector	Light output (max. exp. value)	2×10^6 photons per initial electron
UV lens	Effective solid angle	$2.5 \times 10^{-3}/4\pi$
Photocathode	Quantum efficiency	15.5%
Multichannel plate	Sensitive area	60%
	Gain	350 (at 800 V)
Phosphor screen	Conversion ratio	185 photons/electron
Fibre optics	Transmission loss	65%
CCD	Quantum efficiency	20%
	Noise level	500 electrons
	Full well capacity	5×10^5 electrons
	Pixel size	27 μm × 27 μm

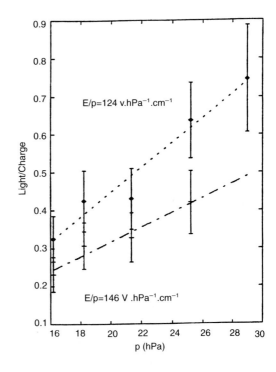

Figure 2. Pressure and reduced field dependence of the light to charge ratio (L/C, photons per avalanche electron) in a parallel amplification element with pure TEA.

gas volume. This corresponds to 16,000 pixels of the CCD, hence on the average about 2 electrons per pixel. Although there is a strong fluctuation in the ionisation density, one can assume that only for a few pixels is the maximum dynamic range exceeded. From these estimates a dynamic range of more than 10^4 is derived.

Based on chamber volume, gas pressure and an average energy of 50 keV deposited in the sensitive volume by a 1 MeV neutron, one derives an absorbed dose of about 1 nGy per event. The CCD camera allows the recording of up to 25 tracks per second, corresponding to a dose rate of 90 μGy.h^{-1}. The rate capability of the chamber itself is much higher (10^4 events per second). If one measures the total energy deposited for this rate and restricts a full analysis to the fraction which can be handled by the optical system, one can achieve a maximum dose rate of 36 mGy.h^{-1}.

Measurement of light gain and photon emission

The absolute light output of the detector was measured at different gas pressures and electric field conditions, using a calibrated photomultiplier. The charge output could be determined by measuring the pulses at the amplification mesh with a calibrated charge-sensitive ADC. Single electrons were produced by irradiating the cathode of the drift cell with a pulsed UV lamp. These were multiplied in the amplification stage by a factor of up to 10^7 (spark limit).

Figure 2 presents the results of these measurements as a function of pressure and for two different reduced field strengths E/p in pure TEA. The light-over-charge ratio slightly decreases with higher E/p values, which can be explained by the shift towards a higher ionisation probability to the expense of molecular excitation[2]. However, this effect is compensated by the increase in total charge.

First results of track recordings

First images of alpha particle tracks from a ^{241}Am source were recorded with our experimental device at a pressure of about 15 hPa TEA. As an example, Figure 3 shows multiple alpha tracks which crossed the chamber within a time interval of about 60 ms. Implementing the gating capability of the image intensifier and the CCD camera will lead to significant improvements in the signal-to-noise performance.

For future analysis of these images, theoretical modelling of the ionisation density distribution along the track was performed. Due to the lack of data for other projectiles and for TEA, only proton tracks in methane could be studied up to now.

The calculation was divided into three steps:

(1) transport of the incident charged particle,
(2) production of delta electrons, and
(3) transport of delta electrons and further ionisation.

For (1) it was assumed

(i) that the energy degradation of a proton can be described by the continuous slowing down approximation,

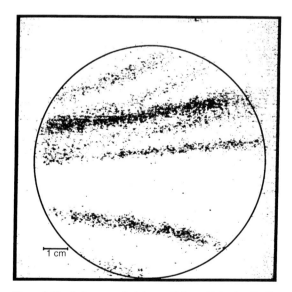

Figure 3. Track images of alpha particles, obtained with the TPC filled with pure TEA at 15 hPa. The circle shows the sensitive area of the TPC.

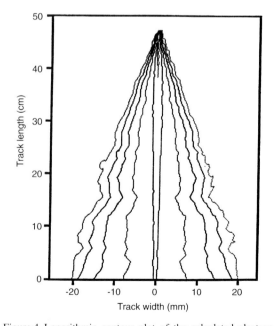

Figure 4. Logarithmic contour plot of the calculated electron density along the track of a 1 MeV proton in 10 hPa CH$_4$. Each line is one order of magnitude in arbitrary units.

(ii) that the angular scattering of the protons can be neglected,
(iii) that the energy loss due to excitation processes is about 50% and
(iv) that the production of ion pairs is exclusively induced by direct ionising interactions, neglecting charge exchange processes[11].

The total proton track was subdivided into track segments 1 mm in length, and the energy loss along each segment was calculated using the total linear stopping power of Ziegler et al[12]. The energy loss fraction due to inelastic proton interactions was determined from the corresponding linear electronic stopping power.

Assuming uniform distribution of inelastic interaction points along each track segment, the spectral distribution of secondary electrons was calculated with the single-differential cross sections of Rudd et al[13] for protons slowed down in methane.

The energy transport of the secondary electrons was handled using a Monte Carlo model[14], assuming that the initial electron flight direction is perpendicular to that of the protons. The simulation of the electron histories took into account elastic electron scattering, impact excitation, and impact ionisation using the same cross sections as summarised by Waibel and Grosswendt[15]. The procedure for simulating the energy degradation of each single electron was stopped as soon as an energy less than or equal to the ionisation threshold of 13.0 eV was reached.

Figure 4 shows the result of these calculations. The lateral width of the track is determined by the energy distribution of the delta electrons which correlates with the projectile velocity. This is why significant differences in the lateral width are expected for different kinds of incident particles of the same stopping power dE/dx. By measuring this width and the dE/dx the incident particle will be identified.

SUMMARY

The principles and first results of measurements with an optically read out time projection chamber for particle tracking are presented. With the first experimental device, light emission was studied in a simple parallel-grid amplification structure with low pressure TEA. It could be shown that with pure TEA at 15 hPa enough light is produced to visualise single electrons, created in α particle tracks. Systematic investigations were performed to quantify the magnitude of light emission per charge in the avalanche after gas amplification.

An improved system, which is expected to be capable of analysing the type, the energy, and the direction of any incident charged particle, in particular also of neutron-induced secondaries, is close to operation. It is ultimately intended to use such an instrument as a reference system for dosimetry and field analysis in complex radiation fields, and for studying interaction mechanisms. For this purpose the chamber walls will be made of tissue-equivalent material, and small quantities of other gases will be added to TEA, in order to approximate its atomic composition to that of tissue.

Due to the large detector volume, the sensitivity is sufficient to measure small dose rates. The kinematics of n-p scattering allow the characterisation of the angular distribution of a neutron field by measuring the directional distribution of recoil protons and application of unfolding procedures. This chamber may also provide a powerful new tool for microdosimetric investigations since the large total volume of the chamber and the expected resolution of below 10 nm allow the large- and small-scale ionisation topology in cellular entities to be studied in detail.

REFERENCES

1. Charpak, G., Fabre, J.-P., Sauli, F., Suzuki, M. and Dominik, W. *An Optical, Proportional, Continuously Operating Avalanche Chamber*. Nucl. Instrum. Methods A**258**, 177–184 (1987).
2. Sauvage, D., Breskin, A. and Chechik, R. *A Systematic Study of the Emission of Light from Electron Avalanches in Low-pressure TEA and TMAE Mixtures*. Nucl. Instrum. Methods A**275**, 351–363 (1989).
3. Breskin, A., Chechik, R., Malamud, G. and Sauvage, D. *Primary Ionization Cluster Counting with Low-pressure Multistep Detectors*. IEEE Trans. Nucl. Sci. **NS-36A**, 316–321 (1989).
4. Malamud, G., Breskin, A. and Chechik, R. *A Systematic Study of Primary Ionization Cluster Counting at Low Gas Pressures*. Nucl. Instrum. Methods A**307**, 83–96 (1991).
5. Pansky, A., Malamud, G., Breskin, A. and Chechik, R. *Detection of X-ray Fluorescence of Light Elements by Electron Counting in a Low-pressure Gaseous Electron Multiplier*. Nucl. Instrum. Methods A**330**, 150 (1993).
6. Hunter, S. R. *Evaluation of a Digital Optical Ionization Ratiation Particle Track Detector*. Nucl. Instrum. Methods A**260**, 469–477 (1987).
7. Turner, J. E., Hunter, S. R., Hamm, R. N., Wright, H. A., Hurst, G. S. and Gibson, W. A. *Development of an Optical Digital Ionization Chamber*. Radiat. Prot. Dosim. **29**, 9–14 (1989).
8. Hunter, S. R., Gibson, W. A., Hurst, G. S., Turner, J. E., Hamm, R. N. and Wright, H. A. *Optical Imaging of Charged Particle Tracks in a Gas*. Radiat. Prot. Dosim. **52**, 323–328 (1994).
9. Marshall, M., Stonell, G. P. and Holt, P. D. *Ionisation Distributions from Proton and other Tracks in Water Vapour*. Radiat. Prot. Dosim. **13**, 41–44 (1985).
10. Peisert, A. and Sauli, F. *Drift and Diffusion of Electrons in Gases, Compilation*. CERN 84-08 (Geneva: CERN) (1984).

11. Grosswendt, B. and Baek, W. Y. *The Energy Dependence of w Values for Protons of Energies up to 10 MeV.* Radiat. Prot. Dosim. **61**, 267–274 (1995).
12. Ziegler, J. F., Biersack, J. P. and Littmark, U. *The Stopping and Range of Ions in Solids.* (New York: Pergamon Press) (1985).
13. Rudd, M. E., Kim, Y.-K., Madison, D. H. and Gay, T. J. *Electron Production in Proton Collisions with Atoms and Molecules: Energy Distributions.* Rev. Mod. Phys. **64**, 441–490 (1992).
14. Grosswendt, B. and Waibel, E. *Transport of Low-energy Electrons in Nitrogen and Air.* Nucl. Instrum. Methods **155**, 145–156 (1978).
15. Waibel, E. and Grosswendt, B. *Spatial Energy Dissipation Profiles, w Values, Backscatter Coefficients, and Ranges for Low-energy Electrons in Methane.* Nucl. Instrum. Methods **211**, 487–498 (1983).

WORKPLACE RADIATION FIELD ANALYSIS

H. Klein
Physikalisch-Technische Bundesanstalt (PTB)
Bundesallee 100, D-38116 Braunschweig
Federal Republic of Germany

INVITED PAPER

Abstract — EURADOS working groups carried out comparison exercises with spectrometers and dosemeters in order to specify the mixed radiation fields at workplaces in nuclear facilities and in practical calibration fields. Carefully specified Bonner sphere spectrometers proved to be best suited to the determination of integral fluence and dose equivalent reference values with uncertainties smaller than 5% and 15%, respectively. Irradiations of personal dosemeters symmetrically mounted on phantoms allowed the directional spectral fluence to be evaluated and ambient and limiting dose equivalent values to be estimated. The readings of commonly used neutron monitors, moderator-type dosemeters and tissue-equivalent proportional counters also showed the rather large deviations from the reference values expected according to their response functions. A survey meter and superheated drop detectors recently developed seem to be better suited for neutron dosimetry in fields with a rather soft energy spectrum. Spectrometry and dosimetry in mixed fields are not as advanced for photons as for neutrons.

INTRODUCTION

This paper deals with the investigation of mixed neutron–photon fields with the major fraction of the dose (equivalent) due to neutrons. Such radiation fields are encountered at workplaces in nuclear power plants, at facilities manufacturing, handling and storing MOX fuel elements or radionuclide neutron sources (the latter being used for calibration purposes, or for inspection in industry), in the transport, intermediate storage and final disposal of spent fuel elements, at high energy accelerators, at fusion plasma experiments, and in aircraft at high altitudes.

The improvement of neutron spectrometry at workplaces was a major task of various cooperation projects supported by the Commission of the European Community within the framework of its Radiation Protection Programme. Different neutron spectrometers, designed and carefully specified for application in radiation protection practice, were employed in four comparison exercises organised, carried out and evaluated by working groups of the European Radiation Dosimetry Group (EURADOS WG7 and WG10). The intercomparisons, which also included survey instruments and personal dosemeters widely used or recently developed, were performed at workplaces in nuclear power plants[1] or in calibration fields replicating practical neutron fields[2-5]. The experience gained from these measurements and the major conclusions will be summarised. More detailed discussions are reported elsewhere[7].

While remarkable progress can be reported for the specification of the neutron component of the mixed field, the simultaneous analysis of the photon fraction still presents serious problems, chiefly due to the scarce knowledge of the neutron sensitivity of photon spectrometers and dosemeters. This topic will be sketched only briefly.

NEUTRON FIELDS AT WORKPLACES

Neutron fields at workplaces generally consist of a few basic components (Figure 1). At nuclear facilities the primary neutrons chiefly originate from spontaneous and n-induced fission or (α,n) reactions which produce rather hard spectra with mean energies of a few MeV and mean ambient dose equivalent conversion factors of 300–500 pSv.cm^2 (also depending on the dose quantity used). Shielding with heavier elements, e.g. Fe, lowers the mean energy of the remaining spectrum due to inelastic scattering, and light shielding materials cause downscattering to an $E^{-\alpha}$ distribution with $0.5 < \alpha < 1.5$ or complete thermalisation. In the case of plasma experiments discharges with a pure deuteron plasma produce 2.5 MeV neutrons and those with mixed deuteron–triton plasmas 14 MeV neutrons which carry the major fraction of the energy released in the fusion reac-

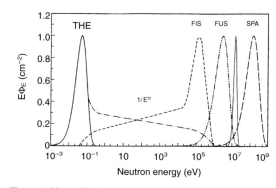

Figure 1. Normalised spectral neutron fluence ϕ_E of primary neutron sources encountered at nuclear power plants (fission FIS), plasma experiments (fusion FUS) and high energy accelerators (spallation SPA), neutron spectra behind shieldings ($1/E^\alpha$) and thermal neutron distributions (THE).

tions. Neutrons much higher in energy are produced by intranuclear cascade reactions of high energy protons in nuclei (e.g. at high energy accelerators and from cosmic radiation). Broad distributions of high energy spallation neutrons will be observed even behind thick shieldings. An extremely wide energy range from thermal to some hundred MeV neutrons must therefore be covered by the instruments used.

According to the various definitions of the dose equivalent (DE) quantities H, different methods and instruments may be used to specify the properties of the neutron component of the mixed field[8]. The method discussed in detail in this paper for determining the (ambient) DE is based on the definition:

$$H = \int h_\phi(E) \, \Phi_E(E) \, dE \quad (1)$$

The fluence-to-DE conversion coefficients h_ϕ are calculated according to the definition of the quantity and the corresponding quality factor, i.e. either maximum DE, H_{MADE}[9], or ambient DE, H*(10), according to ICRU 39[10] or ICRP 60[11].

The spectral neutron fluence Φ_E is determined by the unfolding of vectors with a few measured data or of multichannel pulse height spectra M_i, utilising well-specified fluence response functions $R_\phi(E_j)$ of the spectrometer according to:

$$M_i = \sum_{j=1}^{n} R_\phi^i(E_j) \, \phi_{E_j} \, \Delta E_j \quad \text{for } i = 1,n \quad (2)$$

Another approach results from the definition:

$$H = \int Q(L) \, D_L \, dL \quad (3)$$

and is based on the spectral dose deposition $D_L = dD(L)/dL$ in dependence on the linear energy transfer L (=LET) and the LET-dependent quality factor Q(L), which roughly accounts for the radiobiological effectiveness of the different secondary charged particles produced by the interaction of the neutrons in tissue. The frequency distribution D_y of the lineal energy y deposited in the gas of a tissue-equivalent proportional counter (TEPC) will result in a DE reading M_H if the spectrum is weighted with the y-dependent quality factor q(y) assumed to be proportional to Q(L):

$$M_H = \int q(y) \, D_y \, dy \quad (4)$$

The advantage and limitation of this method have been reviewed elsewhere[12]. Only the data obtained with TEPCs within the framework of the comparisons will therefore be discussed here.

Finally, direct readings of dose-equivalent values can be obtained with area and personal dosemeters according to:

$$M_H = \iint R(E,\Omega) \, \Phi_{E,\Omega} \, dE \, d\Omega \quad (5)$$

provided that the response function of the dosemeter $R(E,\Omega)$ is properly adjusted to the conversion function $h_{\Phi,\Omega}(E,\Omega)$ of the desired quantity, e.g. for the operational quantities ambient DE, h*(E)[10,11] or personal DE $h_p(E,\alpha)$[13].

Most of the dosemeters commonly used show significant deviations from an energy-independent DE reading, which are, however, smoothed out due to the broad energy distributions of practical neutron fields[14]. The remaining deficiencies and the progress achieved with recent developments will be discussed below.

NEUTRON SPECTROMETRY IN MIXED FIELDS

Neutron spectrometers designed and specified for application at workplaces with a significant fraction of the dose equivalent due to neutrons ($H_n > H_\gamma$) must fulfil the following requirements which comprise both general and practical aspects:

(1) Neutron spectrometers must cover a wide dynamic energy range from thermal to some tens (hundreds) of MeV (see Figure 1).
(2) The instruments must be either insensitive to photons or selective to neutrons, i.e. have the capability to distinguish neutron- and photon-induced events. The photon fluence rate can still be larger than the neutron fluence rate and the photon spectra often contain high energy photons (e.g. 6–7 MeV photons from capture reactions).
(3) The dose equivalent rates range from a few $nSv.h^{-1}$ (e.g. neutron background as a result of cosmic radiation) to some $mSv.h^{-1}$ in the vicinity of the neutron source.
(4) An isotropic response is required to determine the (ambient) DE value H_{MADE} or H*. The directional spectral fluence must be separately investigated if personal (H_p) and protection (H_E or E) quantities are required.
(5) The instruments should be (trans)portable and work reliably even under extreme environmental conditions as regards temperature (up to 40°C), humidity (ranging from 25% to 95%), acoustic noise level and electromagnetic interference[1].

Almost all techniques invented in the past 50 years and still suitable for radiation protection purposes were utilised in the comparison exercises carried out by the laboratories collaborating in the EURADOS Working Groups 7 and 10[1–7] (see Table 1).

Bonner Sphere spectrometry (BSS)

Spectrometry with a set of moderating spheres equipped with a thermal neutron detector at their centre was introduced by T.W. Bonner and co-workers[15].

Despite their photon sensitivity, ^{6}LiI scintillators used by the inventor are still in operation[16] because the photon- and neutron-induced events can easily be separated by spectrum analysis. These days, BF_3- and ^{3}He-filled proportional counters (PC) are preferred for their intrinsically better gamma discrimination, and the highest neutron sensitivity is achieved with large volume, spherical ^{3}He PCs[17]. Alternatively, passive thermal neutron detectors (Au activation foils[18], TLD chips[19] may be used if pulsed neutron fields, e.g. at accelerators or in plasma experiments, or fields with high dose rates (>1 mSv.h^{-1}) are to be investigated.

The size and the number of the polythene spheres are selected such that the response functions cover the entire energy range. Bramblett et al[15] had already demonstrated that a careful and detailed experimental calibration of the response is advantageous, but calculations are needed in addition for interpolation purposes. Discrete ordinate neutron transport calculations (ANISN code[20]) proved to be valuable, but the adjustment factors systematically depend on the sphere diameter, which is chiefly due to the simplification necessary in these one-dimensional calculations and the response to thermal neutrons was significantly overestimated, which can mainly be explained by inadequate handling of the thermal neutron transport. Three-dimensional neutron transport calculations (MCNP code) recently performed[16,17,21] succeeded in reproducing the response measured almost independently of the sphere diameter and for all energies from thermal to 15 MeV neutrons[22]. Only the effective neutron detection efficiency of the central detector has to be taken into account by a common adjustment factor.

The normalised count rates of this detector, employed either bare or covered with cadmium or PE spheres with diameters ranging from 5.08 cm (2″) to 45.72 cm (18″), show a smooth dependence on the size, but just the shape indicates the hardness of the neutron spectrum (Figure 2). Unfolding of the few (6–12) measured data into an energy spectrum covering up to 12 decades with at least 3–5 group fluences per decade yields an infinite set of solutions which can be reduced only by appropriate selection of guess spectra or, even better, by means of a priori information from calculations or measurements performed under similar conditions. A comparison exercise with benchmark problems simulated for a given response matrix exhibited excellent agreement between all results obtained with different unfolding algorithms and start spectra[23]. Larger differences in the spectral shape and in particular systematic deviations of the integral DE values from the reference by more than 30%, which were in part observed in these intercomparisons[2,6] pointed to problems with the response matrices used. Indeed, the discrepancies disappeared when calculated response functions were carefully adjusted to the experimental data. Measured and

Table 1. Neutron spectrometers employed in four comparison exercises by different laboratories.

Laboratory	Comparison exercise location			
	RNSF/1	ASPIS	PWR/CLAB	RNSF/2
	Cadarache	Winfrith	Ringhals/Oskarshamn	Cadarache
IPSN-CEA, Fontenay-aux-Roses	BS(F),SC(L) PC(H),MC	—	—	BS(F),SC(L) PC(H),MC
IPSN-CEA, Cadarache	—	—	—	BS(F)
GSF, Neuherberg	BS(L)	BS(L)	BS(L)	BS(L)
IAR, Lausanne	BS(F)*	BS(F)	BS(F)	BS(F)
NPL, Teddington	—	BS(C),SC(L)	BS(A)	BS(C),SC(L)
PTB, Branschweig	BS(C,F),SC(L)	—	BS(C)	BS(C),SC(L) PC(H)
TU, Dresden	—	PC(H),SC(S)	PC(H),SC(S)	PC(H),SC(S)
CMI, Prague	—	—	—	BS(F)
NRI, Rez/MA, Brno	—	—	—	PC(H),SC(S)
BfS, Berlin	BS(C)*	—	—	BS(C)
AEA, Harwell	—	PC(H,He)	—	—
AEA, Winfrith	—	PC(H),SC(L),MC	—	—
AECL, Chalk River	—	—	—	BS(C)
DCMN, Pisa	—	—	—	SDD

*Measured in the course of the second exercise at Cadarache without the water moderator.
(BS(x) = Bonner spheres with spherical (C) or cylindrical (F) ^{3}He proportional counter, LiI (L) scintillator or Au activation foil (A); SC(y) = NE213 liquid (L) or Stilben (S) organic scintillator; PC(z) = hydrogen (H) or ^{4}He (He) filled proportional counters; SDD = superheated drop detector; MC = Monte Carlo calculations).

recalculated readings are statistically compatible (see Figure 2(c)) when the uncertainties of the response functions are carefully evaluated.

For energies higher than 20 MeV, with increasing energy the response functions decrease very similarly for all sphere diameters. Since the responses calculated with different codes and cross section data sets differ significantly, further calibration measurements were needed to select the best extension of the well-established matrix below 20 MeV[24]. Nevertheless, the similarity of the energy dependence prevents unfolding without realistic assumptions about the spectral neutron fluence in this energy region, e.g. taken from Monte Carlo calculations as available for the CERN high energy calibration fields[25]. Additional detectors with a significantly different response for neutron energies above 20 MeV, e.g. Th- and Bi-based fission detectors with thresholds of about 1 MeV and 50 MeV, respectively[19], or modified moderator detectors with heavy metal inlets to increase the high energy response by means of (n,xn) reactions are needed to fix the shape for higher energies[26]. A variety of information from organic scintillation detectors may also be included as additional data points, e.g. the activation due to the ^{12}C(n,2n) reaction (threshold at 10 MeV) and the counting rates for different thresholds[27]. These kinds of detectors are in use at high energy proton accelerators. The corresponding neutron detection efficiencies cannot be precisely calculated and must be determined experimentally.

Recoil proton spectrometry (RPS)

The larger scatter of the integral dose equivalent values compared with the integral fluence, both derived from the same BSS measurements, is chiefly caused by the steep increase in the fluence-to-dose equivalent conversion function in the energy region from 10 keV to 1 MeV. In the case of rather soft spectra which are encountered at nuclear facilities and end at about 1 MeV neutron energy[1], a correct determination of the shape is required because a 20% group fluence between 10 keV and 1 MeV may still contribute 50% to the total DE.

In general, a class of solutions is statistically compatible with the measured rates (Figure 3(a)) and the selection of the 'best' result could be to some extent subjective if no additional information is available. The neutron spectrum can, however, be determined in this energy range with a much higher energy resolution and accuracy if recoil proton detectors are employed (Figure 3(b)). Hydrogen- or methane-filled proportional counters cover the neutron energy range from 50 keV to some MeV[28] (even a lower threshold of 10(1) keV can be achieved if photon-induced events are suppressed by pulse shape analysis[29]). The response functions of cylindrical proportional counters can be reliably calculated if the well-known n-p scattering cross section, ionisation losses due to the finite range of the recoil protons (wall effect) and the influence of the structural material (in- and out-scattering) are considered in Monte Carlo simulations[30]. Recently, the reduced gain at the end

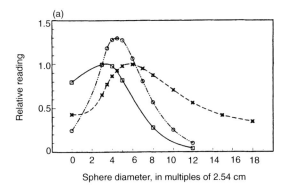

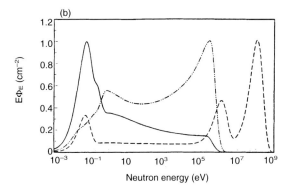

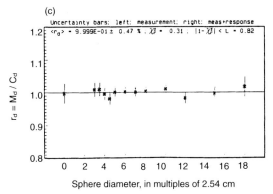

Figure 2. Relative readings of a Bonner spheres (a) in the containment of a pressurised water reactor (full line), in the environment of a transport cask with spent full elements (dashed-dotted) and in the CERN high energy calibration field (dashed) and the corresponding spectral neutron fluence unfolded from these data sets (b). For one example the measured rates M_d are compared with the rates C_d calculated for the spectral fluence and the BSS response matrix demonstrating statistical compatibility (c).

of the wire (electrical effect) of spherical proportional counters could be considered, confirming the empirical corrections needed up to now to adjust the calculated to the measured slope of the pulse height spectra[31]. The practical range of gas-filled proportional counters may be extended to up to about 15 MeV by adding one proportional counter with a ^{4}He filling[32].

Alternatively, organic scintillation detectors (SC) are used to cover the energy range above 1 MeV. Stilben crystals or liquid scintillators NE213 have the capability of separating neutron- and photon-induced events by pulse shape analysis, but in some practical cases it may be a serious drawback that the sensitivity to photons and neutrons is of the same order of magnitude. Provided the light output functions are carefully determined, the response functions can be reliably calculated for neutron energies up to 20(30) MeV because the recoil protons from the well-known n-p scattering dominate the response[33]. For higher energies the contribution of secondary charged particles from the n-C interaction increases and the reaction cross sections are not so well known. Thus the response functions must be determined experimentally[34].

Provided the response matrix is well established by Monte Carlo simulation and experimental calibration, unfolding of measured pulse height spectra results in a unique solution for the spectral neutron fluence. The unfolded neutron spectra are generally in excellent agreement with the spectral fluence derived from neutron time-of-flight spectra measured simultaneously[33]. Discrepancies recently observed in the comparison exercise performed at the IPSN realistic neutron source facility at Cadarache[7], pointed to problems with the response matrices for these systems also. Additional calibration measurements are just being analysed and the results will be considered in the evaluation of the data.

If the data from BSS and recoil proton spectrometers are combined (Figure 4), the resulting neutron spectra exhibit a considerably improved energy resolution for energies higher than 50 keV (PC and SC (Figure 4(a)[35]) or 1.2 MeV (SC only[36] Figure 4(b)) which may reduce the uncertainty of the corresponding DE values.

Bubble or superheated drop detector spectrometers

Recent investigations have shown that bubble[37] or superheated drop[38] detectors may be used for neutron spectrometry because they allow various thresholds between 10 keV and 10 MeV to be achieved. While the

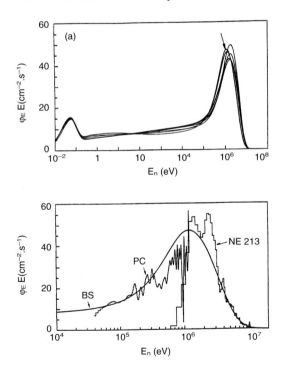

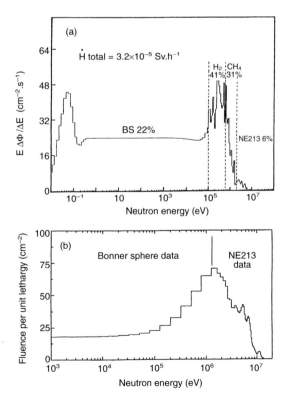

Figure 3. Multiplicity of results (a) obtained for BS measurements in a repository for PuO$_2$ samples[29]. The spectral fluence retained as 'best' result indicated with an arrow in (a) is compared in the overlapping energy region with the high resolution recoil proton data (b) obtained with a liquid scintillator (NE213) and hydrogen filled proportional counters (PC).

Figure 4. Combined analysis of BSS and RPS data, as achieved at (a) IPSN-CEA[36], and (b) NPL[37].

strong temperature dependence of the response of the bubble detectors used was taken into account by appropriate corrections[37], the influence of the temperature in turn was used to vary the threshold of superheated drop detectors in a controlled and reproducible manner[38]. In this way, the neutron emission rate can be determined for effective thresholds. This information may also be helpful in defining the guess spectra for BSS unfolding, but the reliability of the systems has still to be checked under severe environmental conditions as encountered in practice.

Monte Carlo transport calculations

The better the start spectrum selected, the faster the convergence of the few channel unfolding of BS data. Monte Carlo simulations based on a realistic model of the neutron source, the shielding material and/or the environment proved to be very good *a priori* information. For example, the neutron spectra calculated with the MCNP code for the two configurations of the moderator assembly at the IPSN 14 MeV neutron generator at Cadarache[14] were confirmed[2,7] both in shape and in absolute scale which was related to the associated charged particles of the neutron-producing ^{3}H(d,n) reaction. It turned out that the target assembly and the moderator set-up had to be modelled in every detail, an experience already gained from the calculation of the detector response. The simulation of room return neutrons in the PTB bunker with radionuclide sources in its centre was confirmed by BSS measurements[39] as well as the calculation of the high energy calibration fields carried out at CERN to simulate the neutron spectra expected in aircraft[25]. Satisfactory agreement between calculated and measured neutron spectra has recently also been obtained for a mock-up of a transport cask for spent fuel elements, which was equipped with a Cf neutron source instead of fuel elements in order to simulate at least the neutron field of a real transport and storage container. Since the neutron source and the shielding material were well known, it is not surprising that the predicted spectral neutron fluence was confirmed by the BSS data for measurements free-in-air, but the neutrons backscattered from the salt in the drift of a salt mine were also correctly predicted[40].

NEUTRON SURVEY INSTRUMENTS

The practical neutron fields well specified by Monte Carlo calculations and/or spectrometry were used to test the properties of survey instruments in general use and instruments recently developed. Different versions of moderator-type instruments, either of the spherical Leake type[41] or of cylindrical Anderson–Braun design[42], and the various types of TEPCs which were also involved in a former intercomparison[43]. Provided the energy-dependent response of the instrument was known, a consistency check was performed by comparing the reading with the response calculated for the evaluated spectral neutron fluence according to Equation 5.

The results obtained in the course of the four comparison exercises can be summarised as follows:

(1) The Leake-type survey instruments overestimate the DE of soft neutron fields by up to 100%, as also expected because of their response function, if these monitors are calibrated in the 'harder' fields of a Cf or Am-Be source as recommended.

(2) Anderson–Braun-type instruments are expected to show the same deficiencies according to their known response but some models, e.g. Studsvik rem counters, exhibit satisfactory DE readings also in these soft fields.

(3) Both moderator-type instruments fail in high energy fields (under-reading by about a factor of two behind the concrete shield at CERN[25] and also in aircraft) and must be replaced by a modified version with heavy metal inlets which considerably increase the response to neutron energies higher than 20 MeV[26].

(4) A new spherical, moderator-type dosemeter optimised by means of MCNP calculations to measure the ambient dose equivalent according to ICRP 60 and to achieve almost the same calibration factor in the fields of a bare or D_2O-moderated Cf source[44] generally under-reads in soft neutron fields (about 10–15% on the average and at nuclear facilities up to 40% as also predicted).

(5) Two remarkable findings are to be reported for the TEPCs involved. If calibrated with an alpha source for a correct lineal energy scale and checked for accurate determination of photon doses, the TEPCs will generally under-read the neutron DE of the practical soft fields by 40–50%, which could be confirmed by recalculation[6]. In addition, a large scatter of the results obtained with different instruments was observed in all comparisons[1,7], which may be explained in particular by the instability of the portable, handheld instruments (3 of 6 systems used). Neither the Leake-type dosemeters, nor TEPC systems can be recommended for use as a single instrument at workplaces in nuclear facilities. If the measured frequency distribution of the lineal energy is corrected in a sophisticated manner[45] or the lineal energy distribution clearly indicates that a correction factor should be applied in order to comply with the calibration data obtained in a similar neutron field, e.g. that of a D_2O-moderated Cf source or another soft calibration field, the under-reading of the TEPC systems can be reduced. On the other hand, it must be noted that TEPCs are regarded as reference instruments in high energy fields if the major fraction of the neutron dose equivalent is contributed by neutrons with energies higher than 10 MeV[25].

(6) Superheated drop detectors proved to be dosemeters

with sufficient sensitivity and a well-optimised energy dependence. The DE readings obtained in the environment of the transport cask were in agreement with the reference values within 10%[6]. The strong temperature dependence and the sensitivity of the active counting system to acoustic noise prevented this system from being used in the containment building. The systems must be improved in this respect.

DETERMINATION OF PERSONAL DOSE EQUIVALENT AND PROTECTION QUANTITIES

The personal DE, $H_p(10)$, to be recorded and the protection quantities H_E or E depend on the energy and the angle of incidence of the neutrons on the (anthropomorphic) phantom. Both the survey instruments and the spectrometers are designed to have an isotropic response in accordance with the definition of the operational quantity $H^*(10)$. It is, indeed, impossible to build a practicable neutron spectrometer covering an energy range from thermal up to at least 10 MeV at reasonable energy resolution and efficiency and simultaneously being sensitive in a certain solid angle only. The collimator needed would be rather cumbersome and by no means (trans)portable.

In order to check whether the $H^*(10)$ value measured in the particular field is still a conservative estimate for the protection quantity, the directional spectral fluence must be investigated, at least for a rough approach. For this purpose, sets of personal dosemeters were symmetrically mounted on phantoms, either PE spheres[46] or PMMA slabs[47], and irradiated in the point of measurement. PADC- and TLD-based dosemeters were employed. One approach is based on the assumption that the neutron field is composed of a directional and an isotropic fraction having the same spectral fluence as independently measured. An even more sophisticated analysis is possible for multidetector dosemeters with a response separable for thermal, intermediate and high (>70 keV) energy neutrons. Taking into consideration the known angular dependence of the dosemeter response for these energy regions, $H^*(10)$ values can be evaluated, e.g. by means of Monte Carlo unfolding[48]. Finally, protection quantities can be calculated for the various irradiation conditions, e.g. AP, PA, isotropic or rotational.

In general, the personal dose equivalent $H_p(10)$ measured on the phantoms exhibited a rather large scatter, but almost all values were conservative estimates of the highest protection quantity. In the case of an evaluation of $H^*(10)$ values, very good agreement with the reference values was achieved, with deviations less than 20%[6].

PHOTON SPECTROMETRY AND DOSIMETRY

A similar specification of the photon component of a mixed radiation field is still an unsolved problem if the major fraction of the total DE (>50%) is caused by the neutrons. The reason is that the neutron sensitivity of the photon spectrometers and dosemeters is only scarcely known, if at all.

Photon spectrometry

Since photon fields consist of undisturbed primary photons and a continuously distributed background of Compton scattered photons, the measured pulse height spectra must be unfolded. This method has successfully been applied in pure photon fields using Ge, NaI(Tl), BGO spectrometers and, recently, NE213 liquid scintillation detectors[49]. The response to neutrons of the three detector systems with the best energy resolution and highest efficiency for photons (Ge, BGO, NaI) cannot be experimentally separated. Since this response is not known for the entire energy range of neutrons present in the field, if at all, this contribution cannot be calculated using the independently measured neutron spectrum. Further experimental and calculational investigations are required to establish these response matrices. Stilben crystals and NE213 liquid scintillators have the capability of separating signals from secondary charged particles produced by neutron interaction in the scintillator and electrons and positrons from photon interactions. The spectral neutron fluence (at least for $E_n > 1$ MeV) can be evaluated from the neutron-induced pulse height spectrum. Since the photon response functions can be calculated (e.g. EGS4[49]), the photon-induced spectrum may also be unfolded. The resulting spectral photon fluence, however, still contains photons which were produced by the neutrons via inelastic scattering in the casing material or in the scintillator itself. While the scattered neutrons emerge from the detector volume, the subsequent photon has a rather high probability of being detected. In general, the neutron-induced photon events amount to only a few per cent of the neutrons detected, but this fraction increases up to 5(15)% at about 10 MeV neutron energy[50] depending on the detector volume and is chiefly caused by inelastic scattering on the carbon nuclei of the scintillator (4.4 MeV photons). Capture of thermal or moderated intermediate energy neutrons (2.2 MeV photons) gives rise to another extreme of the neutron-induced photon response. Since the third response matrix, i.e. detection of neutron induced photons required for the organic scintillators, has not yet been established, the unfolded photon spectra may be interpreted only with great care.

Photon dosimetry

TEPC systems should be the dosemeters best suited to mixed fields because the photon (low LET) and neutron (high LET) induced fractions of the lineal energy distribution are well separated[24]. Nevertheless, in this detector also, neutron interaction in the wall may produce photons, e.g. by neutron capture in hydrogen, indicating a photon dose although no primary photons are present, e.g. about 10% of the neutron dose in a clean thermal neutron beam[43].

Geiger–Mueller (GM) counters are preferably used as photon survey instruments because the neutron sensitivity is very low ($k_u < 1\%$) for most[51] except thermal and intermediate energies (k_u up to 6%[52]). Since this behaviour is similar for both the GM counter and the TEPC, the 50% higher reading of the GM counters compared with the TEPC result in the containment building of a PWR reactor[1] was not expected and finally attributed to the presence of high energy photons (7 MeV from n-capture in iron) for which commonly used GM counters are not compensated but show an over-reading of up to 100%[53].

Since neutrons are often accompanied by high energy photons, ionisation chambers[53] should be employed, but unfortunately the neutron sensitivity of radiation protection instruments which is needed for corrections is not as well known as for GM counters.

SUMMARY AND CONCLUSIONS

The major intention of EURADOS working groups in performing the comprehensive comparison measurements with neutron spectrometers, and neutron and photon dosemeters was to evaluate the state of the art of dosimetry in mixed fields. The exercise at workplaces in the containment of PWR reactors and in the environment of a transport cask loaded with spent fuel elements[1,6] was most valuable because, besides the practical fields, the extreme measuring conditions (elevated levels of noise and temperatures) had also to be taken into account. Since this was a unique opportunity, the other intercomparisons had to be performed in calibration fields replicating to some extent neutron fields encountered in practice[2–5].

The evaluation of all measurements with neutron spectrometers showed that Bonner sphere systems will be best suited to the determination of the integral fluence and the DE values with a scatter of less than 5% and 15%, respectively, if neutron fields with energies less than 20 MeV are considered. Appropriate detectors must be added in order to extend the dynamic range up to some hundreds of MeV. Well defined response functions are, however, indispensable for reliable unfolding. Since a rigorous uncertainty propagation has not yet been performed, the standard deviation of the scatter of the results obtained with various spectrometers may be regarded as the uncertainty achievable[7].

Proton or helium recoil spectrometry may complement the BS measurements in order to determine the neutron fluence in the energy region with the higher $\bar{Q}$ value ($E_n > 10$ keV) with a considerably improved resolution. The development of reliable spectrometers based on bubble or superheated drop detectors is very much encouraged because these systems are easier to handle.

Realistic Monte Carlo simulations may serve as the ideal *a priori* information, particularly for few-channel unfolding. Provided the neutron source and its environment can be modelled in any detail, the directional spectral neutron fluence calculated with the MCNP code can already be regarded as a reference.

Irradiation of sets of dosemeters on a phantom, if possible using those with spectrometric properties, is the only experimental tool available to complement spectrometric measurements in order to obtain an estimate for the directional fluence. The ambient dose equivalent values evaluated on the basis of these data sets for consistency are in good agreement (within 20%) with the reference values. Considerable improvement can be expected if directly reading detectors with spectrometric properties (electronic dosemeters or SDDs) are used on the phantoms or if the moderator sphere with position-dependent ^{3}He detectors has proved to be applicable[54].

Only a few survey instruments can be recommended for use as single area dosemeters in soft fields at nuclear facilities. Various combinations of two detectors, e.g. Leake-type dosemeters and TEPCs or subsets of Bonner spheres[56], give reasonable results. Special care has to be taken in fields with neutron energies higher than 10 MeV, where TEPC and modified moderator systems are the best choice.

Adequate photon spectrometers and dosemeters are not yet available for application in complex mixed fields. TEPC systems may offer the best approach because the photon-induced response can be separated. The neutron-induced photon response must be determined for those photon dosemeters which should be compensated for the entire energy range up to at least 7 MeV photons.

ACKNOWLEDGEMENTS

The author would like to express his thanks to all members, corresponding members and consultants of EURADOS WG7 and WG10 for the efforts they put into planning, performing and evaluating the comparison exercises and the Commission of the European Community for supporting these activities under contract No FI3P CT920001.

REFERENCES

1. Klein, H. and Lindborg, L. (Eds) *Determination of the Neutron and Photon Dose Equivalent at Workplaces in Nuclear Facilities of Sweden — An SSI-EURADOS Comparison Exercise. Part I: Measurements and Data Analysis.* SSI report 95-15 (Stockholm: SSI) (1995).

2. Klein, H., Thomas, D. J., Chartier, J. L. and Schraube, H. (Eds) *Determination and Realization of Calibration Fields for Neutron Protection Dosimetry as Derived from Spectra Encountered in Routine Surveillance*. EUR report 14927 DE/EN/FR (Luxembourg: CEL) pp. 159–174 (1993).
3. Murphy, M. *A Review of the Neutron Spectrometry Measurements in the ASPIS Reference Fields*. Laboratory report AEA-RS-5548 (Winfrith: AEA) (1994).
4. Chartier, J. L., Kurdjian, J., Paul, D., Audoin, G., Pelcot, G. and Posny, F. *Progress on Calibration Procedures with Realistic Neutron Spectra*. Radiat. Prot. Dosim. **61**, 57–61 (1995).
5. Chartier, J. L., Posny, F. and Buxerolle, M. *Experimental Assembly for the Simulation of Realistic Neutron Spectra*. Radiat. Prot. Dosim. **44**, 125–130 (1992).
6. Lindborg, L., Bartlett, D., Drake, P., Klein, H., Schmitz, Th. and Tichy, M. *Determination of the Neutron and Photon Dose Equivalent at Workplaces in Nuclear Facilities of Sweden — A Joint SSI-EURADOS Comparison Exercise*. Radiat. Prot. Dosim. **61**, 89–100 (1995).
7. Thomas, D. J., Chartier, J. L., Klein, H., Naismith, O. F., Posny, F. and Taylor, G. C. *Results of a Large Scale Neutron Spectrometry and Dosimetry Comparison Exercise at the Cadarache Moderator Assembly*. Radiat. Prot. Dosim. **70**(1–4), 313–322 (This issue) (1997).
8. Schuhmacher, H., Alberts, W. G., Alevra, A. V., Klein, H., Schrewe, U. J. and Siebert, B. R. L. *Characterisation of Photon-Neutron Radiation Fields for Radiation Protection Monitoring and Optimisation*. Radiat. Prot. Dosim. **61**, 81–88 (1995).
9. International Commission on Radiological Protection. *Data for Protection against Radiation from External Sources: Supplement to ICRP publication 14*. Publication 21 (Oxford: Pergamon Press) (1971).
10. Wagner S., Großwendt, B., Harvey, J. R., Mill, A. J., Selbach, H. J. and Siebert, B. R. L. *Unified Conversion Function for the New ICRU Operational Radiation Protection Quantities*. Radiat. Prot. Dosim. **12**, 231–235 (1985).
11. Siebert, B. R. L. and Schuhmacher, H. *Quality Factors, Ambient and Personal Dose Equivalent for Neutrons, Based on the New ICRU Stopping Power Data for Protons and Alpha Particles*. Radiat. Prot. Dosim. **58**, 177–183 (1995).
12. Schuhmacher, H. *Tissue-Equivalent Proportional Counters in Radiation Protection Dosimetry: Expectation and Present State*. Radiat. Prot. Dosim. **44**, 199–206 (1992).
13. Hollnagel, R. *Conversion Functions for Effective Dose E and Effective Dose Equivalent H_E for Neutrons with Energies from Thermal to 20 MeV*. Radiat. Prot. Dosim. **58**, 209–212 (1994) and private communication.
14. Aroua, A., Boschung, M., Cartier, F., Gmür, H., Grecescu, M., Prêtre, S., Valley, J.-F. and Wernli, Ch. *Study of the Response of Two Neutron Monitors in Different Neutron Fields*. Radiat. Prot. Dosim. **44**, 183–187 (1992).
15. Bramblett, R. L., Ewing, R. I. and Bonner, T. W. *A New Type of Neutron Spectrometer*. Nucl. Instrum. Methods **9**, 1–12 (1960).
16. Mares, V. and Schraube, H. *Evaluation of the Response Matrix of a Bonner Sphere Spectrometer with LiI Detector from Thermal Energy to 100 MeV*. Nucl. Instrum. Methods **A337**, 461–473 (1994).
17. Wiegel, B., Alevra, A. V. and Siebert, B. R. L. *Calculations of the Response Functions of Bonner Spheres with a Spherical ^{3}He Proportional Counter using a Realistic Detector Model*. PTB Report N-21 (Braunschweig: PTB) (1994).
18. Bardell, A. G. and Thomas, D. J. *Spectrometry Measurements by NPL at Position A Ringhals Reactor*. In Determination of the Neutron and Photon Dose Equivalent at Workplaces in Neutron Facilities of Sweden, Part I. SSI report 95-15 (Stockholm: SSI) pp. 59–66 (1995).
19. Dinter, H. and Tesch, K. *Determination of Neutron Spectra behind Lateral Shielding of High Energy Proton Accelerators*. Radiat. Prot. Dosim. **42**, 5–10 (1992).
20. Thomas, D. J. *Use of the Program ANISN to Calculate Response Function for a Bonner Sphere Set with ^{3}He Detector*. NPL-Report RSA (EXT) 31 (Teddington, UK: NPL) (1992).
21. Kralik, M. and Novotny, T. *Response Functions of Bonner Spheres with Small Cylindrical ^{3}He Proportional Counters Determined with Monte Carlo Simulations*. Report CMI-GR 2070/95 (Prague: CMI) (1995).
22. Kralik, M., Aroua, A., Grecescu, M., Mares, V., Novotny, T., Schraube, H. and Wiegel, B. *Specification of Bonner Sphere Systems for Neutron Spectrometry*. Radiat. Prot. Dosim. **70**(1–4), 279–284 (This issue) (1997).
23. Alevra, A. V., Siebert, B. R. L., Aroua, A., Buxerolle, M., Grecescu, M., Matzke, M., Mourgues, M., Perks, C. A., Schraube, H., Thomas, D. J. and Zaborowski, H. C. *Unfolding of Bonner Sphere Data. A European Intercomparison of Computer Codes*. Laboratory Report PTB-7.22-90-1 (Braunschweig: PTB) (1990).
24. Alevra, A. V. and Schrewe, U. J. *Measurements with the PTB 'C' Bonner Sphere Spectrometer in the 55 MeV Neutron Field at PSI Villigen for Spectrometric and Calibration Purposes*. Radiat. Prot. Dosim. **70**(1–4), 295–298 (This issue) (1997).
25. Alevra, A. V., Klein, H. and Schrewe, U. J. *Measurements with the PTB Bonner Sphere Spectrometer in High-Energy Neutron Calibration Fields at CERN*. PTB Report N-22, (Braunschweig: PTB) (1994).
26. Birrattari, C., Esposito, A., Ferrari, A., Pelliccioni, M. and Silari, M. *A Neutron Survey Meter with Sensitivity Extended up to 400 MeV*. Radiat. Prot. Dosim. **44**, 193–197 (1992).
27. Aleinikov, V. E. and Timoshenko, G. N. (Unpublished observations).
28. Ing, H., Clifford, T., McLean, T., Webb, W., Cousins, T. and Dhermain, J. *ROSPEC: a Simple Reliable High Resolution Neutron Spectrometer*. Radiat. Prot. Dosim. **70**(1–4), 273–278 (This issue) (1997).

29. Knauf, K., Alevra, A. V., Klein, H. and Wittstock, J. *Neutronenspektrometrie im Strahlenschutz*. PTB-Mitteilungen **99**, 101–106 (1989).
30. Parker, J. B., White, P. H. and Webster, R. J. *The Interpretation of Recoil Proton Spectra*. Nucl. Instrum. Methods **23**, 61–68 (1963).
31. Weise, K., Weyrauch, M. and Knauf, K. *Neutron Response of a Spherical Proton Recoil Proportional Counter*. Nucl. Instrum. Methods **A309**, 287–293 (1991).
32. Birch, R. *An Alpha-Recoil Proportional Counter to Measure Neutron Energy Spectra between 2 MeV and 15 MeV*. Report AERE-R-13002 (Harwell: AERE) (1988).
33. Guldbakke, S., Klein, H., Meister, A., Pulpan, J., Scheler, U., Tichy, M. and Unholzer, S. *Response Matrices of NE213 Scintillation Detectors for Neutrons*. In: Reactor Dosimetry, ASTM STP 1228, Eds H. Farrar IV, E. P. Lippincott, J. G. Williams and D. W. Vehar (American Society for Testing and Materials, Philadelphia) pp. 280–289 (1994).
34. Nolte, R., Schuhmacher, H., Brede, H.J. and Schrewe, U. J. *Measurement of High-Energy Neutron Fluence with Scintillation Detector and Proton Recoil Telescope*. Radiat. Prot. Dosim. **44**, 101–104 (1992).
35. Posny, F., Chartier, J. L. and Buxerolle, M. *Neutron Spectrometry System for Radiation Protection: Measurements at Workplaces and in Calibration Fields*. Radiat. Prot. Dosim. **44**, 239–242 (1992) and Posny, F. private communication.
36. Klein, H., Thomas, D. J., Chartier, J. L., Schraube, H., Kralik, M., Osmera, B. and Grecescu, M. *Realistic Neutron Calibration Fields and Related Dosimetric Quantities*. EUR-report (to be published) and Thomas, D. J. private communication.
37. Rosenstock, W. Schulze, J., Köble, T., Kruzinski, G., Thesing, P., Jannick, G. and Kromholz, H.-L. *Estimation of Neutron Energy Spectra with Bubble Detectors: Potential and Limitations*. Radiat. Prot. Dosim. **61**, 133–136 (1995).
38. d'Errico, F., Alberts, W. G., Curzio, G., Guldbakke, S., Kluge, H. and Matzke, M. *Active Neutron Spectrometry with Superheated Drop (Bubble) Detectors*. Radiat. Prot. Dosim. **61**, 159–162 (1995).
39. Kluge, H., Alevra, A. V., Jetzke, S., Knauf, K., Matzke, M., Weise, K. and Wittstock J. *Scattered Neutron Reference Fields Produced by Radionuclide Sources*. Radiat. Prot. Dosim. **70**(1–4), 327–330 (This issue) (1997).
40. Knauf, K., Alevra, A. V., Wittstock, J., Engelmann, H. J., Khamis, M. and Niehues, N. *Neutron and Photon Spectra and Dose Rates Around a Shielding Cask Placed in a Salt Mine to Simulate a Nuclear Waste Package*. Radiat. Prot. Dosim. **70**(1–4), 251–254 (This issue) (1997).
41. Leake, J. W. *An Improved Spherical Dose Equivalent Neutron Detector*. Nucl. Instrum. Methods **63**, 329–332 (1968).
42. Anderson, I. O. and Braun, J. A. *A Neutron Rem Counter*. Nukleonik **6**, 237–241 (1964).
43. Alberts, W. G., Dietz, E., Guldbakke, S., Kluge, H. and Schuhmacher, H. *International Incomparsion of TEPC Systems Used for Radiation Protection*. Radiat. Prot. Dosim. **29**, 47–53 (1989).
44. Burgkhardt, B., Fieg, G., Klett, A., Plewnia, A. and Siebert, B. R. L. *The Neutron Fluence and H*(10) Response of the New LB 6411 Rem Counter*. Radiat. Prot. Dosim. **70**(1–4), 361–364 (This issue) (1997).
45. Taylor, G. C. *An Analytical Correction for the TEPC Dose Equivalent Response Problem*. Radiat. Prot. Dosim. **61**, 67–70 (1995).
46. Luszik-Bhadra, M. and Matzke, M. *Measurements with PTB Personal Dosemeters*. In: Determination of the Neutron and Photon Dose Equivalent at Workplaces in Nuclear Facility of Sweden, Part I, SSI report 95-15, pp. 104–109 (1995).
47. Bartlett, D. T., Tanner, R. J. and Steele, J.D. *Measurements with NRPB Personal Dosemeters*. In: Determination of the Neutron and Photon Dose Equivalent at Workplaces in Nuclear Facility of Sweden, Part I, SSI report 95-15, pp. 110–112 (1995).
48. Matzke, M., Kluge, H. and Luszik-Bhadra, M. *Directional Information on Neutron Fields*. Radiat. Prot. Dosim. **70**(1–4), 261–264 (This issue) (1997).
49. Büermann, L., Ding, S., Guldbakke, S., Klein, H., Novotny, T. and Tichy, M. *Response of NE213 Liquid Scintillation Detectors to High-Energy Photons ($E_\gamma > 3$ MeV)*. Nucl. Instrum. Methods **A332**, 483–492 (1993).
50. Fowler, J. L., Cookson, J. A., Hussain, M., Schwartz, R. B., Swinhoe, M. T., Wise, C. and Uttley, A. *Efficiency Calibration of Scintillation Detectors in the Neutron Energy Range 1.5–25 MeV by Associated Particle Technique*. Nucl. Instrum. Methods **175**, 449–463 (1980).
51. Guldbakke, S., Jahr, R., Lesiecki, H. and Schölermann, H. *Neutron Response of Geiger-Müller Photon Dosemeters for Neutron Energies between 100 keV and 19 MeV*. Health Phys. **39**, 963–969 (1980).
52. Alberts, W. G., Ambrosi, P. and Kluge, H. *The Response of Some Photon Dosemeters to Slow Neutrons*. In: Proc. IRPA Congress, Compacts Vol. III, pp. 1161–1164 (1984).
53. Büermann, L., Guldbakke, S. and Kramer, H. M. *Referenzstrahlungsfelder zur Kalibrierung von Strahlenschutzdosimetern für Photonen im Energiebereich oberhalb 3 MeV*. In: Strahlenschutz: Physik und Meßtechnik, Eds. W. Koelzer, R. Maushart (Koln: Verlag TÜV Rheinland) Vol. I, pp. 224–229 (1994).
54. Toyokawa, H., Uritani, A., Mori, C. Takeda, N. and Kudo, K. *A Spherical Neutron Counter with an Extended Energy Response for Dosimetry*. Radiat. Prot. Dosim. **70**(1–4), 365–370 (This issue) (1997).
55. Naismith, O., Siebert, B. R. L. and Thomas, D. J. *Response of Neutron Dosemeters in Radiation Protection Environments: an Investigation of Techniques to Improve Estimates of Dose Equivalent*. Radiat. Prot. Dosim. **70**(1–4), 255–260 (This issue) (1997).

NEUTRON DOSE EQUIVALENT RATES CALCULATED FROM MEASURED NEUTRON ANGULAR AND ENERGY DISTRIBUTIONS IN WORKING ENVIRONMENTS

P. Drake† and D. T. Bartlett‡
†Vattenfall A B, Ringhals
S-430 22 Väröbacka, Sweden
‡National Radiological Protection Board
Chilton, Oxon. OX11 0RQ, UK

Abstract — The estimation of non-isotropic neutron dose equivalent quantities in multidirectional fields is complicated. The estimation must be based on the neutron fluence rate as a function of both energy and angle. There are several well developed techniques for measuring neutron energy spectra. However, the directional characteristics have not usually been investigated in detail. The personal dose equivalent inside the PWR stations studied varies by a factor of 4 between AP and ISO irradiation geometries, but only by a factor of 2 between different locations for AP irradiation. This indicates the importance of knowing the directional distribution, in order to interrelate personal dose equivalent, effective dose equivalent (effective dose) and ambient dose equivalent, and to interpret dosemeter and instrument readings. A purpose built phantom has been constructed and used for the separation of the neutron fluence into 18 incident directions. Measurements with the phantom have been made at workplaces at Ringhals Nuclear Power Plant. The phantom consists of a 19 cm diameter boron doped paraffin sphere. Boron was added to decrease the transmission of neutrons through the sphere and to decrease the photon contribution from (n;γ) reactions inside the phantom to the measurement of dose equivalent from incident photons. The phantom has proved to be capable of separating the incident neutrons into 18 directions and effective in reducing (n;γ) reactions. Measured neutron angular distributions with coarse energy information were used with previously measured neutron spectra, to make better estimates of the neutron personal dose equivalent rates at these workplaces in Swedish nuclear facilities. The values of personal dose equivalent calculated with separation into 18 incident directions are slightly lower than those calculated for 2 or 3 directional distributions (AP, ISO and/or ROT). Results of calculations of neutron personal dose equivalent using the new (ICRP Publication 60) Q(L) function and new stopping powers, are also given, with and without resolution into 18 incident directions.

INTRODUCTION

The accurate estimation of non-isotropic neutron dose equivalent quantities in multidirectional fields relies on a knowledge of the spectral distribution of the neutron radiance (the differential distribution of the fluence rate in both energy and angle) in order to interpret or correct the readings of neutron personal dosemeters. The estimation also depends on the orientation of the person exposed to the neutron field. The mean conversion coefficient from fluence to personal dose equivalent inside PWR stations varies by a factor of 4 between AP and ISO irradiation but only by a factor of 2 between different locations for AP irradiation[1]. This indicates the importance of knowing the directional distribution, in order to interpret dosemeter readings and to interrelate personal dose equivalent, effective dose equivalent (effective dose) and ambient dose equivalent. There are several well developed techniques for measuring neutron energy spectra, but there have been only a few investigations of the directional characteristics of neutron radiation fields[1–3].

An international intercomparison was performed at a Swedish pressurised water reactor (Ringhals) and at the Swedish Interim Storage for Spent Nuclear Fuel (CLAB)[1]. In the intercomparison, the personal dose equivalents and effective dose equivalents were calculated from measured neutron energy spectra, and neutron fluence separated into a unidirectional field (AP) and an isotropic field (ISO). The neutron spectra were based on measurements with several different neutron spectrometers. The separation into two directional components was made with different personal dosemeters pointing in six directions on spherical or slab phantoms, but assuming that the neutron spectrum was independent of angle. Using this approach, there were large uncertainties in the estimated values of personal dose equivalent and effective dose equivalent.

New conversion coefficients from neutron fluence to personal dose equivalent and effective dose[4,5] based on ICRP Publication 60[6] and ICRU Report 49[7] have also been applied and a brief comparison of results made. The consequences of the changes will be discussed in more detail elsewhere.

The intercomparison demonstrated the need to investigate further the photon contribution to dosemeter readings, partly because the photon dosemeters' readings depend on the direction and energy of the incident photons and partly because the photons produced by (n;γ) reactions should be excluded from the determination of the photon dose.

MATERIALS AND METHODS

Boron doped phantom

A new phantom has been constructed and used[8] for the separation into 18 incident directions of the neutron fluence at workplaces at Ringhals Nuclear Power Plant inside the containment of a PWR, and near a spent nuclear fuel transport cask. The phantom consists of a 19 cm diameter, boron doped, paraffin sphere. Boron was added to reduce the transmission of neutrons and to reduce the photon contribution to dosemeter readings from $^1H(n;\gamma)\,^2H$ reactions inside the phantom. ^{10}B has a higher thermal neutron capture cross section than hydrogen, and the presence of boron reduces the probability of reactions with hydrogen. The ^{10}B content is 2%. In the sphere surface there are 18 recesses, each 3 cm deep and 19 mm by 19 mm wide, which can hold polymethyl methacrylate (PMMA) inserts with TLD pellets or PADC (poly-allyl diglycol carbonate) foils.

The phantom design was tested using the Monte Carlo code MCNP[8,9] for a simulated ^{252}Cf source and three monoenergetic photon sources (500 keV, 2.2 MeV and 7.6 MeV). To reduce the computation time, the detector elements in each recess were simulated by a slightly larger, single, detector element, (13 × 13 × 2 mm^3). The simulation showed that the phantom had very good properties for separation into 18 incident directions (the dose at 45° was 68% of that at 0°, and decreased rapidly at larger angles) and that both the transmission of neutrons through and reflection from the phantom, and the production of the neutron induced photons was reduced to insignificant levels. The test also showed that the transmission of high energy photons remained significant: using this approach, the directional separation of the photon field will have large uncertainties.

The phantom was then irradiated with a ^{252}Cf source at a hospital, with and without shadow cones. The irradiation with shadow cones was carried out with TLDs only (in the PMMA inserts). Given the complex source configuration, it was not possible completely to shield the primary beam, whilst still allowing as much as was desirable of the scattered radiation to reach the phantom. The estimated fluence at 45° was only about 20% of that measured for the direct beam but this was considered in acceptable agreement taking into account geometrical differences and the large statistical uncertainties for these preliminary calculations.

Measurements in workplace fields

Measurements were performed at Ringhals 4, inside the containment close to the reactor cavity (position **A**, H*(10) rate = 1.46 mSv.h^{-1}) and in the steel airlock entrance into containment (position **L**, H*(10) rate = 0.22 mSv.h^{-1}). Positions **A** and **L** are defined fully in Reference 1.

The irradiation time was 2.5 h in position **A**. Here the boron doped phantom was used with Ringhals TLD (LiF and Li$_2$B$_4$O$_7$ pellets)[10] and Atomic Energy Canada Limited (AECL) track dosemeters[11] in PMMA inserts. The readings of the dosemeters were expected to be large enough to have low statistical uncertainties. The irradiation time in position **L** was 5 h with Ringhals TLD with PMMA inserts, and 29 h with Physikalisch-Technische Bundesanstalt (PTB)[12] and AECL track dosemeters. The readings were lower than at position **A**, resulting in larger measurement uncertainties. Note that for this investigation, the AECL and PTB detectors were cut to smaller dimensions than for normal use.

USE OF CONVERSION COEFFICIENTS

Direct measurements cannot be made of either the protection quantity, effective dose equivalent rate, or the operational quantity, personal dose equivalent rate. Estimates can be derived from measurements of neutron radiance and the application of conversion coefficients. In the investigations reported here, the readings of dosemeters are first converted to neutron radiance using their known energy neutron fluence response characteristics, and the neutron spectrum averaged over all directions determined by means of multispheres. The neutron radiance is then approximately determined for the components of the energy spectrum greater or less than 70 keV, for each of the 18 directions. Conversion coefficients may then be applied to obtain approximate values of H$_p$(10) and H$_E$. Values of conversion coefficients are taken to be exact.

In the case of personal dose equivalent, conversion coefficients for the calibration quantity, H$_p$(10)$_{slab}$, must be used. Values are available for angles of 0, 15, 30, 45, 60, 75 degrees to the normal to the front face about a vertical axis, and ROT and ISO[13]. Values of conversion coefficients for effective dose equivalent at discrete angles have been calculated[14] using a male phantom, for neutrons incident at angles of 0, 40, 60, 75, 90, 120, 140 and 180 degrees about a vertical axis, and for 0, 45 and 90 degrees above and below the horizontal, but only for energies from 1 keV to 20 MeV. In the calculations of effective dose equivalent, the neutron spectra were divided firstly into two energy groups. The directional components below 70 keV were based on TLD values and above 70 keV on track dosemeters. The radiance below 70 keV was further divided into two energy groups: below or above 1 keV. The component below 1 keV was assumed to be either unidirectional or isotropic or a combination of both, and the conversion coefficients from ICRP[15] were used. The conversion coefficients of Stewart et al[14] for 2, 6 or 18 directions were used above 1 keV. This simplification allows the calculation of effective dose equivalent (male) from the results of 18 sets of dosemeters. Comparison was made with the values obtained using only two (AP and ISO)[1] or six directional components.

Calculations have also been carried out of personal dose equivalent using the new recommendations for quality factor[6], and of the new protection quantity, effective dose[6]. Values of the conversion coefficient have been agreed by a joint ICRU/ICRP Task Group and publication is awaited. Here, values were used which were currently available, and not expected to differ greatly from the ICRU/ICRP recommended values. For reasons of consistency, the conversion coefficients calculated by Hollnagel for both quantities were used[4,5]. Those for $H_p(10)_{slab}$ are at angles of 0, 15, 30, 45, 60, 75 degrees to the normal to the front face of a slab phantom about a vertical axis and ROT and ISO. Those for effective dose are for the directions AP, PA, LAT, ROT and ISO. (There are no data available for effective dose for finer angle resolution). The conversion coefficients were not always calculated for exactly the same energy binning or for the same direction as those determined in the measurements. Where the energy and angle bins which were not exactly matched, log/log interpolations or linear interpolations respectively, were performed to calculate appropriate conversion coefficients.

ESTIMATION OF UNCERTAINTIES

The calculations of the dose equivalents were based on several measurements and calculations each with its associated uncertainty. The uncertainty in the calculation of the total dose equivalent for each measurement position can be divided into three main contributions: the uncertainty in estimating the radiance (including its spectral distribution), the uncertainty in combining radiance with the conversion coefficients, and the uncertainty in the orientation of a person in the radiation field. The conversion coefficients are considered, conventionally, to be exact.

The uncertainty in the estimation of the neutron radiance consists of uncertainties in measurements of the neutron spectrum and uncertainties in measuring the direction for different neutron energy intervals. The uncertainty in the measurement of the neutron energy spectra for the measurements in the Swedish intercomparison resulting from uncertainties in input data, counting statistics and unfolding procedures, was estimated to be 10% for the total fluence[1] and somewhat larger for total dose equivalent. The uncertainty in the measurement of the relative neutron fluence in each direction is estimated to be 6% (mean value of the standard deviation in the measured values in this paper). A total uncertainty of 12% (from the sum of the squares) is estimated for the fluence. When the uncertainty in direction is included, an approximate value for the uncertainty in the total dose equivalent from these sources of uncertainty is 20%.

Additional uncertainties in the combining of radiance

Table 1. Calculated energy and angle averaged conversion coefficients with the highest fluence rate incident in the AP, PA or lateral direction. Position A.

	Conversion coefficients from fluence to dose equivalent (pSv.cm²) (Estimated uncertainty 20%)			
	$h_{H_E\phi}$	$h_{E\phi}$	$h_{p\phi slab}$ Old Q(L)	$h_{p\phi slab}$ New Q(L)
Max. on anterior side $H_p(10)$ on anterior side				
83% ISO, 17% AP	4.6	12.1	9.3	12.4
6 directions	4.7	12.0	9.4	11.9
18 directions	4.5	not applicable	8.8	11.4
Max. on posterior side $H_p(10)$ on anterior side				
83% ISO, 17% PA	3.7	10.7	5.4	7.4
6 directions	3.8	10.5	5.0	6.3
18 directions	3.6	not applicable	5.5	7.6
Max. on one lateral side $H_p(10)$ on anterior side				
83% ISO, 17% LAT	3.3	9.2	6.0	8.1
6 directions	3.4	9.0	5.8	7.3
18 directions	3.5	not applicable	7.1	9.5
Reference 1, Max. on anterior side $H_p(10)$ on anterior side				
83% ISO, 17% AP	5.1 (ROT + AP)	11.8	9.3	11.9

$h_{H_E\phi}$ is the conversion coefficient from fluence (ϕ) to effective dose equivalent (H_E), similarly $h_{E\phi}$, etc.

and conversion coefficients for particular dose equivalent quantities consist of (a) uncertainties in matching conversion coefficients for one energy interval and one angle with measured radiance in a similar energy interval and angle when these intervals and angles do not coincide, and (b) uncertainties in estimations or measurements close to 90° or from the rear. The uncertainty due to problems matching neutron energy intervals and directions in the calculation of dose equivalent quantities was estimated to be between 5 and 10%, based on differences obtained when using different matching techniques. Only a small fraction of personal or effective dose equivalent comes from neutrons incident on the rear side of a phantom or person in a field with the highest fluence in the AP direction, therefore the uncertainty in the dose equivalent is little influenced by the uncertainties in interpreting the reading of a dosemeter, or in estimating $H_p(10)$, H_E or E from neutrons incident close to 90°, or from the rear.

The uncertainty in the orientation of a person in the radiation field was not considered in this paper.

The contribution to the uncertainty in estimating the dose equivalents from the uncertainty in total radiance was considered to be of the order of 20%; additional problems of combining conversion coefficients and radiance were considered to contribute on uncertainty of the order of 5 to 10%. The total uncertainty due to these two contributions was estimated to be 20 to 25% (summing the squares of the uncertainties).

RESULTS

The separation of the radiance into 18 directions showed that the radiation fields were more isotropic than the separation into 6 directions had indicated. The results are given for positions **A** and **L**, of calculations of neutron personal dose equivalent (ICRP Publication 15 Q(L)) and effective dose equivalent, with resolution into 6 or 18 incident directions, or based on two directional components. Results of calculations of neutron personal dose equivalent using the new (ICRP Publication 60) Q(L) function and new (ICRU Publication 49) stopping powers, and effective dose are given for the same positions and directional resolution.

The results are presented in Tables 1 and 2 as the calculated conversion coefficients (averaged over energy and angle) from fluence to dose equivalent. The tables also include values taken from Reference 1.

DISCUSSION AND CONCLUSIONS

The investigation with a separation into 18 directions supports previous conclusions that the directional characteristics of radiation fields must be known in

Table 2. Calculated energy and angle averaged conversion coefficients with the highest fluence rate incident in the AP, PA or lateral direction. Position L.

	Conversion coefficients from fluence to dose equivalent (pSv.cm²) (Estimated uncertainty 20%)			
	$h_{H_E\phi}$	$h_{E\phi}$	$h_{p\phi slab}$ Old Q(L)	$h_{p\phi slab}$ New Q(L)
Max. on anterior side $H_p(10)$ on anterior side 80% ISO, 20% AP	5.7	15.8	12.8	17.4
6 directions	5.6	14.6	13.2	16.9
18 directions	4.8	not applicable	11.1	14.6
Max. on posterior side $H_p(10)$ on anterior side 80% ISO, 20% PA	4.4	13.8	6.7	9.5
6 directions	3.7	11.6	4.3	5.5
18 directions	4.0	not applicable	7.0	9.2
Max. on one lateral side $H_p(10)$ on anterior side 80% ISO, 20% LAT	3.8	11.5	7.7	10.7
6 directions	4.2	12.0	8.0	10.2
18 directions	4.2	not applicable	8.0	10.6
Reference 1, Max. on anterior side $H_p(10)$ on anterior side 80% ISO, 20% AP	6.2 (ROT + AP)	15.6	12.8	16.6

$h_{H_E\phi}$ is the conversion coefficient from fluence (ϕ) to effective dose equivalent (H_E), similarly $h_{E\phi}$, etc.

order to make accurate estimates of non-isotropic dose equivalent quantities. The separation into 18 directions instead of 6 directions or 2 directional components (AP and ISO) leads to a slight reduction in the calculated conversion coefficients (averaged over energy and angle) from fluence to dose equivalent; however, this reduction is within the uncertainties of the estimates.

The calculations of energy and angle averaged conversion coefficients have been made not only for the irradiation of persons with the highest intensity being in the AP direction, (corresponding, in general, to the dosemeter being worn on the part of the trunk representative of that most highly exposed), but also for lateral and PA geometry. These field geometries are relevant for inadvertent exposures, and for some realistic working conditions. The results of this investigation show that for old Q(L) values, the readings of dosemeters indicating $H_p(10)_{slab}$, will be conservative for effective dose equivalent (H_E), when worn on the front of the trunk, irrespective of the person's movement in the fields considered. If dosemeters indicate $H_p(10)_{slab}$ with new Q(L) values, then for dosemeters worn on the front of the trunk, effective dose is only not overestimated for irradiations with the highest fluence incident on the anterior surface (AP irradiation). An underestimate of effective dose (E) of about 45% will result from the highest fluence incident in the PA direction and 10% for the lateral direction. The movement of a person in a complex radiation field will normally make isotropic or rotational conversion coefficients better suited for the estimation of personal dose equivalent and effective dose equivalent, and, of course, the characteristics of personal dosemeters need to be considered.

The contribution to the measured photon dose equivalent in mixed neutron–photon fields, of photons from (n;γ) reactions (in a phantom or in an irradiated person), needs to be investigated further, as this can lead to an overestimation of the total dose equivalent.

ACKNOWLEDGEMENTS

The authors wish to express their gratitude to the operations management and the Health Physics Department at Ringhals 4 for permitting and supporting the work inside the containment. The authors also wish to thank their colleagues at AECL, PTB and Ringhals who provided and processed the detectors, and at the Radiation Physics Department (Göteborg) and the Neutron and Reactor Physics Department (Stockholm) who assisted with irradiations and simulations respectively.

REFERENCES

1. Lindborg, L., Bartlett, D. T., Drake, P., Klein, H., Schmitz, T. and Tichy, M. *Determination of Neutron and Photon Dose Equivalent at Work-Places in Sweden. A Joint SSI-EURADOS Comparison Exercise.* Radiat. Prot. Dosim. **61**, 89–100 (1995).
2. Bartlett, D. T. and Bardell, A. G. *Field Measurements of Neutron Energy and Angle Distributions and their Interpretation in Terms of Relevant Radiological Protection Quantities.* In: Proc. 5th. Symp. on Neutron Dosimetry, Neuherberg, 1984, pp. 147–156 (Luxembourg: CEC) (1985).
3. Bartlett, D. T., Britcher, A. R., Bardell, A. G., Thomas, D. J. and Hudson, I. F. *Neutron Spectra, Radiological Quantities and Instrument and Dosemeter Responses at a Magnox Reactor and a Fuel Reprocessing Installation.* Radiat. Prot. Dosim. **44**, 233–238 (1992).
4. Hollnagel, R. A. *Conversion Coefficients for Personal Dose Equivalent Based on Revised Stopping Power Data.* Private communication (1994).
5. Hollnagel, R. A. *Conversion Functions of Effective Dose E and Effective Dose Equivalent H_E for Neutrons with Energies from Thermal to 20 MeV.* Radiat. Prot. Dosim. **54**, 203–212 (1994).
6. ICRP. *1990 Recommendations of the International Commission on Radiological Protection.* Publication 60 (Oxford: Pergamon) (1991).
7. International Commission on Radiation Units and Measurements. *Stopping Powers and Ranges for Protons and Alpha Particles.* Report 49 (Bethesda, MD: ICRU) (1993).
8. Drake, P. and Kirkegard, J. *Construction and Evaluation of a Boron Doped Paraffin Phantom for Measuring the Neutron Angular Distribution.* Report in progress (1995).
9. Briesmaster, J. F. (Ed.) *MCNP — A General Monte Carlo N-Particle Transport Code. Version 4A.* (Radiation Transport Group, Los Alamos National Laboratory) (1993).
10. Drake, P. *Experimental Estimation of Conversion Factors between Air Kerma Free-in-Air and Individual Dose Equivalent Penetrating for 4-7 MeV Photons.* Radiat. Prot. Dosim. **44**(1), 23–39 (1993).
11. Bradley, R. P. and Ryan, F. N. *Results for the 1990 EURADOS-CENDOS Irradiation Programme on Track Etch Detectors.* PTB Report PTB-N-10 (1991).
12. Luszik-Bhadra, M., Alberts, W. G., Dietz, E., Guldbakke, S. and Kluge, H. *A Simple Personal Dosemeter for Thermal, Intermediate and Fast Neutrons Based on CR-39 Etched Track Detectors.* Radiat. Prot. Dosim. **46**, 313–316 (1992).
13. Hollnagel, R. A. *Conversion Coefficients for Personal Dose Equivalent in the ICRU Slab.* Private communication (1993).
14. Stewart, R. D., Tanner, J. E. and Leonowich, J. A. *An Extended Tabulation of Effective Dose Equivalent from Neutrons Incident on a Male Anthropomorphic Phantom.* Health Phys. **65**, 405–413 (1993).
15. International Commission on Radiological Protection. *Data for Use in Protection against External Radiation.* Publication 51 (Oxford: Pergamon) (1987).

A DATABASE OF NEUTRON SPECTRA, INSTRUMENT RESPONSE FUNCTIONS, AND DOSIMETRIC CONVERSION FACTORS FOR RADIATION PROTECTION APPLICATIONS

O. F. Naismith† and B. R. L. Siebert‡
†CIRA, National Physical Laboratory
Teddington, Middlesex, TW11 0LW, UK
‡Physikalisch-Technische Bundesanstalt
Braunschweig, 38023, Germany

Abstract — One of the major problems encountered in dose assessment for neutron radiation protection derives from the imperfect dose equivalent response of the devices used for monitoring. To investigate the performance of such devices in realistic neutron fields and to optimise calibration procedures, knowledge of both the prevalent spectral fluences and the energy response of the dosemeters is required. To facilitate this and similar studies, a database has been developed comprising a catalogue of neutron spectra and energy-dependent response functions together with a software package to manipulate the data in the catalogue. The range of data, features of the programs, and examples for radiation protection applications are described.

INTRODUCTION

Routine monitoring for radiation protection purposes is performed using personal dosemeters or area survey instruments. For neutron radiation, the quantity for assessment is dose equivalent: personal dose equivalent, $H_p(10)$, for personnel measurements, and ambient dose equivalent, $H^*(10)$, for area monitoring[1] The dependence of the dosemeter response on energy and angle of incidence generally does not match that of the dose equivalent quantity which it is intended to measure. Therefore, if a dosemeter is used in a field with a different energy spectrum to the calibration field, the recorded dose equivalent may be significantly different from the true value.

If all workplace spectra were similar, a few measurements would suffice to determine correction factors for dosemeter readings. But the problem is exacerbated by the wide variety of different spectra encountered in working environments. In order to investigate the extent of the problem, as well as approaches to improving measurement accuracy, knowledge of both the prevalent spectral fluences and the energy response of the dosemeters are required. Accordingly a database of neutron spectra and energy-dependent response functions has been developed, together with software to process and analyse the data, which can be installed on a personal computer. In the following, the functions of the programs are outlined, using examples relevant to radiation protection applications to illustrate their use.

THE CATALOGUE

Neutron spectra

There are approximately 500 spectra in the catalogue at present. Almost half of these derive from existing compendia[2,3]; the remainder have been obtained as a result of a questionnaire sent to laboratories worldwide or extracted from the open literature, and some new measurements have also been performed. The data include measured and calculated neutron spectra in the workplace (e.g. nuclear power reactors, fuel reprocessing plants, etc.), calibration fields, and accelerator and reactor-based neutrons. The range of data available is summarised in Table 1.

The reliability of some of these spectra is debatable. Many of the spectra have been measured using Bonner spheres and the earlier data have been unfolded using response functions which have since been improved using modern Monte Carlo calculations. Even a very recent spectrometry intercomparison[4] showed a maximum deviation from the mean of 20% on the spectrum-folded ambient dose equivalent. The spectral data also give no indication of the angular dependency of the field.

Table 1. Summary of contents of catalogue of neutron spectra.

Source type	New	Old[2,3]
Source fabrication	4	
Fuel processing/storage	23	10
Fuel transport containers	12	
Pressurised water reactors	56	22
Boiling water reactors	22	1
Gas-cooled reactors	10	
Accelerators	65	78
Nuclear fusion research	4	5
Radionuclide sources	20	24
Cosmic rays	2	8
Research reactors	3	31
Criticality spectra		49
Monoenergetic on-phantom		40

Response functions

There is a separate library of response functions containing both detector energy responses and dosimetric conversion functions. The majority of the 150 data sets are detector energy responses, including personal dosemeters, area survey instruments and spectrometers. The contents are summarised in Table 2.

Nearly half the data come from an existing catalogue[2]. The new response functions added to the database are likely to provide a more accurate representation of the energy response than the earlier data. This point is especially valid in the case of Bonner sphere response functions which have been continuously refined in the light of intercomparisons and calculations. The library also contains a selection of dosimetric conversion factors, and an additional file of 21 conversion coefficients exists for direct access by the programs which can be modified as and when recommendations change. The data are from numerous sources, based on the recommendations of ICRP 21, ICRP 51, ICRU 39, and ICRP 60, and cover both the operational and protection (limiting) quantities: dose equivalent, effective dose, absorbed dose, radiation weighting factors, etc.

THE PROGRAM PACKAGE

The programs have been written using Microsoft® FORTRAN (version 5.1) and can be installed on any PC with a minimum of 580 kB memory and preferably a 486 processor or above. The package consists of ten programs which are inter-linked by various output files, and has been fully documented[5]. It is named SPKTBIB, an abbreviation for the German SPeKTren BIBliothek — library of spectra. The software has been designed to perform the following functions:

(i) aid installation of new data, and normalise spectra,
(ii) manage the data to facilitate access and retrieval,
(iii) plot spectra and response functions,
(iv) calculate expected readings of detectors by folding response functions with spectra,
(v) derive spectrum-averaged dosimetric quantities,
(vi) order spectra with respect to these quantities or dosemeter response,
(vii) investigate the influence of different bin structures on the values of spectrum-averaged quantities,
(viii) fold response functions over user-specified energy intervals (e.g. thermal, epithermal and fast),

Table 2. Summary of detector response functions in the SPKTBIB database.

Detector type	New	Old[2]
Personal dosemeters		
NTA film	1	1
PADC recoil track detector	14	4
Other recoil track detectors		4
Fission track detectors (e.g. ^{237}Np)	2	5
Albedo dosemeters		4
SHD/bubble detectors	2	
Rem meters	4	3
Long counter		1
Bonner sphere spectrometers	5	3

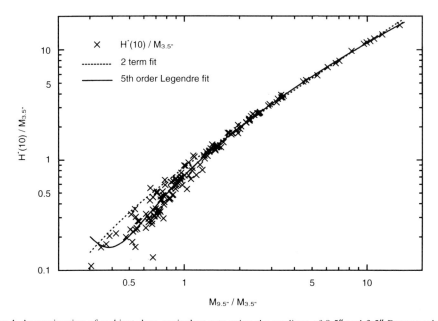

Figure 1. Approximation of ambient dose equivalent rate using the readings of 9.5″ and 3.5″ Bonner spheres.

(ix) combine dosemeter responses to obtain an improved estimate of the dose equivalent response,
(x) combine spectra to mimic another field (e.g. representing a 'realistic' field as a superimposition of calibration spectra).

These features constitute an invaluable tool for a range of applications in radiation protection and metrology, as illustrated in the following examples.

EXAMPLES

The most valuable property of the system is the abundance and diversity of data contained in the catalogue. This will facilitate any study which aspires to investigate characteristics encompassing the complete range of fields encountered throughout the neutron radiation protection environment. The system has already been exploited for a number of such applications including extensive investigations of dosemeter response in the types of spectra typically encountered in the workplace[6–8], and to examine the relationship between the protection and operational dosimetric quantities for neutrons in realistic spectra subsequent to the ICRP 60 recommendations[9]. The system has also been used to investigate methods of 'correcting' dosemeter response[6] using factors based on the type of source environment or a simple measurement in the radiation protection field itself, or by combining dosemeter responses to improve the estimate of dose equivalent.

Application of the 9″ to 3″ Bonner sphere ratio

The use of the ratio of the measured count rates of 9″ and 3″ diameter Bonner spheres to characterise the 'hardness' of a neutron spectrum is well known. The ratio is a monotonically increasing function of neutron energy spanning two orders of magnitude, and as such it may also prove useful in characterising a number of other quantities in dosimetry[10]. In this example, SPKTBIB has been used to fold the Bonner sphere response functions[11] with over 150 fission spectra in the database typical of those encountered in the workplace, and to study the relationship between the sphere ratio and various dosimetric quantities.

The ratio of Bonner spheres of various sizes was first examined to derive the best combination for the purposes stated above. The 9.5″ to 3.5″ ratio was found to provide the best compromise between maximising the range of the values and reducing the scatter on the data. In the spectra investigated the ratio is still a monotonic function of energy, however the range is reduced to about 50.

While the ratio itself may be used to derive information on spectrum hardness or the mean dose equivalent per unit fluence, a more important quantity for radiation protection applications is the dose equivalent rate at a particular location in a field. Either of the 9.5″ and 3.5″ spheres can be used individually to estimate the dose equivalent rate, by calculating a calibration factor using a least squares fit to H*(10)[12]. Neither sphere

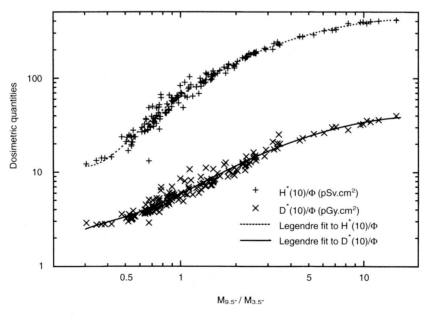

Figure 2. Approximation of the mean neutron fluence conversion factors for H*(10)/Φ and D*(10)/Φ using the readings of 9.5″ and 3.5″ Bonner spheres.

provides a sufficient approximation. However, it may be possible to 'correct' the count rate from one sphere using a function of the 9.5"/3.5" response. A simple linear relation is unsatisfactory. Various alternative expressions were tested for their fit to H*(10), and two in particular gave good results.

A function combining the 9.5" to 3.5" ratio and its square root was derived, using SPKTBIB to calculate the optimum weights for each term in the expression. Another approach is to use Legendre polynomials. This gives theoretically a better fit but is less simple to apply in practice. Both relationships have been plotted in Figure 1.

Some of the coefficients of the polynomial terms are negative. Therefore the propagated uncertainty on the sphere measurements may be relatively large. However, a more realistic estimate of the uncertainty on the calculated dose equivalent rate can be obtained from Figure 1: the fit is correct to within 50% for nearly all the spectra used, and the uncertainty is even lower at the higher end of the scale.

This technique can similarly be applied to other quantities, such as ambient dose, D*(10). The fit to the ambient dose equivalent and dose per unit fluence, using 5th order Legendre polynomials, is shown in Figure 2. These data can subsequently be used to derive the effective quality factor for a field.

A simple ratio of two Bonner sphere measurements can be used to determine some very important properties of a neutron field, which could in turn be used, for example, to determine the type of dosimetry most suited to a particular environment. For example etched track dosemeters for fields with high spectrum-averaged H*(10) or D*(10), and albedo dosemeters, or dosemeters with a thermal radiator, for fields with low mean values.

Simulation of a 'realistic' field as a superimposition of calibration spectra

If 'realistic' calibration fields are to be developed which mimic those found in workplaces in the nuclear industry, one simple technique would be to use a combination of readily available calibration sources. SPKTBIB can be used to optimise such a combination using a least squares fitting procedure on a system of linear equations. For a spectrum with I energy bins one obtains I equations of the form:

$$\Phi_{i,C} = \sum_{j=1}^{J} a_j \Phi_{ij}$$

where i denotes the energy bin, $\Phi_{i,C}$ is the group fluence in the ith bin of the combined calibration field, a_j are the weighting factors, and Φ_{ij} are the group fluences in the ith bin of the Jth calibration spectrum used.

A workplace spectrum can generally be deconvoluted into fast and thermal energy peaks and a 1/E intermediate energy component. The fitting procedure was used to derive a combination of bare and moderated radionuclide neutron source spectra[13] and a Maxwellian thermal energy distribution (e.g. the NPL thermal column) to simulate a field measured in the working environment of a neutron source fabrication facility[14].

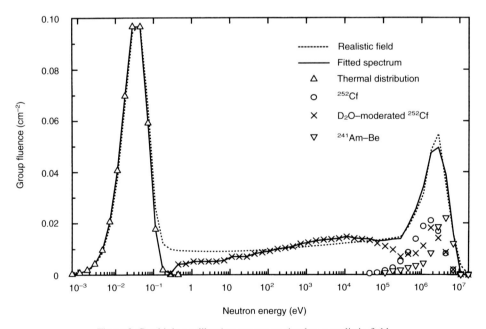

Figure 3. Combining calibration spectra to simulate a realistic field.

The original and fitted spectrum have been plotted in Figure 3.

In view of the success demonstrated in Figure 3, it may be conjectured that appropriate sets of weights a_j can be found for a wide range of 'realistic' calibration fields. (The assumption is made that the instrument response is additive, i.e. the response in the real neutron field can be obtained from a simple addition of the responses in the calibration spectra.) However, this technique may be inapplicable if the high energy peak of the realistic spectrum is in the keV region and accelerator-based fields may have to be developed instead.

These examples illustrate only some of the many features of the software package SPKTBIB. The system has been fully documented and is available on request from either author. (Information via EMail is available from Bernd.Siebert@PTB.DE)

ACKNOWLEDGEMENTS

This work was partly funded by the Commission of the European Communities under contract F13P-CT92 and was supported by EURODOS-CENDOS Working Group IV (Numerical Dosimetry). The authors would like to thank their colleagues who have contributed data to the catalogue and would be grateful to receive communication of any new measurements or calculations.

REFERENCES

1. ICRU. *Quantities and Units for Measurement and Calculation in Radiation Protection*. Report 51 (Bethesda MD: ICRU Publications) (1993).
2. Griffith, R. V., Palfalvi, J. and Madhvanath, U. *Compendium of Neutron Spectra and Detector Responses for Radiation Protection Purposes*. IAEA Technical Report Series No. 318 (Vienna: IAEA) (1990).
3. Ing, H. and Makra, S. *Compendium of Neutron Spectra in Criticality Accident Dosimetry*. IAEA Technical Report Series No. 180 (Vienna: IAEA) (1978).
4. Thomas, D. J., Chartier, J. -L., Klein, H., Naismith, O. F., Posny, F. and Taylor, G. C. *Results of a Large Scale Spectrometry and Dosimetry Comparison Exercise at the Cadarache Moderator Assembly*. Radiat. Prot. Dosim. **70**(1–4), 313–322 (This issue) (1997).
5. Naismith, O. F. and Siebert, B. R. L. *Manual for SPKTBIB: A PC-Based Catalogue of Neutron Spectra*. NPL Report CIRA(EXT) 005 (February 1996).
6. Naismith, O. F., Siebert, B. R. L. and Thomas, D. J. *Response of Neutron Dosmeters in Radiation Protection Environment: An Investigation of Techniques to Improve Estimates of Dose Equivalent*. Radiat. Prot. Dosim. **70**(1–4), 255–260 (This issue) (1997).
7. Burgkhardt, B., Fieg, G., Klett, A., Plewnia, A. and Siebert, B. R. L. *The Neutron Fluence and H*(10) Response of the New LB6411 Rem Counter*. Radiat. Prot. Dosim. **70**(1–4), 361–364 (This issue) (1997).
8. Alberts, W. G., Luszik-Bhadra, M. and Siebert, B. R. L. *Personal Neutron Dosimetry — Conversion Factors, Calibration and Dosimeter Performance*. In: Proc. 4th Conf. on Radiation Protection and Dosimetry, ORNL/TM-12817, pp. 165–174 (Oak Ridge) (1994).
9. Marshall, M., Thomas, D. J., Perks, C. A. and Naismith, O. F. *Radiation Quantities: Significance of the Angular and Energy Distribution of the Radiation Field*. Radiat. Prot. Dosim. **54**(3/4), 239–248 (1994).
10. Aroua, A., Grecescu, M., Pretre, S. and Valley, J.-F. *On the Use of Some Spectrum Quantifiers for Operational Neutron Fields*. Radiat. Prot. Dosim. **54**(2), 99–108 (1994).
11. Wiegel, B., Alevra, A. V. and Siebert, B. R. L. *Calculations of the Response Functions of Bonner Spheres with a Spherical ^{3}He Proportional Counter using a Realistic Detector Model*. Report PTB-N-21 (Braunschweig) (1994).
12. Wagner, S. R., Grosswendt, B., Harvey, J. R., Mill, A. J., Selbach, H. -J. and Siebert, B. R. L. *Unified Conversion Functions for the New ICRU Operational Radiation Protection Quantities*. Radiat. Prot. Dosim. **12**(2), 231–235 (1985).
13. ISO. *Neutron Reference Radiations for Calibrating Neutron Measuring Devices used for Radiation Protection Purposes and for Determining their Response as a Function of Neutron Energy*. ISO 8529 (Geneva; Switzerland: ISO) (1989).
14. Thomas, D. J., Waker, A. J., Hunt, J. B., Bardell, A. G. and More, B. R. *An Intercomparison of Neutron Field Dosimetry Systems*. Radiat. Prot. Dosim. **44**(1/4), 219–222 (1992).

NEUTRON FLUENCE MEASUREMENTS WITH A LIQUID SCINTILLATOR

P. J. Binns, J. H. Hough and B. R. S. Simpson
National Accelerator Centre
PO Box 72, Faure 7131, South Africa

Abstract — An NE213 liquid scintillator operating as a proton recoil spectrometer was utilised to determine the absolute fluence of monoenergetic neutrons in three quasi-monoenergetic fields. Events due to the neutrons of interest were selected by time-of-flight and a portion of the measured response function associated with protons, identified using pulse shape discrimination. Neutron fluence was evaluated by matching a theoretical detector response to the measured data in a region where only events from elastic scattering off hydrogen nuclei contribute to the pulse-height spectrum. The absolute fluence, normalised to unit beam charge, determined for neutrons of energies 25.5, 42.0 and 62.6 MeV was respectively 5.3 ($\pm$0.4), 5.4 ($\pm$0.4) and 5.9 ($\pm$0.5) cm^{-2}.nC^{-1}. The results were corroborated by estimates obtained from measurements of induced activity produced via the ^{12}C(n,2n)^{11}C reaction in the organic compound xylene.

INTRODUCTION

To achieve the desired accuracy in the specification of absorbed dose in neutron therapy, kerma factors for the principal elemental constituents of tissue are needed in the energy range 20–65 MeV[1,2]. Theoretical calculations at these energies are problematic as it is not possible to describe fully all reaction mechanisms in the lighter nuclei and new experimental data can provide integral checks for benchmarking the predictions of various nuclear models and computer codes. The neutron kerma factor is defined as the quotient of kerma and the fluence for neutrons of a particular energy[3] and accordingly this work reports on fluence evaluations of three quasi-monoenergetic neutron fields in which time-resolved kerma assessments have been performed[4]. The absolute fluence for neutrons of the *nominal* energy in each field was determined by analysing pulse-height data measured with an NE213 liquid scintillator. At energies below 20 MeV neutron fluence is readily obtained by accurately matching a theoretical response function to experimental data over the entire pulse height range. Above this energy the contribution of multiparticle break-up from neutron reactions on carbon increases, and nuclear model calculations cannot adequately reproduce the measured line shapes. The method can however be preserved if matching is restricted to a recoil energy region where only events from neutrons scattered elastically off hydrogen nuclei contribute to the pulse height spectrum[5]. The reliability of this evaluation was appraised with fluence estimates from the induced activity produced by the ^{12}C(n,2n)^{11}C reaction in a sample of the organic compound xylene. This evaluation has the advantage that it is unaffected by neutrons with energies below 20 MeV which is the threshold for the activation reaction studied. A comparison of results from the two methods is presented.

EXPERIMENTAL METHOD

Irradiations were undertaken at the neutron time-of-flight (TOF) facility of the National Accelerator Centre (NAC). Collimated neutron fields with nominal energies of 25.5, 42.0 and 62.6 MeV were generated by bombarding a 1.1 mm thick Be target with pulsed protons of incident energies 29.1, 45.0 and 65.3 MeV respectively. Protons of these energies lose only a small fraction of their kinetic energy in traversing the Be target and a dipole magnet positioned downstream from the target assembly sweeps the emergent protons away into a Faraday cup positioned at 15° with respect to the beam direction. Integral beam current readings from the Faraday cup served as the principal beam monitor throughout the experiment. Using electrostatic pulse selection, 4 out of every 5 pulses from the cyclotron were discarded and the time separation between successive beam bursts increased to 441, 375 and 305 ns for the 29, 45 and 65 MeV proton beams respectively.

Measurements were performed with a cylindrical (5 cm long and 5 cm diameter) NE213 liquid scintillator coupled to an RCA 8850 photomultiplier tube[6]. The detector was irradiated end on with the flat surface presented to the beam 6 m from the target. The intensity of the proton beam current on target varied between 20 and 50 nA. Pulse height information (L) was obtained from the dynode chain of the photomultiplier whilst the anode output was fed to a LINK 5010 pulse shape discriminator. This module was modified to provide a fast integral output (F), in addition to the standard outputs for fast timing and pulse shape discrimination (PSD). Each event was chronicled by its time-of-flight (T) relative to the cyclotron RF and was recorded together with L and F using multiparameter data acquisition in event-by-event mode. Corrections for time slewing applied during off-line analysis realised a FWHM of 1.3 ns for the prompt γ peak in the TOF spectrum, giving an intrinsic resolution of ±0.7 ns for the detection system. This corresponds to an energy resolution of ±1.5 MeV for neutrons with an energy of 62.6 MeV.

Induced activation studies were undertaken with a 20 ml sample of xylene $((CH_3)_2C_6H_4)$ sealed in a commercial counting vial. The sample of density 0.864 g.cm^{-3} was positioned in place of the NE213 detector and irradiated for approximately one hour with proton target currents of between 3 and 5 µA. The resulting number of ^{11}C nuclei ($T_{\frac{1}{2}}$ = 20.38 min) produced by the $^{12}C(n,2n)^{11}C$ reaction was determined by measuring the annihilation photons (0.511 MeV) with an HPGe detector. The photopeak efficiency for the measuring geometry was calibrated with a ^{22}Na source of known activity (±0.5%) prepared from a locally standardised solution traceable to the Bureau International des Poids et Mesures, Paris, France. Activity measurements from the HPGe detector were confirmed by $4\pi\beta$-γ liquid scintillation coincidence counting[7] of concurrently activated samples.

The $^{12}C \rightarrow ^{11}C$ transformation probability is considered negligible and the number of ^{12}C nuclei N_c in the sample was assumed constant throughout the irradiation. The number of 0.511 MeV annihilation photons N counted during time t_c is expressed as

$$N = \eta n_e e^{-\lambda t_w} (1 - e^{-\lambda t_c}) \quad (1)$$

where η is the efficiency of the counting system, n_e the number of radioactive nuclei at the end of the irradiation period t_i, λ the decay constant and t_w the time interval between the end of bombardment and the start of the counting period. At an average rate R of ^{11}C formation in the sample, n_e is given by the relationship

$$n_e = (R/\lambda)(1 - e^{-\lambda t_i})$$

Substituting into Equation 1 gives

$$N = \eta(R/\lambda)(1 - e^{-\lambda t_i}) e^{-\lambda t_w} (1 - e^{-\lambda t_c})$$

and

$$R = \frac{N\lambda e^{\lambda t_w}}{\eta(1 - e^{-\lambda t_i})(1 - e^{-\lambda t_c})} = \varphi \overline{\sigma} N_c$$

where φ is the neutron flux density at the irradiation position and $\overline{\sigma}$ the mean cross section for the $^{12}C(n,2n)^{11}C$ reaction.

RESULTS

Neutron TOF spectra corrected for time slewing are shown in Figure 1 for the three quasi-monoenergetic neutron fields. In this representation the time scale is reversed. The distributions are dominated by neutrons emitted by transitions to the ground state of the residual nucleus and each spectrum exhibits a lower energy tail. This continuum arises from both multibody break-up reactions and excitations in the residual nucleus. The electronic PSD was set to allow a small gamma breakthrough and the measured time interval between the

Table 1. The spectral neutron fluence ϕ deduced from the ^{11}C production rate R measured in the activation study. The fluence ϕ'_E (Table 2) corresponding to neutrons of the nominal energy E is given by the product of ϕ and the spectral peak to total ratio.

E (MeV)	R (s^{-1})	$\overline{\sigma}$ ($\times 10^{-27}$ cm^2)	ϕ (>20 MeV) (cm^{-2}.nC^{-1})	Peak to total ratio
25.5	91	10.6	6.1	0.96
42.0	433	27.1	7.1	0.74
62.6	1046	27.9	10.4	0.51

Table 2. Comparison of the integrated neutron fluence normalised to unit beam charge evaluated from measurements in three quasi-monoenergetic fields using an NE213 liquid scintillator (ϕ_E) and activation analysis (ϕ'_E).

E (MeV)	Gate width (MeV)	ϕ_E (cm^{-2}.nC^{-1})	ϕ'_E (cm^{-2}.nC^{-1})
25.5	7.3	5.3 ± 0.4	5.9 ± 0.6
42.0	9.3	5.4 ± 0.4	5.3 ± 0.5
62.6	11.3	5.9 ± 0.5	5.3 ± 0.5

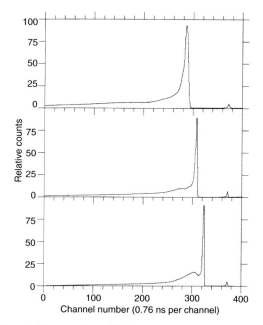

Figure 1. Neutron time-of-flight spectra measured with an NE213 liquid scintillator in three quasi-monoenergetic fields of nominal energies 25.5 MeV (upper panel), 42.0 MeV (middle panel) and 62.6 MeV (lower panel). In this representation the time scale is reversed.

prompt gamma peak on the right and the principal neutron peak is consistent with the relevant neutron energies.

Data were replayed with software gates set to embrace the peak of the nominal neutron energy for each TOF spectrum. During off-line analysis the analogue processing within the LINK module was emulated digitally by computing a pulse shape parameter S for each event based upon the L and F parameters[8]. Figure 2 shows a density plot of pulse height correlated with pulse shape for gated events with a flight time T falling within the principal peak of the TOF spectrum for the p(65) generated field. Separate ridges associated with the different charged recoil particles in the scintillator are observed and those due to protons, deuterons and alpha particles identified. A locus formed by escape protons that exit the detector is also apparent.

The portion of the detector response due to protons arrested inside the scintillator was isolated by placing a two-dimensional window around the appropriate locus on the L vs S display. Projections onto the pulse height axis were then compared with responses calculated using SCINFUL[9], a well established Monte Carlo code. The detector response for neutrons elastically scattered off hydrogen nuclei was modelled and the neutron fluence determined by scaling the calculated response to the measured pulse height data in the region above the threshold for the $^{12}C(n,p)^{12}B$ reaction, 12.6 MeV below the maximum proton recoil energy. The two respective pulse height distributions for 62.6 MeV monoenergetic neutrons are depicted in Figure 3. The calculated response exhibits an abrupt cut-off at large pulse heights as it has not been smoothed for detector and photomultiplier resolution.

In the activation method, the required mean cross section $\bar{\sigma}$ for the production of ^{11}C nuclei in each field was calculated from the above TOF distributions which were converted to energy fluence spectra[10] and then weighted with the appropriate excitation function[9]. Values for R together with $\bar{\sigma}$ and the spectral fluence ϕ determined from the induced activation above 20 MeV are presented in Table 1. The fraction of the total measured activity attributed to neutrons of the nominal energy is given in the last column where the gate widths were those stated in Table 2.

Neutron fluences evaluated from the two methods are compared in Table 2. Fluences were normalised to the beam charge and integrated over gate widths that were chosen to embrace the principal peak of each TOF distribution. Overall uncertainties of 8 and 10% were respectively adopted for the liquid scintillator[11] and activation results. In both instances the uncertainty associated with the relevant reaction cross section was the main contributor to the quoted errors.

CONCLUSION

An NE213 liquid scintillator operating as a proton

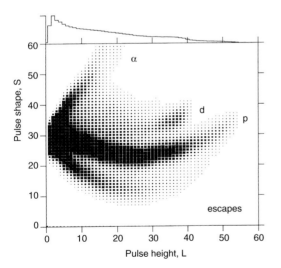

Figure 2. Density plot of pulse height correlated with pulse shape obtained with an NE213 liquid scintillator for monoenergetic neutrons (E = 62.6 MeV). Separate ridges associated with alpha particles (α), deuterons (d), protons (p) and escape protons (escapes) are identified. A projection of all events onto the pulse height axis (top) is also given.

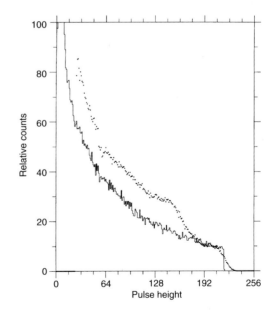

Figure 3. Experimental (···) and predicted (—) pulse height distributions for the proton portion of the scintillator response to 62.6 MeV neutrons. The measured distribution contains all events on the proton locus in Figure 2 including those from the $^{12}C(n,p)^{12}B$ reaction whilst the prediction comprises only those from n-p scattering off hydrogen nuclei. The spectra were normalised in the pulse height region above the threshold for the $^{12}C(n,p)^{12}B$ reaction.

recoil spectrometer was utilised to determine the fluence of monoenergetic neutrons in three well-defined fields. The portion of the measured response function attributed to proton secondaries for each energy was identified by pulse shape discrimination. The absolute fluence, normalised to unit beam charge for neutrons of energies 25.5, 42.0 and 62.6 MeV was respectively 5.3 (±0.4), 5.4 (±0.4) and 5.9 (±0.5) cm^{-2}.nC^{-1}. These results were corroborated by an independent study using activation analysis.

REFERENCES

1. Mijnheer, B. J., Battermann, J. J. and Wambersie, A. *What Degree of Accuracy is Required and can be Achieved in Photon and Neutron Therapy?* Radiother. Oncol. **8**, 237–252 (1987).
2. White, R. M., Broerse, J. J., DeLuca, P. M. Jr., Dietze, G., Haight, R. C., Kawashima, K., Menzel H. G., Olsson, N. and Wambersie, A. *Status of Nuclear Data for Use in Neutron Therapy.* Radiat. Prot. Dosim. **44**, 11–20 (1992).
3. ICRU. *Neutron Fluence, Neutron Spectra and Kerma.* Report 13 (Bethesda MD: ICRU Publications) pp. 25–26 (1969).
4. Binns, P. J. and Hough, J. H. *Kerma Associated with High Energy Neutrons.* Radiat. Prot. Dosim. **52**(1–4), 105–109 (1994).
5. Northcliffe, L. C., Lewis, C. W. and Saylor, D. P. *A Two-parameter Method for the Measurement of Neutron Energy Spectra.* Nucl. Instrum. Methods **83**, 93–100 (1970).
6. Binns, P. J. and Hough, J. H. *Spectral Energy Measurements in a Fast Neutron Therapy Field.* Nucl. Instrum. Methods A**255**, 330–333 (1987).
7. Simpson, B. R. S. and Meyer, B. R. *A Multiple-channel 2- and 3-fold Coincidence Counting System for Radioactivity Standardization.* Nucl. Instrum. Methods A**263**, 436–440 (1988).
8. Smit, F. D. and Brooks, F. D. *Angular Distribution of Neutrons from $^2H(\gamma,n)^1H$ at $E\gamma = 2.75$ MeV.* Nucl. Phys. A**465**, 429–444 (1987).
9. Dickens, J. K. *SCINFUL: A Monte Carlo Based Computer Program to Determine a Scintillator Full Energy Response to Neutron Detection for E_n between 0.1 and 80 MeV: User's Manual and FORTRAN Program Listing.* Oak Ridge National Laboratory Report ORNL-6462 (1988).
10. Jones, D. T. L., Symons, J. E., Fulcher, T. J., Brooks, F. D., Nchodu, M. R., Allie, M. S., Buffler, A. and Oliver, M. J. *Neutron Fluence and Kerma Spectra of a p(66)/Be(40) Clinical Source.* Med. Phys. **19**(5), 1285–1291 (1992).
11. Nolte, R., Schuhmacher, H., Brede, H. J. and Schrewe, U. J. *Measurement of High Energy Neutron Fluence with Scintillation Detector and Proton Recoil Telescope.* Radiat. Prot. Dosim. **44**(1–4), 101–104 (1992).

NEUTRON AND PHOTON SPECTRA AND DOSE RATES AROUND A SHIELDING CASK PLACED IN A SALT MINE TO SIMULATE A NUCLEAR WASTE PACKAGE

K. Knauf†, A.V. Alevra†, J. Wittstock†, H. J. Engelmann‡, M. Khamis‡ and N. Niehues‡
†Physikalisch-Technische Bundesanstalt
Postfach 3345, D-38023 Braunschweig, Germany
‡Deutsche Gesellschaft zum Bau und Betrieb von Endlagern für Abfallstoffe mbH (DBE)
Postfach 1169, D-31201 Peine, Germany

Abstract — The dose equivalent rate around a POLLUX storage container with radioactive waste is increased by neutrons and photons backscattered by the host rock when the waste package is transported to its final position in a repository. This was simulated by a smaller shielding cask with a ^{252}Cf line source. Neutron spectral fluence and dose equivalent rates were measured with Bonner spheres above ground and underground and compared with MCNP calculations. There may be the same quantity of backscattered neutrons as of the unscattered component. The photon spectra were determined with an NE213 scintillation spectrometer above ground. Photon dose equivalent rates are derived and compared with calculated ones and dosemeter readings.

INTRODUCTION

The neutron and photon dose equivalent rates around a container with high-level radioactive waste are increased when the container is transported in an underground repository. Backscattering of neutrons and photons from the surrounding host rock enhances the fluence and thus the dose equivalent rates around the cask.

To forecast the exposure of the personnel in a repository to radiation, the calculations generally carried out have to account for the backscattering effect.

To validate the calculations, an 'Active Handling Experiment with Neutron Sources' (AHE) was performed in a salt mine using a model of the POLLUX transport and storage container reduced in size. The model, a shielding cask, was loaded with a line-shaped ^{252}Cf neutron source in its centre line. Source strength and shielding material were chosen with a view to simulate the radiation field of a POLLUX[1].

The spectral neutron fluences and dose equivalent rates were measured with the shielding cask situated above ground and underground in a transport drift and in a smaller drift specified for emplacement. In all three cases, the axis of the cask was positioned 1.5 m above the floor, and it was deposited at the end of the emplacement drift. The latter situation is encountered in practice when a POLLUX is lowered for final disposition.

The neutron spectrometer used throughout the experiment was a Bonner sphere spectrometer suitable for simultaneous measurements using four detectors[2]. Above ground a liquid scintillation spectrometer (NE213) was additionally applied chiefly to obtain the photon spectra[3]. Also, spherical proton recoil proportional counters were used to achieve a high energy resolution in the range from 15 keV to 1400 keV[3]. Commercial dosemeters (Leake dosemeter for neutrons, TEPC for neutrons and photons, Geiger-Müller counter and ionisation chamber for photons) completed the experimental programme.

At the end, the spectrometric and dosimetric results were compared with the theoretical forecasts obtained by Monte Carlo simulations (MCNP code) of the shielding cask[4,5]. This report focuses on results obtained with Bonner spheres, liquid scintillation counter, Leake dosemeter and photon dosemeters above ground and in the emplacement drift.

EXPERIMENTAL SITUATION

The cask was 91.2 cm in diameter and 135 cm in length, its mass was 6200 kg. The ^{252}Cf line source (60 cm) with a neutron source strength of 1.7×10^8 s^{-1} was surrounded by a stainless steel cylinder which in turn was contained in a polyethylene cylinder. This body was placed in a nodular cast iron container. The cask was positioned with its axis horizontal to the ground.

The detector locations were chosen in the axial and lateral directions (Figure 1). It should be mentioned that the lid and the bottom differed in thickness, therefore measurements were performed at both sides of the cask.

The Bonner sphere spectrometer consisted of spherical proportional counters filled with ^{3}He, which could be fitted into the centres of 11 different polyethylene spheres with diameters from 7.72 cm (3″) to 38.1 cm (15″). The ^{3}He counters were also used bare and shielded with 1 mm cadmium.

The NE213 detector consisted of an aluminium capsule (5.08 cm in diameter and 5.08 cm in length) filled with NE213 scintillator liquid and coupled to a fast photomultiplier tube. The spectrometer was capable of separating neutron- and gamma-induced pulses by pulse shape analysis.

RESULTS

The spectral neutron fluences derived from a Bonner sphere measurement at 2 m distance from the lid of the shielding cask above ground and underground are compared in Figure 2. The MCNP-calculated spectra are also plotted.

The shapes of the calculated and measured spectra are similar, however the calculated shapes are slightly 'softer'. The peak at 24 keV is caused by a dip in the cross-section of the iron, the major shielding material. The peak at 1.5 keV is due to resonance scattering from the sodium nuclei of the salt.

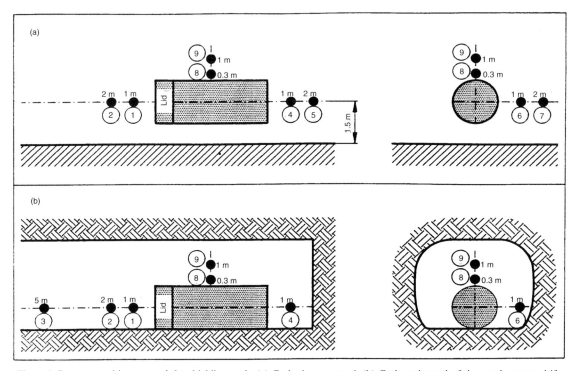

Figure 1. Detector positions around the shielding cask. (a) Cask above ground. (b) Cask at the end of the emplacement drift.

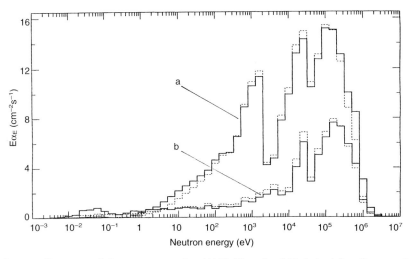

Figure 2. Spectral neutron fluence rate (lethargy representation $d\phi/d(\ln E)$ against $\ln E$) derived from Bonner sphere measurements (solid line) and calculated by MCNP (dotted line) in detector position 2; A, above ground; B, underground.

The measured integral fluences in Figure 2 differ by a factor of 2.75 due to the above-mentioned backscattering, the calculated integral fluence ratio is 2.58. The ratios of the dose equivalent rates (ICRP 21) are 2.09 (measured) and 1.86 (calculated).

The calculated integral fluences deviate from the measured ones by −8% to +20%. The calculated values on the bottom side of the cask are generally slightly higher than those on the opposite side.

Applying the fluence-to-dose-equivalent conversion coefficients (ICRP 21) to the measured and calculated spectral fluences for all positions yields the dose equivalent rates in Table 1. The measured dose equivalent enhancement rises up to a factor of two while the calculation forecast is somewhat less. However, the differences are less than 20% for all detector positions.

Also included in Table 1 are measurements with a Leake counter calibrated using a D_2O-moderated californium source. The enhancement they show is higher as the result of the overestimation of the soft spectrum of the scattered neutrons. Calibration in the moderated field of a californium source is important to obtain comparable results. Calibrations in the field of bare californium or Am-Be sources result in overestimations in these soft neutron fields by up to 100%[2].

The photons emitted from the Cf source were well shielded but the neutrons produced photons in the shielding material and the environment. These photons were measured with the NE213 spectrometer. The spectral fluence unfolded from the photon-induced pulses is plotted for position 8 in Figure 3 together with the spectrum calculated by the MCNP code. The peak at 2.2 MeV is caused by the neutron capture in the hydrogen of the organic scintillator. The peaks at 6 and 7.6 MeV are due to thermal neutron capture in the ^{56}Fe of the cask.

Table 1. Comparison between measured and calculated neutron dose equivalent rates (H_{MADE}) in $\mu Sv.h^{-1}$.

Position	Bonner spheres			Calculation (MCNP)			Leake dosemeter*		
	Above ground	Under- ground	Ratio under/above ground	Above ground	Under- ground	Ratio under/above ground	Above ground	Under- ground	Ratio under/above ground
1	26.0	40.6	1.56	24.9	37.0	1.49	29.5	54.0	1.83
2	8.8	18.4	2.09	8.5	15.8	1.86	10.9	26.2	2.40
3	—	6.4	—	—	5.4	—	—	10.1	—
4	47.4	79.3	1.67	57.4	83.5	1.45	57.4	94.5	1.65
5	16.5	—	—	18.4	—	—	18.4	—	—
6	23.4	50.0	2.14	24.8	45.1	1.82	28.1	62.4	2.22
7	9.4	—	—	9.8	—	—	11.5	—	—
8	67.3	87.0	1.29	67.3	83.4	1.24	81.1	103.4	1.28
9	25.5	41.1	1.61	23.9	38.3	1.60	27.8	51.8	1.86

*Calibrated using a D_2O moderated ^{252}Cf source.

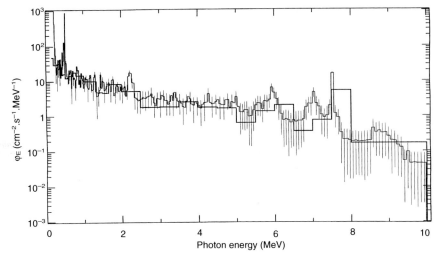

Figure 3. Spectral fluence rate of the photons above ground in detector position 8, unfolded (histogram with uncertainties) and calculated (broad histogram) by the MCNP code.

The integral dose equivalent rate derived from this spectrum was 1.2 μSv.h^{-1} (calculated 1.1 μSv.h^{-1}) which is only 1.8% of the dose equivalent rate of the neutrons. The reading of the photon dosemeters was 1.5 μSv.h^{-1} (ionisation chamber), 1.8 μSv.h^{-1} (TEPC) and 2.4 μSv.h^{-1} (GM counter). The background (<0.2 μSv.h^{-1}) was subtracted.

The high reading of the dosemeters may be due to their neutron sensitivity (all dosemeters) and to an oversensitivity to high-energy photons in the case of the GM counter.

CONCLUSIONS

Scattered neutrons from the surrounding salt make a considerable contribution to the dose equivalent rate even in the vicinity (e.g. at 2 m distance) of a cask with high-level radioactive waste[2]. It may reach the same amount as that of the direct component. From the point of view of radiation protection, this has to be considered when the exposure of personnel handling POLLUX casks in normal operation or in remedial action in a repository is planned.

A total of 43 measurements were made with the Bonner sphere spectrometer[2] and compared with MCNP calculations. The quality of the calculations can be estimated by the fact that the integral fluences predicted deviate from those derived from the measurements by −11% and +20% at most, the mean deviation being +5%. Since the uncertainty of the integral fluence derived from the Bonner sphere data does not exceed 5%, the larger differences should therefore be assigned to deficiencies in the calculations.

On an average the calculated dose equivalent rates are, however, somewhat lower than the measured ones. This is chiefly due to deviations in the shape at the high energy end of the spectra. The calculations predict the measured dose equivalents within ±21%. The uncertainties of the experimental dose equivalent values are less than 15%.

Neutron dosemeters should be calibrated in fields with a spectral characteristic similar to that of the fields encountered[2].

The measured photon dose equivalent rate amounted to only 0.5% to 2% of the measured neutron dose equivalent rate (ICRP 21) depending on the detector position[3]. Nevertheless, the reading of the dosemeter which proved to be best (ionisation chamber) was too high by 75% at most and may be explained by the neutron sensitivity of the instrument (yet unknown). It should thus be possible to measure reliable photon dose equivalent rates around a POLLUX with its higher photon contribution.

ACKNOWLEDGEMENTS

Financial support by the Bundesministerium für Bildung, Wissenschaft, Forschung und Technologie (BMBF contract No 02E84727) and the European Commission (EC contracts No FI2W-CT90–0069 and No. FI3P-920002) is much appreciated. The authors wish to thank K. D. Closs (Forschungszentrum Karlsruhe), the coordinator of the AHE project, for his interest in the work and W. Bernnat from the Universität Stuttgart (KE) and H. Klein from the PTB for valuable discussions. They also thank W. Hobach from the PTB for technical support.

REFERENCES

1. *Active Handling Experiment with Neutron Sources (AHE)*. Final Report, CEC Contract No. FI2W-CT90–0069, A Joint Project of DBE and ANDRA (1995).
2. Alevra, A. V., Knauf, K. and Wittstock, J. *Measurements with PTB Bonner Sphere Spectrometer and Various Dosemeters Around a Model Storage Cask Filled with a ^{252}Cf Source both Free in Air and in a Salt Mine*. PTB Laboratory Report 7.22–95-1 (Braunschweig, FRG: PTB) (1995).
3. Knauf, K., Heimann, C., Novotny, T. and Wittstock, J. *Measurements with a Liquid Scintillation Counter and Proton Recoil Proportional Counters around a Shielding Cask with a ^{252}Cf Line Source — A Contribution to the AHE-Project*. PTB Laboratory Report (to be published).
4. Bernnat, W., Mattes, M. and Pfister, G. *Bestimmung von Neutronen- und Gammaspektren sowie der Dosis in der Umgebung des AHE-Versuchsbehälters auf dem PTB-Gelände*. KE6-FB-73 (Universität Stuttgart, Forschungsinstitut für Kerntechnik und Energiewandlung e.V.) (1995).
5. Bernnat, W. and Mattes, M. *Bestimmung von Neutronen- und Gammaspektren sowie der Dosis in der Umgebung des AHE-Versuchsbehälters innerhalb des Salzbergwerks ASSE*, KE6-FB-74 (Universität Stuttgart, Forschungsinstitut für Kerntechnik und Energiewandlung e.V.) (1995).

RESPONSE OF NEUTRON DOSEMETERS IN RADIATION PROTECTION ENVIRONMENTS: AN INVESTIGATION OF TECHNIQUES TO IMPROVE ESTIMATES OF DOSE EQUIVALENT

O. F. Naismith[†], B. R. L. Siebert[‡] and D. J. Thomas[†]
[†]CIRA, National Physical Laboratory
Teddington, Middlesex, TW11 0LW, UK
[‡]Physikalisch-Technische Bundesanstalt
Braunschweig, 38033 Germany

Abstract — The response of practicable neutron dosemeters for routine use generally does not match the conversion function from fluence for radiation protection quantities such as the ambient dose equivalent. As a consequence, significant errors may be encountered when monitoring in a neutron energy spectrum different from that in which the dosemeter was calibrated, which is almost inevitably the case. A database of neutron energy spectra, detector response functions, and dosimetric conversion factors has been developed, and has been used to investigate the extent of this problem. The paper examines various ways of improving dosemeter response by 'ranking' spectra and deriving correction factors based upon this ordering. In the case of area monitoring, a combination of two responses (e.g. a rem meter and TEPC) may serve to improve the measurement of dose equivalent.

INTRODUCTION

A fundamental problem exists in the calibration of neutron dosemeters which arises from a discrepancy between the energy response of the device and the measurement quantity. Dosemeters for neutron radiation monitoring are designed to measure dose equivalent: personal dose equivalent, $H_p(10)$, for personnel dosimetry and ambient dose equivalent, $H^*(10)$, for area monitoring[1]. In order to measure dose equivalent accurately in any field, the energy response of the dosemeter must match the appropriate fluence-to-dose equivalent conversion function. Unfortunately this is not the case. Area survey instruments have a reasonable dose equivalent response to fast and thermal neutrons but in the intermediate energy region they have a tendency to over-respond. For personal dosemeters various detection mechanisms exist with high sensitivities in the thermal or fast energy region, but problems are encountered at intermediate energies where the capture cross section tends to become insignificant and where the energy of the ions is too low to produce visible tracks. As a consequence, if the energy spectrum of the calibration source is different from the spectrum of the field in which the device is to be used there will be an error in the dosemeter response.

There are two approaches to minimising this problem: one is to develop calibration fields which are more representative of the spectra encountered in the workplace. Chartier *et al*[2] investigate this possibility. However, such a diverse range of fields are encountered that a number of different 'realistic' calibration fields would be necessary as well as a means of determining the most appropriate for each situation. The second approach

Table 1. Responses of four types of personal dosemeter to a variety of workplace spectra after calibration in three ISO-recommended fields.

Spectrum group	Mean energy	$H_p(10)$ (pSv.cm^2)	PADC (with radiator)			PADC (no radiator)			NTA (inc. thermal)			NTA (fast only)		
			Am-Be	^{252}Cf	^{252}Cf (D$_2$O)	Am-Be	^{252}Cf	^{252}Cf (D$_2$O)	Am-Be	^{252}Cf	^{252}Cf (D$_2$O)	Am-Be	^{252}Cf	^{252}Cf (D$_2$O)
GCR	21 keV	31	0.65	0.55	0.65	0.32	0.27	0.33	3.76	5.45	6.51	0.05	0.07	0.09
PWR (low H)	26 keV	33	0.62	0.53	0.63	0.30	0.25	0.30	5.93	8.60	10.3	0.05	0.08	0.09
BWR (low H)	27 keV	31	0.70	0.59	0.71	0.41	0.34	0.41	7.38	10.7	12.8	0.08	0.12	0.14
BWR (med. H)	76 keV	61	0.79	0.67	0.80	0.65	0.53	0.65	1.50	2.17	2.60	0.16	0.22	0.27
PWR (med. H)	110 keV	75	0.87	0.74	0.88	0.74	0.61	0.74	0.91	1.32	1.58	0.20	0.28	0.34
Transport cask	290 keV	180	0.95	0.81	0.96	0.88	0.73	0.89	0.43	0.62	0.74	0.21	0.31	0.37
PWR (high H)	350 keV	160	0.98	0.83	0.99	0.89	0.74	0.90	0.52	0.75	0.90	0.27	0.39	0.46
BWR (high H)	360 keV	190	1.05	0.89	1.06	0.97	0.80	0.98	0.28	0.40	0.48	0.27	0.39	0.46
Reprocessing	1.0 MeV	320	1.17	0.99	1.18	1.15	0.95	1.16	0.64	0.93	1.11	0.51	0.74	0.88
Source fabric'n	1.0 MeV	280	1.19	1.01	1.20	1.18	0.97	1.19	0.90	1.30	1.56	0.60	0.86	1.03

involves applying a correction factor to the dosemeter reading, which is calculated using the known dosemeter energy response and some method of categorising the neutron field.

A database of neutron spectra and detector and dosimetric energy responses has been developed as part of a project to improve calibration methods for neutron dosimetry. The database is accompanied by software to process and analyse the data and is described elsewhere[3,4]. Because of the large number and diversity of spectra in the catalogue, the system is a very powerful tool for examining the problems associated with dosemeter response in a wide range of situations encountered in radiation protection practice and for investigating corrective methods.

DOSEMETER RESPONSE

The database was used to investigate dosemeter response in typical workplace fields after calibration using three ISO-recommended radionuclide sources[5]: ^{241}Am-Be, ^{252}Cf and D_2O-moderated ^{252}Cf. (Response in this context is thus being defined as the quotient of the dosemeter indication by the conventional true value of the particular dose equivalent quantity it is designed to measure, multiplied by the calibration factor for a particular calibration field.) Calculations were performed for four types of personal dosemeter and two rem meters. Two etched track plastic (PADC) responses were used, one incorporating a thermal radiator and the other without, and two NTA film responses, one

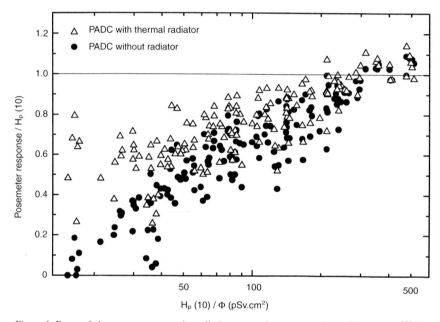

Figure 1. Personal dosemeter response in radiation protection spectra after calibration in ^{252}Cf.

Table 2. Responses of two rem meters to workplace spectra after calibration in ISO-recommended fields.

Spectrum group	$H^*(10)$ (pSv.cm^2)	Harwell 0949			Studsvik		
		Am-Be	^{252}Cf	^{252}Cf (D_2O)	Am-Be	^{252}Cf	^{252}Cf (D_2O)
GCR	19	2.95	2.52	1.87	1.12	1.08	1.01
PWR (low H)	21	2.83	2.41	1.80	1.13	1.09	1.02
BWR (low H)	20	2.72	2.32	1.73	1.16	1.12	1.05
BWR (med. H)	38	2.26	1.93	1.44	1.08	1.04	0.98
PWR (med. H)	50	1.96	1.68	1.25	1.00	0.96	0.90
Transport cask	110	1.52	1.30	0.97	0.92	0.88	0.82
PWR (high H)	100	1.50	1.28	0.96	0.92	0.89	0.83
BWR (high H)	120	1.49	1.27	0.94	0.92	0.89	0.83
Reprocessing	220	1.31	1.12	0.83	1.00	0.96	0.90
Source fabric'n	190	1.27	1.08	0.80	1.00	0.96	0.90

responding only to fast neutrons, the other including a thermal response.

The dosemeter response functions were folded with over 150 spectra from the database, which were divided into ten groups according to source type. The pressurised water (PWR) and boiling water reactor (BWR) fields were each sub-divided into three groups, distinguished by the amount of shielding present between the reactor core and measurement location: spectra with over 2 m of concrete were classified as having a low mean dose equivalent per unit fluence (H); less than 50 cm shielding constituted a 'high H' field. The spectra were also folded with the appropriate dose equivalent quantity in order to compare dosemeter response with the measurement quantity. The under- or over-response of the dosemeter in each field was calculated as:

$$R_H = \left(\frac{I_r}{H_r}\right)\left(\frac{H_c}{I_c}\right)$$

where I_r is the indication in the field, I_c the indication in the calibration field, and H_r and H_c the respective spectrum-averaged dose equivalents. The results are presented in Tables 1 and 2.

The responses, calculated as above using the ^{252}Cf spectrum for 'calibration', have been plotted in Figure 1 for the two PADC dosemeters.

CORRECTION METHODS

Measurement environment

In addition to the dosemeters under-responding in practically every type of field, there is a large amount of scatter in the spectrum-folded data. This must be addressed when considering techniques to improve the response. One possible method is to apply a correction factor based on the type of environment in which the dosemeter is likely to be used. Using the spectrum groups from before, correction factors were calculated by taking the reciprocal of the data from the appropriate columns in Table 1 for personal dosemeters and in Table 2 for area survey meters. The responses of the two ^{252}Cf-calibrated PADC dosemeters corrected in this manner have been plotted in Figure 2.

Classifying spectra using detector response ratios

Correcting dosemeter response using a factor based on the monitoring environment is a relatively straightforward technique. However, although Figure 2 shows an improvement compared with Figure 1, there is still a fairly wide spread in the spectrum-folded data. An alternative method would be to classify the spectra using a simple measurement in the field, on which a correction factor would be based. For example, the relationship between dosemeter response (as a ratio to the spectrum-averaged dose equivalent) and the ratio of Harwell and Studsvik area survey meter responses has been plotted in Figure 3. This particular combination has the advantage of using instruments readily available to health physicists. By fitting a second degree polynomial to the two sets of data a correction function has been derived for each dosemeter.

The spectrum-folded responses of the two PADC

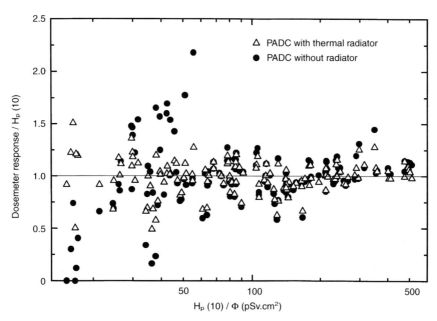

Figure 2. Personal dosemeter response, corrected for measurement environment.

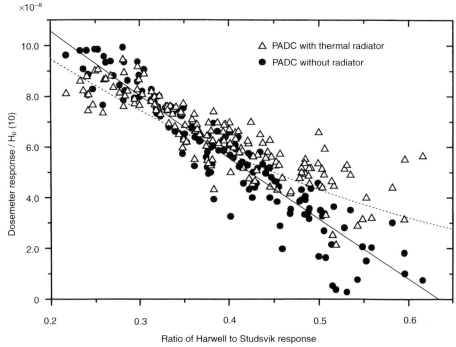

Figure 3. Relationship between dosemeter response and Harwell to Studsvik rem meter response ratio.

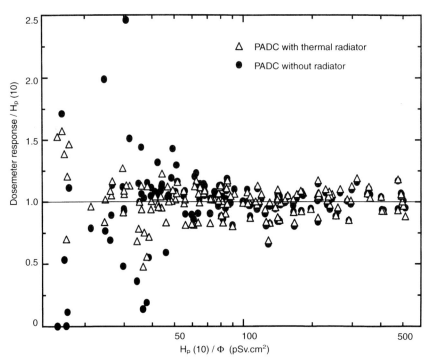

Figure 4. Personal dosemeter response corrected using the Harwell to Studsvik ratio.

dosemeters have been multiplied by factors calculated using these two functions, and the corrected data are plotted in Figure 4. Most of the data now fall within ±20% of the true dose equivalent, compared with a spread of approximately 50% on the data corrected using a factor based on source type alone. This method is, however, limited at present by the accuracy with which the rem meter response functions are known. A more reliable method might be to use a ratio of Bonner sphere responses.

Calibrating dosemeters in the radiation protection field

Each of the techniques described above relies on an initial calibration in an ISO-recommended field. Dosemeters can alternatively be calibrated in the field in which they are to be used if some method exists to determine the dose equivalent rate. A ratio of Bonner sphere responses (e.g. 9.5″/3.5″) can be used for such a purpose, to convert the reading of one of the two spheres to a better approximation of the dose equivalent rate. This technique has been explored in full elsewhere[4].

Area monitoring

So far the paper has concentrated on personal dosimetry. However, the data in Table 2 indicate that the situation for area monitoring is also not ideal. The problem here originates from the over-response of rem meters at intermediate neutron energies. In contrast, an instrument such as a tissue-equivalent proportional counter (TEPC) has an energy response which underestimates dose equivalent in this region. Thus some combination of the two responses might provide an improved dose equivalent response. The software associated with the database was used to optimise a linear combination of the two detector responses by applying a least squares fit to the series of equations

$$H_i^*/\Phi_i = (a_1 R_{i1} + a_2 R_{i2})/\Phi_i$$

where a_j are the weighting factors assigned to each detector, H_i^*/Φ_i is the fluence to ambient dose equivalent conversion factor for the ith energy spectrum, and R_{ij}/Φ_i is the fluence response for the jth detector folded with spectrum i. The combined response is shown in Figure 5, together with the weighted responses of the two instruments. This combination reproduced the spectrum-averaged ambient dose equivalent to within 10% in all cases except for those spectra with a very low mean H*(10). However, even here the dose equivalent response was an improvement over that of either single detector.

ACKNOWLEDGEMENTS

This work was partly funded by the CEC under con-

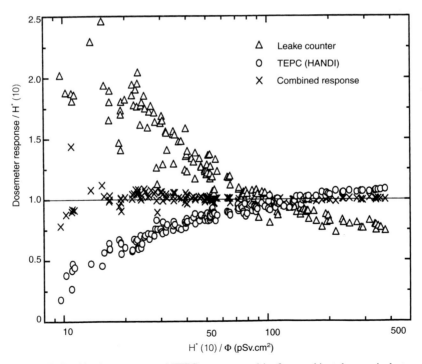

Figure 5. Combined rem meter and TEPC response, and its fit to ambient dose equivalent.

tract F13P-CT92. The authors would like to thank David Bartlett and Rick Tanner at NRPB and Chris Perks at AEA Harwell for their help in providing dosemeter response function data.

REFERENCES

1. ICRU. *Quantities and Units for Measurement and Calculation in Radiation Protection.* Report 51 (Bethesda MD: ICRU Publications) (1993).
2. Chartier, J. L., Posny, F. and Buxerolle, M. *Experimental Assembly for the Simulation of Realistic Neutron Spectra.* Radiat. Prot. Dosim. **44**(1/4), 125–130 (1992).
3. Naismith, O. F. and Siebert, B. R. L. *Manual for SPKTBIB: A PC-Based Catalogue of Neutron Spectra.* NPL Report CIRA(EXT)005, (February 1996).
4. Naismith, O. F. and Siebert, B. R. L. *A Database of Neutron Spectra, Instrument Response Functions, and Dosimetric Conversion Factors for Radiation Protection Applications.* Radiat. Prot. Dosim. **70**(1–4), 241–245, (This issue) (1997).
5. ISO. *Neutron Reference Radiations for Calibrating Neutron Measuring Devices used for Radiation Protection Purposes and for Determining their Response as a Function of Neutron Energy.* ISO 8529 (Geneva, Switzerland: ISO) (1989).
6. Hollnagel, R. A. *Conversion Functions of the Dose Equivalent $H_{sl}(10)$ in the ICRU Slab used for the Calibration of Personal Neutron Dosemeters.* Radiat. Prot. Dosim. **54**(3/4), 227–230 (1994).
7. Wagner, S. R., Grosswendt, B., Harvey, J. R., Mill, A. J., Selbach, H.-J. and Siebert, B. R. L. *Unified Conversion Functions for the New ICRU Operational Radiation Protection Quantities.* Radiat. Prot. Dosim. **12**(2), 231–235 (1985).

DIRECTIONAL INFORMATION ON NEUTRON FIELDS

M. Matzke, H. Kluge and M. Luszik-Bhadra
Physikalisch-Technische Bundesanstalt
Bundesallee 100, D-38116 Braunschweig, Germany

Abstract — An unfolding method is described for determining the directional fluence of neutron fields from the readings of six personal dosemeters mounted on a polyethylene sphere. The method can be applied to every detector system with energy- and angle-dependent response functions, provided a suitable group representation can be found. Results obtained with CR-39 etched track dosemeters in a reactor environment are discussed.

INTRODUCTION

For the estimation of the effective dose produced by neutrons, not only the spectral fluence of the neutron field but also its directional distribution must be known. There is no ideal measuring instrument available for that purpose. The detectors customarily used in neutron dosimetry for the determination of ambient dose equivalent or for spectrometry have a more or less isotropic response. Personal dosemeters mounted on a phantom are better suited, since their response, due to the shielding of the neutrons by the phantom, shows a directional dependence. But since the directional distribution of neutron fields occurring in practice changes with the neutron energy, a dosemeter with spectrometric properties is required which indicates the neutron direction.

The CR-39 etched track personal dosemeter recently developed at the PTB[1] allows a rough spectrometry in three adjacent neutron energy intervals. In each of these intervals, the dose (averaged over the energy distribution in the interval) can be determined separately. It can be assumed that the angular response of these dosemeters does not vary too much in a single interval. Within the uncertainties acceptable for dose determinations, these dosemeters, mounted at different positions on the surface of a spherical phantom, are therefore suitable for approximately determining the directional distribution in each of the intervals.

A general method is described for the evaluation of the directional distribution of neutron fields from measurements with angle-dependent readings. An unfolding algorithm is described, and results for the example of CR-39 track dosemeters irradiated inside a reactor containment are discussed.

FORMALISM

In the following, the example of track detectors is discussed. The method, however, is generally valid and might be applied to every detector system with energy- and angle-dependent response functions. The only requirement is that the energy-dependent response function can be subdivided into a number of energy group responses and that the angular response does not change within such a group.

Six CR-39 track dosemeters mounted on the surface of a 30 cm diameter polyethylene sphere are considered. The coordinate system is chosen in such a way that the origin is positioned in the centre of the sphere, the z axis crosses the detectors mounted at the bottom and at the top, and the other detectors are crossed by the x axis and the y axis, respectively (see Figure 1). The orientation of each detector i is characterised by its normal vector $\mathbf{u}_i = (x_i, y_i, z_i)$, pointing to the centre of the sphere. For each $\mathbf{u}_i$ two of the three vector components are zero.

For the detectors considered here, cylindrical symmetry can be assumed, i.e. the response to incident neutrons of the direction $\mathbf{\Omega} = (\cos\varphi \sin\vartheta, \sin\varphi \sin\vartheta, \cos\vartheta)$ depends only on $\cos\Theta_i = \mathbf{u}_i \cdot \mathbf{\Omega} = x_i \cos\varphi \sin\vartheta + y_i \sin\varphi \sin\vartheta + z_i \cos\vartheta$. It is further assumed that the energy range can be subdivided into three energy groups and that an averaged dose response can be used in each group[1]. With these assumptions and the approximation that the angular response does not vary within such an energy group, the reading M_i of a detector i (calibrated in terms of dose equivalent H* in a unidirectional field with perpendicular incidence) can be represented in this group by the 'model equations':

$$M_i = h_\Phi^* \int d^2\Omega \, R(\mathbf{u}_i, \mathbf{\Omega}) \, \Phi_\Omega(\mathbf{\Omega}) \tag{1}$$

where h_Φ^* is the average fluence to dose equivalent conversion factor taken from Wagner *et al*[2], $\Phi_\Omega(\mathbf{\Omega})$ is the directional fluence (averaged over the energy distribution) in the energy interval considered and $R(\mathbf{u}_i, \mathbf{\Omega})$ is the angle dependent dose equivalent response of the detector i in this interval. The response is taken as normalised to 1 for perpendicular incidence ($\Theta_i = 0$) of neutrons: $R(\cos\Theta_i = 1) = 1$. For isotropic incidence of neutrons, it follows from Equation 1: $M_i = H* \, R_{iso}$ with

$$H* = h_\Phi^* \int d^2\Omega \, \Phi_\Omega(\mathbf{\Omega}) = h_\Phi^* \, \Phi \text{ and}$$

$$R_{iso} = \frac{1}{4\pi} \int d^2\Omega \, R(\mathbf{u}_i, \mathbf{\Omega})$$

$$= \frac{1}{2} \int_{-1}^{+1} d\cos\Theta_i \, R(\cos\Theta_i)$$

SIMPLE MODEL FOR THE DIRECTIONAL DISTRIBUTION

As a simple model for many practical applications it can be assumed that the directional fluence to be measured is a superposition of a monodirectional and an isotropic part:

$$\Phi_\Omega(\Omega) = p \frac{\Phi}{4\pi} + (1-p) \Phi \delta(\Omega - \Omega_0) \quad (2)$$

where p denotes the relative fraction of the isotropic contribution and the delta function selects the incident direction Ω_0. Instead of Equation 1 the readings can be written as

$$M_i = H* p R_{iso} + H*(1-p) R(\mathbf{u}_i \Omega_0) \quad (3)$$

Equation 3 constitutes a system of six equations for the four unknown values $H*$, p, $\cos\vartheta_0$, φ_0, which may be solved by the (non-linear) least squares method minimising the expression:

$$\chi^2 = \sum_i \frac{(M_{0i} - M_i)^2}{\sigma_i^2} \quad (4)$$

where M_{0i} are the measured values with the standard deviations σ_i and the M_i are the readings to be calculated according to Equation 3.

This simple model has been used in a previous paper[3] and led to reliable results in several well-known fields produced by a ^{252}Cf source. The ambient dose equivalent calculated from the dosemeter readings agreed with the reference values within 10% to 20%, and the direction of the monoenergetic component was determined for ϑ_0 and φ_0 within 15°. In addition, the relative isotropic contribution and the fast neutron dose contribution was found to be in satisfactory agreement with the reference values.

IMPROVEMENT: DETERMINATION OF THE DIRECTIONAL DISTRIBUTION BY THE MONTE CARLO METHOD

Instead of parameterising the neutron field according to Equation 2 by the simple model, it can be attempted directly to determine the directional distribution of the fluence by standard methods as used in neutron spectrum unfolding.

Using a suitable group representation for $\Phi_\Omega(\Omega)$ with angular elements $\Delta^2\Omega_k$ (k = 1,...,N) in the energy group considered, the integral of Equation 1 can be replaced by a sum:

$$M_i = h*_\Phi \sum_{k=1}^{N} \Delta^2\Omega_k R(\mathbf{u}_i \Omega_k) \Phi_\Omega(\Omega_k) \quad (5)$$

Since a value of N of the order of 1000 must be used for a suitable group representation valid for distributions occurring in practice, the linear least squares method minimising a χ^2 expression according to Equation 4 with M_i from Equation 5 cannot be applied. The system is considerably underdetermined. However, setting $f_k = h*_\Phi \Phi_\Omega(\Omega_k)\Delta^2\Omega_k$ and considering $\chi^2(\mathbf{f})$ as a function of the N unknown elements $\mathbf{f} = (f_1,...f_k,...f_N)$, the method

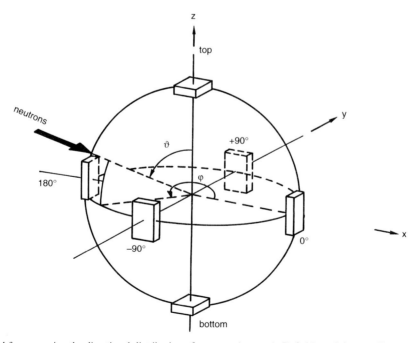

Figure 1. Method for measuring the directional distribution of neutrons (see text). Definition of the coordinate system and angles.

of maximum entropy can be used to construct a multivariate probability density for $\mathbf{f}$: $P(\mathbf{f}) = \text{const} \times \exp(-\beta \chi^2(\mathbf{f}))^{(4,5)}$ with $P(\mathbf{f}) = 0$, if one $f_k < 0$. A unique, most probable value of $\mathbf{f}$ might not exist, but it is possible to calculate its expectation value $\langle \mathbf{f} \rangle$ and the covariance matrix of $\mathbf{f}$ from the probability density. The parameter β can be determined from the constraint that the expectation value of χ^2 should be equal to the number of degrees of freedom involved, which is equal to the number 6 of measured readings. The Monte Carlo code MIEKE[6] used in neutron spectrometry was modified to calculate the expectation values $\langle f_k \rangle$ and their uncertainties, which are given by the covariance matrix $\langle f_k f_m \rangle - \langle f_k \rangle \langle f_m \rangle$ (see Reference 6). This new program allows not only the expectation values $h_{\Phi}^* \langle \Phi_\Omega(\Omega_k) \Delta^2 \Omega_k \rangle$ but also interesting integrals like the ambient dose H* or the current through a fixed plane to be calculated. Also, values relevant for personal dosimetry like the (averaged) directional dose or the effective dose can be determined from the integrals:

$$h_{\Phi}^* \int d^2\Omega \, W(\Omega) \, \Phi_\Omega(\Omega)$$

where $W(\Omega)$ is the corresponding weighting function.

The Monte Carlo program based on the algorithm of the MIEKE code is rather time-consuming for large numbers N of groups. The code recently developed by Weise[7] (SPECAN program with the essential unfolding subroutine UNFANA) contains an analytical approach to this Monte Carlo unfolding; it is much faster than the MIEKE code. This code was included in the new program in its present version. The new program allows all relevant quantities for N = 1600 angle groups to be calculated in a few minutes.

Information on the directional distribution is required for workplaces. As an example, measurements were performed in the reactor containment of the Ringhals 4 reactor in Sweden. The position of the sphere with the six dosemeters was at the entrance level inside the containment, denoted by position A in a previous report on intercomparison measurements[8]. The readings of Table 1 are the results of a new measurement with a slightly changed dosemeter design[1] compared to that of the previous intercomparison[8]. In the table the read-

Table 1. Experimentally obtained dosemeter readings for thermal (M_{th}), intermediate (M_{Int}) and fast neutrons (M_{Fast}) for the irradiations at position A of the Ringhals reactor and unfolded results with the Monte Carlo code MIEKE[6]. The intermediate energy range is between 50 meV and 70 keV[3]. ϑ_{MAX}, φ_{MAX}: values for maximum of the distribution.

Position of the dosemeter	M_{th} (μSv.h^{-1})	M_{Int} (μSv.h^{-1})	M_{Fast} (μSv.h^{-1})
0°	62 ± 40	499 ± 50	1119 ± 112
+90°	54 ± 40	315 ± 40	165 ± 40
−90°	81 ± 40	360 ± 40	269 ± 40
top	96 ± 40	404 ± 40	515 ± 52
bottom	57 ± 40	231 ± 40	46 ± 40
180°	46 ± 40	469 ± 47	162 ± 40
H*	583 ± 91	1167 ± 53	1565 ± 103
ϑ_{MAX}	14°	70°	70°
φ_{MAX}	−52°	−6°	−6°
Isotropic fraction	94%	88%	19%

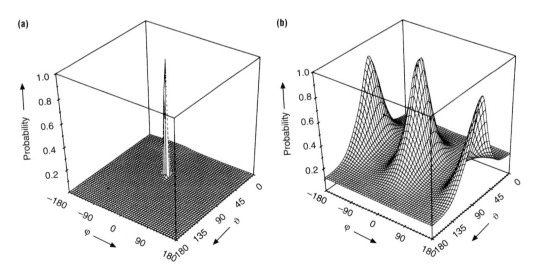

Figure 2. Directional distribution (as seen by an observer in the centre of the sphere) obtained by the SPECAN (UNFANA)[7] code for the irradiations performed at the Ringhals reactor. (a) fast neutrons; (b) intermediate neutrons. Coordinate ranges: $0 \leq \vartheta \leq 180°$, $-180° \leq \varphi \leq 180°$.

ings of the dosemeters experimentally obtained for the thermal, the intermediate and the fast neutron energy group are given. It is seen that the intermediate part is more isotropic than the fast one. The readings were first analysed by the MIEKE code, leading to the results given in the last rows of the table. The uncertainties quoted are obtained from the elements $\langle f_k f_m \rangle - \langle f_k \rangle \langle f_m \rangle$ of the covariance matrix. The angles for the maximum values of the distribution entering the sphere as seen by an observer in the centre of the sphere are also given. The integral $\int d^2\Omega\, \Omega\, \Phi_\Omega(\Omega)$ corresponds to the time-integrated current density and vanishes for isotropic fields. Its absolute value divided by the fluence is between 0 and 1 and can be defined as the relative non-isotropic fraction of the distribution (see Table 1). The whole distribution of the fluence entering the sphere as seen by an observer in the centre of the sphere is shown in Figure 2. It was evaluated by the SPECAN (UNFANA) code. Again the more isotropic behaviour of the intermediate neutrons is seen.

CONCLUDING REMARKS

A method has been developed to measure the directional distribution of neutrons in different energy intervals. The method can be applied to every detector system with energy- and angle-dependent response functions, provided that a suitable energy group representation can be found. The directional distribution in each energy group can be calculated for a number of up to 1600 angular groups. Using the method described, measurements have already been performed in various neutron fields at reactors, accelerators and fuel containers, and will be described in a forthcoming paper.

ACKNOWLEDGEMENTS

The authors thank Prof. Dr K. Weise for providing the SPECAN (UNFANA) code and Dr P. Drake for performing the irradiations at Ringhals 4.

REFERENCES

1. Luszik-Bhadra, M., Alberts, W. G., d'Errico, F., Dietz, E., Guldbakke, S. and Matzke, M. *A CR-39 Track Dosemeter for Routine Individual Neutron Monitoring.* Radiat. Prot. Dosim. **55**, 285–293 (1994).
2. Wagner, S. R., Grosswendt, B., Harvey, J. R., Mill, A. J., Selbach, H.-J. and Siebert, B. R. L. *Unified Conversion Functions for the New ICRU Operational Radiation Protection Quantities.* Radiat. Prot. Dosim. **12**, 231–235 (1985).
3. Luszik-Bhadra, M., Kluge, H. and Matzke, M. *Measurement of the Directional Distribution of Neutrons with Personal Neutron Dosemeters.* Radiat. Prot. Dosim. **66**, 335–338 (1996).
4. Weise, K. and Matzke, M. *A Priori Distributions from the Principle of Maximum Entropy for the Monte Carlo Unfolding of Particle Energy Spectra.* Nucl. Instrum. Methods **A280**, 324–330 (1989).
5. Matzke, M. *Unfolding and Radiation Damage Prediction with and without a-priori Information.* In: Reactor Dosimetry: Methods, Applications, and Standardization. ASTM STP 1001 (American Society for Testing and Materials, Philadelphia), pp. 425–433 (1989).
6. Matzke, M. *Unfolding of Pulse Height Spectra: The HEPRO Program System.* (Physikalisch-Technische Bundesanstalt, Braunschweig) Report PTB-N-19 (1994).
7. Weise, K. *Mathematical Foundation of an Analytical Approach to Bayesian-Statistical Monte Carlo Spectrum Unfolding.* (Physikalisch-Technische Bundesanstalt, Braunschweig) Report PTB-N-24 (1995).
8. Klein, H. and Lindborg, L. (eds) *Determination of the Neutron and Photon Dose Equivalent at Work Places in Nuclear Facilities of Sweden* (Swedish Radiation Protection Institute, Stockholm) Report SSI-rapport 95-15 (1995).

APPLICATION OF NEUTRON SPECTROMETRY IN THE DRPS NEUTRON DOSIMETRY SERVICE

K. Barlow[†], M. Jackson[†], A. French[†] and J. R. Harvey[‡]
[†]Defence Radiological Protection Service (DRPS), Institute of Naval Medicine
Alverstoke, Hampshire PO12 2DL, UK
[‡]Dosimetry Consultant, Dursley, Gloucestershire, GL11 4LW, UK

Abstract — A method for improving the accuracy of dose assessments undertaken using CR-39 etched track dosemeters is described. This method compensates for the limited energy and angular response of the dosemeter. This work was undertaken following a review of operational neutron fields which indicated that, without this correction, the neutron dose could be underestimated by up to a factor of at least 2. These correction factors are derived from a detailed knowledge of the neutron fields in which the dosemeter is worn and calibrated, together with the energy and angular response of the dosemeter. The occupational location correction factors enable a more accurate assessment of neutron dose to be obtained for both AP and ROT fields, using the dose quantity effective dose equivalent H_E.

INTRODUCTION

The DRPS planar neutron dosemeter consists of two CR-39 elements side by side. These elements are chemically etched prior to assessment on an automatic reader[1]. The DRPS neutron dosimetry service is approved by the UK Health and Safety Executive (HSE) and currently issues dosemeters to both Armed Forces and civilian personnel who are exposed to a wide range of neutron sources and associated neutron fields. Examples of these include ^{241}Am-Be and ^{252}Cf sealed sources, linear accelerators and pressurised water reactors in nuclear submarines.

The dosemeter is calibrated using the field from an ^{241}Am-Be source which is usually different from the neutron field in which the dosemeter is worn occupationally. This will give rise to an error in the dose assessment stemming from the imperfect energy and angular response of the dosemeter. To improve the accuracy of the dose assessment location correction factors (LCFs) have been evaluated for both anterior-posterior (AP) and rotational (ROT) fields, in terms of the effective dose equivalent H_E.

METHOD

LCFs are derived from knowledge of the neutron fields in which the dosemeter is worn and calibrated, together with data related to the energy and angular response of the dosemeter.

The calibration and operational neutron fields have been characterised using a transportable neutron spectrometer (TNS)[2], although, in some instances, published spectra have been used for some of the occupational fields[3]. The TNS uses six detectors to measure neutrons from thermal energies up to 10 MeV. Two BF_3 detectors, one covered with a cadmium sheath, are used to measure thermal and epithermal neutrons. Three spherical proportional proton recoil detectors, containing hydrogen at different pressures, detect neutrons with energies ranging from 30 keV to 1.4 MeV and a hydrogenous liquid scintillator is used to detect neutrons in the range 1–10 MeV. Analysis of the data provides a differential energy spectrum for 42 energy groups, over three decades of energy, in addition to thermal and epithermal fluence rates.

The energy and angular dependence of the dosemeter has been determined as part of the type testing requirements for the HSE approval of the neutron dosimetry system. The energy response of the dosemeter has been assessed from 100 keV to 66 MeV and the energy response, relative to ambient dose equivalent $H^*(10)$[4,5], is shown in Figure 1. The angular response of the dosemeter has been assessed, over the energy range 144 keV to 15.1 MeV, for angles of incidence up to 85°[5]. This is shown in Figure 2.

Energy correction factors

Energy correction factors (E), given in Table 1, are used in the calculation of the LCF to correct for the energy response of the dosemeter. Energy correction factors are calculated by converting the energy response of the dosemeter relative to $H^*(10)$, as shown in Figure 1, to an energy response relative to $H_E(AP)$.

$$E = E_{H^*(10)} \frac{H_E(AP)/\phi}{H^*(10)/\phi} \quad (1)$$

where $E_{H^*(10)}$ is the energy response relative to $H^*(10)$, $H_E(AP)/\phi$ is the effective dose equivalent per unit fluence and $H^*(10)/\phi$ is the ambient dose equivalent per unit fluence.

Fractional angular response

The angular response data for the dosemeter, shown in Figure 2, has been extrapolated to obtain an estimate of the angular response of the dosemeter at 7.5°, 115° and 157.5°. The fractional angular response (F) data used in the calculation of the LCF are dependent on the field geometry and as a consequence, F for an AP field differs significantly from that for a ROT or ISO field. Equations 2 to 4 have been derived to enable the calculation of F for AP, ROT and ISO fields. Equation 2 indicates that the value of F which should be used in AP neutron radiation fields is unity.

$$F_{(AP)} = 1.0 \quad (2)$$

$$F_{(ROT)} = \frac{\frac{1}{\pi} \int_0^\pi (R_\alpha^*/R_0^*) \, d\alpha}{H_E(ROT)/H_E(AP)} \quad (3)$$

$$F_{(ISO)} = \frac{\frac{1}{2} \int_0^\pi (R_\alpha^*/R_0^*) \sin\alpha \, d\alpha}{H_E(ISO)/H_E(AP)} \quad (4)$$

where R_α^* is the $H^*(10)$ angular response for radiation incident at α degrees.

Table 2 details the experimental and extrapolated data for seven neutron energies together with the fractional angular response for the rotational irradiation geometries, $F_{(ROT)}$, evaluated using Equation 3. The value of F for isotropic radiation geometry could not be evaluated since values of $H_E(ISO)$ were not available.

CALCULATION OF THE LOCATION CORRECTION FACTOR

The neutron spectra obtained from the TNS measurements are produced in terms of fluence rates for 44 energy bands. Each fluence rate spectrum is converted to a neutron dose equivalent rate spectrum using effective dose equivalent per unit fluence conversion factors for the mean neutron energy[6]. The neutron dose rates

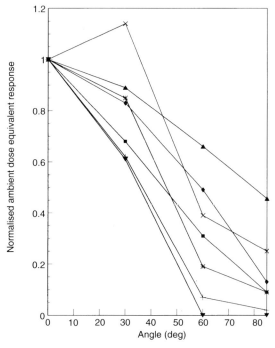

Figure 2. Normalised ambient dose equivalent response of the planar neutron dosemeter as a function of angle of incidence. (▼) 144 keV, (+) 240 keV), (∗) 565 keV, (■) 1.2 MeV, (×) 2.1 MeV, (◆) 5.3 MeV, (▲) 15.1 MeV.

Table 1. Energy correction factors for the DRPS planar dosemeter.

Neutron energy (MeV)	Energy correction factor (E)
0.151	0.20 ± 0.05
0.250	0.78 ± 0.01
0.565	0.95 ± 0.04
1.2	1.04 ± 0.01
2.1	0.66 ± 0.10
5.3	0.65 ± 0.16
14.0	0.46 ± 0.12

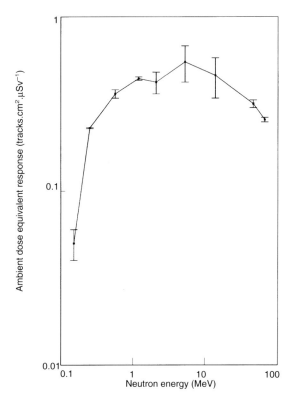

Figure 1. Ambient dose equivalent response of the planar neutron dosemeter as a function of energy.

for each of the 44 energy points are then normalised and consolidated into seven larger energy bands, where the logarithmic mid point of each energy band represents each energy for which the energy correction factor and the fractional angular response has been determined. Neutron spectra obtained from published data are also consolidated into the same energy bands.

The location correction factor for each operational neutron field can then be found by combining the normalised dose for each energy band, the energy correction factor and the fractional angular response, as shown in Equation 5:

$$LCF = \frac{\sum_{i=1}^{i=7} (P_i)(E_i) \text{ (calibration field)}}{\sum_{i=1}^{i=7} (P_i)(E_i)(F_i) \text{ (occupational field)}} \quad (5)$$

where P_i is the fraction of $H_E(AP)$ dose in the ith energy band, E_i is the energy correction factor in the ith energy band and F_i is the fractional angular response for the ith energy band. The neutron dose can then be calculated by multiplying the assessed dose by the LCF.

The value of F is dependent on the field geometry and the LCF for a ROT field is consequently different from that for an AP field.

RESULTS

The values of the location correction factor derived for a range of individual neutron fields are given in Table 3. The errors from the TNS neutron spectra measurements, from the energy and angular response of the planar dosemeter, and those from the conversion coefficients are combined to determine an absolute error for the LCF. The standard deviation of the LCFs given in Table 2 is 20%.

For the AP fields the LCF varies between 0.90 and 2.12 whereas for ROT fields the LCF varies between 1.32 and 3.42. The ROT values are all greater than the AP values due to the fall-off in the angular response of the dosemeter with the increasing angle of the incident neutron radiation.

The final stage of the LCF derivation is to establish whether the operational neutron field is AP, ROT, or ISO or a combination of the three. This is currently being assessed by undertaking various surveys using isotropic and planar dosemeters on a PMMA slab phantom, or by considering the working practices and environment of personnel wearing the dosemeter.

DISCUSSION

The DRPS has considered using a multi-element isotropic CR-39 dosemeter which has an improved energy and angular response[1]. The isotropic CR-39 dosemeter, however, contains a minimum of three elements which

Table 3. Values of the LCF for various AP and ROT neutron fields.

Location number	LCF-AP	LCF-ROT
1	1.13	1.90
2	1.42	2.34
3	1.23	2.05
4	1.07	1.82
5	0.87	1.54
6	1.24	2.01
7	1.38	2.33
8	1.91	3.22
9	2.12	3.42
10	1.27	2.14
11	0.96	1.46
12	0.90	1.32
13	1.00	1.93

Table 2. Angular response of the DRPS planar neutron dosemeter.

Mean angle (degrees)	7.5	30	60	80	115	157.5	
Range (degrees)	0–15	15–45	45–75	75–85	85–135	135–180	
Neutron energy (MeV)			Angular response normalised to 1 at 0°				$F_{(ROT)}$
0.144	0.90*	0.61	0	0	0*	0*	0.418
0.250	0.90	0.62	0.07	0.02	0*	0*	0.465
0.565	0.96*	0.85	0.19	0.09	0*	0*	0.663
1.2	0.92*	0.68	0.31	0.09	0*	0*	0.576
2.1	1.03*	1.14	0.39	0.25	0.08*	0*	0.727
5.3	0.96*	0.83	0.49	0.13	0*	0*	0.461
15.1	0.97*	0.89	0.66	0.45	0.19*	0*	0.594

*Estimated data from Figure 2.

would have serious implications on the operating costs and efficiency of the DRPS dosimetry service. Isotropic CR-39 dosemeters could be useful for small dosimetry services which do not have access to neutron spectrometry systems.

CONCLUSIONS

The DRPS data for several operational neutron fields indicates that corrections must be made for CR-39 planar etched track dosemeters. If these correction factors are not applied then the dose assessment could give rise, in the case of rotational fields, to an underestimate of the neutron dose by a factor of at least two. Having developed the methodology for calculating location correction factors their calculation simply relies on gaining knowledge of the operational neutron spectra. The DRPS has used a transportable neutron spectrometer to determine the spectra although other methods, for example Bonner sphere measurements and bubble spectrometers, could also be used.

ACKNOWLEDGEMENTS

The authors would wish to thank AEA Winfrith for assistance provided in calibrating the TNS detectors. The authors also wish to thank Physikalisch Technische Bundesanstalt, Germany and the GSF-Munchen, Institut für Strahlenschutz, Neuherberg, Germany, for energy and angular calibrations performed and sponsored by Working Group 8 of EURADOS. Additionally the authors would like to thank the National Physical Laboratory, the Defence Radiological Standards Centre (DRaSTaC), AWE, Aldermaston, UK, and Berkeley Technology Centre, Berkeley, UK, for other irradiations provided.

REFERENCES

1. Harvey, J. R., French, A. P., Jackson, M., Renouf, M. C. and Weeks, A. R. *An Automated Neutron Dosimetry System Based on the Chemical Etch of CR-39*. Radiat. Prot. Dosim. **70**(1-4), 149–152 (This issue) (1997).
2. Armishaw, M. J., Curl, I. J. and McCracken, A. K. *The Development of a Transportable Neutron Spectrometer TNS2, (November 1987)*. UKAEA Report RP&SG/IJC/P(87)93 (United Kingdom Atomic Energy Authority (now AEA Technology) Winfrith Technology Centre, Dorchester, Dorset, DT2 9DH, UK) (1987).
3. IAEA. *Compendium of Neutron Spectra and Detector Responses for Radiation Protection Purposes*. STI/DOC/10/318 (Vienna: IAEA) (1990).
4. Jackson, M., Harvey, J. R. and Weeks A. R. *Low Energy Neutron Response — Planar and Isotropic Neutron Dosemeters*. CR Neutron Dosimetry Report 9/94 (Defence Radiological Protection Service, Institute of Naval Medicine, Alverstoke, Gosport, Hants, PO12 2DL, UK) (1994).
5. Weeks, A. R., Ford, T. D. and Harvey, J. R. *An Assessment of Two Types of Personal Neutron Dosemeter which utilise the AUTOSCAN 60 Autoscan Etch-Track Reading System*. UK Nuclear Electric, Technology Division Report TEPZ/REP/0084/93 (1993).
6. ICRU. *Determination of Dose Equivalents Resulting from External Radiation Sources — Part 2*. Report 43 (Bethesda, MD: ICRU Publications) (1988).

A NEW APPROACH TO LOW LEVEL MONITORING IN MIXED RADIATION FIELDS

S. Pszona
Soltan Institute for Nuclear Studies
05–400 Otwock-Swierk, Poland

Abstract — The new method for measuring ambient dose equivalent in mixed neutron–gamma fields has been proposed. It has been shown that the moderator technique, up to now used for neutron monitoring only, can be adjusted for monitoring both gamma and neutron radiation. The fluence response of a device consisting of a ^{3}He proportional counter and 203 mm diameter polythene sphere has been evaluated and compared with the ambient dose equivalent conversion factors for photons.

INTRODUCTION

A thermal neutron detector at the centre of a polythene moderator is the most widespread technique used for ambient dose equivalent measurements in neutron fields. The substantial improvement of this technique is reported to be achieved in two ways, namely by optimising the moderator response by doping with special materials[1], and by using a high pressure ^{3}He proportional counter and active spectrometry[2] for signal analysis. During the experiments with the ^{3}He proportional counters, it has been observed that gamma radiation which is registered at the lower part of pulse-height spectrum is well separated from the pulse-height spectrum caused by thermal neutrons. These observations are illustrated on the Figures 1 to 2 in this paper.

The separation of the two components, i.e. gamma and neutrons seen in the pulse-height spectrum, justifies asking the question 'Can a moderator type instrument based on use of a selected ^{3}He proportional counter, be adjusted for determination of ambient dose equivalent for photons and neutrons in a single measuring run?' For this purpose a study of the photon responses of a ^{3}He counter in a polythene moderator of 203 mm diameter has been undertaken to reveal the possible advantages as well as the limitations of the method.

MATERIALS AND METHODS

A spherical proportional counter (PC) of 40 mm diameter, filled with 160 kPa of ^{3}He and 53 kPa of Xe in a stainless steel envelope is placed inside the 203 mm diameter polythene (PE) sphere than has been chosen for this study. There are two approaches for selection of a ^{3}He detector for the purpose of this work. Firstly, a detector is filled with ^{3}He at relatively low pressure such that a substantial part of pulse height spectrum due to thermal neutrons is due to the wall effects, with the characteristic low energy edge caused by ^{3}H recoils. Secondly, a counter is filled to a very high pressure of ^{3}He such that the wall effects are negligible compared with the 764 keV absorption peak. The second approach will not be discussed in this paper.

The pulse-height spectra from a 160 kPa ^{3}He detector placed in a PE sphere when irradiated in standardised neutron–gamma fields of Cf and ^{137}Cs[3] have been measured and the radiation components have been evaluated.

The fluence responses to gamma radiation of the applied PC in the polythene sphere of 203 mm diameter have been calculated using the ANISN code[4] with the 21 groups gamma library type DLC-31. The parallel monoenergetic photon beam has been considered.

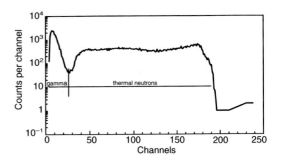

Figure 1. 160 kPa proportional counter pulse-height spectrum. Counter in 203 mm diam. polythene sphere irradiated by paraffin shielded Cf neutron source.

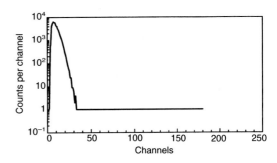

Figure 2. 160 kPa proportional counter pulse-height spectrum. Counter in 203 mm diam. polythene sphere irradiated by ^{137}Cs photon source.

RESPONSE TO PHOTONS

Generally the readings, M(E), of a system, both to neutrons and photons, are related to the incoming particle fluence $\phi(E)$ through the following relation:

$$M(E) = k\, F(E)\, \phi(E)$$

where k is a proportionality coefficient and F(E) is a filtration function for the given moderator shape. The filtration functions F(E) for monoenergetic neutrons are well known for PE spheres of different diameters while for photons there are no published data.

The difference between fluence response of the studied system and an 'ideal' response represented by the $H^*(10)/\phi$ ratio can be written as follows:

$$|\, K\, M(E)/\phi(E) - H^*(10)/\phi(E)\, | < \epsilon(E)$$

and should be within the acceptable limits, expressed numerically by $\epsilon(E)$. K is a normalisation coefficient.

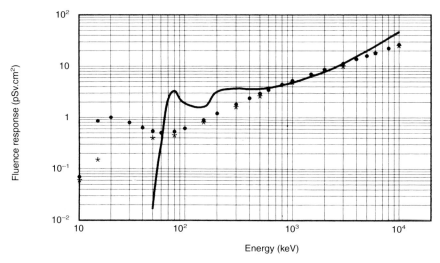

Figure 3. Fluence response of 160 kPa counter in 203 mm diam. polythene sphere. First extreme case: Xe in counter gas considered only. Solid line, K M(E)/ϕ. (●) $H^*(10)/\phi$. (*) H_E/ϕ. Counter wrapped in 1 mm thick Pb foil.

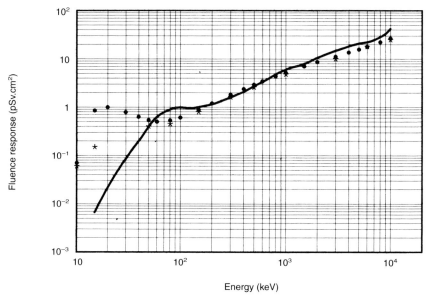

Figure 4. Fluence response of 160 kPa counter in 203 mm diam. polythene sphere. Second extreme case: Fe wall considered only. Solid line, K M(E)/ϕ. (●) $H^*(10)/\phi$. (*) H_E/ϕ. Counter wrapped in 0.6 mm thick Pb foil.

The aim of this work is to estimate $\epsilon(E)$ and thus define to what extent a moderator based detecting system used for neutrons can be applied for estimating both gamma and neutron ambient dose equivalent for penetrative radiation.

RESULTS AND DISCUSSION

The pulse height spectrum as shown on Figure 1 is obtained when the counter–moderator system was exposed to a Cf neutron source placed in a 250 mm diameter paraffin sphere, the case with the highest gamma contribution from the neutron source. As seen from Figure 1, the left part of this spectrum, below the characteristic edge due to ^{3}H recoils, i.e. below channel 25, consists of the pulses from photon interactions with the walls and gas of the detector. This is confirmed by the pulse-height spectrum in a pure ^{137}Cs gamma source as shown on Figure 2.

Photons generated by captured thermal neutrons, by hydrogen mainly, contribute to 'gamma channels' in the pulse-height spectrum. This effect has been experimentally estimated as proportional to the number of counts in the 'neutron channels'. It has been found that gamma channel counts due to gamma capture effects are equivalent to 1.6% of the counts in the neutron channels. Therefore the true readings in the gamma channels can be easily determined. The effect of He recoils due to elastic collisions with fast neutrons is negligibly small.

RESULTS OF CALCULATIONS

The fluence response of the PC surrounded by a 203 mm polythene sphere is determined by the attenuation and scattering effects on the PE sphere and by the counting efficiency of the bare detector. The latter is difficult to calculate and at present the calculations were performed for two extreme assumptions, namely

(i) that the process of energy absorption of photons is determined by Xe in the counter, and
(ii) that the process of energy absorption is determined solely by interactions of photons with the stainless steel wall of the counter.

In both cases, due to high Z elements in the bare counter, it has to be wrapped in Pb foils to avoid the over-response of the system at lower energies. The results are shown in Figure 3 and Figure 4 and are compared with $H^*(10)/\phi$. The envelope values of effective dose equivalent per unit fluence, H_E/ϕ, were included in the figures to show the direction of the deviation of the fluence responses of the studied system. The true response is situated between the extreme cases taken in the calculations. It seems that at energies higher than 300 keV the response is determined by the counter wall and at low energies by the Xe content of the PC gas. Generally, the results of the calculations, which can be assumed as preliminary, are encouraging for further studies on the proposed method for ambient dose equivalent measurements in neutron–gamma fields.

The experimentally determined numerical responses of the studied assembly in the fields of standard sources give the following results: for Pu-Be neutrons 1.3×10^{10} Sv^{-1} and 1.44×10^{10} Sv^{-1} for Cs photons. These results indicate that the method is specially suitable for low level monitoring.

CONCLUSIONS

The results of the experiments and the calculations have shown that the moderator based measuring system, till now solely used for neutron ambient dose equivalent measurements, can be used for gamma and neutron ambient dose equivalent determinations in a single measurement run. The numerical values of the responses for photons and neutrons indicate that this method is suitable for low level monitoring of mixed neutron–gamma radiation fields.

ACKNOWLEDGEMENTS

The author is very indebted to K. Wincel and B. Zaremba for performing calculations of the fluence spectra for the polythene spheres. Thanks are also due to J. Kula and A. Dudzinski for help in preparation of a manuscript and a conference poster. This work was supported by Grant No 209519101 from Polish State Committee for Scientific Research.

REFERENCES

1. Hsu, H. H., Casson, W. H., Olsher, R. H. and Vasilik, D. G. *Design of a New High Energy Neutron Dose Equivalent Meter by Monte Carlo Simulations.* Radiat. Prot. Dosim. **61**(1–4), 163–166 (1995).
2. Pszona, S. *Low-Level Neutron Monitoring Using High Pressure ^{3}He Detector.* Radiat. Prot. Dosim. **61**(1–4), 129–132 (1995).
3. Józefowicz, K., Golnik, N. and Zielczyński, M. *Dosimetric Parameters of Simple Neutron + Gamma Fields for Calibration of Radiation Protection Instruments.* Radiat. Prot. Dosim. **44**(1–4), 139–142 (1992).
4. Engle, W. W. *User Manual for ANISN One-Dimentional Discrete Ordinate Transport Code with Anisotropic Scattering.* RSIC Report K-1963 (1967).

ROSPEC — A SIMPLE RELIABLE HIGH RESOLUTION NEUTRON SPECTROMETER

H. Ing†, T. Clifford†, T. McLean†, W. Webb†, T. Cousins‡ and J. Dhermain§
†Bubble Technology Industries, Chalk River, Ontario, Canada
‡Space Systems & Technology Section, Defence Research Establishment Ottawa, Ottawa, Ontario, Canada
§Establishement Technique Central De L'Armement, Ministere de la Defense, Cedex, France

Abstract — A first prototype of ROSPEC (Rotating Spectrometer) was constructed over 10 years ago with the aim of achieving a high resolution neutron spectrometer for radiation protection applications around sources of fission neutrons. Studies have confirmed that the main contribution to dose equivalent from such neutrons involve those over the energy range 50 keV to 4.5 MeV. Thus, a neutron spectrometer spanning this energy range would be adequate for the intended purpose. The selected sensors for ROSPEC were four spherical gas counters filled with different pressures of hydrogenous gas. The pressures were selected to cover four separate energy segments within 50 keV to 4.5 MeV with generous overlap between adjacent energy segments. The sensors rely on hydrogen recoil from elastic neutron scattering. Low energy recoil protons whose pulse heights were contaminated with gamma-induced signals were not used. Large signals from recoil protons whose ranges were comparable to the detector size were also rejected because of distortion of the theoretical response function. The first ROSPEC prototype was successfully used to determine the neutron spectra for several radiation fields of dosimetric interest. However, the analyses of the data required spectroscopic expertise. In recent years, the enormous advances in electronic miniaturisation and computer power has allowed the re-design of the prototype spectrometer into a compact, reliable, user-friendly instrument. Data acquisition and spectral unfolding can now be done with simple computer commands, transforming ROSPEC from a specialist tool to an everyday routine instrument for radiation protection. Typically, a reliable spectral measurement can be done over periods of hours, e.g. overnight. Over the past few years, several ROSPECs have been built for various groups who have used them to measure an enormous number of neutron spectra of interest to military and nuclear power applications. These spectra are providing a detailed understanding of important neutron interaction processes and a scientific basis for choosing personal neutron dosemeters for a variety of radiation environments. ROSPEC has already been accepted as the reference neutron spectrometer for NATO experimental studies. It is now rapidly being adopted as a reference secondary standard for neutron dosimetry in nuclear institutions.

INTRODUCTION

The use of hydrogen-filled proportional counters for neutron detection dates back to the 1940s. However, since monoenergetic neutrons produce a pulse-height distribution spanning from 0 to the maximum proton energy for n-p scattering, neutron spectroscopy using such an approach requires 'spectral unfolding'. This made the technique unattractive until the widespread use of computers in the 1960s.

During the 1970s, one of the authors (H. I.) started to investigate the application of hydrogenous counters for spectral measurements of neutrons of interest to radiation protection as part of a larger spectral dosimetry programme. Some of this work has already been summarised in the literature[1,2].

Of particular importance to the current work was the development of a system of four spherical hydrogenous gas counters, filled to different pressures, to act as a composite spectrometer for measurement of fission and degraded fission neutrons. The counters were mounted on a slowly rotating platform to ensure that they all sweep the same volume in space so that the measured fluences are absolutely correlated to avoid normalisation requirements among the counters. Due to the rotation of the counters, the system was called ROSPEC for ROtational SPECtrometer. For historic interest, a picture of the original ROSPEC is reproduced in Figure 1.

The original ROSPEC consisted of three small (2 cm radius) counters filled with 1, 4 and 10 atm of hydrogen and one larger counter (6" in diam.) filled with 0.5 MPa (5 atm) Ar–methane. This choice allowed ROSPEC to span the energy range 50 keV to 4.5 MeV, where the bulk of neutron doses is found in essentially all practical environments where fast neutrons from fission are of radiological concern. The electronic components for the original ROSPEC were mainly modular, commercial units which were readily available. The pulse-height outputs from ROSPEC were stored on magnetic cassette (home-built) and transferred to a large central computer for spectral unfolding. More details of the original ROSPEC along with results obtained using the instrument in a variety of applications can be found in the earlier publications.

EVOLUTION OF ORIGINAL ROSPEC

The advent of inexpensive, powerful, personal computers (PC) in the 1980s permitted a significant technological improvement to ROSPEC operation. It was a natural extension to replace the hardwired, multichannel analyser with a PC-based multichannel analyser. With some changes to the unfolding code to allow execution on the PC, ROSPEC became a self-contained instrument yielding spectral results minutes after data acquisition.

Such a version of ROSPEC was used to measure neutron spectra of military interest in the late 1980s associated with fast-burst reactors operated in the steady-state mode.

Figure 2 shows some typical spectra at close range to a fast-burst reactor. It is interesting to see the intermediate energy neutrons build up as the distance from the reactor increases.

Figure 3 shows the measured spectrum at 400 m from the reactor. The spectral shape is now typical of a well-moderated fission spectrum. The structure in the spectrum is associated with resonances in the oxygen of the air.

Figure 4 shows the spectrum at 400 m filtered by 10 cm of iron. The softening of the spectrum is in accordance with theoretical predictions[3].

EMERGENCE OF MODERN ROSPEC

Requirements by the military of several NATO countries[4] for a rugged, simple-to-use version of ROSPEC for field use led to a re-design of the spectroscopic system. The original, mostly commercial and

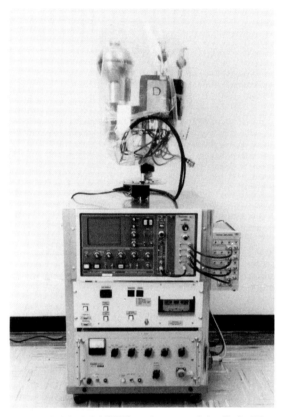

Figure 1. Original ROSPEC as it appeared in Chalk River Nuclear Laboratories in the 1970s and early 1980s.

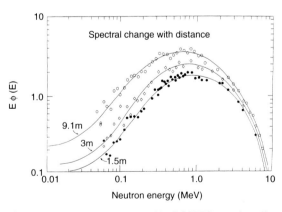

Figure 2. Neutron spectra measured by ROSPEC at various distances in the vicinity of a fast-burst reactor. The lines are drawn as guides for the eye.

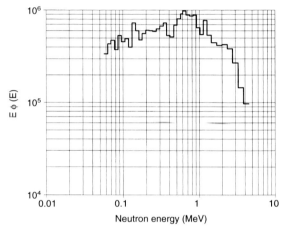

Figure 3. The neutron spectrum at 400 m from the fast-burst reactor.

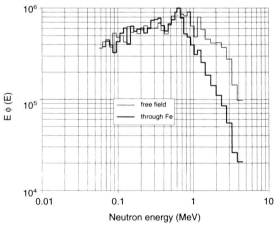

Figure 4. The effect of a 10 cm slab of iron on the neutron spectrum at 400 m.

modular electronics were replaced by specially designed circuits using the latest components which reduce the size and weight of the system considerably. The use of powerful lap-top computers for data analysis and presentation along with macro-programming allows complex data processing to be properly carried out at the push of a button, with minimal input or spectroscopic experience on the part of the operator.

Figure 5 shows a modern ROSPEC with the lid (which protects the counter assembly) removed. The preamps are housed inside the tubular handle below the counters. The smaller counters have been increased slightly in size (2″ diameter) relative to the earlier counters, but the large counter remained unchanged. The size increase allows for a more generous overlap between the high energy limit of the 1 MPa (10 atm) counter with the low energy limit of the large counter. Four separate high voltage power supplies (visible in the figure) are used to operate the proportional counters.

The counter assembly is mounted on a platform which rotates slowly and continuously when ROSPEC is turned on. The rotation is gear driven on a toothed circular track with self-lubricating components for lasting operation. Mounted on the underside of the rotating platform are numerous circuits boards which contain four amplifiers, four separate ADCs, low voltage power supplies, and a microcomputer. The microcomputer controls the data processing of the signals from the four ADCs, identifying the pulse-height information with the appropriate counter, storing the data into a buffer and transferring the data from the buffer to the data acquisition computer through a rotary link and by use of a RS422 serial interface. The data acquisition computer can be located next to the 'head unit' or can be located up to 1 km from it.

When the lid is in place, the head unit is a sealed cylinder, whose contents can be viewed through a circular transparent port on the top of the lid. There are no accessible electronic controls pertaining to spectroscopic performance.

To use ROSPEC, one connects the data transmission cable between the 'head unit' and the data acquisition computer, turns the power on for both units and pushes a key on the computer to start data acquisition. During measurement, one can peer through the viewport to see the counters rotating slowly, watch the pulse-height spectra accumulate simultaneously in all four counters (or singly) or examine various quantities of particular interest — e.g. dead time, count rate, elapsed time, etc. At the end of the measurement, data acquisition is halted at the push of a key and the pulse-height distributions unfolded on command.

The outputs of the measurement are the unfolded spectrum — tabulated and plotted — along with computed dose equivalent, kerma and respective rates. Typical measurement in fields of interest take a matter of hours to get high quality spectra. No adjustments, calibrations or routine checks are needed to use ROSPEC; it is designed for simple, maintenance-free operation.

Figure 5. A modern version of ROSPEC. The head unit is shown without its lid to reveal inner components. The lap-top shown does not represent the latest models used.

RESULTS WITH ROSPEC

The final stage of ROSPEC production involves calibration and testing. ROSPEC is calibrated using monoenergetic neutrons produced by the p-^{7}Li, p-^{3}H and d-d reactions using a 3 MV Van de Graaff accelerator at Defence Research Establishment, Ottawa. Its absolute detection efficiency (which is determined theoretically from known gas volumes and pressures) is checked using a long counter during accelerator calibrations and with a Cf neutron source traceable to national standards of several countries — including Canada and the United States.

Figure 6 shows typical results from monoenergetic neutron calibrations and gives some indication of spectrometer resolution.

Figure 7 shows a typical spectrum from a Cf standard. The light curve is the theoretical uncollided fluence spectrum at the point of measurement (50 cm from source). The measured spectrum shows more neutrons in the region below 1.5 MeV, which is due to room-scattered neutrons. The agreement in total fluence between the measured and theoretical spectra from numerous experiments with many ROSPECs is always within ±5%.

Figure 8 shows a measured spectrum of an Am-Be source at the Royal Military College in Kingston, Ontario, using ROSPEC in combination with a special scintillation spectrometer called SSS[5]. The SSS spec-

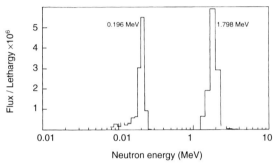

Figure 6. Typical results from monoenergetic neutron calibration using the p-^{7}Li and p-T reaction produced by a Van de Graaff accelerator.

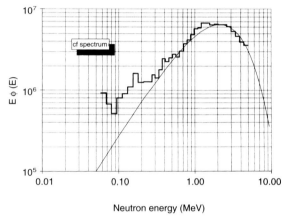

Figure 7. Measured spectrum at 50 cm from Cf standard. The lighter curve is the theoretical uncollided fluence at the point of measurement.

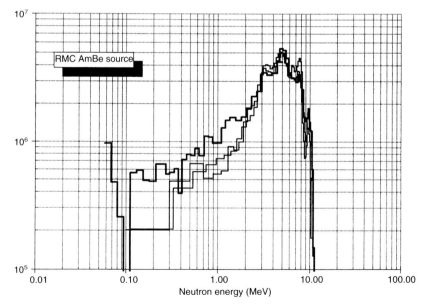

Figure 8. Measured spectrum of an Am-Be neutron source. The light curves are theoretical spectra[6,7].

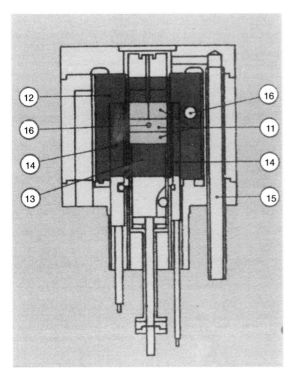

Figure 9. Simplified schematic of the Harmonie reactor at Cadarache, France. Only part of the steel reflector is shown. Legend: 11, slab reactor core; 12, uranium reflector; 13, security block; 14, control rod; 15, core ventilation; 16, central and tangential experimental channels.

trometer consists of an array of 16 3 × 3 mm cylinders of plastic scintillators mounted on a single photomultiplier to act as a composite scintillator. The size of these scintillators was chosen to eliminate gamma contamination of the recoil proton pulse-height distribution above 4.0 MeV. SSS was made to complement ROSPEC to allow measurement of neutrons up to 20 MeV and intended for simple operation — the avoidance of the usual n/γ pulse-shaped discrimination electronics. Theoretical Am-Be spectra[6,7] are also shown in the figure (light lines). The agreement between measurement and theory is good in view of the uncertainty of the theoretical spectrum. The measured spectrum is richer in neutrons below about 2 MeV mainly due to room scattering.

Figure 9 shows a simplified schematic diagram of the Harmonie reactor in Cadarache, France. For the ROSPEC measurement, the central canal was filled with depleted uranium and steel in order to make the reactor as homogeneous as possible. The measurement was made 1 m from the reactor face for a period of 1800 s with a reactor power of 7 mW.

Calculations of the neutron fluence at the location of measurement were made using MCNP4 using a mock-up of the assembly but without the inclusion of the floor.

Figure 10 compares the ROSPEC results with the calculated spectra. Within the accuracy of the mock-up, the spectral agreement is regarded as good. The total measured fluence agrees with the calculated value to with 1% (which is probably fortuitous).

CONCLUDING REMARKS

Many (modern) ROSPECs are now in use throughout

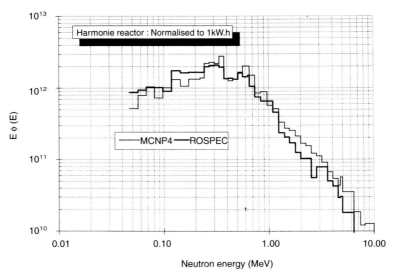

Figure 10. Measured and calculated spectra from Harmonie at 1 m from the reactor surface.

the world. Experience by various groups indicates that the spectrometer is accurate and surprisingly simple to use — as already indicated in the literature[8]. Years of measurements by military scientists have proved the reliability of ROSPEC under a variety of environmental conditions and the instrument has been accepted[4] as a reference neutron spectrometer for NATO military studies.

The ease of use of ROSPEC has caught the attention of scientists in non-military nuclear institutions where recognised neutron hazards exist. Rather than focusing on high quality neutron spectra, these groups are mainly concerned with inadequacies of personal neutron dosemeters. Their need is to have some idea of the energies of neutrons responsible for radiation exposures and to have an accurate measurement of the neutron dose rates at locations of concern. Such groups are now beginning to use ROSPEC as a secondary standard and to regard the measured dose rates at such locations as calibrations for their personal neutron dosemeters. It can be expected that the increasing use of such an accurate spectrometer for such applications will undoubtedly yield a wealth of spectral data for diverse operational environments which will finally allow health physicists truly to understand the general features of neutron spectra around working environments which has eluded them for decades.

REFERENCES

1. Cross, W. G., and Ing, H. *Neutron Spectroscopy*. In: The Dosimetry of Ionizing Radiation, Vol. II, ch. 2. Eds K. R. Kase, B. E. Bjarngard and F. H. Attix. (New York: Academic Press) (1987).
2. Ing, H., Cross, W. G. and Bunge, P. J. *Spectrometers For Radiation Protection at Chalk River Nuclear Laboratories*. Radiat. Prot. Dosim. **10**, 137–145 (1985).
3. Ing, H. and Makra, S. *Compendium of Neutron Spectra In Criticality Accident Dosimetry*. IAEA Technical Report Series No. 180, p. 116 (Vienna: IAEA) (1978).
4. Stanka, M. B. *The ROSPEC Neutron Spectrometer, A NATO Collaborative Success*. In: Proc. Workshop on Biomedical Aspects of Nuclear Defence, NATO Technical Proceeding AC/243 (Panel 8) Tp/6 B3.1-B3.8 (1995).
5. Bubble Technology Industries. *Simple Scintillation Spectrometer, BTI-SSS for Neutron Spectroscopy to 15MeV, Manual*. BTI report 6 May 1994; revised 3 November 1994.
6. Geiger, K. W. and Vander Zwan, L. *Radioactive Neutron Source Spectra From ^9Be (α, n) Cross Section Data*. Nucl. Instrum. and Methods, **131**, 315–321 (1975).
7. Burger, G. and Schwartz, R. B. *Guidelines on Calibration of Neutron Measuring Devices*. IAEA Technical Report Series No. 285 (Vienna: IAEA) (1988).
8. Schwartz, R. B. and Eisenhauer, C. M. *Test of A Neutron Spectrometer in NIST Standard Fields*. Radiat. Prot. Dosim. **55**, 99–105 (1994).

SPECIFICATION OF BONNER SPHERE SYSTEMS FOR NEUTRON SPECTROMETRY

M. Kralik†, A. Aroua‡, M. Grecescu‡, V. Mares§, T. Novotny†, H. Schraube§ and B. Wiegel∥
†Czech Metrological Institute (CMI)
Radiova 1, CZ-10200 Prague 10, Czech Republic
‡Institut de Radiophysique Appliquée (IRA), CH-1015 Lausanne, Switzerland
§GSF-Forschungszentrum für Umwelt und Gesundheit
Neuherberg, D-85758 Oberschleissheim, Germany
∥Physikalisch-Technische Bundesanstalt (PTB)
D-38116 Braunschweig, Germany

Abstract — The response functions of four widely used Bonner sphere spectrometers (BSS) with an LiI scintillator and different ^{3}He detectors were calculated by means of the three-dimensional Monte Carlo neutron transport code, MCNP, taking into consideration a detailed description of the detector set-up; they were then compared with experimental calibration data. The calculated response functions reproduce the calibration data in shape. A common adjustment factor is only needed to take into account the neutron detection efficiency of the central detector. An interpolation method proposed to correct the BSS response functions to another polyethylene density was verified.

INTRODUCTION

Since its introduction in 1960[1], a Bonner sphere spectrometer (BSS) has been the only instrument which allows the spectral neutron fluence to be measured in a wide range of energies from thermal up to at least 20 MeV. It is therefore often used for specifying neutron fields at workplaces and in different fields obtained in the laboratory for the calibration of area and personal dosemeters. The desired ambient dose equivalent value is then calculated from the neutron spectrum applying the corresponding fluence-to-dose-equivalent conversion function. Compliance with the recent recommendations of ICRP 60[2] concerning the operational and limiting dose equivalent quantities requires an improved sensitivity and accuracy in spectral neutron fluence measurements with BSS. The response functions of the BSS must, therefore, be determined with a reasonable accuracy in the entire neutron energy range. Since experimental calibrations can be performed at selected

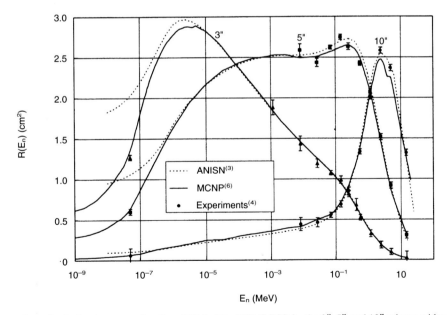

Figure 1. Comparison of calculated response functions $R(E_n)$ of the PTB-C BSS for the 3″, 5″, and 10″ spheres with corresponding experimental data.

energies only, reliable calculations are needed, at least for interpolation purposes.

Previous calculations of BSS response functions performed with one-dimensional codes, e.g. with the ANISN code[3], could not take into consideration all the influences on the response such as a non-spherical detector geometry. In order to obtain agreement between the calculated response functions and the measured calibration data[4], different scaling factors had to be applied to the response functions calculated for the various sphere diameters, and even the shape had to be adjusted in order to fit the thermal neutron response measured (see Figure 1). Better results were obtained by means of the three-dimensional neutron transport Monte Carlo code, MCNP, although details of the detector set-up were still neglected[5]. Many doubts concerning the influence which the various approximations might exert on the calculated responses were circumvented by using a realistic model of the detector set-up with all relevant details[6]. As illustrated in Figures 1 and 2, this approach proved to be the best of those available to date.

In this paper the response functions of different BSS calculated by the MCNP code are discussed, including detailed descriptions of either an LiI scintillator[7] or, of spherical[6] or cylindrical[8,9] ^{3}He proportional counters, and the results compared with the experimental data obtained with monoenergetic[4,9] and thermal[10] neutrons.

EXPERIMENTAL SET-UP

BSS currently in use, chiefly differ in the polyethylene density and as regards the central detector for thermal neutrons chosen, namely:

(i) the GSF system[7] with a polyethylene density of 0.95 g.cm^{-3} and a cylindrical LiI(Eu) crystal scintillator of 4 mm diameter and 4 mm length;
(ii) the PTB-C system[6] with a polyethylene density of 0.946 g.cm^{-3} and a ^{3}He-filled spherical proportional counter of type SP90 with 32 mm diameter (Centronics);
(iii) the PTB-F system[8] with the same polyethylene density as the C system and a ^{3}He-filled cylindrical proportional counter of type 0.5 NH 1/1K of 8 mm diameter and 10 mm length (LCC France);
(iv) the IRA system[7] with the polyethylene density of 0.918 g.cm^{-3} and the same cylindrical proportional counter as the PTB-F system.

All these BSS were calibrated with monoenergetic neutrons at the PTB accelerator facility and the thermal beam at NPL (except for the IRA system). Corrections for room return neutrons (shadow cone measurements) and target scattered neutrons (Monte Carlo calculations) were generally applied. Depending on the neutron energy, proton recoil or telescope detectors served as reference systems for neutron fluence measurements.

RESULTS AND DISCUSSION

For all neutron calibration energies E and all Bonner sphere diameters d, the ratios $r_{d,E} = R_c(d,E)/R_e(d,E)$ of calculated $R_c(d,E)$ and experimental $R_e(d,E)$ response values were determined. The evaluation procedure

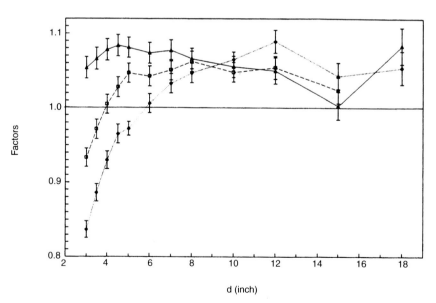

Figure 2. Factors which fit calculated responses of the PTB-C BSS to measured responses as a function of sphere diameter d. MCNP simulations were performed either with (□) a simplified[5] or (▲) a detailed geometry model[6]. (◆) ANISN[3]. (1 inch = 2.54 cm).

described elsewhere[6] was applied for calibration data of all four BSS in order to calculate weighted mean values of the ratios $r_{d,E}$ for individual detector systems averaged either over all spheres $r_E = \langle r_{d,E} \rangle_E$ or over all calibration energies $r_d = \langle r_{d,E} \rangle_d$ taking into account their uncertainties. The experimental data were taken from the literature[4,9,10] and also the calculated data[6-8]. The values of r_E and r_d for individual BSS are represented in Figures 3 and 4. To ensure uniform presentation, the mean values of r_E and r_d were adjusted to unity. The uncertainties given comprise all correlations due to the experimental calibration and those associated with the fluence reference data[6]. Since the measurements in the thermal beam at NPL were not done for all systems and no detailed analysis has yet been performed to explain the rather outlying points at thermal energies[6], the thermal data were not included in the further analysis.

Figure 3 shows that there is no significant trend in the dependence of the mean ratio r_E on the neutron energy. The very similar scatter of PTB and GSF data is not

Table 1. Adjustment factors f and their relative standard uncertainty u for the individual BSS under consideration.

System	GSF	IRA	PTB-F	PTB-C
f	0.742	0.856	1.042	1.070
u(%)	1.3	2.0	1.8	1.8

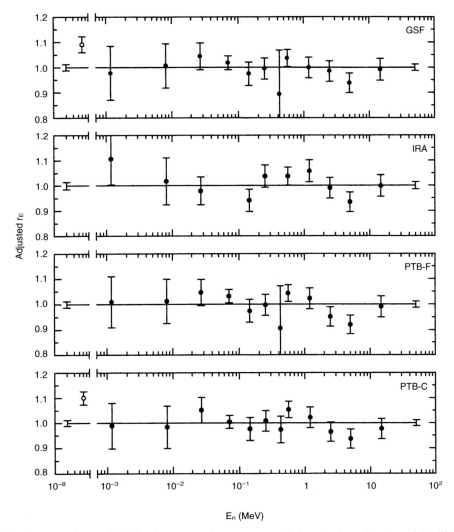

Figure 3. Adjusted mean ratios r_E of calculated to measured responses considering all sphere diameters of the different BSS at a given calibration energy E_n; the bars represent their standard uncertainty. The empty symbols depict measurements in the thermal beam.

surprising because all three systems were simultaneously calibrated and are correlated due to the same fluence reference data. The ratios r_d (Figure 4) still show a slight increase with the diameter d up to the 12″ sphere, which is most pronounced for the system with the LiI detector, but it still seems justified to average the mean ratios to obtain a single common ratio for every BSS. Its reciprocal value is used as an adjustment factor to scale the calculated response matrix properly (Table 1).

INTERPOLATION OF RESPONSE FUNCTIONS

Since there are many BSS of equal design but different polyethylene density, and since the calculation of a complete response matrix is very time-consuming, an interpolation method has been suggested[6]. This method is based on the fact that the relative variation of the polyethylene density, $\Delta\rho/\rho$, has the same effect on the response as has a relative variation of the wall thickness, $\Delta t/t$, of the polyethylene sphere. The reliability of this method was checked using the responses $R_{0.946}$ and $R_{0.921}$ calculated for the arrangement of the PTB-F BSS assuming polyethylene densities of $\rho = 0.946$ g.cm^{-3} and 0.921 g.cm^{-3}, respectively. The directly calculated responses $R_{0.921}$ were compared with those obtained from interpolation of the responses calculated for $\rho = 0.946$ g.cm^{-3}, denoted as $R^i_{0.921}$[8]. The ratio

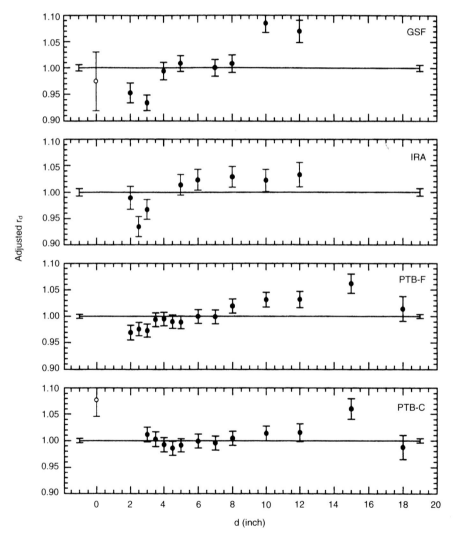

Figure 4. Adjusted mean ratios r_d of calculated to measured responses considering all calibration energies for a given sphere diameter d of each BSS; the bars represent their standard uncertainty. The empty symbols depict measurements in the thermal beam. (1 inch = 2.54 cm).

$R_{0.946}/R_{0.921}$ along with the ratio $R^i_{0.921}/R_{0.921}$ are shown in Figure 5 for the 2.5″ and 15″ spheres.

Since the ratio of interpolated to directly calculated responses is nearly constant and equal to unity for all spheres, the interpolation method may be used as a reasonable approach for calculating the response matrix of the BSS which is of identical design but of another polyethylene density.

CONCLUSIONS

It has been shown that the BSS response functions calculated by means of the three-dimensional Monte Carlo neutron transport code, MCNP, using a realistic model of the central detector reproduce the available calibration data in shape. If calibration data with monoenergetic neutrons are available, the calculations should be adjusted separately to each sphere diameter. If the calibration can be performed only with standard radionuclide neutron sources, a common adjustment factor can be applied in order to take into account the neutron detection efficiency of the central detector.

The interpolation method used to correct the BSS response functions to another polyethylene moderator density or intermediate sphere diameter was verified by additional calculations.

Properly specified response functions proved to be indispensable to achieve satisfactory agreement of the unfolded neutron spectra and the corresponding integral fluence and dose equivalent values obtained in comparison exercises[11-13].

ACKNOWLEDGEMENTS

This work was supported by the European Commission under contract FI3P CT92-0002 and by the Swiss Federal Office for Education and Science under contract No 94.0039. The authors are very grateful to Dr Horst Klein, co-ordinator of the CEC project, for valuable discussions and support.

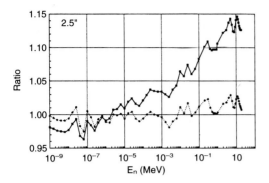

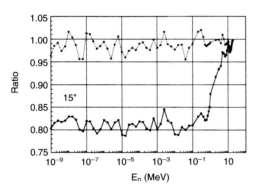

Figure 5. Ratios $R^i_{0.921}/R_{0.921}$ (dashed line) and $R_{0.946}/R_{0.921}$ (solid line) of calculated responses for the PTB-F BSS arrangement.

REFERENCES

1. Bramblett, R. L., Ewing, R. I. and Bonner, T. W. *A New Type of Neutron Spectrometer.* Nucl. Instrum. Methods **9**, 1–12 (1960).
2. ICRP. *1990 Recommendations of the International Commission on Radiological Protection.* Publication 60. Ann. ICRP **21**, (1–3) Oxford: Pergamon Press (1991).
3. Thomas, D. J. *Use of the Program ANISN to Calculate Response Functions for a Bonner Sphere Set with ^{3}He Detector.* NPL Report RSA(EXT) 31 (Teddington: NPL) (March 1992).
4. Alevra, A. V., Cosack, M., Hunt, J. B., Thomas, D. J. and Schraube, H. *Experimental Determination of the Response of Four Bonner Sphere Sets to Monoenergetic Neutrons (II).* Radiat. Prot. Dosim. **40**, 91–102 (1992).
5. Mares, V., Schraube, G. and Schraube, H. *Calculated Neutron Response of a Bonner Sphere Spectrometer with ^{3}He Counter.* Nucl. Instrum. Methods A **307**, 398–412 (1991).
6. Wiegel, B., Alevra, A. V. and Siebert, B. R. L. *Calculations of the Response Functions of Bonner Spheres with a Spherical ^{3}He Proportional Counter using a Realistic Detector Model.* Report PTB-N-21 (Braunschweig: PTB) (1994).
7. Mares, V. and Schraube, H. *Evaluation of the Response Matrix of a Bonner Sphere Spectrometer with LiI Detector from Thermal Energy to 100 MeV.* Nucl. Instrum. Methods A **337** 461–473 (1994).
8. Kralik, M. and Novotny, T. *Response Functions of Bonner Spheres with Small Cylindrical ^{3}He Proportional Counter Determined by Monte Carlo Calculations.* Report CMI-GR2070/95 (1995).
9. Aroua, A., Grecescu, M., Lerch, P., Valley, J.-F. and Vylet, V. *Evaluation and Test of the Response Matrix of a Multisphere Neutron Spectrometer in a Wide Energy Range Part I. Calibration.* Nucl. Instrum. Methods A **321**, 298–304 (1992). See also:
Aroua, A. and Grecescu, M. *MCNP Calculations of the Response Matrix of the IRA Bonner Spheres vs Experimental Calibration Data.* IRA Internal Report, Lausanne (July 1995).

10. Thomas, D. J., Alevra, A. V., Hunt, J. B. and Schraube, H. *Experimental Determination of the Response of Four Bonner Sphere Sets to Thermal Neutrons*. Radiat. Prot. Dosim. **54**, 25–31 (1994).
11. Lindborg, L., Bartlett, D., Drake, P., Klein, H., Schmitz, Th. and Tichy, M. *Determination of Neutron and Photon Dose Equivalent at Workplaces in Nuclear Facilities in Sweden, A Joint SSI-EURADOS Comparison Exercise*. Radiat. Prot. Dosim. **61**, 89–100 (1995).
12. Thomas, D. J., Chartier, J.-L., Klein, H., Naismith, O. F., Posny, F. and Taylor, G. C. *Results of a Large Scale Neutron Spectrometry and Dosimetry Comparison Exercise at the Cadarache Moderator Assembly*. Radiat. Prot. Dosim. **70**(1–4), 313–322 (This issue) (1997).
13. Klein, H. *Workplace Radiation Field Analysis*. Radiat. Prot. Dosim. **70**(1–4), 225–234 (This issue) (1997).

IMPROVED NEUTRON SPECTROMETER BASED ON BONNER SPHERES

A. Aroua†, M. Grecescu†, S. Prêtre‡ and J.-F. Valley†
†Institute for Applied Radiophysics, CH-1015 Lausanne, Switzerland
‡Swiss Nuclear Safety Inspectorate, CH-5232 Villigen-HSK, Switzerland

Abstract — A modification of the Bonner spheres neutron spectrometer is proposed in order to improve its energy resolution at low and intermediate energies. The small diameter polyethylene spheres are surrounded by layers of neutron absorbers with suitable composition and thickness. The system design is based on response function calculations with the neutron transport code ANISN. Several absorbers were investigated and, eventually, boron with natural isotopic composition has been retained. The optimal set of detectors consists of the 2″ and 3″ spheres surrounded by boron layers of 0.0628, 0.837 and 2.76 g.cm^{-2}. Narrower response functions have been obtained with one maximum per decade in the energy range 1 eV–100 keV. An experimental calibration has been performed with neutron beams of 0.186, 2, 24 and 144 keV. The computed response functions have been adjusted to the calibration points. Simulations with numeric neutron spectra confirmed the improved energy resolution of the new system.

INTRODUCTION

In recent years, an improved characterisation of neutron fields in the epithermal and intermediate energy range was required in various applications: boron neutron capture therapy (BNCT)[1,2], radiation field measurements inside nuclear power plants[3,4], and production of reference neutron fields for the calibration of neutron monitors[5].

The Bonner spheres neutron spectrometer is suited for radiation protection measurements at nuclear facilities; however, it suffers from its low energy resolution and lack of response functions with maxima at intermediate energies over 4 decades. A better resolution at low energies and a better coverge of the epithermal and intermediate energy range with peaked response functions will improve the accuracy of the system with respect to: the analysis of spectra containing structures in this energy range; the definition of the spectral shape in the region between 100 keV and 1 MeV where the fluence to dose conversion factor h_Φ varies rapidly; and the evaluation of dosimetric quantities for radiation protection.

Wang and Blue[1] developed a neutron spectrometer similar to the Bonner spheres system, designed with an improved resolution in the energy range 1 eV–10 keV. The spectrometer consists of a spherical ^{3}He proportional counter and a set of paraffin spheres surrounded by hemispherical shells loaded with ^{10}B. As this system was developed for a special application (BNCT), it has a limited energy range and an anisotropic response, being unsuitable for general radiation protection use.

Based on the same principle, but using a cadmium shell between two layers of polyethylene, Kryuchkov and Semenova published a set of response functions with improved energy resolution[6].

The present work describes a neutron spectrometer with improved performance in the epithermal and intermediate energy range based on the Bonner spheres system of the Institute of Applied Radiophysics (IAR), Lausanne. The improvement is based on the principle proposed by Wang and Blue[1], namely the use of a suitable neutron absorber around a few spheres, but the previously mentioned limitations have been eliminated by careful design. The system has an isotropic response. Its response functions could be reliably calculated as the spheres are made of polyethylene with a well-defined chemical composition, in contrast to paraffin. The new response functions display one maximum per decade between 1 eV and 100 keV. The modified system has been calibrated with monoenergetic neutron beams.

MATERIAL AND METHODS

Design principles

The IAR Bonner spheres system consists of a set of polyethylene spheres with diameters 2, 2.5, 3, 4.2, 5, 6, 8, 9, 10, 12 and 15 inch. The neutrons moderated in the sphere are detected by a ^{3}He cylindrical proportional counter (1 cm diam. × 1 cm) located at the centre of each sphere and coupled to a suitable electronic system. A detailed description of the construction and performance of this system has been published[7].

It is possible to change the shape of the response functions in the low and intermediate energy region by surrounding the small spheres with a layer of absorber having a neutron capture cross section proportional to 1/v such as ^{10}B or ^{6}Li. The effect of the absorber is mainly to cut the low energy tail of the response function, thereby reducing its width and displacing the position of the maximum on the energy scale. For a given sphere, the maximum position can be controlled by varying the absorber thickness; however, a compromise must be achieved with the unavoidable reduction in sensitivity. The procedure is effective for the spheres with diameters up to 6″, whose response functions are predominating in the energy region of interest.

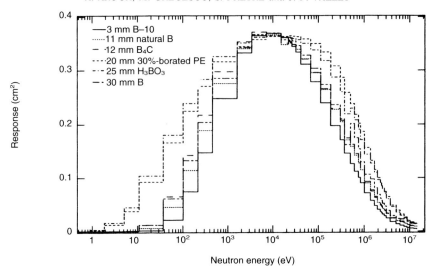

Figure 1. Response functions of the 2″ Bonner sphere covered with spherical shells of different materials containing the same amount of ^{10}B.

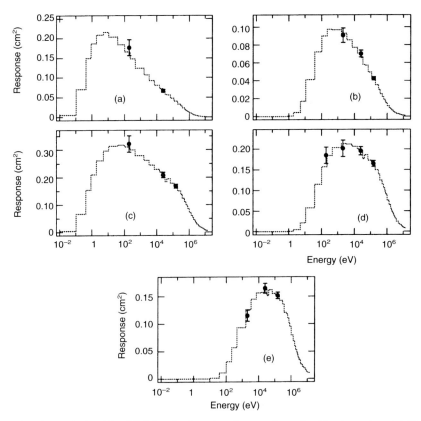

Figure 2. Response functions calculated by ANISN and adjusted to the experimental calibration points: (a) 2″ + 0.0628 g.cm^{-2} boron; (b) 2″ + 0.837 g.cm^{-2} boron; (c) 3″ + 0.0628 g.cm^{-2} boron; (d) 3″ + 0.837 g.cm^{-2} boron; (e) 3″ + 2.76 g.cm^{-2} boron.

The system design is based on the calculation of the response functions with various combinations of materials and geometries, in order to choose the best solution in terms of energy resolution, sensitivity, cost and dimensions.

Calculation of the response matrix

The calculation of the response matrix was performed using the one-dimensional transport code ANISN[8] which had been used previously for calculating the response functions of the IAR Bonner spheres system[9]. The cylindrical proportional counter is filled with ^{3}He at a nominal pressure of 8×10^5 Pa according to the manufacturer, and the polyethylene density is 0.916 ± 0.003 g.cm^{-3}. The effective pressure in the proportional counter may be different from the nominal one and an adjustment of the calculated response functions to experimental calibration points is necessary.

The neutron absorber has the shape of a spherical shell surrounding a polyethylene sphere. In order to determine its optimal composition, simulations with several materials were performed. The ^{10}B has been preferred to ^{6}Li and cadmium because of the monotonous 1/v shape and the higher value of its neutron absorption cross section. Calculations showed that the design goals could be reached with suitable ^{10}B thicknesses. However, highly enriched ^{10}B is expensive and subsequent simulations were performed with various chemical compounds containing boron with natural isotopic composition (20% ^{10}B and 80% ^{11}B): elemental boron, B_4C, B_2O_3, H_3BO_3, borated polyethylene (30% B). Equivalent thicknesses containing the same amount of ^{10}B have been used. Typical results are presented in Figure 1, showing that elemental boron gives the closest response function to that obtained with ^{10}B. The response functions obtained with hydrogen-containing compounds (borated polyethylene and boric acid) are significantly wider.

An additional investigation was carried out using a 3" diameter sphere made entirely of borated PE. Its response function is similar to that of the 3" polyethylene sphere covered with 4.5 mm boron, but its response is 50 times lower, which precludes the use of this material.

Eventually boron with natural isotopic composition and in powder form was chosen for the absorber shells. Different metals were tested for the walls of the powder container (1 mm of aluminium, steel or copper); their effects on the response functions are negligible and aluminium was retained for mechanical reasons.

The response functions were calculated for the small diameter spheres (2" to 6") covered with different thicknesses of natural boron between two sheets of 1 mm aluminium. An optimal set of sphere–absorber combinations was established as a compromise between the following factors: reduced width of the response func-

Table 1. Characteristics of the boron layers.

Thickness (mm)	Volume (cm^3)	Mass (g)	Density (g.cm^{-3})	Mass thickness (g.cm^{-2})
1	31.8	19.97	0.628	6.28×10^{-2}
11	442	336	0.761	8.37×10^{-1}
35	2288	1803	0.788	2.76

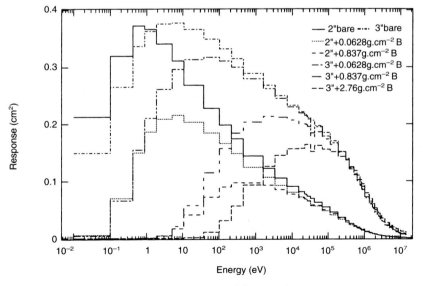

Figure 3. Response matrix of the improved Bonner spheres spectrometer.

tions, their uniform distribution (one peak per decade in the energy range 1 eV to 100 keV), acceptable sensitivity reduction, reasonable complexity (number of combinations) and cost. The set consists of the 2″ and 3″ polyethylene spheres used together with three boron absorbers of different thicknesses.

RESULTS

Response functions

For practical reasons, the absorbers were manufactured as cylindrical aluminium boxes filled with boron powder (purity 97%, maximum grain size 1 μm). The inner dimensions of the boxes are 82 mm in diameter and height, suitable for accommodating the 3″ sphere. The necessary boron layer thickness was determined by taking into account the density of compacted boron powder. The effective boron quantity was eventually obtained by weighing the manufactured boxes (Table 1). A definitive set of response functions was computed with the experimental boron mass thickness by taking into account the impurities specified by the chemical analysis.

An adjustment of the ANISN calculations to experimental calibration points is necessary to correct for whatever detail may be missed in the simulation, which is based upon the assumption that the detector has a perfect spherical symmetry and upon the value of the ^{3}He pressure in the proportional counter specified by the manufacturer. The system was calibrated with quasi-monoenergetic neutron beams (0.186, 2, 24 and 144 keV) at the research reactor of the Physikalisch-Technische Bundesanstalt (PTB), Braunschweig[10].

The calculated response functions have been adjusted to the experimental calibration points, using a least squares fit. The results of the adjustment are presented in Figure 2. It is remarkable that the individual adjustment factors are quite close to each other (within ±2.3%), the mean value being 0.577. This points to a common origin of the difference between the calculated and the experimental responses, undoubtedly the true value of the ^{3}He density.

The adjusted matrix of the manufactured spectrometer is represented in Figure 3 and shows that the design goals have been achieved. The maxima of the response functions are evenly spread (roughly one per decade) between 1 eV and 100 keV, reduced widths have been obtained for the response functions of the new detectors and their reduction in sensitivity does not exceed a factor of 4.

A comparison between the measurement duration for the set of 7 detectors considered in Figure 3 and a set of 6 conventional Bonner spheres (2″, 2.5″, 3″, 4.2″, 5″ and 6″) covering the same energy range has been performed. In a typical neutron field encountered in radiation protection, with a moderate dose equivalent rate of 100 μSv.h^{-1}, the effective counting time required to have less than 1% statistical error is of the order of 74 min for the first set and 19 min for the second one. The increase of the measuring time by a factor of about 4 is still acceptable.

Spectrometric performance

The spectrometric performance of the new system has been evaluated by simulation using four synthetic neutron spectra consisting of a single peak in the epithermal energy region, situated at 1, 10, 100 and 1000 eV respectively. For each of these spectra the counting rate

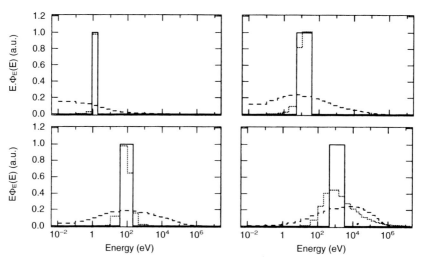

Figure 4. Comparison of results obtained on test spectra with a standard Bonner spheres set and with the improved spectrometer: (———) test spectrum; (– – – –) unfolded spectrum with standard matrix; (.....) unfolded spectrum with improved matrix. For details, see text.

expected from a given detector has been determined by convolution of the spectrum with the response function of this detector. The two sets of detectors previously considered have been used. The unfolding has been performed with a version of the SAND code[11] using 47 energy groups. For all the unfolding trials a 1/E spectrum over the whole energy range was used as a 'first guess', without any *a priori* information about the presence of a peak.

The spectrometric results obtained with the two sets of detectors are compared in Figure 4. They show a significant improvement of the resolution. Clearly the boron-based set gives better spectrometric information in the epithermal region; indeed, it always correctly locates the peak (in most cases also with a correct amplitude), while the spectrum determined by the conventional spectrometer shows only a very broad structure.

CONCLUSIONS

An improved version of the multisphere neutron spectrometer has been developed. The simulations showed the possibility of improving the spectrometer performance (energy resolution, more uniform distribution of the response functions) in the epithermal and intermediate energy region by using neutron absorbers of different thicknesses around the small spheres. They led to the choice of natural boron and allowed the selection of an optimised set of sphere diameter–boron thickness combinations. The prototype boron-based spectrometer, with practical dimensions and weight, was calibrated with monoenergetic neutron beams between 0.186 and 144 keV and its response matrix was determined by adjusting the computed response functions to the calibrated points. The investigation of its performance by simulation confirmed the improvement of the energy resolution in the epithermal region, with no dramatic decrease in sensitivity.

ACKNOWLEDGEMENTS

The present work has been performed under contract No 94.0039 with the Swiss Federal Office for Education and Science; the authors are grateful for this financial support. The cooperation of Dr E. Dietz (PTB) during the experimental calibration of the system is gratefully acknowledged.

REFERENCES

1. Wang, C. K. and Blue, T. E. *A Neutron Spectrometer for Neutrons with Energies between 1 eV and 10 keV*. Nucl. Instrum. Methods A**290**, 237–241 (1990).
2. Perks, C. A. and Gibson, J. A. B. *Neutron Spectrometry and Dosimetry for Boron Neutron Capture Therapy*. Radiat. Prot. Dosim. **44**(1/4), 425–428 (1992).
3. Aroua, A., Boschung, M., Cartier, F., Grecescu M., Prêtre, S., Valley, J.-F. and Wernli, C. *Characterisation of the Mixed Neutron–Gamma Fields Inside the Swiss Nuclear Power Plants by Different Active Systems*. Radiat. Prot. Dosim. **51**(1), 17–25 (1994).
4. Lindborg, L. and Klein, H. (eds) *Determination of Neutron and Photon Dose Equivalent at Work Places in Nuclear Facilities of Sweden: An SSI-EURADOS Comparison Exercise*. SSI-Report 95-15 (Stockholm: SSI) (1995).
5. Chartier, J.L., Posny, F. and Buxerolle, M. *Experimental Assembly for the Simulation of Realistic Neutron Spectra*. Radiat. Prot. Dosim. **44**(1/4), 125–130 (1992).
6. Kryuchkov, V. P. and Semenova, G. I. *Neutron Responses for Sphere Spectrometer*. (in Russian) IHEP Reprint 87-191 (Serpukhov: IHEP) (1987).
7. Aroua, A., Grecescu, M., Lerch, P., Prêtre, S, Valley, J.-F. and Vylet, V. *Evaluation and Test of the Response Matrix of a Multisphere Neutron Spectrometer in a Wide Energy Range, Part 1. Calibration*. Nucl. Instrum. Methods A**321**, 298–304 (1992).
8. Engle, W. W., Jr *A User Manual for ANISN, A One Dimensional Discrete Ordinates Transport Code with Anisotropic Scattering*. AEC Research and Development Report K 1693 (1967).
9. Aroua, A., Grecescu, M., Lanfranchi, M., Lerch, P., Prêtre, S. and Valley, J.-F. *Evaluation and Test of the Response Matrix of a Multisphere Neutron Spectrometer in a Wide Energy Range, Part 2. Simulation*, Nucl. Instrum. Methods A**321**, 305–311 (1992).
10. Guldbakke S., Dietz, E., Kluge, H. and Schlegel, D. *PTB Neutron Fields for the Calibration of Neutron Sensitive Devices*. In: Strahlenschutz: Physik und Messtechnik, Band I, pp. 240–247 (Fachverband für Strahlenschutz e.V.) (1994)
11. McElroy, W. N., Berg, S., Crockett, T. and Hawkins, R. *A Computer-Automated Iterative Method for Neutron Flux Spectra Determined by Foil Activation*. Report AFWL-TR-67-41, Vols I–IV (US Air Force Weapons Lab., Kirtland, AFB, New Mexico) (1967).

HIGH ENERGY RESPONSE FUNCTIONS OF BONNER SPECTROMETERS

A. V. Sannikov*, V. Mares and H. Schraube
GSF — Forschungszentrum für Umwelt und Gesundheit
Institut für Strahlenschutz, Neuherberg
Postfach 1129, D-85758 Oberschleißheim, Germany

Abstract — Response functions of Bonner sphere spectrometers were calculated for neutron energies from 10 MeV to 1.5 GeV by the Monte Carlo high energy transport code HADRON. Calculations were made for two types of thermal neutron detector inside the Bonner spheres: ^{3}He proportional counter and ^{6}LiI scintillation detector. The results obtained are compared with calculations using the MCNP and the LAHET Monte Carlo codes. An analysis of some possible error sources in the calculation of high energy response functions and in the interpretation of the experimental data in high energy neutron fields is presented.

INTRODUCTION

The problem of radiation safety of air crews arising after ICRP Publication 60[1] has created fresh interest in the response functions of active and passive dosimetric devices for high energy radiations. This is related first of all to neutron measuring devices due to the large contribution of high energy neutrons above 20 MeV to the total dose equivalent at the altitudes of civil air flights[2].

The present paper is devoted to Monte Carlo calculations of the response functions of Bonner sphere spectrometers (BSS) in the neutron energy range from 10 MeV to 1.5 GeV. Such calculations performed earlier by different multigroup numerical programmes up to 400 MeV have shown large discrepancies between them. Three different Monte Carlo codes were therefore used to try to understand the sources of systematic errors. Another aim was to estimate the additional contribution of secondary charged particles produced by high energy hadrons in the detectors for thermal neutrons and in the surrounding volume.

COMPUTER CODES

The calculations were made by the MCNP4A[3], HADRON[4,5] and LAHET[6] Monte Carlo computer codes. The MCNP code was extended up to 100 MeV using the neutron cross sections for hydrogen and carbon from the LA100 library[7].

The high energy transport code HADRON is based on the cascade-exciton model of nuclear reactions. This model includes a cascade stage (modified version of the Dubna cascade model), a pre-equilibrium stage (exciton model) and an equilibrium evaporation stage. Low energy neutrons are transported by the Monte Carlo code FANEUT. The results for low neutron energies responses of the ^{3}He BSS were found to be in a good agreement with the MCNP data[8].

The LAHET code, developed at LANL in Los Alamos, is based on the HETC high energy transport code of ORNL in Oak Ridge. This programme has different options for simulation of inelastic interactions of hadrons with nuclei: Bertini and the ISABEL cascade models. At the de-excitation stage of nuclear reaction, two versions of the pre-equilibrium exciton model or the Fermi break-up model may be used. The transport of the low energy neutrons is calculated by the code HMCNP which is a modified version of MCNP.

RESULTS AND DISCUSSION

Response comparison

The ^{3}He Bonner sphere spectrometer (BSS) considered in this paper is based on the spherical proportional counter manufactured by Centronic Ltd (Type SP90), 32 mm in diameter, filled with ^{3}He gas of the nominal pressure of 172 kPa and krypton of 100 kPa. The helium number density was taken to be 4.25×10^{19} (cm^{-3}) at 293 K, as in the calculations[8] for low energy neutrons. The counter wall is made from 0.51 mm thick stainless steel. The reaction of interest is ^{3}He(n,p)^{3}H with Q = 0.765 MeV. A typical pulse-height spectrum for low energy neutrons, characterised by a main peak corresponding to the total absorption of protons and tritons in a gas, and by a wide plateau due to wall effects, ranges approximately from 0.2 to 0.9 MeV.

The low energy neutron cross sections for ^{3}He nuclei were taken from the ENDF/B-VI library[9] and extrapolated above 20 MeV. The results of calculations performed by the three Monte Carlo codes are shown in Figure 1 for several Bonner spheres. The data obtained by the MCNP[10] and HADRON programmes agree within the limits of 15% below 100 MeV. This is not the case for the LAHET data which are much higher as a rule. Calculations with LAHET were made using the standard option that includes the Bertini cascade model and the Fermi break-up model.

---

*On leave from the Institute of High Energy Physics, 142284 Protvino, Moscow Region, Russia.

There are different possible reasons for such discrepancies. One of them is that the total inelastic cross sections of hadron interactions with nuclei are calculated by the LAHET code itself in the frame of the cascade model. Conversely, MCNP and HADRON use evaluations of experimental data. The more important source of errors was found in the comparison of calculated secondary neutron spectra from the reaction C(p,xn) at 113 and 256 MeV with the experimental data[11,12]. The HADRON data are in better agreement with experiment at low neutron energies where LAHET, in the version employed, gives an overestimation up to a factor of 2.

The more pronounced differences for small spheres in Figure 1 may be explained by the large contribution of low energy neutrons from the first interaction of neutrons with carbon nuclei. In the case of large spheres, this effect is masked by averaging the processes of neutron production and absorption over several collisions.

Nevertheless, for the present times we cannot draw any final conclusion about the LAHET code because of insufficient experience of its use. However, it is strongly recommended that the LAHET calculations be performed using the more recent version of the code, including the new physics models like the ISABEL intranuclear cascade model and the multistage pre-equilibrium exciton model. Unfortunately, this version of the code is not yet available in the public domain at this time.

HADRON calculations

Calculations with HADRON were made simultaneously with and without taking into account the contribution of secondary protons and pions to a hadron cascade in polyethylene. For this purpose, each neutron had an additional weight that was equal to unity if all the parents of this neutron were neutrons. In other cases it was set to zero. The obtained contribution of secondary charged hadrons to the total response of the ^{3}He BSS is presented in Table 1. It may be seen that this effect, not considered in earlier multigroup numerical calculations, becomes important at neutron energies above 200 MeV and for large Bonner spheres.

The full set of the ^{3}He BSS smoothed response functions is shown in Figure 2 by solid curves from 1 MeV to 1500 MeV (MCNP below 10 MeV and HADRON above 10 MeV). These responses include only ^{3}He(n,p)^{3}H events with energy depositions from 0.2 to 0.9 MeV. The dashed curves in Figure 2 show the total responses including all events above 0.2 MeV.

Table 1. Per cent contribution of neutrons, produced by secondary protons and pions in polyethylene, to the total response of the ^{3}He BSS with diameters d.

d (in)	Neutron energy (MeV)		
	200	500	1000
5	1.4	8.8	15.1
8	1.0	10.2	19.9
12	1.2	10.1	21.5
18	1.7	16.8	28.1

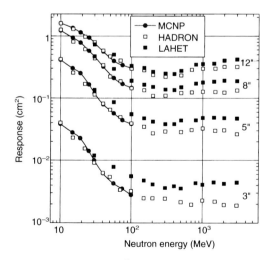

Figure 1. Comparison of the ^{3}He BSS responses calculated by the Monte Carlo codes MCNP, HADRON and LAHET.

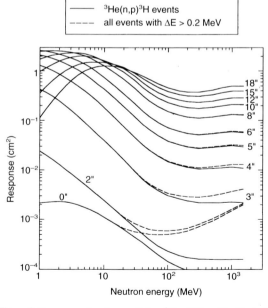

Figure 2. Response functions of the ^{3}He BSS with (dashed curves) and without (solid curves) contribution of secondary charged particles from high energy hadrons calculated by the codes MCNP (below 10 MeV) and HADRON (above 10 MeV).

These responses have been calculated by the HADRON code taking into account the additional contribution of secondary charged particles produced by high energy hadrons in a counter and surrounding volume. The main source of such events are protons created by hadrons in the counter wall. Although this effect increases considerably the responses of small spheres to high energy neutrons, it is not assumed to be important for broad high energy neutron spectra, e.g. outside shields of high energy sources.

This situation is drastically changed for the BSS with a ^{6}LiI thermal neutron detector due to its much higher density. The calculations have been performed for the 4 mm diameter × 4 mm ^{6}LiI(Eu) scintillation crystal in the geometry considered recently[13]. A typical experimental pulse-height spectrum for low energy neutrons is compared in Figure 3 with HADRON calculations for monoenergetic high energy neutrons. The latter event spectra are much wider than the spectrum of the ^{6}Li(n,α)^{3}H reaction and range up to 100 MeV and more for 1 GeV neutrons.

The response functions of the ^{6}LiI BSS calculated with and without taking into account secondary charged particles produced by high energy hadrons in the detector are shown in Figure 4. It can be seen that the high energy hadron events provide a considerable contribution to the responses of all Bonner spheres. This factor makes more difficult the problem of interpretation of pulse-height spectra in high energy neutron fields compared to the low energy ones. This is related first of all to the background subtraction. One of possible ways is to try to extract a gaussian-like ^{6}Li(n,α)^{3}H peak from the event spectra and to consider all other events as a background. The response functions shown by solid curves in Figure 4 must be used in this case in the unfolding problem.

CONCLUSIONS

Comparison of the ^{3}He BSS response functions calculated by three Monte Carlo codes has shown good agreement of the MCNP and HADRON data up to 100 MeV. These are, however, inconsistent with the LAHET results above approximately 30 MeV. The main reason for such discrepancies is, in our opinion, the overestimation of the low energy tail of secondary neutron spectra from inelastic collisions of high energy neutrons with carbon nuclei calculated by the LAHET code. The contribution of secondary charged particles produced by high energy hadrons in a detector and surrounding volume was found to be not very important for the ^{3}He BSS. In the case of ^{6}LiI BSS, this factor considerably changes the response functions in the high energy range and makes the problem of interpretation of pulse-height spectra in high energy neutron fields much more difficult.

ACKNOWLEDGEMENTS

This work was supported by the Commission of the European Union under contract FI3P CT92-0026.

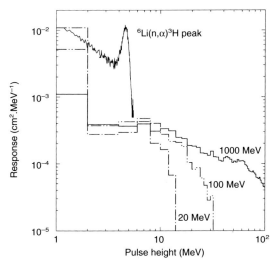

Figure 3. Calculated event spectra from secondary charged particles produced by high energy neutrons in the bare ^{6}LiI detector (4 mm diameter × 4 mm) in comparison with the experimental pulse-height spectrum for low energy neutrons (relative units).

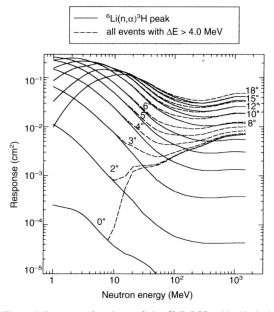

Figure 4. Response functions of the ^{6}LiI BSS with (dashed curves) and without (solid curves) contribution of secondary charged particles from high energy hadrons calculated by the codes MCNP (below 10 MeV) and HADRON (above 10 MeV).

REFERENCES

1. ICRP. *1990 Recommendations of the International Commission on Radiological Protection.* Publication 60 Ann. ICRP, **21**(1–3) (Oxford: Pergamon Press) (1991).
2. *Radiation Exposure of Civil Aircrew.* Proc. Workshop, Luxembourg, 1991. Radiat. Prot. Dosim. **48**(1) (1993).
3. Briesmeister, J. F. (Ed.) *MCNP — A General Monte Carlo N-Particle Transport Code, Version 4A.* LA-12625-M (Los Alamos National Laboratory) (1993).
4. Savitskaya, E. N. and Sannikov, A. V. *High Energy Neutron and Proton Kerma Factors for Different Elements.* Radiat. Prot. Dosim. **60**, 135–146 (1995).
5. Sannikov, A. V. and Savitskaya, E. N. *Ambient Dose Equivalent Conversion Factors for High Energy Neutrons Based on the ICRP 60 Recommendations.* Radiat. Prot. Dosim. **70**(1–4) 383–386 (This issue) (1997).
6. Prael, R. E. and Lichtenstein, H. *User Guide to LCS: The LAHET Code System.* LANL Report LA-UR-89-3014 (Los Alamos National Laboratory) (1989).
7. Young, P. G. et al. *Transport Data Libraries for Incident Proton and Neutron Energies to 100 MeV.* LANL Report LA-11753-MS (Los Alamos) (1990).
8. Mares, V., Schraube, G. and Schraube, H. *Calculated Neutron Response of a Bonner Sphere Spectrometer with ^{3}He Counter.* Nucl. Instrum. Methods, A**307**, 398–412 (1991).
9. Hendricks, J. S., Frankle, S. C. and Court, J. D. *ENDF/B-VI Data for MCNP.* LANL Report LA-12891 (Los Alamos) (1994).
10. Mares, V. and Schraube, H. *Improved Response Matrices of Bonner Sphere Spectrometers with ^{6}LiI Scintillation Detector and ^{3}He Proportional Counter between 15 and 100 MeV.* Nucl. Instrum. Methods, A**366**, 203–206 (1995).
11. Meier, M. M. et al. *Differential Neutron Production Cross Sections and Neutron Yields from Stopping-length Targets for 113-MeV Protons.* LANL Report LA-11518-MS (Los Alamos) (1989).
12. Meier, M. M. et al. *Differential Neutron Production Cross Sections and Neutron Yields from Stopping-length Targets for 256-MeV Protons.* LANL Report LA-11656-MS (Los Alamos) (1989).
13. Mares, V. and Schraube, H. *Evaluation of the Response Matrix of a Bonner Sphere Spectrometer with LiI Detector from Thermal Energy to 100 MeV.* Nucl. Instrum. Methods, A**337**, 461–473 (1994).

MEASUREMENTS WITH THE PTB 'C' BONNER SPHERE SPECTROMETER IN THE PSI VILLIGEN 55 MeV NEUTRON FIELD FOR SPECTROMETRY AND CALIBRATION PURPOSES

A. V. Alevra and U. J. Schrewe
Physikalisch-Technische Bundesanstalt
Bundesallee 100, D-38116 Braunschweig, Germany

Abstract — The PTB Bonner sphere spectrometer was used for measurements in the 55 MeV quasi-monoenergetic neutron beam at PSI, Villigen. The spectral fluence of the neutron background was determined using the fluence response matrix well specified for energies below 20 MeV. The spectral fluence of the neutron beam was analysed, with the background subtracted, by testing five different high energy extensions of the existing response matrix, and compared with time-of-flight spectral and telescope integral fluence measurements taken as reference. The high energy response shapes which agreed best with the references were used to extend our response matrix.

INTRODUCTION

For some time now there has been an increase in human activity in environments where high energy neutrons (from tens of MeV to several GeV) are present in significant quantities. The radiation fields around high energy particle accelerators, at neutron or proton therapy facilities, or the fields induced by cosmic rays in flight altitudes of modern civil aircraft, are typical examples of such environments. As a consequence, there is an increasing interest in reliable dosimetry and, by implication, in the spectrometry of high energy neutron fields.

Among the existing instruments usually utilised in neutron field spectrometry below 20 MeV, the Bonner spheres (BS) could possibly be used at higher energies, provided that their fluence responses in the extended energy range can be properly established.

The measurement reported in this paper was performed in the 55 MeV quasi-monoenergetic neutron field at the Paul Scherrer Institute (PSI) in Villigen, Switzerland. The considerable contribution of the neutron background could be determined using the already existing fluence response matrix of our BS spectrometer (BSS) for neutron energies below 20 MeV. For the analysis of the high energy component of the field, the response matrix was extended using calculated high energy response shapes taken from the literature.

THE NEUTRON FIELD

Intense quasi-monoenergetic neutron beams were produced by bombarding a 2 mm thick beryllium target with 60.9 MeV protons. The ^{9}Be(p,n)^{9}B reaction strongly feeds the ground state and the lowest excitation level of ^{9}B, producing an almost monoenergetic neutron peak at 55 MeV (a 12% half-width peak followed by a lower energy distribution as shown in Figure 4, TOF).

The proton beam passing the Be target is bent by a magnet and stopped in a Faraday cup, while the neutrons from the target are collimated in the initial proton beam direction, 0°. The collimation is axially symmetric; the collimator, which is made of brass, has a circular aperture which first decreases with the distance from the target, reaches its minimum of 13.5 mm in diameter at 1520 mm from the target, then increases continuously, reaching a diameter of 25.5 mm at the end of the collimator, at 2720 mm from target[1].

The actual spectral neutron fluence and the real shape of the beam profile are influenced by weak transitions into states of higher excitation of ^{9}B, other reaction channels, energy loss of the incident protons in the target, the finite size of the beam spot, and neutron scattering from various parts, e.g. target, collimator walls, air. All these together cause, for the spectral neutron fluence, a significant energy spreading up to a broad low energy tail, and for the beam profile a broadening and smooth edges.

The beam profiles at various distances from the target were measured with various small neutron fluence sensors (pin diode, NE102 scintillator, low pressure proportional counter) moved in the plane perpendicular to the neutron beam axis in the 0° (horizontal), 90° (vertical) and 45° directions. The measured beam profiles can be described in a common way as a superposition of Gaussian distributions:

$$\frac{\phi(x,\rho)}{\phi(x,\rho=0)} = \frac{1}{\sigma\sqrt{2\pi}} \int_{-\omega}^{+\omega} \exp\left[-\frac{1}{2}\left(\frac{\xi-\rho}{\sigma}\right)^2\right] d\xi \qquad (1)$$

where $\phi(x,\rho)$ denotes the fluence at the point defined by x, i.e. the distance from the target, and ρ, the distance to the beam axis. The least squares fit of different beam profiles measured at various distances x produced the values $(\omega/x) = 4.2 \times 10^{-3}$ and $(\sigma/x) = 8 \times 10^{-4}$ for the parameters ω (describing the width of the beam profile) and σ (describing the smoothness of its edge) respectively. The solid angle of the neutron beam determined from these parameters is 5.5×10^{-5} sr.

The spectral neutron fluence was measured with a

cylindrical NE213 scintillation detector, 51 mm in diameter and 102 mm in length, by applying time-of-flight (TOF) techniques for neutron energy determination and pulse shape analysis for neutron–photon discrimination. The limited length of the flight paths, the time resolution of the scintillator and the repetition rate of the pulsed proton beam restricted the application of the TOF techniques to neutron energies greater than about 4 MeV[2].

The neutron fluence of the collimated beam was monitored by means of (i) the proton beam charge collected in the Faraday cup (proton monitor), and (ii) the number of counted events produced in a 2 mm thick NE102 plastic scintillator which was positioned behind the neutron collimator (neutron monitor). By continuous on-line controlling of the monitor ratios, the drifts of the proton beam transport system could be detected and corrected, guaranteeing in this way stable beam profiles in all measurements.

The absolute neutron fluence was measured with a proton recoil telescope (PRT)[3]. A polyethylene radiator foil was placed at the centre of the beam, and the recoil protons emitted from the (n,p) scattering in a small range of angles around zero degrees were detected and separated from other reaction products using two proportional counters and two silicon detectors in quadruple coincidence. Because of the small range of proton scattering angles accepted by the PRT, the proton energy was proportional to the energy of incident neutrons, and by applying proton energy discrimination the absolute neutron fluence of the collimated neutrons with energies greater than 48 MeV, ϕ_0, could be determined. In the measurement position, the area with spatially almost constant neutron fluence was larger than the radiator area so that the measured fluence is about equal to $\phi_0(x,\rho=0)$ described in Equation 1. At a distance of 10.6 m from the target, where the BSS measurements were made, the neutron fluence was found to be 5.14 ± 8.0% $cm^{-2}.nC^{-1}$ relative to the proton beam charge and 0.2436 ± 8.0% cm^{-2} per monitor count relative to the NE102 monitor. The total number of neutrons in the beam, N_0, was obtained by integration of Equation 1:

$$\frac{N_0}{Q} = \frac{2\pi}{Q} \int_0^\infty \rho\, \phi_0(x,\rho)\, d\rho \qquad (2)$$

THE MEASUREMENTS

The PTB 'C' BSS and the unfolding procedure used in this work are briefly described in Reference 4 and the references therein.

The majority of the spheres involved in the measurements have diameters considerably larger than the neutron beam cross section at 10.6 m distance. In order to irradiate the spheres with a spatially constant neutron fluence covering their whole cross section, a scanning procedure was applied. The spheres were moved in the plane perpendicular to the neutron beam within a square with a lateral length of 2a. If we describe the position of a sphere in the plane at the time t by the Cartesian coordinates y(t) and z(t), a path covering almost homogeneously the whole square can be described by $y(t) = (-1)^n \dot{y}t$ and $z(t) = (-1)^m \dot{z}t$. $\dot{y}$ and $\dot{z}$ denote the velocities. The exponents n and m are integer numbers which are incremented by 1 at the moments when $|y(t)| = a$ and $|z(t)| = a$, respectively, while the velocities are chosen such that $\dot{z} = (k/l)\, \dot{y}$, where k and l are integer numbers without any common factor except unity, their ratio being, however, close to unity. The velocities in both directions were thus chosen such that they were about the same, $0.1\, m.s^{-1}$, so that the cycle time for the periodical movement in each direction was of the order of 10 s, much shorter than the typical irradiation time of 500 s to 1000 s. The time-dependent fluctuations of the neutron beam intensities were therefore averaged, and the mean fluence per proton beam charge valid for any point of the square could be obtained from Equation 2 as $\phi_0^{SCAN}/Q = (N_0/a^2)/Q$ (and in a similar way when related to the NE102 neutron monitor).

BSs with diameters up to 10 inches were measured with a 30 cm × 30 cm scanning area, the 10, 12 and 15 inch spheres with 50 cm × 50 cm scanning area and the 15 and 18 inch spheres at 60 cm × 60 cm. The measurements at two different scanning areas performed with the 10 inch and the 15 inch spheres served to verify the procedure of fluence determination.

During scanning, the BSs detected not only collimated neutrons from the target, but also background consisting of neutrons scattered by the walls and the air of the experimental hall. The background was measured with the BSs placed outside the scanned area. Figure 1 shows the relative BS readings due to the background and to the direct (collimated) neutrons (i.e. the readings obtained by scanning minus the background readings) as a function of the sphere diameter. The zero diameter indicates the bare counter. The uncertainty bars reflect

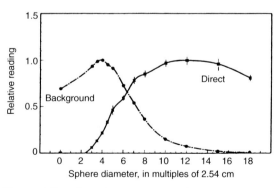

Figure 1. BS readings due to background and direct neutrons (×) Direct relative to 12″ (0.5859 counts per monitor count). (○) Background relative to 4″ (0.5198 counts per monitor count).

the statistical standard deviations of the monitor and BS readings. The deviations of the curves from smooth shapes are compatible with the statistical uncertainties.

THE HIGH ENERGY RESPONSES OF THE BONNER SPHERES

For the unfolding of the direct spectrum, fluence response values of up to about 100 MeV are necessary. From the literature five different high energy BS responses were selected, calculated in most cases for systems differing from ours. Assuming that the high energy shapes of these responses, especially those of the large spheres which are of interest in this case, do not depend essentially on the system details (e.g. central detector geometry), all five variants were tested. These are:

(i) the IAR-L responses[5],
(ii) the HASL responses[6],
(iii) the GSF-1 responses[7],
(iv) the GSF-2 responses, a revised version of the previous ones[8], and
(v) the LANL responses[9].

In fact, from the mentioned references only the high-energy shapes were used, which were normalised to our response values at energies of about 10 MeV in order to extend the responses of our BSs up to several hundreds of MeV. Figure 2 shows as examples the extended responses of the 12 and 18 inch spheres. The literature data were reduced to the energy groups used by us (logarithmically equidistant, five per order of magnitude) and completed by means of cubic spline interpolations for sphere diameters used in our BSs. In one case, for the IAR-L shapes, the response functions were extrapolated as shown in Figure 2, because no 18 inch calculated data were given.

RESULTS

Figure 1 suggests that the direct spectrum consists chiefly of high energy neutrons, while the background has strong thermal and epithermal components but a negligible contribution at high energies. The background neutron spectrum could therefore be unfolded from the measured data using the unextended response matrix of our BSs, which is well specified from thermal to 20 MeV neutron energies. The result is shown in Figure 4.

The direct spectrum is already known from TOF measurements for energies above 5 MeV. This known part, rebinned in our five energy groups per order of magnitude and completed at lower energies with various shapes which were tested, was used as *a priori* information in the unfolding procedure. With the five different high energy extensions of the response matrix, five different solutions were obtained for the direct neutron spectrum which are shown in Figure 3.

From the five high energy extensions of the responses, the IAR-L reproduces best the TOF spectrum shape and the absolute fluence determined with the PRT within the limits of one standard deviation of ±8% of the measurement. For this reason, the IAR-L high energy response shapes were retained for the extension of our response matrix up to 398 MeV. Finally, the resulting matrix was extrapolated up to 1 GeV, keeping constant the negative slopes of the response functions in a log-lin representation as in Figure 2.

Figure 4 presents the results obtained with the BSs in

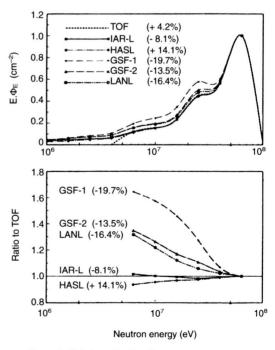

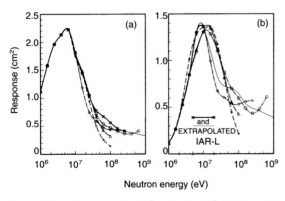

Figure 2. Responses of (a) 12″ and (b) 18″ spheres. (×) IAR-L, (◇) HASL, (□) GSF-1, (△) GSF-2, (○) LANL.

Figure 3. Solutions for the direct neutron spectrum.

this work compared with the TOF neutron spectrum shown in its original energy binning. The total spectrum as measured in the scanned area can practically be divided in two regions: at high energies, the direct neutrons from the target and, at low energies, background neutrons. Only in the energy range from 10 keV to 4 MeV does each of the two components contribute to the total spectrum with more than 10%. Altogether, the background represents about a third of the total neutron fluence.

CONCLUSIONS

Using the Bonner sphere spectrometer, knowledge of the PSI 55 MeV quasi-monoenergetic neutron field has improved by determining the spectral fluence distribution of the background neutrons and that of the collimated neutrons for energies below 5 MeV. The already known high energy part of the neutron spectrum could be used to obtain an experimentally based high energy extension of the Bonner sphere fluence response matrix. Our extrapolation of the response matrix up to 1 GeV has no experimental support from the present work and is not in agreement with the LANL data which indicate important response increases above 250 MeV. As the available data (LANL) are given only in few energy points and do not contain the 15 inch BS, more detailed calculations are necessary.

The extended response matrix obtained in this work has already been successfully used to investigate high energy neutron fields at CERN-Genève[4], where neutrons with energies above 250 MeV made no remarkable contributions.

ACKNOWLEDGEMENTS

The authors express their gratitude to Dr H Klein who contributed to the preparation of this work by valuable discussions and suggestions, and Drs V Dangendorf and W. Newhauser for their help during the measurements. This work was partially supported by the Commission of the European Communities within the framework of the project FI3P-CT92-0002.

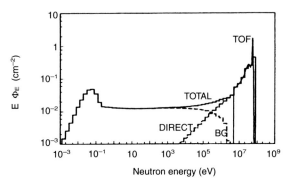

Figure 4. Comparison of this work and TOF spectrum.

REFERENCES

1. Henneck, R., Gysin, C., Hammans, M., Jourdan, J., Lorenzon, W., Pickar, M. A., Sick, I., Burzynski, S. and Stammbach, T. *A Facility for Monoenergetic Polarized Neutrons of 30–70 MeV*. Nucl. Instrum. Methods Phys. Res. **A259**, 329–340 (1987).
2. Nolte, R., Schuhmacher, H., Brede, H. J. and Schrewe, U. J. *Measurements of High Energy Neutron Fluence with Scintillation Detector and Proton Recoil Telescopes*. Radiat. Prot. Dosim. **44**, 101–104 (1992).
3. Schuhmacher, H., Siebert, B. R. L. and Brede, H. J. *Measurement of Neutron Fluence for Energies Between 20 MeV and 65 MeV Using a Proton Recoil Telescope*. In: Proc. NEANDC Specialists' Meeting on Neutron Cross Section Standards for the Energy Region above 20 MeV, Uppsala 1991. OECD: NEANDC-305, pp. 123–134 (1991).
4. Alevra, A. V., Klein, H. and Schrewe, U. J. *Measurements with the PTB Bonner Sphere Spectrometer in High-Energy Neutron Calibration Fields at CERN*. Report PTB-N-22 (Braunschweig, Germany) (December 1994).
5. Aroua, A., Grecescu, M., Lanfranchi, M., Lerch, P., Prêtre, S. and Valley, J.-F. *Evaluation and Test of the Response Matrix of a Multisphere Neutron Spectrometer in a Wide Energy Range. Part II. Simulation*. Nucl. Instrum. Methods **A321**, 305–311 (1992).
6. Sanna, R. S. *Thirty-one Group Response Matrices for the Multisphere Neutron Spectrometer over the Energy Range Thermal to 400 MeV*. USAEC, HASL-267 (March 1973).
7. Mares, V. and Schraube, H. *Evaluation of the Response Matrix of a Bonner Sphere Spectrometer with LiI Detector from Thermal Energy to 100 MeV*. Nucl. Instrum. Methods Phys. Res. **A337**, 461–473 (1994).
8. Schraube, H. Private Communication. GSF, Neuherberg (1995).
9. Hsu, H. H., Alvar, K. R. and Vasilik, D. G. *A New Bonner Sphere Set for High Energy Neutron Measurements: Monte Carlo Simulation*. In: Proc. IEEE Nucl. Sci. Symp., Oct–Nov. 1993, pp. 845–850 (1994).

CALIBRATION OF A NEUTRON SPECTROMETER IN THE ENERGY RANGE 144 keV TO 14.8 MeV WITH ISO ENERGIES

W. Rosenstock, T. Köble, G. Kruziniski and G. Jaunich
Fraunhofer-Institut für Naturwissenschaftlich-Technische Trendanalysen (INT)
PO Box 1491, D-53864 Euskirchen, Germany

Abstract — For calibration purposes measurements were performed with the neutron spectrometer system ROSPEC (gas proportional counters) and SSS (plastic scintillation detector) at ISO energies from 144 keV to 14.8 MeV. The spectrometer ROSPEC consists of four spherical proportional counters filled at different pressures to cover the complete energy range from 50 keV to 4.5 MeV. The SSS extends the energy range from 4 MeV to 15 MeV. The experimental data prove that both instruments are able to reproduce the nominal neutron energies within an accuracy of less than 8%. The fluence measurements show in the energy range from 50 keV to 2.5 MeV a systematic overestimation of approx. 6% of the absolute neutron fluences. Between 4.2 MeV and 14 MeV these systematic deviations are below 8% and above 14 MeV those deviations are completely within the statistical error of less than 3%.

INTRODUCTION AND AIMS

The equivalent dose of neutrons depends strongly on the neutron energy for the same fluence. For optimising radiation shielding and for neutron dose measurements the neutron energy spectra must be well known in order to get satisfactory results. In the case of pulsed neutron sources these spectra may be obtained by time-of-flight spectroscopy. If non-pulsed neutron sources such as spontaneous fission sources or radioactive waste are considered, the situation is much more difficult. In this case the neutron spectrum can be obtained by, for example, utilising the (n,p) reaction (detecting the recoiling charged particle) and unfolding the resultant pulse height spectrum. The spectrometer system discussed here makes use of this proton recoil and unfolding technique. In order to get data on the precision and reliability of the system measurements were made with this spectrometer system by exposing it to energy and intensity calibrated neutron beams at the Physikalisch Technische Bundesanstalt (PTB) in Braunschweig, Germany.

DESCRIPTION OF THE NEUTRON SPECTROMETERS ROSPEC AND SSS

The Rotating Neutron Spectrometer (ROSPEC)* consists of four spherical gas proportional counters. Three have a diameter of 5.08 cm and are filled with pure hydrogen at pressures of 76 kPa, 405 kPa and 1.01 MPa (0.75, 4 and 10 atm). The fourth has a diameter of 15.24 cm and is filled with a mixture of argon and methane at a pressure of 507 kPa (5 atm). Each detector covers a well-defined energy range, overlapping with the energy range of the detector for the next higher energy. In all, a neutron spectrum can be measured from 50 keV to 4.5 MeV. The neutrons are detected by the measurement of the recoiling protons produced in the gas proportional counters. The four detectors are mounted on a circular aluminium platform which rotates about a vertical axis to average out local variations in the neutron field. Therefore the neutron field is determined in the centre of the rotation circle.

The complete electronics for the detectors consisting of pre-amplifiers, shaping amplifiers and pulse height analysers are also mounted on or under this circular plate. The whole assembly is located in an aluminium cylinder with a diameter of 40.5 cm and a height of 49 cm and weighs 20 kg. It is fed by a systems power supply and connected for control and data acquisition with a PC notebook computer via a RS422-Interface[1,2].

The Simple Scintillation Spectrometer (SSS) is designed to extend the upper limit of the energy range of ROSPEC, so that with both instruments measurements can be made up to neutron energies of 15 MeV. The unfolding of the pulse height spectra from both spectrometers is done by a modified version of the SPEC-4 CODE[3] on the PC notebook.

The SSS consists of two parts: a probe containing the neutron detector, its photomultiplier and a high voltage supply; and an analyser module, which includes a shaping amplifier, analogue-to-digital converter, display and systems power supply. The active detector element is an array of small plastic scintillators, which are 3 mm in diameter and 3 mm long. Gammas produce only small pulses in scintillators with these dimensions, which are rejected by a discriminator.

THE MEASURING PROCEDURE AT THE PTB

At the accelerator facility of the PTB monoenergetic neutrons are produced by various nuclear reactions[4,5].

Measurements were performed at the following ISO (International Organization for Standardization)

---
*ROSPEC and SSS are manufactured by Bubble Technology Industries, Chalk River, Ontario, Canada.

energies: 0.144 MeV, 0.565 MeV, 1.2 MeV, 2.5 MeV, 5.0 MeV and 14.8 MeV[6]. In addition measurements were made at 0.100 MeV, 0.842 MeV, 4.2 MeV, 13.89 MeV and 14.3 MeV.

After producing neutrons of a definite energy by means of the corresponding reaction, first an energy measurement was done by the PTB, for which a time-of-flight spectrometer was used. The flux measurement of the monoenergetic neutrons was performed by the PTB using hydrogen-filled proportional counters and a recoil proton telescope depending on the neutron energy[4].

The contribution of neutrons scattered in the air, at the detector support and from the thick surrounding concrete walls to the measured flux must be taken into account. This is achieved by an additional measurement placing an appropriate shadow cone between the neutron source and the detector. This shadow cone is a truncated cone, which consists of steel on the side facing the neutron source and of polyethylene on the other side. This background flux is subtracted from the preceding flux measurement in order to get the neutron flux caused only by neutrons, which directly reached the detector.

THE RESULTS OF THE CALIBRATION MEASUREMENTS

The energy calibration

Figures 1–3 show some of the unfolded neutron spectra measured with the ROSPEC and SSS at the various neutron energies. The measured neutron energy E is determined by calculating the centre of the measured energy distribution in the following way:

$$E = \frac{\sum_i E_i \Delta E_i F_i}{\sum_i \Delta E_i F_i},$$

with
E_i : mean energy of channel i
ΔE_i : energy width of channel i
F_i : neutron fluence in channel i

In Table 1 are listed the nominal irradiation energy and the measured neutron energy E with the corresponding peak width. It can be clearly seen that the measured neutron energies agree well with the nominal energies within the accuracy of the spectrometers. This accuracy is given by the peak width (FWHM). This peak width, which principally consists of the resolution effects caused by the instruments, the neutron beam width and the distance from the source to the detector, is primarily determined by the resolution of the detectors. The energy resolution was determined from the original pulse height spectra, because the energy channel width of the unfolded spectra was chosen in the order of magnitude of the energy resolution of the instruments by the manufacturer.

At the neutron energy of 14.8 MeV it was not possible to state the energy resolution, because the measured energy distribution is not symmetric due to electronic pulse height limiting.

In case of the SSS the peak width is independent of

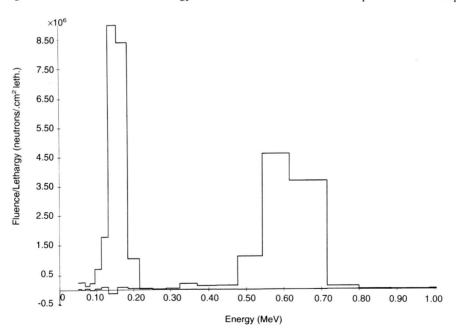

Figure 1. Spectra of monoenergetic neutrons with nominal energies of 100 keV and 565 keV.

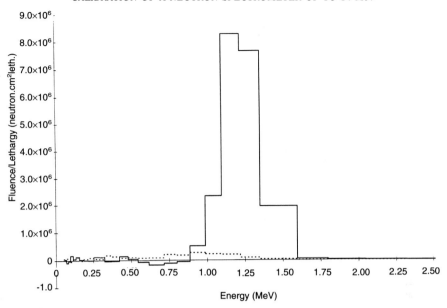

Figure 2. Spectrum of monoenergetic neutrons with a nominal energy of 1.2 MeV and the corresponding background measurement (dashed line).

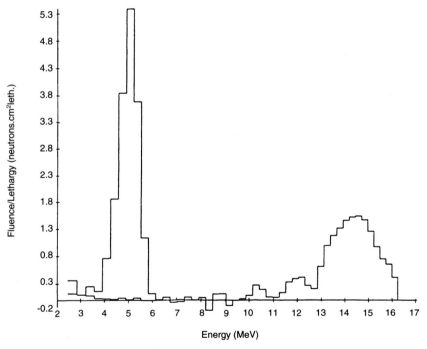

Figure 3. Spectra of monoenergetic neutrons with nominal energies of 5.0 MeV and 14.3 MeV.

the energy and around 19%, while for the ROSPEC measurements it is around 10%, except for 100 keV, where it is 31% (Table 1).

The flux calibration

Table 2 shows the results of the measured neutron flux. As indicated in Table 2 the flux measured with the spectrometer ROSPEC is systematically higher by approx. 6% in reference to the calibration standard (PTB) except for the energy of 4.2 MeV, where the flux is determined by the composition of the spectra of the two instruments. In the case of the flux measurements with the SSS the measured flux agrees with the calibration standard within the given accuracy.

The shadow cone measurements show that the background flux is between 2% and 11% of the total neutron flux for the ROSPEC measurements and between 1% and 2% for the SSS measurements. The higher background of the ROSPEC measurements mainly arises from the larger dimensions of the ROSPEC detectors and therefore a larger volume, into which neutrons may be scattered. The dashed line in Figure 2 shows the intensity of the backscattered neutrons at the energy 1.2 MeV measured with the shadow cone. At that energy the highest background (11.2% of the total neutron flux) was detected.

In order to examine the statistical errors in the neutron flux repeated measurements were performed at 0.100 MeV, 0.144 MeV, 0.565 MeV and 13.89 MeV. These errors are between 1% and 3%. To achieve these small errors a neutron fluence of greater than 10^5 n.cm^{-2} is needed. This corresponds to a measuring time of about 10 min according to the neutron source strengths at the PTB.

CONCLUSION

The neutron spectrometer system ROSPEC plus SSS represents a reliable and versatile system for neutron spectroscopy in the energy range from 100 keV up to 14.3 MeV. Its strong point is its ability to reproduce correctly the neutron energies with a good accuracy. For all neutron energies the deviation of the measured energy from the nominal energy is less than 8%. Except for the energy of 100 keV the energy resolution of the ROSPEC is in the order of 10%. For the scintillation spectrometer (SSS) the energy resolution is in the order of 20%.

The measured neutron fluence is slightly too high for energies lower than 4.2 MeV. In general the measured flux agrees with the calibration standard within 8% for all energies.

To get statistically good results (error less than 3%) a neutron fluence of some 10^5 n.cm^{-2} is needed in the

Table 1. Measured neutron energy and peak width of the measured energy spectra.

Spectrometer	Nominal energy (MeV)	Measured energy, E (MeV)	Peak width (FWHM) (%)
ROSPEC	0.100	0.10	31
ROSPEC	0.144*	0.15	11
ROSPEC	0.565*	0.61	10
ROSPEC	0.842	0.85	12
ROSPEC	1.20*	1.18	10
ROSPEC	2.50*	2.56	5.5
ROSPEC + SSS	4.20	4.04	14
SSS	5.00*	4.91	20
SSS	13.89	13.8	18
SSS	14.31	14.1	19

*ISO energies.

Table 2. Comparison of the measured neutron flux with the actual flux.

Nominal neutron energy (MeV)	Distance (source to detector) (cm)	Neutron flux (our measurements) (n.cm^{-2}.s^{-1})	Neutron flux (PTB measurements) (n.cm^{-2}.s^{-1})	Difference between the neutron flux measurements (%)
0.100	100	372.7 ± 7.7	345.0 ± 20.3	8.0
0.144	100	456.7 ± 3.9	428.0 ± 21.8	6.7
0.565	220	230.8 ± 1.4	218.3 ± 11.4	5.7
0.842	220	252.3	247.8 ± 14.5*	1.8
1.2	220	408.7	391.0 ± 15.9	4.5
2.5	220	912.3	861.9 ± 30.9	5.9
4.2	220	82.3	89.2 ± 3.5	−7.7
5.0	100	2172.5	2079.5 ± 79.2	4.5
13.89	103	4250.8 ± 85.6	4361.9 ± 100.0	−2.6
14.3	103	4866.7	5021.4 ± 247.0*	−3.1
14.8	100	5111.2	5140.5 ± 180.4	−0.6

*Not background-corrected.

case of a nearly monoenergetic energy distribution, and for a broader neutron spectrum (e.g. fission neutrons) the neutron fluence should exceed at least 10^6 n.cm^{-2}.

REFERENCES

1. Ing, H. *Using the ROSPEC Spectrometer*. Presented at NIST, Gaithersburg (1992).
2. Ing, H., Cross, W. G. and Bunge, P. J. *Spectrometers for Radiation Protection at Chalk River Nuclear Laboratories*. Radiat. Prot. Dosim. **10**(1–4), 137–145 (1985).
3. Benjamin, P. W., Kemshall, C. D. and Brickstock, A. *The Analysis of Recoil Proton Spectra*. AWRE Report (1968).
4. Strzelczyk, H. *Experimentelle Einrichtungen für Kalibrierungen am Dosimetriemeβplatz der Gruppe 6.5 "Neutronendosimetrie" der Physikalisch-Technischen Bundesanstalt* (Bericht ND-27) (Braunschweig: PTB) (1986).
5. Guldbakke, S., Dietze, E., Kluge, H. and Schlegel, D. *PTB Neutron Fields for the Calibration of Neutron Sensitive Devices*. Strahlenschutz: Physik und Meβtechnik, Vol. I, pp. 240–247 (Köln: Verlag TÜV Rheinland) (1994).
6. International Organization for Standardization (ISO). *Neutron Reference Radiations for Calibrating Neutron Measuring Devices used for Radiation Protection Purposes and for Determining their Response as a Function of Neutron Energy*. International Standard ISO 8529 (Geneva: ISO) (1989).

RECENT DEVELOPMENTS IN THE SPECIFICATION AND ACHIEVEMENT OF REALISTIC NEUTRON CALIBRATION FIELDS

J. L. Chartier†, B. Jansky‡, H. Kluge§, H. Schraube∥ and B. Wiegel§
†Institut de Protection et de Sûreté Nucléaire, DPHD/SDOS/LRDE
BP no 6, F-92265 Fontenay-aux-Roses Cedex, France
‡Nuclear Research Institute Rez. plc., Czech Republic
§Physikalisch-Technische Bundesanstalt, Postfach 33 45
D-38023 Braunschweig, Germany
∥GSF-Forschungszentrum-Neuherberg-Institut für Strahlenschutz
Ingolstädter Landstr. 1, D-85758 Oberschleißheim, Germany

Abstract — In order to calibrate more accurately the neutron dosemeters involved in radiation protection, the concept of 'Realistic Neutron Calibration Fields' is considered as an appropriate alternative solution, making necessary new irradiation facilities which generate well-characterised neutron fields with energy and angular distributions replicating more closely practical workplace conditions. Several experienced laboratories have collaborated on a European project and proposed various approaches which are reviewed in this paper. A short description of the facilities currently in operation is given as well as a few characteristics of the available radiation fields. This description of the state of art is followed by a discussion of the problems to be solved for using such facilities for calibration purposes according to well-specified calibration procedures.

INTRODUCTION

For the past ten years, the spectral neutron fluences of mixed neutron–photon fields encountered at many workplaces in nuclear installations have been measured and the results compiled in well-documented publications[1,2]. As a result, a large variety of neutron spectral distributions have been identified. Due to the characteristics of neutron dosimetric devices currently in operation, which are far from responding in terms of dose equivalent in radiation fields covering an energy range of 9 to 10 decades, and because the anisotropic distribution of these fields must be allowed for, radiation protection monitoring in such fields is a difficult problem not yet satisfactorily solved. In addition, the last ICRP Recommendations[3] require an improvement of the sensitivity of neutron dosemeters as a consequence of a substantial decrease of the primary dose limits. Much data have been derived from measurements in nuclear power plants, in factories manufacturing radionuclide sources, in nuclear plants involved in the assembling and the reprocessing of fuel elements as well as in the environments of transport containers. All these radiation protection situations can be modelled by a fission-like source surrounded by various biological shieldings yielding leakage spectra at workplaces which exhibit three main components, namely:

(i) a 'shifted' fission spectrum, because of absorption and scattering in the shielding materials;
(ii) A down-scattered part proportional to $E^{-\alpha}$, with α ranging from 0.5 to 1.5;
(iii) a 'thermal' distribution.

The relative fractions of these contributions vary strongly according to the initial neutron spectrum and the shielding configurations as shown in Figure 1 where four typical samples A, B, C, D, are examples of spectra with increasing hardness, i.e. with increasing mean energy $\overline{E}_\phi$ from 7.7 keV up to 0.94 MeV, derived from measurements performed in power plants[4]. For comparison, the ^{252}Cf fission spectrum is shown, to illustrate and quantify the energy shift of realistic spectra.

With respect to the characteristics of current radiation protection devices, special care has to be given to the determination of their calibration factor, which consequently depends strongly on the field spectrum involved for this purpose. Conventional ISO sources (^{241}Am-Be, ^{252}Cf, ^{252}Cf(D$_2$O)[5]), which are generally used for calibrating dosemeters, belong to the group of hard spectra (see spectrum D in Figure 1), and it is necessary to investigate alternative solutions. From those different types of spectra the concept of 'Realistic Neutron Calibration Fields' (RNCF) has been put forward. The aim is the calibration of dosimetric monitors in neutron fields replicating radiation fields encountered at workplaces in the laboratory. These fields can be characterised in terms of the relevant operational quantities.

In this paper, it is intended to present the state of the art of the different approaches currently proposed by several laboratories in designing and achieving RNCF, to review the characteristics of the available radiation fields, and to address the question of calibrating dosemeters in such facilities.

REQUIRED CHARACTERISTICS OF RNCF

In order to fulfill the requirements implied in the calibration of instruments, realistic neutron spectra facilities must meet mandatory conditions as listed in the following.

Calibrations shall be performed at one (or several) well-specified location(s) or calibration zone(s), where the properties of an expanded field are compatible with the overall dimensions of the instrument to be calibrated (area dosemeters) or those of the calibration phantom (personal dosemeters). To assess the required 'homogeneity' of the field in the calibration area, its characteristics, i.e. the spectral fluence or the angular fluence distribution with respect to neutron energy, briefly called spectrum, have to be investigated in order to define the reference neutron spectrum. In addition, the accompanying photon spectrum must be determined. Taking into account the diversity of situations faced in practical radiation protection, the flexibility of an RNCF is also a highly appreciated property. For a given facility, this feature concerns both its capability of replicating several practical neutron fields, and, for a given spectrum, if necessary, the wide range of dose equivalent rates covered. From a practical standpoint, flexibility assumes that rather minor modifications of the experimental set-up are carried out to change the configuration and consequently the calibration field. The last step deals with the practical use of such radiation fields for which agreed calibration procedures are to be established and tested before being incorporated in an international standard.

STATE OF THE ART OF RECENT RNCF

Several approaches have been proposed to replicate realistic neutron spectra. Based on combinations of computational simulations and experimental work, they either took advantage of already existing installations or, in other cases, they gave rise to the development of new assemblies. In this section the principles of the currently available facilities are summarised whereas more detailed descriptions are given elsewhere[6–11]. Regarding the primary neutron source type involved in their development, those facilities can be arranged in two groups: the fission-based neutron fields and the accelerator-based neutron fields.

Fission-based neutron fields

In the first method, bare or moderated fission radionuclide neutron sources (^{252}Cf, ^{252}Cf(D_2O)) are selected but their initial emission spectrum is modified by shielding materials or/and by scattering on the environment (walls, air, surroundings). Accordingly, the resulting neutron spectrum is laboratory-specific.

At NRI Rez laboratory[6], several reference spectra have been set up on the basis of a ^{252}Cf fission spectrum altered by different moderator configurations either in spherical geometry (Fe; radii 10, 15, 25 cm) or in slab

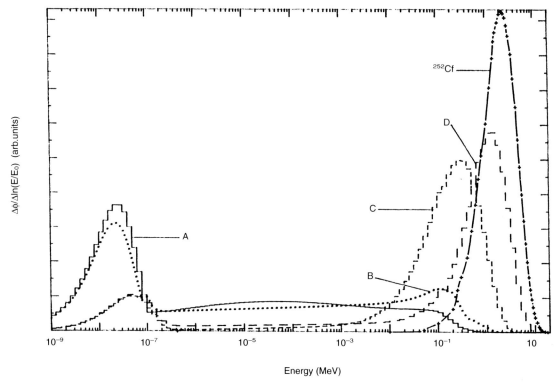

Figure 1. Examples of realistic neutron fields measured at workplaces in lethargy representation $\Delta\phi/\Delta\ln(E/E_0)$ as a function of neutron energy E. $\Delta\phi$ is the neutron group fluence in the interval $\Delta\ln(E/E_0)$, E is the neutron energy with $E_0 = 1$ MeV.

geometry. In the latter case, six iron–polyethylene assemblies combine three Fe layers (5, 15, 25 cm thick) and one $(CH_2)_n$ layer 10 cm thick. Spectrometric measurements (proton recoil and Bonner sphere techniques), performed at distances of 50 cm from the slab set-up and 100 cm from the source of the $^{252}Cf(Fe)$ or $^{252}Cf(D_2O)$ arrangement, agree quite well with previous calculations and ensure a reliable characterisation of the neutron field at the calibration locations. In spherical configuration, the radiation field is considered to be directional (room-scattered neutrons subtracted by using a shadow cone), and in the slab configuration, the calibration field is multidirectional because the room-scattered contribution is included. Numerical data are given in Table 1 and some neutron spectra referring to the slab configuration are shown in Figure 2.

At PTB Braunschweig[7], by intercepting the 'direct' neutrons of ISO sources (^{252}Cf or $^{252}Cf(D_2O)$) by a shadow bar, the room-return contribution at a distance of 170 cm from the source centre has been investigated. Bonner sphere measurements carried out at that point are in good agreement with calculations using the Monte

Table 1. General characteristics of current realistic neutron spectra facilities.

Institute (method)	Primary source	Moderating assembly Geometry/Material			Neutron field geometry	$\bar{E}_\phi$ (MeV)	$\bar{E}_H^*$ (MeV)	$\bar{h}_\phi^*$ (pSv.cm²)	Calibration distance (cm)
NRI Rez (exp.)	^{252}Cf	Spherical r = 15 cm r = 10;15;25 cm	⇒ ⇒	D_2O Fe	*Directional *(With shadow bar)	Recoil proton measurements Comparison with calculations			100
	^{252}Cf	Slab 5;15;25 cm 10 cm	⇒ ⇒	Fe $(CH_2)_n$	Multidirectional	0.14 to 1.33	0.82 to 2.15	49 to 275	50 cm from surface
PTB (I) (exp.;calc.)	$^{252}Cf + D_2O$ ^{252}Cf Am-Be	Shadow bar + Room scattering			≈ Isotropic	0.094 0.42 0.68	0.97 1.26 1.94	33 115 132	170 170 170
	$^{252}Cf + D_2O$ no Cd shell	Room scattering			Multidirectional	0.31	1.89	61	170
IPSN-SDOS (exp.; calc.)	^{238}U fission (14.6 MeV neutrons)	Shell Duct Slab	⇒ ⇒ ⇒	Fe $(CH_2)_n$ H_2O	Directional	0.07 to 0.185	0.80 to 1.10	28 to 69	30 cm from the duct exit
	^{238}U fission (3 MeV neutrons)	Duct Slab	⇒ ⇒	$(CH_2)_n$ H_2O	Directional	0.10 to 0.27	0.91 to 0.97	35 to 87	
PTB (II) (exp.; calc.)	Li (p, n) Be (2.5 MeV protons)	Spherical (r = 5 cm) + Shell (10 cm)	⇒ ⇒	$(CH_2)_n$ C	Directional	0.004 to 0.022	0.033 to 0.195	11 to 20	100
	Li (p, n) Be (3.3 MeV protons)	Spherical (r = 5 cm) + Shell (10 cm)	⇒ ⇒	$(CH_2)_n$ C	Directional	0.021 to 0.095	0.203 to 0.575	19 to 46	100
GSF (exp.;calc.)	Be (d,n) B (2.8 MeV deuterons)	*Spherical *Spherical *Slab *Slab *+Cavity scattering	⇒ ⇒ ⇒ ⇒	D_2O D_2O Fe $(CH_2)_n$	Multidirectional	0.21 0.38 0.74 0.77	0.76 1.22 1.24 1.97	83 105 204 147	47 107 47 47

$\bar{E}_\phi$: mean fluence energy.
$\bar{E}_H^*$: mean ambient dose equivalent energy.
$\bar{h}_\phi^*$: mean fluence-to-ambient dose equivalent conversion coefficient[24].
Calibration distance: distance from the neutron source to the point of test if not specified otherwise.
'Multidirectional', 'directional' or isotropic' refers to a general and qualitative description of the radiation field.
'Method' refers to the manner in which the spectral information was obtained, either experimentally, or by calculation, or both.

Carlo N-Particle Transport code MCNP4A[12]. The calculated spectra were used as initial information in a modified STAY'SL unfolding procedure[13]. The angular distribution of the field is quasi-isotropic[14]. The same technique[14] can been applied with bare ^{252}Cf and ^{241}Am-Be sources. This option may be interpreted as an extension of the use of ISO sources requiring only a shadow bar as additional equipment (Table 1 and Figure 3).

Another practical field at PTB is produced by combination of the direct and room-scattered neutrons from a ^{252}Cf(D$_2$O) source without Cd shell. In this case, the resulting neutron spectrum, which is multidirectional, varies strongly with respect to the distance from the source.

At the IPSN/SDOS laboratory in Cadarache, the fission-induced technique is applied by inserting an intense neutron source (14.6 MeV, DT reaction; 3 MeV, DD reaction) at the centre of a ^{238}U converter[8,9]. The induced fission spectrum is modified by additional shields (iron, water) and a polyethylene duct contributes to the slowing down and the thermalisation of neutrons. Monitoring of the accelerator is performed by the associated particle technique. A calibration zone (30 cm × 30 cm × 30 cm) has been defined on the set-up's symmetry axis. A computational study[15] of the spectral fluence and the angular distribution of the radiation field in a cross section of the directional field has suggested further modifications of the facility to better approach a quasi-parallel beam geometry. The flexibility of this facility is quite satisfactory and has been demonstrated in the frame of international intercomparisons[16,17]. The various experimental set-ups are well controlled by MCNP calculations and have been verified by joint spectrometric measurements (Table 1 and Figure 4). Nevertheless, additional investigations are necessary to have a better knowledge of neutron spectra which should be accurately replicated with that facility. A tentative calibration procedure combining the predicted spectrum and an environmental integral cross-checking has been successfully implemented[18] for the calibration of area survey instruments.

Accelerator-based neutron fields

Instead of a fission spectrum as primary source, the performances of ^{9}Be(d,n)^{10}B and ^{7}Li(p,n)^{7}Be nuclear

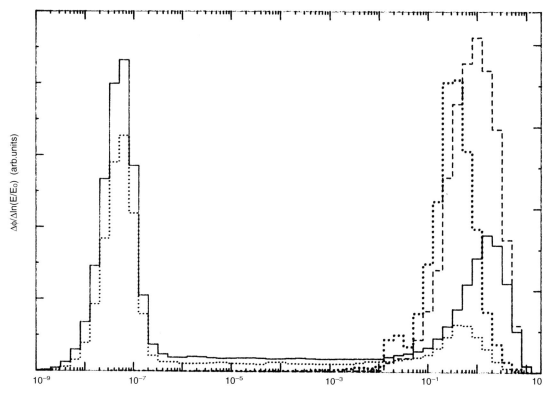

Figure 2. Examples of realistic neutron field spectra at the NRI Rez facilities with different moderator configurations in the slab geometry. (———) ^{252}Cf (Fe 5 cm, (CH$_2$)$_n$ 10 cm); (. . . .) ^{252}Cf (Fe 25 cm, (CH$_2$)$_n$ 10 cm); (– – – –) ^{252}Cf (Fe 5 cm, no (CH$_2$)$_n$); (· · · ·) ^{252}Cf (Fe 25 cm, no (CH$_2$)$_n$).

reactions have been evaluated at GSF[10] and PTB[11] respectively.

At GSF, the neutron field produced by 2.8 MeV deuterons impinging on a Be target is degraded by a D_2O moderator 30 cm thick and an iron or polyethylene slab in front of the target, respectively, and scattered by the concrete walls of a cavity (1.1 m × 1.3 m × 1.7 m) surrounding the target assembly. Bonner sphere measurements of the neutron field at different positions along the beam axis show a low varying thermal contribution which simulates rather hard spectra encountered in some locations at power plants[19] (Table 1 and Figure 5). Further investigations are still necessary to study the abilities of this prototype facility with respect to its application in the calibration of instruments by using other different moderating arrangements.

Both reactions, Li(p,n)Be and Li(p,n)Be*, were investigated computationally at PTB in order to predict the performances of a thick metallic Li target embedded in various moderator assemblies combining polyethylene and carbon. Several configurations were considered in terms of neutron fluence, mean energy $\bar{E}_\phi$ and mean fluence-to-dose equivalent conversion coefficients $\bar{h}_\phi^*$, flexibility and spectral photon contributions for two incident proton energies of 2.5 MeV and 3.3 MeV. As a conclusion from recent calculations, using the MCNP4A code, the association of $(CH_2)_n$ and C shells, 5 and 10 cm in thickness respectively, appears to be a good compromise to simulate rather soft spectra ($\bar{h}_\phi^*$ in the range from 11 pSv.cm² to 46 pSv.cm²) like those measured at workplaces behind heavy radiological shielding. The neutron field is directional, and changes of the mean neutron energy $\bar{E}_\phi$ of the neutron spectrum can be effected by acting on the energy of the incident protons or by taking into consideration the neutron field emerging at different angles with respect to the direction of the incident charged particles. In Figure 6, neutron spectra are represented which correspond to the general energy distribution produced by this technique. In addition, this technique requires a low scattering experimental environment to optimise flexibility. Spectrometric measurements in these fields have been performed at PTB. First results will be available in the near future.

CALIBRATION OF DOSIMETRIC INSTRUMENTS WITH RNCF

The use of conventional ISO sources for calibrating dosimetric devices follows recommended methods and procedures which are thoroughly described in several documents[20,21]. A similar process should be applied for RNCF, taking into account their specifications and those of area or personal dosemeters.

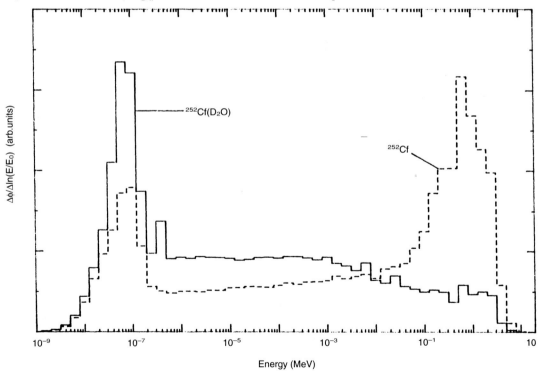

Figure 3. Realistic neutron field spectra at the PTB irradiation facility with radionuclide reference neutron sources. Energy spectra of scattered neutron produced behind a shadow object by a bare and a D_2O-moderated ^{252}Cf source both at the geometric centre of the irradiation room. Results of the Bonner sphere measurements using the STAY'SL code for unfolding[6].

Calibration of area survey meters

The calibration factor is given by:

$$N_{H^*} = \frac{H^*(10)}{M}$$

where $H^*(10)$ is determined by calculation from the spectral fluence distribution of the field under consideration and M stands for the reading of the instrument.

Knowledge of the neutron spectrum is generally derived from two methods: either neutron spectrometry (Bonner spheres and recoil proton counters) or radiation transport calculations combined, for instance, with an integral experimental verification. The latter technique also applies for the photon spectrum characterisation. If a neutron area survey instrument can be considered to have an isotropic response, it is not necessary to know the angular distribution of the neutron field and, in principle any facility listed in Table 1 is appropriate for calibration. Nevertheless, the homogeneity of the field in the calibration area has to be verified in order to ensure that an expanded field exists over a volume occupied by the whole instrument. If such a condition is not satisfied, additional investigations have to be performed to quantify the 'volume dependence' of the reference spectrum or to increase the relevant uncertainties accordingly.

Calibration of personal dosemeters

Personal dosemeters are intended to give an indication M which is representative of $H_p(10)$, the personal dose equivalent, an operational quantity defined in the body. According to ICRU[22], calibration of dosemeters worn on the trunk should be performed on an ICRU slab phantom, in terms of the dose equivalent at 10 mm depth in the phantom. That quantity depends indirectly on both the neutron spectrum and the angular distribution of the radiation field. Furthermore, when taking into account the response of a personal dosemeter, which is also energy- and angle-dependent, its indication M will be influenced by the corresponding field characteristics. Two limiting cases are the following:

(1) In a directional calibration field, a calibration factor N_{H_p} can be determined:

$$N_{H_p} = \frac{H_p(10, \theta_0)}{M(\theta_0)}$$

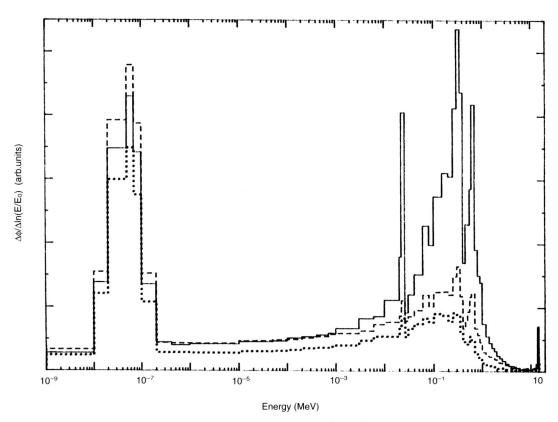

Figure 4. Realistic neutron field spectra at the Cadarache IPSN-SDOS facility (^{238}U-induced fission by 14.6 MeV neutrons) obtained by modifying the additional water shield thickness (MCNP-4A calculations). (——) no additional H$_2$O shield, (– – –) 5 cm H$_2$O shield, (· · ·) 20 cm H$_2$O shield.

where θ_0 is the angle between the direction of the field and the normal vector at the centre of the dosemeter, and $M(\theta_0)$ the angular response of the dosemeter in that field.

(2) In a multidirectional calibration field, a calibration factor is obtained which includes the folding of the energy and angle dependences of the neutron spectrum with those of the dosemeter response function in terms of the same variables, resulting in the reading M. A special case is the PTB facility[7] where the following calibration procedure is applied on a routine base. Due to the removal of the 'direct' neutrons from the source by the shadow object, the measured reading is simply attributed to the fluence rate or dose equivalent rate produced by the scattered neutrons, resulting in a calibration factor for the limiting case of the irradiation in a isotropic neutron field.

As a reminder, on-phantom calibration requires irradiation conditions for which the radiation field is considered as expanded over a volume sufficiently large to contain the whole phantom. This condition may be difficult to comply with, and results of calibration performed in multidirectional RNCF need therefore to be carefully analysed.

CONCLUSIONS

Several options have been proposed to implement calibration facilities replicating typical neutron spectra like those encountered in practical radiation protection situations. An investigation of the database associated

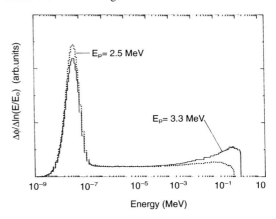

Figure 6. Realistic neutron field spectra at PTB based on the neutron emission of a thick Li target (Li(p,n)Be reaction) embedded in a combined polyethylene–carbon moderating assembly.

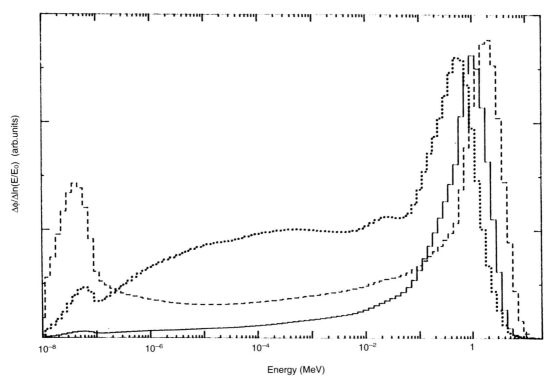

Figure 5. Realistic neutron field spectra at GSF (GRENF facility) for three different shielding arrangements. (· · ·) D_2O 30 cm sphere, (———) 10 cm iron slab, (– – –) 10 cm $(CH_2)_n$ slab.

to the SPKTBIB$^{(2,23)}$ program package shows that the majority of neutron spectra exhibit $\bar{h}^*_\Phi$ values lower than 300 pSv.cm^2 with more than 50% of data below 100 pSv.cm^2. On the assumption that this coefficient is a reliable indicator of the hardness of a neutron spectrum, it appears that current RNCF should be able to provide appropriate simulated spectra. Nevertheless, further optimisation of the set-ups is necessary to improve the similarity between the workplace and laboratory spectra with the help of additional calculations performed on a more systematic basis. Furthermore, the calibration of dosimetric instruments (area and personal dosemeters) requires additional work to finalise these prototype assemblies in order that they reach the level of operational calibration facilities, with their own calibration procedures and standards.

ACKNOWLEDGEMENTS

This work was partly supported by the European Communities in the frame of the Contract FI3P CT92 0002.

REFERENCES

1. Griffith, R. V., Palfalvi, J. and Madhvanath, U. *Compendium of Neutron Spectra and Detector Responses for Radiation Protection Purposes.* IAEA Technical Reports Series No. 318 (Vienna: IAEA) (1990).
2. Naismith, O. F. and Siebert, B. R. L. *A Database of Neutron Spectra, Instrument Response Functions and Dosimetric Conversion Factors for Radiation Protection Applications.* Radiat. Prot. Dosim. **70**(1–4), 241-245 (This issue) (1997).
3. ICRP. *1990 Recommendations of the International Commission on Radiological Protection.* Publication 60, (Oxford: Pergamon) (1991).
4. Aroua, A., Boschung, M., Cartier, F., Grecescu, M., Prêtre, S, Valley, J. F. and Wernli, C. *Characterisation of the Mixed Neutron–Gamma Fields Inside the Swiss Nuclear Power Plants by Different Active Systems.* Radiat. Prot. Dosim. **51**, 17–25 (1994).
5. ISO. *Neutron Reference Radiations for Calibrating Neutron Measuring Devices Used for Radiation Protection Purposes and for Determining their Response as a Function of Neutron Energy.* International Standard ISO 8529 (International Organization for Standardization, Genève, Switzerland) (1989).
6. Jansky, B., Turzik, Z. and Marek, M. *Reference Neutron Spectra Based on ^{252}Cf Sources in NRI Rez.* NRI Report UJV 10368 R.D (1994).
7. Kluge, H., Alevra, A. V., Jetzke, S., Knauf, K., Matzke, M., Weise, K. and Wittstock, J. *Scattered Neutron Reference Fields Produced by Radionuclides Sources.* Radiat. Prot. Dosim. **70**(1–4), 327-330 (This issue) (1997).
8. Chartier, J. L., Posny, F. and Buxerolle, M. *Experimental Assembly for the Simulation of Realistic Neutron Spectra.* Radiat. Prot. Dosim. **44**, 125–130 (1992).
9. Paul, D., Pelcot, G. and Itié, C. *Advances in Realistic Neutron Spectra: Progress in Fluence Monitoring of the DD Reaction.* Radiat. Prot. Dosim. **70**(1–4), 331–335 (This issue) (1997).
10. Schraube, H., Hietel, B., Jakes, J., Mares, V., Schraube, G. and Weitzenegger, E. *GRENF: The GSF-realistic Neutron Field Facility.* Radiat. Prot. Dosim. **70**(1–4), 337–340 (This issue) (1997).
11. Final report on the CEC project FI3P CT92 0002 for the period of July 1, 1992 to June 30, 1995. PTB contribution (1995).
12. Briesmeister, J. F. (Ed.) *MCNP — A General Monte Carlo N-Particle Transport Code, Version 4A.* LA-12625 (1993).
13. Matzke, M. *Unfolding of Pulse Height Spectra. The HEPRO Program System.* Report PTB-N-19 (Braunschweig: PTB) (1994).
14. Luszik-Bhadra, M., Kluge, H. and Matzke, M. *Measurement of the Directional Distribution of Neutrons with Personal Neutron Dosemeters.* Radiat. Prot. Dosim. **66**(1–4), 335–338 (1996).
15. Final report on the CEC project FI3P CT92 0002 for the period of July 1, 1992 to June 30, 1995. IPSN/SDOS contribution (1995).
16. Final report on the CEC project Bi7-031 for the period of July 1990 to June 1992. IPSN/SDOS contribution (1992).
17. Thomas, D. J., Chartier, J. L., Klein, H., Naismith, O. F., Posny, F. and Taylor, G. C. *Results of a Large Scale Neutron Spectrometry and Dosimetry Comparison Exercise at the Cadarache Moderator Assembly.* Radiat. Prot. Dosim. **70**(1–4), 313–322 (This issue) (1997).
18. Chartier, J. L., Kurkdjian, J., Paul, D., Itié, C., Audoin, G., Pelcot, G. and Posny, F. *Progress on Calibration Procedures with Realistic Neutron Spectra.* Radiat. Prot. Dosim. **61**, 57–61 (1995).
19. Final report on the CEC project FI3P CT92 0002 for the period of July 1, 1992 to June 30, 1995. GSF contribution (1995).
20. ISO. *Reference Neutron Radiations: Dosimetry Fundamentals Related to the Basic Quantities Characterizing the Radiation Field.* Committee Draft ISO/TC85/SC2/WG2 8529-2 (International Organization for Standardization, Genève, Switzerland) (1996).
21. ISO. *Reference Neutron Radiations: Calibration of Area and Personal Dosimeters and the Determination of their Response as a Function of Neutron Energy and Angle of Incidence.* Committee Draft 8529-3 (International Organization for Standardization, Genève, Switzerland) (1995).
22. ICRU. *Measurement of Dose Equivalents from External Photon and Electron Radiations.* Report 47 (Bethesda, MD: ICRU Publication) (1992).
23. Naismith, O. F., Siebert, B. R. L. and Thomas, D. J. *Response of Neutron Dosemeters in Radiation Protection Environments an Investigation of Techniques to Improve Estimates of Dose Equivalents.* Radiat. Prot. Dosim. **70**(1–4), 255–260 (This issue) (1997).
24. Wagner, S. R., Grosswendt, B., Harvey, J. R., Mill, A. J., Selbach, H. J. and Siebert, B. R. L. *Unified Conversion Functions for the New ICRU Operational Radiation Protection Quantities.* Radiat. Prot. Dosim. **12**(2), 231–235 (1985).

RESULTS OF A LARGE SCALE NEUTRON SPECTROMETRY AND DOSIMETRY COMPARISON EXERCISE AT THE CADARACHE MODERATOR ASSEMBLY

D. J. Thomas†, J.-L. Chartier‡, H. Klein§, O. F. Naismith†, F. Posny‡, and G. C. Taylor†
† National Physical Laboratory, Teddington, Middlesex, TW11 0LW, UK
‡ Institut de Protection et de Sûreté Nucléaire, DPHD/SDOS/LRDE
Fontenay-aux-Roses, France
§ Physikalisch-Technische Bundesanstalt
Postfach 33 45, D-38023 Braunschweig, Germany

Abstract — Eurados Working Group 7 recently organised a large-scale comparison of neutron spectrometry and dosimetry measurements at the IPSN/SDOS laboratory of the CEA Cadarache Research Centre in France. A large number of participants took part with a range of instruments including spectrometers, tissue-equivalent proportional counters, personal dosemeters, and survey instruments. The neutron field used for the exercise was a primarily low energy neutron spectrum similar to those which have been measured recently around nuclear facilities. This paper presents the results of the measurements and attempts to draw conclusions about the accuracy attainable with the various devices, their advantages and drawbacks, and potential problems.

INTRODUCTION

Since neutron dosemeters, both area survey instruments and personal dosemeters, do not have response functions which enable them to measure the required operational quantities (ambient and personal dose equivalent respectively) at all energies, their readings are unreliable. In an unknown neutron spectrum it is difficult even to estimate the possible errors in their readings. For this reason neutron spectrometry in areas around nuclear facilities where workers are subject to neutron doses has assumed increasing importance over recent years. With spectrometric information the ambient dose equivalent can be deduced using tabulated fluence to dose equivalent conversion coefficients, and if angular information about the fields is available, personal dose equivalent (and effective dose equivalent) can also be determined. Correction factors for dosemeters can then be assigned. One question which still needs to be addressed is whether the accuracy of available spectrometers is adequate, particularly since the uncertainties in spectrometric techniques are very difficult to quantify from first principles. One of the best methods of assessing the uncertainties is via comparison exercises, preferably in spectra which are well known from calculation.

Working Group 7 (WG7) of the European Radiation Dosimetry Group (Eurados) has now organised four intercomparisons of spectrometers and dosemeters[1–3]. The fourth, and most comprehensive, was undertaken during 1994 at the IPSN/SDOS laboratory of the CEA Cadarache Research Centre in France. It involved the largest number of participants and the widest range of instruments to date. These are listed in Table 1. The instruments included not only spectrometers, but also survey instruments, personal dosemeters, and various tissue-equivalent proportional counters (TEPCs). The latter provide an alternative method of characterising the field, in some respects complementary to the spectrometry. This paper represents a first attempt to evaluate the wealth of data which came out of the exercise.

FACILITY

The facility used for the comparison has been described in detail in other papers[4,5] and only a brief outline is given here. Neutrons are generated in the first instance at about 14 MeV using the d-T reaction. These then bombard a ^{238}U converter which produces fission neutrons, which in turn are moderated by a hemispherical layer of iron (10 cm), and a layer of water (10 cm). Finally the path to the reference position for the measurements is via a hollow polyethylene duct which produces additional low energy scattered neutrons. (Neutron capture in the hydrogen of the polyethylene produces large numbers of 2.2 MeV photons which provide a challenge for neutron spectrometers which use n/γ discrimination).

Since the geometry of the facility, the energy of the primary neutrons, and the relevant nuclear cross sections were all known, it was possible to calculate the spectrum using Monte Carlo transport techniques. The calculation was performed[5] using the MCNP-3A[6] code, and this provided important benchmark data for assessing the results. Although the final spectrum shows a small residual 14 MeV peak this accounts for only 0.2% of the fluence. The spectrum is essentially a low energy one with the majority of neutrons below 1 MeV and a thermal component amounting to about 50% of the total fluence. It is similar in type to spectra which have been measured in recent years around nuclear facilities.

All the measurements were performed during three time periods in 1994. Each participant was assigned a number of days, based on their anticipated requirements, and during these the facility was run solely for them. Monitoring of the neutron production was provided by three charged particle detectors which counted associated alpha particles from the ^{3}H(d,n)^{4}He reaction. This provided the essential links between measurements by different participants, and within any series of measurements made by a group. The final results are presented as fluence or dose equivalent per average monitor count, α, taking a mean of the three monitors. Investigation of the variation of the counts relative to their mean indicated a maximum uncertainty from monitor variations of about ±3%.

RESULTS

Broad range spectrometry results

Nine measurements of the full energy range spectrum were submitted, and these are plotted in Figure 1 where they are compared with the MCNP calculations. The data for fluence and ambient dose equivalent for the total spectra and for five broad energy groups are detailed in Table 2. All these spectra are based essentially on Bonner sphere (BS) measurements although the results of IPSN/SDOS Fontenay and NPL also include data from NE213 scintillators and in the case of Fontenay also from proton recoil counters.

The MCNP calculated spectrum was not made available to the participants before they submitted their results and so various types of *a priori* information were used when unfolding the BS measurements. Because of their low resolution BSs on their own can not hope to pick out the 14 MeV peak in the spectrum, so those spectra which do include this feature have either included it specifically in the *a priori* information because its presence can be inferred from the way the spectrum has been produced, or in the case of IPSN/SDOS Fontenay and NPL, measured it using additional high resolution spectrometers.

The first conclusion to be drawn from Figure 1 is that all the participants measured very similar gross features, i.e. a large thermal peak, and a significant intermediate energy component which extended up in energy to a broad 'bump' feature in the 10 keV to 1 MeV region. Closer inspection reveals some differences in the magnitude and exact position of the 'bump' in the keV region which have repercussions for the derived dose equivalent.

Figure 2 illustrates the degree of agreement between the measurements for integral fluence (a), and ambient dose equivalent (b). For total fluence the standard deviation for all results, excluding the calculation, was 6.6%. However, one value, that from CMI, stood out as being 15% higher than the others. In their report on the measurements the CMI group had noted instrumental difficulties which resulted in uncertainty about which of two different measured response sets to take: either those from a pulse-height analyser measuring the spectrum from the ^{3}He proportional counter used in the BS set, or those from a scaler system counting events above a fixed discriminator threshold. Shortage of time at the measurement site had meant that it had been impossible to resolve the differences. The submitted results were

Table 1. Instruments and participants in the 1994 Cadarache spectrometry and dosimetry comparison. (Key: C — Canada, CH — Switzerland, CZ — Czech Republic, D — Germany, F — France, GB — United Kingdom, I — Italy, S — Sweden.)

Bonner spheres	Scintillators	Proton recoil counters	TEPCs	Area survey instruments	Personal dosemeters	SDDs (Bubble detectors)
IPSN-FAR (F)	Univ. Dresden (D)	Univ. Dresden (D)	SSI (S)	NPL (GB)	Ringhals (S)	Univ. Pisa* (I)
NPL (GB)	IPSN-FAR (F)	IPSN-FAR (F)	ZFU (D)	PTB (D)	PTB (D)	NPI (CZ)
PTB (D)	NPL (GB)	PTB (D)	IPSN-FAR (F)	IPSN-FAR (F)	AECL (C)	
CMI (CZ)	PTB (D)	NRI (CZ)	NPL (GB)	KfZ (D)	COGEMA (F)	
GSF (D)	NRI (CZ)		GSF (D)	NPI (CZ)	IPSN-FAR (F)	
IRA (CH)			PTB (B)		NRPB (GB)	
AECL (C)			CEA-Grenoble (F)		IRA (CH)	
BfS (D)			AECL (C)		NPI (CZ)	
IPSN-Cadarache (F)			CERN (CH)			

*The superheated drop detectors of the University of Pisa were used both as dosemeters and as a spectrometer for the region above 100 keV.
Key: AECL, Atomic Energy of Canada Ltd.; BfS, Bundesamt für Strahlenschutz; CEA, Commissariat à l'Energie Atomique; CERN, European Laboratory for Particle Physics; CMI, Czech Metrological Institute; COGEMA, Compagnie Général Matières Nucléaires; GSF, Forschungzentrum für Umwelt und Gesundheit GmbH; IPSN, Institut de Protection et de Sûrete Nucléaire (FAR, Fontenay aux Roses); IRA, Institute de Radiophysique Appliquée; KfZ, Forschungzentrum Karlsruhe; NPI, Nuclear Physics Institute; NPL, National Physical Laboratory; NRI, Nuclear Research Institute; NRPB, National Radiological Protection Board; PTB, Physikalisch- Technische Bundesanstalt; Ringhals, Health Physics Group, Ringhals reactor; SSI, Swedish Radiation Protection Institute; ZFU, Zentrum für Umweltforschung.

based on the pulse-height analyser data. When all the results were compared, and it became apparent that the CMI value was higher than all the others, the alternative set of responses was tried and found to give much better agreement with the other participants. Using the revised CMI datum, the standard deviation for all nine results became 3.4%. This represents remarkably good agreement.

The variation of the results for dose equivalent, is illustrated in part (b) of Figure 2. (The quantity used is ambient dose equivalent, derived using the fluence to dose equivalent conversion coefficients of Wagner et al[7]. The spread of results here is greater than for fluence which illustrates the classic weakness of BS measurements. Although they provide accurate data for the total fluence, their low resolution and the uncertainty in the unfolding, mean that they cannot localise the fluence very accurately. This fact is illustrated in Table 2 by the larger variances in the results for fluence in broad groups than for the total fluence, as well as by the greater spread for the total ambient dose equivalent values. Dose equivalent is a sensitive measure of the shapes of the spectra in the keV region. The results for total ambient dose equivalent had a standard deviation of 19%, a value which dropped to 13% when the revised CMI datum was used. Inspection of Figure 1 brings out reasons for the greater spread of total dose equivalent compared to fluence. For example, the slightly higher energy for the 'bump' in the AECL spectrum results in a high dose equivalent, whereas the lower fluence in the BfS spectrum in this region results in a low value. The low value for IRA appears to result from low fluences in all the broad energy groups of Table 2 except 10 to 100 keV.

The standard deviations of the total fluence and dose equivalent derived from the data set with the corrected CMI datum represent a measure of the state-of-the-art for presently available BSs. This is reasonably satisfactory. However, the need to correct one of the results highlights an important issue. Had it not been part of a comparison exercise, there would have been no way of knowing that the original CMI data were high. It is important that instruments are tested and calibrated in advance, and that sufficient time is made available at a measurement site for technical problems to be recognised and rectified.

High resolution spectrometers

Four laboratories made high resolution spectrometry measurements of the upper energy region with both scintillators (either NE213 or stilbene) and proton recoil counters. Combining the data from these instruments enabled participants to cover the region from 20 MeV to 100 keV or lower (10 keV in the case of one participant). Figure 3 shows the results compared to the calculated spectrum. No fundamental conclusions are immediately apparent, but several points can be made about the data.

All four measurements see similar features to a greater or lesser extent. (The discontinuity in the NRI data just above 1 MeV is, in part at least, believed to be an instrumental effect.) The PTB measurement is the only one which extends low enough in energy to see the 24 keV iron window, however, the extent of the structure seen by this group, in the 350 keV region for example, is questionable, and may be an artefact of the unfolding. For most of the energy region the measured results scatter on either side of the calculated spectrum. In the opinion of the participants better agreement can be achieved, but this requires improvements to the instrument response functions. Efforts are under way in this area. Nevertheless, even at the present state-of-the-art, high resolution results can be extremely useful, particularly in determining whether the BS spectrum unfolded in the upper energy region is correct[1]. If, for example, BfS and AECL had had access to high resol-

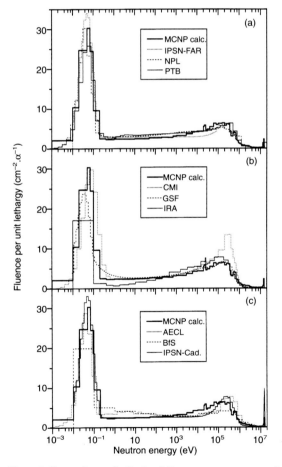

Figure 1. Comparison of all the full energy range spectral results, as originally submitted by the participants, with each other and with the MCNP calculation.

ution data in the 100 keV to 1 MeV region it would have given some indication that the BS unfolding was not assigning the fluence to exactly the right region.

Tissue-equivalent proportional counters

Nine groups participated with TEPCs using a variety of different systems involving different instrument designs (spherical and cylindrical), different wall thicknesses (from 1 mm to 6.3 mm), different simulated diameters (from 0.5 μm to 3 μm), different gas fillings, and, perhaps most important of all, different methods of deriving best estimates of the dose equivalents. The comparison is thus not of 'like with like', but of a set of instruments where the common feature is the measurement of lineal energy. Table 3 summarises the results, and the participants' best estimates of the dosimetric quantities are illustrated in Figure 4. Since the MCNP results represent a reasonable average of the spectral data, they are included in the table and figure to provide a cross reference to the spectrometry. The TEPC neutron dose values are compared to calculated kerma in ICRU tissue derived using the conversion coefficients of Caswell et al[8]. The TEPC neutron dose equivalent results are quoted for the Q(L) relationship of ICRP Publication 21[9] and are compared to ambient dose equivalent calculated using the conversion coefficients of Wagner et al[7]. (The latter conversion coefficients provide estimates of the ICRU operational quantities[10] as calculated with the ICRP 21 Q(L) relationship. Although values would change if the Q(L) relationship recommended in ICRP Publication 60[11] were used, the overall conclusions would not change significantly.)

Two features of the data are immediately obvious from Figure 4. Firstly, the large variation in the results, and secondly, the neutron dose equivalent values are significantly lower than the MCNP calculation while the gamma dose equivalent is on average higher. (The gamma results are discussed later in connection with gamma dosemeter results.)

The low neutron dose equivalent results were to a large extent expected. TEPCs are known from measurements of their response functions[12,13] to underestimate neutron dose equivalent, particularly at low energies. Since users of these instruments are aware of this fact, some of them attempt to correct for this, either by using a correction factor derived from measurements in a soft neutron field, for example heavy water moderated ^{252}Cf, or using other approaches[14]. These correction techniques provide multiplicative factors which increase the neutron dose equivalent above the value obtained from the TEPC if used simply with a lineal energy calibration using an alpha particle source or the 'proton edge' feature in the spectrum.

In an attempt to compare just the uncorrected TEPC results the correction factors have been removed in those cases where their values were given in the reports to the evaluators. Figure 5, which plots the neutron dose equivalent against detector type, includes both the corrected (signified by the 'a' added to the laboratory name) and uncorrected values for CEA-Grenoble, NPL,

Table 2. Results of full energy range spectrometry measurements, mainly Bonner spheres.

Energy range (eV)	Participating laboratory											Mean**	Stand. dev.** σ
	Calc. MCNP	IPSN Cad.	BsF	AECL	IRA	GSF	CMI*		PTB	NPL	IPSN FAR		
	Absolute fluence (cm^{-2}.α^{-1})												
<0.4	58.9	60.6	52.8	54.1	50.05	47.9	51.1	(62.9)	53.4	55.7	62.5	54.3	8.7%
0.4–10^4	28.0	26.4	36.2	33.7	26.8	32.9	24.2	(29.0)	36.1	35.3	30.0	31.3	12%
10^4–10^5	11.9	8.5	8.0	8.4	15.4	12.1	12.1	(15.1)	10.5	10.9	8.2	10.4	24%
10^5–10^6	11.4	11.3	8.3	14.4	12.1	11.8	16.8	(21.6)	12.2	10.0	10.3	11.9	21%
>10^6	1.5	1.5	2.0	1.7	0.4	0.6	0.60	(0.74)	1.4	1.7	1.2	1.25	47%
Total	111.6	108.3	107.3	112.3	105.0	105.4	104.9	(129.2)	113.7	113.7	112.1	109.2	3.4%
	Absolute ambient dose equivalent (nSv.α^{-1})												
<0.4	0.52	0.54	0.46	0.47	0.43	0.42	0.47	(0.57)	0.47	0.48	0.56	0.48	9.3%
0.4–10^4	0.23	0.22	0.31	0.28	0.20	0.27	0.20	(0.23)	0.29	0.29	0.25	0.26	16%
10^4–10^5	0.35	0.27	0.22	0.25	0.46	0.35	0.36	(0.44)	0.31	0.32	0.25	0.31	23%
10^5–10^6	1.92	2.23	1.51	2.75	1.90	2.01	2.99	(3.81)	2.22	1.79	1.75	2.13	23%
>10^6	0.59	0.74	0.77	0.64	0.12	0.23	0.23	(0.28)	0.64	0.66	0.48	0.50	49%
Total	3.61	3.99	3.27	4.40	3.11	3.28	4.24	(5.34)	3.94	3.54	3.27	3.67	13%

*Both sets of data from CMI (see text) are represented in this table, the revised results first, followed by the originals in brackets.
**Calculated using the revised CMI data.

PTB, and ZFU. Table 3 also includes both sets of values. The correction factors ranged from 1.32 to 1.75. It is important to note, however, that the factors removed were those reported to the evaluators. Some of the other results may also contain corrections, but insufficient information was given for the values to be determined. The effects of removing the correction factors were to reduce further the values of neutron dose equivalent relative to the MCNP calculation, from a ratio of 0.69 to 0.58, and also to reduce the standard deviation from 26% to 22%.

In an attempt to identify trends in the results with parameters such as wall thickness, simulated diameter, etc., the data were plotted as functions of these parameters, but no obvious conclusions became apparent. Probably the most revealing plot was that against detector type, shown in Figure 5. With the correction factors removed there is good agreement between the neutron

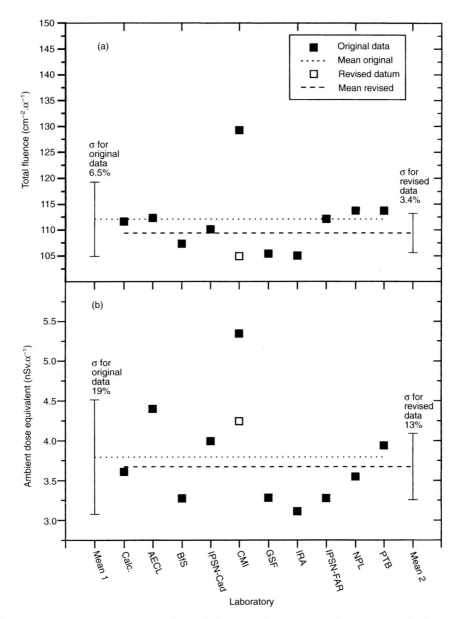

Figure 2. The total fluence and dose equivalent values and the standard deviations, σ, for these values for the participants who measured the total spectrum.

dose equivalent results for the 2″ and 5″ Far West Technology (FWT) counters and the IPSN/SDOS Fontenay 5 cm right cylindrical counter. The standard deviation of these is only about 6%. However, these are the lowest values of all with an average ratio of 0.47 to the MCNP results. One slightly surprising aspect of this is that the IPSN/SDOS-Fontenay result was reported to be based on a heavy water moderated ^{252}Cf calibration, and as such might be expected to be closer to the MCNP calculation.

The remaining data are all higher, with a mean ratio of 0.68 relative to the MCNP result, and as a group

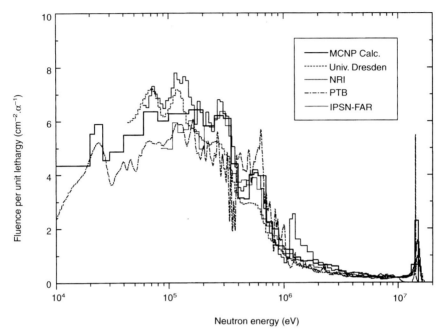

Figure 3. High resolution spectrometry results from scintillators and proton recoil counters compared against the MCNP calculation.

Table 3. Summary of gamma, neutron, and total dose and dose equivalent values from the TEPC measurements.

	\multicolumn{10}{c	}{Participating laboratory}	Mean	Stand. dev.	Calc. MCNP	Mean/ MCNP
	AECL	IPSN FAR	CEA Gren.	CERN HANDI	CERN REM500	GSF	NPL	PTB	SSI	ZFU				
\multicolumn{15}{c}{Derived dose/kerma values (nGy.α^{-1})}
Gamma	0.36	0.50	0.22	0.45		0.35	0.44	0.45	0.48	0.37	0.40	22%		
Neutron	0.16	0.21	0.24	0.20		0.15	0.19	0.26	0.12	0.26	0.20	25%	0.23	0.88
Total	0.52	0.71	0.45	0.64		0.50	0.62	0.70	0.60	0.63	0.60	15%		
\multicolumn{15}{l}{Derived dose equivalent values including correction factors used by some participants to improve neutron dose equivalent results (nSv.α^{-1})}
Gamma	0.41	0.50	0.25	0.51		0.55	0.49	0.45	0.51	0.37	0.45	21%	0.28	1.60
Neutron	1.62	1.72	2.71	2.12	2.86	1.79	2.84	3.47	2.45	3.22	2.48	26%	3.61	0.69
Total	2.02	2.22	2.96	2.62		2.34	3.33	3.92	2.96	3.58	2.88	22%	3.90	0.74
\multicolumn{15}{l}{Derived dose equivalent values excluding correction factors used by some participants to improve neutron dose equivalent results (nSv.α^{-1})}
Neutron	1.62	1.72	1.55	2.12	2.86	1.79	1.80	2.46	2.45	2.45	2.08	22%	3.61	0.58
Total	2.02	2.22	1.80	2.62		2.34	2.29	2.91	2.96	2.82	2.44	17%	3.90	0.63

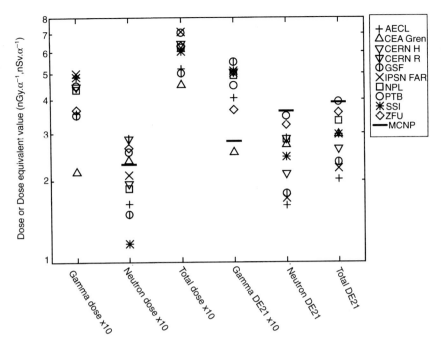

Figure 4. Variation of the TEPC results for neutron and gamma dose, and for dose equivalent (DE) calculated using the Q(L) relationship of ICRP 21.

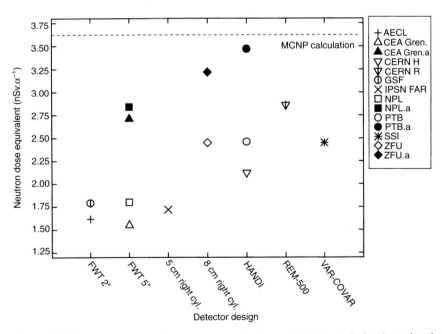

Figure 5. Dependence of TEPC neutron dose equivalent results on the type of detector used, showing values both with (solid symbols) and without (open symbols) the correction factors introduced by several participants to improve the dose equivalent estimate.

exhibit a larger scatter. Even after the correction factor has been removed, the ZFU result obtained with an 8 cm right cylindrical counter is higher than the results from the FWT and IPSN/SDOS-Fontenay instruments. Two of the TEPCs, the HANDI and the REM-500, are commercial instruments with built-in analysis making it difficult to know or make adjustment for quantities such as the lineal energy calibration. The variance–covariance technique of SSI differs from the others and depends on calibration in known neutron fields and on determination of the photon dose equivalent fraction by measurement with an independent detector, e.g. a Geiger-Müller counter.

For the TEPC results the overall impression remains of large variations, even when known correction factors are removed, and of a significant underestimate of the neutron dose equivalent unless fairly substantial correction factors are applied.

Neutron survey instruments

Table 4 summarises the neutron survey instrument results. These instruments were designed at different times to measure different quantities, either maximum dose equivalent[9], H_{MADE}, or ambient dose equivalent, $H^*(10)$, derived with the Q(L) relationship of either ICRP 21 or ICRP 60. Regardless of the dose equivalent quantity they were designed to measure, they can be calibrated to give the correct response for any of these quantities in any standard calibration field. The results in Table 4 are thus quoted for some or all of these three quantities. Also given in brackets are the ratios to these quantities as determined from the MCNP calculations. The column in which the result for any instrument is tabulated corresponds to the quantity for which it had been calibrated.

The Harwell 0949 instrument was originally designed to measure H_{MADE}. There is a recommended technique for setting up the instrument involving an Am-Be source and a test jig. This does not correspond exactly to a free-field calibration with an Am-Be source, but for H_{MADE} gives a similar calibration. This instrument over-reads for all quantities, by up to a factor of two for H_{MADE}, although there was a noticeable improvement if calibrated in a heavy water moderated ^{252}Cf field, particularly for $H^*(10)_{ICRP\,60}$. The Harwell N91 is a new version of the 0949. It has essentially the same response function and gave similar results.

The Studsvik 2202, the Cramal 21, and the Dineutron gave good results in this field. Results for the Cramal 20 were low by about 25% compared to the MCNP calculation. The bubble detector results of the University of Pisa and NPI were high by 17% and low by 14% respectively. Finally the Nuclear Enterprises NM2 reading was 46% high while the recently developed EG&G Berthold LB6411 moderating-type instrument[15] designed to measure $H^*(10)_{ICRP\,60}$ gave readings which were low by 25 to 30%.

The overall impression is one of large variations. For the Harwell instruments, the Anderson-Braun type NM2, and the LB6411, the high or low readings can be anticipated from the known response functions and the

Table 4. Results for area survey instruments and, in brackets, ratio to MCNP calculation.

Participating laboratory	Instrument	Remarks	H_{MADE} (nSv.α^{-1})	$H^*(10)_{ICRP21}$ (nSv.α^{-1})	$H^*(10)_{ICRP60}$ (nSv.α^{-1})	Calibration field where known
IPSN-FAR	MCNP calc.	Reference	3.39	3.62	5.34	
PTB	Harwell 0949	Leake-type detector with spherical ^{3}He counter	6.63 (1.96)	6.77 (1.87)	7.94 (1.49)	Am-Be
			5.57 (1.64)	5.65 (1.56)	7.11 (1.33)	^{252}Cf (bare)
			3.99 (1.18)	4.07 (1.12)	5.37 (1.01)	^{252}Cf(D$_2$O moderated)
NPL	Harwell 0949	"	6.90 (2.04)			Manufacturer's recommended set-up
IPSN-FAR and IPSN-Cad.	Harwell N91	Updated Harwell 0949		6.12 (1.69)		Am-Be
	Studsvik 2202	Modified Anderson-Braun		3.71 (1.02)		Am-Be
	CRAMAL 21	^{3}He counter + perforated Cd		3.94 (1.09)		
	CRAMAL 20	^{3}He counter + Cd		2.68 (0.74)		
	DINEUTRON	Uses 2 moderating spheres		3.98 (1.10)		^{252}Cf
IAR	Studsvik 2202	Modified Anderson-Braun		3.40 (0.94)		^{252}Cf (D$_2$O moderated)
NPI	NE-NM2	Anderson-Braun type		4.95 (1.46)		^{252}Cf (bare)
	BD-100 R	'Bubble' detector		2.90 (0.86)		^{252}Cf (bare)
Univ. Pisa	SDD	Superheated drop detector		4.25 (1.17)		Monoenergetic neutrons
KFZ	EG&G LB6411	Spherical moderator with cylindrical ^{3}He counter.			4.00 (0.75)	Am-Be
	Prototype 2				3.68 (0.69)	Am-Be

spectrum. The good results for the Stuksvik imply that it cannot have the same response function as the similar Anderson-Braun instrument.

Personal neutron dosemeters

In general, personal dosemeter measurements were reported in terms of personal dose equivalent, $H_p(10)$, or directional dose equivalent, $H'(10)$. Some groups, however, irradiated sets of dosemeters at different points around a phantom to investigate the directional dependence of the fluence. Taking into consideration the energy and angle dependence of the detector response it was possible to derive an estimate of $H^*(10)$. Only these data are compiled in Table 5, and for these the agreement with the reference data is remarkably good. (Not all data for personal dosemeters was received by the deadline).

Photon dose results

Although not specifically a photon dosimetry intercomparison, several participants made photon measurements, and the TEPC results also provided photon information. The photon dosemeter results of NPI, SSI, and IRA indicated a gamma dose equivalent of around 0.6 nSv.α^{-1}. This can be compared with the average value from the TEPCs of 0.45 nSv.α^{-1}. The MNCP-3A calculation, which gave a value of 0.28 nSv.α^{-1}, thus appears to underestimate the gamma dose equivalent. (Note: recent calculations with MCNP-4A and new cross sections give a calculated answer about 20% higher in better agreement with the measurements.)

SUMMARY AND CONCLUSIONS

The exercise described here represents one of the most comprehensive comparisons of neutron spectrometric and dosimetric instrumentation ever undertaken. The neutron spectra measured with BSs agreed well with each other and with the MCNP calculation confirming that properly characterised BS systems are well suited to determining dose equivalent reference values in workplace neutron fields. High resolution spectrometry can add valuable information in the high energy region. Survey instruments gave variable results and need to be used with caution. Personal dosemeter results were surprisingly good although this may reflect the competence and care taken deriving $H^*(10)$ rather than the general state of the art.

TEPC results were characterised by the large range of values over which they extended and a significant under-read for neutron dose equivalent. This under-read was consistent with the known response functions of these devices and several users employed techniques to try to correct for this. It is recommended that a clear statement of exactly what correction factors have been applied be given on all occasions, and where the TEPC analysis is under the full control of the user, i.e. excluding commercial devices with built in analysis, some uniformity of approach would be very desirable.

Where device response functions are known their readings can be calculated from the spectrum and compared with the measured values. Comparison of measurement and calculation for the Harwell instruments, the EG&G Berthold LB6411, and the HANDI TEPC have already shown agreement, thus giving confidence in both the measured spectrum and the response functions for these instruments. Spectrometric information derived from the SDD measurements with detectors of different thresholds is still being analysed but the results with this novel technique look encouraging.

The evaluation of this exercise will continue. A summary of all the available results together with the calculated spectrum and angular dependence of the field has been distributed to all participants who will use the data to check their analyses and investigate discrepancies. Extended reports on the exercise will be produced in due course.

ACKNOWLEDGEMENTS

The authors would like to thank all the participants for making their data available, IPSN/SDOS for the use of their excellent facilities at Cadarache and the effort put into the organisation of the measurements, and Eurados, and hence the CEC, for support of WG7.

Table 5. Results of ambient dose equivalent measurements with personal dosemeters on phantoms, and (in brackets) ratio to value estimated from the MCNP calculation.

Participating laboratory	Dosemeter type	Phantom type	Calibration field	Dose equivalent (nSv.α^{-1})
NRPB	PADC*	PMMA** slab	^{252}Cf (bare)	2.91 (0.80)
Ringhals	TLD	PMMA slab	PWR field	4.22 (1.17)
	TLD albedo	PMMA slab	PWR field	4.15 (1.15)
PTB	PADC	CH$_2$ sphere	^{252}Cf (bare)	3.19 (0.88)
COGEMA	TLD albedo	PMMA slab	Realistic spec.	3.80 (1.05)

*Etched track detector using the polymer, poly allyl diglycol carbonate, (PADC), often referred to as CR-39.
**Polymethyl methacrylate, commonly known as Perspex.

REFERENCES

1. Klein, H. *Workplace Radiation Field Analysis*. Radiat. Prot. Dosim. **70**(1–4), 225–234 (This issue) (1997).
2. Klein, H. and Lindborg, L. (Eds) *Determination of Neutron and Photon Dose Equivalent at Work Places in Nuclear Facilities of Sweden. An SSI – Eurados Comparison Exercise: Part 1: Measurements and Data Analysis*. SSI-report 95-15 (Stockholm: SSI) (1995).
3. Lindborg, L., Bartlett, D. T., Drake, P., Klein, H., Schmitz, Th. and Tichy, M. *Determination of Neutron and Photon Dose Equivalent at Workplaces in Nuclear Facilities in Sweden*. Radiat. Prot. Dosim. **61**(1–3), 89–100 (1995).
4. Chartier, J.-L., Posny, F. and Buxerolle, M. *Experimental Assembly for the Simulation of Realistic Neutron Spectra*. Radiat. Prot. Dosim. **44**(1–4), 125–130 (1992).
5. Chartier, J.-L., Kurdjian, J., Paul, D., Itié, C., Audoin, G., Pelcot, G. and Posny, F. *Progress on Calibration Procedures with Realistic Neutron Spectra*. Radiat. Prot. Dosim. **61**(1–3), 57–61 (1995).
6. Briesmeister, J. F. (Ed.) *MCNP — A General Monte Carlo Code for Neutron and Photon Transport, Version 3A*. Los Alamos Report LA-7396 (1986).
7. Wagner, S. R., Groβwendt, B., Harvey, J. R., Mill, A. J., Selbach, H. -J. and Siebert, B. R. L. *Unified Conversion Functions for the New ICRU Operational Radiation Protection Quantities*. Radiat. Prot. Dosim. **12**, 231–235 (1985).
8. Caswell, R. S., Coyne, J. J. and Randolph, M. L. *Kerma Factors for Neutron Energies Below 30 MeV*. Radiat. Res. **83**, 217–254 (1980).
9. ICRP. *Data for Protection against Ionizing Radiation from External Sources: Supplement to ICRP Publication 15*. Publication 21 (Oxford: Pergamon Press) (1973).
10. ICRU. *Determination of Dose Equivalents Resulting from External Radiation Sources*. Report 39 (Bethesda, MD; ICRU Publications) (1985).
11. ICRP. *1990 Recommendations of the International Commission on Radiological Protection*. Publication 60 (Oxford: Pergamon Press) (1991).
12. Alberts, W. G., Dietz, E., Guldbakke, S., Kluge, H. and Schuhmacher, H. *International Intercomparison of TEPC Systems used for Radiation Protection*. Radiat. Prot. Dosim. **29**(1–2), 47–53 (1989).
13. Menzel, H. G., Lindborg, L., Schmitz, Th., Schuhmacher, H. and Waker, A. J. *Intercomparison of Dose Equivalent Meters based on Microdosimetric Techniques: Detailed Analysis and Conclusions*. Radiat. Prot. Dosim. **29**(1–2), 55–68 (1989).
14. Taylor, G. C. *An Analytical Correction for the TEPC Dose Equivalent Response Problem*. Radiat. Prot. Dosim. **61**(1–3), 67–70 (1995).
15. Burgkhardt, B., Fieg, G., Klett, A., Plewnia, A. and Siebert, B. R. L. *The Neutron Fluence and H*(10) Response of the New LB6411 Rem Counter*. Radiat. Prot. Dosim. **70**(1–4), 361–364 (This issue) (1997).

CURRENT STATUS OF AN ISO WORKING DOCUMENT ON REFERENCE RADIATIONS: CHARACTERISTICS AND METHODS OF PRODUCTION OF SIMULATED PRACTICAL NEUTRON FIELDS

J. C. McDonald[1], W. G. Alberts[2], D. T. Bartlett[3], J.-L. Chartier[4], C. M. Eisenhauer[5], H. Schraube[6], R. B. Schwartz[5] and D. J. Thomas[7]
[1]Pacific Northwest National Laboratory, Richland WA 99352, USA
[2]Physikalisch-Technische Bundesanstalt, 38116 Braunschweig, Germany
[3]National Radiological Protection Board, Didcot, Oxon, OX11 0RQ, UK
[4]Institut de Protection et Sûreté Nucléaire, Commissariat à l'Energie Atomique 92265 Fontenay-aux-Roses, France
[5]National Institute of Standards and Technology, Gaithersburg MD 20899, USA
[6]GSF-Forschungszentrum Neuherberg, 85758 Oberschleissheim, Germany
[7]National Physical Laboratory, Teddington, Middlesex, TW11 0LW, UK

Abstract — The International Organization for Standardization (ISO) TC85/SC2/WG2 has convened a working group to develop a standard dealing with the production, characterisation and use of reference neutron radiation fields simulating those found in the workplace. This standard is being developed to provide guidance to the laboratories developing simulated practical reference neutron fields, and to the users of these fields. The ISO currently recommends four reference neutron spectra for the calibration of radiation protection instruments and dosemeters. However, due to the energy dependence in the dose equivalent responses of these devices, it is useful to develop additional neutron spectra for calibration purposes. This standard will describe the methods of production and the expected characteristics of simulated practical neutron fields. The types of facilities considered include those using radionuclide neutron sources, accelerators and reactors. Facilities producing high energy neutrons for the calibration of instruments and dosemeters used at accelerator laboratories and in aircraft for in-flight monitoring are also considered. The status of this standard is being presented now in order to provide a progress report and to elicit comments and suggestions before the standard is finalised.

INTRODUCTION

Several ISO standards deal with the production, characterisation and use of neutron fields for the calibration of personal dosemeters and survey meters used for radiation protection purposes[1-3]. These standards describe reference radiations with neutron energy spectra that are well defined and well suited for use in the calibration laboratory. However, the neutron spectra commonly encountered in routine radiation protection situations are, in many cases, quite different from the neutron spectra of the reference radiations specified in the ISO standards. Specifically, the reference radiations described in ISO 8529[1] have dose equivalent mean energies above 2 MeV and do not have thermal neutrons present, whereas most workplace neutron spectra contain thermal neutrons and many have dose equivalent mean energies below 2 MeV. Many workplace fields have spectrum averaged fluence-to-dose equivalent conversion coefficients $\langle h_\Phi^*(10)\rangle$ that are significantly less than the coefficients for the ISO 8529 spectra. Therefore, it is useful to specify additional reference radiations for the calibration of personal neutron dosemeters and survey meters. The ISO working draft standard described herein specifies the methods for producing and characterising neutron fields that more closely resemble those found in practical workplace situations.

It has long been recognised that, because of the energy dependence in their dose equivalent responses, neutron personal dosemeters and survey meters calibrated using reference radiations such as unmoderated ^{252}Cf or ^{241}Am-Be may yield incorrect measurements in a workplace neutron field if the neutron energy spectrum of this field is significantly different from that of the reference radiation[4-8]. Several possibilities have been suggested for improving this situation. First, the workplace neutron energy spectrum can be measured and a correction factor for the detector can be calculated. This method requires measurements of the energy spectrum and knowledge of the dose equivalent response of the detector. The technique involves exposure of dosemeters at the time of the spectrometry measurements. Unfortunately, this approach is often impractical because a suitable location in the workplace may not be available for such calibration measurements. Another approach is to construct a laboratory facility designed to produce a neutron field with an energy spectrum that simulates the spectrum found in the workplace. When this field has been properly characterised, it can then be used for the direct calibration of personal neutron dosemeters and survey meters that will be used in a particular workplace.

This latter approach has the advantage of providing a neutron source with an appropriate, reproducible energy spectrum and the possibility of having a variable intensity. Within the neutron source facility, it is possible to control the scattering geometry as well as such environmental effects as temperature and humidity. For these reasons, the technique has been employed at a number of calibration laboratories[9–12], and this standard gives guidance for establishing and characterising simulated practical neutron spectra that can be used as reference calibration sources for radiation protection purposes.

The ISO draft standard discussed in this report is being developed mainly for primary and secondary calibration laboratories that need to develop facilities to produce reference radiations having simulated practical neutron energy spectra. It is also expected that the standard will be of interest to potential users of such reference radiations.

SUMMARY OF CONTENTS OF THE WORKING DOCUMENT

The ISO working draft standard document discusses the production of simulated practical neutron spectra, and several examples of facilities generating such spectra are given. Only those facilities that are generally available to outside users, and for which complete characterisations have been performed, were considered. Although it is recognised that many laboratories have constructed specialised neutron calibration sources, it was necessary to limit the number of facilities described in this standard to those satisfying the aforementioned criteria.

The types of facilities considered are those using radionuclide neutron sources[10], accelerators[9,12] and reactors[11]. In each of these cases, a simulated practical neutron calibration spectrum was generated by placing a variety of absorbing and scattering materials between the primary neutron source and the detector. The draft standard discusses the advantages and disadvantages of each type of facility used to produce the neutron spectra.

An example of the assembly used to produce a simulated practical neutron spectrum is the neutron source assembly developed at the Institut de Protection et de Sûreté Nucléaire, Commissariat à l'Energie Atomique, Cadarache Laboratory in France[13]. The neutron spectrum generated at this facility contains a high energy component, representing the primary source of neutrons, a scattered component with an approximately $1/E_n$ dependence and a thermal neutron component. These general characteristics are common to many neutron spectra encountered in the workplace.

The draft standard contains guidelines for the characterisation of simulated practical neutron fields. The methods suggested for the characterisation of such fields include Monte Carlo calculations and measurements of the neutron energy spectra. The limitations and expected uncertainties associated with these methods are discussed. It is recommended that both approaches be used in parallel, and it is also recommended that independent measurements and calculations by different laboratories be performed at the facility under study in order to verify the dosimetric parameters.

Characterisation of the simulated practical neutron field must include a determination of the type and quantity of contaminating radiations present. Since neutron fields are nearly always accompanied by gamma rays, it is necessary to determine the contribution to dose equivalent delivered by this particular radiation. Recent results of intercomparison measurements with various neutron spectrometers at the IPSN-CEA Cadarache facility were reported by D. Thomas et al during this symposium[14]. The draft standard also requires that the intensity of those neutron fields expected to vary quickly with time be monitored using methods appropriate to the particular neutron production mechanism. These methods include: ionisation chambers, proportional counters, and solid state charged particle detectors.

The draft standard discusses the sources of uncertainties present in calibration measurements using simulated practical neutron fields. The evaluation and reporting of uncertainties follow recommendations given in ISO/IEC Guidelines to the Expression of Uncertainty in Measurement[15].

CONCLUSIONS

In recent years, it has become clear that additional neutron reference radiations were needed for the calibration of personal neutron dosemeters and survey meters. This results from the fact that neutron dosemeters and survey meters are generally quite energy dependent in their dose equivalent response.

Several calibration laboratories have developed simulated practical neutron sources in order to provide reference neutron radiations with energy spectra that more closely resemble spectra found in the workplace. In order to provide guidance on the production of simulated practical neutron fields and requirements for their characterisation, the ISO draft standard described in this report was developed. The draft standard contains descriptions of several types of neutron sources, indicates how these sources are to be characterised and how they are to be used.

This report was prepared for the purposes of describing the contents of the standard at its present state of development, and eliciting comments and suggestions from potential users of the standard. These comments should be directed to members of the writing group who are also the authors of this report.

ACKNOWLEDGEMENT

This work was supported in part by the US Department of Energy under Contract DE-AC06-76RLO-1830.

REFERENCES

1. ISO 8529. *Neutron Reference Radiations for Calibrating Neutron-Measuring Devices.* (International Organization for Standardization (ISO), Geneva, Switzerland) (1989).
2. ISO DIS 10647. *Procedures for Calibrating and Determining the Response of Neutron-Measuring Devices Used for Radiation Protection* (International Organization for Standardization (ISO), Geneva, Switzerland) (1994).
3. ISO CD 8529-Part 3. *Reference Neutron Radiations: Calibration of Area and Personal Dosimeters and the Determination of Their Response as a Function of Neutron Energy and Angle of Incidence.* (International Organization for Standardization (ISO), Geneva, Switzerland) (1995).
4. Piesch, F. *Calibration Techniques for Personal Dosemeters in Stray Neutron Fields.* Radiat. Prot. Dosim. **10**(1–4), 159–173 (1985).
5. Naismith, O. F. and Siebert, B. R. L. *A Database of Neutron Spectra, Instrument Response Functions, and Dosimetric Conversion Factors for Radiation Protection Applications.* Radiat. Prot. Dosim. **70**(1–4), 241–245 (This issue) (1997).
6. Aroua, A., Boschung, M., Cartier, F., Grecescu, M., Prêtre, S., Valley, J.-F. and Wernli, Ch. *Characterisation of the Mixed Neutron-Gamma Fields Inside the Swiss Nuclear Power Plants by Different Active Systems.* Radiat. Prot. Dosim. **51**(1), 17–25 (1994).
7. Lindborg, L., Bartlett, D., Drake, P., Klein, H., Schmitz, Th. and Tichy, M. *Determination of Neutron and Photon Dose Equivalent at Workplaces in Nuclear Facilities in Sweden — A Joint SSI-EURADOS Comparison Exercise.* Radiat. Prot. Dosim. **61**(1–3), 89–100 (1995).
8. Naismith, O. F., Siebert, B. R. L. and Thomas, D. J. *Response of Neutron Dosemeters in Radiation Protection Environments: an Investigation of Techniques to Improve Estimates of Dose Equivalent.* Radiat. Prot. Dosim. **70**(1–4), 255–260 (This issue) (1997).
9. Chartier, J. L., Posny, F. and Buxerolle, M. *Experimental Assembly for the Simulation of Realistic Neutron Spectra.* Radiat. Prot. Dosim. **44**(1–4), 125–130 (1992).
10. Kluge, H., Alevra, A. V., Jetzke, S., Knauf, K., Matzke, M., Weise, K. and Wittstock, J. *Scattered Neutron Reference Fields Produced by Radionuclide Sources.* Radiat. Prot. Dosim. **70**(1–4), 327–330 (This issue) (1997).
11. Medioni, R. and Delafield, H. J. *Criticality Accident Dosimetry: An International Intercomparison at the SILENE Reactor.* Radiat. Prot. Dosim. **70**(1–4), 445–454 (This issue) (1997).
12. Aroua, A., Hofert, M., Sannikov, A. V. and Stevenson, G. *Reference High Energy Neutron Fields at CERN.* CERN Report CERN/TIS-RP/TM/94-12 (CERN, Geneva) (1994).
13. Chartier, J. L., Jansky, B., Kluge, H., Schraube, H. and Weigel, B. *Recent Developments in the Specification and Achievement of Realistic Neutron Calibration Fields.* Radiat. Prot. Dosim. **70**(1–4), 305–312 (This issue) (1997).
14. Thomas, D. J., Chartier, J. L., Klein, H., Naismith, O. J., Posny, F. and Taylor, G. G. *Results of a Large Scale Neutron Spectrometry and Dosimetry Comparison Exercise at the Cadarache Moderator Assembly.* Radiat. Prot. Dosim. **70**(1–4), 313–322 (This issue) (1997).
15. ISO. *Guide to the Expression of Uncertainty in Measurement* (International Organization for Standardization (ISO), Geneva, Switzerland) (1993).

SCATTERED NEUTRON REFERENCE FIELDS PRODUCED BY RADIONUCLIDE SOURCES

H. Kluge†, A. V. Alevra†, S. Jetzke‡, K. Knauf†, M. Matzke†, K. Weise† and J. Wittstock†
†Physikalisch-Technische Bundesanstalt
Bundesallee 100, D-38116 Braunschweig, Germany
‡Fachhochschule Braunschweig/Wolfenbüttel, Salzdahlumer Str. 46–48
D-38302 Wolfenbüttel, Germany

Abstract — Realistic fields similar to those occurring in practice can be approximately realised using the ISO-recommended radionuclide sources if the neutrons scattered by the concrete walls of a medium sized irradiation room are used. At the PTB, such fields were established at the irradiation facility with radionuclide reference neutron sources. In order to determine the spectral fluence rate at the point of test as the basis of calibrations, the different spectra were calculated using the MCNP code, and measured by means of a Bonner sphere spectrometer. The measured count rates were unfolded using different codes. The total fluence rate can be determined with a relative standard uncertainty of about 4%. The relative standard uncertainty of the ambient dose equivalent is found to be about 5%.

INTRODUCTION

At present, there is no neutron dosemeter which allows the specific quantity of dose equivalent used in area monitoring or in individual monitoring to be measured independently of the energy and the direction of incident neutrons. For this reason, instruments should be calibrated in neutron fields which, as far as energy and directional distribution are concerned, are similar to those in which the devices are to be used.

The International Organization for Standardization (ISO) recommends radionuclide sources (unmoderated and D_2O-moderated ^{252}Cf, ^{241}Am-B(α,n) and ^{241}Am-Be(α,n)) for the calibration of neutron measuring devices used for radiation protection purposes[1]. Calibrations are normally performed using the 'direct' neutrons reaching the measuring device without additional scattering from the environment. The contribution of scattered neutrons to the instrument reading is measured separately. This can be done, for instance, by means of a shadow object of negligible neutron transmission, which is positioned between the source and the point of measurement. From among these sources, the neutron energy spectrum of the D_2O-moderated ^{252}Cf source (spherical moderator 30 cm in diameter, covered with cadmium) covers the energy range from the Cd cut-off energy (0.5 eV) to several MeV. However, the important thermal component is lacking, and the field is nearly unidirectional at the point of measurement for calibration distances normally used. In contrast, most of the fields found in practice contain a large portion of scattered neutrons, with a considerable portion in the range of thermal energies and, in general, a broad range of angles of incidence.

Such fields were approximately simulated using the above-mentioned radionuclide sources and taking into account the neutrons scattered by the concrete walls of a medium sized room. A special field is that of the D_2O-moderated ^{252}Cf source if the cadmium shielding is removed and the 'direct' and scattered neutrons are used. The field follows from the superposition of a quasi-aligned and a quasi-isotropic component, the ratio of which can be changed if the distances between them are changed. Calibrations can be carried out in these fields if the energy and the angular distribution of the neutrons are well known. At the PTB, corresponding fields have now been established at an existing irradiation facility with radionuclide reference neutron sources[2].

IRRADIATION FACILITY

The facility is installed in a 7 m × 7 m × 6.5 m bunker room whose walls and ceiling are made of reinforced concrete 1 m thick. Five bare sources, three ^{252}Cf sources of different source strengths, a ^{241}Am-Be(α,n) source and a ^{137}Cs source for testing, can be located near the geometric centre of the room. The D_2O moderator assembly can be positioned exactly in the centre of the room. An additional adjustable thermal component of the 'direct' neutrons was achieved at the moderator assembly by covering half the D_2O moderator with cadmium and by making it possible to rotate the moderator about its vertical axis. Due to the dimensions of the room and the thickness of the walls, a large number of wall-scattered neutrons with rather low energies and practically isotropic directional distribution are present at the standard calibration distance of 170 cm. These neutrons can either be used in addition to the 'direct' neutrons from the source or separately. In the latter case, the only thing to do is to eliminate the 'direct' neutrons by arranging an appropriate shadow object between the source and the point of test.

MCNP CALCULATIONS

For comparison, the neutron energy spectrum of the wall-scattered neutrons was simulated using the MCNP Code[3]. In order to simulate reality as closely as possible, the composition of the walls was investigated taking samples at different depths and locations. The samples were analysed at the PTB and the following composition, which was practically independent of the depth, was found and used for the calculations (percentages by weight): H: 1.2; O: 51.8; Na: 0.5; Mg: 0.8; Al: 6.2; Si: 28.7; S: 0.5; K: 1.4; Ca: 6.5 and Fe: 2.4. The walls and the ceiling are heavily reinforced by an iron meshwork 18 mm in thickness and with a nominal mesh width of 120 mm × 140 mm at a depth of about 5 cm below the surface. For the simulation, this meshwork was replaced by a thin homogeneous layer of iron of the same mass applied to the inner surfaces of the bunker room. In addition, the air, the main parts of the detector positioning system and the shadow objects were simulated (the latter being a cone composed of 20 cm iron with a front diameter of 3.2 cm and 30 cm polyethylene with a back diameter of 12 cm, and a block of polyethylene with 10% boron content, dimensions: 30 cm × 30 cm × 50 cm). 'Point detectors' were used to calculate the neutron energy spectra at the point of test behind the shadow objects. The D_2O moderator assembly of the PTB covered by a Cd half-shell was completely simulated using a Maxwell distribution with a spectrum parameter of T = 1.42 MeV for the ^{252}Cf source at the moderator centre. The same distribution was used for the bare ^{252}Cf source spectrum. For the ^{241}Am-Be source, the measured spectrum[4] was taken as the initial spectrum.

MEASUREMENTS

The main problem in determining the spectral neutron fluence or fluence rate is the large energy range of the scattered neutrons reaching from thermal energies to several MeV. The only possibility of getting experimental information about the neutron energy spectrum over such a large energy range with one spectrometer is the Bonner sphere method. Corresponding measurements were carried out using the PTB's Bonner sphere system[5] with a spherical ^3He proportional counter as the central detector. Ten spheres with diameters ranging from 7.62 cm (3″) to 30.48 cm (12″) and, in addition, the ^3He counter, bare and covered with Cd, were used. All measurements were performed at a centre-to-centre distance of 170 cm between the source (moderator) and the Bonner spheres. The distance between the centre of the source and the front face of the shadow object used was 2.5 cm for the ^{241}Am-Be(α,n) source, 9.5 cm for the bare and 65.2 cm for the D_2O moderated ^{252}Cf source.

DATA EVALUATION

Three different procedures were applied in order to unfold the measured data and to derive the neutron energy spectrum and the resulting integral quantities: the thermal fluence rate, the total fluence rate and the total dose equivalent rate. The first is a modified STAY'SL code[6] which allows a neutron energy spectrum given as pre-information to be adjusted within the range determined by the uncertainties. This code was used with the information resulting from the MCNP calculations. As the MCNP calculations are expected to be more uncertain in the range of thermal energies, a relative standard uncertainty of 60% was assumed for the

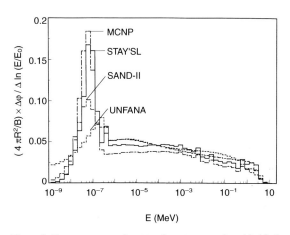

Figure 1. Energy spectra of scattered neutrons produced behind a shadow object by a D_2O-moderated ^{252}Cf source at the geometric centre of the irradiation room: intercomparison of the Bonner sphere measurements using different unfolding codes. Here and in the following figure, the corresponding spectrum resulting from the MCNP calculations is given for comparison. $\Delta\varphi$ is the neutron group fluence rate in the interval $\Delta\ln(E/E_0)$ at distance R, E is the neutron energy with E_0 = 1 MeV and B is the neutron source strength.

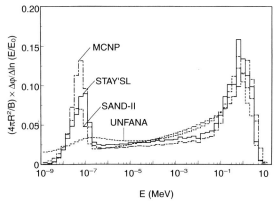

Figure 2. Energy spectra of scattered neutrons produced behind a shadow object by a ^{252}Cf source. For notations, see Figure 1.

spectral fluence rates in this energy range and one of 20% for energies above 0.5 eV. The second procedure, UNFANA, is based on a new analytical unfolding method[7]. Here, the spectrum probability distribution which, according to Bayesian statistics and the principle of maximum entropy, is most appropriate to the data, is replaced by a suitable exponential distribution from which an estimate of the desired spectrum can be analytically calculated. The third procedure is a modified SAND-II code where an initial spectrum is needed for the iteration procedure, which was assumed to be a superposition of a Maxwell distribution for thermal neutrons, a $1/E^{\alpha}$ slowing down part (E neutron energy) and a fission neutron spectrum for the fast neutrons[8].

RESULTS

As a representative example of the results of the different unfolding procedures, those obtained for the ^{252}Cf sources (bare and moderated) are presented in Figures 1 and 2. The spectra resulting from the MCNP calculations, which were used as pre-information for the STAY'SL code, are shown for comparison. Results for the bare ^{252}Cf source using the SAND-II code were given earlier[9] and have been included for comparison. When the unfolded spectra were folded again with the response functions of the Bonner spheres, the resulting count rates were consistent with those measured for all spectra.

Integral quantities characterising the different fields such as the mean neutron energy, $\bar{E}$, and the ambient dose equivalent averaged mean neutron energy, $\tilde{E}^*$, and quantities relevant to calibrations such as the total neutron fluence rate, φ, normalised to the neutron source strength, B, and mean neutron fluence-to-ambient dose equivalent conversion factors averaged over the neutron energy spectra, $\bar{h}^*_\varphi$, together with the fractional contribution of thermal neutrons to the total fluence rate, φ_{th}/φ, are compiled in Table 1 for all fields investigated. Here and in the following, $\bar{h}^*_\varphi$ stands for $\bar{h}^*_\varphi(10)$. $\tilde{E}^*$ and $\bar{h}^*_\varphi$ were calculated using the conversion factors for monoenergetic neutrons given by Wagner et al[10], based on the Q(L) relationship according to ICRP 21[11], and by Siebert and Schuhmacher[12], which are based on ICRP 60[13] taking into account new stopping power data for protons and alpha particles[14].

DISCUSSION

Although the spectra resulting from the different unfolding procedures show a certain ambiguity within the uncertainties, the integral quantities differ only by a few per cent due to negative correlations. It is difficult to state realistic uncertainties for the values, as the input

Table 1. Intercomparison of the results of different unfolding procedures. The corresponding values resulting from the MCNP calculations are given for comparison.

Source	Procedure	Field type	$\bar{E}$ (MeV)	$\varphi/B \times 10^6$ (cm^{-2})	φ_{th}/φ (%)	$\tilde{E}^*_{21}$ (MeV)	$\bar{h}^*_{\varphi,21}$ (pSv.cm^2)	$\tilde{E}^*_{60}$ (MeV)	$\bar{h}^*_{\varphi,60}$ (pSv.cm^2)
^{252}Cf(D$_2$O)	MCNP	Scattered	0.079	2.40	44.4	0.94	28.5	0.86	34.7
	STAY'SL	Scattered	0.094	2.63	37.6	0.97	32.5	0.90	39.5
	UNFANA	Scattered	0.104	2.55	39.7	1.12	32.0	0.99	38.6
	SAND-II	Scattered	0.095	2.57	33.9	0.92	34.2	0.84	43.6
^{252}Cf	MCNP	Scattered	0.39	2.56	28.6	1.25	107	1.17	128
	STAY'SL	Scattered	0.42	2.71	21.2	1.26	115	1.18	139
	UNFANA	Scattered	0.52	2.79	17.4	1.75	110	1.57	129
	SAND-II	Scattered	0.53	2.77	18.4	1.70	115	1.55	135
^{241}Am-Be	MCNP	Scattered	0.65	2.15	28.4	1.96	126	1.80	144
	STAY'SL	Scattered	0.68	2.24	19.6	1.94	132	1.79	153
	UNFANA	Scattered	0.62	2.19	16.5	1.84	125	1.69	146
	SAND-II	Scattered	0.72	2.21	17.5	2.07	133	1.89	154
^{252}Cf(D$_2$O)	MCNP	Scatt. + dir.	0.29	5.22	26.4	1.93	57.6	1.75	66.2
	STAY'SL	Scatt. + dir.	0.31	5.54	22.5	1.89	61.4	1.72	70.7
	UNFANA	Scatt. + dir.	0.28	5.48	21.6	1.59	63.7	1.46	74.2
	SAND-II	Scatt. + dir.	0.29	5.46	20.3	1.73	63.9	1.58	77.6

The mean neutron energy, $\bar{E}$, the ratio of total neutron fluence rate, φ, and neutron source strength of the source used, B, the relative contribution of the fluence rate of thermalised neutrons, φ_{th}, to the total fluence rate, the ambient dose equivalent averaged mean neutron energy, $\tilde{E}^*$, and mean neutron fluence-to-ambient dose equivalent conversion factors, $\bar{h}^*_\varphi$, are given. $\bar{h}^*_{\varphi,21}$ refers to the conversion factors given by Wagner *et al*[10], based on the Q(L) relationship according to ICRP 21[11], and $\bar{h}^*_{\varphi,60}$ to those given by Siebert and Schuhmacher[12], based on ICRP 60[13], using new stopping power data and ranges of protons and α particles according to ICRU 49[14].

quantities and the results, too, are correlated. However, the standard deviations of the mean values for φ/B and φ_{th}/φ stated in the table and for the product $\bar{h}^*_\Phi\varphi/B$, relevant to the determination of the ambient dose equivalent rate, may be taken as a clue to the inherent uncertainties attributable to the different procedures.

In addition, contributions to the uncertainties (relative standard deviations), including the correlations, must be taken into account, which follow from the source strength determination (1.3% to 1.6%) and the determination of a common scaling factor for the response functions of the Bonner spheres (3%).

An overall relative standard uncertainty of about 4% for the total fluence rate, of about 8% for the fluence rate of neutrons in the range of thermal energies and of about 5% for the ambient dose equivalent rate is therefore stated.

REFERENCES

1. International Organization for Standardization (ISO). *Neutron Reference Radiations for Calibrating Neutron Measuring Devices used for Radiation Protection Purposes and for Determining their Response as a Function of Neutron Energy*. International Standard ISO 8529 (1989).
2. Guldbakke, S., Dietz, E., Kluge, H. and Schlegel, D. *PTB Neutron Fields for the Calibration of Neutron Sensitive Devices*. In: Strahlenschutz: Physik und Meßtechnik (Köln: Verlag TÜV Rheinland GmbH) Band I, pp. 240–247 (1994).
3. Briesmeister, J. F. (ed.) *MCNP — A General Monte Carlo N Particle Transport Code — Version 4A*. LA-12625-M, (1993).
4. Kluge, H. and Weise, K. *The Neutron Energy Spectrum of a ^{241}Am-$Be(\alpha,n)$ Source and Resulting Mean Fluence to Dose Equivalent Conversion Factors*. Radiat. Prot. Dosim. **2**(2), 95–93 (1982).
5. Alevra, A. V., Cosack, M., Klein, H., Knauf, K., Matzke, M., Plewnia, A. and Siebert, B. R. L. *Development of Neutron Spectrometers for Radiation Protection Practice*. EUR-Report 1326B, Vol. 1, pp. 132–142 (Luxembourg: CEC) (1991).
6. Matzke, M. *Unfolding of Pulse Height Spectra: The HEPRO Program System*. Report PTB-N-19, Braunschweig (1994).
7. Weise, K. *Mathematical Foundation of an Analytical Approach to Bayesian-Statistical Monte Carlo Spectrum Unfolding*. Report PTB-N-24, Braunschweig (1995).
8. Alevra, A. V. *Measurements with the PTB Bonner Sphere Spectrometer and a Leake Rem Counter*. In: Determination of Neutron and Photon Dose Equivalent at Work Places in Nuclear Facilities of Sweden. SSI-Report 95-15, pp. 42–58 (1995).
9. Alevra, A. V., Klein, H., Knauf, K. and Wittstock, J. *Neutron Field Spectrometry for Radiation Protection Dosimetry Purposes*. Radiat. Prot. Dosim. **44**(1/4), 223–226 (1992).
10. Wagner, S., Großwendt, B., Harvey, I. R., Mill, A. J., Selbach, H. J. and Siebert, B. R. L. *Unified Conversion Functions for the New ICRU Operational Radiation Protection Quantities*. Radiat. Prot. Dosim. **12**(2), 231–235 (1985).
11. ICRP. *Data for Protection against Ionizing Radiation from External Sources*. Publication 21 (Oxford: Pergamon) (1976).
12. Siebert, B. R. L. and Schuhmacher, H. *Quality Factors, Ambient and Personal Dose Equivalent for Neutrons, Based on the New ICRU Stopping Power Data for Protons and Alpha Particles*. Radiat. Prot. Dosim. **58**(3), 177–183 (1995).
13. ICRP. *1990 Recommendations of the International Commission on Radiological Protection*. Publication 60, Ann. ICRP **21**(1–3) (Oxford: Pergamon) (1991).
14. ICRU. *Stopping Powers and Ranges for Protons and Alpha Particles*. Report 49 (Bethesda, MD: ICRU) (1993).

ADVANCES IN REALISTIC NEUTRON SPECTRA: PROGRESS IN FLUENCE MONITORING OF THE DD REACTION

D. Paul, G. Pelcot and C. Itié
Institut de Protection et de Sûreté Nucléaire (IPSN)
Département de Protection de la santé de l'Homme et de Dosimétrie, Service de Dosimétrie
B.P. n°6, F-92265 Fontenay-aux-Roses Cedex, France

Abstract — As part of a programme on simulation of realistic neutron spectra at workplaces, a new facility based on the ^{2}H(d,n)^{3}He reaction (abbrev.: DD) which yields a 3.3 MeV neutron field, is being developed at the IPSN/SDOS Laboratory in Cadarache. Additional shields at the exit of an accelerator target can provide, in the laboratory, a replication of some spectral conditions encountered in practice. Spectra resulting from this reaction will complete those obtained from the DT process in the same laboratory. Reasons for using the DD reaction and the associated particle counting technique as a monitor of neutron flux are presented. Instead of detecting the alpha particle from the DT reaction, protons are counted from the competitive reaction ^{2}H(d,p)^{3}H with respect to the DD reaction. The fact that the two last reactions are involved in the monitoring and are markedly anisotropic in the CM system, means that a more complex situation has to be dealt with to calculate the neutron yield per proton measured in a solid angle. A conversion factor between neutrons and protons must be calculated and compared with reference measurements performed with several instruments. Based on the new assembly, a 'realistic' spectrum calculated with the MCNP code is proposed. Results from the MCNP code are multiplied by a (n,p) factor calculated by another program, MONITOR, which takes into account the kinematics, the cross sections and the energies involved in the reactions. The theoretical fluence predicted at a reference point in this spectrum is validated directly with the area monitor previously calibrated through a specific methodology. This work provides a new tool of investigation to obtain realistic neutron spectra to be used in the calibration of radiation protection instruments that should fulfil the ICRP recommendations, as given in ICRP Publication 60.

INTRODUCTION

The concept of realistic neutron spectra used for the calibration of dosimetric instruments in conditions similar to their practical use has already been described[1-4]. In order to approach realistic spectral conditions[5], several laboratories have decided to work together within the framework of a CEC contract[3]. The IPSN/SDOS laboratory in Fontenay-aux-Roses and Cadarache initiated this objective in 1988 with a first generator, i.e. an accelerator-based neutron source producing 14.6 MeV neutrons from the ^{3}H(d,n) reaction (SAMES J25).

The present work was carried out to provide a new prototype facility for the laboratory which will complete the first assembly. This new tool is of particular importance in solving the problem of calibration of radiation protection instruments to fulfil the ICRP recommendations as specified in ICRP Publication 60[6].

The set-up is patterned on the the SAMES J25 accelerator with two important modifications: the value of the primary neutron energy produced by a SAMES T400 accelerator using the ^{2}H(d,n) reaction and the initial shielding of the assembly, without an iron shell. The monitoring system is more complex and based on the competitive reaction ^{2}H(d,p) for the detection of the associated proton particle. The calculations and measurements relative to fluence monitoring and the characteristics of new spectra obtained with the new facility are presented.

THE BARE NEUTRON SOURCE

The facility is based on a SAMES T400 accelerator (max HV, 400 kV). Deuteron ions are accelerated with H.V. 330 kV. The present engineering (mechanics, cooling system) allows for only a small current, around 150 µA. The DD process (^{2}H(d,n)^{3}He reaction) is performed by using deuterated targets from CEA Valduc (target thickness: 1352 µg.cm^{-2}). The reaction at HV 330 kV yields a 3.3 MeV neutron field. The competitive reaction: ^{2}H(d,p)^{3}H, produces 2.5 MeV protons used to monitor the neutron emission of the target (Table 1).

Due to neutron scattering in the target assembly, the bare neutron source consists of a spectrum which has been simulated (simplified model) with MCNP Monte Carlo code[7] at a distance of 100 cm from the target (Figure 1) and normalised to 1 neutron emitted in 4π sr. The study of this 'bare' configuration A allows the scattered contribution (20% in fluence) and the absorp-

Table 1. Data processing, DD reaction.

Deuteron energy, E_D (MeV)	0.330
Neutron energy, E_n (MeV)	3.268
Proton energy, E_p (MeV)	2.473
Anisotropy factor, K	0.789
Ratio (σ_n/σ_p), R	1.263
Total number of neutrons per proton detected y_n (4π sr)	0.697×10^6

tion correction in the target backing ($\approx 14\%$ in fluence) to be taken into account.

THE CONFIGURATIONS SIMULATING 'REALISTIC' SPECTRA

The technique uses a ^{238}U fission-induced neutron field[1] from a 3 MeV neutron source (configuration A). The target of the accelerator is placed at the centre of a quasi-spherical ^{238}U converter to give the first realistic neutron spectrum on a T400 accelerator (configuration B). The original fission spectrum can be modified by an additional water shield (10 cm) and a polyethylene duct (thermalisation), in order to obtain a spectrum with a rather low mean energy, as is usually encountered behind biological shields (configuration C).

In Figure 2, the main characteristics of the configurations (B) and (C) are presented.

The study of neutron spectra in the calibration zone ($30 \times 30 \times 30$ cm^3) at a distance of 30 cm from the duct was also performed by the Monte Carlo simulation (MCNP) for configurations B (Figure 3) and C (Figure 4). Furthermore, the main characteristics of neutron spectra shown in Figures 3 and 4 are summarised in Table 2 and in Table 3 in fluence and dose equivalent terms for all configurations.

MONITORING: ASSOCIATED PARTICLE SYSTEM

The MONITOR program has been employed to evaluate the characteristics of the monitoring system based on the competitive reaction ^{2}H(d,p)^{3}H. The calculation was performed by the same methodology as the ^{3}H(d,n)^{4}He reaction on the SAMES 150 kV J25 type accelerator currently in operation in this laboratory[1].

Due to the fact that two nuclear reactions are involved in the DD processs and that both reactions are markedly anisotropic in the centre of mass system (CM), a more complex situation has to be considered in a solid angle $\Delta\Omega_p$ (proton detection).

The protons are measured by three semiconductor

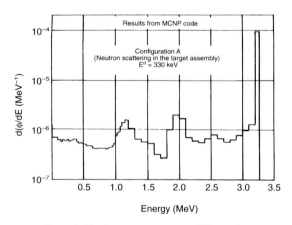

Figure 1. The bare neutron source, DD reaction.

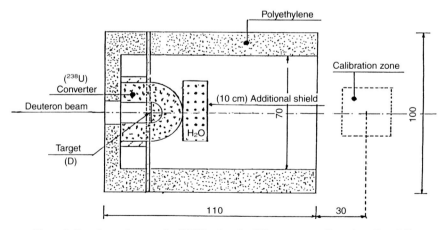

Figure 2. Experimental set-up for RNSF using the DD process, configurations B and C.

detectors (silicon diodes) placed at an angle of 176° to the incident 330 keV deuteron beam. The spectrum of the proton detector can be observed in Figure 5.

A check source (^{241}Am) has also been installed to test the three diodes and to control the reliability of the monitoring system (based on the mean value of the three detector counts).

From the expression:

$$Y_n = (4\pi/\Delta\Omega_p) K_p (\sigma_n/\sigma_p)$$

the total neutron yield per measured proton can be derived. In this formula, the symbols have the following meanings:

$\Delta\Omega_p$ = mechanical solid angle of measurement: 18×10^{-6} sr

K_p = anisotropy factor of the ^{2}H(d,p) reaction

$$K_p = [\sigma_p/4\pi]/[(d\sigma_p/d\omega')(d\omega'/d\omega)]$$

where

$(d\omega'/d\omega)$ = ratio of the differential solid angle $d\omega'$ in the CM system and the corresponding solid angle $d\omega$ in the laboratory,

$(d\sigma_p/d\omega')$ = differential cross section of proton yield in the CM system,

$(\sigma_n/\sigma_p = R)$ = ratio of total neutron and proton production cross sections.

The kinematics of nuclear reactions were calculated using the information from Marion and Young[8] in the case of a thin target (reactions occur only at the bombarding deuteron energy).

Results for K_p, R and Y_n calculations are summarised in Table 1 for a deuteron energy $E_D = 0.33$ MeV. The results of our program (MONITOR) have been compared with those of Ruby and Crawford[9] then with Hertel and Wehring[10] simulating their conditions for the associated proton detection. The difference is less than 2% with Ref. 9 and 6% with Ref. 10. Ruby and Crawford comment that, in practice, uncertainties in cross sections, target loading ratio and uniformity of target loading limit the accuracy to a few per cent.

CALCULATED AND MEASURED NEUTRON FLUENCES

Predicted fluences for configurations A, B, and C have been determined combining Y_n with the MCNP calculated fluence spectrum (normalised to 1 neutron emitted in 4π sr). The calculation takes into account the scattered contribution and the absorption correction in the target backing.

Fluence measurements have been carried out for the three configurations of the set-up at the calibration point in the beam. The following instruments were used: a calibrated Harwell N91[11] and a ^{3}He counter with an 8 inch sphere, Cadarache Bonner spheres with MORELPA code[1]. The N91 instrument comprises a 208 mm diameter polyethylene sphere, containing a spherical ^{3}He proportional counter Type SP9, enclosed in a perforated Cd liner. The moderator/detector assembly, denoted Neutron detector 0075, is housed in a moulded

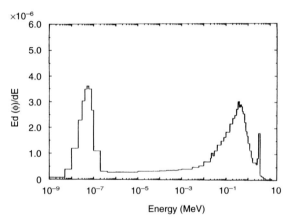

Figure 3. RNSF, configuration B. 3 MeV + ^{238}U + (CH$_2$)$_n$ duct.

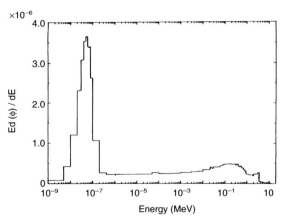

Figure 4. RNSF, configuration C. 3 MeV + ^{238}U + (CH$_2$)$_n$ duct + H$_2$O (10 cm).

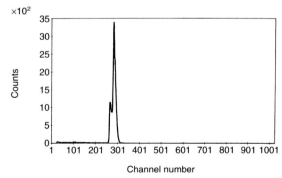

Figure 5. Spectrum of the proton detector (silicon diode).

case having two lateral compartments for electronic circuitry and batteries. The Cadarache counter coupled with the 8 inch sphere is a French cylindrical counter type 0.5 NH 1/1 KF. The MORELPA code is a homemade code developed in the Cadarache laboratory. The calibration coefficient for N91 and the 8 inch sphere with the French counter were calculated from the fluence response curve folded with the MCNP calculated spectrum[13]. These instruments were also calibrated in our wide reference spectra using procedures of ISO/CD 10647. Results are presented for configurations A, B and C in Table 3.

Predicted and experimental values are in better agreement for configurations B and C than for configuration A. The fluence ratio (experiment/calculation) for the monoenergetic configuration A is 1.14 with the 8 inch sphere and 1.27 with N91. The 8 inch result is lower than those obtained by Hertel and Wehring[10], who have reported 1.18 for a thin target, 1.37 for a thick target!

There may be several explanations for the difference between theory and experiment in configuration A. The first hypothesis deals with the determination of the 3.3 MeV fluence derived from the energy response of the instruments and the MCNP calculated spectrum. Table 3 exhibits the results of the 8 inch and N91 obtained with two different response functions in the region of interest (3 MeV). Thus, the precision of the response is essential in this calculation. The other hypothesis concerns the simulation of interactions in the target (thick) to evaluate with precision the monoenergetic fluence in the measured neutron field. Finally, the double peak observed on the diode spectrum (Figure 5) may be suspicious of the presence of d_2^+ (165 keV) in the deuteron beam. This energy of incident deuterons should be integrated in the MCNP simulation. Calculations and experiments will be performed in the future to check these two last hypotheses.

CONCLUSION

A new model of the Realistic Neutron Spectra Facility (RNSF) was established and simulated at the IPSN/SDOS Laboratory. This new tool provides complementary solutions and more spectra available for calibrating area and individual neutron dosemeters in practical situations at workplaces. A ^{238}U fission-induced neutron field based on the DD reaction offers numerous advantages over existing techniques: a more 'realistic' model regarding the initial energy (3 MeV) and obviously a greater flexibility since it avoids the tritide target. Nevertheless, fluence monitoring of the DD process is more complex than for the DT reaction, due to the anisotropy of the two reactions involved. Further calcu-

Table 2. Characteristics of RNSF, configurations B and C.

Configuration	Mean energy (fluence spect.) E_ϕ (MeV)	Thermal neutrons $E \leq 0.4$ eV Th. (%)	Intermediate neutrons 0.4 eV $< E \leq 10$ keV Int. (%)	Fast neutrons $E > 10$ keV HE (%)	Mean conversion factor h_ϕ^* (pSv.cm^2)
B: [3.3 MeV + ^{238}U + PE duct]	0.27	35.3	17.9	46.8	89
C: [B + H$_2$O (10 cm)]	0.10	59.8	21.4	18.8	35

Table 3. Fluence and dose equivalent determined with several detectors for RNSF configurations A, B, and C.

	Fluence (cm^{-2})	H*(10) (nSv)
Configuration A	Monoenergetic neutrons 3.3 MeV	
Calculation (MCNP) × y_n	6.6*	2.7*
Harwell N91	8.4	3.4
^{3}He counter + 8" sphere	7.5	2.9
Configuration B	[3.3. MeV + ^{238}U + PE duct]	
Calculation (MCNP) × y_n	13.8*	1.3*
Harwell N91	12.7	1.2
Bonner spheres	12.3	0.9
Configuration C	[3.3 MeV + ^{238}U + PE duct + H$_2$O (10 cm)]	
Calculation (MCNP) × y_n	8.1*	0.3*
Harwell N91	7.8	0.3

* Normalised to 1 monitor count.

lations and experiments still have to be performed, especially for the monoenergetic configuration. Additional spectrometric characterisation of the available neutron fields are needed in order for RNSF to be used as a calibration facility. This will be investigated in the future by studying the homogeneity and angular distribution of the reference spectrum.

for his participation in making MCNP calculations, J. Kurkdjian and H. Muller (SDOS-CAD) for their assistance.

This work was partly supported by the Commission of European Communities (Contract FI3P-CT92–0002).

ACKNOWLEDGEMENTS

The authors wish to thank J. L. Chartier (SDOS-FAR)

REFERENCES

1. Chartier, J. L., Posny, F. and Buxerolle, M. *Experimental Assembly for the Simulation of Realistic Neutron Spectra.* Radiat. Prot. Dosim. **44**(1/4), 125–130 (1992).
2. Klein, H., Thomas, D., Chartier, J. L. and Schraube, H. *Determination and Realization of Calibration Fields for Neutron Protection Dosimetry as Derived from Spectra Encountered in Routine Surveillance.* EUR report 14927 DE/EN/FR (Luxembourg: CEC) pp. 159–174 (1993).
3. CEC Contract FI3P-CT92–0002 (1995).
4. Chartier, J. L., Jansky, B., Kluge, H., Schraube, H. and Wiegel, B. *Recent Developments in the Specification and Achievement of Realistic Neutron Calibration Fields.* Radiat. Prot. Dosim. **70**(1–4), 305–312 (This issue) (1997).
5. Siebert, B. R. L., Schraube, H. and Thomas, D. J. *A Computer Library of Neutron Spectra for Radiation Protection Environments.* Radiat. Prot. Dosim. **4**(1/4), 135–137 (1992).
6. ICRP, *1990 Recommendations of the International Commission on Radiological Protection.* Publication 60 (Oxford: Pergamon) (1991).
7. Briesmeister, J. F. *MCNP-A General Monte Carlo N-particle Transport Code version 4A.* LA-1265-M (Los Alamos) (1993).
8. Marion, J. B. and Young, F. C. *Nuclear Reaction Analysis, Graphs and Tables.* Neutronendosimetrie-PTB-6.5 Report (1968).
9. Ruby, L. and Crawford, R. B. *Anisotropy Factors for the Determination of Total Neutron Yield from the $D(d,n)^3He$ and $T(d,n)^4He$ reactions.* Nucl. Instrum. Methods **24**, 413–417 (1963).
10. Hertel, N. E. and Wehring, B. W. *Absolute Monitoring of DD and DT Neutron Fluences using the Associated-particle Technique.* Nucl. Instrum. Methods **172**, 501–506 (1980).
11. Harwell Instruments, AEA Technology. *Technical Specification and User Guide* (1993).
12. Buxerolle, M. *Private Communication* (1990).
13. Chartier, J. L., Kurkdjian, J., Paul, D., Itié, C., Audoin, G., Pelcot, G. and Posny, F. *Progress on Calibration Procedures with Realistic Neutron Spectra.* Radiat. Prot. Dosim. **61**, 57–62 (1995).

GRENF — THE GSF REALISTIC NEUTRON FIELD FACILITY

H. Schraube, B. Hietel, J. Jakes, V. Mares, G. Schraube and E. Weitzenegger
GSF-Forschungszentrum für Umwelt und Gesundheit mbH.,
Institut für Strahlenschutz Neuherberg, D-85758 Oberschleißheim, Germany

Abstract — At the GSF, an irradiation facility has been established which allows one to test and calibrate neutron measuring devices under realistic neutron field conditions. It consists of an accelerator neutron source, an irradiation cavity, and a variable moderating assembly. As a first approach to spectra observed in reactor environments, a simple set-up of heavy water, polyethylene and iron shields was installed. The spectral neutron fluences were measured with a Bonner spectrometer and partly verified by Monte Carlo calculation.

INTRODUCTION

None of the presently available devices for determining individual or ambient neutron dose equivalent measures sufficiently exactly the respective operational quantity defined by the International Commission on Radiation Measurements and Units (ICRU). Instead, their energy dependence of response deviates considerably from the required one. In a wide range between thermal energy and a few hundred keV, there are only a few monoenergetic neutron sources available to study the response, but their installation and use is accompanied with great technical effort. Therefore, broad spectra which are similar to those encountered in radiation protection practice, may serve to study the response by comparing the integral reading in these fields with results obtained from folding the assumed responses with well defined spectral data.

There are already several approaches available for the simulation of realistic fields, though with slightly different aims. Only the following examples should be mentioned: The first one was certainly a ^{252}Cf spontaneous fission source, surrounded by a 30 cm diameter heavy water sphere[1]. A more elaborate installation employing a small accelerator with a few hundreds of keV was achieved by Chartier et al[2]. For the ^{3}H(d,n) reaction the tritium target was surrounded by a ^{238}U shell to convert 14 MeV neutrons into fission neutrons. An installation at a reactor with a thermal neutron converter and a small irradiation cavity was established in Winfrith[3]. For the study of emergency dosemeters, the SILENE reactor with an open core and variable shielding is operating in Valduc[4]. The common characteristics of all these facilities is that the spectral source strength is essentially a fission spectrum. The same is done in this work, but making use of a different technique.

This work is part of the European joint effort to provide neutron sources for radiation protection purposes which allow the simulation of neutron spectra as encountered in the nuclear fuel cycle. The main purpose of the facility is to test dosimetric devices whose energy dependence of response is based on complex physical processes, by a combination of several processes, or by combination of different devices, so that a verification of predicted and/or calculated responses in a wide energy range is necessary.

THE GRENF FACILITY

For the design of the facility, only simple considerations were made: (i) the primary neutron source should have energies comparably to those of fission neutrons; (ii) the cavity should be large enough to place any kind of portable device, but also phantoms with individual dosemeters fixed to the surface; (iii) the source strength should be sufficiently high for delivering an acceptable dose rate at the place where dosemeters are to be irradiated; (iv) the dose rate should be variable without changing much any other parameters, e.g. the average spectral neutron yield; (v) the facility should be well shielded, and no radiation protection problems should appear in adjacent laboratories; and (vi) there should be sufficient space provided to place shielding material in front of the neutron source for modelling of a variety of spectral conditions. Essential parts of these technical provisions were already available at the GSF, so the facility could be developed without major problems.

Neutrons are produced by the reaction Be(d,n) with deuterons accelerated to 2.8 MeV by a Van-de-Graaff generator. The Be target was thick enough to stop the accelerated particles totally. The average energy of the neutrons emitted in the forward direction is approximately 2 MeV which corresponds well to the energy of fission neutrons. The target is positioned inside an irradiation cavity with the dimensions 130 cm length, 170 cm height and 116 cm width (Figure 1). In front of the target the moderating assembly can be placed, in the figure the 30 cm D_2O sphere. The wall of the cavity consists of 25 cm thick concrete and may be modified by additional scatter material on the inner side of the cavity. Here, a 5 cm thick iron scatterer is mounted. Monitoring is established by a tissue-equivalent disc type transmission chamber fixed directly in front of the target[5].

NEUTRON SPECTRA TO BE SIMULATED

The spectra to be simulated had, in the first instance,

been chosen from measurements which were done inside a 440 MWe pressurised water reactor[6] in Dukovany, Czech Republic, and near a thin walled transport container with fresh MOX fuel elements (Figure 2). They exhibit the type of spectra observed in environments of fission sources with a comparably small amount of moderating and scattering material. At the MOX element container only a small scatter component from the floor is observed where the container rested. The two positions inside the PWR are (1) above the pressure vessel with the shield effect of the vessel water and the iron, and (2) behind a thicker concrete shield. The spectra contain an increasing intermediate component while the peak energies are shifted towards energies below 1 MeV. The thermal neutron component appears to be small.

MC CALCULATIONS OF THE EXPECTED SPECTRA

In order to simulate a comparably large intermediate spectral component, D_2O was used as moderator, because of its low neutron capture capabilities. The spherical shape was not chosen purposely. This D_2O moderator was already available and a transfusion into a differently shaped container was not done to avoid impurities from the heavy water.

The Monte Carlo calculations were performed using the MCNP code version 4A[7] and cross section data of the ENDF/B-VI library[8]. The total GRENF arrangement was modelled very closely on the physical reality. The neutron fluences were averaged over small spherical volumes with a diameter of 20 mm at the points A, B, and C and normalised to one starting particle. The corresponding spectra are given in Figure 3.

The only problem arose from the proper modelling of the neutron source when the neutron field is produced by the reaction $^9Be(d,n)^{10}B$ in a thick target bombarded by 2.8 MeV deuterons. For the calculation of the spectral and angular neutron yield, the (d,n) neutron production cross section data for the slowing down deuteron energies in the target layer are required. Because of insufficient knowledge of these data, the neutron source was approximated by 2 MeV monoenergetic neutrons isotropically emitted from the disc surface of 10 mm diameter in the front hemisphere. This approximation corresponds fairly well with the observations of Inada et al[9] who used the time-of-flight technique to measure energy and angular distributions of fast neutrons from the same neutron source.

EXPERIMENTAL PROCEDURE

The spectral distribution of the neutron fluence at the

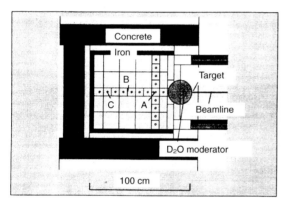

Figure 1. Irradiation facility consisting of an irradiation cavity with concrete walls and iron shield, accelerator target, exchangeable moderator assembly (here, the 30 cm D_2O sphere) and mounting positions for the detector devices (top view).

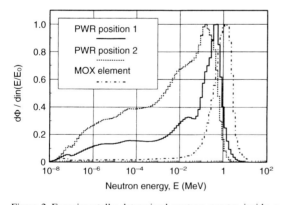

Figure 2. Experimentally determined neutron spectra inside a PWR at two different experimental positions (1) and (2), and close to a MOX element container. The spectra are presented as neutron fluence per log energy interval, i.e. equal areas exhibit equal neutron fluences. The spectra here are normalised to unity at the most probable energy.

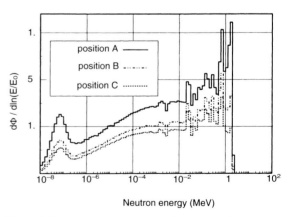

Figure 3. Calculated spectral neutron fluence per log energy interval normalised to equal target neutron yield using the MCNP Monte Carlo code. The source term was simplified by 2 MeV monoenergetic neutrons emitted into the front hemisphere with the D_2O moderator in place.

dosemeter irradiation positions was measured employing a 9-channel Bonner sphere spectrometer (BSS)[10] equipped with a 4 mm × 4 mm LiI(EU) scintillator in the centre of the Bonner spheres with sizes 2″ to 12″. Subtraction of the background count rate due to photons was done by an appropriate non-linear fit to the experimental pulse height distributions[11]. The relative background count rate in the region of the full energy peak was around 5% for the medium size spheres and did not exceed 8% for the large spheres and the bare counter. This low sensitive BSS was chosen as it also allows measurement of relatively high neutron fluence rates.

The first measurements with the BSS were performed when a 30 cm heavy water sphere moderator was in place. The purpose — beyond the reasons mentioned already — was to verify the experimental results by calculations. Further experiments were done by replacing the heavy water sphere by 10 cm thick and 40 cm × 40 cm wide slabs of polyethylene and iron. The centre of the front surface of the slab was placed at the point where the centre of the sphere had been positioned (Figure 1). The fluence distribution was scanned with a 5″ BS. For the D_2O arrangement, the fluence rate decreased along the beam axis at point C to approximately 60% of the value at point A.

DATA UNFOLDING

The spectra were unfolded using the SANDII code[12] which calculates the result from the BS response matrix, the count rate vector and a first guess spectrum by an iterative mathematical procedure[13]. Three different guess spectra were used as start spectra (a) a thermal Maxwellian and 1/E contribution, (b) an additional fast fission spectrum, and (c) applying the result of the MC calculation in Figure 3 spectrum A.

RESULTS AND DISCUSSION

Figure 4 presents the result of the unfolding when the three different start spectra were used. It is recognized that (i) all three results are very similar in shape and absolute fluence rate, (ii) even when a fast fission neutron peak is not included into the guess spectrum, the result contains this peak though with a somewhat worse resolution, (iii) the iron resonance at 24 keV is only present when it was introduced from the MC calculated spectrum into the guess spectrum. But also the spectra without any *a priori* information on this energy, give back a small bump at this energy. In Figure 5 the resulting spectra at position A are shown when the iron and polyethylene shields are in position. To normalise the spectra to the same peak height, the spectrum with the polyethylene shield had to be multiplied by 2.5, at the same neutron yield from the target. The shapes now appear different from the previous results. For both situations, the intermediate spectral contribution is small, and the thermal one is small for the iron shield and significantly higher for the polyethylene shield. It should be emphasised, however, that the spectral results are only valid for a detector system whose directional response is constant. As the different spectral components at the point of interest originate from different directions, e.g. scattered from the walls or transmitted through the moderators, the response of a non-isotropic device has to be studied in more detail. This is especially true for individual dosemeters fixed to the front of a phantom.

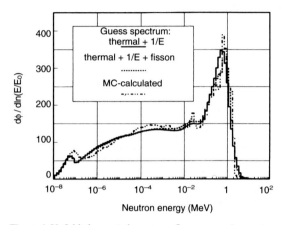

Figure 4. Unfolded spectral neutron fluence per log energy interval at the experimental position C when the spherical D_2O moderator is placed, using three different guess spectra. The spectra are based on the same experimental data set and given for an arbitrary target yield.

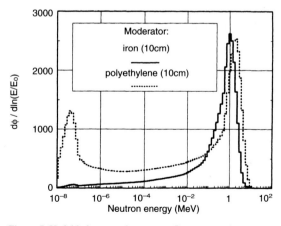

Figure 5. Unfolded spectral neutron fluence per log energy interval at the experimental position A when 10 cm thick moderators of iron and polyethylene are placed. The spectra are normalised to the same peak height for an arbitrary target neutron yield.

CONCLUSIONS

The realistic neutron field facility presented here exhibits a versatile installation for testing of individual and ambient dose equivalent meters. If the direction and energy-dependent response of the device under consideration is known in the fast and thermal energy range from experiments or calculations, it is possible to determine the integral response in the intermediate range. The modelling of the arrangement by Monte Carlo calculation was satisfying, although the source term could only be approximated. Though only three simple moderator arrangements have been tested so far, the observed spectral distributions are essentially different with respect to their intermediate and thermal neutron fluence contribution. Even small shielding thicknesses in combination with the scattered neutron component do have the effect that much larger moderating arrangements, e.g. inside a PWR, may be simulated.

ACKNOWLEDGEMENTS

It is gratefully appreciated that the Commission of the European Union DGXII supported the work under contract FI3P-CT92-0002.

REFERENCES

1. Schwartz, R. B., Eisenhauer, C. M. and Grundl, J. A. *Experimental Verification of the Neutron Spectrum from the NBS D_2O-Moderated ^{252}Cf-Source*. NUREG/CR-3399 (1983).
2. Chartier, J. L., Posny, F. and Buxerolle, M. *Experimental Assembly for the Simulation of Realistic Neutron Spectra*. Radiat. Prot. Dosim. **44**(1–4), 124–130 (1992).
3. Murphy, M. F. *A review of Neutron Spectrometry Measurements in the ASPIS Reference Field*. UKAEA Report, AEA-RS-5548, (1993)
4. Medioni, R. and Delafield, H. J. *An International Intercomparison of Critical Accident Dosimetry Systems at the SILENE Reactor, Valduc, France. Part I: Reactor and Reference Dosimetry of Radiation Fields*. AEA and IPSN-CEA joint report HPS/TR/H/1 (1995).
5. Goodman, L. J. *Neutron Dosimetry at the Radiological Research Accelerator Facility*. In: Proc. First Symp. on Neutron Dosimetry. (Eds G. Burger, H. Schraube and H. G. Ebert) EUR 4896, Vol. I, pp. 177–206 (1972).
6. Jakes, J., Schraube, G., Weitzenegger, E. and Schraube, H. (Unpublished observations).
7. Briesmeister, J. F. (Ed) *MCNP — A General Monte Carlo N-Particle Transport Code, Version 4A*. LA-12625-M (Los Alamos National Laboratory) (1993).
8. Hendricks, J. S., Frankle, S. C. and Court, J. D. *ENDF/B-VI Data for MCNP*. LANL Report LA-12891 (Los Alamos) (1994).
9. Inada, T., Kawachi, K. and Hiramoto, T. *Neutrons from Thick Target Beryllium (d,n) Reactions at 1.0 MeV to 3.0 MeV*. J. Nucl. Sci. Technol. **5**(1), 22–29 (1968).
10. Mares, V. and Schraube, H. *Evaluation of the Response Matrix of a Bonner Sphere Spectrometer with LiI Dector from Thermal Energy to 1 MeV*. Nucl. Instrum. Methods Phys. Res. A**337**, 461–473 (1994).
11. Thomas, D. J., Alevra, A. V., Hunt, J. B. and Schraube, H. *Experimental Determination of the Response of Four Bonner Sphere Sets to Thermal Neutrons*. Radiat. Prot. Dosim. **54**(1), 25–31 (1994).
12. McElroy, W. N., Berg, S., Crockett, T. and Hawkins, R. G. *Spectra Unfolding*. AFWL-TR-67-41, Vol. 1–4 (1967).
13. Alevra, A. V., Siebert, B. R. L., Aroua, A., Buxerolle, M., Grecescu, M., Matzke, M., Mourgues, M., Perks, C. A. and Schraube, H. *Unfolding Bonner-Sphere Data: European Intercomparison of Computer Codes*. PTB Laboratory Report 7.22-90-1 (1990).

DOSIMETRIC CHARACTERISTICS OF THE IHEP NEUTRON REFERENCE FIELDS

A. G. Alekseev and S. A. Kharlampiev
Institute for High Energy Physics
Protvino, Moscow Region, Russia 142284

Abstract — The Institute of High Energy Physics (IHEP) reference radiation fields for calibration include radionuclide neutron sources and high energy fields behind shielding. The ambient dose equivalent rates have been measured using instruments developed specially for measurements in a field of mixed radiation having unknown neutron spectra. The problem of adequate simulation by the IHEP reference neutron fields for user's fields is discussed.

INTRODUCTION

Interpretation of neutron detector readings in terms of dose equivalent is an important problem. Since the response of a neutron detector has usually an energy dependence which does not adequately simulate the energy dependence of dose equivalent, its use as a dosemeter in the real neutron fields met in nuclear physics after a simple calibration in a neutron field of a radionuclide source (such as ^{252}Cf) leads to considerable errors. There are various approaches to solve this problem. One method[1,2] of dosemeter calibration in workplace fields is available for personal and area dosemeters. Another method[3] is the calibration at a neutron field having a spectrum simulating the user's field. The method of superposition of calibration fields[4] to obtain a weighted neutron response is a further step in this direction. All of these approaches in one way or another are concerned with the application of neutron reference fields for these goals.

Requirements for a neutron field to be a 'reference' field are the following. First of all it should have a neutron spectrum which simulates typical workplace neutron spectra. Its main characteristics, such as neutron spectrum, dose equivalent rate, angular distribution, etc., must be well established. It should be reproducible and sufficiently intense to provide the possibility of quick calibration.

At the Institute for High Energy Physics (IHEP), the workplaces concerned with personal exposure are concentrated mostly at the experimental hall of the IHEP 70 GeV proton synchrotron. The radiation fields behind the shielding of the accelerator have very complicated component composition where neutrons in a wide energy range from thermal energies up to the primary proton beam's energy of 70 GeV are prevalent. Numerous investigations of real radiation fields behind the accelerator's shielding[2,5–7] show that they can be separated into three main groups: (1) high energy fields behind the upper shielding (320–350 cm of concrete); (2) low energy neutron fields behind the side shielding (~ 600 cm of concrete); (3) radiation fields with a large contribution of charged particles near to the charged particle transport beams.

The IHEP neutron reference fields include the high energy reference field and four radionuclide reference neutron fields[8] formed by filtered and direct radiation

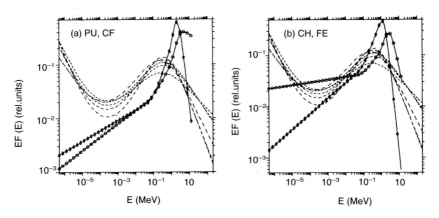

Figure 1. Neutron spectra of the radionuclide reference fields PU, CF (a) and CH, FE (b) in comparison with typical workplace neutron spectra behind the side shielding of 70 GeV IHEP accelerator (various dashed lines). All spectra correspond to total fluence of 1 cm^{-2}. (○) PU in (a), CH in (b). (◇) CF in (a), FE in (b).

of ^{252}Cf and ^{239}Pu - Be radionuclide sources: ^{239}Pu - Be in free air (PU), ^{252}Cf in free air (CF), ^{252}Cf at the centre of a polyethylene sphere of 30 cm diam. (CH) and ^{252}Cf at the centre of iron sphere of 30 cm diam. (FE). The radionuclide reference neutron fields are reproduced at the middle of a special calibration room having dimensions of 5.4 m × 13 m and 4 m height. The distance between a radionuclide source creating the field and a reference point is the same for all the four fields (0.75 m) as well as the distance from a floor (1.65 m).

The high energy reference field (HEF) is reproduced at 1 m height outside the upper shielding in the experimental hall of the IHEP 70 GeV proton accelerator. Inside a volume of 0.5 × 0.6 × 0.3 m³ with the centre at the reference point the intensity of radiation varies in less than 10%.

MEASUREMENT TECHNIQUES

The dosemeters being used in measurements are listed in Table 1. The detailed description of measurement procedures may be found in the original papers. Briefly, the tissue-equivalent proportional counter TEPC[9] allows one to obtain dose equivalent and absorbed dose of radiation of any kind by simultaneously measuring the charge produced in a cavity and the event spectrum.

The analogue component rem meter ACR[10] includes three high-pressure ionisation chambers (IC): an argon-filled IC with an aluminium wall (γ and charged particles absorbed dose); tissue-equivalent IC (total absorbed dose); and ^{3}He-filled IC in a 10″ diameter polyethylene moderator (neutron dose equivalent).

Other instruments are: multisphere spectrometer (MS) with ^{6}LiI(Eu) detector (being used as a dosemeter according to the method[8] of linear combination of six spheres readings), thermoluminescence detector IKS (TLD on base of alumophosphate glass)[11], and argon-filled proportional counter SLETCP with aluminium wall.

For neutron spectra measurements in the IHEP reference fields the IHEP multisphere spectrometer was used[12]. Cadmium-covered polyethylene spheres of 2″, 3″, 5″, 8″, 10″ and 12″ diameter with a ^{6}LiI(Eu) diam. 10 mm × 10 mm thermal detector have been used in measurements[8] in the radionuclide neutron reference fields. For spectrum measurements in high energy neutron fields a 18″ diameter sphere and a carbon activation scintillation detector (^{12}C(x,xn)^{11}C) were involved. Another spectrometer available in IHEP, a passive neutron dosemeter-spectrometer (PDSN)[7] including a set of fissile radionuclide detectors, has also been used for spectra measurements in high energy neutron fields.

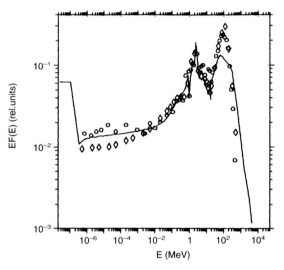

Figure 2. Neutron spectra of the IHEP high energy reference field, ($\diamond$) measured by multisphere spectrometer, ($\circ$) compared with calculations by FLUKA for top concrete shield[13], and (——) by ROZ6H for 220 cm concrete, small angle incidence. All spectra correspond to total fluence of 1 cm^{-2}.

Table 1. The instruments used in dose equivalent measurements in the IHEP reference fields.

Instrument	Refs	Measuring components	Scopes, limitations	Errors	Certification
TEPC	9	photons and charged particles, neutrons	$\bar{E}_n > 0.8$ MeV	9% 9%	VNIIFTRI, State Standard of Russia
ACR	10	photons and charged particles, neutrons	$H_n/H_\gamma < 100$ $H_n/H_\gamma > 0.1$	10% 18%	″ ″
MS	8	neutrons		15%	VNIIM, State Standard of Russia
TLD IKS	11	photons	$H_\gamma = 1 - 100$ mSv	15%	″
SLETCP	—	photons		10%	—

RESULTS AND DISCUSSION

The results of neutron and photon ambient dose equivalent measurements for the IHEP radionuclide reference neutron fields are given in Table 2. They are given per one neutron from a radionuclide source, for convenience of further applications. The results are the expert estimation of the values usually taken as an average between TEPC, ACR and MS results except in the cases when the error obtained after treatment of instrument readings exceeds the permissible level for the dosemeter — here we use the data for IKS and SLETCP.

The neutron spectra of radionuclide reference neutron fields are shown in Figure 1 in comparison with accelerator low energy spectra outside the side shield. It can be seen that the radionuclide fields do not closely simulate the real spectra from the accelerator, but nevertheless have a wider neutron spectrum like real ones and a shape of separate parts, rather similar to that of a real spectrum, which could be interesting from the point of view of a detector response. Then, in Table 3 the components of a neutron spectrum as well as the main spectral characteristics are given for the reference fields in comparison with the same values for the real neutron fields behind the IHEP accelerator shielding, measured by MS[5] and by PDSN[7]. The real neutron spectra are subdivided into several groups according to spectrum hardness. As can be seen from the table, the range of ambient dose equivalent conversion factors (193–322 pSv·cm^2) and the mean neutron energies in the spectrum (0.76–3.32 MeV) as well as the ratio between separate parts of the neutron spectra cover most parts of real neutron spectra behind the side shielding of the IHEP accelerator.

Table 2. Measured ambient dose equivalents (expert estimation) and some important integral characteristics of the IHEP reference fields.

Reference field	H*(10) (fSv per source neutron)		$h_n^*(10)$ (pSv·cm^2)	$\bar{E}_n$ (MeV)	$N_{10''}/N_{5''}$ (rel.units)
	Neutrons	γ + ch.part			
Radionuclide reference fields					
PU	4.9 ± 9%	0.18 ± 20%	322	3.32	1.32
CF	4.5 ± 12%	0.22 ± 15%	284	1.83	1.01
CH	0.63 ± 10%	0.25 ± 10%	193	1.36	0.73
FE	3.4 ± 9%	0.023 ± 15%	216	0.76	0.61
High energy reference field					
HEF	4.0 ± 13% (μSv per cycle)	0.8 ± 15% (μSv per cycle)	282	49	0.86

Table 3. Comparison of the main characteristics of neutron spectra of the reference and real workplace fields.

Type	Spectra	Parts of F_n (rel.units)			$h_n^*(10)$ (pSv·cm^2)	$\bar{E}$ (MeV)
		intermed.	fast	>20 MeV		
	Radionuclide reference fields					
	PU	0.11	0.89	0	322	3.32
	CF	0.13	0.87	0	284	1.83
	CH	0.41	0.59	0	193	1.36
	FE	0.20	0.80	0	216	0.76
	HEF	0.20	0.36	0.44	282	49.1
	Measured workplace neutron fields					
Soft,	MS[5]	0.79–0.84	0.16–0.21	0	41–52	0.11–0.14
iron	PDSN[7]	0.51–0.73	0.24–0.49	0.–0.01	75–137	0.6–5.9
Side	MS[5]	0.45–0.67	0.28–0.47	0.01–0.04	106–157	0.7–3.5
shielding	PDSN[7]	0.37–0.60	0.37–0.58	0.05–0.16	156–222	5.7–47
Upper	PDSN[7]	0.16–0.31	0.17–0.44	0.37–0.65	254–374	77–297
shielding	MS	0.20	0.36	0.44	282	49
	ROZ6H[14] (calc.)	0.22	0.39	0.38	253	71

The main advantage of the radionuclide neutron reference fields is a constant spectrum and component composition, good reproducibility and intensity (30–200 nSv.s^{-1}) which cannot be reached in real fields behind the accelerator shielding.

The neutron spectrum at the high energy reference field presented in Figure 2 is measured by the IHEP multisphere spectrometer using *a priori* information. It is compared with a theoretical spectrum calculated by FLUKA[13] and ROZ6H[14] computer codes. The presence of two peaks, in the evaporation region and near 100 MeV, is confirmed by all authors. Such behaviour of a high energy neutron spectrum also appears in the measured cosmic neutron spectrum[15].

CONCLUSION

Investigations of dosimetric characteristics of the IHEP reference neutron fields have been carried out by different dosemeters specially developed for measurements in mixed radiation fields. The ambient dose equivalent data were obtained with an accuracy of 10–20%. The main spectral characteristics of this set of reference fields is also determined and compared with real neutron fields. It is shown that some reference fields (CH, HF) allow one to calibrate routine dosemeters directly; and the full set of the reference fields can be used for dosemeter investigations, dosemeter intercomparisons and other applications.

REFERENCES

1. Piesch, E. *Calibration Techniques for Personnel Dosemeters in Stray Radiation Fields*. Radiat. Prot. Dosim. **10**, 159–173 (1985).
2. Bystrov, Yu., Golovachik, V. and Kharlampiev, S. *Calibration of Rhodium Monitors in Terms of Total Equivalent Dose*. Internal Report RRD IHEP (1989).
3. Schwartz, R. and Eisenhauer, C. USNRC Report NUREG/GR-1204 (1980).
4. Britvich, G. and Chumakov, A. *Reference Fields Superposition Method for Dosimetric Apparatus Calibration to Operate within the Neutron Radiation Fields*. Preprint IHEP 93-66, Protvino (1993).
5. Belogorlov, E., Britvich, G., Krupny, G., Lebedev, V., Lukanin, V., Makagonov, A., Peleshko, V. and Rastsvetalov, Ya. *Neutron Low Energy Spectra Behind IHEP Accelerator Shielding*. Preprint IHEP 85-148, Serpukhov (1985).
6. Krupny, G., Lebedev, V., Peleshko, V. and Rastsvetalov, Ya. *Spectral-Angular Distributions of Charged Particles Outside Biological Shielding of 70 GeV Serpukhov Accelerator*. IHEP Preprint 88-85, Serpukhov (1988).
7. Sannikov, A. *A Passive Neutron Dosemeter-Spectrometer for High Energy Accelerators*. IHEP Preprint 90-133, Protvino (1990).
8. Britvich, G., Volkov, V., Kolevatov, Yu., Kremenetsky, A., Lebedev, V., Mayorov, V., Rastsvetalov, Ya., Trykov, L. and Chumakov, A. *Spectra and Integral Values of Reference Neutron Fields from Radionuclide Neutron Sources*. IHEP Preprint 90-48, Protvino (1990).
9. Abrosimov, A., Alekseev, A. and Golovachik, V. *Tissue-Equivalent Radiation Monitor for Dose Equivalent Measurements of Radiation Behind the Shielding of a Proton Accelerator*. Kernenergie **34**(3), 108–111 (1991).
10. Abrosimov, A., et al. *Dose Characteristics of IHEP Reference Neutron Fields*. IHEP Preprint 93-43, Protvino, (1993). Abrosimov, A., et al. Geratebasis fur die metrologische Sicherstellung der Strahlungsuber wachung am Synchrophasotron Serpuchow. Kernenergie **31**(5), 214 (1988).
11. Bochvar, N. A., et al. *Method of IKS Dosimetry*. (Atomizdat, Moscow) (1977). Baranenkov, N. N., et al. IHEP Preprint 89-122, Serpukhov (1989).
12. Belogorlov, E., Britvich, G., Krupny G. et al. IHEP Preprint 85-3, Serpukhov, (1985).
13. Roesler, S. and Stevenson, G. *July 1993 CERN-CEC Experiments: Calculations of Hadron Energy Spectra from Track-Length Distributions Using FLUKA*. CERN internal report CERN/TIS-RP/IR/93-47 (1993).
14. Averin, A., et al. In: Proc. Topical Meeting on Advances in Mathematical Computing and Reactor Physics, Pittsburgh, USA. Vol. 5 (1991). Also Gorbatkov D., et al. In: Proc. Fourth European Particle Accelerator Conf., EPAC'94, London, 1994 Vol 3, p. 2588.
15. Schraube, H., Jakes, J., Sannikov, A., Weitzenegger, E., Roesler, S. and Heinrich, W. *The Cosmic Ray Induced Neutron Spectrum at the Summit of the Zugspitze (2963 m)*. Radiat. Prot. Dosim. **70**(1–4), 405–408 (This issue) (1997).

SILENE, A TOOL FOR NEUTRON DOSIMETRY

B. Tournier, F. Barbry and B. Verrey
Institut de Protection et de Sûreté Nucléaire
IPSN/CEA Valduc, 21120 Is sur Tille, France

Abstract — The SILENE reactor today is a reference source able to meet the expectation of a great number of researchers in the fields of physics, neutronics, biology and dosimetry. The latest technical developments and its multiple operating possibilities enable this experimental reactor of the French Nuclear Protection and Safety Institute (IPSN) to provide a large range of calibrated neutron and gamma radiation fields which can be reproduced very easily.

INTRODUCTION

Originally designed to study criticality accidents[1], the reactor named SILENE, a French acronym for 'Free Neutron Evolution Irradiation Source', the facility of the French Nuclear Protection and Safety Institute (IPSN) is a high-performance solution-fueled reactor of great operating flexibility, able to meet the requirements of experimental research in many fields, such as those of physics, neutronics, dosimetry and biology as observed by the latest international intercomparison of criticality accident dosimetry systems organised under the sponsorship of the European Communities (EC) and the International Atomic Energy Agency (IAEA)[2].

THE SILENE SOURCE

This source is highly suitable for studying the effects of both low and high doses. In any case, SILENE is extremely useful for calibrating a very wide range of methods and means of radiation measurements applied in a mixed field (neutron + gamma).

The use of shields of different sizes and materials, the intensity of the emitted gamma and neutron radiation as well as the various possible operating modes make SILENE a power research source.

A power excursion or 'divergence' is produced by withdrawing the control rod from the core at various predetermined levels according to the experimental requirements and the desired effects.

The core

The reactor core is a small annular tank located in the centre of a large concrete cell, $19 \times 12 \times 10$ m^3. The fissile solution (71 g.l^{-1} of uranyl nitrate with the uranium 93% enriched in ^{235}U) is prepared in a laboratory under the cell, far from the reactor core.

To carry out an experiment, the uranium solution is pumped into the core up to a predetermined supercritical level, with the control rod inserted to prevent divergence. This control rod is placed in the core in a 7 cm

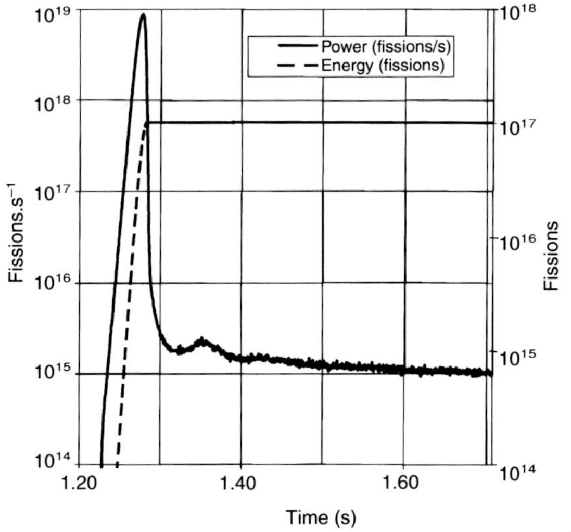

Figure 1. Typical PULSE.

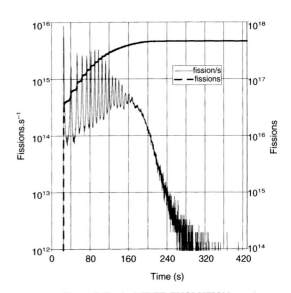

Figure 2. Typical FREE EVOLUTION.

Table 1. Total doses and dose rates at 4 m from the core[3].

Modes	Total fissions $\times 10^{17}$	Total dose (Gy)	Ratio (gamma/neutron)	Total dose rate (Gy.s^{-1})
Without shielding				
PULSE	≤3.0	≤12.0	≅1.5	≤1200
FREE EVOLUTION	≤5.0	≤21.0	≅1.5	≤8.3
STEADY STATE	≤5.0	≤21.0	≅1.5	40×10^{-9} to 12×10^{-3}
Lead shielding				
PULSE	≤3.6	≤5.30	≅0.2	≤515
FREE EVOLUTION	≤4.0	≤5.90	≅0.2	≤2.9
STEADY STATE	≤4.0	≤5.90	≅0.2	15×10^{-9} to 4×10^{-3}
Polyethylene shielding				
PULSE	≤3.8	≤5.70	≅10	≤530
FREE EVOLUTION	≤5.0	≤7.50	≅10	≤0.75
STEADY STATE	≤5.0	≤7.50	≅10	15×10^{-9} to 4.5×10^{-3}
Steel shielding				
PULSE	≤4.1	≤3.70	≅0.4	≤290
FREE EVOLUTION	≤4.0	≤3.70	≅0.4	≤0.5
STEADY STATE	≤4.10	≤3.70	≅0.4	9×10^{-9} to 2.7×10^{-3}

diameter channel. Various control rods (boron and cadmium rods) can be used, but they are all hollow so as to allow the introduction of test capsules into the channel which then serves as an irradiation cavity. Power excursion is achieved by withdrawing the control rod which moves vertically in the central cavity.

Once the experiment is over, the fissile solution containing the radioactive fission products is drained into a special tank, located in a shielded room. This allows rapid access to the reactor cell.

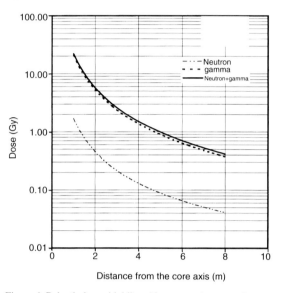

Figure 4. Polyethylene shielding. Neutron and gamma doses at different distances from the core axis for 10^{17} fissions[3].

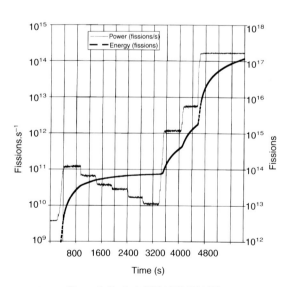

Figure 3. Typical STEADY STATE.

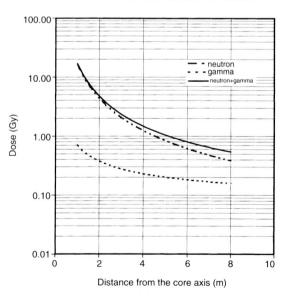

Figure 5. Lead shielding. Neutron and gamma doses at different distances from the core axis for 10^{17} fissions[3].

Operating modes

Depending on various parameters, such as core reactivity, the control rod withdrawal rate, and the presence or absence of an auxilliary neutron source, SILENE can be operated in three different modes, called 'PULSE', 'FREE EVOLUTION' and 'STEADY-STATE'.

PULSE mode is obtained by the rapid withdrawal of the control rod, with or without an additional neutron source. The power excursion is very brief (a few milliseconds) and a very high peak power can be obtained (up to 1000 MW). The energy released does not exceed 3.6×10^{17} fissions (in an unshielded reactor configuration) (Figure 1).

FREE EVOLUTION is achieved by slowly removing the control rod fitted with an auxiliary neutron source; reactor power oscillates during the experiment. This mode allows high energy generation in a few minutes (5×10^{17} fissions in an unshielded reactor configuration) (Figure 2).

STEADY STATE mode allows low or high energy to be released (5×10^{17} fissions) at a constant power ranging from a few milliwatts to 10 kW. The flexibility of this type of operation is appropriate for special dosimetry techniques implying current measurements (Figure 3).

Shields

Cylindrical shields can be placed around the source to modify leakage radiation characteristics, for instance to degrade the neutron spectrum and modify the neutron/gamma dose ratio.

Since 1993, the reactor has been able to operate in four configurations: bare source or source shielded by lead, steel or polyethylene.

The space surrounding the reactor core (230 m² floor area) allows equal dose irradiation of large objects in arcs corresponding to isodoses. Extra shields (concrete walls, etc) can be placed in the reactor room in order to simulate specific situations. By means of suitable materials, it is also possible to modify significantly the radiation fluences and spectra emitted in the irradiation cavity located in the centre of the reactor.

CHARACTERISTICS OF THE EMITTED RADIATION

The characterisation of the leakage radiation fields was carried with different techniques:

(1) For the neutron spectrum: activation detectors, Bonner spheres, proton recoil spectrometer (associated with organic scintillator).
(2) For the neutron doses: direct measurements were performed with ionisation chambers (tissue-equivalent and aluminium) and diodes and results obtained have been compared with values deduced from the spectrum measurements.
(3) For the gamma doses: thermoluminescence dosemeters, energy-compensated Geiger-Müller counters, ionisation chambers.

Dose and dose rate

Figures 4, 5 and 6 show the variation of the doses[3] (normalised to 1×10^{17} fissions) measured at different distances from the reactor axis for different configurations. Table 1 give the main characteristics of the mixed field at a distance of 4 m from the reactor axis.

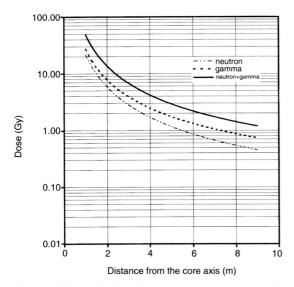

Figure 6. Without shielding. Neutron and gamma doses at different distances from the core axis for 10^{17} fissions[3].

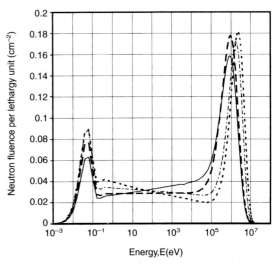

Figure 7. Neutron spectrum for the different shields[5]. (— — — —) without shields, (———) lead shields, (— — — —) polyethylene shields, (———) steel shields.

The neutron and gamma doses at different distances from the core with steel shielding[4] are still being characterised.

Neutron fluence and neutron spectrum

Figure 7 shows the neutron spectrum (normalised to unit fluence) for the different shields[5].

REFERENCES

1. Barbry, F. *A Review of the Silene Criticality Excursions Experiments*. IPSN Valduc Report SRSC 93.220 (1993).
2. Medioni, R. *International Intercomparison of Criticality Accident Dosimetry Systems at the Silene Reactor*. IPSN Fontenay-aux Roses Report on contract No 92-PR 007 (1993).
3. Serbat, A. *Doses Neutrons et Gamma mesurées autour de Réacteur SILENE sans Écran ou entouré d'Écrans de Plomb ou de Polyéthylène*. DPEN/DE Centre d'Etudes du Bouchet. Private communication (1994).
4. Verrey, B. *Cartographie des Rayonnements Neutrons et Gamma*. IPSN Valduc Note technique SRSC 93–03 (1993).
5. Kurdkdjian, J. *Spectres due Réacteur Silène*. IPSN/DPHD/CE Cadarache. Private communication (1995).

DETERMINATION OF NEUTRON ROOM SCATTERING CORRECTIONS IN RCL'S CALIBRATION FACILITY AT KAERI

S. C. Yoon, S. Y. Chang, J. S. Kim and J. L. Kim
Health Physics Department
Korea Atomic Energy Research Institute
PO Box 105, Yusung, Taejon, 305-600, Korea

Abstract — Neutron room scattering corrections that should be made when neutron detectors are calibrated with a D_2O moderated ^{252}Cf neutron source in the centre of a calibration room are considered. Such room scattering corrections are dependent on specific neutron source type, detector type, calibration distance, and calibration room configuration. Room scattering corrections for the responses of thermoluminescence dosemeter and two different types of spherical detectors to a neutron source in the Radiation Calibration Laboratory (RCL) neutron calibration facility at the Korea Atomic Energy Research Institute (KAERI) were experimentally determined and are presented. The measured room scattering results are then compared with theoretical results calculated by predicting room scattering effects in terms of parameters related to the specific configuration. Agreement between measured and calculated scattering correction is generally about 10% for three kinds of detectors in the calibration facility.

INTRODUCTION

Neutron room scattering corrections that should be made when neutron dosemeters and measuring devices are calibrated with a D_2O-moderated ^{252}Cf neutron source in the centre of a calibration room are considered. Such room scattering corrections are dependent on specific neutron source type, detector type, calibration distance, and calibration room configuration. Unfortunately, these effects are sometimes ignored, giving rise to faulty calibrations[1]. When neutron room scattering is considered, data of a complete neutron spectral measurement as a function of distance in the calibration room would be used to evaluate the effect of the neutrons reflected from the surfaces of a calibration room. However, this kind of approach is not always feasible due to time or instrumentation constraints. Monte Carlo calculations based on various simplifying assumptions are more commonly used to determine the neutron scattering corrections[2,3].

The Radiation Calibration Laboratory (RCL) has been operated for almost two decades in the Korea Atomic Energy Research Institute (KAERI) as one of the secondary standard dosimetry calibration laboratories, and the old RCL has moved and finished expanding recently. The neutron calibration room which is considered here is a new RCL neutron calibration facility, which is located in the basement of a newly constructed radiation application building at KAERI. The calibration room is 8 m long, 6 m wide and 6 m high. The source and detector are placed at a height of 2.9 m near the centre of the room. The room is entirely enclosed with concrete.

Room scattering corrections for the responses of a thermoluminescence dosemeter (TLD) and two different types of spherical detectors with the neutron source in the RCL neutron calibration facility at KAERI were experimentally determined and are presented. The measured room scattering results are then compared with theoretical results calculated by predicting room scattering effects in terms of parameters related to the specific configuration[2,4].

THEORETICAL CONSIDERATIONS

Scattering from the room walls, ceiling, and floor, or 'room scattering', is not a new problem and has been investigated at many laboratories in the past. Room scattering in the case of a completely enclosed concrete room is a much more serious problem. When we consider the reflected fluence of neutrons in the ideal case of a neutron source at the centre of a spherical cavity, the singly reflected neutron fluence in the cavity is everywhere constant and isotropic, as shown by Savinskii and Filyushkin[5].

The relative response of a detector to reflected and source neutrons, M_s/M_o, depends on the two spectra, as well as on the relative fluence. Therefore, the relative response can be expressed as

$$M_s/M_o = 4.5g\ (R_\Phi^s/R_\Phi^0)(r/r_c)^2 \qquad (1)$$

where R_Φ^s and R_Φ^0 are the spectrum-averaged responses for the reflected and source neutrons, respectively. The predicted room scattering correction S divided by r^2 from Equation 1 is given by:

$$S = 4.5g\ (R_\Phi^s/R_\Phi^0)(r/r_c)^2 \qquad (2)$$

These calculations were made for a spherical cavity. In applying them to a real room which has the form of a rectangular parallelopiped, one needs to find the radius of a spherical room which has the same surface area as the real room. This suggests that the effective radius r_c of an equivalent spherical room can be obtained from

$$4\pi\ r_c^2 = \Sigma\ A_i \qquad (3)$$

where A_i is the area of the ith surface of the room.

Finally, the room scattering correction S from the rectangular parallelopiped shaped room can be calculated using equation 2 after finding the effective radius r_c of an equivalent spherical room from Equation 3.

EXPERIMENTAL RESULTS

Measurements have been made for three kinds of neutron detectors as a function of distance from a D_2O-moderated ^{252}Cf neutron source at KAERI. The neutron detectors included an Eberline 9 inch diam. spherical rem meter (9″ sphere), an Eberline 3 inch cadmium-covered detector (3″ sphere), and an Albedo TLD (Teledyne TLD: PB-3 Badge). These instruments were chosen because the 9″ spherical rem meters and the TLD are commonly used in nuclear power plants and industry and because the ratio of the response of the 9″ and 3″ spherical detectors can sometimes be used as an index to derive dose calibration factors for an albedo TLD in various neutron fields[6]. Griffith et al[7] found that the dose calibration factor for albedo neutron dosemeters in various neutron spectra could be related to the ratio of responses of the 9″ and 3″ spherical detectors.

Measurements of the TLD were carried out on a phantom. Measurements of the response D of the 9″ and 3″ spheres and the TLD were first corrected for air scatter by subtracting a contribution of 2.3% and 4.5% per metre for 9″ and 3″ spheres and 3.0% per metre for the TLD, respectively. These recommended air scatter corrections were derived from McCall's air scatter calculations[8] and Hankins measurements of detector response functions[9]. These corrected responses, multiplied by the square of the source–detector r in metres, are plotted against r^2 in Figures 1 and 2. These points lie approximately on a straight line which can be expressed as

$$M r^2 = M_0(1 + S r^2) \qquad (4)$$

The linear equations from least squares fit are shown in the figures. The intercept of such a plot at r = 0 gives the response M_0 at 1 m from neutron source and the slope S gives information on the relative response from reflected neutrons, namely, room scattering corrections. In Figure 1 the measured response ratio of 9″ and 3″ spheres at 1 m, which includes room-reflected neutrons, is 4.46 and the slopes are S = 0.17 m^{-2} and S = 0.33 m^{-2}, respectively for the 9″ and the 3″ spheres. Thus the relative effect of reflected neutrons for the 3″ sphere is about 2 times greater than for the 9″ sphere. In the case of the TLD, the slope is determined to be S = 0.115 m^{-2} as shown in Figure 2.

COMPARISON OF THEORETICAL AND EXPERIMENTAL RESULTS

Theoretical room scattering correction can be estimated by calculation using Equation 2. A comparison of the measured and the calculated room scattering correction is shown in Table 1. The calculated $g(R_\Phi^S/R_\Phi^0)$ values for the D_2O-moderated ^{252}Cf neutron source were 0.86, 1.4 and 0.58 for 9″ sphere, 3″ sphere and albedo TLD, respectively[10]. The effective radius of the RCL calibration room calculated from Equation 3 was 4.58 m. Agreement between measured and calculated

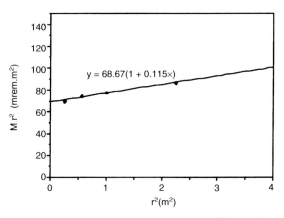

Figure 2. Response D of the TLD as a function of distance from a D_2O-moderated ^{252}Cf neutron source in the RCL calibration facility.

Table 1. Comparison of the measured and the calculated neutron room scattering corrections of detector responses for the D_2O-moderated ^{252}Cf neutron source at KAERI.

Detector	Measured S (m^{-2})	Calculated S (m^{-2})	Meas./Calc. S (% at 1 m)
9″ sphere	0.17	0.185	91.9
3″ sphere	0.33	0.30	110.0
TLD	0.115	0.124	92.7

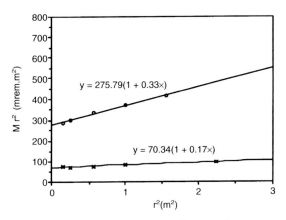

Figure 1. Response D of the Eberline 9″(x) and 3″(o) spheres as a function of distance from a D_2O-moderated ^{252}Cf neutron source in the RCL calibration facility.

scattering correction is generally about 10% for three kinds of detectors in the calibration room. The fact that both measured corrections for 9″ and 3″ spheres in the RCL calibration room are slightly different from calculated corrections suggests that perhaps prediction for the calibration room is complicated because it does not have perfect concrete-enclosed six walls, but has a 12 cm thick concrete slab structure (3.5 m long × 6 m high) in front of the door which is intended to prevent neutrons from leaking out of the door. In addition, there is some concern that the 3″ sphere used for measuring the scattering correction in the calibration room was almost 10 years old. However, no problems have been shown so far.

CONCLUSIONS

Because of the fact that the agreement in Table 1 is good for 9″ sphere, 3″ sphere and albedo TLD and the measured points have good least squares fits (Figures 1 and 2), it can be concluded that the formula (Equation 2) is reliable for the calculation of the room scattering conrrections for detector calibration at the RCL neutron calibration facility. Further scattering correction measurements will be performed for 9″ cylindrical rem meter and Bonner spheres at the RCL calibration facility. These measurements and their interpretation will be the subject of a further paper.

REFERENCES

1. Liu, J. C., Sims, C. S., Casson, W. H., Murakami, H. and Francis, C. *Neutron Scattering in ORNL Calibration Facility.* Radiat. Prot. Dosim. **35**(1), 13–21 (1991).
2. Jenkins, T. M. *Simple Recipe for Ground Scattering in Neutron Detector Calibration.* Health Phys. **39**, 41–47 (1987).
3. Eisenhauer, C. M. *Review of Scattering Corrections for Calibration of Neutron Instruments.* Radiat. Prot. Dosim. **28**(4), 253–262 (1989).
4. Schwartz, R. B. and Eisenhauer, C. M. *Procedures for Calibrating Neutron Personnel Dosimeters.* NBS Special Publication 633 (National Institute of Standards and Technology) (1989).
5. Savinskii, A. K. and Filyushkin, I. V. *An Estimate of the Contribution by Neutrons Scattered in Enclosed Rooms to a Total Radiation Dose.* In: Proc. Symp. on Neutron Monitoring for Radiation Protection Purposes, UCRL-Trans-10771 (1972).
6. Eisenhauer, C. M. and Schwartz, R. B. *Measurement of Neutrons Reflected from the Surfaces of a Calibration Room.* Health Phys. **42**, 489–495 (1982).
7. Griffith, R. V., Hankin, D. E., Gammage, R. B., Tommasino, L. and Wheller, R. V. *Recent Developments on Personnel Neutron Dosimeters — a Review.* Health Phys. **36**, 235–260 (1979).
8. McCall, R. C., Jenkins, T. M. and Shore, R. A. *Transport of Accelerator Produced Neutrons in a Concrete Room.* SLAC-PUB-2214 (Stanford Linear Accelerator Center) (1978).
9. Hankins, D. E. *Energy Dependence Measurements of Remmeters and Albedo Neutron Dosimeters at Neutron Energies of Thermal and Between 2 keV and 5.67 MeV.* In: Proc. Int. Radiation Protection Assn IV International Congress, Paris, France, 24–30 April 1977. pp. 553–556 (1977).
10. Eisenhauer, C. M., Hunt J. B. and Schwartz, R. B. *Calibration Techniques for Neutron Personal Dosimetry.* Radiat. Prot. Dosim. **10**(4), 45–57 (1985).

TESTS OF INSTRUMENTS FOR NEUTRON AND GAMMA RAY MEASUREMENTS AT SEVERAL RADIONUCLIDE NEUTRON SOURCES

F. Spurný†, I. Votočková†, J.-F. Bottollier-Depois‡, J. Kurkdjian‡ and D. Paul‡
†Department of Radiation Dosimetry, Nuclear Physics Institute
Czech Academy of Sciences Na Truhlářce 39/64, 180 86 Praha 8, Czech Republic
‡Institut de Protection et de Sureté Nucléaire
Départment de Protection de la Santé d'Homme et de Dosimétrie
BP No 6, 92265 Fontenay-aux-Roses Cedex, France

Abstract — Neutron radiation protection instruments have been tested and gamma ray field characteristics have been studied for several radionuclide neutron sources (Am-Li, Am-F, Am-Be, Pu-Be, Cf, and Cf/D_2O/Cd) of different construction available at two laboratories. Several active (TEPC, moderator-type rem meters, scintillator- and GM-based dose rate meters, Si- and GM-based individual electronic dosemeters) and passive detectors (TLDs, bubble) have been used. Particular attention has been paid to the contribution of the gamma component to the total dose equivalent. The influence of Pb shielding of the sources has been characterised for all sources mentioned above, our values are generally a little lower than some of those recommended in the literature. The present studies also revealed the importance of particular features of some of the devices used (sensitivity to fast neutrons of some photon detectors, overestimation of high energy photons).

INTRODUCTION

Neutron detectors and dosemeters can also be sensitive to the photons always present in the field of a neutron source (and vice versa for photon detectors). The data concerning the photon dose component are, for most neutron fields, less complete and less specified than for neutrons. In some cases these data are practically non-existent.

The basic goal of these studies was to bring additional and/or new data on the photon component in the radiation fields for most of the known radionuclide neutron sources. The neutron component has also been studied by means of some radiation protection instruments.

The results obtained are discussed and analysed, some general conclusions concerning both the field characteristics as well as detector responses to neutrons and photons are formulated.

MATERIALS AND METHODS

The irradiation geometry and the radionuclide neutron sources are characterised in Table 1. Neutron and photon detectors used are described in Tables 2 and 3, their calibration has been performed by means of ^{252}Cf neutrons and ^{60}Co photons, respectively, in terms of H*(10), employing ICRP 21 quality factors*[1].

* ICRP 60 recommendations have not been fully officially adopted, as well as corresponding reference values for the sources studied. For this reason ICRP 21 was preferred. The passage to ICRP 60 will not change the conclusions.

RESULTS AND DISCUSSION

Cadarache neutron sources

Typical sets of direct readings in μSv of the H*(10) calibration sources are presented in Table 4. Comments on the responses of different instruments and detectors follow.

Neutron detectors

(a) There is a reasonable agreement between the readings of most of the detectors used and the reference values.

Table 1. Neutron sources studied: geometry of irradiation.

Neutron source[a]	Distance (m) to Detector Floor		Irradiation hall (m^3)	Emission rate[b] (s^{-1})	Conversion factor[c] ($Sv.cm^2$)
Am-Li (3)	0.50	1.5	∅7×4	2.85×10^5	3.0×10^{-10}
Am-F (1)	0.50	1.5	10×6×4	1.60×10^5	3.3×10^{-10}
Cf (1)	0.50	1.5	10×6×4	2.56×10^6	3.4×10^{-10}
Cf (2)	0.75	4.0	25×12×9	1.60×10^8	3.4×10^{-10}
Pu-Be (1)	0.50	1.5	10×6×4	1.03×10^7	3.8×10^{-10}
Pu-Be (3)	0.50	1.5	∅7×4	2.14×10^7	3.8×10^{-10}
Am-Be (3)	0.50	1.5	∅7×4	1.09×10^7	3.8×10^{-10}
Am-Be (2)	0.75	4.0	25×12×9	3.53×10^7	3.8×10^{-10}
Am-Be (3)	0.50	1.5	∅7×4	1.85×10^6	3.8×10^{-10}
Am-Be (1)	0.50	1.5	10×6×4	2.09×10^6	3.8×10^{-10}

[a](1) Prague; (2) Cadarache; (3) Fontenay-aux-Roses.
Cf(2) used also inside the D_2O sphere (d = 300 mm) covered with Cd (1.5 mm).
[b]On 1 June 1994.
[c]ICRP 21 quality factors[1].

(b) It seems, nevertheless, that some particular features can be stated:
  (i) The energy dependence of the responses of two rem meters used seems to be different, NM2 being relatively more sensitive to higher energies and, also, to intermediate neutrons.
  (ii) Overestimation of intermediate neutrons by moderator type rem meters is re-confirmed (compare Cf and Cf/D$_2$O/Cd).
  (iii) The relative response of BDND is probably not entirely flat in the region of 1 to 2 MeV (see reading for Cf).

Photon detectors

(a) Direct readings are generally lower than reference values.
(b) These variations can be explained when individual characteristics of different instruments are taken into account. For example, over-readings of NB 3201 can be attributed to the sensitivity of the plastic scintillator to fast neutrons (they are relatively more important for the Am-Be source).

TEPC

It should be remembered that the lower limit of the high LET (neutron) contribution is taken to be 3.5 keV.μm^{-1}. This means that some of the neutron-induced events are considered as a photon contribution and vice versa. When the total values of dose equivalent are taken into account, their agreement with reference values is satisfactory for Cf sources. An overestimation is seen for the Am-Be source (+18%); this is mainly due to the high LET part and is on the limit of statistical reliability.

GENERAL REMARKS ON THE PHOTON CONTRIBUTIONS AT VARIOUS SOURCES

The sets of direct readings presented in Table 4 have been gathered for all sources and fields given in Table 1, for both bare as well as shielded (1.5 mm of Pb) sources. The general behaviour of responses corresponded well to remarks formulated in the case of Cadarache sources. This permits the following formulation of some general conclusions concerning the photon contribution.

Table 2. Neutron detectors.

Detector	Measuring principle	Remarks
NAUSICAA (Steel, France)	Tissue-equivalent proportional counter; wall 10 mm tissue	Neutron events are taken to be above 3.5 keV.μm^{-1}[2]
N91 rem meter (Harwell, UK)	Moderator-type device; ^{3}H proportional counter	Simulates roughly a 8 inch diameter sphere[3]
NM2 rem meter (Nuclear Enterprises)	Moderator-type device; BF$_3$ counter	Simulates roughly a 10 inch diameter sphere[4]
Bubble neutron detectors (BDND-BTT, Canada)	Superheated liquid in a viscous medium	Energy threshold 0.1 MeV; sensitivity: 1 bubble per 1 μSv[5]

Table 3. Photon detectors.

Detector	Measuring principle	Remarks
NAUSICAA (Steel, France)	Tissue-equivalent proportional counter; wall 10 mm tissue	Photon events are taken to be below 3.5 keV.μm^{-1}[2]
NB 3201 (Tesla, Czech Rep.)	Plastic scintillator with small NaI:Tl	Designed for environmental photon dose measurements[6]
RP 114 (ZMA, Czech Rep.)	Set of GM counters covered with 2 mm of Fe	Designed for environmental photon dose measurements[7]
DMC 90 (Mérlin-Gérin, France)	Si diode	Photon electronic individual dosemeter[8]
TLDs	^{7}LiF, AlP glasses, CaSO$_4$:Dy Al$_2$O$_3$	Responses to thermal and fast neutrons known[9]

Influence of Pb shield on photon contribution

For ^{252}Cf and Pu-Be sources the Pb-shield diminishes the photon dose equivalent, on average by about 10%. This influence is generally more important for sources with ^{241}Am, emitting 59.5 keV photons. Its extent can be appreciated from the results presented in Table 5. The particular features of different instruments are also reflected in these results:

(i) the decrease is clearly most important for the DMC 90 instrument having the highest response to 60 keV photons;
(ii) the decrease is generally less for the fast neutron sensitive NB 3201 than for GM-based RP 114.

It should be also emphasised that, as mentioned in ISO 8529[10], the influence of Pb shielding can vary with source construction (see Am-Be sources). It is preferable always to use ^{241}Am neutron sources with a Pb shield.

Photon contribution to the dose equivalent

Taking into account the particular features of different photon detectors, the estimated relative photon contributions to the dose equivalent in the fields of radionuclide neutron sources studied are presented in Table 6. Mainly TLDs and DMC 90 have been considered in this estimation due to knowledge of their responses to neutrons of all energies[9] and their small dimensions. These contributions are systematically a little lower than values recommended.

MICRODOSIMETRIC CHARACTERISTICS OF FIELDS STUDIED

Microdosimetric characteristics of the fields studied at radionuclide neutron sources are presented in Table 7.

Table 4. Dose equivalents as measured at Cadarache neutron sources with different detectors and instruments.

Detector		Dose equivalent rate* (μSv.h^{-1})		
		Am-Be/Pb	Cf	Cf/D$_2$O/Cd
Reference value**		703	2772	660
N 91		719	2995	818
NM 2	Neutrons	820	3130	956
BDND		680	3630	700
NAUSICAA		846	2990	687
Reference value**		35$^{2)}$	132	110
NB 3201		39	153	86
RP 114	Photons	28	122	81
DMC 90		24	138	85
TLDs		25	116	82
NAUSICAA		30	141	87

In H(10) of reference radiations; statistical uncertainties (1 σ_{rel}) about 10% for BDND and neutron contribution measured with NAUSICAA, about 5% for all other detectors. **ISO 8529; reference value for photons and Am-Be/Pb taken as 5% of value for neutrons.

Table 5. Influence of Pb shield on the photon contribution.

Neutron source*		Ratio of photon dose equivalent with and without 1.5 mm of Pb as measured by means of		
		NB 3201	RP 114	DMC 90
Am-Li	(3)	0.095**	0.15	0.10
Am-F	(1)	0.040	0.046	0.020
Am-Be	(1)	0.48	0.50	—
Am-Be	(2)	0.072	0.096	0.035
Am-Be	(3)	0.43	0.50	0.29

*See Table 1 for sources.
**1σ_{rel} ~10% for all values presented.

Table 6. Relative photon contribution to the dose equivalent at radionuclide neutron sources.

Neutron source	Relative H*(10)
Am-Li/Pb	0.34 ± 0.05*
Am-F/Pb	0.087 ± 0.015
Cf/D$_2$O/Cd	0.12 ± 0.02
Cf	0.038 ± 0.005
Am-Be/Pb	0.033 ± 0.005
Pu-Be	0.030 ± 0.004

*2σ

Table 7. Microdosimetric characteristics of radiation fields at radionuclide neutron sources.

Neutron source		Characteristics measured with NAUSICAA (t = 1h)				
		H$_{low}$*	H$_{high}$*	D$_{high}$	Q$_{high}$	Q
		(μSv)		(μGy)		
Am-F/Pb	(1)	0.68**	5.4	0.70	7.7	3.0
Cf	(1)	5.1	99	12	8.5	6.2
Cf	(2)	141	2988	303	9.8	7.0
Cf/D$_2$O/Cd	(2)	89	687	81	8.4	4.5
Am-Be/Pb	(1)	4.9	90.3	12.0	7.5	5.6
Am-Be/Pb	(2)	30.3	846	101	8.4	6.7
Am-Be/Pb	(3)	33.9	540	66.9	8.1	5.7
Pu-Be	(1)	16.1	493	61.1	8.1	6.6
Pu-Be	(3)	54.1	1187	143	8.3	6.3

*low, below 3.5 keV.μm^{-1}; high, above 3.5 keV.μm^{-1}.
**Relative uncertainties at 20 min data accumulation typically: for H$_{low}$, Q$_{high}$ and Q: 1σ_{rel} <10%; for H$_{high}$ and D$_{high}$: 1σ_{rel} 10 to 15%.

(1) There is a reasonable agreement for the same sources studied in different laboratories; the scattered radiation contribution seems to be lower, as expected, in Cadarache as compared to two other laboratories (compare Q values for Cf and Am-Be sources).
(2) Only small differences exist in the high LET part of event spectra (>3.5 keV.μm^{-1}) for all these fields.

fields mentioned. The Pb shield influence has been quantitatively characterised. Our results are systematically lower (~15–20%) than the values recommended in some previous documents[10].
(3) The high LET part of microdosimetric spectra is roughly the same for all neutron sources studied. It should be taken into account in some applications, for example in radiobiology.

CONCLUSIONS

(1) The radiation protection instruments used give reasonable information on the neutron component characteristics of the fields at radionuclide neutron sources. Some differences observed can be well attributed to the intrinsic properties of a device.
(2) A fairly complete set of data has been gathered on the photon contribution to the dose equivalent in the

ACKNOWLEDGEMENTS

The authors are much obliged to Dr A. Rannou for his support of the present studies and to Dr J.-L. Chartier for stimulating discussions of the results and their presentation, their thanks are also due to G. Pelcot and H. Muller for their technical assistance.

This work was partially supported in the framework of CEC Project CT93-0072.

REFERENCES

1. ICRP. *Data for Protection against External Radiation*. Publication 21 (Oxford: Pergamon Press) (1973).
2. Bousset, P., et al. *Description de l'Instrument NAUSICAA conçu pour Effecteur des Mesures Instantanées de H, D, et du Spéctre T.E.L. en Champs Complexes*. In: Proc. IRPA8, Montréal, May 1992, pp. 463–466 (1992).
3. Technical Specification and User Guide. *N 91 Portable Neutron Monitor* (AEA Technology, Harwell Instruments) Product code 0949/6 (1993).
4. Anderson, I. O. and Braun, J. A. *A Neutron Rem Counter*. Nukleonik **6**, 237–241 (1964).
5. Ing, H. *Bubble Technology Industries Report* (Chalk River: AECL) (1991).
6. Vierebl, L., Nováková, O. and Jursová, L. *Combination Scintillation Detector for Gamma Dose Rate Measurements*. Radiat. Prot. Dosim. **20**, 272–276 (1990).
7. *Radiometer RP 114; Description Manual* (ZMA Ostrov nad Ohří, Czech Republic) (1992).
8. *DMC90 — Manuel d'utilisation* (Mérlin-Gérin, France) (1991).
9. Spurný, F. *Methods of Dosimetry for the External Exposure and their Use*. DSc. thesis, Prague (July 1984).
10. ISO 8529. *Rayonnements neutroniques de référence destinés a l'étalonnage des instruments de mesures des neutrons utilisés en radioprotection et a la détermination de leur réponse en fonction de l'énergic des neutrons.* M 60-516 (Décembre 1989).

A NEW EXPRESSION FOR DETERMINATION OF FLUENCES FROM A SPHERICAL MODERATOR NEUTRON SOURCE FOR THE CALIBRATION OF SPHERICAL NEUTRON MEASURING DEVICES

M. Khoshnoodi and M. Sohrabi
National Radiation Protection Department
Atomic Energy Organization of Iran
P.O. Box 14155-4494, Tehran, Islamic Republic of Iran

Abstract — A new expression modifying the inverse square law for determination of neutron fluences from spherical moderator neutron sources is reported. The formalism is based on the neutron fluence at a point outside the moderator as the summation of fluxes of two groups of neutrons: direct neutrons from the central region of the moderator, and moderated neutrons which, to a first approximation, are scattered from the outermost layers of the spherical moderator. The expression has been further developed for spherical neutron measuring devices with an appropriate geometry factor which corrects the reading of the device for non-uniform irradiation of the detector. The combination of the new fluence function and those of the air and room scattered components introduce a calibration model. The fluence relationship obtained for moderated sources may conveniently be used for calculating the more rapid change of neutron dose at close distances than that which is based on the inverse square dependence.

INTRODUCTION

During the past two decades many laboratories have utilised small neutron sources such as ^{252}Cf, Am-Be etc. for the calibration of neutron measuring devices. The neutron emission by these sources is rather isotropic[1] and this makes it possible to implement the inverse square law to explain the relation between the induced response at the position of a small detector and the strength of the source.

The expanded use of moderated neutron sources, in particular the D_2O-moderated ^{252}Cf[2], for the calibration of neutron area monitoring devices has made many calibration laboratories investigate the correction factors[3,4] which, on the basis of detector and source volume effects, evaluate the deviation of the fluence response according to the inverse square distance variations of the neutron fluence.

It is postulated here that the fluence rate of slowed down neutrons outside a spherical moderated source is produced by a converging field at the point of interest, where a small neutron detector is assumed to be situated. The neutron fluence rate obtained using this method will then be expanded over the surface of a spherical neutron measuring device in order to predict the deviation arising because of the use of a large spherical detector instead of a point detector.

FLUENCE RATE DERIVATION

In a homogeneous spherical moderating medium containing a small neutron source at the centre, the fast neutrons emitted by the central source are slowed down through many interactions with the nuclei, in particular, with the hydrogen nucleus. At a point outside the sphere moderator, the fluence of the slowed down neutrons can be determined using four considerations:

(1) The slowing down process is isotropic in a homogeneous non-absorbing medium.
(2) The slowed down neutrons reaching the surface of a spherical moderator are uniformly distributed.
(3) The emerging moderated neutrons, to a first approximation, originate in the outermost layers of the sphere moderator.
(4) The surface element is considered as a point source with anisotropy behaviour which, for the moderated neutrons, is defined by the cosine of the angle between the direction of the emission and the normal to the surface element.

According to point (2) the fluence rate σ_s of slowed neutrons on the surface of a spherical source can be given by:

$$\sigma_s = Q\alpha/(4\pi r_s^2) \qquad (1)$$

where Q is the strength of the central source; r_s is the moderator radius and α is a constant accounting for the relative numbers of source neutrons undergoing the slowing down process in the moderator. In non-absorbing media this parameter can be defined by the relation:

$$\alpha = 1 - \exp(-\Sigma_t r_s)$$

Where Σ_t is the macroscopic removal cross section for source neutrons.

The surface element (Figure 1) may also be assumed to be a source of direct neutrons and those make a few interactions very close to the central source. These neutrons introduce a flux density which, on the surface of the moderator, is defined by:

$$\sigma_o = Q\exp(-\Sigma_t r_s)/(4\pi r_s^2) \qquad (2)$$

The differential flux density, $d\Phi$, introduced by the surface element, ds, at a distance, d_o, can be determined by:

$$d\Phi = \frac{[\sigma_s F(\beta) + \sigma_o \delta(\cos\beta - \cos\theta)]ds}{2\pi x^2} \quad (3)$$

with

$$x^2 = d_o^2 + r_s^2 - 2d_o r_s \cos\theta \quad (4)$$

$$ds = 2\pi r_s^2 \sin\theta \, d\theta \quad (5)$$

The anisotropy factor[1], $f(\beta)$, is shown to be:

$$F(\beta) = 2\cos\beta \quad (6)$$

The integration of the term which deals with the delta function may be performed using the following identity:

$$f(z) = \int \delta(z - y) f(y) dy \quad (7)$$

Hence, overall integration of Equation 3 will result in the relation:

$$\Phi = (Q\alpha/2\pi) f_s + (Q/4\pi) \exp(-\Sigma_t r_s) f_c \quad (8)$$

with

$$f_s = [1 + \sqrt{(1 - r_s^2/d_o^2)}]^{-1} \quad (8a)$$

$$f_c = (d_o - r_s)^{-2} \quad (8b)$$

where the functions f_s and f_c are related to the fluence of slowed down and direct neutrons, respectively.

In practical situations, where measurements and calibrations are performed in a closed room, the detector reading usually exceeds that expected according to Equation 8. The room reflected and air inscattered neutrons may have significant roles in this increase. It has been shown that in the central region of an experimental room, the component of reflected neutrons by the room walls is constant[5] and the component of air inscattered neutrons is inversely proportional to the distance between source and detector[6]. These results have been deduced from the fact that the scattered neutrons are originally emitted by a very small neutron source[5], nevertheless, they also may be applied for moderation sources as large as D_2O-moderated $^{252}Cf^{(3)}$.

The outscattering component, which takes into account the attenuation of source neutrons by air, is introduced by an exponential factor multiplied by Equation 8. By considering the effects mentioned above, the count rate, $c(d_o)$, of a small thermal detector, irradiated by a moderation neutron source, can be expressed by:

$$c(d_o) = K\{e^{-\Sigma d}[f_s + R d_o^2 f_c] + A d_o + S d_o^2\}/d_o^2 \quad (9)$$

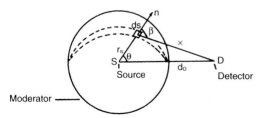

Figure 1. The surface element geometry.

where Σ is the air linear attenuation coefficient which, for thermal neutrons, amounts to $0.0256 \, m^{-1}$ and d is a distance as measured from the front face of the source to that of the detector. The length, d, is however replaced by d_o, as in short distances the outscattering effect is negligible and at greater separation distances it approaches d_o. A and S respectively are air inscattered and room scattered fractional components at one metre. K and R are defined as follows:

$$K = Q\alpha\epsilon/(2\pi) \quad (10)$$

$$R = [\tau \exp(-\Sigma_t r_s)]/2 = \tau(\sigma_s/\sigma_o)/2 \quad (11)$$

K is known as the source–detector characteristic constant and R is introduced as the ratio of responses induced by two groups of neutrons; neutrons of the first group leave the moderator from its central region probably with not more than one interaction, while, those of the second are scattered out of the moderator surface after slowing down in the moderator. ϵ is the detector efficiency for thermal neutrons and τ is a ratio of the spectrum-averaged responses of the detector for the intermediate and source neutrons energy, respectively.

DETECTOR GEOMETRY CORRECTION

In order that the fluence response of the spherical detector conforms with Equation 8, the readings must be corrected for the geometric condition of the device in a neutron field.

Axton's theory[7] for calculation of the geometric correction factor may be developed into the more complicated situation where a stray field of neutrons is incident from a moderated source such as D_2O-moderated ^{252}Cf. In this situation the following equations are intended to define the desired correction factors:

$$F_1(d_o) = 1 + ag(d_o) \quad (12)$$

$$F_2(d_o) = 1 + bh(d_o) \quad (13)$$

Here a and b are known as the average neutron effectiveness parameters[7] to be discussed later. The correc-

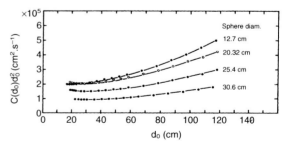

Figure 2. The plotted points represent the product of the measured count rate and the square of the separation distance between the centres of the source and BF_3 detector, $c(d_o)d_o^2$, as a function of d_o. The curves are the calculated product obtained from Equation 9 and the constants values given in Table 1. The uncertainties are too small to be shown.

tion functions, $g(d_o)$ and $h(d_o)$, are determined according to the two-group neutrons classification, which is shown analytically by the two-term fluence expression, i.e. Equation 8. The first term, the fluence rate of the slowed down neutrons, may be used to obtain the following relation:

$$g(d_o) = \frac{d_o(1+\sqrt{(1-r_s^2/d_o^2)}\,[f(x_2)-f(x_1)]}{(r_s r_d)^2} - 1 \quad (14)$$

where

$$f(x) = x_2^2[x - \sqrt{(x^2 - r_s^2)} + r_s \mathrm{acos}(r_s/x)]$$
$$\quad - 1/3[x^3 - (x^2 - r_s^2)^{3/2}] \quad (14a)$$
$$x_1^2 = (d_o - r_d)^2 \quad (14b)$$
$$x_2^2 = d_o^2 - r_d^2 \quad (14c)$$

The correction function for the second term, which purely indicates the inverse square distance dependence, is given by:

$$h(d_o) = \frac{x_3^2[f'(x_2) - f'(x_1)]}{d_o r_d^2} - 1 \quad (15)$$

where

$$f'(x) = \frac{(r_s^2 + r_d^2 - d_o^2)}{(x - r_s)} - (x - r_s) - r_s \ln(x - r_s) \quad (15a)$$
$$x_3^2 = (d_o - r_s)^2 \quad (15b)$$

Equation 9 may be rewritten as a generalised calibration model if the first and second terms of Equation 9 are multiplied by the geometry correction factors given by Equations 12 and 13, respectively. This model is represented by the relation:

$$c(d_o)d_o^2 = K\{\exp(-\Sigma d_o)\,[1 + ag(d_o)]f_s + R\exp(-\Sigma_s d_o) [1 + bh(d_o)]f_c + Ad_o + Sd_o^2\} \quad (16)$$

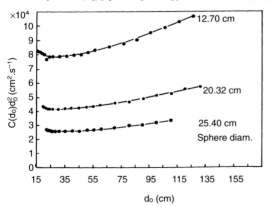

Figure 3. The plotted points represent the product of the measured count rate and the square of the separation distance between the centres of the source and the spherical dose rate meter, $c(d_o)d_o^2$, as a function of d_o. The curves are the calculated product obtained from Equation 16 and constant values given in Table 2.

In this equation a and b determine the effectiveness of these geometric effects, represented by Equations 14 and 15, in producing a response in the detector; Σ and Σ_s are the air linear attenuation coefficients for thermal and the source neutrons.

The least squares technique may be used to optimise the model parameters. In order to reach sufficient accuracy in the optimisation, the number of data points used in the fitting procedure must be more than the square number of the parameters introduced in Equations 9 and/or 16.

MEASUREMENTS

Two series of measurements were made using a cylindrical bare BF_3 counter (h = 10 cm, d = 3 cm), manufactured by Studvik, and a spherical dose equivalent rate meter of Leake's type[8]. The series of counts read by the BF_3 detector and the dose equivalent rate meter, were respectively used for optimising the constants of Equations 9 and 16. For irradiation of the detectors four different polyethylene-moderated ^{252}Cf sources were prepared by loading the Bonner spheres of diameters 12.7, 20.32, 25.4 and 30.6 cm with a small ^{252}Cf neutron source at the centre. The axis of the BF_3 detector was taken to be normal to the line joining the centres of the source and detector. In this situation the BF_3 detector was assumed to behave as a point detector, because, at a distance, greater than 15 cm between the centre of the source and that of the detector, the neutron fluence does not change significantly on the surface of the detector. For the measurements with the BF_3 detector, the closest distances between the centre of the detector and the centres of the moderating spheres of the quoted diameters were 17.5, 18.1, 18.9 and 22.3 cm, respectively. In each range of separation distance around 20 sets of counts were measured at four different incremental steps of 1, 2, 5 and 10 cm. In order to take the data from the spherical dose equivalent rate meter, the same arrangement for the range of separation distance between the centre of the moderated source and that of the rate meter was used. In these measurements the closest distance between source moderator and detector moderator surfaces was when they were touching.

DATA ANALYSIS AND RESULTS

Tables 1 and 2 show the results obtained with the calibration models, i.e. Equations 9 and 16, and the least squares fitting technique. The tables include characteristic constants, K, for different source–detector combinations. These constants represent the detector response due to the contribution of thermal and intermediate energy neutrons only. It is clear that the maximum response will occur when the moderated source of 20.32 cm diameter is employed for irradiation of the BF_3 thermal detector (see Table 1). This table also shows the response ratio, R, changing with the diameter of the moderating sphere. Except for the largest moderating sphere, the response ratio variations appear to be a decreasing function of r_s within the quoted uncertainties. This behaviour may arise due to a shift of neutron energy from a higher to lower energy spectrum. Table 2 shows the value of constants, excluding b, from the generalised calibration model represented by Equation 16.

Attempts were made to fit the data sets to this equation for optimisation of all the parameters, but were unsuccessful because of the lack of high output source strength, from the larger moderating spheres, in providing more numbers of accurate measurements at greater separation distances. However, in this situation, a small contribution of direct neutrons to the fluence response strongly suggested re-evaluating the model by neglecting the geometric correction term associated with the fluence of direct neutrons.

Another set of data, provided by J. B. Hunt from NPL, taken from the same dose rate meter but irradiated by a D_2O-moderated ^{252}Cf square, were fitted to Equation 16: the results are given for consideration in Table 2.

It is clearly seen from Figure 2 that the experimental data are in good agreement with the curves representing the four-parameters calibration model, Equation 9. Figure 3 shows the curves were plotted using the generalised calibration model with the parameters optimised for different sets of data. In all cases the quoted uncertainties are standard deviations and they were calculated using the least squares technique.

ACKNOWLEDGEMENT

The authors would like to thank Dr J. B. Hunt from NPL in England for his valuable contribution to this study.

Table 1. Intercomparison of the constants obtained using the data taken by a bare BF_3 thermal detector as irradiated by spherical moderator neutron sources.

Sphere diam. (cm)	$K \times 10^{-5}$ ($cm^2.s^{-1}$)	R	$A \times 10^4$ (cm^{-1})	$S \times 10^6$ (cm^{-2})
12.70	3.527 ± 0.030	0.008 ± 0.002	8.94 ± 0.61	60.13 ± 0.67
20.32	3.628 ± 0.061	0.0030 ± 0.0005	7.91 ± 0.46	42.87 ± 0.55
25.40	2.686 ± 0.011	0.0010 ± 0.0002	6.17 ± 0.51	39.13 ± 0.37
30.60	1.602 ± 0.013	0.0030 ± 0.0004	10.00 ± 0.88	37.99 ± 0.61

Table 2. Intercomporison of the constants obtained using the data taken with the spherical dose euqivalent rate meter as irradiated by three spherical moderator neutron sources and the D_2O-moderated ^{252}Cf source.

Sphere diam. (cm)	$K \times 10^{-5}$ ($cm^2.s^{-1}$)	a	R	$A \times 10^4$ (cm^{-1})	$S \times 10^6$ (cm^{-2})
12.7	1.179 ± 0.034	−0.086 ± 0.041	0.068 ± 0.008	20.00 ± 1.37	3.17 ± 0.74
20.32	0.668 ± 0.036	−0.034 ± 0.148	0.024 ± 0.011	20.00 ± 2.70	5.56 ± 1.27
25.40	0.435 ± 0.024	−0.055 ± 0.111	0.009 ± 0.008	10.00 ± 4.25	11.75 ± 2.52
30.6 D_2O-mod.	30.33 ± 0.026	−0.106 ± 0.009	0.0010 ± 0.0002	7.12 ± 0.01	0.618 ± 0.002

REFERENCES

1. Hunt, J. B. *The Calibration and Use of a Long Counter for the Accurate Measurement of Neutron Flux Density*. NPL Report RS5 (April 1976).
2. Schwartz, R. B. and Eisenhauer, C. M. *The Design and Construction of a D_2O-Moderated ^{252}Cf Source for Calibrating Neutron Personnel Dosimeters Used at Nuclear Power Reactors*. NUREG/CR-1204 (January, 1980).
3. Kluge, H., Weise, K. and Hunt, J. B. *Calibration of Neutron Sensitive Spherical Devices with Bare and D_2O-Moderated ^{252}Cf Sources in Rooms of Different Sizes*. Radiat. Prot. Dosim. **32**, 233–244 (1990).
4. Khoshnoodi, M. and Sohrabi, M. *The Application of a New Geometry Correction Function for the Calibration of Neutron Spherical Measuring Devices Using Large Volume Neutron Sources*. Radiat. Prot. Dosim. **44**(1/4), 121–124 (1992).
5. Eisenhauer, C. M. and Schwartz, R. B. *The Effect of Room-Scattered Neutrons on the Calibration of Radiation Protection Instruments*. In: Proc. Fourth Symp. on Neutron Dosimetry, Neuherberg, Munchen, 1981. EUR-7445 (Luxembourg: CEC), Vol. I, p. 421–430 (1981).
6. Eisenhauer, C. M. *A Study of the Angular and Energy Distribution of Radiation at Small Distances from a Point Source of Gamma Rays or Neutrons*. Nucl. Sci. Eng. **27**, 240–251 (1967).
7. Axton, E. J. *The Effective Centre of a Moderating Sphere When Used as an Instrument for Fast Neutron Flux Measurement*. J. Nucl. Energ. **26**, 581–583 (1972).
8. Leake, J. W. *Spherical Dose-Equivalent Neutron Detector. Type 0075 — A Recommendation for Change in Sensitivity*. Nucl. Instrum. Methods, **170**, 287–288 (1980).

THE NEUTRON FLUENCE AND H*(10) RESPONSE OF THE NEW LB 6411 REM COUNTER

B. Burgkhardt†, G. Fieg†, A. Klett‡, A. Plewnia§ and B. R. L. Siebert§
†Forschungszentrum Karlsruhe, POB 3640, D 76021 Karlsruhe, Germany
‡EG&G Berthold, POB 100163, D 75312 Bad Wildbad, Germany
§Physikalisch-Technische Bundesanstalt, POB 3345, D 38023 Braunschweig, Germany

Abstract — A new rem counter with a response tailored to match the shape of H*(10)/Φ as defined by ICRU and ICRP has been developed recently in the frame of a Technology Transfer Project of the Research Centre Karlsruhe with the industrial partner EG&G Berthold, Wildbad. In this paper, a carefully established, consistent fluence and ambient dose equivalent response function is provided for this detector as well as the effective response calculated for more than 500 neutron spectra from a catalogue.

INTRODUCTION

Rem counters are used for routine area surveillance. Most are based on a neutron moderator with a central detector. In general, the moderator is a polyethylene sphere or cylinder and the cental detector is commonly most sensitive to thermalised neutrons such as a ^{3}He proportional counter which allows discrimination against photons by setting an appropriate threshold. One of the main advantages of rem counters is their reliability and their simple handling in practical use.

A new rem counter, LB 6411, with a response tailored to match the shape of H*(10)/Φ as defined by ICRP[1] and ICRU[2] has recently been developed in the frame of a Technology Transfer Project of the Research Centre Karlsruhe with the industrial partner EG&G Berthold, Wildbad[3]. In this paper a carefully established, fluence and ambient dose equivalent response function is provided for this detector as well as the effective response calculated for more than 500 neutron spectra from a catalogue[4]. The value of H*(10)/Φ as a function of neutron energy are taken from Ref. 5. The performance of the LB 6411 is compared with that of the Leake counter as described in Ref. 6.

INSTRUMENT

The rem counter LB 6411 consists of a polyethylene moderator sphere with a diameter of 25 cm, a central cylindrical ^{3}He proportional counter and internal Cd absorbers and perforations. The instrument has an integrated high voltage supply and signal processing and is connected to a microprocessor-controlled portable datalogger. The design of the counter is shown in Figure 1 and its essential features are summarised in Table 1.

The optimisation of the H*(10) response function has been performed by extensive calculations with the Monte Carlo transport program MCNP[7] using detailed modelling and the latest version of the European Fusion cross section File EFF. Many parameters have been varied in the search for an optimised response: the diameter of the moderator, and the material, position and perforated fraction of absorbing layers (Table 1). The resulting configuration has been confirmed by calibration measurements at the Physikalisch-Technische Bundesanstalt (PTB) using monoenergetic neutrons with energies from thermal up to 19 MeV[8]. A single fit factor close to 1 was sufficient to adjust the calculated to the absolute experimental response. This factor accounts for the effective number of ^{3}He nuclei in the counter tube and the selected threshold.

EVALUATION OF FLUENCE AND DOSE EQUIVALENT RESPONSE FUNCTION

An interactive homemade computer program based on a least squares spline interpolation has been used for the fitting procedure. This program achieves a sound compromise between the closeness to the data and the smoothness of the fit. The experimental data were given weights in accordance with their experimental variances

Table 1. Essential features of the LB 6411.

Electronics	Integrated high voltage supply and signal processing, connected to a microprocessor-controlled portable datalogger.
Physical basis	^{3}He proportional counter, moderating polyethylene sphere of 25 cm diameter and Cd absorbers
Optimisation	Search for optimal physical instrument parameters using detailed MCNP modelling and the newest cross sections (European Fusion File, EFF).
Experimental verification	Monoenergetic neutrons from thermal through 19 MeV at the PTB
New quantity indication	Ambient dose equivalent H*(10) for neutrons according to ICRP 60 from thermal to 20 MeV neutrons
Low energy dependence	+10% to −30% in the energy range 50 keV up to 10 MeV
High sensitivity	≈3 counts per nSv

361

and the values calculated with MCNP were given lesser weights and used for a well justified interpolation. Resonances in the fluence response due to carbon are not represented, as rem counters are designed for use in broad spectra. Therefore, a set of only 47 energy points and a cubic Lagrange interpolation are sufficient to describe all relevant structures of the response function. The interpolation formula for the cubic Lagrange interpolation is given by

$$P(x) = \sum_{i=1}^{4} y_i \left[\prod_{j=1, j \neq i}^{4} (x - x_j) \bigg/ \prod_{j=1, j \neq i}^{4} (x_i - x_j) \right] \quad (1)$$

where y_i are the values of the responses and x_j the logarithms of the neutron energies and where $x_2 \leq x \leq x_3$, except for the beginning and the end of the defined energy range ($E_{n\text{-min}}$, $E_{n\text{-max}}$). The original data and the fitting results are shown in Figure 2 which demonstrates the quality of the fitted procedure, especially in the relevant high neutron energy range.

This fluence response has then been used to establish a numerical calibration factor for a broad calibration field, N, defined as the ratio of the ambient dose equivalent, $H^*(10)$, and the value indicated by the rem counter in the calibration field, M, by integration:

$$H^*(10) = \int_{E_{n\text{-min}}}^{E_{n\text{-max}}} dE_n\, h^*_\phi(E_n)\, \Phi_{E_n}(E_n) \quad (2)$$

$$M = \int_{E_{n\text{-min}}}^{E_{n\text{-max}}} dE_n\, R_\phi(E_n)\, \Phi_{E_n}(E_n) \quad (3)$$

where E_n, R_ϕ and Φ_{E_n} are the neutron energy, the evaluated fluence response and the spectrum, respectively. $E_{n\text{-min}}$ and $E_{n\text{-max}}$ are taken as 1 meV and 20 MeV, respectively.

Table 2 shows neutron fluence and ambient dose equivalent response for the LB 6411 and fluence-to-ambient dose equivalent conversion coefficients at 47 Lagrange points.

The spectrum from the bare ^{252}Cf neutron source (ISO 8529) has been used as the calibration spectrum.

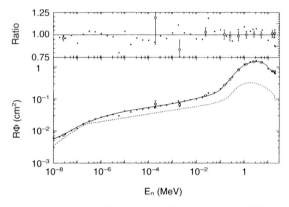

Figure 2. Lower part: Experimental (circles), calculated (dots) and evaluated neutron fluence responses for the LB 6411 (solid line) and the Leake counter (dashed line) as a function of neutron energy. Upper part: Response ratios of experimental and calculated values to the fitted values.

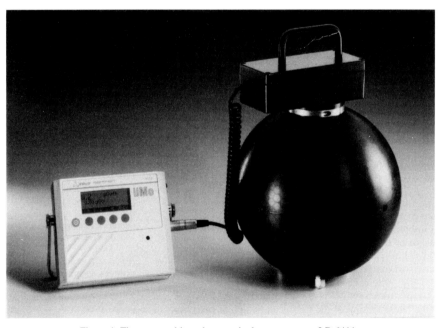

Figure 1. The new ambient dose equivalent rate meter LB 6411.

The numerical calibration factors, N, obtained are 324.34 pSv and 1372.0 pSv per count for the LB 6411 and the Leake counter, respectively. Figure 3 shows the H*(10) response of the LB 6411 and the Leake counter numerically calibrated in the bare ^{252}Cf spectrum. The LB 6411 shows a three times lower over-response for intermediate neutrons and provides a four times higher sensitivity than the Leake counter. The agreement between the numerical and the measured calibration factor for the LB 411 using a bare ^{252}Cf neutron source was found to be better than 10%. Table 3 shows, for often used calibration spectra, the fluence weighted ambient dose equivalent readings of the LB 6411, NM/Φ, and the mean neutron fluence to ambient dose equivalent conversion factor, H*(10)/Φ. These data support the assumption that the LB 6411 rem counter will work reliably in a wide range of spectra.

PERFORMANCE AT WORKPLACES

The quality and usefulness of a rem counter is determined by its ability to measure the ambient dose equivalent in spectra found at workplaces. A reliable test of rem counters with known response functions can be carried out numerically by computing the ratio of the ambient dose equivalent indicated, i.e. N M, to H*(10) in spectra found in workplaces[4]. The values for M and H*(10) are obtained from Equations 2 and 3. The mean neutron fluence to ambient dose equivalent conversion factor, H*(10)/Φ, is used as an indicator of the hardness of a spectrum. The results of this check on the performance are shown in Figure 4 for about 500 spectra found

Table 2. Neutron fluence and ambient dose equivalent response for the LB 6411 and fluence-to-ambient dose equivalent conversion coefficients at 47 Lagrange points.

E_n (MeV)	R_Φ (cm^2)	$R_{H^*(10)}$ (pSv.cm^2)	$h^*_\Phi(10)$ (pSv.cm^2)
1.00×10^{-9}	4.00×10^{-3}	1.30	6.60
1.00×10^{-8}	5.57×10^{-3}	1.81	9.00
2.53×10^{-8}	7.67×10^{-3}	2.49	1.06×10
1.00×10^{-7}	1.31×10^{-2}	4.25	1.29×10
2.00×10^{-7}	1.70×10^{-2}	5.51	1.35×10
5.00×10^{-7}	2.22×10^{-2}	7.20	1.36×10
1.00×10^{-6}	2.65×10^{-2}	8.60	1.33×10
2.00×10^{-6}	3.08×10^{-2}	9.99	1.29×10
5.00×10^{-6}	3.65×10^{-2}	1.18×10	1.20×10
1.00×10^{-5}	4.07×10^{-2}	1.32×10	1.13×10
2.00×10^{-5}	4.49×10^{-2}	1.46×10	1.06×10
5.00×10^{-5}	5.04×10^{-2}	1.64×10	9.90
1.00×10^{-4}	5.46×10^{-2}	1.77×10	9.40
2.00×10^{-4}	5.91×10^{-2}	1.92×10	8.90
5.00×10^{-4}	6.58×10^{-2}	2.13×10	8.30
1.00×10^{-3}	7.18×10^{-2}	2.33×10	7.90
2.00×10^{-3}	7.90×10^{-2}	2.56×10	7.70
5.00×10^{-3}	9.14×10^{-2}	2.96×10	8.00
1.00×10^{-2}	1.04×10^{-1}	3.37×10	1.05×10
2.00×10^{-2}	1.20×10^{-1}	3.89×10	1.66×10
3.00×10^{-2}	1.28×10^{-1}	4.15×10	2.37×10
5.00×10^{-2}	1.46×10^{-1}	4.73×10	4.11×10
7.00×10^{-2}	1.65×10^{-1}	5.35×10	6.00×10
1.00×10^{-1}	1.96×10^{-1}	6.36×10	8.80×10
1.50×10^{-1}	2.79×10^{-1}	9.05×10	1.32×10^2
2.00×10^{-1}	3.40×10^{-1}	1.10×10^2	1.70×10^2
3.00×10^{-1}	5.00×10^{-1}	1.62×10^2	2.33×10^2
5.00×10^{-1}	7.45×10^{-1}	2.42×10^2	3.22×10^2
7.00×10^{-1}	9.37×10^{-1}	3.04×10^2	3.75×10^2
9.00×10^{-1}	1.07	3.47×10^2	4.00×10^2
1.00	1.12	3.63×10^2	4.16×10^2
1.20	1.21	3.93×10^2	4.25×10^2
2.00	1.40	4.54×10^2	4.20×10^2
3.00	1.47	4.77×10^2	4.12×10^2
4.00	1.46	4.74×10^2	4.08×10^2
5.00	1.40	5.54×10^2	4.05×10^2
6.00	1.33	4.31×10^2	4.00×10^2
7.00	1.25	4.05×10^2	4.05×10^2
8.00	1.16	3.76×10^2	4.09×10^2
9.00	1.07	3.47×10^2	4.20×10^2
1.00×10	9.50×10^{-1}	3.08×10^2	4.40×10^2
1.20×10	8.47×10^{-1}	2.75×10^2	4.80×10^2
1.40×10	7.61×10^{-1}	2.47×10^2	5.20×10^2
1.50×10	7.34×10^{-1}	2.38×10^2	5.40×10^2
1.60×10	7.28×10^{-1}	2.36×10^2	5.55×10^2
1.80×10	7.20×10^{-1}	2.34×10^2	5.70×10^2
2.00×10	7.15×10^{-1}	2.32×10^2	6.00×10^2

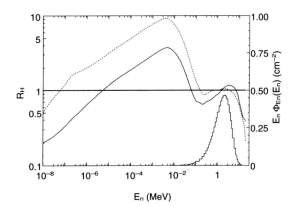

Figure 3. Ambient dose equivalent response as a function of neutron energy for the LB 6411 (solid line) and the Leake Counter (dashed line) numerically calibrated (Equations 2 and 3) in a bare ^{252}Cf spectrum (histogram).

Table 3. Fluence weighted ambient dose equivalent readings for the LB 6411 and ambient dose equivalent in calibration spectra.

	^{252}Cf-D$_2$O mod.	^{252}Cf bare	^{241}Am-Be	^{241}Am-B
NM/φ (pSv.cm^2)	111	**385**	407	458
(H*(10)/φ)(pSv.cm^2)	105	**385**	391	408
NM/H*(10)	1.06	**1.00**	1.04	1.12

in a catalogue[4] and for a sub-set of over 100 spectra measured in reactor environments. The dashed and dash-dotted lines in the upper part of Figure 4 enclose the allowed domain of R_H(rel.), namely, $2/3 \leq R_H$(rel.) $\leq 3/2$. In particular, at workplaces in reactors (cf. upper part of Figure 4), the LB 6411 indicates $H*(10)$ within +10% and −30% for the whole range. For the other spectra the performance of LB 6411 is also seen to be much better than that of the Leake counter.

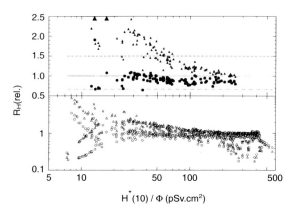

Figure 4. Calculated relative ambient dose equivalent response, R_H(rel.) for the LB 6411 (circles) and the Leake counter (triangles) as a function $H*(10)/\Phi$. Bottom part: In about 500 spectra found in a catalogue[4]. Upper part: In a sub-set of over 100 spectra found in reactor environments, the values for the two enlarged triangles are 4.72 and 2.53.

CONCLUSION

The LB 6411 rem counter has been established as a useful and sufficiently accurate survey instrument for the measurement of $H*(10)$ in routine monitoring. The use of the consistent fluence and ambient dose equivalent response provided here allows even more accurate measurements.

DEDICATION

The authors wish to dedicate this work to their recently retired colleague Ernst Piesch, an eminent scientist in the field of dosimetry and much respected in this capacity and as a person.

REFERENCES

1. ICRP. *1990 Recommendation of the International Commission on Radiological Protection*. Publication 60 (Oxford: Pergamon Press) (1990).
2. ICRU. *Quantities and Units for Use in Radiation Protection Dosimetry*. Report 51 (Bethesda, MD: ICRU Publications) (1993).
3. Klett, A., Maushart, R., Burgkhardt, B., Fieg, G. and Piesch, E. *A New Neutron Dose Equivalent Rate Meter with Improved Energy Response in Terms of H*(10)*. In: Proc. 4th Conf. on Radiation and Protection Dosimetry, Orlando, FL, USA, 23–27 October 1994. ORNL/TM-12817, p. 227 (1994).
4. Naismith, O. F. and Siebert, B. R. L. *A Database of Neutron Spectra, Instrument Response Functions, and Dosimetric Conversion Factors for Radiation Protection Applications*. Radiat. Prot. Dosim. **70**(1-4), 241–245 (This issue) (1997).
5. Siebert, B. R. L. and Schuhmacher, H. *Quality Factors, Ambient and Personal Dose Equivalent for Neutrons, Based on the New ICRU Stopping Power Data for Protons and Alpha Particles*. Radiat. Prot. Dosim. **58**, 177–183 (1995).
6. Griffith, R. V., Palfalvi, J. and Madhvanath, U. *Compendium of Neutron Spectra and Detector Responses for Radiation Protecton Purposes*. Techn. Report Series No. 318, (Vienna: IAEA) (1990).
7. Briesmeister, J. (Ed.) *MCNP — A General Monte Carlo Code for Neutron and Photon Transport*. Report LA 7396-M, Rev. 2 (Los Alamos, USA) (1986, Revised 1991).
8. Guldbakke, S., Dietz, S., Kluge, H. and Schlegel, D. *PTB Neutron Fields for the Calibration of Neutron Sensitive Devices*. In: Strahlenschutz: Physik und Meβtechnik. Eds W. Koelzer and R. Maushart (Verlab TÜV Rheinland, Köln) Vol. I, pp. 240–247 (1994).

A SPHERICAL NEUTRON COUNTER WITH AN EXTENDED ENERGY RESPONSE FOR DOSIMETRY

H. Toyokawa†, A. Uritani†, C. Mori†, N. Takeda‡ and K. Kudo‡
†Department of Nuclear Engineering
Nagoya University, Nagoya 464-01, Japan
‡Electrotechnical Laboratory, Tsukuba 305, Japan

Abstract — A neutron counter which is applicable to spectrometry and dosimetry over a wide energy range has been developed. It gives energy spectra, integral fluences and dose equivalent of incident neutrons with energies from thermal to 15 MeV. The counter consists of a spherical polyethylene moderator and three slender ^{3}He position-sensitive proportional counters inserted into the moderator. The position-sensitive proportional counters give detection position profiles of neutrons, slowed down to thermal energies in the spherical moderator, which give the above mentioned information. The responses of the counter as a function of the neutron energy were calculated with Monte Carlo simulations. The feasibility of the counter for applications to neutron spectrometry and dosimetry was examined with an unfolding method. It is shown that the counter is applicable to the spectrometry and dosimetry of neutrons over a wide energy range.

INTRODUCTION

Although Bonner spheres[1] were developed more than three decades ago, they still play an important role in neutron spectrometry, because their fluence responses extend from thermal energies to more than 20 MeV. The responses of the spheres are functions of the neutron energy and moderator diameter. The larger the sphere diameter, the higher the neutron energy where the response reaches its maximum. Therefore, using a set of spheres with different diameters, a set of count rate data is obtained from which an energy spectrum is deduced. The responses of the Bonner spheres have been studied by many researchers both experimentally and theoretically. Alevra[2] showed that the accuracy of the measurements of the neutron fluence with Bonner spheres can be improved to within 4%. A theoretical investigation to optimise the measurements with Bonner spheres has been reported[3].

The present work reports a study of an application to neutron spectrometry and dosimetry of a neutron counter, with a spherical polyethylene moderator and three slender position-sensitive proportional counters inserted into the moderator. The method for obtaining an energy spectrum and integral fluence from the counter response is similar to that of the Bonner spheres. However, only a single measurement with a single counter system is needed, rather than multiple measurements with a set of the Bonner spheres.

DESCRIPTION OF THE COUNTER

A schematic drawing of the counter and the set-up used for response calculations is shown in Figure 1. The diameter of the spherical moderator is 26 cm. It was estimated using Monte Carlo simulations so that the responses extend up to 15 MeV. The outer diameter and the inner diameter of a position-sensitive proportional counter are 1 cm and 0.9 cm, respectively. The counting gas is a mixture of ^{3}He (101 kPa) and CF_4 (70 kPa). Because the addition of the CF_4 gas shortens the ranges of the proton and the triton produced in the ^{3}He(n,p)^{3}H reaction[4], the proportional counter has a good position resolution which was evaluated to be about 0.7 cm in FWHM when the applied voltage is 1500 V. The position of the neutron detection was calculated by the charge division method.

In the response calculations, a point source geometry was considered. The source was placed on the X-axis 2 m away from the centre of the spherical moderator (Figure 1). The source–detector set-up used for the response calculations was chosen such that the geometry met an experimental set-up at the standard neutron field of the Electrotechnical Laboratory.

Because the cross section of the ^{3}He(n,p)^{3}H reaction is extremely large for low energy neutrons, most of the neutrons detected with the porportional counters are thermal neutrons. Therefore the detection position profiles obtained with the proportional counters can be assumed to be the distribution of the thermalised neutrons in the moderator. The distribution is sensitive to the energy of the incident neutrons. As the energy of the incident neutrons increases, the thermal neutrons are distributed more broadly and deeply in the spherical moderator. This is mainly due to the fact that the mean free path of the incident fast neutrons, with energies of about 1 MeV or more, is much larger than that of the slower neutrons. Therefore, the probability of neutron detection with the proportional counters in the deeper region of the sphere increased. In other words, the sensitivity of the neutron counter in the 'central region' of the sphere is high for fast neutrons. Conversely, the sensitivity near the surface of the sphere is high for slow neutrons, because they can be thermalised with fewer collisions. Moreover the detection probability in the central region is significantly low for slow neutrons, because these neutrons have a high probability of escaping from the moderator surface without producing any detection signal.

The response of a Bonner sphere corresponds approximately to the density of the thermalised neutrons around the thermal neutron detector, placed at the centre of the spherical moderator, as a function of the incident neutron energy. The response to fast neutrons of a large sphere is higher than that of a smaller sphere. The reason is that incident fast neutrons can undergo multiple collisions in the spherical moderator with a large diameter and can be thermalised enough in order to be detected. Moreover, the probability that they escape from the counter is low because the neutron moderation takes place in a deep region of the sphere. On the other hand, the response of a small sphere to fast neutrons is low, compared to that of a larger sphere, because the probability of escaping from the sphere prior to thermalisation is rather high for these neutrons.

Therefore the responses of the Bonner spheres as a function of the neutron energy and the moderator thickness are similar to those of the new neutron counter described here. The two neutron detection systems are, in principle, based on the same technique for sensing the energy of the incident neutrons. The use of position-sensitive proportional counters in a single moderator sphere, instead of a set of moderator spheres, significantly reduces time and effort in practical measurements.

CALCULATION OF RESPONSE FUNCTIONS

A Monte Carlo simulation code was developed to investigate the counter responses. The cross section data were from the JENDL-3 nuclear data file[5]. In the simulations, neutrons were generated isotropically from the point source and incident on the counter. The history of each of the neutrons incident on the counter was followed until the neutron escaped from the moderator or it interacted with a ^{3}He nucleus, producing a proton–triton pair from the ^{3}He(n,p)^{3}H reaction in the proportional counters. When the latter interaction took place, the interaction position in the proportional counter was calculated.

Figure 2 shows the calculated position profiles (or the counter responses as a function of the axial position) for various monoenergetic neutrons. In each calculation, 10^7 neutrons, which corresponds to the neutron fluence of 1.9×10^4 cm^{-2} at the centre of the moderator (when the moderator is removed) were incident of the counter.

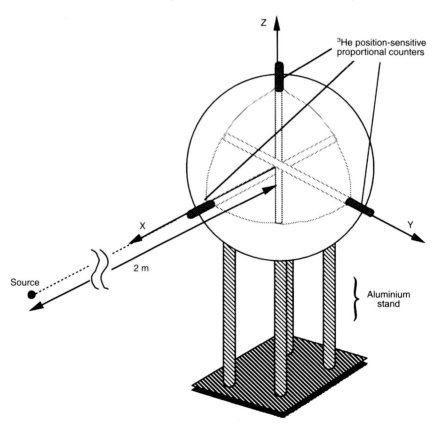

Figure 1. A schematic drawing of the counter and the source-detector geometry.

The abscissa of Figure 2 shows the axial position in the position-sensitive proportional counter placed along the X-axis (Figure 1). Therefore the neutrons were incident on the counter from the right-hand direction in Figure 2. As the energy of the incident neutrons increases, the position profiles spread to the deep region, which is about within ±5 cm of the axial position. As stated before, the position profiles approximately show the distribution of the thermalised neutrons. It is shown in Figure 2 that the distribution is sensitive to the energy of the incident neutrons. In the first step of the feasibility study of the counter, thought of as a neutron spectrometer to be used under various geometric conditions of neutron irradiation, only the responses obtained for the X-proportional counter for the point source geometry were examined.

The numbers shown at the top of Figure 2 are position groups which were introduced to calculate the response at a certain depth from the moderator surface along the X-axis as a function of the neutron energy. Therefore, the increment of the number corresponds to the decrease of the diameter of the Bonner spheres. Because the responses were not optimized as a function of the position-division, the groups were chosen so that the position-division would be uniform: (1) −13 cm to −7.8 cm, (2) −7.8 cm to −2.6 cm, (3) −2.6 cm to 2.6 cm, (4) 2.6 cm to 7.8 cm, and (5) 7.8 cm to 13 cm, respectively. Because the responses of the counter are 'directional', which means that the distribution of the thermalised neutrons varies significantly as the source–detector geometry changes[6], the response calculation should be done for each of the practical conditions in which the counter was used.

Figure 3 shows the counter responses as a function of the incident neutron energy obtained for the X-proportional counter. Each of the curves shows the detection counts at one of the five position groups, per unit fluence at the centre of the moderator, as a function of the incident neutron energy. The energy resolution seems to be rather poor because the response curves overlap considerably. The advantages of the new neutron counter are, however, the extended neutron fluence response from thermal energies to more than 10 MeV

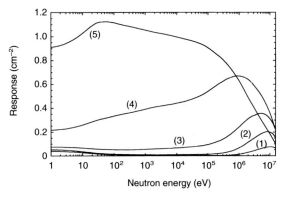

Figure 3. Calculated responses for the five position groups shown in Figure 2 for the X-proportional counter as a function of the incident neutron energy.

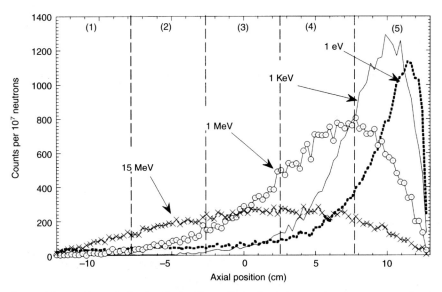

Figure 2. Detection position profiles calculated with the Monte Carlo simulations or the responses as a function of the axial position for various monoenergetic neutrons. The position profiles are shown only for the X-proportional counter. The numbers indicate the position groups: (1) −13 cm to −7.8 cm, (2) −7.8 cm to −2.6 cm, (3) −2.6 cm to 2.6 cm, (4) 2.6 cm to 7.8 cm, and (5) 7.8 cm to 13 cm, respectively.

and the ability to reduce time and effort significantly in practical measurements compared with the conventional methods.

FEASIBILITY TEST FOR NEUTRON SPECTROMETRY

Detection position profiles for neutrons with various energy spectra were calculated by Monte Carlo simulations. A feasibility test for neutron spectrometry was done with a spectrum unfolding method using the calculated position profiles. The response matrix used for the test had 5 position bins and 20 energy bins. The unfolding code used here was SAND-II[7]. At first, the response to monoenergetic neutrons was examined. The energy spectra obtained for (a) 50 eV and (b) 1 MeV monoenergetic neutrons are shown in Figure 4. From the figures, it can be seen that the estimated spectra represent the original ones well. The initial guess spectra used for the unfolding were white spectra, i.e. no *a priori* information was used. Figure 5 shows the estimated energy spectrum (histogram) of incident neutrons with Gaussian distribution, peaked at 200 keV with 50% FWHM (dashed line). Figure 6 shows the estimated energy spectrum (histogram) and the original Maxwellian spectrum (dashed line) described by the following equation:

$$S(E) = \frac{1}{\sqrt{(2\pi)}} T^{-3/2} \sqrt{E} \exp(-E/T) \quad (T = 1.42 \text{ MeV})$$

where E is the neutron energy in MeV. The equation describes quite well the neutron energy spectrum from the spontaneous fission of ^{252}Cf. Although the energy resolution of the counter seems to be rather poor, the estimated spectra are in good agreement with the original ones.

Further discussions may be necessary for the practical application of the counter to neutron spectrometry and dosimetry, because there were slight differences between the calculated position profiles and those obtained experimentally. The differences are mainly due to some simplification of the models used in the simulation. Careful adjustments of some of the parameters used would improve the accuracy of the simulation, and would make it possible to apply the counter in practice.

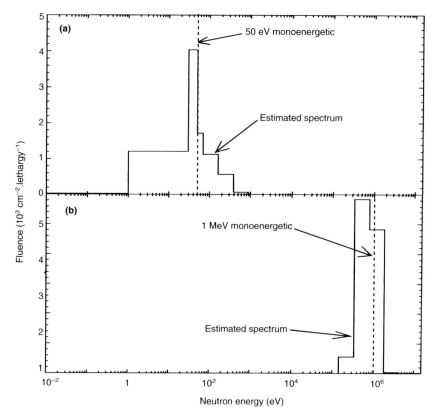

Figure 4. Estimated energy spectra with the SAND-II code (histogram) for (a) 50 eV and (b) 1 MeV monoenergetic neutrons (dashed line). The initial guesses were white spectra. Original fluence = 9.42×10^3 cm^{-2}. Estimated fluence: (a) 8.44×10^3 cm^{-2}, (b) 9.57×10^3 cm^{-2}.

CONCLUSIONS

A study on the estimation of the energy spectra of the incident neutrons using the new spherical neutron counter was carried out. The response functions of the counter were calculated with Monte Carlo simulations. The applicability of the counter to spectrometry and dosimetry was examined with simulations. Spectrum unfolding with the SAND-II code was applied to three characteristic neutron fields with monoenergetic, Gaussian, and Maxwellian spectral distributions. The response matrix used for the spectrum unfolding had 5 rows and 20 columns. The estimated spectra agreed well with the original ones. Rough estimations of neutron spectra in a wide energy range were possible. It was concluded that it is possible to apply the new neutron counter to spectrometry and dosimetry in a wide energy range.

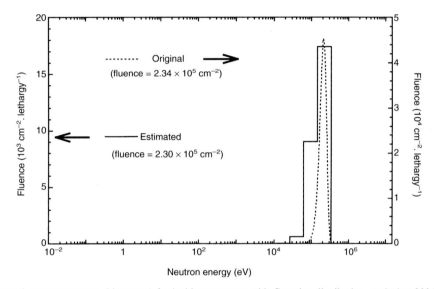

Figure 5. Estimated energy spectrum (histogram) for incident neutrons with Gaussian distribution, peaked at 200 keV with 50% FWHM (dashed line).

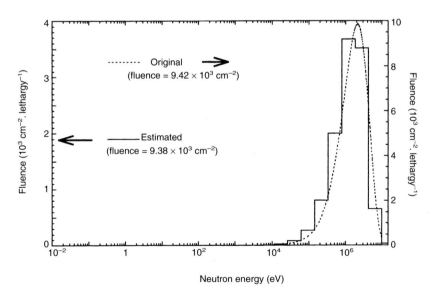

Figure 6. Maxwellian neutron energy spectrum from the spontaneous fission of ^{252}Cf (dashed line) and the corresponding spectrum estimated with the SAND-II code (histogram).

ACKNOWLEDGEMENTS

The authors wish to express their thanks to Dr Hiroshi Miyahara and Mr Shiko Ito, of Department of Nuclear Engineering, Nagoya University, for their valuable discussions.

REFERENCES

1. Bramblett, R. L., Ewing, R. I. and Bonner, T. W. *A New Type of Neutron Spectrometer*. Nucl. Instrum Methods **9**, 1–12 (1969).
2. Alevra, A. V. *Accurate Neutron Fluence Measurements Using Bonner Spheres*. In: Proc. Eighth ASTM-EURATOM Symp. on Reactor Dosimetry, Vail, Colorado, 1993. ASTM STP 1228, pp. 290–299 (1993).
3. Weise, K. *Optimisation in Neutron Spectrometry and Dosimetry with Bonner Spheres Using a General Measure of Quality for Experiments*. Radiat. Prot. Dosim. **37**(3), 157–164 (1991).
4. Kopp, M. K., Valentine, K. H., Chritophorou, L. G. and Carter, J. P. *New Gas Mixture Improves Performance of ^{3}He Neutron Counters*. Nucl. Instrum Methods **201**, 395–401 (1982).
5. JENDL-3 Compilation Group, Japanese Nuclear Data Committee. *Japanese Evaluated Nuclear Data Library, Version-3-JENDL-3*. JAERI 1319 (1990).
6. Toyokawa, H., Uritani, A., Mori, C., Takeda, N. and Kudo, K. *A Multipurpose Spherical Neutron Counter*. IEEE Trans. Nucl. Sci. **NS-42**(4), 644–648 (1995).
7. McElroy, W. N., Berg, S., Crockett, T. and Hawkins, R. AFWL-TR-67-41, Vol. 1 (Air Force Weapons Laboratory, Kirkland) (1967).
8. ICRP. *Data for Use in Protection Against External Radiation*. ICRP publication 51 (Oxford: Pergamon Press) pp. 74 (1987).

COMPUTATIONAL DOSIMETRY

B. R. L. Siebert† and R. H. Thomas‡
†Physikalisch-Technische Bundesanstalt
POB 3345 — D 38023 Braunschweig, Germany
‡Lawrence Livermore National Laboratory, University of California
7000 East Avenue, L-382, Livermore, CA 94550 and Berkeley, USA

INVITED PAPER

"All the mathematical sciences are founded on relations between physical laws and the laws of numbers, so that the aim of exact science is to reduce the problems of nature to the determination of quantities by operations with numbers".

James Clerk Maxwell in '*On Faraday's Lines of Force*' (1856)

Abstract — A definition of the term 'Computational Dosimetry' is presented that is interpreted as the sub-discipline of computational physics which is devoted to radiation metrology. It is shown that computational dosimetry is more than a mere collection of computational methods. Computational simulations directed at basic understanding and modelling are important tools provided by computational dosimetry, while another very important application is the support that it can give to the design, optimisation and analysis of experiments. However, the primary task of computational dosimetry is to reduce the variance in the determination of absorbed dose (and its related quantities), for example in the disciplines of radiological protection and radiation therapy. In this paper emphasis is given to the discussion of potential pitfalls in the applications of computational dosimetry and recommendations are given for their avoidance. The need for comparison of calculated and experimental data whenever possible is strongly stressed.

INTRODUCTION

The term 'Numerical Dosimetry' is puzzling and its meaning requires analysis. That this recently coined term has found its way into general circulation affords yet another validation of Gresham's Law.* The term is of dubious etymological parentage but is nevertheless widely used. Indeed the authors were asked to present a paper on the topic of 'Numerical Dosimetry' to this Symposium and wrote an abstract to this title. However, upon reflection the term 'Computational Dosimetry' seems a more precise description of the topics covered.

First we must define measurement. Fortunately there is general agreement, supported by dictionaries, that 'to measure' is synonymous with 'to estimate' and thus, following Maxwell, we infer that dosimetry — the measurement of absorbed dose — includes the idea of its calculation. The terminal '-metry' is derived from the Greek μετρον and has the 'general sense of action, process or art of measuring (something specified by the initial element)'.**

Nevertheless, the attraction of the term 'numerical

---

* 'Bad money drives out good'. Sir Thomas Gresham (1558). Some scholars attribute the law to Copernicus.
** The Oxford English Dictionary adds the explanatory comment — 'Most of the subjects in -meter have correlative words in -metry, denoting specifically the process of measuring by the instrument called "-meter"'. Thus presumably dosimetry is the determination of dose with a dosemeter.

dosimetry' needs explanation before it is dismissed. The authors of this paper speculate that, of the several options offered by the dictionary, those who coined the term would choose either 'by numbers' or 'expressed by a number or numbers'. However, since all determinations of dose, whether by measurement or calculation, are expressed by a number (of units of the quantity) it would seem likely that it is the first of these two options that is intended. We therefore infer that 'Numerical Dosimetry' is presumably then the determination of absorbed dose 'by numbers'. If our speculation is correct this interpretation is indeed most apt because the detailed mechanisms of most of the radiation transport codes now used for the calculation of dose use random number generators.

The authors prefer to describe these Monte Carlo techniques, and indeed any mathematical solutions to radiation transport problems — which are nowadays basic tools used in the dosimetry of ionising radiation, by the term 'Computational Dosimetry'.

Computational dosimetry is the sub-discipline of computational physics devoted to radiation metrology. 'To compute' derives from the Latin root *putare* (to clear up or settle) — and from the etymology one can then infer that to compute does not merely mean to put numbers together but rather to place them in order. In that sense, computational dosimetry may be defined as the process of connecting and ordering of known data, by means of relations based on theory or established models, in order to create new data and to reveal new insights. The main goal of this paper is to elucidate this process of acquiring new data by means of computational dosimetry and to question the validity of the knowledge obtained in this way.

It is the contention of the authors that computational dosimetry is more than a mere collection of computational methods. This viewpoint is demonstrated by an analysis of its widespread use in the papers presented to this Symposium. About half of the papers presented use methods of computational dosimetry. The heavy reliance on Monte Carlo methods is seen from the fact that about 40 papers use transport codes.

This paper addresses the need of basic and derived data as input for computations, shows the role of theory and models in, and discusses some subjects of, computational dosimetry. The potential pitfalls of computational dosimetry are summarised in the appendix.

BASIC AND DERIVED DATA

The basic data used in dosimetry are atomic and nuclear cross sections for the interactions of interest. In principle, data such as kerma coefficients, stopping powers or W values may be computed from fundamental cross section data and are therefore often referred to as 'derived data'. In practice, however, the fundamental cross sections are sometimes not known well enough to allow a sufficiently accurate calculation of these derived data and it is often necessary to determine their values by direct measurement.

The need for basic cross section data is fundamental and much remains to be done in this area[1]. Unfortunately the measurement of basic data is tedious, difficult and, in general, expensive. In the present economic climate financial support is dwindling for such measurements and the dearth of information is likely to be with us for some time.

Fortunately several compilations and evaluations of those available cross section data which are of special interest for neutron dosimetry have been published, together with programs to utilise these evaluated data. Prominent examples are ENDF[2] and NJOY[3], both available from the Radiation Shielding Information Center (RSIC)*, Oak Ridge National Laboratory or the Nuclear Energy Agency, (NEA)**. Both these organisations distribute not only data but also computer codes suitable for dosimetric applications. So as to ensure the quality assurance of dosimetric determinations it is highly recommended that the assistance of these organisations be sought for information as a means of practising quality assurance of dosimetric determinations. Data for nuclear reactions used in the production of neutron fields are contained in an IAEA Technical Report 273[4]. Kerma values are published by the International Commission on Radiation Units and Measurements ICRU[5,6], stopping power values by ICRU[7] and Ziegler et al[8] and W values by the ICRU[9]. In the field of radiation therapy special attention is drawn to a new IAEA technical document on Atomic and Molecular Data[10]. A special source of interaction data are so-called 'generators' for high energy particle transport codes. These generators are usually needed as integral parts of such codes, (see, for example, References 11–13), because detailed differential cross section data are still scarce[14].

ON THEORIES AND MODELS

There are two basic assumptions in the science of radiation dosimetry: first that the fundamental laws of physics apply and second that the absorption of energy by living organisms is the beginning of a complex chain of reactions which are bio-chemico-physical in nature and which may be ultimately either beneficial (e.g. in the case of radiation therapy) or detrimental (e.g. in the case of carcinogenesis).

A theory is a formulation of apparent relationships or underlying principles of certain observed phenomena which has been verified to some degree. A theory therefore serves two purposes: it 'explains' observed data and, by the process of induction, predicts the outcome of potential experiments.

The goal of a complete theory of radiation dosimetry would be to predict the reaction of living organisms and/or the organs of an animal irradiated by ionising radiation. Such a 'dosimetric theory' does not yet exist. Nevertheless, computational dosimetry is based on sound theoretical fundamentals. Theoretical input from mathematics, atomic and nuclear physics is used.

Because Monte Carlo techniques are extensively used in computational dosimetry one of the most important theoretical considerations is to ensure that the underlying model is physically correct and free from any divergence, even under a variety of complex simulations. The proper application of statistical methods will then permit reliable estimates of the variance of the quantity being calculated, even when a large number of trials are needed to estimate even a small numerical value.

Models possess some features of a theory but are usually limited in scope. They attempt to describe known data rather than to predict new information: models are not founded on first principles but are merely descriptive. For instance, target theory[15] as applied to the photographic action of X rays is indeed a theory, however, its numerous offspring applied to biological cells are better termed models.

In the context of this paper the term model is used for the computational image of an experiment, i.e. as a general term for all input data and the algorithms used for simulating an experiment. It must be borne in mind

---
*RSIC: Internet: pdc@ornl.gov and WWW: htp://epicws.epm.ornl.gov.
**NEA: Le Seine St. Germain 12, Boulevard des Iles 92130 Issy-les-Moulineaux, France; Internet: sartori@nea.fr

that such an image is always an approximation to reality. Apart from technical restrictions such as finite computing resources there are limitations in describing the experimental situation. For instance, in depth dose calculations for therapeutic applications there is an uncertainty in anatomical geometries imposed by the finite resolution of the tomographic process. As another example, in calculating response functions of sensors assumptions need to be made on the homogeneity of materials used in its construction which usually cannot be checked by any feasible measurement. As a specific example in this latter regard, small changes in the hydrogen content of materials (caused by adsorption) or slight changes in the content of neutron absorbing substances (*e.g.* in borated materials), can have a strong influence on the neutron moderating properties of the materials comprising the sensor.

In view of these limitations it is clear that all such models are imperfect to some degree. It is necessary to provide a rigorous and balanced assessment of both the merits and limits of any model, but even the most careful discussion can never replace experimental verification. Nevertheless, even bearing these limitations in mind, such models are versatile tools of computational dosimetry. They generate new insights, provide basic understanding and give strong support to the design, optimisation and analysis of experiments. In some sense, such models may be regarded as replacing experiments provided they are based on sound assumptions, use correct input data, are carefully checked and are applied within their scope. A simple example which exemplifies these constraints is the calculation of a response function of a radiation detector: if the simulation agrees satisfactorily with measurements in monoenergetic calibration fields, then use of the function at other energies, within the calibration range, is justified.

SOME TOPICS OF COMPUTATIONAL DOSIMETRY

The primary objective of computational dosimetry is to enhance the accuracy of the determination of absorbed dose. Computational analyses and simulations are the main tools used to achieve these objectives.

In this paper we first give examples of the applications of computational dosimetry to radiation therapy; to the calculation of particle spectra; the calculation of dose distributions; the calculation of instrument response functions and the calculation of conversion coefficients. Following these discussions an example of the optimisation of experiments and the tools used to analyse radiation environments are then discussed.

Newcomers to computational dosimetry will find the detailed discussion of the MCNP code in Ref. 16 (available from RSIC or NEA) and the text on Monte Carlo techniques by Lux and Koblinger[17], which contains an excellent bibliography. Many useful sub-routines for numerical mathematics are to be found in a book by Press *et al*[18]. An excellent general text has been written by Paic[19] and, finally, the book by Knoll[20] on radiation sensors is of great value, especially if response functions of detectors are to be calculated.

For the casual user, it will usually be most economic and efficient to use standard program packages such as MCNP. However, it is vital that some workers continue to develop new codes so that existing codes are challenged, improvements in accuracy are achieved and 'user-friendliness' is improved. Furthermore, there is no substitute for the experience and insight gained by writing one's own programs.

Provided adequate computing capacity is available, the reduction of the inherent statistical variance in simulations is nowadays a rather small problem. In any computation a rigorous analysis of variance should always be undertaken. Considerable effort is needed to study the sensitivity of the endpoint being calculated with variations of the input parameters. There is no generally agreed method as how best to address the very difficult and complicated problem of the treatment of the variances associated with the input data such as cross sections and values for derived data. For these reasons it is once again stressed that experiments are indispensable to check simulations.

Radiotherapy

The medical uses of radiation are the largest source of man-made exposure to ionising radiations. The current annual average worldwide effective dose *per capita* from medical exposures is about 0.6 mSv, half of which comes from therapy and the remainder from diagnostic procedures[21]. Many diagnostic methods, particularly the generation of images using tomography[22] and therapy planning[23] rely on highly sophisticated computational methods. Although it is beyond the scope of this paper to give a detailed account of these methods, some general remarks seem appropriate.

As reported in this conference (*e.g.* by White *et al*[23]), modern radiotherapy planning is a prominent example of the attempt to find a substitute for measurements. Clearly there are many measurements that are impracticable in human patients, no matter how important the data required. The alternative, but necessary, task is formidable because an enormous effort in diagnostic imaging is required to describe adequately the anatomical details, with respect to geometry and material properties, of any specific patient. Even when and if this problem is adequately solved there remains the very basic problem of the verification of such procedures, *i.e.* the question whether one can design mock-ups of the situations encountered in patients which are accessible to measurement and calculation.

The question of the relative biological effectiveness to be applied to the absorbed dose in different organs or tissues is another area for improvement. In order that existing models may be tested, and possibly improved, it will first be necessary to improve the accuracy of calculated particle spectra inside a patient and to complement these with studies on a micrometre, or better nanometre, dosimetric scale which would give insight on track structures and provide a basis for their interaction with the structures of the cells affected by radiation. Here an interdisciplinary approach is required in view of the complex problems encountered in radiation chemistry and biology (*e.g.* see References 24 and 25).

Spectra and dose distributions

Spectra and dose distributions are needed not only in therapy planning but also in general radiation protection. A knowledge of spectra in the workplace is needed so that appropriate detectors may be selected to perform radiation monitoring and that the instruments may be properly calibrated.

As a practical example most instruments used for surveying neutron fields are far from ideal. Before attempting a measurement it would be helpful to consult catalogues of typical neutron spectra[26,27] but ultimately the spectrum in the actual site of measurement must be determined.

Spectra for use in calibration and for the determination of the response of neutron-measuring devices are provided by the International Standards Organization (ISO)[28]. It is highly recommended that these data be used so as to provide a consistent basis for measurement.

Even small changes in neutron spectra may produce significant changes in instrument readings, thereby increasing the variance of the determination. An example demonstrating this is shown in Figure 1 where two estimates of the fission spectrum from bare (unmoderated) ^{252}Cf are shown. In the MCNP code[16] this spectrum is simulated by the two-parameter Watt spectrum. The differences between the two spectra seem very small but, as is shown in Table 1, if these spectra are used to compute the mean neutron energy; or fluence to ambient dose equivalent conversion coefficients, $H^*(10)/F$; or the response of Bonner spheres[29] marked differences are encountered.

The use of calculated spectra and the resulting doses to optimise shielding can have significant economic impact. As a rule of thumb, about 20% of the total construction cost for large accelerators is due to shielding. Excessive shielding which may be necessitated by uncertainties associated with dose estimates can be very costly at these large installations and it is worthwhile making considerable efforts in refining these calculations. Furthermore, in most countries accurate calculations are required in order to assess the environmental impact of such installations before a permit to construct the new facility will be granted. Last, but not least, spectra provide physical insight which may be helpful or even needed in many applications.

Details on the calculation of absorbed dose distribution can be found in a recent NCRP report[30] and some additional remarks are given below in context with conversion functions.

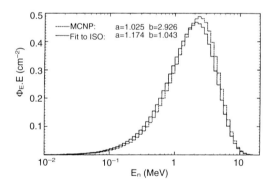

Figure 1. The spectrum from a bare ^{252}Cf neutron fission source as given by ISO[30] (solid line) and in MCNP[16] (dotted line) as a function of neutron energy.

Response and conversion functions

Nowadays, it is good practice to accompany the development of a new detector with a computational simulation of its performance. In this procedure computational images are used to optimise the design of the instrument and then the model is used to compute the response of the detector in several calibration fields in which measurements with the instrument being designed have been made. When good agreement between these calculations and measurements is obtained one can be confident that further calculations

Table 1. The influence of different representations of the fission spectrum from a bare ^{252}Cf neutron source on the neutron fluence averaged energy, $\langle E_n \rangle$, and ambient dose equivalent, $H^*(10)/\Phi^{(37)}$, and on the reading of Bonner spheres[29].

	$\langle E \rangle$ (MeV)	$H^*(10)/\Phi$ (pSvcm^2)	3" (cm^2)	5" (cm^2)	8" (cm^2)	10" (cm^2)	15" (cm^2)
ISO[28]	2.13	385	0.309	1.66	2.30	1.88	0.701
MCNP[16]	2.31	390	0.286	1.66	2.30	1.91	0.753
Δ(%)	8.5	1.3	−7.4	0.0	0.0	1.6	7.4

of the instrument response may be trusted over the entire range of energy covered by the calibration fields. However, attention must be given to resonances in cross sections of the materials of the detector which may produce anomalous readings in limited energy regions; see, for example Ref. 29. To illustrate the use of this technique we select two from the many examples presented in this conference: a new rem counter[31] and a study on the use of a silicon diode as personal dosemeter[32]. Both examples clearly demonstrate the benefits of using the methods of computational dosimetry.

So that the numerical values of the protection quantities and operational quantities resulting from irradiation from external sources of ionising radiation may be compared it is necessary to know the distribution of the absorbed dose throughout the actual human body or computational model. When this distribution of the absorbed dose is known, the appropriate protection and operational quantities may be determined.

Considerable effort is required for calculations of the absorbed dose distribution in mathematical computational models. However, modern computing methods (particularly those that use Monte Carlo techniques) have the advantage in that they can deal with complex irradiation conditions and provide absorbed dose distributions for irradiation in a wide variety of radiation fields. Excellent statistical precision may usually be obtained from such computations but, in addition to the statistical uncertainties inherent in such calculations, other uncertainties may arise from several sources (*e.g.* the data used to simulate radiation interactions with tissue; differences in anatomical modelling; and the inherent variation in human anatomy). In addition, in practical situations, large uncertainties in the estimation of the absorbed dose distributions may arise because of a lack of detailed knowledge of the external radiation field. Although great precision or accuracy is not required in these calculations for the purpose of radiological protection *per se* it is nevertheless most important to compare the results obtained by several types of calculation — particularly those made in differing computational models and using different computational techniques — so that possible systematic errors may be detected. Such comparisons require appropriate precision in the calculation, even if this precision is not strictly needed for radiological protection.

The experimental approach to the determination of these quantities requires construction of a realistic phantom of the human body using tissue-equivalent material and an extensive measurement programme. Accurate measurements require considerable effort and resources and this is particularly true in the case of neutrons. Such resources are generally beyond the means of all but the largest laboratories[33]. Consequently, the number of systematic experimental determinations* of the absorbed dose reported in the scientific literature is limited and resort is usually made to the alternative approach of calculation. The inter-relation between the different quantities used in radiation protection needs a careful analysis[34]. Attention to several influences must be given, particularly so in the calculation of conversion coefficients for neutrons[35]. The currently recommended conversion factors[36] are expected to be soon superseded by new data based on ICRP 60 recommendations. However, new $H^*(10)$ data for neutrons are available[37] and in view of the increasing demand for conversion coefficients for $H^*(10)$ for neutrons with high energies, we refer to a contribution to this conference[38].

Design and optimisation of experiments

The practical realisation of realistic neutron calibration fields is of great practical importance[39]. As an example we discuss one example based on a well-designed moderator around a lithium target bombarded by monoenergetic protons. The optimum design requires a compromise between the need to remove heat dissipated in the lithium target and the distortion of the neutron spectrum resulting from moderation by the cooling water. Two designs were explored. In the one, the coolant was led through an O-ring forming an annulus of coolant around the circumference of the target and in the other the water was led so that it cooled the entire backing of the target.

Table 2 shows that the change in spectra by neutron moderation has a strong influence both on the fluence averaged energies and the ambient dose equivalents.

Analysis of radiation environments

Finally, we turn to one of the most important tasks of computational dosimetry, which is to analyse radiation environments and find methods to enhance the accuracy

Table 2. Neutron fluence-averaged energy, $\langle E_n \rangle$, and ambient dose equivalent, $H^*(10)/\Phi^{(37)}$, at a distance of 1 m from a Li(p,n) target using a cooling disc or a cooling ring (see text).

Angle	Disc $\langle E \rangle$ (keV)	Ring $\langle E \rangle$ (keV)	Disc $H^*(10)/\Phi$ (pSv.cm^2)	Ring $H^*(10)/\Phi$ (pSv.cm^2)
0-blank	469	521	285	312
90-blank	205	210	160	165
0-mod.	16.4	19.0	21.9	23.6
90-mod.	6.36	7.26	15.5	16.1

---

*The term "experimental determination" is used here to distinguish it from the shorter term "measurement" because throughout this paper the term measurement is taken to include calculation.

of the determination of absorbed dose and dose equivalent.

The program Spektren Bibliothek (SPKTBIB), presented to this conference[27] provides techniques both to improve estimates of dose equivalent[40] and to optimise existing multi-element detectors[41].

SPKTBIB carries out simulations based on catalogued radiation spectra of various environments and predicts the response of dosemeters by calculating sets of conversion coefficients. This technique permits the selection of detectors appropriate for measurement in a given radiation environment with no tedious or expensive experimental effort. However, it must be emphasised that the assumptions used both for the spectra and response functions used in the simulation must be based on experimentally verified data. In that sense SPKTBIB provides a classical example of what computational dosimetry is 'all about': Create a computational image relying on experimentally verified pieces of information (spectra, response functions) and substitute experiments by less expensive and faster calculations. The advantage of calculations is, that verified pieces of input can be combined with great ease, to correspond to a multitude of practical situations, whereas an experiment very often provides answers to only one situation at a time.

CONCLUSIONS

Computational dosimetry is the sub-discipline of computational physics devoted to radiation metrology. It is essential for the analysis and organisation of the complex body of data which is needed in dosimetry and it provides models, evaluations and concepts for this purpose.

The ability to use numerical means to simulate practical situations is one of the most powerful tools of computational dosimetry: experimental or clinical set-ups may be represented; particle spectra and dose distributions may be determined; response functions (or matrices) for instruments and conversion coefficients may be calculated. Computational dosimetry greatly facilitates the optimisation of experiments and supplements the analysis of the data. In appropriate cases computational dosimetry can even replace experiments.

These great benefits are not free — thorough training and experience are needed if they are to be obtained. It is necessary fully to understand and study the influence of the variance of input data (cross sections and geometry) although this may involve considerable effort when using standard Monte Carlo codes. An important task for the future is to develop standardised procedures to this end.

In summary, computational dosimetry greatly enhances the overall quality of radiation dosimetry in therapy and radiation protection. Nevertheless it is important to remember that despite this great utility appropriate experimental verification of any calculated data is nevertheless indispensable!

ACKNOWLEDGEMENTS

The European Communities have provided generous support to EURADOS and its Working Group IV on Numerical Dosimetry over many years and thus have greatly contributed to the growing use of and experience with the techniques provided by computational dosimetry.

One of us (RHT) is grateful to the support of the Hazards Control Department, Lawrence Livermore National Laboratory, University of California, for their gracious hospitality and support.

APPENDIX

Caveat emptor!

General comments on the use of computational dosimetry

(1) Random number generators are used as the backbone of any simulation, but they are never ideal and the statistics of large numbers do not hold for imperfect random generators. So as to minimise the possibility of error the minimum that should be done with any calculation is to use different initial 'seeds' and to perform a sequential analysis at least once.

(2) The convergence of Monte Carlo calculations must be carefully examined — especially in the case of small calculated endpoints. Internal counters should be used in order to obtain some information as to whether they are appropriately sampled. If possible, analytical checks should be performed, e.g. in computing absorbed doses one should also calculate the first collision dose by analytical means.

(3) Variance reduction techniques are sometimes unavoidable, e.g. in deep penetration problems. However, one must keep in mind that these techniques only redistribute the variance. The use of point detectors is recommended for detectors. It may be helpful to check the normalisation by additional track length or surface fluence estimation in a volume around or a surface of at least one point detector.

(4) In the specification of variances there are two conflicting requirements. On the one hand variances should be given as small as can be justified so that the differences in predictions between alternative procedures may become apparent. On the other hand, if safety is at stake prudence demands that additional safety factors be applied when giving variances.

(5) The scope and limits of a given model (computational image) with respect to the reliable ranges of arguments and parameters should be discussed, e.g. nuclear resonances, if not properly

taken into account, limit the accuracy of interpolations between experimental points using the model.
(6) Resonances in cross sections need to be carefully considered. In some cases it is important that they be taken into account using the best energy resolution available, e.g. in computing the transmission through an iron filter at a reactor. It is not usually helpful to specify conversion coefficients at resonance energies.
(7) Reference data such as provided by IAEA[4], ICRP[36], ICRU[7] or ISO[28] should be used. If other data seem more accurate it is most important to point out the differences between the data used and reference data and to explain the use of other data. The same holds true for the use of evaluated cross sections.
(8) As a minimum a rudimentary sensitivity analysis should be performed by varying some input data, e.g. the radius of a sphere used as estimator for the dose equivalent at a point in the determination of H*(10) or by selecting cross section data for a given nuclear reaction from different evaluations.
(9) A variation of input parameters and data is very helpful in gaining more insight and as consistency check; e.g. in studying a moderator assembly the mass density of one or more cells or regions may be varied (including considering them as voids).
(10) If at all possible, important problems should be treated using more than one code. It is most helpful to participate in benchmark problems so that codes may be validated.
(11) Finally, documentation should present not only the final results. A strong plea is made for sufficient details of the methods and algorithms used, the input data and any intermediate results so that the reader may be able to reproduce the published data; e.g. it is helpful to specify the spectral fluence and not merely spectrum weighted data and for instance documenting intermediate steps in an optimisation may help the reader.

REFERENCES

1. Chadwick, M. B., DeLuca, Jr, P. M. and Haight, R. C. *Nuclear Data Needs for Radiation Protection and Therapy Dosimetry.* Radiat. Prot. Dosim. **70**(1–4), 1–12 (This issue) (1997).
2. Rose, P. F. and Dunford, C. L. (eds) *ENDF-102 Data Formats and Procedures for the Evaluated Nuclear Data File ENDF-6*. BNL-NCS-44945 (National Nuclear Data Center, Brookhaven National Laboratory, Upton, NY, USA (1990). For newest versions turn to RSIC or NEA.
3. MacFarlane, R. E., Muir, D. W. and Boicourt, R. M. *The NJOY Nuclear Data Processing System.* Los Alamos National Laboratory, LA-9303-M, Vol. I, II (1982), Vol. III (1987), Vol. IV (1985). For newest versions turn to RSIC or NEA.
4. IAEA. *Handbook on Nuclear Activation Data.* Technical Reports Series No. 273 (Vienna: International Atomic Energy Agency) (1987).
5. ICRU. *Neutron Dosimetry for Biology and Medicine.* Report 26 (Bethesda, MD: International Commission on Radiation Units and Measurements) (1977).
6. ICRU. *Photon, Electron, Proton and Neutron Interaction Data for Body Tissues.* Report 46 (Bethesda, MD: ICRU Publications) (1992).
7. ICRU. *Stopping Power and Ranges for Protons and Alpha Particles.* Report 49 (Bethesda, MD: ICRU Publications) (1993).
8. Ziegler, J. F., Biersack, J. P. and Littmark, U. *The Stopping of Ions in Solids* (Oxford: Pergamon Press) (1985).
9. ICRU. *Average Energy Required to Produce an Ion Pair.* Report 31 (Bethesda, MD: ICRU Publications) (1979).
10. IAEA. *Atomic and Molecular Data for Radiotherapy and Radiation Research.* IAEA-TECDOC-799 (Vienna: IAEA) (1995).
11. Fassò, A., Ferrari, A., Ranft, S. R. P. J., Stevenson, G. R. and Zazula, J. M. *FLUKA92.* In: Proc. Workshop on *Simulating Accelerator Radiation Environments, SARE.* Santa Fe, USA. Report LA-12835-C, Ed. A. Palounek, (1994).
12. Prael, R. E. and Lichtenstein, H. *User Guide to LCS: The Lahet Code System.* Los Alamos National Laboratory, LA-UR-89-3014 (1989).
13. Mokhov, N. *MARS12 Code System.* In: Proc. SARE Workshop, Santa Fe (1993) and *The MARS Code System User's Guide Version 13(95).* Report FN-628 (Fermi National Accelerator Laboratory) (1995).
14. Konig, A. J. *Review of High Energy Data and Model Codes for Accelerator-Based Transmutation.* Nuclear Energy Agency, NEA/NSC/DOC(92)12 (1992).
15. Blau, M. and Altenburger, K. *Über einige Wirkungen von Strahlen, II.* Z. Phys. **12**, 315–329 (1922).
16. Briesmeister, J. (Ed.) *MCNP — A General Monte Carlo Code for Neutron and Photon Transport.* Report: LA 7396-M (Los Alamos, USA) (1986, revised 1991).
17. Lux, I. and Koblinger, L. *Monte Carlo Particle Transport Methods: Neutron and Photon Calculations.* (Boca Raton: CRC Press) (1991).
18. Press, W. H., Flannery, B. P., Teukolsky, S. A. and Vetterling, W. T. *Numerical Recipes* (Cambridge University Press) (1989).
19. Paic, G. (Ed.) *Ionizing Radiation: Protection and Dosimetry.* (Boca Raton: CRC Press) (1988).
20. Knoll, G. F. *Radiation Detection and Measurement* (New York: John Wiley) (1979).

21. Valentin, J. *UNSCEAR Data Collections on Medical Radiation Exposures: Trends and Consequences*. Radiat. Prot. Dosim. **57**, 85–90 (1995).
22. ICRU. *Use of Computers in External Beam Radiotherapy Procedures with High-Energy Photons and Electrons*. Report 42 (Bethesda, MD: ICRU Publications) (1987).
23. White, R. M., Chadwick, M. B., Chandler, W. P., Hartmann Siantar, C. L., Resler, D. A. and Weaver, K. A. (Unpublished observations).
24. Watt, D. E. and Hill, S. J. A. *An Empirical Model for the Induction of Double Strand Breaks in DNA by the 'Indirect' Action of Radiation*. Radiat. Prot. Dosim. **52**, 17–20 (1994).
25. Watt, D. E. *A Unified System of Bio-Effectiveness and its Consequences in Practical Applications*. Radiat. Prot. Dosim. **70**(1–4), 529–536 (This issue) (1997).
26. Griffith, R. V., Palfalvi, J. and Madhvanath, U. *Compendium of Neutron Spectra and Detector Responses for Radiation Protection Purposes*. Techn. Report Series No. 318 (Vienna: IAEA) (1990).
27. Naismith, O. F. and Siebert, B. R. L. *A Database of Neutron Spectra, Instrument Response Functions, and Dosimetric Conversion Factors for Radiation Protection Applications*. Radiat. Prot. Dosim. **70**(1–4), 241–245 (This issue) (1997).
28. ISO. *Neutron Reference Radiations for Calibrating Neutron-Measuring Devices Used for Radiation Protection Purposes and the Determination of their Response as a Function of Neutron Energy*. ISO 8529 (International Standards Organization, Geneva, Switzerland) (1989).
29. Wiegel, B., Alevra, A. V. and Siebert, B. R. L. *Calculations of the Response Functions of Bonner Spheres with a Spherical ^{3}He Proportional Counter Using a Realistic Detector Model*. (Physikalisch-Technische Bundesanstalt, Braunschweig) PTB Report N-21 (1994).
30. NCRP. *Conceptual Basis for Calculations of Absorbed-Dose Distributions*. NCRP Report No. 108 (Bethesda, MD: National Council on Radiation Protection and Measurements) (1991).
31. Burgkhardt, B., Fieg, G., Klett, A., Plewnia, A. and Siebert, B. R. L. *The Neutron Fluences and H*(10) Response of the New LB 6411 Rem counter*. Radiat. Prot. Dosim. **70**(1–4), 361–364 (This issue) (1997).
32. Luszik-Bhadra, M., Alberts, W. G., Dietz, E. and Siebert, B. R. L. *Feasibility Study on an Individual Electronic Neutron Dosemeter*. Radiat. Prot. Dosim. **70**(1–4), 97–102 (This issue) (1997).
33. Tanner, J. E., Piper,. R. K., Leonowich, J. A. and Faust, I. G. *Verification of an Effective Dose Equivalent Model for Neutrons*. Radiat. Prot. Dosim. **44**, 171–174 (1992).
34. Siebert, B. R. L. *Radiation Quantities: Their Inter-Relationship*. Radiat. Prot. Dosim. **54**(3–4), 193–202 (1994).
35. Schuhmacher, H., Hollnagel, R. A. and Siebert, B. R. L. *Sensitivity Study of Parameters Influencing Calculations of Fluence-to-Ambient Dose Equivalent Conversion Coefficient for Neutrons*. Radiat. Prot. Dosim. **54**, 221–225 (1994).
36. ICRP. *Data for Use in Protection Against External Radiation*. Publication 51. Ann. ICRP **17**(2,3) (Oxford: Pergamon Press) (1987).
37. Siebert, B. R. L. and Schuhmacher, H. *Quality Factors, Ambient and Personal Dose Equivalent for Neutrons, Based on the New ICRU Stopping Power Data for Protons and Alpha Particles*. Radiat. Prot. Dosim. **58**, 177–183 (1995).
38. Sannikov, A. V. and Savitskaya, E. N. *Ambient Dose Equivalent Conversion Factors for High Energy Neutrons Based on the ICRP 60 Recommendations*. Radiat. Prot. Dosim. **70**(1–4), 383–386 (This issue) (1997).
39. Chartier, J. L., Jansky, B., Kluge, H., Schraube, H. and Wiegel, B. *Recent Developments in the Specification and Achievement of Realistic Neutron Calibration Fields*. Radiat. Prot. Dosim. **70**(1–4), 305–312 (This issue) (1997).
40. Naismith, O. F., Siebert, B. R. L. and Thomas, D. J. *Response of Neutron Dosemeters in Radiation Protection Environments: an Investigation of Techniques to Improve Estimates of Dose Equivalent*. Radiat. Prot. Dosim. **70**(1–4), 255–260 (This issue) (1997).
41. Alberts, W. G., Dörschel, B. and Siebert, B. R. L. *Methodological Studies on the Optimisation of Multi-Element Dosemeters in Neutron Fields*. Radiat. Prot. Dosim. **70**(1–4), 117–120 (This issue) (1997).

ON THE CONSERVATIVITY OF $H_P(10)$

G. Leuthold, V. Mares and H. Schraube
GSF-Forschungszentrum für Umwelt und Gesundheit
Institut für Strahlenschutz, Neuherberg
D-85758 Oberschleißheim, Germany

Abstract — For neutrons, numerical values for the Personal Dose Equivalent, $H_p(10,\alpha)/\Phi$, in the ICRU TE slab phantom were recently recommended in an international standard (IEC-1323, 1995) and are intended for calibrating individual dosemeters fixed to the front of the phantom, with respect to energy and angle of incidence. These recommended conversion factors are compared with the present MCNP calculations of conversion coefficients $h_{p\Phi}(10;E,\alpha)$, and with calculations of Effective Dose, E (ICRP 60), and Effective Dose Equivalent, H_E (ICRP 26), in the anthropomorphic ADAM phantom for monoenergetic and parallel neutron radiation (expanded field) with an angle of incidence varying from 0° (A-P) up to 75° to the left side of the phantom in the energy range from thermal up to 19 MeV. For the calculation of the conversion coefficients $h_{p\Phi}(10;E,\alpha)$ and the Effective Dose Equivalent, H_E, in the phantom, a quality factor is applied which takes into account the revised Q(L) relationship of ICRP 60 and the new stopping power data for protons and alpha particles of ICRU Publication 49 (1993). For the calculation of Effective Dose, E, the smooth radiation weighting factor w_R of ICRP 60 is used.

INTRODUCTION

The ICRU has introduced the quantity Personal Dose Equivalent, $H_p(10)$, as an estimator for the Effective Dose, E, as the primary limiting quantity, for all conditions of penetrating radiation, i.e. radiation type, energy and angle of incidence. It is defined as the dose equivalent in 10 mm depth of an individual. The definition of Personal Dose Equivalent, $H_p(10)$, does not lead to a unique value because it depends both on the individual and on the site chosen in which to obtain $H_p(10)$.

For the purpose of development, type testing and calibration of individual dosemeters, the quantity was extended by the ISO[1] to the TE slab phantom. In this case $H_p(10)$ has a unique value for a given radiation field. It serves as some kind of 'estimator for the estimator', and shall be sufficiently conservative compared with the Effective Dose, at least for an irradiation incidence which is restricted to the frontal half space.

MONTE CARLO CALCULATIONS

Monte Carlo calculations were performed using the MCNP code version 4A[2]. This code is capable of transporting neutrons and photons as well as electrons. The MCNP4 code was implemented to run under a UNIX operation system on the GSF CONVEX C3840 supercomputer with four CPUs and one GFLOP peak vector speed. The MCNP code allows modelling of three-dimensional configurations defined in a Cartesian coordinate system. The volumes are built up by surfaces using the intersection and union operators.

Calculations were performed for parallel monoenergetic beams of neutrons. The neutron cross section data library EFF1LIB[3] used in the present calculation is based on the ENDF/B-5. This library contains cross section data for 69 isotopes in the range from 10^{-5} eV to 20 MeV. The photon interaction tables used were based on evaluated data from Hubbell et al[4]. To model the thermal neutron scattering by molecules at room temperatures, the $S(\alpha,\beta)$ cross section tables for light water were used.

PHANTOMS

The phantoms used for the calculations are the ICRU slab phantom (30 cm × 30 cm × 15 cm) consisting of 4-element ICRU tissue, and the male anthropomorphic phantom ADAM. The shapes of the body and internal organs of ADAM are defined by equations of surfaces of cylinders, cones, ellipsoids, planes and tori. Internal organs are considered to be homogeneous in composition and density. Different densities and compositions are used for the lungs, skeleton, skin and the bulk of the body. The angle of irradiation incidence for both phantoms varies from 0° (A-P) up to 75° in steps of 15° to the left. The space between source and phantoms was assumed to be vacuum.

BASIC DATA

For the calculation of the Personal Dose Equivalent in the ICRU slab phantom, a neutron quality factor was used which takes into account the revised Q(L) relationship of ICRP 60[5] and the new stopping power data for protons and alpha particles in water by ICRU 49[6]. The kerma values applied are taken from ICRU 26[7]. Above 10 MeV neutron energy, the recommended kerma values of White et al[8] are used. Figure 1 shows the quality factor as function of neutron energy. The method of calculating the quality factor may be found elsewhere[9]. There is a very good agreement with a calculation by Siebert and Schuhmacher[10] except in the energy range above 5 MeV. The reason is that in their calculation the kerma values of Caswell et al[11] are used which show differences, especially for carbon and oxygen, in comparison with ICRU 26[7], and the more recent data of White et al[8]. As kerma values are required both for the calculation of absorbed dose *and* for the calculation of the quality factor, Personal Dose Equivalent will be influenced twice. It is therefore important which kerma values are applied. Though the differences are not large in respect of practical requirements, the calculational origin should be based on more recent data which appear more accurate, as $H_p(10,\alpha)$ serves as operational quantity for calibration purposes.

RESULTS

Calculations of Personal Dose Equivalent, $H_p(10;\alpha)$, in the ICRU slab phantom were not performed for all energies given in IEC-1323[12]. The energies chosen are 0.025 eV, 2 keV, 565 keV and the high energies 14.8 and 19 MeV in order to see the influence of the above mentioned difference in the kerma values and quality factor. The values are given in Table 1. Figure 2 shows the ratio of Personal Dose Equivalent, $H_p(10;\alpha)$, given by IEC[12] to $H_p(10)$ obtained in this work for the five selected neutron energies as a function of the angle of incidence. The differences at low neutron energies may be attributed to the use of the $S(\alpha,\beta)$ cross section tables for light water whereas for the calculation of the IEC data the $S(\alpha,\beta)$ cross sections of polyethelene were applied[10]. The difference at high energies due to the different kerma values used exceeds 20%.

The calculation of the Effective Dose, E, and the Effective Dose Equivalent, H_E (ICRP 26)[13], in the ADAM phantom were performed for all energies given in IEC-1323 Table A.2[12]. Figure 3 shows the result for the Effective Dose, E, as function of neutron energy for different angles of incidence to the left side of the phantom. Additional values for 0° (A-P) are given which form an upper limit and allow an interpolation for neutron energies in between. The smooth radiation weighting factor function w_R of ICRP 60[5] was used.

Figure 4 gives the ratio of Personal Dose Equivalent to Effective Dose, $H_p(10;\alpha)/E$. For the Personal Dose Equivalent the IEC values are used except at 14.8 and 19 MeV. Over a wide range the ratio is smaller than one, i.e. Personal Dose Equivalent is not conservative in relation to E. The ratios of Personal Dose Equivalent to Effective Dose Equivalent, $H_p(10;\alpha)/H_E$, however,

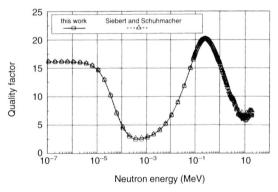

Figure 1. Quality factor as function of neutron energy in an ICRU tissue mass element. Differences in the high energy region come from different kerma values used.

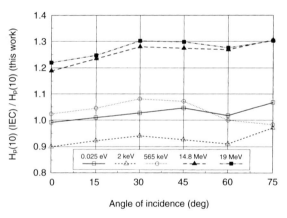

Figure 2. Ratio of Personal Dose Equivalent, $H_p(10)$, in the ICRU slab phantom given by IEC[12] to $H_p(10)$ obtained in this work for five neutron energies as function of the angle of incidence.

Table 1. Personal Dose Equivalent, $H_p(10,\alpha)$, in the ICRU slab phantom for five neutron energies at different angles of incidence.

E_n(MeV)	$H_p(10,\alpha)/\Phi(pSv.cm^2)$					
	0°	15°	30°	45°	60°	75°
2.5×10^{-8}	11.50	10.50	8.87	6.32	3.97	1.62
2.0×10^{-3}	9.69	8.91	7.73	5.86	3.80	1.72
0.565	347	334	321	292	245	117
14.8	472	456	454	449	454	396
19.0	492	478	477	473	486	436

given in Figure 5, show conservative behaviour over nearly the whole range of energies and angles.

CONCLUSION

Calculations of quantities such as the Personal Dose Equivalent or the Ambient Dose Equivalent which are determined by the application of the neutron quality factor should take into account recommended stopping power and kerma values. The introduction of Effective Dose, E, via the radiation weighting factor, w_R leads to a non-conservativity in the low and medium neutron energy range which is strongly pronounced at higher angles of incidence. Two of the reasons are qualitatively the following:

(i) Originally, the radiation weighting factor w_R[5] was derived from calculations in the ICRU sphere, i.e. it was taken from the ratio of dose equivalent to energy dose ('quality factor') in 1 cm depth of the sphere. This procedure was equivalent to a concentration of all organs into an organ '1 cm depth in the sphere'. The change of stopping powers in ICRU Report 49[6] reduced the numerical values of the conversion factor $H_p(10)/\Phi$ to some extent, but no more the radiation weighting factor.

(ii) With increasing angle α, $H_p(10)$ is essentially influenced by the oblique mass of material which the radiation has to pass through. This mass is, for a certain angle α, larger in the case of a rectangular slab phantom than for a phantom with curved surface.

The use of the older concept of Effective Dose Equivalent, H_E, which takes into account the ICRP 60 recommendations establishes conservativity over a wide range of neutron energy and angle of incidence. If the radiation weighting factor concept should be retained a change of the numerical values of w_R could be a solution[14].

Of course, the overall effect of 'non-conservativity' depends essentially on the spectral distribution of the incident neutron field, and may be unimportant for pure fast neutron spectra.

ACKNOWLEDGEMENTS

The work was supported by the Commission of the European Union DGXII under contract FI3P-CT92–0002. This is gratefully acknowledged.

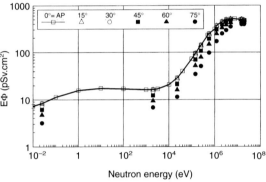

Figure 3. Effective Dose, E, in the ADAM phantom as function of neutron energy for six angles of incidence to the left side of the phantom.

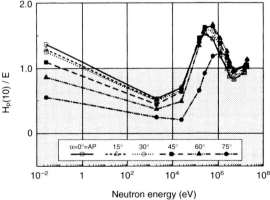

Figure 4. Ratio of Personal Dose Equivalent, $H_p(10)$, in the ICRU slab phantom to Effective Dose, E, in the ADAM phantom for six angles of incidence. The shaded area exhibits the region of non-conservativity.

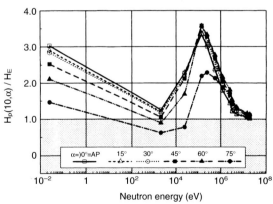

Figure 5. Ratio of Personal Dose Equivalent, $H_p(10)$, in the ICRU slab phantom to Effective Dose Equivalent, H_E, in the ADAM phantom for six angles of incidence. The shaded area exhibits the region of non-conservativity.

REFERENCES

1. ISO. *Reference Photon Radiation: Calibration of Area and Personal Dosemeters and the Determination of their Response as a Function of Photon Energy and Angle of Incidence.* ISO-DIS 4037-3. Draft International Standard (1996).
2. Briesmeister, J. F. (Ed.) *MCNP — A General Monte Carlo N-Particle Transport Code, Version 4A.* Report LA-12625-M (Los Alamos National Laoratory) (1993).
3. Vontobel, P. *EFF1LIB: A NJOY Generated Neutron Data Library Based on EFF-1 for the Continuous Energy Monte Carlo Code MCNP.* PSI Report Nr. 107, (Villigen: PSI) (1991).
4. Hubbell, J. H., Veigele, W. J., Briggs, E. A., Brown, R. T., Cramer, D. T. and Howerton, R. J. *Atomic Form Factors, Incoherent Scattering Functions, and Photon Scattering Cross Sections.* J. Phys. Chem. Ref. Data **4**, 471 (1975).
5. ICRP. *1990 Recommendations of the International Commission on Radiological Protection.* Publication 60 (Oxford: Pergamon Press) (1991).
6. ICRU. *Stopping Powers and Ranges for Protons and Alpha Particles.* Report 49 (Bethesda, MD: ICRU Publications) (1993).
7. ICRU. *Neutron Dosimetry in Biology and Medicine.* Report 26 (Bethesda MD: ICRU Publications) (1977).
8. White, R. M., Broerse, J. J., DeLuca Jr, P. M., Dietze, G., Haight, R. C., Kawashima, K., Menzel, H. G., Olsson, N. and Wambersie, A. *Status of Nuclear Data for Use in Neutron Therapy.* Radiat. Prot. Dosim. **44**(1/4), 11–20 (1992).
9. Leuthold, G., Mares, V. and Schraube, H. *Calculation of the Neutron Ambient Dose Equivalent on the Basis of the ICRP Revised Quality Factors.* Radiat. Prot. Dosim. **40**(2), 77–84 (1992).
10. Siebert, B. R. L. and Schuhmacher, H. *Quality Factors, Ambient and Personal Dose Equivalent for Neutrons, Based on the New ICRU Stopping Power Data for Protons and Alpha Particles.* Radiat. Prot. Dosim. **58**(3), 177–183 (1995).
11. Caswell, R. S., Coyne, J. J. and Randolph, M. L. *Kerma Factors for Neutron Energies below 30 MeV.* Radiat. Res. **83**, 217–254 (1980).
12. IEC. *Radiation Protection Instrumentation — Neutron Radiation — Direct Reading Personal Dose Equivalent and/or Dose Equivalent Rate Monitors.* Norme International/International Standard CEI/IEC-1323 (1995).
13. ICRP. *Recommendations of the International Commission on Radiological Protection.* Publication 26 (Oxford: Pergamon Press) (1977).
14. Leuthold, G. and Schraube, H. *Critical Analysis of the ICRP 60 Proposals for Neutron Radiation and a Possible Solution.* Radiat. Prot. Dosim. **54**(3/4), 217–220 (1994).

AMBIENT DOSE EQUIVALENT CONVERSION FACTORS FOR HIGH ENERGY NEUTRONS BASED ON THE ICRP 60 RECOMMENDATIONS

A. V. Sannikov and E. N. Savitskaya
Institute for High Energy Physics
142284 Protvino, Moscow Region, Russia

Abstract — Calculations of the ambient dose equivalent conversion factors for neutron energies from 20 MeV to 5 GeV have been performed using the Monte Carlo high energy transport code HADRON. The kerma approximation extended to the high energy region was applied in the calculations. The results obtained are compared with the previous data for an infinite slab 30 cm thick as well as with the recent calculations for the ICRU sphere. The influence of the phantom shape and dimension is expected to be fairly unimportant for high energy radiations. The discrepancies in dose equivalent data are explained mainly by incorrect quality factors of secondary charged particles used in earlier calculations. Another conclusion is that the ambient dose equivalent H*(10) satisfies the requirement of conservativity for high energy radiations in practical cases of radiation fields outside shields.

INTRODUCTION

The ICRP recommended in Publication 60[1] that the quality factor, Q, be redefined; this involves revision of the dosimetric quantities used in radiation measurements. In neutron dosimetry, the main operational quantity for area monitoring is the ambient dose equivalent, H*(10). Calculations of the corresponding conversion factors h*(10), for the new Q(L) dependence have been performed recently by several groups for neutron energies below 20 MeV[2-4] and by Nabelssi and Hertel[5] in the neutron energy range from 30 to 180 MeV. The present work is devoted to calculation of the ambient dose equivalent conversion factors for neutron energies from 20 MeV to 5 GeV. The calculations were performed in the geometry of a normal irradiation of an infinite tissue-equivalent slab 30 cm thick. In spite of the fact that this phantom differs from the 30 cm diameter tissue-equivalent sphere recommended by the ICRU, such a choice seems to be appropriate at high energies where the high anisotropy of secondary radiation reduces the effect of different shapes near the phantom surface.

METHOD

The calculations were performed by the Monte Carlo high energy transport code HADRON[6,7] based on the cascade–exciton model of nuclear interactions. The tissue slab was divided for calculation purposes into 31 layers separated at the depths of 0.8 cm, 1.2 cm, 2 cm, 3 cm etc. up to 30 cm. The doses in the second layer were related to corresponding ambient doses. Primary and secondary neutrons, protons and charged pions were transported through the phantom. Kinetic energy of charged particles heavier than a proton, such as d,t,^{3}He,α and recoil nuclei having short enough ranges, was assumed to be deposited locally. Low LET particles (photons, electrons, positrons and muons) were not transported. This component was taken from Alsmiller's data[8] which were described in detail and seem to be reliable. The total ambient dose equivalent was calculated as a sum of the following components; (a) heavy particles (A > 1); (b) protons; (c) charged pions; (d) low energy neutrons; (e) photons, electrons, muons.

The kerma concept extended by us recently to all ionising radiations[7,9] was applied in the calculations. In the frame of this approach, the kerma factor for ionising particles, k, is considered as a sum of the electromagnetic, k_e, and nuclear, k_n, components where k_e for charged paticles is equal to L/ρ. The kerma factor k_n is separated into the partial kerma factors for heavy (A > 1) and light (p,π) secondary charged particles: $k = k_e + k_h + k_l$. The kerma approximation is applied to the first two components related to secondary charged particles with short ranges. The component k_l is due to protons and pions transported through the phantom and is considered separately in the calculations. The extensive discussion of the method as well as the basic equations used in Monte Carlo calculations are given elsewhere[9].

The partial kerma factors and quality factors for high energy hadrons were calculated beforehand. First of all, the quality factors $\overline{Q}_e(E)$ for all possible secondary charged particles from pions to ^{16}O nuclei averaged over their ranges were calculated as a function of kinetic energy:

$$\overline{Q}_e(E_i) = \frac{1}{E_i} \int_0^{E_i} Q(L_i(E))dE \qquad (1)$$

The ICRU 49 recommendations[10] on stopping power of protons and α particles in liquid water were used. In other cases the semi-empirical formulae of Anderson and Ziegler[11] were applied below 100 MeV per nucleon and the well-known Bethe formula with density effect correction above this energy. The calculated energy dependences $\overline{Q}_e(E)$ for five different charged particles are shown in Figure 1. It can be seen that the ICRP 60 quality factors exceed in most cases those of ICRP 21. Exceptions are recoil nuclei (A > 4) with L > 200 keV.μm^{-1} and high energy particles having L < 30 keV.μm^{-1}. The partial kerma factors $k_h(E)$ and the corresponding quality factors $\overline{Q}_h(E)$ for neutrons, protons and charged pions were calculated using the HADRON code and $\overline{Q}_e(E)$ dependences for heavy particles. The obtained data for nucleons are presented in Figure 2. One important feature of these results is that the ICRP 60 quality factors $\overline{Q}_h$ for nucleons below 150 MeV are lower than the ICRP 21 ones. The explanation of this effect may be found in the comparison of the quality factors $\overline{Q}_e(E)$ for carbon and oxygen nuclei (Figure 1) which are nearly half as low for the ICRP 60 Q(L) dependence compared to the ICRP 21 value in the practically important energy range of 1–20 MeV. The neutron kerma factors $k_n(E)$ and quality factors $\overline{Q}_n(E)$ (ICRP 60 Q(L)) below 20 MeV were taken from Siebert and Schuhmacher[4]. In the case of ICRP 21 Q(L) dependence, the neutron quality factors calculated by Leuthold et al[2] were used.

RESULTS AND DISCUSSION

The calculated ambient absorbed dose and dose equivalent (ICRP 21 Q(L)) are shown in Figure 3 in comparison with previous calculations[2,3,5,8,12–17]. Our results for absorbed dose agree on an average with earlier calculations. Some differences between the present data and our old results[17] are explained by the better description of secondary particle spectra at energies of 20–60 MeV and by taking into account the decrease of

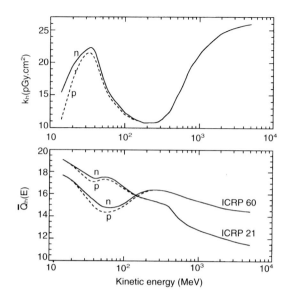

Figure 2. Partial kerma factors, $k_h(E)$, and average quality factors, $\overline{Q}_h(E)$, of heavy particles (A > 1) produced by nucleons in the ICRU tissue.

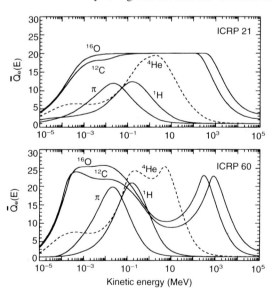

Figure 1. Quality factors of charged particles averaged over their ranges in the ICRU tissue for two Q(L) dependences recommended by the ICRP.

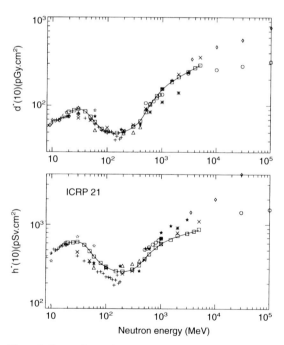

Figure 3. Comparison of calculated data for neutron ambient absorbed dose and ambient dose equivalent (ICRP 21 Q(L)). Solid curve shows the smoothed data of present calculations, dashed curve, the ICRP 51 recommendation. Key to References: (△) 12, (☆) 13, (∗) 14, (★) 8, (◇) 15, (○) 16, (✢) 2, (×) 17, (+) 3, 5, (□) present data.

nucleon density in nuclei during the cascade stage above 1 GeV. Another reason is the higher accuracy of the present calculations. Statistical errors of ambient absorbed doses were less than 2% and for ambient dose equivalent they were as a rule lower than 1%. The scatter of different data for dose equivalent is much higher than for absorbed dose. In the high energy region above 400 MeV, our calculations give lower values of ambient dose equivalent compared to other data. This is explained by the overestimated quality factor of 20 attributed to heavy particles in most previous calculations. It can be seen from Figure 2 that with more acccurate consideration these values lie between 11 and 14 at hadron energies above 400 MeV. The ICRP 21 dose equivalents calculated by Nabelssi and Hertel[5] above 20 MeV are 20–30% lower than our results. These differences are mainly explained by large discrepancies in quality factors. It should be mentioned that the 30% jump of quality factor[3,5] between 20 and 30 MeV cannot be justified by any physical reason. From the analysis of the method used in calculations of the quality factors by Nabelssi and Hertel[5], the conclusion can be drawn that slowing down of charged α particles was not, probably, taken into account. Comprison with the depth dose distributions for 180 MeV[5,8] and 3 GeV neutrons[8] made previously[9] confirms this suggestion as well as the small sensitivity of depth dose distributions to the phantom shape.

The above presented dose equivalents are related to the ICRP 21 Q(L) relationship. The ambient dose equivalents calculated by different groups using the ICRP 60 Q(L) dependence are shown in Figure 4. Our results at 20 MeV agree well with the data calculated by the MCNP code[2–4]. The discrepancies with the data[5] above 20 MeV are less than 20%. Such a small difference is somewhat strange taking into account the large divergence for the ICRP 21 dose equivalent. As expected, differences between ambient dose equivalents for the new and old Q(L) relationships are not large and do not exceed 10%.

It is a well-known fact that the ambient dose equivalent H*(10) is not a conservative quantity for high energy monoenergetic particles. For that reason, there are various proposals to change the depth of measurement of ambient dose equivalent in the high energy region. For instance, Nabelssi and Hertel[5] proposed using the quantity H*(120) for neutrons above 30 MeV. In our opinion, the common shortcoming of such proposals is that they are based on the unrealistic assumption of exposure by *monoenergetic* high energy particles that never takes place in practice. In all the practical situations, exposure of personal and public takes place in the conditions of equilibrium spectra outside shields where high energy particles are accompanied by large numbers of low energy secondaries. As a consequence, the maximum of the dose equivalent for equilibrium high energy spectra lies near the phantom surface. An additional confirmation may be found in Aroua et al[18] devoted to the recent study of build-up effects in high energy radiation fields at CERN. The practical conclusion of this consideration is that there is no necessity to introduce any innovations in the definition of ambient dose equivalent H*(10) which seems to us to be successful and stable.

Table 1. Ambient dose and ambient dose equivalent conversion factors for high energy neutrons calculated using the ICRP 21 and ICRP 60 Q(L) relationships.

E_n (MeV)	$d^*(10)$ (pGy.cm^2)	$h^*(10)$ (pSv.cm^2)	
		ICRP 21	ICRP 60
20	85.1	606	591
25	88.4	613	586
30	91.0	620	586
40	85.7	573	528
50	74.7	488	440
60	65.0	419	377
80	54.8	345	320
100	50.9	313	300
150	49.1	280	285
200	51.0	275	285
300	57.6	293	306
400	68.1	334	349
500	84.8	393	420
600	102	450	487
800	130	529	580
1000	153	591	647
1500	188	677	733
2000	211	728	789
3000	245	791	862
4000	270	837	915
5000	290	870	951

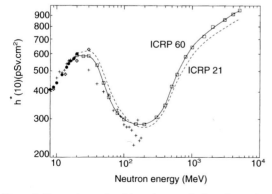

Figure 4. Comparison of ambient dose equivalent data calculated using the ICRP 60 Q(L) dependence. The present results for the ICRP 21 Q(L) dependence are shown for illustration. Key to References: (+) 2, (+) 3, 5, (●) 4, (□) present data.

CONCLUSIONS

The new approach based on the extension of the kerma approximation to the high energy region was applied in calculations of the ambient dose equivalent for high energy neutrons. The ambient absorbed dose and ambient dose equivalent conversion factors obtained for two Q(L) dependences are presented in Table 1. Comparison with the data of previous calculations has shown that the discrepancies are explained mainly by the approximate methods of estimating the quality factors of secondary charged particles produced by high energy hadrons. Another reason is large statistical errors in most earlier calculations. Consideration of applicability of the ambient dose equivalent H*(10) for high energy particles enables us to draw the conclusion that in practical situations of high energy radiation fields outside shields this quantity provides a conservative estimation for effective dose equivalent.

REFERENCES

1. ICRP. *1990 Recommendations of the International Commission on Radiological Protection*. Publication 60 (Oxford: Pergamon Press) (1991).
2. Leuthold, G., Mares, V. and Schraube, H. *Calculation of the Neutron Ambient Dose Equivalent on the Basis of the ICRP Revised Quality Factors*. Radiat. Prot. Dosim. **40**, 77–84 (1991).
3. Nabelssi, B. K. and Hertel, N. E. *Ambient Dose Equivalents, Effective Dose Equivalents and Effective Doses from 10 to 20 MeV*. Radiat. Prot. Dosim. **48**, 145–151 (1993).
4. Siebert, B. R. L. and Schuhmacher, H. *Quality Factors, Ambient and Personal Dose Equivalent for Neutrons Based on the New ICRU Stopping Power Data for Protons and Alpha Particles*. Radiat. Prot. Dosim. **58**, 177–183 (1995).
5. Nabelssi, B. K. and Hertel, N. E. *Ambient Dose Equivalent, Deep Dose Equivalent Index and ICRU Sphere Depth-Dose Calculations for Neutrons from 30 to 180 MeV*. Radiat. Prot. Dosim. **51**, 169–182 (1994).
6. Golovachik, V. T., Kustarjov, V. N., Savitskaya, E. N. and Sannikov, A. V. *Absorbed Dose and Dose Equivalent Depth Distributions for Protons with Energies from 2 to 600 MeV*. Radiat. Prot. Dosim. **28**, 189–199 (1989).
7. Savitskaya, E. N. and Sannikov, A. V. *High Energy Neutron and Proton Kerma Factors for Different Elements*. Radiat. Prot. Dosim. **60**, 135–146 (1995).
8. Alsmiller, R. G., Armstrong, T. W. and Coleman, W. A. *The Absorbed Dose and Dose Equivalent from Neutrons in the Energy Range 60 to 3000 MeV and Protons in the Energy Range 400 to 3000 MeV*. ORNL-TM-2924R (Oak Ridge) (1970).
9. Sannikov, A. V. and Savitskaya, E. N. *Ambient Dose Equivalent Conversion Factors for High Energy Neutrons Based on the New ICRP Recommendations*. IHEP Preprint 95–98 (Protvino) (1995).
10. ICRU. *Stopping Powers and Ranges for Protons and Alpha Particles*. ICRU Report 49 (Bethesda, MD: ICRU Publications) (1993).
11. Andersen, H. H. and Ziegler, J. F. *The Stopping and Ranges of Ions in Matter* (New York: Pergamon Press) (1977).
12. Zerby, C. D. and Kinney, W. E. *Calculated Tissue Current-to-Dose Conversion Factors for Nucleons below 400 MeV*. Nucl. Instrum. Methods **36**, 125–140 (1965).
13. Irving, D. C., Alsmiller, R. G. and Moran, H. S. *Tissue Current-to-Dose Conversion Factors for Neutrons with Energies from 0.5 to 60 MeV*. Nucl. Instrum. Methods **51**, 129–135 (1967).
14. Wright, H. A., Anderson, V. E., Turner, J. E., Neufeld, J. and Snyder, W. S. *Calculation of Radiation Dose due to Protons and Neutrons with Energies from 0.4 to 2.4 GeV*. Health Phys. **16**, 13–31 (1969).
15. Armstrong, T. W. and Chandler, K. C. *Calculations of the Absorbed Dose and Dose Equivalent from Neutrons and Protons in the Energy Range from 3.5 GeV to 1.0 TeV*. Health Phys. **24**, 277–286 (1972).
16. Golovachik, V. T., Potjomkin, E. L., Lebedev, V. N. and Frolov, V. V. *Doses from High Energy Particles in Tissue Equivalent Phantoms*. Preprint IHEP 74-58 (Serpukhov) (1974).
17. Sannikov, A. V. and Savitskaya, E. N. *Ambient Dose and Ambient Dose Equivalent Conversion Factors for High Energy Neutrons*. CERN Preprint CERN/TIS-RP/93-14 (1993).
18. Aroua, A., Höfert, M., Sannikov, A. V., Stevenson, G. R. and Vaerman, C. *An Investigation of Build-up Effects in High-Energy Radiation Fields using a HANDI TEPC*. Radiat. Prot. Dosim. **62**(3), 171–173 (1995).

CALCULATION OF THE PERSONAL DOSE EQUIVALENT, $H_P(10)$, FOR NEUTRONS IN THE MIRD PHANTOM

R. A. Hollnagel
Physikalisch-Technische Bundesanstalt
D-38116 Braunschweig, Germany

Abstract — The personal dose equivalent, $H_p(10)$, was calculated in the anthropomorphic MIRD phantom. $H_p(10)$ is compared with the primarily limited quantities, the effective dose, E, and the effective dose equivalent, H_E, computed in the same Monte Carlo run. The operational quantity, $H_{sl}(10)$, in the ICRU slab was found to approximate $H_p(10)$ for multidirectional incidences. All quantities were obtained with the same database. Also the correlation with the organ dose equivalents in the testes and thymus are given graphically. The dependence of $H_p(10)$ on the location in the trunk is discussed briefly.

INTRODUCTION

The personal dose equivalent, $H_p(d)$, as defined in ICRU Report 51[1], is the dose equivalent (DE) in ICRU soft tissue, at an appropriate depth, d, below a specified point on the body. For strongly penetrating radiation a depth of 10 mm is frequently employed, ICRU states, in which case it is denoted as $H_p(10)$. The ICRU annotated in its Report 39[2], that the quantity $H_p(10)$, at a given location on the anterior portion of the trunk, can be related to the effective dose equivalent[3], H_E, received by the trunk for radiation incident from anteriorly to laterally on the body.

The values of $H_p(10)$ are expected to be more closely connected than those of other protection quantities with the reading of a personal dosemeter above the point specified. Meanwhile H_E was replaced by the ICRP with the effective dose E[4,5]. However, both H_E and E are mentioned in ICRU Report 51: $H_p(d)$ and H_E among others belong to a 'coherent system of quantities', therefore H_E is also considered here. This paper deals with $H_p(10)$ itself and comparisons of its values with the effective dose equivalent H_E, with the effective dose E and with the $H_{sl}(10)$[6,7] in the ICRU slab, but not with dosemeter properties. With respect to the last quantity ICRU Report 51 refers the reader to ICRU Report 47[6], noting that the calibration of the dosemeter is generally performed on the ICRU slab.

Being defined in the exposed individual, $H_p(10)$ depends on the individual's shape and the location specified on the anterior trunk. Thus, it is a multivalued quantity. For the calibration of personal dosemeters, the ICRU slab is a suitable phantom and the DE in the slab, $H_{sl}(10)$, below the dosemeter to be calibrated is taken as the calibration quantity in substitution for $H_p(10)$ itself. Whereas $H_{sl}(10)$ is obviously not practical for pure posterior or lateral exposure for the neutron energy range considered here, i.e. thermal to 20 MeV, because of the different transmission properties of the slab and the trunk, one may expect for multidirectional incidences, such as rotational or isotropic, that, with the associated averaging of deviations, the values of $H_{sl}(10)$ and $H_p(10)$ turn out to be rather close together. Values of $H_p(10)$ in anthropomorphic phantoms have so far been lacking. The following questions are addressed: How good is $H_p(10)$ as a substitute for H_E or E to serve as a limiting quantity? Is $H_{sl}(10)$ sufficiently closely related to $H_p(10)$ in order to be a suitable surrogate for $H_p(10)$? And is $H_p(10)$ easier to estimate than E or H_E in different irradiation geometries? Also the question of how the DE of superficial organs, e.g. the testes and thymus, compare with $H_p(10)$, is addressed, because many calculations of organ doses in anthropomorphic phantoms are already available and the DE of certain superficial organs might suit as approximations to $H_p(10)$.

CALCULATIONAL METHOD

In all computations the Q(L) relationship from ICRP 60[4] and the mass stopping power values, dE/dx, from ICRU Report 49[8] were used. The Monte Carlo program utilised in calculating the doses in the volumes was HL_PH[5]. Organ doses H_T were estimated simultaneously during a computer run. The irradiation geometries were AP (anterior-posterior), PA (reverse of AP), Lat (lateral from both sides), Rot (rotational) and Iso (isotropic).

The anthropomorphic phantom in which $H_p(10)$ was calculated, is the male version of the MIRD phantom, ADAM[5]. In the MIRD model the organs or tissues are described as combinations of simple geometric bodies in the form of mathematical inequalities. Organ or tissue doses are usually obtained from Monte Carlo (MC) estimates in those geometrical regions of the phantom. MC programs built in combinatorial geometry, allow input of these sets of inequalities in a versatile manner. The implementation of supplementary regions in the ADAM phantom, in which to estimate $H_p(10)$, is somewhat arbitrary and not possible in a straightforward way. Therefore their construction is described in detail for those readers who want to verify the results:

Twenty volumes — arranged in five vertical columns and four horizontal rows on the anterior trunk — lie between the surfaces of elliptical cylinders with half axes (a,b) = (18.9, 8.9) cm and (19.1, 9.1) cm with a thickness of 2 mm and a depth of about 1 cm in the trunk of ADAM. Additionally, vertical planes with their normals directed laterally and horizontal planes bound those volumes.

One column lies in the middle of the anterior trunk, two are shifted sideways by ±8 cm and another two by ±15 cm, to the left and the right. Volumes symmetric about the vertical central line are treated as one single region for the estimation of DE, just like symmetric organs, resulting in 12 regions instead of 20 volumes. The centroids of the volumes are at 17, 40, 51 and 62 cm from the trunk bottom. The dimensions of the regions amounted to (8×10) cm^2 each, in plan view. The regions, apart from the lowest and outermost ones, contain bone material from the rib cage; whereas it was considered for the radiation transport calculations, dose deposition was only recorded in the soft tissue portion of the volumes, which was less than half of the respective volumes. The main reason for the exclusion was, that ICRU soft tissue ($H_p(d)$ is the DE in that material) is similar in elemental composition and density to that of soft tissue, in which the H_p volumes were lying, but not to bone. In particular, the different densities of ICRU soft tissue (1 g.cm^{-3}) and of bone (1.486 g.cm^{-3}) would have introduced an artificial discontinuity in the H_p regions. The volume contents which are needed when normalising the MC results, were determined separately with an uncertainty less than 0.5% by means of test rays penetrating the volumes. In the figures, except

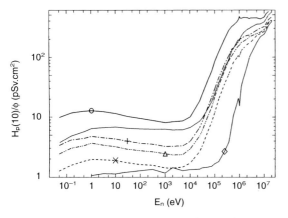

Figure 1. The personal dose equivalent (DE), $H_p(10)$, in the trunk of the male MIRD phantom for five irradiation geometries (○) AP, (◇) PA, (×) Lat, (+) Rot, (△) Iso. The dotted curve not marked by a symbol represents the effective DE, H_E, at AP in the same phantom.

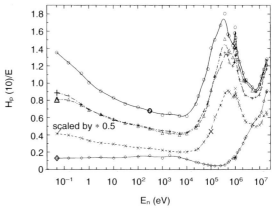

Figure 3. Ratio of the personal DE, $H_p(10)$, to the effective dose E (ICRP 60), both in the phantom 'ADAM' for five neutron incidences as indicated. Symbols as Figure 1.

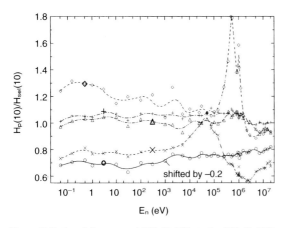

Figure 2. Ratio of the personal DE, $H_p(10)$, to the DE, $H_{sl}(10)$, in the ICRU slab under five irradiation geometries, identical for the MIRD phantom and the ICRU slab, except for Lat which was under α = 75° instead of 90° to the normal of the slab face over the site specified for $H_{sl}(10)$. Symbols as Figure 1.

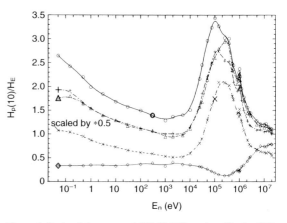

Figure 4. Ratio of the personal DE, $H_p(10)$, to the effective DE, H_E, (ICRU Report 51), both in the phantom 'ADAM' for five neutron incidences as indicated. Symbols as Figure 1.

for the last one, $H_p(10)$ was taken as the average over the twelve frontal H_p regions.

RESULTS AND DISCUSSION

The symbols in Figures 2–7 represent the values calculated. In some figures (2,5,6,7) the results scatter due partially to high statistical uncertainty, so that the curves obtained by a spline fitting procedure do not run through them but serve as a guide to the eye. The shape of the $H_p(10)$ curves in Figure 1 is roughly similar to that of other conversion functions for neutrons; the PA curve is rather low, as expected. The curve for H_E at AP incidence is also displayed showing that $H_p(10)$ is a conservative estimate for H_E at AP geometry. The ratio of $H_p(10)$ to $H_{sl}(10)$ in the ICRU slab (Figure 2) is very near unity for rotational (Rot) and isotropic (Iso) incidences, but deviates more for the unilateral irradiations AP, PA and Lat. The peak for PA around 500 keV neutron energy corresponds to clear deviations between the results for the central column of H_p volumes and the ones shifted sideways at PA exposure, not shown in the figure. This apparently is caused by the gradient of the depth DE curves at the relevant depths.

Figure 3 shows that the ratio $H_p(10)/E$ ranges between 0.4 and 1.8 except for the PA incidence, which is, anyway, an exceptional situation for H_p. In general, $H_p(10)$ underestimates the effective dose E below 50 keV neutron energy. Whether this involves $H_p(10)$ not being a suitable limiting quantity, can only be answered correctly for the spectrum of the working place in question, i.e. instead of the above ratio the respective conversion functions must be folded with that spectrum in numerator and denominator. Then it can be expected that in most working fields a ratio close to unity would result. In Figure 4 the ratio $H_p(10)/H_E$ exhibits a spread between 1 and 3.4, again except at PA. Consequently, $H_p(10)$ could replace the effective DE H_E as a limiting quantity. But there exists no advantage in calculating $H_p(10)$ instead of H_E or E, because the degree of complexity of the computations is the same for each quantity.

The ratio of $H_p(10)$ to the DE in the genitalia region of the MIRD phantom, which contains the testes (Figure 5), is not far from unity except for lateral incidence (Lat), where the shielding of the testes is higher. The corresponding ratio for the thymus (Figure 6) is rather dissimilar and higher at intermediate energies. The organ of choice to make estimates of $H_p(10)$ may be the testes. The dependence of $H_p(10)$ on the site within the

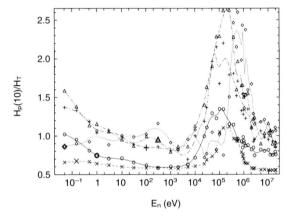

Figure 6. Ratio of the personal DE, $H_p(10)$, to the organ DE H_T in the thymus, both calculated in 'ADAM', for five neutron incidences. The thymus lies at a depth in ADAM of about 23 mm. The curve for Lat was scaled with a factor of 0.05 and 0.5 added. (○) AP, scaled by × 0.8; (◇) PA, (×) Lat, (+) Rot, (△) Iso.

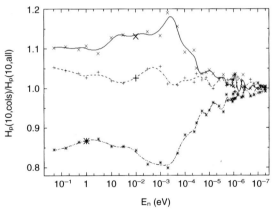

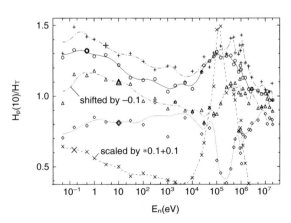

Figure 5. Ratio of the personal DE, $H_p(10)$, to the organ DE H_T in the testes, both calculated in 'ADAM', for five neutron incidences. The same symbols as in Figure 4 are assigned. The depth of the testes varies from about 5 to 35 mm.

Figure 7. Ratio of the personal DE, $H_p(10)$, averaged over four regions arranged vertically in columns ($H_p(10,cols.)$), one column in the middle of the trunk (×), two shifted sideways by 8 cm (+), the last two by 15 cm (∗) to the mean value of $H_p(10)$ over all twelve regions for AP incidence ($H_p(10,all)$).

trunk is noticeable only for lateral positions (columns) and not for the vertical ones. For AP incidence (Figure 7), $H_p(10)$ at lateral positions deviates from the average by $\pm 15\%$ up to $E_n = 50$ keV.

CONCLUSIONS

Values of the personal dose equivalent, $H_p(10)$, were calculated for neutrons in anthropomorphic phantoms. These are similar in neutron energy dependence to other conversion functions in phantoms. As $H_p(10)$ is assigned to an individual, and moreover the site for $H_p(10)$ is not uniquely defined, it has multiple values since there is some degree of arbitrariness in choosing the location in the anterior trunk.

The computation of $H_p(10)$ is of comparable complexity to that of E or H_E. The calculation of $H_p(10)$ is not requisite, however, for personal dosimetry because the DE $H_{sl}(10)$ in the ICRU slab is a suitable surrogate for $H_p(10)$ for radiation from the frontal half space or for multidirectional fields such as rotational and isotropic incidences. Though $H_p(10)$ underestimates the effective dose E below 50 keV neutron energy, in realistic spectra of working places, it might well be used instead of E or H_E as the limiting quantity, provided that the exposure is not predominantly from the back. That remains to be investigated in more detail. At least for one organ, the testes, there is a fairly close correlation which permits estimation of $H_p(10)$ to about 30%. The ratio $H_p(10)/H_T$ for the thymus shows much higher deviations.

REFERENCES

1. ICRU. *Quantities and Units in Radiation Protection Dosimetry*. ICRU Report 51 (Bethesda, MD: ICRU Publications) (1993).
2. ICRU. *Determination of Dose Equivalent Resulting from External Radiation Sources*. ICRU Report 39 (Bethesda, MD: ICRU Publications) (1985).
3. ICRP. *Recommendations of the International Commission on Radiological Protection*. ICRP Publication 26 (Oxford: Pergamon Press) (1977).
4. ICRP. *1990 Recommendations of the International Commission on Radiological Protection*. ICRP Publication 60 (Oxford: Pergamon Press) (1991).
5. Hollnagel, R. A. *Conversion Functions of Effective Dose E and Effective Dose Equivalent H_E for Neutrons with Energies from Thermal to 20 MeV*. Radiat. Prot. Dosim. **54**, 209–212 (1994).
6. ICRU. *Measurement of Dose Equivalents from External Photon and Electron Radiations*. ICRU Report 47 (Bethesda, MD: ICRU Publications) (1992).
7. Hollnagel, R. A. *Conversion Functions of the Dose Equivalent $H_{sl}(10)$ in the ICRU Slab used for the Calibration of Personal Neutron Dosemeters*. Radiat. Prot. Dosim. **54**, 227–230 (1994).
8. ICRU. *Stopping Powers and Ranges for Protons and Alpha Particles*. ICRU Report 49 (Bethesda, MD: ICRU Publications) (1993).

ORGAN DOSES AND DOSE EQUIVALENTS FOR NEUTRONS ABOVE 20 MeV

V. Mares, G. Leuthold and H. Schraube
GSF-Forschungszentrum für Umwelt und Gesundheit Institut für Strahlenschutz, Neuherberg
Postfach 1129, D-85758 Oberschleißheim, Germany

Abstract — MCNP Monte Carlo calculations with an extended cross section data set have been performed for an anthropomorphic MIRD (ADAM) phantom in the neutron energy range between 20 and 100 MeV. The irradiation conditions vary from 0° (frontal incidence) up to 75° to the left side of the MIRD phantom. From the fluence calculations, organ doses and dose equivalents are derived using the kerma approximation. The quality factor as function of neutron energy for the calculation of effective dose equivalent, H_E, is derived from initial charged particle spectra produced in an ICRU tissue element by primary neutrons. Above 20 MeV an approximation was applied. Resulting organ doses and dose equivalents are compared with a data set published by Nabelssi and Hertel (1993).

INTRODUCTION

The dosimetry of neutrons and other charged particles at aviation altitudes became of interest since ICRP 60[1] recommended that air crews be considered as occupationally exposed people. The spectral distribution of neutrons at these altitudes covers a wide range between thermal energy and some 100 MeV. Below 20 MeV organ doses, dose equivalents and operational quantities are sufficiently known from Monte Carlo calculations in anthropomorphic phantoms and in ICRU recommended phantoms, respectively, whereas above 20 MeV only a few data are available, especially for organ doses and dose equivalents. Therefore, MCNP Monte Carlo code calculations have been performed for an anthropomorphic MIRD (ADAM) phantom.

MCNP CALCULATIONS

Geometrical and physical conditions

Calculations of neutron and neutron-induced photon fluences in organs were performed using the MCNP-4A[2] Monte Carlo code developed in Los Alamos. The parallel monoenergetic beam of neutrons uniformly irradiated the phantoms. Neutron histories were started from the surface of a disc source centred on and perpendicular to the source–phantom axis. The angles of incidence varied from 0° to 75° to the left side of the phantom in 15° steps. The space between source and phantom was assumed to be vacuum. The neutron and photon fluences were averaged over every organ in the phantoms and normalised to be per one starting particle per source area.

Anthropomorphic phantom

The male anthropomorphic phantom used in this work represents a modified version of the mathematical hermaphrodite MIRD-5 phantom as developed by Kramer et al[3]. The body and internal organs of ADAM are defined by equations of surfaces of cylinders, cones, ellipsoids, hyperboloids and tori were implemented in the MCNP geometrical input. The oesophagus was added to the phantom using the model reported by Zankl et al[4]. Internal organs were considered to be homogeneous in composition and density. Different densities and compositions were used for the lungs, skeleton, skin and the bulk of the body. The composition description of these four tissues were limited to the 14 elements H, C, N, O, Na, Mg, P, S, Cl, K, Ca, Fe, Zr and Pb. Other elements accounting for less than 5×10^{-3} by weight were neglected in this work. The asymmetric volumes and areas of the organs which cannot be calculated automatically by MCNP itself were established stochastically using a ray tracing procedure.

Cross section data

The MCNP calculation was extended up to 100 MeV using the neutron cross sections from the library LA100[5]. This library contains neutron cross section data for nine elements ^{1}H, ^{9}Be, ^{12}C, ^{16}O, ^{27}Al, ^{28}Si, Fe, W, ^{238}U evaluated up to 100 MeV by matching calculations with experimental data using the PSR-125/GNASH statistical/pre-equilibrium/fission theory code. For the remaining elements of the phantom tissue composition the code was allowed to extrapolate the cross section data of the ENDF/B-VI library[6] between 20 and 100 MeV with constant value. To model the thermal neutron scattering by molecules at room temperatures, the $S(\alpha,\beta)$ cross section tables for light water were used for all four tissues. The photon interaction tables used in the Monte Carlo calculations were based on evaluated data from Hubbell et al[7].

BASIC DATA

For the conversion of neutron fluence to absorbed dose the kerma approximation was applied. The kerma approximation is used even for high neutron energies, i.e. it is assumed that the range of the secondary charged particles is small compared with the size of the organs. This seems justified for radiation protection purposes because it simulates a situation where secondary charged particle equilibrium is established by the surrounding material, an assumption which is closer to actual practice than assuming a vacuum around the phantom.

The elemental kerma values were taken from ICRU 26[8]. Above 10 MeV the recommended kerma values of White et al were used[9]. For secondary photons the kerma factors were derived from elemental energy absorption coefficients given by Hubbell[10]. Above 20 MeV data from Storm and Israel were used[11].

For the calculation of organ dose equivalent $H_{T,q}$ (the index q indicates that the quality factor is applied) a quality factor q_n for neutrons was applied which takes into account the revised Q(L) relationship of ICRP 60[1] and the new stopping power data for protons and alpha particles in water of ICRU 49[12].

$$H_{T,q}(E) = \int q_n(E') k_n(E') \Phi_{T,n}(E',E) dE' + \int k_{ph}(E'') \Phi_{T,ph}(E'',E) dE''$$

where:

$q_n(E')$ is the quality factor as function of neutron energy E',

$k_n(E')$ is the fluence to kerma conversion factors as function of neutron energy E' depending on the elemental composition of the tissue in organ T,

$k_{ph}(E'')$ is the fluence to kerma conversion factor as function of photon energy E'',

$\Phi_{T,n}(E',E)$ and $\Phi_{T,ph}(E'',E)$ are the neutron and photon fluences in the organ.

For the calculation of the equivalent organ dose H_T

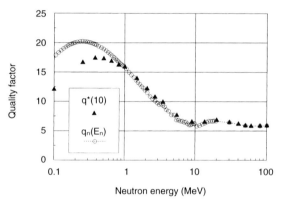

Figure 1. Neutron quality factor and q*(10) in the ICRU sphere as function of neutron energy. Above 20 MeV data for q*(10) of Sannikov et al[13] are used as an approximation for the quality factor q_n.

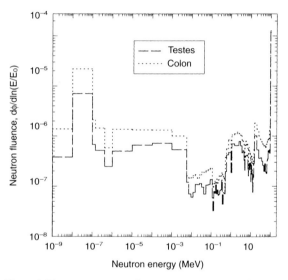

Figure 2. The neutron spectra in testes and colon for 100 MeV neutrons for front incidence on the anthropomorphic MIRD (ADAM) phantom, using the MCNP-4A Monte Carlo code.

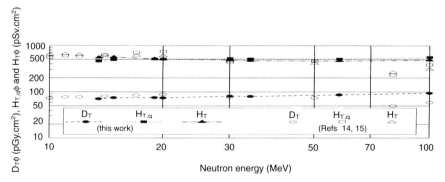

Figure 3. Organ absorbed dose D_T, organ dose equivalent $H_{T,q}$ and equivalent organ dose H_T in the high energy range for testes. Closed symbols refer to this work, open symbols to data from the literature[14,15].

392

(according to ICRP 60) the smooth radiation weighting factor function w_R was applied:

$$H_T(E) = w_R(E) \left[\int k_n(E') \Phi_{T,n}(E',E)dE' + \int k_{ph}(E'') \Phi_{T,ph}(E'',E)dE'' \right]$$

Effective dose E and effective dose equivalent H_E were calculated by use of the new organ weighting factors w_T of ICRP 60[1] where

$$E = \sum_T w_T H_T$$

and

$$H_E = \sum_T w_T H_{T,q}$$

Figure 1 shows the quality factor q_n and the ratio $q^*(10) = H^*(10)/D^*(10)$ at 10 mm in the ICRU sphere as function of neutron energy. As there is a good agreement between both quantities in the MeV region up to 20 MeV, the assumption was made that the quality factor up to 100 MeV can be approximated by values of $q^*(10)$. Data for $q^*(10)$ above 20 MeV were taken from Sannikov et al[13].

RESULTS

In Figure 2, as an example, the neutron fluences in testes and colon are shown for 100 MeV neutrons and angle of incidence equal to 0°, i.e. A-P irradiation. It can be seen that neutron spectra in testes and colon differ in absolute scale and also slightly in the shape. This effect can be explained by the different position of organs in the body of ADAM. Testes belong to the surface organs while the colon to deeper lying organs.

Figure 3 and 4 show the absorbed dose D_T, the organ

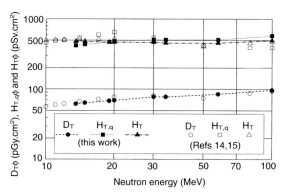

Figure 4. Organ absorbed dose D_T, organ dose equivalent $H_{T,q}$ and equivalent organ dose H_T in the high energy range for colon. Closed symbols refer to this work, open symbols to data from the literature[14,15].

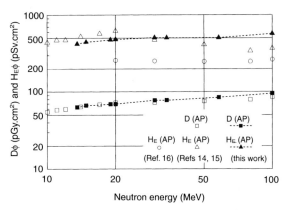

Figure 6. Comparison of absorbed dose D and effective dose equivalent H_E for the ADAM phantom in the high energy range. Closed symbols refer to this work, open symbols to data from the literature[14–16].

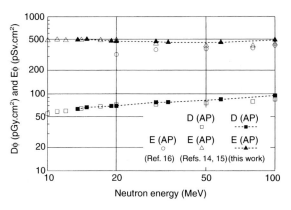

Figure 5. Comparison of absorbed dose D and effective dose E for the ADAM phantom in the high energy range. Closed symbols refer to this work, open symbols to data from the literature[14–16].

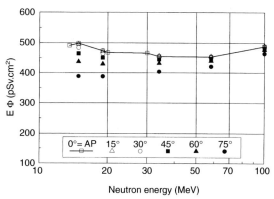

Figure 7. Effective dose E as function of neutron energy in the energy range up to 100 MeV for different angles of incidence on the ADAM phantom.

dose equivalent H_{T_q} and the equivalent organ dose H_T for the testes and for the colon for AP irradiation in comparison with data of Nabelssi and Hertel[14,15]. Differences can be seen for small organs (e.g. testes) near the front above 50 MeV whereas deeper lying organs (e.g. colon) are in good agreement with the equivalent organ dose H_T.

Figure 5 and 6 show results for the total absorbed dose D, the effective dose E and the effective dose equivalent H_E for AP irradiation, respectively. For the absorbed dose and the effective dose there is a good agreement with the data of Nabelssi and Hertel[14,15]. Differences in the organ dose equivalents $H_{T,q}$ and the effective dose equivalent H_E are due to different procedures[14-16] to take into account the neutron quality factor.

In Figure 7, the effective dose E is depicted for the six selected angles of incidence to the phantom. It is recognised that with increasing energy the kerma distribution inside the phantom becomes more uniform; hence the influence of the angle of incidence on the effective dose decreases, because the kerma approximation is used when determining the dose.

CONCLUSIONS

For deeper lying organs and for the total ADAM phantom the equivalent organ dose and the effective dose are in good agreement with other calculations up to 100 MeV neutron energy. It can be stated that the kerma approximation is a conservative estimation of the total absorbed dose and the effective dose and exceeds the real values obtained from calculations with realistic ranges of secondary charged particles to some extent.

For small organs close to the surface differences are more evident, although some observed differences from the published data are attributed to statistical uncertainties rather than to systematic trends.

ACKNOWLEDGEMENTS

The work was supported by the Commission of the European Union DGXII under contracts FI3P-CT92-0026 and 0002.

REFERENCES

1. ICRP. *1990 Recommendations of the International Commission on Radiological Protection*. Publication 60 (Oxford: Pergamon Press) (1991).
2. Briesmeister, J. F. (Ed) *MCNP — A General Monte Carlo N-Particle Transport Code, Version 4A*. LA-12625-M, (Los Alamos National Laboratory) (1993).
3. Kramer, R., Zankl, M., Williams, G. and Drexler, G. *The Calculation of Dose from External Photon Exposures Using Reference Human Phantoms and Monte Carlo Methods, Part I: The Male (Adam) and Female (Eva) Adult Mathematical Phantoms*. GSF-Bericht S-885 (1982).
4. Zankl, M., Petoussi, N. and Drexler, G. *Effective Dose and Effective Dose Equivalent — The Impact of the New ICRP Definition for External Photon Irradiation*. Health Phys. **62**(5), 395–399 (1992).
5. Young, P. G. et al. *Transport Data Libraries for Incident Proton and Neutron Energies to 100 MeV*. LANL Report LA-11753-MS, (Los Alamos) (1990).
6. Hendricks, J. S., Frankle, S. C. and Court, J. D. *ENDF/B-VI Data for MCNP*. LANL Report LA-12891, (Los Alamos) (1994).
7. Hubbell, J. H., Veigele, W. J., Briggs, E. A., Brown, R. T., Cramer, D. T. and Howerton, R. J. *Atomic Form Factors, Incoherent Scattering Functions, and Photon Scattering Cross Sections*. J. Phys. Chem. Ref. Data **4**, 471 (1975).
8. ICRU. *Neutron Dosimetry in Biology and Medicine*. Report 26 (Bethesda MD: ICRU Publications) (1977).
9. White, R. M., Broerse, J. J., DeLuca, P. M. Jr, Dietze, G., Haight, R. C., Kawashima, K., Menzel, H. G., Olsson, N. and Wambersie, A. *Status of Nuclear Data for Use in Neutron Therapy*. Radiat. Prot. Dosim. **44**(1/4), 11–20 (1992).
10. Hubbell, J. H. *Photon Mass Attenuation and Energy–absorption Coefficients from 1 keV to 20 MeV*. Int. J. Appl. Radiat. Isot. **33**, 1269–1290 (1982).
11. Storm, E. and Israel, H. I. *Photon Cross Sections from 1 keV to 100 MeV for Elements Z = 1 to Z = 100*. Nucl. Data Tables A7, 565–681 (1970).
12. ICRU. *Stopping Powers and Ranges for Protons and Alpha Particles*. Report 49 (Bethesda, MD: ICRU Publications) (1993).
13. Sannikov, A. V. and Savitskaya, E. N. *Ambient Dose Equivalent Conversion Factors for High Energy Neutrons Based on the ICRP 60 Recommendations*. Radiat. Prot. Dosim. **70**(1–4), 383–386 (This issue) (1997).
14. Nabelssi, B. K. and Hertel, N. E. *Ambient Dose Equivalents, Effective Dose Equivalents, and Effective Doses for Neutrons from 10 to 20 MeV*. Radiat. Prot. Dosim. **48**(2), 153–168 (1993).
15. Nabelssi, B. K. and Hertel, N. E. *Effective Dose Equivalents and Effective Doses for Neutrons from 30 to 180 MeV*. Radiat. Prot. Dosim. **48**(3), 227–243 (1993).
16. Iwai, S., Uehara, T., Sato, O., Yoshizawa, N., Furihata, S., Takagi, S., Tanaka, S. and Sakamoto, Y. *Evaluation of Fluence to Dose Equivalent Conversion Coefficients for High Energy Neutrons — Calculation of Effective Dose Equivalent and Effective Dose*. In: Proc. 2nd Specialists' Meeting on Shielding Aspects of Accelerators, Targets and Irradiation Facilities (SATIF2), CERN, Geneva 12–13 October 1995.

DOSIMETRY FOR OCCUPATIONAL EXPOSURE TO COSMIC RADIATION

D. T. Bartlett[1], I. R. McAulay[2], U. J. Schrewe[3], K. Schnuer[4], H.-G. Menzel[5], J.-F. Bottollier-Depois[6], G. Dietze[3], K. Gmür[7], R. E. Grillmaeir[8], W. Heinrich[9], T. Lim[8], L. Lindborg[10], G. Reitz[11], H. Schraube[12], F. Spurný[13] and L. Tommasino[14]
[1]National Radiological Protection Board, Chilton, Oxon, OX11 0RQ, UK
[2]Trinity College, Dublin 2, Ireland
[3]Physikalisch-Technische Bundesanstalt, PO Box 3345, D-38023 Braunschweig, Germany
[4]European Commission, DGXI, Bât. Wagner, L-2920 Luxembourg
[5]European Commission, DG XII-F.6, T61-1/31, Rue de la Loi 200, B-1049 Brussels, Belgium
[6]Institut de Protection et de Sûreté Nucléaire, F-92265 Fontenay-aux-Roses Cedex, France
[7]Paul Scherrer Institut, CH-5232 Villigen PSI, Switzerland
[8]Universität des Saarlandes, D-6602 Saarbrucken-Dudweiler, Germany
[9]Universität Gesamthochschule, D-57068 Siegen, Germany
[10]Statens Strålskyddsinstitut, S-17116 Stockholm, Sweden
[11]Deutsche Forschungsanstalt für Luft-und Raumfahrt e.V., Linder Höhe, D-51147 Köln, Germany
[12]GSF-Forschungszentrum Neuherberg, D-85375 Oberschleissheim, Germany
[13]Nuclear Physics Institute, Na Truhlářce 39, CZ 18086 Prague 8, Czech Republic
[14]Agenzia Nazionale per la Protezione dell'Ambiente, Via V. Brancati, 48, I-00144 Roma, Italy

INVITED PAPER

Abstract — In the course of their work, aircraft crew and frequent flyers are exposed to elevated levels of cosmic radiation of galactic and solar origin and secondary radiation produced in the atmosphere, aircraft structure, etc. This has been recognised for some time and estimates of the exposure of aircraft crew have been made previously and included in, for example, UNSCEAR (United Nations Scientific Committee on the Effects of Atomic Radiation) publications. The recent increased interest has been brought about by several factors — the consideration that the relative biological effectiveness of the neutron component was being underestimated; the trend towards higher cruising altitudes for subsonic commercial aircraft and business jet aircraft; and, most importantly, the recommendations of the International Commission on Radiological Protection (ICRP) in Publication 60, and the revision of the Euratom Basic Safety Standards Directive (BSS). In 1992, the European Dosimetry Group (EURADOS) established a Working Group to consider the exposure to cosmic radiation of aircraft crew, and the scientific and technical problems associated with radiation protection dosimetry for this occupational group. The Working Group was composed of fifteen scientists (plus a corresponding member) involved in this field of study and with knowledge of radiation measurement at aviation altitudes. This paper is based on the findings of this Working Group. Where arrangements are made to take account of the exposure of aircraft crew to cosmic radiation, dose estimation procedures will not be necessary for persons for whom total annual doses are not liable to exceed 1 mSv, and therefore, in general, for crew on aircraft not routinely flying above 8 km. Where estimates of effective dose and, in the case of female staff who are pregnant, equivalent dose to the embryo or fetus, are required (for regulatory or other purposes), it was concluded that the preferred procedure was to determine route doses and fold these with data on staff rostering.

INTRODUCTION

In the course of their work, aircraft crew and frequent flyers are exposed to elevated levels of cosmic radiation of galactic and solar origin and secondary radiation produced in the atmosphere, aircraft structure, etc. This has been recognised for some time and estimates of the exposure of aircraft crew have been made previously and included in, for example, UNSCEAR (United Nations Scientific Committee on the Effects of Atomic Radiation) publications[1].

In 1992, the European Dosimetry Group (EURADOS) decided to establish a working group to be concerned with the exposure to cosmic radiation of aircraft crew, and with the scientific and technical problems associated with radiation protection dosimetry for this occupational group. The Working Group was composed of fifteen scientists (plus a corresponding member) nominated by virtue of their involvement in this field of study and their knowledge of radiation measurement at the altitudes associated with aviation. A report has been prepared and is to be published.

The primary reasons for the establishment of the Working Group were the recommendation by the International Commission on Radiological Protection in ICRP Publication 60[2] that exposure of aircraft crew to cosmic radiation in jet aircraft should explicitly be considered as occupational exposure and because of the overall reduction in dose limits recommended in the same publication, and the revision of the European

Basic Safety Standards Directive (BSS)[3]. The assessment of exposure to cosmic radiation might be required for all aircraft crew who were liable to receive effective doses in excess of 1 mSv per year. There were also the considerations that the relative biological effectiveness of the neutron component was being underestimated and of the capability of attaining higher cruising altitudes for subsonic commercial aircraft and business jet aircraft.

Three sub-groups were set up to prepare draft reports on various aspects. Sub-Group 1 was to describe the radiation environment at altitudes relevant to civil aviation, and to assess experimental studies carried out to characterise, and computational codes used to calculate the cosmic radiation fields. Sub-Group 2 was to survey measurement procedures and quantities and to consider appropriate instrumentation. Sub-Group 3 was to report on techniques and problems associated with calibration, reference instruments and reference fields.

This paper presents a summary of the findings of the Working Group, with particular emphasis on dosimetry.

THE RADIATION ENVIRONMENT AT AIRCRAFT ALTITUDES: MEASUREMENTS AND CALCULATIONS

A full description of the radiation environment at radiation altitudes may be found in the paper of Reitz[4] given at the Luxembourg workshop[5] organised by the Commission of the European Communities. A brief description only is given here.

The radiation environment in the atmosphere is a result of the interaction of the primary cosmic radiation with the magnetosphere and the atmosphere of the earth. The intensities and the components of the radiation field change with magnetic latitude and altitude and with the solar activity. With the exception of very rare solar flares, the composition of the radiation field inside the atmosphere is determined by the galactic cosmic radiation, but modulated by the effect of solar activity. The galactic cosmic radiation arises from sources outside the solar system but within our galaxy, and is incident continuously on the Earth. Cosmic radiation is produced by stellar flares, stellar coronal mass ejection, super-nova explosions, pulsar acceleration and explosions of galactic nuclei. It is contained by the interstellar magnetic field ($B \sim 10^{-9}$ T) and its mean residence time is about 200 million years and average fluence rate 10 $cm^{-2}.s^{-1}$. It consists of about 85% protons, 12% helium ions, 1% heavier ions (all known chemical elements are present), and 2% electrons. The radiation field in free space is isotropic. For high energies the number of particles, N, as a function of their energy, E, are represented by a function ($N(E) \sim E^{-y}$ with $y \approx 2.5$). Particles with energies greater than 10^{20} eV have been detected. There is a decrease in intensity for low energies owing to the deflection within the interplanetary space by the magnetic fields of the solar wind. The intensity of the solar wind, which consists mainly of ionised hydrogen and an equal number of electrons, varies during the 11-year cycle of solar activity. The low energy solar wind particles do not contribute to the radiation environment inside the atmosphere. During high solar acitivity, the solar wind is stronger and so therefore are the magnetic fields associated with it, resulting in a decrease of the cosmic ray intensity. The converse is true when the solar activity is low.

A cosmic ray particle has to penetrate the Earth's magnetic field in order to enter the atmosphere. This means that at the equator only particles of high energy can reach the atmosphere. As latitudes increase towards the poles particles of lower energies can do so, and at the poles, particles of all energies can enter in the direction of the magnetic field. This effect is reduced at civil aviation altitudes by absorption of low energy particles in the higher atmosphere, with the result that at geomagnetic latitudes greater than about 50°, no further increase occurs and a plateau in the particle flux is reached.

A particle penetrating into the atmosphere loses most of its energy by ionising atoms and molecules. At high energies nuclear interactions also take place. Through fragmentation, the number of particles with high atomic number decreases strongly through the atmosphere and is therefore of minor importance at civil aviation altitudes. Protons are mainly responsible for the production of secondaries at these altitudes and large numbers of particles may be created in successive interactions by the protons and by subsequent interactions of the secondaries. The principal reaction that takes place is the production of neutrons, secondary protons, and the pion triplet by an interaction of a primary proton, or a neutron, with air. The charged pions decay to muons and neutrinos. Most of the muons penetrate the atmosphere down to sea level, although some decay to electrons and neutrions. The neutral pions decay into gamma rays. For a given magnetic latitude and phase of solar cycle the particle fluence rate is fairly constant from 150 km to 50 km altitude being about 1 (proton) cm^{-2}. At 50 km it increases owing to the production of secondary particles and reaches a maximum at about 20 km, known as the Pfotzer maximum (approximately 10 $cm^{-2}.s^{-1}$) comprised mainly of electrons, neutrons and protons, and decreases after that due to absorption. In most of the interactions the incoming particle ejects nucleons out of the target nucleus and continues its path still as a highly energetic particle. The highly excited target nucleus cools down by releasing some additional nucleons and ends up as either a stable or radioactive nuclide. In such nuclear disintegrations, there are events called 'stars', described in this way because of their appearance in nuclear emulsions. Stars consist mainly of protons, neutrons and alpha particles with mean energies of less than 10 MeV.

Neutrons are produced by two processes in two separate energy ranges. Low energy neutrons are evaporated from highly excited nuclei similar to the neutron emission from fission fragments. The spectral neutron fluence has a Maxwellian (Watt) distribution with a mean energy of about 2 MeV and a maximum of about 1 MeV. Their angular distribution is fairly isotropic. High energy neutrons originate as knock-on particles in peripheral collisions or in charge exchange reactions of high energy protons. They show a non-isotropic angular distribution, since they are emitted preferentially in the forward direction. Neutrons do not enter the atmosphere from outside but leak from it to the exosphere, since unlike charged particles they are not affected by the geomagnetic field and can thus easily escape from the atmosphere. Most of such neutrons decay into protons and electrons which will be trapped in the so-called radiation belts. In the upper part of the atmosphere the neutron fluence rate in the downward direction is higher than that in the upward direction and has an angular distribution with a maximum in the vertically downward direction. At minimum solar activity more primaries enter the atmosphere thereby producing more neutrons. In polar regions the neutron production rate is therefore enhanced considerably compared to solar maximum conditions. The variation due to the solar cycle, of the neutron intensity at the equator is only about 5%, since low energy primaries are excluded from this high geomagnetic cut-off area and high energy primaries undergo only a slight solar modulation. Large fluxes of neutrons are produced in the atmosphere when a strong solar flare hits the earth. The neutron fluence rate depends in a complex way on geomagnetic latitude and altitude, decreasing with an effective attenuation length from about 220 to 160 g.cm^{-2} going from equatorial latitudes to about 60° geomagnetic latitude. The mean energy of the primaries which induce cascades in low geomagnetic latitudes is higher than of those in polar regions. More neutrons per interaction are produced and the secondaries have a higher mean energy and are able to penetrate deeper into the atmosphere. For these reasons the maximum of the neutron intensity is located at lower altitudes in the equatorial than in the polar regions. Near the pole the maximum neutron flux appears at about 18 km, at the equator at about 15 km. The spectra of neutrons have been measured in several experiments. There remains some uncertainty, and further measurements and calculations are being carried out.

SOLAR FLARES

A phase of high solar activity corresponds to an occurrence of discrete, local perturbations on the solar surface with sunspots being the most remarkable manifestations. The solar cosmic radiation of concern at altitudes relevant to civil aviation is that produced by solar flares — sudden sporadic eruptions of the chromosphere of the sun. At maximum solar activity up to 10 significant flares per year occur. During the years of minimum solar activity, on average only one event can be observed. The largest eruptions with more than 10^{10} particles.cm^{-2} take place at the end of the period of maximum solar activity. The rise in dose rates associated with a flare is quite rapid, usually of the order of minutes and the duration of a flare may be of hours or longer. Only a small fraction (about 3%) of flares produce significant fluences of high energy particles and only a small fraction of these cause an increased intensity of cosmic radiation[6,7].

A few solar flares give rise to high instantaneous dose equivalent rates — of the order of mSv.h^{-1}, but are of short duration leading to a maximum integrated dose equivalent of a few mSv at aviation altitudes. At altitudes greater than 15 km the chance of higher dose equivalent rates from solar flares is greater than at subsonic levels resulting from the lower energies of the solar radiation protons and therefore the shorter penetration of both primary and secondary particles (note also magnetic latitude effects). During the February 1956 flare the upper limit of the calculated dose equivalent rate was about 50 mSv.h^{-1} at 20 km altitude, and over 10 mSv.h^{-1} at 10 km[8]. (There was no direct measurement in the atmosphere during the maximum of the February 1956 event.) The dose rates for flares observed since 1956, including the large flares of August 1972 and August 1989, have been much smaller. The Space Environment Service Centre (SESC) calculated a dose equivalent rate of 1.1 mSv.h^{-1} at an altitude of 19.8 km over the North Pole for the September 1989 event[9].

Major solar flares are rare. O'Brien has recently calculated[10] the additional contribution to dose equivalent for regular polar flights over the period February 1984 to July 1992, during which 14 periods of energetic solar activity were observed. At 12 km the additional contribution was calculated to be 3%, and at 18 km 7%. A long-term forecast of solar flares is not possible, but there have been some improvements in short-term forecasting in recent years.

MEASUREMENTS

Many different types of active and passive measuring devices have been used to determine the characteristics of radiation fields on board aircraft. Tissue-equivalent proportional counters or ionisation chambers have been used to measure dose and dose equivalent for all components of the field. High pressure (Ar) ionisation chambers, GM-counters, scintillators, silicon diodes, thermoluminescence detectors, photographic films, rem meters, Bonner spheres, organic scintillators, proton recoil counters, activation detectors, nuclear emulsions, fission foils, superheated drop detectors and solid state nuclear track detectors have been used to estimate the directly ionising, photon and neutron components. Relatively few measurements, with nuclear emulsions and nuclear track detectors, have been carried out to characterise the high Z component of the radiation field.

Many data were obtained between 1965 and 1977 by American and Soviet scientists. They were reviewed and presented during the Luxembourg workshop[5] and together with some representative results obtained by other authors in the period 1990–1993 are presented as a function of flight altitude and latitude in Figure 1[11]. For simplicity only values of total dose equivalent are presented in the figure. Data for the same phase of the solar cycle agree well, within 20–30%. Generally, relative uncertainties given by different authors for total dose equivalent rate values are estimated to be about 20% (1σ). For the low LET component they are judged to be better (±10%) than for the high LET component (±30%). Because of the complexity of the radiation field on board aircraft, interpretation may also have varied from one author to another. The uncertainties given reflect only statistical, not systematic uncertainties, and should therefore be taken as lower limits.

The following conclusions can be drawn: location within an aircraft does not affect the total dose equivalent to more than 5–10%; measured dose equivalent rates depend on geographical latitude, increasing from the equator toward either pole up to a latitude of about 50°, remaining approximately constant at higher latitudes (the increase is greater for the high LET component (a factor of 3 to 5) than for the low LET component (a factor of 1.5 to 2.5)); total dose equivalent rates increase with flight altitude for all latitudes, being roughly twice as great at 12 km as at 9 km; the measured values of total dose equivalent rates correlate with the variation in cosmic radiation intensity due to the solar cycle of about 11 years, being about 30% higher in the period 1974–76 (solar minimum) than in 1991 (solar maximum); at latitudes of 50° and greater, and at normal flight altitudes, the relative contributions of the high and low LET components of dose equivalent are broadly similar; the average quality factor value is about 2.

Concorde operation has permitted the accumulation of many data at typical supersonic flight altitudes. The average total dose equivalent rate in the period 1976–83 was 11 μSv.h^{-1} for altitudes of about 18 km, and average values reported for 1988, 1989, 1990 were similar[12]. Values measured by Russian scientists in 1977 for Russian supersonic aircraft are in agreement[13]. Relative contributions of the neutron and low LET components are the same as for subsonic flight altitudes. The average quality factor for total dose equivalent is also comparable[14].

CALCULATIONS

In nuclear collisions of cosmic ray protons in the atmosphere, hadrons (protons, neutrons, pions and kaons), and leptons (muons and electrons), are generated. These secondary particles undergo further interactions or decay into other particles. Based on these interactions and decays, a hadron component, an electron–photon component and a muon component develop in the atmosphere. A program that allows the calculation of energy spectra and dose contributions by primary particles and secondaries produced in the atmosphere is the package LUIN developed by O'Brien more than twenty years ago[15]. This code calculates the one-dimensional hadronic and electromagnetic cascade and the muon component in the atmosphere starting with a cosmic ray primary proton spectrum at the top of the atmosphere. Solar modulation and geomagnetic shielding are considered in detail. Contributions by heavy nuclei to the atmospheric cascades are included by considering their nucleons as free particles. Fluxes of heavy nuclei in the atmosphere are not described by this code.

LUIN describes high energy nucleon–nucleus collisions by a phenomenological model which can be applied to analytical transport theory. Atmospheric cosmic ray fluxes are obtained as analytic solutions to an approximation of the Boltzmann equations. This model has been developed to calculate spectra above 100 MeV[16]. In order to account for neutron fluence rates at lower energies, the cosmic ray neutron spectrum reported by Hess[17] is used as an extrapolation. From the spectral fluence for the particles, values of dose equivalent are obtained by folding with conversion coefficients (depth dose data) for the maxima for bilateral irradiation of a semi-infinite 30 cm slab of tissue-equivalent material using ICRP Publication 15 Q(L). The results of calculations by O'Brien of the components as a function of altitude are shown in Figure 2.

LUIN has continued to be under development and several updated verions have been released. O'Brien provided calculations of some dose equivalent rates given in Table 1 using the 1994 version of LUIN. There is broad agreement with the experimental data shown in Figure 1. A program for use on personal computers, CARI-2, which is based on the LUIN package, has been released[18].

Computer codes describing the production of particle cascades in more detail have been developed at the large accelerator centres. These Monte Carlo programs are used to model interactions in detector set-ups and to perform shielding calculations for high energy particle beams. Two programs of this type have been developed

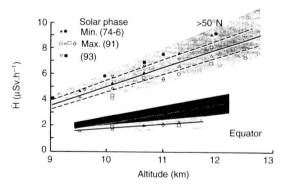

Figure 1. Representative values of measured dose equivalent rates.

at CERN and the University of Milan. The program GEANT[19] describes cascades. It contains a package called GHEISHA to calculate hadronic cascades. The second code FLUKA[20] has been developed to describe hadronic cascades including lower energies. These codes could be used to perform more detailed calculations of secondary particle spectra generated by the cosmic radiation in the atmosphere. This implies that the models mentioned above have to be extended to include cosmic ray spectra, solar modulation geomagnetic cut-off and exponential density distribution of the atmosphere. As a test, proton and neutron spectra behind a slab of air have been calculated using LUIN, GHEISHA and FLUKA for the same proton spectrum and different areal densities up to 500 g.cm^{-2}. The general observation is that fluence rates predicted by LUIN are smaller (tens of per cent) than those predicted by the two other models. In a recent study[21], the FLUKA code was employed in a validation experiment at a low altitude (3000 m, 700 g.cm^{-2}), on a mountain-top, near solar minimum. The evaluated neutron spectrum shows a distinct peak around 100 MeV due to intranuclear cascade processes, in agreement with earlier calculations[22]. These data may be substituted for the Hess spectrum in an extrapolation of LUIN calculations.

The energy spectra of cosmic ray nuclei in the atmosphere from incident heavy ions can be calculated considering the attenuation of the particle flux by fragmentation in collisions with air nuclei and the energy loss by ionisation. Calculations and measurements were performed in the early seventies[23,24]. The results of the calculations were in good agreement with experimental data which had been taken with plastic nuclear track detectors in several balloon flights at altitudes between 20 and 25 km. An improved version of this heavy ion transport code was later successfully used to calculate cosmic ray LET-spectra for different space missions[25]. A code which has been developed by other authors for the purpose of space dosimetry gives comparable results[26]. Based on more precise fragmentation cross section data which have been measured during recent years, this heavy ion transport code can be used for calculations of cosmic ray nuclei fluxes in the atmosphere.

DOSIMETRY AND MONITORING PROCEDURES

Dosimetry

The quantity required to be assessed in order to demonstrate compliance with any regulations which may be introduced, is effective dose (with the additional particular specification of equivalent dose to the embryo or fetus). The protection quantities are not directly measurable and in guidance from competent authorities, including national governments, it is generally accepted that certain operational quantities for personal and area monitoring may substitute for the protection quantities in routine monitoring and usually those values are entered in records as if they were protection quantities. Where limits or investigation levels are approached, it may be necessary to make more accurate estimates of the protection quantities. If one has detailed knowledge of, or can make assumptions about the radiation field (particle type, energy and angle distributions), the protection quantities may be calculated directly. With knowledge of dosemeter and instrument response characteristics, and information on, or assumptions of the orientation of the persons exposed to the field, the readings of personal dosemeters or of area monitoring instruments may be interpreted directly in terms of the protection quantities. For example, the assumption may be made that the field is rendered isotropic by movement of the exposed persons or that a person is exposed in the orientation which would give the largest value of effective dose or organ equivalent dose. The orientation of a person relative to a defined direction in a non-isotropic field is also important in the interpretation of readings of personal dosemeters. Alternatively the readings of dosemeters and instruments (and, indeed calculations) may be related to operational quantities, personal dose equivalent or ambient dose equivalent. Considerations of the most suitable approach must include what monitoring procedures will be necessary,

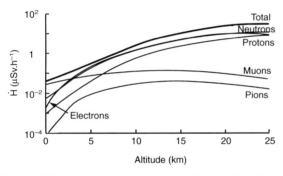

Figure 2. Major components of cosmic radiation, as a function of altitude, for solar radiation minimum (latitude 55°N) (from Reference 13).

Table 1. Dose equivalent rate in μSv.h^{-1}, calculated using LUIN 94, at 0° geographical longitude for different latitude, solar modulation and altitude (heliocentric potential: 100 MV for solar minimum and 1000 MV for solar maximum).

Altitude (shielding)		Geographical latitude		
		0°	50°	70°
12 km (200 g.cm^{-2})	sol. min.	2.6	6.1	9.0
	sol. max.	2.5	4.5	5.1
9 km (300 g.cm^{-2})	sol. min.	1.4	2.9	4.0
	sol. max.	1.3	2.2	2.4

for example, whether calculations supported by area monitoring will be sufficient, or if some personal monitoring will be required.

Knowledge of the complex radiation field in aircraft at operational altitudes is incomplete. Further calculations and measurements are being undertaken or are planned by a number of laboratories. Calculations can be of particle radiance for each particle type, which data can then be folded with conversion coefficients to obtain dose quantities. Calculations by O'Brien[15] of the components of dose equivalent (maximum for bilateral irradiation of a slab) as a function of altitude using this approach are shown in Figures 2 and 3. Detailed information on the field components is necessary in order to interpret and relate measurements made by different instruments, and to relate different measurement/assessment quantities. Several types of device may be used to measure with reasonable accuracy the absorbed dose to a small mass of tissue covered by an attenuating layer for the directly ionising components and photons. Tissue-equivalent proportional counters measure the absorbed dose to a small mass of tissue-like material similarly covered. Dose equivalent to this small mass may be estimated by appropriately weighting the dose distribution in lineal energy. For some components of the radiation field (generated by the cascade process), a state of quasi-equilibrium can be considered to exist, thus allowing approximations to be made as to the relationships of quantities, or allowing the readings of instruments and dosemeters whose response characteristics have been determined in terms of one dose quantity, to be used to assess another. Dose equivalent in tissue shielded by 1 cm of tissue, for example, may be equated to ambient dose equivalent or effective dose. This consideration may not apply to the neutron component; the aircraft contains much hydrogeneous material (fuel and passengers for example), but this may not produce an approximation to radiation equilibrium for the proton and neutron fields. The later calculations of O'Brien[16] give depth dose and depth dose equivalent (using ICRP Publication 60 Q(L) distributions for bilateral isotropic irradiation of a 30 cm slab of tissue. For the data for 12 km, the depth dose distribution is essentially flat (at about $3\ \mu Gy.h^{-1}$.) The depth dose equivalent shows a 10% lower value at 15 cm compared to the maximum near the surface (a value of about $8\ \mu Sv.h^{-1}$). This indicates a significant decrease in neutron dose equivalent with depth in tissue. This is consistent with calculations of ratios of ambient dose equivalent to effective dose for both calculated and measured neutron spectra. Discrimination in most detectors between neutron events and proton events is impossible without using coincidence techniques. Although the (p,p) cross section is still a factor of 3 lower at 100 MeV than the (n,p) cross section, proton tissue 'kerma' is over a factor of 10 higher than neutron tissue 'kerma' (see Figure 4 from Savitskaya and Sannikov[27]), and differences in effective and equivalent doses remain to be calculated.

The dose equivalent from neutron radiation becomes the dominant component for altitudes above about 10 km at latitudes above about 50°. Neutron energies range up to the maximum of the incident protons. The evidence is that there are two major components in the spectral distribution, between 1 and 10 MeV, and between 50 and 150 MeV, but there remains uncertainty as to the spectral distribution. Ambient dose equivalent will be an adequate estimator of effective dose for neutron energies between 1 and 20 MeV, for isotropic fields the overestimate of E by H*(10) decreases from a factor of 4 at 1 MeV down to 1.8 at 20 MeV. At neutron energies above 35 MeV, ambient dose equivalent will slightly underestimate effective dose. However it is important to average spectrally the ratio of H*(10) to E, and to consider the field geometry. The ratio of H* to E is lower for AP and higher for rotational and isotropic incidence. It is concluded that even for the hardest spectral distributions considered, ambient dose equivalent remains a conservative estimator for the protection quantity, effective dose. It is reasonable to assume that aircraft crew members are exposed isotropically rather than from the front. Lacking precise data and ignoring the dependence of neutron spectrum on latitude, a conversion factor from H* to E equal to 0.7 ± 0.2 has been calculated for the neutron dose equivalent component

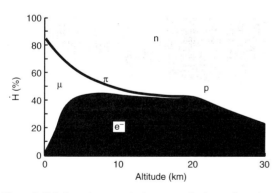

Figure 3. Relative dose equivalent contribution of major components of cosmic radiation, as a function of altitude, for solar minimum (latitude 55°N) (from Reference 13).

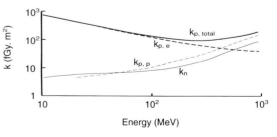

Figure 4. Neutron and proton tissue kerma, as a function of neutron and proton energy (from Reference 27).

independent of flight altitude. (It should be emphasised that, for the energy range between 18 MeV and 1 GeV, the calculations were made using constant fluence to dose conversion coefficients, on the assumption of approximately constant ratios of coefficients in this energy range. It is considered that the resulting inaccuracies are not significant in the context of all other uncertainties.)

The energy of primary galactic protons ranges up to 10^{14} MeV with a modal energy (fluence distribution) of 300 MeV, and of primary solar cosmic rays up to 400 MeV, but most in the range 1–100 MeV. The energies of secondary protons are in the same range, with dose rate dominated by particles of energies below 100 MeV. The range of 10 mm in tissue corresponds to a proton energy of about 35 MeV. For protons of energy sufficient to penetrate to deep-lying organs (80 MeV protons penetrate 50 mm in tissue), the application of a radiation weighting factor of 5 may be inappropriate. Protons of energies from some tens to a few hundreds of MeV incident on the body can be characterised as low LET radiation for most of their range. In a recent publication[28], the NCRP have recommended a value of 2 for the radiation weighting factor. (Eye lens dose and skin dose may need to be considered also.) Spectral information is important.

Procedures

To estimate effective dose to aircraft crew or frequent flyers, four procedures may be envisaged — (i) no dose estimates required; (ii) determination of route doses and occupancy factors; (iii) on-board area monitoring and occupancy factors; (iv) partial or full issue of personal dosemeters.

(i) For short flights with little time spent at altitudes above 8 km, it may be shown that combining route doses (which may be only a few μSv) with the staff flying hours, total annual doses will not exceed 1 mSv for groups of staff restricted to such a work programme. Staff would not be considered to be exposed workers. Some measurements may be required to verify the assessment.

(ii) In the case of exposed workers for whom records are required, estimates of individual doses would be obtained from route doses (for whatever quantity is selected) and records of staff rostering. These estimates would be placed in the dosimetry records of staff as assessed values of the protection quantities. The route doses may be based on measured values, preferably with a reference instrument system or from dose rates calculated from flight parameters and solar cycle phase. Confirmation by measurement of the calculated dose rates should be made periodically (not more frequently than annually) for representative routes. This option is the most generally applicable. The calculational method is still under development and further calculations are to be made. Occupational doses for regulatory purposes can be derived from the results of these calculations, staff roster information and the flight log of altitude, latitude and time. It is understood that in the USA, the Federal Aviation Administration has accepted mathematical modelling of radiation exposure in aircraft as an alternative to measurement. The calculated doses can be directly effective dose and organ equivalent dose (calculated for adult anthropoid phantoms).

(iii) Commercial aircraft which fly above 15 km already carry monitoring devices but these are primarily intended as warning devices in case of a steep increase in dose rates due to solar flares. The increase in cruising altitudes of some aircraft of recent design (particularly in the case of private aircraft) results in dose rates in these aircraft approaching 10 μSv.h^{-1} and on-board radiation monitoring equipment may be desirable. This option might be considered for general application in future designs of commercial and business aircraft when it could be designed in.

(iv) Simple designs of thermoluminescence dosemeter to measure the photon plus directly ionising component, and an etched track neutron dosemeter may be sufficient for personal dosimetry, if this is required, subject to corrections applied to their readings to take account of their energy and directional dependence of response, and the difficulties of measuring low doses. Electronic dosemeters of much greater sensitivity and very low minimum measurable doses are becoming available. Good control would need to be exercised over the personal dosemeters and the location specific background controls. It is not considered that this option will need to be taken up other than to supplement or verify the estimates of dose obtained from other methods based on area monitoring.

Couriers and other specialist occupational groups who fly more often than other passengers on commercial aircraft, and aircraft crew on private aircraft may need to have their exposure assessed by their employers. For flights on commercial aircraft, it would seem reasonable that arrangements could be made with the airlines to obtain estimates of route doses. For the assessment of doses to persons travelling frequently on private aircraft, employers may need, independently, to consider the options outlined above.

CALIBRATION AND REFERENCE FIELDS

Calibration

The calibration of an instrument comprises the process of comparing under equivalent conditions the reading of the instrument with the reference value obtained from a primary or secondary standard device for the respective quantity. The ratio of these readings gives the calibration factor of the instrument. In principle, this calibration factor can only be applied correctly if the actual measuring conditions are equal to the calibration conditions. Deviations from the calibration conditions usually lead to errors of the instrument reading which depend on the amount of deviation and on the instrument properties. This requirement, apparently trivial, is a particularly problematic one for the calibration of radiation dosemeters. Radiation fields may consist of many different elementary particles which interact with matter with vastly different effects depending on particle energy and type. In particular, the cosmic radiation field at typical flight altitudes contains particles from the complex cascade reactions of high energy galactic radiation with the upper atmosphere. Given the complexity of the cascade reactions, it is doubtful whether a complete understanding is ever possible, particularly since the particle energies extend well above those which can be achieved with current accelerator technology. In order to calibrate measurement devices for use in aircraft, three different approaches are conceivable:

(i) The calibration can be performed in the radiation field of interest itself, i.e. in aircraft. This would require an accompanying reference instrument which measures the quantity correctly for all different particles present in cosmic radiation fields. The limitations and uncertainties of this reference instrument should be well described.

(ii) A reference calibration field (on the ground) could be used as a primary standard, provided that this reference field contains the more important components present in the cosmic radiation field and that either the relative dose fractions or the spectral fluences of the various components are known and are being monitored. Alternatively, a device may be used which truly measures the relevant dosimetric quantity.

(iii) The response characteristics of a device may be determined for the major, or most important part of the radiation field to be measured.

Reference Fields

The radiation field at flight altitudes contains a complex mixture of the primary cosmic radiation and their secondary reaction products produced in the upper part of the atmosphere. So far, there are no reference fields on the ground for the routine calibration of dose equivalent meters including the same radiation components of identical composition as the cosmic radiation field at flight altitudes. Various attempts have been made to use the radiation fields behind the shielding walls of high energy particle accelerators for calibration purposes. It has been shown that the radiation components which produce the major dose fractions in the cosmic radiation fields, i.e. photons, leptons, and neutrons of energies up to, and exceeding 100 MeV, are present in these fields and that dose rates similar to those at flight altitudes can be achieved. Mixed high energy particle fields have been recently set up at the super proton synchrotron (SPS) at CERN[29–31]. These stray radiation fields are created by beams of high energy protons and pions with momenta of 120 and 205 GeV/c impinging on a 0.5 m long copper target. The target area was surrounded by a massive shielding and the calibration fields were located outside of the shielding. The radiation produced at large angles to the beam passes either through a 0.4 m thick iron shield or 0.8 and 1.6 m thick concrete shields. Measurement positions were on the roof and on the side of the shielding. The roof shields produce almost uniform fields over an area of 2 m by 2 m. The dose equivalent rates in the various position varied between 0.8 $mSv.h^{-1}$ and 6 $mSv.h^{-1}$. The monitoring of the high energy hadron beam was by means of an air-filled ionisation chamber in the primary beam together with two rem meters on the roof shield. The radiation field in each calibration position was calculated by using the Monte Carlo code FLUKA. The investigations of the CERN reference fields started in 1992. Since 1993, these fields have been available for testing and calibration of detectors and equipment from other European laboratories, and various measurements have already been performed on the initiative of EURADOS and the CERN radiation protection group. Three campaigns of measurements were performed in 1993 and 1994 with 19 institutions participating, using 25 different techniques (including multisphere systems, rem meters, activation detectors, proton-recoil spectrometers, etched track detectors, bubble damage detectors, albedo dosemeters, GM counters, scintillators, TLDs and several types of TEPC based dose equivalent meters).

The preliminary conclusions from the joint studies were that the high energy neutron spectral fluences contained features similar to those aboard aircraft and that reasonable agreement was found between the results of Monte Carlo calculations and the experimental spectral fluence, integral fluence and integral dose measurements. Detailed analyses are in progress.

CONCLUSIONS

It is concluded that most of the experimental data relating to aircraft crew exposure to cosmic radiation can be reconciled in spite of the rather high uncertainties attached to the measurements. The agreement of these data with theoretical estimations is not fully satisfactory and there are some areas in which improved data would reduce the uncertainties. More precise specification of the detailed contributions by different particles, especially neutrons and protons, to the total dose equivalent would permit a more correct interpretation of data acquired with different instruments and detectors.

For dose assessment and recording purposes where

required (and possibly also for ALARA and for epidemiological purposes, or for legal reasons) estimates of effective dose and/or equivalent dose to the embryo or fetus (or surface of the abdomen), may be needed. Measurements may be made of ambient dose equivalent and used as estimates of the protection quantity, effective dose, with or without corrections based on knowledge of the radiation field, or estimates of effective dose and organ equivalent dose made directly from instrument readings, interpreted using *a priori* knowledge of the radiation field. It is considered desirable that where there is a knowledge of the radiation field, this should be applied to obtain the best estimate of the protection quantity. Aircraft crew are one of the most highly exposed occupational groups.

Dose estimation procedures should not be necessary for persons for whom total annual doses will not exceed 1 mSv and therefore, in general, for aircraft not routinely flying above 8 km. In general, the preferred procedure in order to estimate doses to aircraft crew or frequent flyers is to determine route doses and fold these with data on staff rostering. Route doses are likely to be obtained from calculations of the radiation field as a function of aircraft altitude and latitude (and solar cycle phase), with validation by measurement. It is considered unlikely that personal dosemeters will be required.

The provision of a sound metrological basis for measurements at aviation altitudes is important. A suitable well specified reference field needs to be available for this purpose.

Full investigation of the solar cycle influence on the exposure level is desirable. There are only small discrepancies between older and newer data, but the extent of solar cycle modulation on the exposure level is not adequately known. Experimental studies should be prolonged at least up to the next solar maximum, which will occur in the year 2000, and the data acquired correlated with neutron monitor network data and with improved calculated data.

A significant improvement of model predictions is possible using interaction codes such as those developed at CERN and the University of Milan (FLUKA). These predictions would also contribute to the information about field composition. Other codes under development may in the future contribute further improvements.

ACKNOWLEDGEMENTS

The authors would like to acknowledge their many colleagues for their contributions to the work reported. Meetings and coordination work of EURADOS Working Group 11 'The Radiation Exposure and Monitoring of Air Crew' were supported by the European Commission, Directorate General XI, Environment, Nuclear Safety and Civil Protection, Unit for Radiation Protection. The research work of a number of members of the Working Group was supported under Contract F13P-CT92-0026 by the European Commission, Radiation Protection Research Action.

REFERENCES

1. United Nations Scientific Committee on the Effects of Atomic Radiation. *Sources, Effects and Risks of Ionising Radiation.* UNSCEAR 1993. Report to the UN General Assembly with Scientific Annexes. New York: United Nations (1993).
2. International Commission of Radiological Protection. *1990 Recommendations of the International Commission on Radiological Protection.* ICRP Publication 60 (Oxford: Pergamon Press) (1991).
3. Council Directive 96/29/EURATOM of 13 May 1996 *Laying Down the Basic Safety Standards for Protection of the Health of Workers and the General Public Against the Dangers Arising from Ionizing Radiation.* Official Journal of the European Communities **39**, L159 (29 June 1996).
4. Reitz, G. *Radiation Environment in the Stratosphere.* Radiat. Prot. Dosim. **48**, 5–20 (1993).
5. Reitz, G., Schnuer, K. and Shaw, K. (Eds) *Radiation Exposure of Civil Aircrew.* Proceedings of a Workshop, Luxembourg, 25–27th June 1991. Radiat. Prot. Dosim. **48** (1993).
6. Lantos, P. *The Sun and its Effects on the Terrestrial Environment.* Radiat. Prot. Dosim. **48**, 27–32 (1993).
7. Wardman, P. *Solar Flares.* Paper presented to the Society of Radiological Protection at a meeting at Imperial College, London (October 1972).
8. Armstrong, T. W., Alsmiller, R. G., Jr and Barish, R. *Calculation of the Radiation Hazard at Supersonic Aircraft Altitudes Produced by an Energetic Solar Flare.* Nucl. Sci. Eng. **37**, 337–342 (1969).
9. Barish, R. *Health Physics Concerns in Commercial Aviation.* Health Phys. **59**, 199–204 (1990).
10. O'Brien, K., Sauer, H. H. and Friedberg, W. *Atmospheric Cosmic Rays and Solar Energetic Particles at Aircraft Altitudes.* In: Proc. Conf. on Natural Radiation Environment, (in press) (1995).
11. EURADOS Working Group 11, Eurados Report 1996-01. *The Radiation Exposure and Monitoring of Air Crew.* European Commission publication Radiation Protection 85 (Luxembourg: EC) (1996).
12. Davies, D. M. *Cosmic Radiation in Concorde Operations and the Impact of New ICRP Recommendations on Commercial Aviation.* Radiat. Prot. Dosim. **48**, 121–124 (1993).
13. Akatov, Yu. A. *Some Results of Dose Measurements Along Civil Airways in the USSR.* Radiat. Prot. Dosim. **48**, 59–64 (1993).
14. Nguyen, V. D., Bouisset, P., Kerlau, G., Parmentier, N., Akatov, Yu. A., Archangelsky, V. V., Smirenniy, L. N. and Siegris, M. *A New Experimental Approach in Real Time Determination of the Total Quality Factor in the Stratosphere.* Radiat. Prot. Dosim. **48**, 41–46 (1993).

15. O'Brien, K. *The Cosmic Ray Field at Ground Level.* In: The Natural Radiation Environment II, Proc. 2nd Int. Conf. On Natural Radiation Environment, August 1972, Houston, USA. Ed. J. A. S. Adams, Conf-720805-PC, pp 15–54 (US Dept. of Commerce, National Technical Information Service, Springfield, VA) (1972).
16. O'Brien, K. and Friedberg, W. *Atmospheric Cosmic Rays at Aircraft Altitudes.* Environ. Int. **20**(5), 645–663 (1994).
17. Hess, W. N., Patterson, H. W., Wallace, R. and Chupp, G. L. *Cosmic-ray Neutron Energy Spectrum.* Phys. Rev. **116**(2), 446–457 (1959).
18. Friedberg, W., Duke, F. E., Snyder, L., O'Brien, K., Parker, D. E., Shea, M. A. and Smart, D. F. *Computer Program CARI-2.* (US Dept of Commerce, National Technical Information Service, Springfield, VA) (1994).
19. *GEANT Detector Description and Simulation Toll.* CERN Program Library Office, (Geneva: CERN) (undated).
20. Fasso, A., Ferrari, A., Ranft, J. and Sala, P. -R. *FLUKA: Present Status and Future Developments.* In: Proc. IVth. Int. Conf. on Calorimetry in High Energy Physics. p. 493, (World Scientific) (1994).
21. Schraube, H., Jakes, J., Sannikov, A., Weitznegger, E., Rösler, S. and Heinrich, W. *The Cosmic Ray Induced Neutron Spectrum at the Summit of the Zugspitze (2963 m).* Radiat. Prot. Dosim. **70**(1–4), 405–408 (This issue) (1997).
22. Merker, M. *The Contribution of Galactic Cosmic Rays to the Atmospheric Neutron Maximum Dose Equivalent as a Function of Neutron Energy and Altitude.* Health Phys. **25**, 524–527 (1973).
23. Allkofer, O. C. and Heinrich, W. *Attention of Cosmic Rays Heavy Ion Fluxes in the Upper Atmosphere by Fragmentation.* Nucl. Phys. B**71**, 429–438 (1974).
24. Allkofer, O. C. and Heinrich, W. *Measurements of Cosmic Ray Heavy Nuclei at Supersonic Transport Altitudes and Their Dosimetric Significance.* Health Phys. **27**, 543–551 (1974).
25. Heinrich, W. *Calculation of LET-spectra of Heavy Cosmic Ray Nuclei at Various Absorber Depths.* Radiat. Effects **34**, 143–148 (1977).
26. Adams, J. H., Tylka, A. J. and Stiller, B. *LET Spectra in Low Earth Orbit.* IEEE. Trans. Nucl. Sci. **NS-33**, 1386–1389 (1986).
27. Savitskaya, E. N. and Sannikov, A. V. *High Energy Neutron and Proton Kerma Factors for Different Elements.* Radiat. Prot. Dosim. **60**(2), 135–146 (1995).
28. National Committee on Radiation Protection. *Radiation Exposure and High-altitude Flight.* NCRP Commentary No. 12 (Bethesda, MD: NCRP) (1995).
29. Höfert, M. and Stevenson, G. R. *The CERN-CEC High-Energy Reference Field Facility.* CERN Report TIS-RP/94-02/CF (Geneva: CERN) (1994).
30. Aroua, A., Höfert, M., Sannikov, A. V. and Stevenson, G. R. *Reference High-Energy Radiation Fields at CERN.* CERN Report TIS-RP/94-12/CF (Geneva: CERN) (1994).
31. Höfert, M., Stevenson, G. R. and Alberts, W. G. *Measurement of Radiation Fields and Dosimetry in the Stratosphere.* In Proc. Conf. Strahlenschutz: Physik und Meßtechnik of the Fachverband für Strahlenschutz e.V. at Karlsruhe, 24–26 May 1994 Proc. der 26 Jahrestagung des FS, p. 541 (FZK: Karlsruhe) (1994).

THE COSMIC RAY INDUCED NEUTRON SPECTRUM AT THE SUMMIT OF THE ZUGSPITZE (2963m)

H. Schraube†, J. Jakes†, A. Sannikov†*, E. Weitzenegger†, S. Roesler‡ and W. Heinrich‡
†GSF-Forschungszentrum für Umwelt und Gesundheit, Institut für Strahlenschutz
Neuherberg, D-85758 Oberschleißheim, Germany
‡Universität Gesamthochschule Siegen, Fachbereich Physik
D-57068 Siegen, Germany

Abstract — This paper describes a neutron spectrometry experiment at the summit of the mountain Zugspitze. The measured spectral neutron fluence rate is compared with results of particle transport calculations of cosmic primaries and secondaries down to the depth of 700 g.cm^{-2} in the atmosphere. The results may serve as a basis for the estimation of neutron exposure to air crews and other persons flying frequently.

INTRODUCTION

The study of the radiation to which air crews and other frequently flying individuals are exposed became an important interdisciplinary task in health physics. An essential contribution to the radiation risk is caused by secondary neutrons. Therefore, the determination of the spectral distribution of neutrons at civil flight levels is the primary basis for the derivation of relevant radiation protection quantities.

Radiation transport calculations may provide spectral fluences of all particles created in cascades in the atmosphere when a high energy cosmic particle impinges on the top of the atmosphere. The benefit of the calculations with respect to a full description of the radiation fields at any altitude and any geographical position at any time depends essentially on the knowledge of the primary cosmic radiation on top of the atmosphere, all possible interaction processes and the transport models used. On the other hand, airborne experiments suffer from inconvenient equipment, non-stationary experimental positions for measurements in aeroplanes, insufficient knowledge of the instrument responses, and also from poor statistics.

This paper describes the attempt to investigate the neutron spectra though not at flight altitudes, but at a level with increased fluence rate. The aim of the experiment was (i) to test the spectrometry system for air borne experiments, (ii) to collect more reliable data, and (iii) to verify the radiation transport calculations in the atmosphere.

EXPERIMENTAL CONDITIONS

The measurements were done on the roof of an atmospheric observatory which is positioned at the summit of the Zugspitze, the highest mountain in Germany (2963 m) at the geographical position 47°25'N and 11°E.

Bonner sphere spectrometer

The BS spectrometer employed consisted of a set of 14 homogeneous polyethylene spheres (density 0.95 g.cm^{-3}) with a spherical ^{3}He proportional counter placed in the centre as thermal neutron detector. The total set included spheres with diameters of 2.5, 3, 3.5, 4, 5, 6, 7, 8, 9, 10, 11, 12, and 15″. In addition, a sphere of 9″ diameter was used with an internal lead layer which replaced the polyethylene between 3″ and 4″ diameter. This layer serves as a high energy neutron converter by (n,xn′) nuclear processes. The sphere exhibits the important measuring channel for high energy neutrons. The ^{3}He proportional counter (SP90 of Centronic Ltd) of 32 mm inner diameter had a nominal pressure of 172 kPa ^{3}He gas and 100 kPa krypton.

The response characteristics of the system with the stated nominal counter gas filling had been well studied experimentally[1,2] in the range between thermal neutron energy and 15 MeV, and by calculations[3,4] up to 100 MeV. The responses to high energy neutrons in excess of 40 MeV were taken from calculations with the code HADRON[5] which were in good agreement with the MCNP calculations between 10 and 100 MeV. In the case of the 9″ sphere with lead, the responses were obtained using our data for the 9″ sphere without lead multiplied by the ratio of responses for this sphere with and without lead[6].

The spectrometer was calibrated against another BS spectrometer with ^{7}Li(EU) scintillator[7] to account for deviations of the actual counter pressure from the nominal pressure for which the calculation had been performed[3]. Pulse height analysis was provided to ensure proper noise subtraction.

The BS spectrometer was placed in the open air in a thin aluminium container whose temperature could be kept within 20 ± 3°C. The total experiment time was 144 h. The counting time for each sphere was chosen

---

* On leave from the Institute for High Energy Physics, 142284 Protvino, Moscow Region, Russia.

in such a way that the uncertainty due to counting statistics was slightly less than 2.5%.

Monitoring

To account for changes in the neutron fluence rate throughout the experiment due to changing atmospheric conditions, simple neutron fluence monitoring was established by employing a 5″ Bonner sphere.

Data unfolding

For the processing of the experimental data sets, two different unfolding codes were employed: the SAND II code[8] calculates a count rate vector from a first guess spectrum and improves the solution by an iterative process[9].

The BON95 code[10] is a modified version of the BON94 program[11]. It uses an *a priori* by parametrisation of the neutron spectrum which is described by a superposition of a thermal Maxwellian peak, an E^{-b} tail of epithermal neutrons, an 1/E tail of intermediate neutrons, a fast neutron peak with variable temperature and width and high energy peak with a most probable energy of 100 MeV. The solution obtained is used further as a first guess spectrum for an iterative procedure.

COSMIC RAY TRANSPORT CALCULATION

The cosmic ray induced neutron fluence rate in the atmosphere has been calculated using the Monte Carlo (MC) code FLUKA[12–14]. FLUKA simulates the whole high energy hadronic and electromagnetic cascade from TeV to thermal energies. Hadronic interactions in FLUKA are simulated based on different models depending on the energy of the primary particles. The Dual Parton model is applied at energies above 5 GeV whereas MC implementations of models for resonance production and decay and for an intranuclear cascade plus pre-equilibrium plus evaporation sequence describe inelastic hadronic collisions at intermediate energies[15]. These models were themselves carefully tested against measurements and describe successfully the main aspects of high and low energy particle production. Neutrons below 20 MeV are transported using a neutron cross section data set[16].

The calculations were performed in the depth of the atmosphere down to the summit of the Zugspitze (2963 m) at its geographical position for the time of the experimental verification, based on primary cosmic ray spectra and considering the effects of solar modulation and geomagnetic shielding. The amount of solar modulation of the cosmic ray flux density depends on the general level of solar activity which typically exhibits an 11-year activity cycle. It effectively results in a decrease or increase of the primary flux density below 10 GeV. This has been taken into consideration based on ground neutron monitor data using the sinusoidal modulation model[17]. The atmosphere was modelled using the combinatorial geometry package of the FLUKA code. A plain geometry was applied subdividing the atmosphere down to the depth of the Zugspitze (700 g.cm^{-2}) into 14 slabs of equal depth and a standard chemical composition of air of 76% nitrogen and 24% oxygen was used. Note that results of the calculations depend on the assumptions made, i.e. on the primary flux density and the model of the atmosphere, and may give rise to uncertainties in addition to the statistical uncertainties. The interactions of charged secondary hadrons with air were simulated down to 10 keV whereas the interactions of secondary neutrons were treated down to thermal neutron energies. In total 120 000 primaries were followed with energies up to 1300 GeV and with different uniformly distributed direction cosines against the normal. The spectra were scored on a logarithmic basis in energy above 20 MeV and according to the HILO neutron multigroup structure[16] below this energy at each boundary between two atmospheric slabs of different densities.

RESULTS

Monitor data

During the course of the experiment a strong atmospheric depression field crossed the Alps. This caused a considerable change in the air pressure and hence in the count rate of the monitor. To account for the non-constant neutron fluence rate, it was assumed that it depends exponentially on the air pressure. From a least squares fit through the experimental data an attenuation factor of 0.82 ± 0.16% per mbar was determined. Figure 1 shows that the exponential law is fulfilled for all data on the 99% confidence level, with one exception.

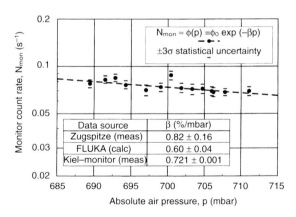

Figure 1. Count rate of the 5″ Bonner sphere neutron monitor obtained at the summit of the Zugspitze plotted against the atmospheric air pressure. The attenuation data in the table are additionally given for the FLUKA calculated fluence rates and for the Kiel monitor[18].

Spectral data

In Figure 2 the resulting experimental spectra are depicted which were obtained by the two unfolding codes. From the FLUKA calculations as well as from much earlier calculations[19] it was reasonable to assume that a spectral component around 100 MeV due to intranuclear cascade processes was present even at the low altitude. It was therefore modelled into the BON95 code as described above, but also used in the first estimated spectrum for the SAND II code. If the start spectrum was in first instance taken as a simple 1/E distribution, the resulting spectra developed the thermal and intermediate range well, and the fast evaporation peak, but the high energy contribution was shifted to the upper end of the full range, i.e. around 1 GeV. The spectra of the parametric solution in Figure 2 are presented with an estimation of the uncertainty. The solutions of both unfolding agree well up to 10 MeV, for the 100 MeV peak the BON95 unfolding delivers a somewhat higher solution.

In Figure 3 three calculated spectral distributions of the neutron fluence are presented, as obtained by the two codes FLUKA and LUIN[20], and taken from the literature[19]. While the results of FLUKA are in reasonable agreement with the data of Merker, the LUIN results differ in shape and energy integrated fluence rate.

In Figure 4 the experimentally determined spectrum (SAND II solution) is compared with the calculated results. It is observed that the shapes of both results are rather similar. While in the calculations the peak around 100 MeV is only a little higher than the experimental findings, the calculated total fluence rate is almost a factor of 2 higher than the experimental one.

It has to be considered, however, that the statistical uncertainties of the MC calculations are of the order of 40%, and that the experimental result may depend on the unfolding procedure applied.

CONCLUSIONS

The neutron spectra derived by two different unfolding methods from the experimental data at the summit of the Zugspitze are in good agreement within estimated error limits. The spectra obtained from radiation transport calculation from the top of the atmos-

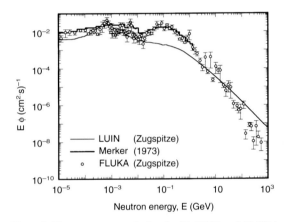

Figure 3. Neutron spectra calculated with LUIN and FLUKA codes compared with calculations of Merker[19] for the same depth (700 g.cm^{-2}) in the atmosphere.

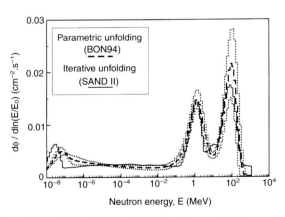

Figure 2. Spectral neutron fluence per logarithmic energy interval (E_O = 1000 MeV) obtained from the measurements by two different unfolding procedures (full line: SAND, broken lines: BON95 with upper and lower confidence limits).

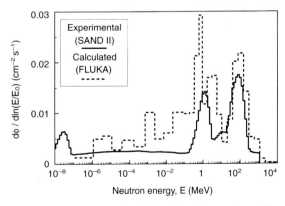

Figure 4. Comparison of calculated and experimentally determined neutron spectra at 700 g.cm^{-2} depth in the atmosphere which is approximately equivalent to 2963 m altitude (E_0 = 1000 MeV).

phere down to a depth equivalent to the same position deliver a similar shape, a considerably higher total neutron fluence rate, and a cascade peak at 100 MeV which is in good agreement with the experiment. It is concluded that the modelling of the atmosphere, the statistics of experiments and calculations, and the unfolding procedures require further studies and improvements.

ACKNOWLEDGEMENTS

The authors are grateful to Prof. Dr K. Schäfer of the Fraunhofer Institut für atmosphärische Umweltforschung, Garmisch, for his support and hospitality during the measurements. Two of us (W.H. and S.R.) are obliged to Dr A. Ferrari for providing the FLUKA code. The financial support of the European Union under contract FI3P-CT92-0026 is highly appreciated.

REFERENCES

1. Alevra, A. V., Cosack, M., Hunt, J. B., Thomas, D. J. and Schraube, H. *Experimental Determination of the Response of Four Bonner Sphere Sets to Monoenergetic Neutrons (II)*. Radiat. Prot. Dosim. **40**, 91–102 (1992).
2. Thomas, D. J., Alevra, A. V., Hunt, J. B. and Schraube, H. *Experimental Determination of the Response of Four Bonner Sphere Sets to Thermal Neutrons*. Radiat. Prot. Dosim. **54**, 25–31 (1994).
3. Mares, V., Schraube, G. and Schraube, H. *Calculated Neutron Response of a Bonner Sphere Spectrometer*. Nucl. Instrum. Methods Phys. Res. A**307**, 398–412 (1991).
4. Mares, V. and Schraube, H. *Improved Response Matrices of Bonner Sphere Spectrometers with ^{6}LiI Scintillation Detector and ^{3}He Proportional Counter between 15 and 100 MeV Neutron Energy*. Nucl. Instrum. Methods Phys. Res. A**366**, 203–206 (1995).
5. Sannikov, A., Mares, V. and Schraube, H. *High Energy Response Functions of Bonner Spectrometers*. Radiat. Prot. Dosim. **70**(1–4), 291–294 (this issue) (1997).
6. Hsu, H. H., Alvar, K. R. and Vasilik, D. G. *A New Bonner-Sphere Set for High Energy Neutron Measurements — Monte-Carlo Simulation*. IEEE Trans. Nucl. Sci. **NS-41**, 938–940 (1994).
7. Mares, V. and Schraube, H. *Evaluation of the Response Matrix of a Bonner Sphere Spectrometer with LiI Detector from Thermal Energy to 100 MeV*. Nucl. Instrum. Methods Phys. Res. A**337**, 461–473 (1994).
8. McElroy, W. N., Berg, S., Crockett, T. and Hawkins, R. G. *Spectra Unfolding*. AFWL-TR-67-41, Vol. 1–4 (1967).
9. Alevra, A. V., Siebert, B. R. L., Aroua, A., Buxerolle, M., Grecescu, M., Matzke, M., Mourgues, M., Perks, C. A. and Schraube, H. *Unfolding Bonner-sphere Data: A European Intercomparison of Computer Codes*. PTB Laboratory Report 7.22-90-1 (1990).
10. Sannikov, A. V. *BON95, a Universal User-independent Unfolding Code for Low Informative Neutron Spectrometers* (to be published).
11. Sannikov, A. V. *Bon94 Code for Neutron Spectra Unfolding from Bonner Spectrometer Data*. CERN Internal Report, CERN/TIS-RP/IR/94-16 (1994).
12. Fasso, A., Ferrari, A., Ranft, J. and Sala, P.-R. *FLUKA: Present Status and Future Developments*. In: Proc IVth Int. Conf. on Calorimetry in High Energy Physics, La-Biodola (Elba), Italy, 1993. Eds A. Menzione and A. Scribano (World Scientific) p. 493 (1994).
13. Fasso, A., Ferrari, A., Ranft, J. and Sala, P.-R. *FLUKA: Performances and Applications in the Intermediate Energy Range*. In: Proc. Specialists Meeting on Shielding Aspects of Accelerators, Targets and Irradiation Facilities, Arlington, USA, 1994 (OECD/NEA) p. 287 (1994).
14. Aarnio, P.-A. In: Proc. Int. Conf. on Monte Carlo Simulation in High Energy and Nuclear Physics, MC'93, Tallahassee, USA, 1993. Eds P. Dragovitsch, S. L. Linn and M. Burbank (World Scientific) p. 88 (1994).
15. Fasso, A., Rerrari, A., Ranft, J., Sala, J., Stevenson, G. R. and Zazula, J. M. *FLUKA92*. In: Proc. Workshop on Simulating Accelerator Radiation Environments, SARE, Ed. A. Palounek. Santa Fe, USA. Los Alamos LA-12835-C p. 134 (1994).
16. Cuccoli, E., Ferrari, A. and Panini, G. C. *A Group Library from JEF 1.1 for Flux Calculations in the LHC Machine Detectors*. ENEA-Bologna Report JEF-DOC-340 (1991).
17. Adams, J. H., Silberberg, R. and Tsao, C. H. *Cosmic Ray Effects in Microelectronics, Part I: The Near Earth Particle Environment*. Naval Research Laboratory Report 4506 (1973).
18. Röhrs, K. Private communication (1995).
19. Merker, M. *The Contribution of Galactic Cosmic Rays to the Atmospheric Neutron Maximum Dose Equivalent as a Function of Neutron Energy and Altitude*. Health Phys. **25**, 524–527 (1973).
20. O'Brien, K. *LUIN, A Code for the Calculation of Cosmic Ray Propagation in the Atmosphere* (update of HASL-275) US Department of Energy Report EML-338 (1978).

EXPERIMENTAL APPROACH TO THE EXPOSURE OF AIRCREW TO COSMIC RADIATION

F. Spurný
Department of Radiation Dosimetry, Nuclear Physics Institute
Czech Academy of Sciences
Na Truhlářce 39/64, 180 86 Praha 8, Czech Republic

Abstract — Results of measurements made on board subsonic aircraft since 1991 are presented and discussed. Several types of active devices as well as passive detectors have been used. The same types of instruments have also been tested in high energy reference fields at CERN and at JINR Dubna. This has permitted the correction of the readings of neutron dosemeters for the probable spectrum on board. The correction factor for low LET radiation measuring instruments has been estimated on the basis of available particles (electrons, protons) spectra. It has been found that the exposure level depends on parameters known from theoretical considerations (geomagnetic position, flight altitude, etc.), some concrete examples are given. The measured data have also been compared with the data available from the world network of cosmic radiation monitors. It has been found that the correlation is satisfactory. Nevertheless, the studies have to continue until after the next solar minimum (~2007) to improve the reliability of the correlation. Some comments on radiation protection aspects of the topic are also made.

INTRODUCTION

ICRP Publication 60 recommends including aircrew among occupationally exposed persons[1]. Since 1991 many studies have been started to characterise the level of aircrew exposure with more precision, both experimentally and theoretically.

Efforts here have been concentrated on the experimental approach: this contribution summarises the results obtained in the period 1991–1995. The results of measurements performed on board aircraft as well as during on-Earth studies behind the shielding of high energy accelerators are presented and discussed. Several different instruments and detectors have been tested and used for these purposes.

Some comments on radiation protection aspects of the topic are also given.

EXPERIMENTAL METHODS

Radiation fields on board aircraft can be divided into the neutron (contributing mainly to the high LET component) and directly ionising (mainly low LET) components (mostly electrons, high energy protons, photons)[2]. Many different instruments and detectors have been tested and used since 1991: they are described in more detail in previous publications[3,4]. They are listed as follows:

All radiation

NAUSICAA equipment with a tissue-equivalent proportional counter[5]

Neutron (high LET) component

Moderator-type neutron rem counter NM2 (Nuclear Enterprises)

Bubble damage neutron detectors (BDND-BTI Chalk River)
etched track detectors (TED-Pershore, UK)

Directly ionising (low LET) component

High pressure Ar-filled ionisation chamber RSS 112 (Reuter-Stokes)
Scintillation-based environmental radiation dose rate meter NB 3201 (Tesla-Czech Rep.)[6]
GM-based environmental radiation dose rate meter RP 114 (ZMA Ostrov, Czech Rep.)
Two individual electronic dosemeters: DCM 90 (Merlin-Gerin, France) based on Si-diode; and D222 (ZMA Ostrov, Czech Rep.) based on miniature GM-counters
Thermoluminescence detectors (TLDs) — $CaSO_4$, Al_2O_3:C

Neutron (high LET) detectors were calibrated with a ^{252}Cf neutron source, directly ionising (low LET) detectors with ^{60}Co photons, in both cases in free-in-air geometry and in terms of $H^*(10)$ with ICRP 21 quality factors[7].

It should be mentioned that the radiation field composition on board is different from calibration radiation for both components. The correction factors to be applied due to the basic difference in measured and calibration field composition are discussed further.

Most of the equipment used could also be sensitive to other radiation components than that for which they are designated. 'Neutron' detectors could be sensitive to high energy protons through the secondary particles appearing during neutron–nuclear interactions, low LET detectors could be sensitive to neutrons particularly of energies above few tens of MeV[8]. The extent of the contribution of other components to the response of an instrument is very difficult to estimate quantitatively.

Nevertheless, it is considered that due to the low proton contribution and the higher QF for high energy neutrons[2] these contributions do not increase the reading more than 5 to 10% in either case.

RESULTS AND DISCUSSION

On board measurements

From April 1991 up to the end of June 1995 the measurements on board aircraft were performed during about 20 return commercial flights, mostly on board A310–300 and TU154M aircraft. Flight routes extended from the equator (Singapore, 1.3°N) up to more northern regions (north Atlantic routes, up to 65°N). Flying altitudes varied from 27,000 feet (8.2 km) to 41,000 feet (12.5 km), the most typical levels were between 33,000 and 37,000 feet. The most complete studies were made during long haul return flights, Prague–New York, Prague–Montreal, Prague–Singapore and Prague–Bangkok. Some of the results obtained are presented in Tables 1 to 3, in terms of H*(10) of calibration sources, i.e. ^{252}Cf neutrons and ^{60}Co photons. The results obtained during the whole period mentioned exhibit some general rules. They can be formulated in the following way:

(1) Measured values are practically constant for geographical latitudes above 50–55°N, they decrease when going towards the equator. For the flights to south-east Asia this decrease is practically complete at about 25°N, beyond this the level is again constant. The decrease is more important for the neutron than for low LET component.
(2) Dose rate increases smoothly with flight altitude, at 12 km being roughly twice that at 9 km.
(3) The values measured after the beginning of 1993 are about 15% higher than those measured in April 1991.
(4) Values measured with NAUSICAA TEPC are, in the high LET (neutron) region, systematically higher than apparent dose equivalent values deduced from the readings of rem counter and/or bubble detectors. The average ratio is about 2 for the region above 50°N, a little lower when approaching to the equator. As far as the low LET region is concerned, NAUSICAA values above 50°N are systematically about 25% higher than for the most of other detectors; when approaching the equator all readings are closely similar.
(5) Si-based individual dosemeters of low LET radiation give values systematically about 30% lower than for other detectors. The same effect has been observed in other high energy particle fields and may be attributed to differences in high energy particles energy transfer coefficients[9].

During two long-haul return flights on Atlantic routes (PRG-NYC; PRG-YMC) large surface etched track detectors (PADC) (CR-39) were also exposed. A detector of the same type and origin has been kept in the laboratory for background measurement. In both cases statistically reliable differences could be registered (total of 11,367 particle tracks in on-board exposed samples, 7411 in the background; total surface more than 100 cm^2) corresponding to a net track density about 34 cm^{-2}. Of that, about 6 cm^{-2} can be attributed to fast neutrons, about 1–2 cm^{-2} to high energy protons. However the residual track density is too high to be explained by primary heavier charged particles. Their origin could be connected with so-called stars[10], the possibility of such an explanation is being studied.

Table 1. Average readings of dose equivalent rate measuring instruments; February 1995.

Flight route	Altitude (km)	Apparent dose equivalent rate (μSv.h^{-1})			
		Low LET component		High LET component	
		NAUSICAA	RP114	NAUSICAA	NM2
PRG-YMC[a]	9.5	1.99 ± 0.09[d]	1.7 ± 0.1	2.24 ± 0.32	1.0 ± 0.1
YMC-PRG	10.1	2.41 ± 0.05	2.0 ± 0.1	2.38 ± 0.25	1.3 ± 0.1
PRG-YMC	10.7	2.90 ± 0.09	2.3 ± 0.1	3.98 ± 0.42	1.6 ± 0.2
YMC-PRG	11.3	3.29 ± 0.14	2.6 ± 0.2	3.35 ± 0.33	1.8 ± 0.2
PRG-ABH[c]	11.3	1.62	1.40	1.91	0.99
ABH-PRG[c]	10.0[b]	1.27	1.23	1.00	0.60
ABH-BKK	11.3	0.91 ± 0.04	1.08 ± 0.07	0.84 ± 0.10	0.48 ± 0.08
BKK-ABH	10.1	0.72 ± 0.04	0.89 ± 0.07	0.52 ± 0.08	0.29 ± 0.04

[a]PRG-Prague, YMC-Montreal, ABH-Abu Dhabi, BKK-Bangkok.
[b]Average value from three levels (10.1, 11.3 and 9.5 m).
[c]For PRG-ABH and ABH-PRG the actual values varied during the flight.
[d]Here for all instruments and in all Tables 1 to 3 — only statistical uncertainties.

ON-EARTH REFERENCE FIELDS; CORRECTION FACTORS

Apparent dose equivalent values do not give H*(10) values on board because of the different energy spectrum (neutrons) resp. depth dose distributions (low LET component). The appropriate correction factors could be obtained through the analysis of spectral data. In the case of low LET radiation it has been found that the apparent H*(10) values should be multiplied by a factor of 1.25[11], this is close to the ratio experimentally observed for the region above 50°N comparing NAUSICAA values (wall thickness 10 mm) with the data from other detectors.

For the neutron component, available spectra differed too much to estimate the correction factor properly by means of a theoretical approach. Reference fields have recently been formed behind the shielding of high energy accelerators at Dubna[12], and CERN[13]. Several years ago it was established that the appropriate correction factor for rem counters and/or etched track detectors in Dubna high energy fields is about 2[14]. In a much more comprehensive way this correction factor has been confirmed during a series of experiments performed at CERN reference fields behind the concrete shielding[2,13]. Typical results obtained by us are shown in Table 4. The average values of the ratio of NAUSICAA and other detectors' readings are close to 2 as well as in the case of on-board measurements (see Table 3 — average ratio equal to 1.88 ± 0.08 for NM2 and 1.88 ± 0.14 for BDNDs).

CONCLUSIONS

(1) After correction factors are applied, the level of aircraft crew exposure can be estimated at $\geq 50°N$ to be about 6 to 8 μSv per hour at typical flight altitudes (11 km). Low and high LET components contribute to this figure roughly equally, the average quality factor measured with NAUSICAA TEPC is equal to 1.80 ± 0.12. When going towards the equator these figures decrease by a factor of about 2 to 3 for low LET radiation, 4 to 5 for the high LET component. The overall uncertainty is estimated to be about ±15%.

(2) The figures given correspond to the heliocentric potential between 300 and 400 MV, minimum and maximum values can differ by a factor up to 1.6[15]. The studies described in this contribution should therefore be continued to improve knowledge of the correlation between solar cycle phase and the level of exposure (15% difference between 1995 and April 1991 corresponds well to that calculated for the different helocentric potentials).

(3) Such studies should also be continued to help in further development of calculations[2].

(4) If 500 h of flight above 10 km is taken as a typical value, annual exposure would be about 4 mSv with ICRP 21 quality factors, about 5 mSv when ICRP 60 recommendations[1] are adopted. It exceeds values typical for medicine (1 mSv) or industry (1.6 mSv)[16]. Particular problems can arise for pregnant women. Regarding the dosimetry to be performed, the main conclusions are formulated in the review paper presented at this symposium[2]. It

Table 3. Integral dose equivalent readings of high LET instruments; February 1995.

Flight route*	Apparent high LET equivalent (μSv)			
	NAUSICAA	NM2	Bubble damage detectors	
			PND	BD100R
PRG-YMC	27.3 ± 3.1	12.3 ± 2.0	12.8 ± 2.0	12.3 ± 2.0
YMC-PRG	17.0 ± 1.7	9.2 ± 1.2	10.3 ± 1.8	6.7 ± 1.4
PRG-ABH	9.4 ± 0.4	4.8 ± 0.8	3.6 ± 0.7	4.2 ± 0.9
ABH-BKK	4.1 ± 0.8	2.3 ± 0.4	1.5 ± 0.5	2.5 ± 0.7
BKK-ABH	3.1 ± 0.5	1.7 ± 0.3	1.9 ± 0.8	2.0 ± 0.8
ABH-PRG	5.8 ± 1.0	3.5 ± 0.5	4.7 ± 1.3	3.5 ± 0.8

*For flight altitude abbreviations, see Table 1.

Table 2. Integral dose equivalent readings of low LET equipment; February 1995.

Flight route*	Apparent low LET dose equivalent (μSv)				
	NAUSICAA	RP114	D222	DMC90	TLDs
PRG-YMC	20.9 ± 0.7	16.9 ± 0.8	18.0 ± 2.0	12.0 ± 1.3	—
YMC-PRG	17.5 ± 0.5	13.8 ± 1.0	13.8 ± 1.5	9.2 ± 1.5	34 ± 10**
PRG-ABH	7.9 ± 0.8	6.9 ± 0.7	7.5 ± 1.1	5.8 ± 1.3	—
ABH-BKK	4.5 ± 0.7	5.3 ± 0.8	5.0 ± 0.7	3.8 ± 0.5	—
BKK-ABH	4.2 ± 0.6	4.2 ± 0.6	4.5 ± 0.7	3.5 ± 0.6	—
ABH-PRG	7.4 ± 0.9	7.1 ± 1.1	7.4 ± 1.1	5.0 ± 0.7	24 ± 8**

*For flight altitude and abbreviations see Table 1.
**For the total route from Prague to Prague.

should only be emphasised that our experience demonstrated the difficulties in measuring doses to aircraft crews using existing passive individual dosemeters.

ACKNOWLEDGEMENTS

The author is much obliged to A. Rannou and J.-F. Bottollier-Depois for their support in measurements with NAUSICAA equipment.

This work was partially supported in the framework of CEC Project CT92–0026.

Table 4. Comparison of readings of high LET (neutron) radiation measuring instruments at CERN reference fields behind the concrete shielding.

Position[13]	Read values of H*(10)* (pSv) by		
	NAUSICAA	NM2 rem counter	BDND
Top concrete — T6	524 ± 25**	267 ± 7	270 ± 20
Side concrete — S2	644 ± 60	303 ± 7	310 ± 20

*In terms of ^{252}Cf neutrons; per 1 count of PIC monitor; $E(p + \pi^+) \sim 205$ GeV/c.
**Only statistical uncertainties.

REFERENCES

1. ICRP. *1990 Recommendations of the International Commission on Radiological Protection*. ICRP Publication 60 (Oxford: Pergamon Press) (1991).
2. Bartlett, D. et al. *Dosimetry for Occupational Exposure to Cosmic Radiation*. Radiat. Prot. Dosim. **70**(1–4), 395–404 (This issue) (1997).
3. Spurný, F. et al. *Dosimetric Characteristics of Radiation Fields on Board CSA Aircraft as Measured with Different Active and Passive Detectors*. Radiat. Prot. Dosim. **48**, 73–78, (1993).
4. Spurný, F., Votočková, I. and Bottolier-Depois J.-F. *Aircrew Exposure on Board a Subsonic Aircraft Studied with Complex Set of Dosimetric Methods*. Radioprotection (France), (to be published).
5. Bousset P. et al. *Description de l'Instrument NAUSICAA concu pour effectuer des Mesures Instantanées de H., D. et T.E.L. en Champs Complex*. In: Proc. IRPA 8, Montreal, May 1992, Acte de Congrés, pp. 463–466.
6. Viererbl, L., Nováková, O. and Jursová, L. *Combined Scintillation Detector for Gamma Dose Rate Measurements*. Radiat. Prot. Dosim. **20**, 272–276 (1990).
7. ICRP. *Data for Protection against Ionizing Radiation from External Sources*. Publication 21 (Oxford: Pergamon Press) (1971).
8. Spurný, F. *Methods of Dosimetry for External Radiation and their Use*. DSc Thesis (in Czech), Prague, July 1984.
9. Bartlett, D. private communication, 1995.
10. Wilson J. W. et al. *Transport Methods and Interactions for Space Radiation*. NASA Reference Publication 1257 (December 1991).
11. Michalik, V., Pernicka, F., Spurny, F. and Nguyen, V. D. *Some Aspects of the Exposure of Aircrew Members to Cosmic Radiation*. Radiat. Prot. Dosim. **54**, 255–258, (1994).
12. Aleinikov V. E., Bamblevskij, V. P., Komochkov, M. M., Krylov, A. R., Mokvev, Ya. V. and Timoshenko, G. N. *Reference Neutron Fields from Metrology of Radiation Monitoring*. Radiat. Prot. Dosim. **54**, 57–60 (1994).
13. Höfert M. and Stevenson, G. R. *The CERN-CEC High Energy Reference Field Facility*. Report CERN/TIS-RP/94-02/CF (Geneva: CERN) (January, 1994).
14. Spurný, F. and Bamblevskij, V. P. *Response of Some Detectors to Neutrons of Reference Fields at the JINR, Dubna*. Report IRD CAS 248/88 (in Russian) (1988).
15. O'Brien, K. et al. *Atmosphere Cosmic Rays and Solar Energetic Particles at Aircraft Altitudes*. In: Proc. NRE VI, Montreal, June 1995.
16. Prouza, Z., Spurný, F., Klener, V., Fojtikova, I., Fojtik, P. and Podskubkova, J. *Occupational Radiation Exposure in the Czech Republic*. Radiat. Prot. Dosim. **54**, 333–336 (1994).

RESULTS OF DOSIMETRIC MEASUREMENTS IN SPACE MISSIONS

G. Reitz†, R. Beaujean‡, J. Kopp‡, M. Leicher‡, K. Strauch†, and C. Heilmann§
†DLR, Institut für Luft- und Raumfahrtmedizin
Abt. Strahlenbiologie, D-51140 Köln, Germany
‡Universität Kiel, Institut für Kernphysik
Olshausenstr. 40/60, D-24118 Kiel, Germany
§Centre de Recherches Nucleaires, GADVI
23 rue du Loess, F-67037 Strasbourg, France

Abstract — Detector packages consisting of thermoluminescence detectors (TLDs), nuclear emulsions and plastic nuclear track detectors were exposed in different locations inside spacecraft. The detector systems, which supplement each other in their registration characteristics, allow the recording of biologically relevant portions of the radiation field independently. Results are presented and compared with calculations. Dose equivalents for the astronauts have been calculated based on the measurements; they lie between 190 $\mu Sv.d^{-1}$ and 860 $\mu Sv.d^{-1}$.

INTRODUCTION

The radiation field in near earth orbit is composed of electromagnetic radiation and charged particles of solar and galactic origin, as well as of albedo neutrons, produced by interactions of the primary galactic radiation with the Earth's atmosphere. Photons and particles with energies between some eV and up to 10^{18} eV are present. In orbits between 200 and 600 km the main absorbed dose inside spacecraft is contributed by protons trapped in the Earth's magnetic field in the region of the South Atlantic Anomaly (SAA), an area where the radiation belt comes closer to the Earth's surface due to a displacement of the magnetic dipole axes from the Earth's centre. Outside spacecraft the dose is dominated by the electrons of the horns of the radiation belt located at about 60° latitude in the polar regions. The particle fluxes inside the belts are modulated by the 11-year solar cycle. With increasing solar activity proton flux decreases while electron flux increases and vice versa. The cosmic ray flux is highest at minimum solar activity and nearly independent of the altitude in the above mentioned region, but it depends strongly on the orbit inclination. The flux increases from low to high latitudes. Only cosmic ray particles of very high energies are able to penetrate the magnetic field to low latitudes. At 100 MeV per nucleon the cosmic ray fluxes differ by a factor of about 10 between maximum and minimum solar activity conditions, whereas at about 4 GeV only a variation of about 20% is observed. Secondary particles are produced by interactions of the galactic cosmic rays and the protons of the radiation belt with the nuclei of the shielding material and with those of the bodies of the astronauts. In most of such interactions the incoming particle ejects nucleons out of the target nucleus and continues its path as a still highly energetic particle. The highly excited target nucleus releases some additional nucleons and ends up as either a stable or radioactive nuclide. In such nuclear disintegration 'stars' mainly protons, neutrons and alpha particles are produced.

Since the beginning of manned spaceflight the problem of radiation protection from the multiple sources of ionising radiation of the space environment has been a permanent topic of experimental biomedical research in nearly all spaceflight missions. So far, it has not been possible to provide a quantitative description of the radiation field, especially at the edge of the radiation belt. Dosemeters were therefore part of each manned mission.

In this work the sparsely and the densely ionising components of the radiation field have been measured with TLDs, plastic nuclear track detectors and nuclear emulsions. Based on the obtained data equivalent doses are determined.

METHODS

The contribution of the sparsely ionising component of the radiation field was determined with lithium fluoride TLDs. In the energy range of interest, the sensitivity of this material can be considered constant. The fluence of heavy charged particles and their linear energy transfer (LET) spectrum was determined with plastic track detectors of different LET thresholds for particle registration. The detector systems used are diallylglycol carbonate (CR-39), cellulose nitrate-Kodak (CN_K), cellulose nitrate-Daicel (CN_D) and polycarbonate (Lexan). Since the registration range of no single plastic detector covers the relevant LET portion of the space radiation spectrum, only such a combination of the matched spectra from different detector systems can generate LET spectra adequate for dosimetric calculations. The density of nuclear disintegration stars was determined in nuclear emulsions. The absorbed dose deposited by neutrons has been estimated from the differences between doses recorded in lithium fluoride detector materials differing in their relative contents of the isotopes ^{6}Li and ^{7}Li. For this purpose Harshaw lithium fluoride detector materials TLD-600 and TLD-700 were used.

Nuclear track detector packages were located at different places inside MIR station during the German MIR92 mission and the ESA EUROMIR94 and in several Spacelab missions. During D2 in addition each Payload Specialist (PS) was equipped with three track detector packages. These dosemeters were permanently attached to the waist, the neck and the leg of the PS.

To obtain the dose equivalent contributed by the different radiation types the absorbed dose has to be multiplied by a weighting or quality factor. This factor takes into account that the radiation effect is not only dependent on the absorbed dose, but also on the type and energy of the radiation causing that dose. For example, heavy charged particles with higher energy loss per unit track length (LET) are more effective in producing biological effects than electrons or fast protons with lower rates of energy loss. Since the quality factor is related to LET, the determination of the contribution of the densely ionising radiation to the equivalent dose requires a measurement of the fluence rate as a function of LET. This was done only for heavy charged particles. For the other components the absorbed dose was converted into dose equivalent by means of mean quality factors determined as spectral averages from known or postulated energy spectra of the constituents.

The dose equivalent contributed by nuclear disintegration stars has been calculated with the assumption of a mean number of particles per disintegration (only charged particles) of 3.75 (2.59 protons and 1.6 alpha particles) and a mean particle energy of 14.2 MeV[1]. The mean Q has been determined to be 7.7 using the Q(L) defined in ICRP 60[2].

The number of low energetic neutrons measured is correlated to the high energetic neutron flux. Therefore the number of thermal neutrons, which are calculated from the difference of the TLD-600 and TLD-700 readings, may serve as a lower flux limit for fast neutrons. The mean energy of the neutrons produced in nuclear disintegration stars is taken to be about 7.9 MeV[1]. The energy spectrum of the albedo neutrons coming from the Earth's atmosphere shows two maxima, one at 10 MeV and one at about 100 MeV[3]. Therefore the calculation of the dose equivalent was done choosing the mean neutron energy of 10 MeV and a mean weighting factor of 10, which slightly overestimates the neutron equivalent dose.

The sparsely ionising radiation dose is the dose measured in TLD-700 corrected by the contribution of the stars and by the heavy particles with LET values higher than 10 keV. μm^{-1} using the LET dependence of the TL signal after Horrowitz[4]. The dose equivalent is obtained by multiplying this dose with a mean Q of 1.3 which was chosen based on the energy spectra of the protons present in the radiation belt[5] (see also Ref. 6),

since the protons are the dominant contributors to dose inside a spacecraft. There is only a minor difference using ICRP 21[7] instead of ICRP 60 values in the dose equivalent determination for the sparsely ionising component.

DOSIMETRIC RESULTS AND DISCUSSION

The flight parameters of the missions under investigation are given in Table 1. There were two low inclination missions, D2 and IML2, all others were high inclination missions.

In Figure 1 total absorbed doses determined from the highest records of the used detector systems are compared to measurements in TLD-700 chips and to calculated doses for 10 g.cm^{-2} and 20 g.cm^{-2} shielding. Calculations are done with the AP8 model[8] and the Kiel transport code for cosmic rays[9] and are the sum of the trapped proton dose and the galactic cosmic ray dose. Doses determined from TLD-700 readings are lower than the total absorbed doses, since TLD-700 has a response to high LET particles substantially lower than 1. Since the mean shielding inside Spacelab is assumed to be at least 20 g.cm^{-2}, considerable differences are observed between measurements and calculations. For lower altitudes the model underestimates the dose considerably, for higher altitudes it comes closer to the measurements. In addition, it is well known that for solar minimum conditions the model tends to overesti-

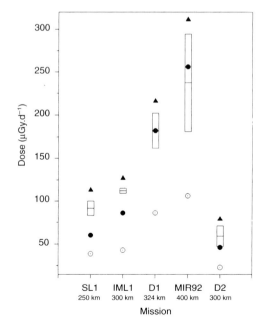

Figure 1. Absorbed doses determined from the whole data set of the detector packages (△) compared with readings from TL700 chips (□) and with dose calculations for 10 g.cm^{-2} (●) and for 20 g.cm^{-2} aluminium shielding (○).

mate the dose. The inclusion of the contribution of secondary products to the dose in the model may help to improve the situation.

The complete set of measurements is compiled in Table 2. Variations in absorbed doses less than 25% are observed in most of the Spacelab missions. The exception was IML2 in which the absorbed dose differs by about 60%, variations of the same order were observed in both MIR missions. These variations are due to the different shielding distributions around the dosemeters. Those distributions are only approximately known and depends on the arrangement of the equipment inside the spacecraft. Absorbed doses varied over the human body by about 20% during D2. During SL1, IML1 and MIR92 the solar activity was distinctly increased compared to the remaining missions which were close to solar minimum conditions. Comparing the data an increase of absorbed dose with altitude and with decreasing solar activity is clearly shown.

For a given inclination the flux of heavy particles inside a spacecraft in low earth orbits depends only on the solar activity and the mass shielding around the dosemeters. Looking at the data of the high inclination missions one may observe for most of the data an increase of flux with decreasing solar activity, but by no means so clearly expressed as for the absorbed dose data. Variations in fluxes up to a factor of 2 are observed and even in one mission, IML2, a variation by a factor of 6. The reason for that and also for the variations in flux inside is the strong influence of the shielding distributions around the single dosemeter positions inside the spacecraft. The spread of measured data from different positions emphasises the importance of actual measurements. The strong influence of inclination on the heavy ion flux is clearly expressed in the highly reduced flux for high LET (low energy) particles for the low inclination missions. This is demonstrated in Figure 2 in which measurements of heavy ion fluxes are shown as a function of LET together with calculations for the missions MIR92, IML1 and D2[10].

Table 1. Mission parameters.

Mission	Inclination (grad)	Altitude (km)	Mission duration (h)	Launch date
SL1	57	250	247.5	28 November 1983
D1	57	324	168	25 October 1985
IML1	57	348	194	22 January 1992
MIR92	51.5	400	190	17 March 1992
D2	28.5	296	240	26 April 1993
IML2	28.5	296	353	8 July 1994
EUROMIR94	51.5	400	756	4 October 1994

Table 2. Dosimetric measurements in recent manned missions.

Mission	Absorbed dose ($\mu Gy.d^{-1}$)	Stars ($cm^{-3} d^{-1}$)	Neutrons ($\mu Gy.d^{-1}$) (estimate)	Total planar particle fluence rate with LET > *** ($cm^{-2}.d^{-1}$)	
				1.7 $GeV.cm^{-1}$	1.3 $GeV.cm^{-1}$
SL1	100 ± 10	730 ± 14	*	0.55 ± 0.04	0.95 ± 0.05
	83 ± 8	470 ± 9		0.32 ± 0.04	0.64 ± 0.07
D1	202 ± 8	503 ± 21	5.2	0.53 ± 0.07	2.58 ± 0.15
	162 ± 7	436 ± 18		0.34 ± 0.05	2.09 ± 0.13
IML1	115 ± 1	743 ± 28	3.6	0.43 ± 0.03	1.33 ± 0.09
	109 ± 2			0.21 ± 0.03	0.73 ± 0.08
MIR92	294 ± 13	612 ± 21	6.8	0.30 ± 0.05	1.78 ± 0.11
	178 ± 6			0.19 ± 0.03	1.29 ± 0.08
EUROMIR94	380 ± 7	*	4.6	0.78 ± 0.04	2.24 ± 0.06
	234 ± 7			0.48 ± 0.03	0.96 ± 0.04
D2**	71 ± 4	383 ± 22	3.9	<0.07	0.67 ± 0.11
	58 ± 3	54 ± 7			0.46 ± 0.11
IML2	114 ± 11	651 ± 28	3.4	<0.07	1.82 ± 0.10
	73 ± 4	491 ± 16			0.29 ± 0.04

*Not determined.
**D2 data are from personal dosemeters, only stars are determined from environmental dosemeters.
***Number of particles passing through a unit area of a planar surface. Measurements in CN-Daicel and CN-Kodak.

The contribution of the secondaries, stars and neutrons, which are produced mainly in interaction of the galactic cosmic ray and belt protons with the nuclei of the spacecraft material or the human body, to the radiation environment is not well known. The measured numbers of stars in the high inclination missions do not show the expected influence of the solar activity. The D2 measurements are in the right direction, but the great difference in the measurements is difficult to understand. Our own estimates of the fast neutron flux range from 0.7 to 1.3 $cm^{-2}.s^{-1}$, which seems to be in good agreement with measurements of Haskins[11] for a shuttle flight at solar maximum at an altitude of 300 km and an inclination of 57° showing fluxes ranging from 0.68 to 1.06 $cm^{-2}.s^{-1}$ behind different shieldings. On the other hand, the measured neutron flux of EUROMIR94 is too low compared to the MIR92 flux. To improve the determination of the fast neutron flux, the CR-39 track dosemeter of Luszik-Bhadra[12] was recently included in our detector packages for the running EUROMIR95 mission and is planned to be used in future spaceflights.

Table 3 shows the contribution of the different radiations to the total absorbed dose and to the dose equivalent for the MIR92 (high inclination) and for the D2 (low inclination) mission. Although minor in their contribution to the absorbed dose, the densely ionising components always contribute about 50% to the dose equivalent or even more. A compilation of absorbed dose rates, quality factors and dose equivalent rates and mission dose equivalents is shown in Table 4. The mean quality factor of the radiation field varies from 2.1 to 4.1. Mission dose equivalents range from about 3 to 27 mSv.

The annual mean effective dose to the public from all natural sources is about 2.5 mSv. Recently, the annual radiation limit was reduced from 50 mSv to 20 mSv for occupationally exposed persons on earth by the International Commission on Radiation Protection (ICRP) partly based on a re-evaluation of the Hiroshima and Nagasaki data[2]. Crews of civil aircraft are considered in this recommendation as occupationally exposed. Their exposure to cosmic rays can reach 10 mSv per year. For spaceflight activities NASA uses guidelines[13] which recommend an annual maximum limit of 500 mSv and a career limit ranging from 1 Sv to 4 Sv for different ages and sexes. The mission dose equivalents for most of the European astronauts remained well below the mean yearly limit of 20 mSv recommended by ICRP as limit for occupationally exposed persons on earth. The radiation load added to the astronauts is in the range of the annual mean effective dose. Only during the 30 day mission of EUROMIR94 was the mean yearly limit exceeded.

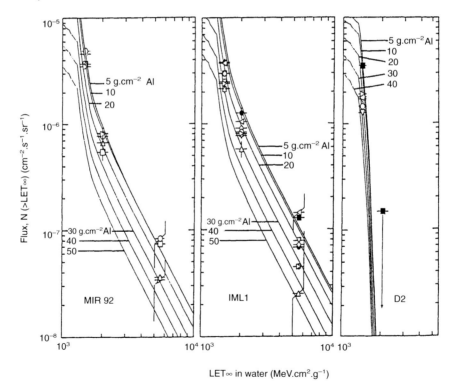

Figure 2. Measured and calculated LET spectra for MIR92, IML1, and D2 (same symbols for same dosemeter). For a certain LET value L_x the spectrum quotes the number of particles with LET higher than L_x.

CONCLUSIONS

Actual measurements at different locations inside spacecraft are the only way to obtain really confident information about the radiations present and are therefore indispensable for radiation protection measures. The reason for is that the absorbed dose is dependent on a variety of parameters which preclude at this time an accurate prediction of the dose in future missions.

The calculation of the dose equivalent as presented is the best that can be done at present. The two major uncertainties are the only rudimentary knowledge of biological effectiveness of the densely ionising rdiations and the potential influence of microgravity on the manifestation of radiation damage. Investigations have to be intensified on the ground and in space to get significant results.

ACKNOWLEDGEMENTS

The work at Kiel is supported by DARA under grants 50 QV 8566 and 50 WB 9418.

Table 3. Energy and dose equivalent rates of the sparsely ionising and the three densely ionising components of the radiation field in the MIR92 and D2 missions including the mean Q for these missions. Calculations using the highest recorded measurements.

Mission	Radiation	Dose rate ($\mu Gy.d^{-1}$)	Q	Dose equivalent rate ($\mu Sv.d^{-1}$)
MIR92	Sparsely ionising radiation	286	1.3	372
	Stars	5.2	7.7	40
	Heavy ions	7.7	20.8	160
	Neutrons	6.8	10	68
	Total	306	2.1	640
D2	Sparsely ionising radiation	64	1.3	83
	Stars	3.3	7.7	25
	Heavy ions	2.9	13.1	38
	Neutrons	4.6	10	46
	Total	75	2.6	192

Table 4. Dose rate, mean quality factor, dose equivalent rate and mission dose equivalent in manned missions.

Mission	Dose rate ($\mu Gy.d^{-1}$)	Mean quality factor	Dose equivalent rates ($\mu Sv.d^{-1}$)	Mission dose equivalent (mSv)
SL1	113	4.1	459	4.7
D1	216	2.2	478	3.3
IML1	126	3.0	377	3.0
MIR92	306	2.1	640	5.1
D2	75	2.6	192	1.9
IML2	120	2.2	260	3.8
EUROMIR94			≈860	≈27

REFERENCES

1. Schäfer, H. J. *Nuclear Emulsion Measurements of the Dose Contribution from Tissue Disintegration Stars on the Apollo-Soyuz Mission*. University of West Florida, Report No. 105-860/2 (1977).
2. ICRP. *1990 Recommendations of ICRP*. Report 60. (Oxford: Pergamon Press) (1991).
3. Armstrong, T. W. Chandler, K. C. and Barish, J. *Calculations of Neutron Flux Spectra Induced in the Earth's Atmosphere by Galactic Cosmic Rays*. J. Geophys. Res. **78**(16), 2715–2726 (1973).
4. Horowitz, Y. S. *Thermoluminescence and Thermoluminescent Dosimetry*, Vol. II (Boca Raton: CRC Press) (1984).
5. NCRP. *Guidance on Radiation Received in Space Activities*, Report 98 (Bethesda: MD: NCRP) (1989).
6. Haffner, J. W. *RBE for Protons and Alpha Particles*. NASA SP-71, pp. 513–525 (1964).
7. ICRP. *Recommendations of ICRP*. Report 26. (Oxford; Pergamon Press) (1977).

8. Sawyer, D. M. and Vette, J. I. *AP8 Trapped Proton Environment for Solar Maximum and Solar Minimum* (National Space Science Data Center, Goddard Space Flight Center) NSSDC/WDC-A-R&S 76-06 (1976).
9. Leicher, M. *Energieverlustspektren schwerer Ionen der kosmischen Strahlung innerhalb des Weltraumlabors IML1.* Diplomarbeit, Universität Kiel (1993).
10. Beaujean, R., Kopp, J., Leicher, M. and Reitz, G. *HZE-Dosimetry on MIR92, IML1 and D2.* Adv. Space Res. (To be published).
11. Haskins, P. S., McKisson, J. E., Weisenberger, A. G., Ely, D. W., Ballard, T. A., Dyer, C. S., Truscott, P. R., Piercey, R. B., Ramayya, A. V. and Camp, D. C. *Effects of Increased Shielding on Gamma-radiation Levels within Spacecraft.* Adv. Space Res. **12** (2–3), (2)461-(2)464 (1992).
12. Luszik-Bhadra, M., Alberts, W. G., d'Errico, F., Dietz, E., Guldbakke, S. and Matzke, M. *A CR-39 Track Dosemeter for Routine Individual Neutron Monitoring*, Radiat. Prot. Dosim. **55**, 285–293 (1994).
13. NCRP. *Guidance on Radiation Received in Space Activities.* Report 98 (Bethesda, MD: NCRP) (1989).

DOSIMETRY OF HIGH ENERGY NEUTRONS AND PROTONS BY ^{209}Bi FISSION

W. G. Cross† and L. Tommasino‡
†AECL, Chalk River Laboratories, Chalk River, Ontario, Canada K0J 1J0
‡ANPA, Via V. Brancati 48, 00144 Roma, Italy

Abstract — Fission of Bi by neutrons and protons of energies above 50 MeV provides a detector for the high energy part of cosmic ray neutron and proton doses received by personnel in commercial aircraft. A stack of damage track detectors and Bi radiators, of large total area, has been tested and has adequate sensitivity. Knowledge of the fission cross sections for neutrons and protons is critical for such measurements. Contrary to a common assumption that these are nearly equal, both old and recent measurements and recent calculations show that $\sigma_{Bi}(p,f)$ is more than twice $\sigma_{Bi}(n,f)$. Calculations indicate that Bi fission can probably determine dose equivalents from high energy cosmic ray neutrons to within ±30%, and from cosmic ray protons to within ± 25%. The response is nearly the same for equal dose equivalents of cosmic ray neutrons and protons. The accuracy for neutrons could be improved if cosmic ray neutron spectra were better determined.

INTRODUCTION

There is considerable current interest in determining radiation doses to the crews of high-flying aircraft. At the typical altitudes and latitudes of most commercial long distance flights, neutrons produce roughly half the dose equivalent, while protons produce around 10%. The neutron energy spectrum extends to at least several GeV and a substantial part of the neutron dose equivalent comes from energies above 50 MeV. At such energies, dose rate meters and dosemeters conventionally used in the MeV range have a poor energy response or are not sufficiently sensitive to measure accurately the low dose rates encountered.

A detector that has its principal neutron response at energies above 50 MeV is provided by the fission of ^{209}Bi. Bismuth fission chambers have been used since 1948 for measuring high energy neutrons around accelerators and in cosmic rays[1-3]. The practical threshold of the Bi(n,f) reaction is around 40–50 MeV. Damage track detectors that register the fission fragments from this reaction have recently been used for measurements of the high energy part of neutron dose equivalents in high flying aircraft[4], as outlined in the next section.

A critical requirement for such a dosemeter is knowledge of the fission cross section as a function of neutron energy. Old measurements of $\sigma_{Bi}(n,f)$ are not very accurate, but they indicate that at least up to 400 MeV the fission cross section for neutrons is two to three times smaller than that for protons, $\sigma_{Bi}(p,f)$, which has been determined much more accurately. Nevertheless, for over 30 years many authors (e.g. Hess et al[5] Wollenberg and Smith[6], Swanson and Thomas[7], Dinter and Tesch[8]) have assumed that the neutron and proton fission cross sections of Bi are nearly equal. Recent measurements and calculations confirm the early measured cross sections. The main aim of the present paper is to derive more reliable values of $\sigma_{Bi}(n,f)$ and to consider the capabilities of Bi fission for determining high energy neutron and proton dose equivalents in cosmic rays.

SENSITIVE CHAMBERS AND DAMAGE TRACK DETECTORS USING Bi FISSION

To have sufficient sensitivity to measure the small fluxes of high energy neutrons in cosmic rays, a Bi fission detector must have a very large area. Hess et al[5] described a fission chamber with an effective area of 65,000 cm^2, of 43 parallel plates coated with 1 mg.cm^{-2} of Bi. The very limited data on cosmic ray neutrons obtained with this chamber have been described[2]. Hewitt et al[9] tested an improved chamber[10] but reported no measurements. For measurements in commercial aircraft, such chambers have the disadvantages of complexity, bulkiness and a high sensitivity to microphonics.

To facilitate measurements in aircraft, a Bi fission detector based on the registration of damage tracks has

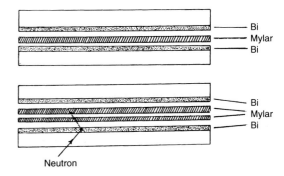

Figure 1. Arrangements of Bi radiators and Mylar detector foils in Bi fission damage track detectors. In the upper arrangement only one detector foil is surrounded by Bi, while in the lower, there are two Mylar foils in which a single fission fragment may produce two adjacent tracks.

been recently developed[4,11]. It is shown schematically in Figure 1. Large areas can be easily scanned by spark counting the etched-through holes induced by fission fragments in thin Mylar films. To increase the detector sensitivity, the plastic film is sandwiched between two Bi radiators, as shown in the upper part of Figure 1. This geometry also removes the front–back asymmetry of the detector. The Bi radiators are formed by a layer about 25 mg.cm^{-2} thick deposited on a 100 μm thick Mylar film. Because of the simplicity and low weight of such Bi radiators, large total detector areas (up to 1.5 m^2) are now conveniently obtained with a compact and light weight stack of similar detectors.

The detector background can be improved by using adjacent paired, thin detector films (lower part of Figure 1) and by requiring that a fission track be observed at the same position in both films. Coincidence events can be easily recorded as paired holes on the two geometrically-matched Al foil replicas[12]. Ultra low level counting can thus be achieved.

Table 1 shows the registration efficiencies, for 100 MeV protons, of stacks having 64 identical detectors with a total area of 14,400 cm^2, made of 4 or 6 μm detector films or of paired 4 μm films[4]. At the same proton energy, Hess et al[2] obtained an efficiency for their Bi chamber of 0.015 counts per proton.

In comparison with the Bi chamber, this compact detector stack, in addition to providing a similar registration efficiency, is insensitive to microphonics and other electronic noise. Because of the low background, adequate sensitivity for measuring typical, average dose equivalent rates in intercontinental flights (e.g. a few μSv.h^{-1}) can be achieved by using sufficiently long exposure times. From experience accumulated to date, the most practical detector consists of 6 μm films between two Bi radiators. This is dictated by a trade-off between simplicity and sensitivity.

DISTRIBUTION OF DOSE EQUIVALENT WITH NEUTRON ENERGY

The authors are aware of only three groups that have measured the fluence spectrum of high energy cosmic ray neutrons at altitudes typical of commercial aircraft routes[2,9,13]. All these measurements used thermal neutron detectors under various thicknesses of hydrogenous moderators (e.g. multispheres) as the main spectrometer, sometimes supplemented by various threshold detectors[2]. Although spectra up to 400 MeV were derived from multisphere data, these high energy results are very crude and probably unreliable.

These spectra, φ(E), are often plotted as fluence against MeV on a logarithmic energy scale. Such plots give the misleading impression that there are relatively few neutrons at high energies. By plotting fluence per unit logarithmic energy interval (lethargy) or, what is the same thing, E φ(E), and using a linear ordinate scale, the fluence between any two energies is proportional to the area under the curve between these energies.

Figure 2 shows two examples of measured cosmic ray neutron spectra (Hess et al[2], Nakamura et al[13]). A third set of three measured spectra (Hewitt et al[9,13]) shows very large differences in shape among themselves

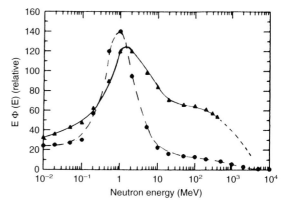

Figure 2. Measured spectra of cosmic ray neutrons. That of Nakamura et al[14] is for 11.28 km and 24° N latitude (▲), while that of Hess et al[2] is for various altitudes up to 12.2 km and at 44° N (●). Spectra are plotted per unit lethargy.

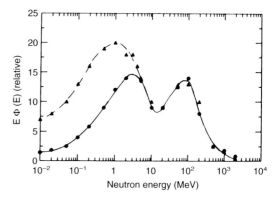

Figure 3. Cosmic ray neutron spectra at altitudes of (▲) 5 and (●) 17.5 km and geomagnetic latitude 42°, calculated by Merker[16].

Table 1. Registration efficiencies (recorded tracks per 100 100-MeV protons) for different detector film thicknesses.

Film thickness (μm)	Registration efficiency	
	Measured	Calculated
4	1.12 ± 0.2	1.2 ± 0.12
6	0.47 ± 0.10	0.80 ± 0.08
4 + 4	0.33 ± 0.14	0.47 ± 0.05

and from those of Figure 2. These data have recently been re-analysed[15]. Spectra calculated by Monte Carlo[16,17], of which two are shown in Figure 3, have a totally different shape, showing a strong peak of cascade neutrons at energies centred near 100 MeV. Similar cascade peaks are seen in recent measurements (e.g. Dinter and Tesch[8]) and Monte Carlo calculations[18] of spectra of neutrons leaking from shielding around high energy accelerators. While differences in spectral shape do occur at different latitudes, the large differences between the measured shapes are probably due mainly to experimental errors and illustrate how poorly cosmic ray neutron spectra are known.

The dose equivalent at any energy is given by multiplying the fluence by the appropriate conversion coefficient. Up to 200 MeV the recent conversion coefficients of Siebert and Schuhmacher[19] were used between $H^*(10)$ and fluence, ϕ. At higher energies the values of Sannikov and Savitskaya[20] were used. Above 400 MeV, $H^*(10)/\phi$ increases rapidly with energy and considerably increases the dose contribution of high energy neutrons.

The resulting distribution of ambient dose equivalent (per unit lethargy) for the spectrum of Nakamura et al[14] is shown in Figure 4. The Nakamura spectrum was extrapolated to 3 GeV as shown by the dotted curve in Figure 2. For this extended spectrum about 32% of the dose equivalent comes from neutrons above 50 MeV.

CROSS SECTIONS FOR FISSION OF ^{209}Bi BY PROTONS AND NEUTRONS

Protons

Measurements of proton fission cross sections are relatively extensive and, up to 500 MeV, reasonably accurate. These measurements, shown in Figure 5, have been evaluated by Fukahori and Pearlstein[21] who also calculated $\sigma_{Bi}(p,f)$ using the ALICE-P Monte Carlo code, obtaining reasonable agreement with measured data. A choice of parameters in the code leads to the three curves in this figure. The short-and-long-dashed curve was considered to give the best fit to the data. It is fitted by the empirical expression

$$\sigma_{Bi}(p,f) = 217\,[1-\exp(-0.00782\,(E_p-36.6))] \quad (1)$$

where the proton energy E_p is in MeV and cross sections are in millibarns. Measurements also extend to 30 GeV but it is not clear whether the cross section remains flat or decreases.

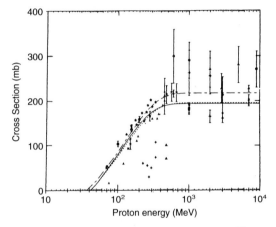

Figure 5. Comparison of measured values of the ^{209}Bi(p,f) cross section with those calculated with the ALICE-P code. The three curves correspond to different choices of parameters in the calculation. The short-and-long-dashed curve is considered to give the best fit and is fitted by Equation 1 (Fukahori and Pearlstein[21]).

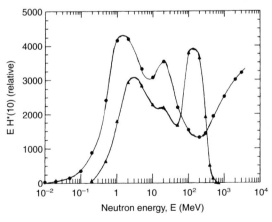

Figure 4. Distributions of $H^*(10)$ (per unit lethargy) for the cosmic ray neutron spectrum of Nakamura et al[14] (circles) and for a calculated spectrum outside the shielding of the CERN SPS accelerator[29] (triangles).

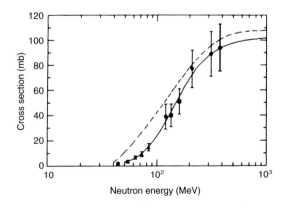

Figure 6. Cross sections for ^{209}Bi(n,f) measured by (▲) Kelly and Wiegand[22], (●) Goldanskii et al[24] and (■) Eismont et al[25]. The solid curve is an empirical fit to the data, given by Equation 2. The dashed curve is Fukahori and Pearlstein's fit to the ^{209}Bi (p,f) cross section[21] divided by 2.

Neutrons

At energies above 40 MeV, we are aware of only the measured data shown in Figure 6. Kelly and Wiegand[22] derived cross sections relative to the fission cross section of ^{232}Th. More recent measurements of $\sigma_{Th}(n,f)$ up to 400 MeV[23] permit absolute values of $\sigma_{Bi}(n,f)$ to be derived from the Kelly and Wiegand data. Goldanskii et al[24] measured cross sections from 120 MeV to 380 MeV. Both the above groups used neutrons from deuteron stripping, which resulted in considerable spreads in energy. The horizontal error bars in Figure 6 indicate the estimated full width at half maximum of the neutron beam — not the uncertainty of its mean energy.

Measurements at 135 and 160 MeV were made recently by Eismont et al[25]. These are considerably more reliable, because the neutron beam was nearly monoenergetic, and they confirm the older measurements. These authors also calculated the neutron cross sections using the LAHET Monte Carlo code, obtaining results in reasonable agreement with measurements.

The solid curve in Figure 6 is our empirical fit to the measured data, given by the smoothed step function

$$\sigma_{Bi}(n,f) = 103 - 103/[1+ (\log (E_n)/5.03)^{13.81}] \quad (2)$$

where E_n is the neutron energy in MeV and cross sections are in millibarns. At energies above 400 MeV it is very probable that the ratio of the Bi(n,f) and Bi(p,f) cross sections is nearly constant and this is assumed in subsequent calculations.

Fukahori and Pearlstein[21], noting that Bi(n,f) cross sections are roughly half those of protons, estimated the energy variation of neutron cross sections by dividing the values of Equation 1 by 2. This is shown by the dashed curve in Figure 6. At lower energies it significantly overestimates the experimental data.

Equations 1 and 2 appear to provide the most reliable cross sections for the Bi(p,f) and Bi(n,f) reactions currently available. The relative values for neutrons and protons are particularly important when Bi fission detectors are calibrated by a known fluence of high energy protons and then used to measure neutrons. This is the only practical method of calibration, since sources of nearly monoenergetic neutrons at energies above 100 MeV are rare.

Measurements of neutron spectra using Bi fission detectors, made with the assumption that the (n,f) and (p,f) cross sections are equal (e.g. Hess et al[2], Dinter and Tesch[8]) will have underestimated the high energy part of the spectrum by a factor of approximately 2.

DETERMINATION OF NEUTRON DOSE EQUIVALENTS

The relative response of a Bi fission detector for measuring H*(10) from monoenergetic neutrons of different energies is proportional to the quotient of $\sigma_{Bi}(n,f)$ and the conversion coefficient H*(10)/φ. Combining $\sigma_{Bi}(n,f)$ from Equation 2 with H*(10)/φ yields the variation of response with energy shown by the solid line in Figure 7. Only above about 100 MeV is the response really significant.

The variation of relative response from one broadly distributed spectrum to another will be very much smaller than for monoenergetic neutrons. For example, the average 'relative responses', given by $\sigma_{Bi}(n,f)\phi/H^*(10)$ are 176, 167, 182 and 174 mb.nSv^{-1}.cm^{-2} for the four widely varying spectra of Figures 2 and 3. Because the detector can be calibrated to read nearly correctly in any assumed spectrum, the error in measuring cosmic ray doses will depend on how closely this matches the actual spectrum. While the shape is not expected to vary much over the range of altitudes of importance, there is little knowledge of variations of shape with latitude as well as very considerable uncertainty in the general shape. It is estimated that dose equivalents from cosmic ray neutrons could be determined within ±30%.

DETERMINATION OF PROTON DOSE EQUIVALENTS

Spectra of cosmic ray protons have been determined in a number of reasonably consistent measurements. Figure 8 shows one example of a measured spectrum[26] and one of a calculated spectrum[27], both for a minimum of the solar cycle. The energy of the peak shifts with solar activity.

Values of conversion coefficients for protons were taken from ICRP 51[28]. They apply to $H_p(10)/\phi$ in a tissue slab 30 cm thick. The relative response of Bi fission for measuring dose equivalents from monoenergetic protons of different energies, $\sigma_{Bi}(p,f)\phi /H^*(10)$, is shown by the dashed curve in Figure 7. The mean value

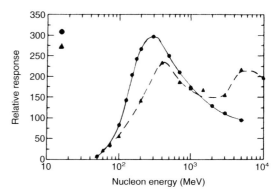

Figure 7. The relative reading of a Bi fission detector per unit dose equivalent of monoenergetic (●) neutrons or (▲) protons. This is proportional to the ratio of the Bi fission cross section to the appropriate conversion coefficient, shown here in mb.nSv^{-1}.cm^{-2}. In the ratio for protons, the sharp increase near 4 GeV is probably not real.

of this quantity, averaged over the calculated cosmic ray proton spectrum of Figure 8, is 176 mb.nSv^{-1}. cm^{-2}, in good agreement with the neutron value. Thus this detector has an approximately equal response to dose equivalents from cosmic ray neutrons and protons and it is not necessary to separate neutron and proton events. In contrast, for most other detectors such separation requires additional coincidence counters.

Combining the calculated spectrum of Figure 8 with these conversion coefficients gives the distribution of $H_p(10)$ with proton energy shown in Figure 9. Only a small part of this proton dose equivalent is below 200 MeV. Above this energy the proton efficiency of a Bi fission detector varies by only ± 25% even for monoenergetic protons, as shown in Figure 7. Hence such a detector can give a reasonably accurate measure of cosmic ray proton dose equivalents.

CONCLUSIONS

There are large discrepancies among different measurements, and between measurements and calculations, of the spectra of cosmic ray neutrons, which must be regarded as poorly known. Measurements and calculations of $\sigma_{Bi}(n,f)$ are consistent, although not very precise, and the values are less than half those of $\sigma_{Bi}(p,f)$. At energies above 100 MeV, Bi fission detectors can measure dose equivalents from monoenergetic neutrons within a factor of 2. For high energy cosmic ray neutrons, their potential accuracy is probably within ±30%, and would be improved by more accurate determinations of the spectrum and its variations with latitude. Dose equivalents from cosmic ray protons can probably be measured within ±25%. The Bi fission reaction has an approximately equal response to dose equivalents from cosmic ray neutrons and protons.

ACKNOWLEDGEMENTS

The contribution to this work of one of the authors (L.T.) has been made possible thanks to the Radiation Protection Research contract F13P-CT 92–0020 of the Commission of the European Communities. The authors are grateful to Dr W. Heinrich for providing unpublished spectra.

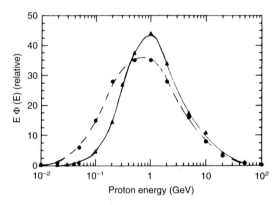

Figure 8. Examples of the cosmic ray proton spectra at solar minimum. The intensities shown are arbitrary (Meyer et al[26], Heinrich[27]). (▲) measured; (●) calculated. 10 km.

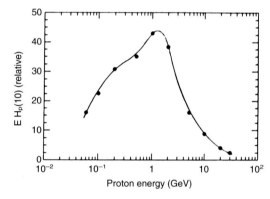

Figure 9. Distribution with energy of $H_p(10)$ in a 30 cm thick tissue slab, for the calculated proton spectrum shown in Figure 8. Dose equivalents are per unit lethargy.

REFERENCES

1. Wiegand, C. *High Energy Neutron Detector.* Rev. Sci. Instrum. **19**, 790–792 (1948).
2. Hess, W. N., Patterson, H. W., Wallace, R. and Chupp, E. L. *Cosmic Ray Neutron Energy Spectrum.* Phys. Rev. **116**, 445–457 (1959).
3. Smith, A. R. *Threshold Detector Applications to Neutron Spectroscopy at the Berkeley Accelerators.* In: Proc. First Symp. on Accelerator Radiation Dosimetry and Experience, Brookhaven. pp. 224–260 (1966).
4. Zhou, D., Cavaioli, M., Spurny, F., Teodori, K. and Tommasino, L. *Bismuth-fission Detector for High Energy Cosmic-ray Nucleons* (To be published) (1996).
5. Hess, W. N., Patterson, H. W., Wallace, R. and Chupp, E. L. *Delay-line Chamber has Large Area, Low Capacitance.* Nucleonics **15**(3), 74 (1957).
6. Wollenberg, H. A. and Smith, A. R. *Energy and Flux Determinations of High Energy Nucleons.* In: Proc. Second Int. Conf. on Accelerator Dosimetry and Experience, CONF 691101 (Technical Information Center, US Department of Energy) pp. 586–594 (1969).
7. Swanson, W. P. and Thomas, R. H. *Dosimetry for Radiological Protection at High Energy Particle Accelerators.* In: The Dosimetry of Ionizing Particles, Eds K. R. Kase, B. E. Bjarngard and F. H. Attix. San Diego, Academic Press. Vol. 3, pp.1–161 (1990).

8. Dinter, H. and Tesch, K. *Determination of Neutron Spectra Behind Lateral Neutron Shielding of High Energy Proton Accelerators.* Radiat. Prot. Dosim. **42**, 5–10 (1992).
9. Hewitt, J. E., Hughes, L., McCaslin, J. B., Smith, A. R., Stephens, L. D., Syvertson, C. A., Thomas, R. H. and Tucker, A. B. *Exposures to Cosmic Ray Neutrons at Commercial Jet Aircraft Altitudes.* In: Natural Radiation Environment III, Vol. 2, CONF 780422 (Technical Information Center, US Department of Energy) pp. 855–881 (1980).
10. McCaslin, J. B., Patterson, H. W., Smith, A. R. and Stephens, L. D. *Some Recent Developments in Technique for Monitoring High Energy Radiations.* In: Proc. First Int. Cong. on Radiation Protection, (Oxford: Pergamon Press) pp. 1131–1137 (1968).
11. Tommasino, L., Caggiati, F., Cavaioli, M., Notaro, M., Teodori, R., Torri, G., Zhou, D., Byrne, J. and O'Sullivan, D. *From a Complex Cosmic-ray Stack to a Simple Dosimeter System for Aircrew Exposure.* In: Proc. Int. Conf. on Natural Radiation Environment, In press (1996).
12. Geisler, F. H., Philips, P. R. and Walker, P. H. *Search for the Superheavy Element in Natural and Proton Irradiated Materials*, Nature **224**, 428–429 (1973).
13. Hewitt, J. E., Hughes, L., Baum, J. W., Kuehner, A. V., McCaslin, J. B., Rindi, A., Smith, A. R., Stephens, L. D., Thomas, R. H,. Griffith, R. V. and Welles, C. G. *Ames Collaborative Study of Cosmic Ray Neutrons: Mid-latitude Flights.* Health Phys. **34**, 375–384 (1978).
14. Nakamura, T., Uwamino, Y., Ohkubo, T. and Hara, A. *Altitude Variation of Cosmic Ray Neutrons.* Health Phys. **53**, 509–517 (1987).
15. Hajnal, F. and Wilson, J. *High-altitude Cosmic Ray Neutrons: Probable Source for the High Energy Protons of the Earth's Radiations Belts.* In: Proc. 8th Congress Int. Radiat. Prot. Assoc., Montreal. p. 1620 (1992).
16. Merker, M. *Contribution of Galactic Cosmic Rays to the Atmospheric Neutron Maximum Dose Equivalent as a Function of Neutron Energy and Altitude.* Health Phys. **25**, 524–527 (1973).
17. Armstrong, T. W., Chandler, K. C. and Barish, J. *Calculations of Neutron Flux Spectra Produced in the Earth's Atmosphere by Galactic Cosmic Rays.* J. Geophys. Res. **78**, 2715–2726 (1973).
18. Zazula, J. M. and Tesch, K. *Analysis of the Transverse Shielding Problem at Proton Accelerators using a Hadronic Cascade Code with Low Energy Particle Modules.* Nucl. Instrum. Methods **A286**, 279–294 (1990).
19. Siebert, B. R. L. and Schuhmacher, H. *Quality Factors, Ambient and Personal Dose Equivalent for Neutrons Based on the New ICRU Stopping Power Data for Protons and Alpha Particles.* Radiat. Prot. Dosim. **58**, 171–183 (1995).
20. Sannikov, A. V. and Savitskaya, E. N. *Ambient Dose and Ambient Dose Equivalent Conversion Factors for High Energy Neutrons.* Report CERN/TIS-RP/TM/93–38 (Geneva: CERN) (1993).
21. Fukahori, T. and Pearlstein, S. *Evaluation at the Medium Energy Region for Pb-208 and Bi-209.* In: Intermediate Energy Nuclear Data for Applications. INDC(NDS)-245, (Vienna: IAEA) pp. 93–128 (1991).
22. Kelly, E. L. and Wiegand, C., *Fission of Elements from Pt to Bi by High Energy Neutrons.* Phys. Rev. **73**, 1135–1139 (1948).
23. Lisowski, P. W. et al. *Neutron Induced Fission Cross Sections for ^{232}Th, ^{235,238}U, ^{237}Np and ^{239}Pu.* In: 50 Years with Nuclear Fission (Am. Nuclear Society, La Grange Park, Illinois) pp. 443–448 (1989).
24. Gol'danskii, V. I., Pen'kina, V. S. and Tarumov, E. Z. *Fission of Heavy Nuclei by High Energy Neutrons.* Soviet Physics JETP **2**, 677–687 (1956).
25. Eismont, V. P., Prokovyev, A. V., Rimski-Korsakov, A. A. and Smirnoff, A. N. *Neutron Induced Fission Cross Sections of ^{209}Bi and ^{238}U in the Intermediate Energy Region.* In: Proc. Int. Conf. on Nuclear Data for Science and Technology, Gatlinburg, Tenn. USA. pp. 360–362 (1994).
26. Meyer, P., Ramaty, R. and Webber, W. R. *Cosmic Rays — Astronomy with Energetic Particles.* Physics Today **27**, 23–33 (1974).
27. Heinrich, W. Private communication. Universität Gesamthochschule, Siegen, Germany (1995).
28. ICRP. International Commission on Radiological Protection. *Data for use in Protection against External Radiation.* ICRP Publication 51, (Oxford: Pergamon Press) (1987).
29. Roesler, J. J. and Stevenson, G. R. *July, 1993 CERN-CEC Experiments: Calculation of Hadron Energy Spectra from Track Length Distribution using FLUKA.* Report CERN/TIS-RP/IR/933–47, (1993).

MEASUREMENTS OF NEUTRON SPECTRA AT THE STANFORD LINEAR ACCELERATOR CENTER

V. Vylet, J. C. Liu, S. H. Rokni and L.-X. Thai
Stanford Linear Accelerator Center
PO Box 4349, M.S. 48
Stanford, CA 94309, USA

Abstract — The purpose of the present measurements was to determine neutron spectra outside the shielding of high energy electron accelerators, with a special emphasis on the high energy (>20 MeV) portion of the spectra. The spectrometer system used consists of a standard set of Bonner spheres augmented by a custom made 30.48 cm (12″) sphere with a 1 cm inner layer of lead, which increases its response in the high energy region. The system was further complemented by a plastic scintillator used for the measurement of ^{11}C activation. A pair of rem meters was also used: one standard Anderson-Braun, and one with an inner layer of lead which has an extended high energy response. Measured spectra and relevant dosimetric quantities outside shielding at three different electron beam energies are presented.

INTRODUCTION

In neutron fields outside the shielding of high energy accelerators, a significant fraction of the total dose equivalent may be generated by neutrons with high energies (above 20 MeV). Although the existence of this high energy component is undisputed, there are limited detailed measurements, especially around electron machines. Depending on the neutron production mechanism and the energy of the interacting photon, photo neutrons can be divided into three groups: giant resonance (from a few MeV to about 30 MeV), pseudo-deuteron (about 30-300 MeV) and photopion (above 140 MeV) production. Our measurements include spectra resulting from three different electron energies: 46.6 GeV at the Final Focus Test Beam Facility (FFTB), 2.3 GeV at the Booster-Synchrotron of the Stanford Synchrotron Radiation Laboratory (SSRL), and 120 MeV at the SSRL Linac. The above production processes will therefore contribute in variable proportions to the measured spectra.

INSTRUMENTATION

Our regular Bonner sphere system consists of five moderators with diameters of 5.08, 7.62, 12.7, 20.32, 25.4 and 30.48 cm (or 2, 3, 5, 10 and 12 in). The thermal neutron detector is 4×4 mm^2 LiI crystal. All the moderators and the scintillation probe were purchased from Ludlum Measurements, Inc. The counting system consists of a portable MCA, Nomad from ORTEC, and a laptop computer. Adopting the same approach as Biratari et al[1] for their LINUS rem meter, a 1 cm thick inner lead layer was added to a 12″ custom sphere to extend its energy response, by taking advantage of the (n, xn) reaction in lead at higher neutron energies. Whenever possible, the Bonner sphere system was complemented with ^{11}C activation, which has a reaction threshold of 20 MeV, using a 12.7×12.7 cm^2 (5 × 5 in^2) cylindrical plastic scintillator.

In addition to the spectrometric system, a pair of rem meters were also used: one standard Anderson-Braun, and one custom made with an inner layer of lead, following the LINUS design. These rem meters were used with standard NIM electronics. To monitor the stability of the measured radiation fields, a moderated BF$_3$ counter was used at a fixed location during measurements at each of the three facilities. In some instances, readings of a fixed rem meter or charge on a Faraday cup were also used for this purpose.

DATA ANALYSIS

Detector responses were normalised to a selected count rate of the reference monitor. Bonner sphere and carbon activation data were processed using the BUNKI[2] code to unfold neutron spectra. A 'flat' 1/E first guess spectrum was assumed over all but the last three energy groups, where a 1/E^2 or 1/E^3 tail was applied above 134 MeV. The ^{11}C response function was integrated in the multisphere response matrix in terms of saturated activity per unit fluence rate, which enables inclusion of the plastic scintillator as just another 'sphere' in the unfolding process. Fluence-to-dose equivalent conversion factors h_ϕ from 10CFR835[3] were used. The energy response of the custom 12″ sphere was calculated using the LAHET[4] code system, while Sanna's data were used for the other spheres. Energy responses of the custom and regular 12″ spheres are shown in Figure 1. The absolute sensitivities of the Bonner sphere system and the two rem meters were determined using a calibrated ^{252}Cf source and are described in detail by Thai[5]. In low duty factor accelerators at SLAC the dead time effect in the measurements was mitigated by time spread of the thermal neutron signal due to moderation. The dead time correction was estimated to be less than 4% at SSRL SPEAR, 3% at SSRL Linac and 1% at FFTB.

INTERCOMPARISON

In order to verify the validity of our response matrix, notably its high energy end, it was very desirable to perform measurements in well characterised high energy neutron spectra. The CERN-CEC high energy reference field facility[6] has been used in the past for a series of CERN-CEC intercomparison experiments. The available neutron spectra and relevant field quantities were investigated by a number of experimental teams using Bonner spheres, ^{11}C activation, low pressure tissue-equivalent proportional counters and activation detectors. In addition, these spectra and field quantities were also calculated using the FLUKA[6] code. Our participation in the fourth intercomparison therefore provided us with the opportunity to test our system.

The radiation fields were created by a beam of 205 GeV.c^{-1} protons and pions incident on a 50 cm long copper target. These fields were measured behind four different shielding configurations. The two measurements reported here were done on the roof of the beamline at 90° from the target, one above a 40 cm thick iron shield and one above 80 cm of concrete. The general shape of two neutron spectra measured with Bonner spheres (BS), presented in Figure 2, is in good agreement with FLUKA calculations[6], although the measured low energy portion is slightly higher. A comparison of measured against calculated fluence is presented in Table 1. Both these data and spectra in Figure 2 are normalised to 10^6 counts of the PIC (Precision Ionisation Chamber) monitor, assuming 2.34×10^4 beam particles per PIC count. Table 1 indicates a better agreement for the concrete shielding, but the measurements seem systematically to underestimate the high energy fluence and overestimate the low energy part, when compared with calculations. The same type of discrepancies was also observed by other experimenters.

MEASUREMENTS AT SLAC

FFTB

The first series of measurements was performed around the heavy shielding of the 46.6 GeV beam dump of the FFTB facility, illustrated in Figure 3. In a preliminary survey the highest dose equivalent rates were found

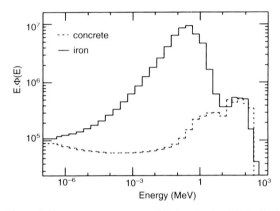

Figure 2. Neutron spectra measured during the 4th CERN-CEC intercomparison.

Table 1. Comparison of measured fluence with FLUKA calculations at the CERN-CEC facility for two shielding configurations.

	Iron			Concrete		
	$\Phi < 20$	$\Phi > 20$	Φ total	$\Phi < 20$	$\Phi > 20$	Φ total
FLUKA	11.44	0.66	12.10	0.83	0.62	1.45
BS	16.51	0.50	17.01	1.08	0.49	1.57
Ratio	1.44	0.78	1.41	1.30	0.79	1.08

The three columns in each case indicate fluence (cm^{-2} per 10^6 PIC counts — see text) below 20 MeV, above 20 MeV, and total.

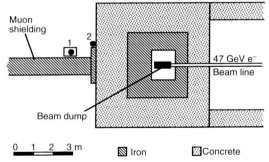

Figure 3. Schematic drawing of the FFTB dump area. Measurement locations are indicated with numbered round bullets.

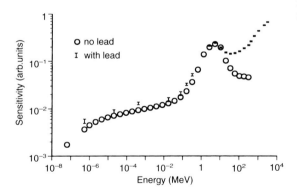

Figure 1. Comparison of response functions for a 12″ Bonner sphere with and without a 1 cm inner lead layer, located at 4 cm from the centre.

at location 2. The neutron spectrum measured at this location is shown in Figure 4; the spectrum at location 1 is very similar. Both of these spectra exhibit a two-peak structure, with a first peak around 1 MeV and the second slightly below 100 MeV, resembling the spectrum behind concrete shielding measured at CERN. The presence of high energy neutrons was also confirmed by measurements with the pair of rem meters. The relevant quantities from FFTB measurements are summarised in Table 2: dose equivalent rates measured with Anderson-Braun (AB) rem meter, LINUS and Bonner spheres (BS), average energy E_{av}, and fluence-to-dose equivalent conversion factor h_ϕ. The quantity in the last column, H26, is the fraction of the total dose equivalent from neutrons with energies above 26 MeV. The choice of this particular limit was somewhat dictated by the energy group structure used in the BUNKI code. The response of the LINUS rem meter was consistently higher than the standard Anderson-Braun and agreed with the Bonner spheres to 20%.

SSRL SPEAR

SPEAR is an electron storage ring used for synchrotron radiation research. The electron beam originates in a 120 MeV linac and is further accelerated to 2.3 GeV in a booster synchrotron. The beam is then injected into the SPEAR ring. Our measurements were performed on the roof of SPEAR, at 90° above a Faraday cup, which is at the end of the injection line from the booster synchrotron. The Faraday cup is shielded by 10.2 cm (4″) of lead and the concrete roof blocks are 61 cm (2 ft) thick. The neutron spectrum measured at this location is shown in Figure 5. Note that the spectrum has only a small high energy fraction and a single peak at 1 MeV. This difference with the FFTB spectra could be explained by the combined effect of lower electron energies and lateral lower high energy neutron yields at 90°. Relevant dosimetric quantities for this location are also summarised in Table 2. Due to equipment failure, ^{11}C activation could not be measured, and the 12″ sphere with lead was the only indicator in this energy region. Therefore, the H26 value of 5.7% might be subject to a large uncertainty from the unfolding procedure.

SSRL Linac

When not being injected into the booster synchrotron, the 120 MeV beam from the linac can be sent into a Faraday cup for diagnostic purposes. Our measurements were performed in the diagnostic room, outside the shielding cave housing the Faraday cup, as shown in Figure 6. No photopion production is expected at this electron energy, which is below the photopion production threshold. A ^{11}C activation measurement yielded results indistinguishable from the background. As in the case of SPEAR, the measured neutron spectrum, also presented in Figure 5, exhibits an even smaller high energy portion. This high energy tail comes from the quasi-deuteron process. The spectrum is mostly a giant resonance spectrum degraded by the cave shielding and strong scatter contribution inside the diagnostic room.

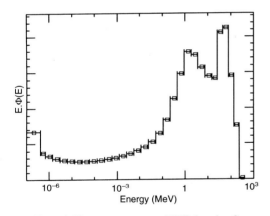

Figure 4. Neutron spectrum at FFTB location 2.

Table 2. Summary of measurement results at SLAC.

Location	Dose equivalent rate (μSv.h^{-1})			E_{av} (MeV)	h_ϕ (Sv.cm^2)	H26 (%)
	AB	LINUS	BS			
FFTB-1	48	62	77	13.8	2.3×10^{-10}	31.6
FFTB-2	133	220	188	19.3	2.4×10^{-10}	38.9
SSRL SPEAR	—	—	189	2.0	7.4×10^{-11}	5.7
SSRL Linac	—	—	88	0.9	8.5×10^{-11}	1.0

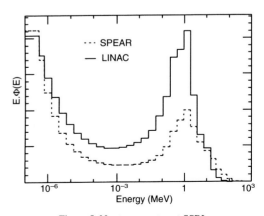

Figure 5. Neutron spectra at SSRL.

CONCLUSION

Comparison with CERN reference spectra shows an overall agreement in spectral shape, and a fluence agreement better than 50%. As expected, a large high energy neutron component was measured in the forward direction at a very energetic (46.6 GeV) electron beam at FFTB. Due to the thick shielding, this is an 'equilibrium' neutron spectrum, similar to the CERN reference spectrum outside the concrete shield. In this case, dose equivalent rates measured by LINUS agree to 20% with the Bonner spheres, while the standard Anderson-Braun rem meter systematically underestimates by up to 30%. On the other hand, the two spectra at SSRL are much softer, due to lower beam energies, 90° production angle and thinner shield. Both spectra exhibit a single peak around 1 MeV and fluence-to-dose equivalent conversion factors lower by a factor of 3 compared to FFTB. This is also confirmed by the H26 values in Table 2.

The preliminary results reported here are the first in series of ongoing measurements. The shape of the high energy spectra is based only on data from a 12″ sphere with lead and ^{11}C activation. The response functions of these two detectors are quite similar, which does not improve the energy resolution of the system. Additional detectors sensitive in the high energy region would be desirable. Our calculations indicate that Sanna's response function for the standard 12″ sphere is too high above 77 MeV. Calculation of an updated response matrix and additional measurements are in progress.

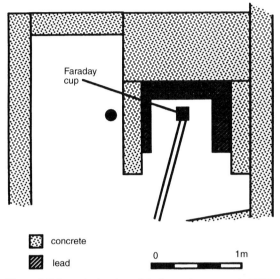

Figure 6. Schematic drawing of the beam dump cave in SSRL Linac. The measurement location is indicated by a round bullet.

ACKNOWLEDGEMENT

This work was supported by the US Department of Energy under contract DE-AC-03-76SF00515.

REFERENCES

1. Birattari, C., Ferrari, A., Nuccetelli, C., Pelliccioni, M. and Silari, M. *An Extended Range Neutron Rem Meter*. Nucl. Instrum. Methods Phys. Res. **A297**, 250–257 (1990).
2. Lowry, K. A. and Johnson, T. L. *Modifications to Iterative Recursion Unfolding Algorithms and Computer Codes to Find more Appropriate Neutron Spectra*. Memo. Rep. 5340 (Naval Research Laboratory, Washington, DC (1983).
3. Title 10 Code of Federal Regulations Part 835, **58**(238), Reg. #65458 (1993).
4. Prael, R. E. and Lichtenstein, H. *User Guide to LCS: The LAHET Code System*. LANL Report LA-UR-89-3014 (Los Alamos) (1989).
5. Thai, L.-X. *MS Thesis*, San Jose State University (to be published).
6. Hofert, M. and Stevenson, G. R. *The CER-CEC High-energy Reference Field Facility*. In: Proc. 8th Int. Conf. on Radiation Shielding, Arlington, Texas, USA (American Nuclear Society) pp. 635–642 (1994).

SHIELDING MEASUREMENTS FOR A PROTONTHERAPY BEAM OF 200 MeV: PRELIMINARY RESULTS

A. Mazal[1], K. Gall[2], J. F. Bottollier-Depois[3], S. Michaud[1], D. Delacroix[4], P. Fracas[4], F. Clapier[5], S. Delacroix[1], C. Nauraye[1], R. Ferrand[1], M. Louis[1] and J. L. Habrand[1]
[1]Centre de Protonthérapie d'Orsay, BP 65, 91402, Orsay Cédex, France
[2]Massachuset General Hospital and Harvard Cyclotron Unit, 44 Oxford Street, Cambridge Massachusetts, 02138 USA
[3]Institut de Protection et de Sûreté Nucléaire, BP6, F-92265, Fontenay aux Roses, Cedex, France
[4]UGSP/SPR Commissariat à l'Energie Atomique Saclay, F-91191, Gif sur Yvette, France
[5]Institut de Physique Nucléaire, F-91406, Orsay, Cedex, France

Abstract — The preliminary results are presented of a series of measurements performed at the Centre de Protonthérapie d'Orsay to evaluate the neutron yield when a 200 MeV proton beam interacts with aluminium and water, the attenuation of different thicknesses of concrete at different angles and the response of different detectors. The effect of shielding has been fitted to exponential functions with first estimations of attenuation lengths for both targets and angles from 0° to 90°. The attenuation length for aluminium changes from 900 kg.m^{-2} at 0° to 680° kg.m^{-2} at 90°, in agreement with published data. Similar values have been obtained for water targets. The source term seems to be underestimated, in spite of using an extended range rem counter. Microdosimetric spectra have been obtained simulating the conditions used to extend the energy range or rem counters.

INTRODUCTION

Radiation shielding required for protontherapy facilities is dominated by the neutron yield[1,2] produced by the proton beam when interacting with different materials in the beam line and with the patient. Concrete is the most common material used for shielding as a good compromise between attenuation, mechanical characteristics and cost. The problem of measuring neutron dose equivalents at energies higher than 10 MeV is not easily solved with conventional methods like rem counters. The complexity of the radiation environment in these facilities usually requires the use of different detectors to have a better knowledge of the radiation field to cover a broad energy spectrum, minimising the uncertainties of measurements. The use of tissue-equivalent proportional counters for neutron and gamma measurements in complex environments can be considered as a reference provided the necessary environmental conditions are taken into account when performing and interpreting the results. The preliminary results of a series of measurements performed at the Centre de Protonthérapie d'Orsay are presented here. The neutron yield when a 201 MeV proton beam interacts with aluminium and water, the attenuation of different thicknesses of concrete at different angles, and the response of different detectors are all evaluated.

METHODS AND MATERIALS

Beam and shielding

The 201 ± 1 MeV proton beam is produced by a synchrocyclotron. The energy dispersion ΔE/E is 0.7%, and the maximum external current available is 2.4 μA (proton flux 1.5×10^{13} s^{-1})[3]. The beam current used for these measurements has been limited to 10 nA (proton flux 6.2×10^{10} s^{-1}). The beam is conducted towards one treatment room and one experimental area[4]. In the latter, a bunker has been built with blocks of concrete (density = 2.2 g.cm^{-3}, Figure 1). The total thickness of concrete was changed during the measurements from 3 m to 0 m (Figure 2). The shielding on the top of the bunker has been kept constant (1 m of concrete). The beam current has been further reduced to 5 nA for measurements with no shielding around the target. Beam monitoring has been performed with a Faraday cup and with a parallel plate transmission ion chamber.

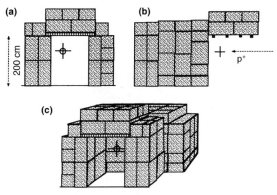

Figure 1. General layout for the shielding measurements: (a) frontal view; (b) lateral view; (c) perspective view.

Targets and positions of measurements

Two targets have been used: aluminium and water, to represent the patient and some beam modifier devices like modulators and absorbers. The aluminium target was 90 mm in diameter and 150 mm length. The water target was 200 × 200 mm² and 320 mm length. In both cases, the target was thick enough to stop the beam.

Measurements have been performed behind the concrete walls at angles from 0° to 90° from the beam direction, at 4.5 m from the target. Additional measurements have been done for specific problems like build-up effects and distance to the target.

Detectors

Different detectors have been used:

(a) an ion chamber, TE wall, 515 cm³, (Nardeux, France), used to evaluate the gamma component of the dose equivalent;
(b) a tissue-equivalent proportional counter (TEPC, Far West Technology, USA)[2], outer housing of 0.8 mm of stainless steel, wall material 2.46 mm of Shonka A-150 tissue-equivalent plastic, internal diameter 56.9 mm, single wire, filled with a propane-based tissue-equivalent gas at low pressure, calibrated by an internal ^{244}Cm alpha source;
(c) 'NAUSICAA'[5] tissue-equivalent proportional counter (IPSN, France), outer housing of 0.3 mm Ni, with a sensitive volume of 5 cm length and 5 cm diameter, single wire, filled with a propane-based tissue-equivalent gas at low pressure simulating a 3 μm biological target under 1 cm of tissue-equivalent wall, internal calibration by a ^{244}Cm alpha source, to get microdosimetric spectra (dose weighted lineal energy single-event distributions) of the complex field;
(d) AE 2202 D Neutron Dose Rate meter (Studsvik, Sweden), with and without 2 cm lead around as a 'moderator' to measure the dose equivalent;
(e) an extended range rem counter, with lead within the moderator, following the theoretical calculations and practical measurements performed by Birattari et al[6,7];
(f) LiF thermoluminescence detectors within moderators and films in albedo, like those used for personal monitoring[8].

RESULTS

Microdosimetric spectra

Microdosimetric dose weighted lineal energy single-event distributions have been obtained for different situations in order to evaluate changes in the spectra behind the shielding and the effect of different materials around a detector. In Figure 3, the spectra with no shielding at 0° and 90° are presented for three configurations: the 'Nausicaa' TEPC is free-in-air, or surrounded by 10 cm of water or by 2 cm of lead. The spectra have been normalised to absorbed dose of the 'free air' response at each direction for comparative purposes. The water

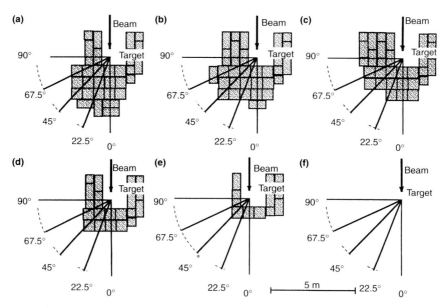

Figure 2. Concrete configurations for the shielding measurements: the thickness of concrete has been changed from (a) to (f) during the measurements. The reason for such combinations is to cover most of the desired thicknesses at each angle with a limited number of available blocks. The lines at 0°, 22.5°, 45°, 67.5 and 90° represent the measuring angles. The detectors have been placed at 4.5 m from the target.

thickness corresponds to a range of protons with energies even larger than 100 MeV, and simulates a position in the centre of a human body, while the lead was studied to evaluate a practical approach to get an extended range neutron survey meter[6,7].

Spectra for the two targets at different directions and for different shieldings are presented elsewhere[5].

Shielding

Figure 4 shows the effect of concrete shielding measured with a conventional rem counter at different angles from the direction of the incident beam, for a water target.

In practice, the attenuation of the dose equivalent through shielding is expressed with a fit to experimental data through an exponential equation[1,2,9–11]:

$$H(t) = H_0 \exp(-t/\lambda)/r^2 \quad (1)$$

where $H(t)$ (in Sv) is the dose equivalent behind a thickness t of shielding, in $g.cm^{-2}$, H_0 is the 'source term' (the dose equivalent at 1 m with no shielding, or extrapolated to 0 shielding thickness) in $Sv.m^2$; λ is the attenuation length for dose equivalent through the shield, in $g.cm^{-2}$, and r is the distance in m from the source to the shield surface, or point of interest outside the shield.

With our particular set-up, a large background of dose equivalent has been measured. The results presented in this paper correspond to measurements of dose equivalent made at a constant distance (4.5 m) to the source. In consequence, the preliminary fit of our results has been obtained with the following analytical expression:

$$H(t) = (H(0) - B) \exp(C t) + B \quad (2)$$

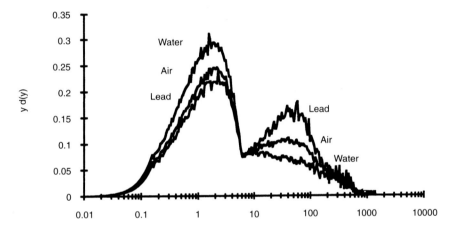

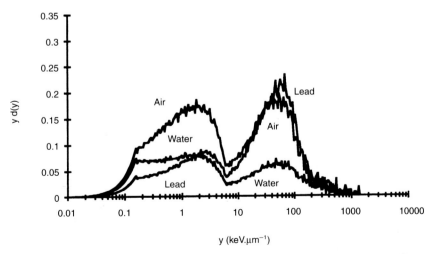

Figure 3. Microdosimetric spectra (dose weighted lineal energy single-event distributions plotted against linear energy) with no shielding, (a) at 0° and (b) 90° with the tissue-equivalent proportional counter free-in-air, surrounded by 10 cm of water, and surrounded by 2 cm of lead. Figures are normalised to the spectra 'in air'.

where $H(0)$ is the dose equivalent in Sv, measured with no shielding, or extrapolated from attenuation curves, t is the thickness of concrete in m, B is the dose equivalent background in Sv and C is the attenuation coefficient, the inverse of the attenuation length in m^{-1}. The original coefficients fitted with this approach have been converted into H_0, λ and B (Figure 5) in order to compare the first two terms with data from the literature (Table 1).

Directional effects and comparisons between detectors

Figures 6 and 7 and Table 2 show the effect of the angle and the shielding on the measurements performed with different detectors. All the figures represent dose equivalents in Sv per proton on the target, while Table 2 compares total dose equivalents integrated during specific measurements.

DISCUSSION

In Figure 5(a), the values of the 'source' term (H_0) show that with a conventional rem counter there is a factor of two between aluminium and water targets, and a nearly constant value at all angles, specially for the aluminium target. This can be assumed to result from the neutron energy range of response of this detector, which falls off steeply above 10 MeV, sensing the neutron yield coming from an isotropic evaporation process. This process is expected to be more important for the aluminium target. H_0 has been estimated including the measurement of the dose equivalents when no shielding was present (configuration 6, 'with zero' in Figure 5). When the measurements are fitted using only values when an 'equilibrium' of the neutron spectra has been obtained (after around 60 cm of concrete, values 'without zero' in Figure 5), the aluminium and the water mesurements of H_0 are closer and diminish with the emission angle.

The attenuation coefficient 'lambda' measured with the same detector (Figure 5(b)) shows a similar tendency: small differences between water and aluminium, with higher values for small angles where the neutron spectrum is the hardest. Finally, the dose equivalent background (Figure 5(c)) is also higher for the aluminium target. The increase of this background for the large angles is related to the shielding configuration (opened towards the back of the target) and the short distances to the external concrete walls in the experimental room.

Figure 5 (d, e and f) show the difference in the parameters H_0, λ and B of the model when the measurements are done with the extended range rem counter ('modified'), and the fit is done without the measurements with no shielding ('without zero'). The 'source' term H_0 and the attenuation coefficient 'lambda' are rather similar for both targets and show lower values for large angles. This confirms a better response of the detector to a hard spectra, and that a very important low energy component has been eliminated with the first shielding. No differences are seen for the background, already related with low energy neutrons after passing through a large thickness of shielding.

In Table 1, it is shown that the attenuation length has in general a very good agreement with other measured

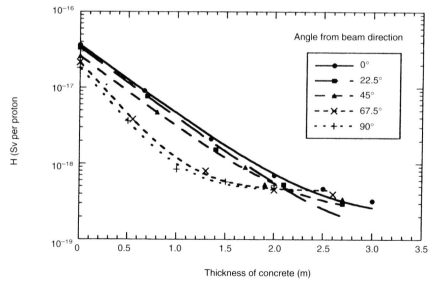

Figure 4. Attenuation of the dose equivalent by concrete at different angles from the beam direction, measured with a conventional rem counter. The experimental error (1 SD) is estimated at 2.5%, for simplicity it is not included in the graph.

and calculated data[1,2,12,13]. For the source term, no data are available for water. For an aluminium target, our values are in good agreement for large angles, but they are a factor 5 to 10 times lower than published data for small angles. The data for H_0 from Hagan et al[13] cannot be extrapolated to zero thickness of concrete, so our comparisons were performed with a shield of 1 m of concrete, with similar conclusions. The difference in source terms for our data and those previously published may result from several factors. Firstly, the Hagan data were derived from Monte Carlo calculations that are known to overpredict dose equivalent in the forward direction[2]. This difference can also be related to the response of the detector: in spite of the extended range,

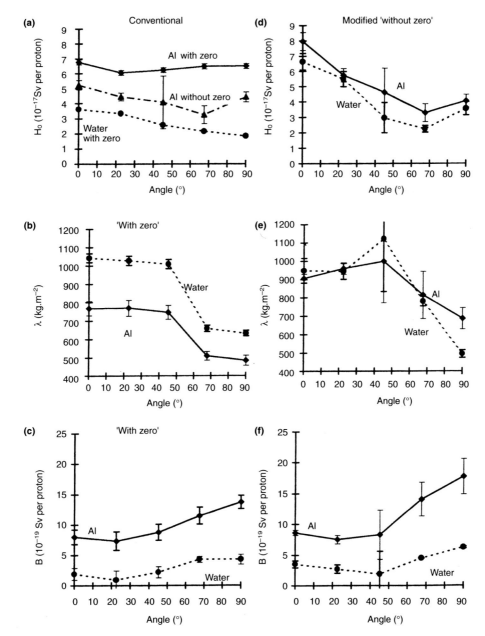

Figure 5. Source term H_0, attenuation length λ and background B: (a,b,c) conventional rem counter, including measurements of dose equivalent with no shielding ('with zero'); (d,e,f) modified rem counter, data for no shielding have not been taken into account ('without zero'). The lines are not a fit, they are presented to 'guide the eye'.

Table 1. Comparison of the attenuation length and source term with measured and calculated data from the literature.

Angle (deg)	Attenuation length λ (kg/m^2)					
	Water target Measured	Aluminium target Measured	Aluminium target Measured[2]	Aluminium targed Calculated[12]	Iron target Calculated[13] at 3 to 4 m depth	Copper target Calculated[1]
0	950 ± 40	900 ± 30	900 ± 20	1050	1020	470 (for 50 MeV neutron) 880 (for 200 MeV neutron)
22			880 ± 30		980	
22.5	940 ± 40	960 ± 30			980	
45	1120 ± 290	1000 ± 230	750 ± 20	943	920	
67.5	780 ± 30	820 ± 130			670	800
90	490 ± 20	680 ± 60	506 ± 8	570	730	230 (for 50 MeV neutron) 690 (for 200 MeV neutron)

Angle (deg)	H_0 at 1 m (10^{-15} Sv.m^2 per proton)				
	Water target Measured	Aluminium target Measured	Aluminium target Measured[2]	Aluminium target Calculated[12]	Copper target Calculated[1]
0	1.3 ± 0.1	1.6 ± 0.1	7.9 ± 0.5	10	1.4 (for 50 MeV neutron) 7.4 (for 200 MeV neutron)
22			4.7 ± 0.5		
22.5	1.1 ± 0.1	1.15 ± 0.08			
45	0.6 ± 0.2	0.9 ± 0.1	2.72 ± 0.18	4.5	
67.5	0.44 ± 0.04	0.67 ± 0.1		1.5	
90	0.71 ± 0.08	0.89 ± 0.04	0.89 ± 0.05	0.2	0.18 (for 50 MeV neutron) 1.3 (for 200 MeV neutron)

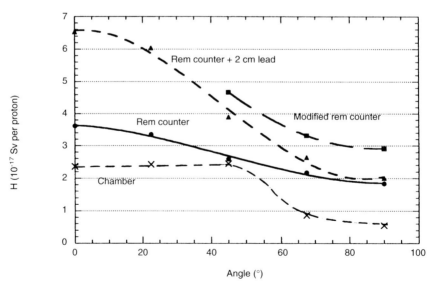

Figure 6. Dose equivalent at different angles, measured with different detectors, with a water target and no shielding. The experimental error (1 SD) has been estimated at 2.5%. The lines are presented to 'guide the eye'.

the modified rem counter may still underpredict the dose equivalent from high energy neutrons. Further investigation of the cause of the discrepancy will be done by more extensive analysis.

The B coefficient can be used in conditions similar to our particular set-up, where the shielding is not completely closed, and where additional walls can produce a background of low energy neutrons, like near the entrance of labyrinths. For calculations around fully shielded rooms, this term should not be included.

The particular problem of build-up attenuation and moderation around TEPC, and moderators around rem counters[6,7] has been identified with the spectra analysis in Figure 3, and it can be used for further applications. The low energy component of the neutron spectra (low energy protons, high LET), is attenuated drastically, at 0° and 90°, by the water around the detector, as well as for the gamma components at 90°. At 0°, the presence of high energy neutrons, the effect of the proton build-up and the gamma production by low energy neutrons near the core of the detector is observed at the low LET region. The inelastic processes effect of lead signifies an increase in the response of detectors sensitive to low energy neutrons (protons), as is shown for the high LET component at 0°, but this effect seems not to be complete as already stated. No significant change of the spectra is obtained at 90° for the low energy neutrons. For both angles there is an obvious attenuation of the gamma components by lead.

There is still further work to be done to compare the response of all the detectors used in this study, but Figures 6 and 7 already show the important change in the response of the modified rem counter at low angles compared with the conventional one. Table 2 shows a systematic shift of 20% in the response of FLi, used as described by one of the authors[8], compared with the rem counter.

CONCLUSIONS

A full set of data has been obtained including microdosimetric spectra, absorbed doses and dose equivalents with different detectors, for different directions (0°–90°), behind different concrete thicknesses (0–3 m, 2.2 g.cm^{-3}) and for two targets (Al and water) of usual application in facilities devoted to protontherapy at 200 MeV. For the aluminium target, the attenuation length is in good agreement with published data, while the source terms may be underestimated by the extended range rem counter. The background introduced in our model should be used only for measurements inside experimental areas.

Rem counters can be used for area monitoring, but they need to be modified to take into account the energies higher than 10 MeV as encountered in facilities devoted to clinical applications of proton beams. The results of conventional methods (films, LiF) for staff dosimetry can be related to the real dose equivalent under known conditions (spectra estimation behind thick shielding). This information can be obtained using tissue-equivalent proportional counters as reference detectors.

ACKNOWLEDGEMENTS

This work was supported by the Ministère de la Santé, the Caisse Régionale d'Assurance Maladie de l'Ile de France, the Ligue de Lutte Contre le Cancer (France), and by US NIH grant CA 21239.

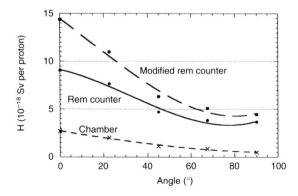

Figure 7. Dose equivalent at different angles, measured with different detectors, with a water target and a shield of 0.6 m of concrete. The experimental error (1 SD) has been estimated at 2.5%. The lines are presented to 'guide the eye'.

Table 2. Comparison of LiF and rem counter measurements for some of the configurations studied.

Set-up (Figure 2)	Target	Angle (deg)	Concrete (m)	Rem counter (mSv)	LiF (mSv)	LiF/Rem
A	Al	0.0	3	0.153	0.12	0.78
A	Al	90.0	1	0.434	0.40	0.92
A	Al	22.5	2.7	0.262	0.19	0.73
A	Al	45.0	2.7	0.217	0.17	0.78
D	water	90.0	1	0.68	0.54	0.79
D	water	22.5	1.4	1.05	0.83	0.79
D	water	170.0	0	19	13.59	0.72
F	Al	22.5	0	1.7	1.44	0.85
F	Al	45.0	0	1.83	1.30	0.71
					Mean	0.79
					SD	0.07

REFERENCES

1. Thomas, H. and Stevenson, R. *Radiological Safety Aspects of the Operation of Proton Accelerators*. Tech. Report Series 283 (Vienna, IAEA) (1988).
2. Siebers, J. V. *Shielding Measurements for a 230 MeV Proton Beam*. Thesis, University of Wisconsin-Madison (1990).
3. Debray, P. *List of Synchrocyclotrons*. Entry FM4 In: Proc. 9th Int. Conf. on Cyclotrons and Applications. Ed. G. Gendreau (Les Éditions de Physique, Caen, France) p. 858 (1981).
4. Mazal, A. and Habrand, J. L. *La Protonthérapie: le Centre d'Orsay*. Path. Biol. **41**(1), 122–125 (1993).
5. Bottollier-Depois, J. F., Plawinski, L., Spurný, F. and Mazal, A. *Microdosimetric Investigations in Realistic Fields*. Radiat. Prot. Dosim. **70**(1–4), 203–206 (This issue) (1997).
6. Birattari, C., Ferrari, A., Nuccetelli, C., Pelliccioni, M. and Silari, M. *An Extended Range Neutron Rem Counter*. Nucl. Instrum. Methods Phys. Res. **A297**, 250–257 (1990).
7. Birattari, C., Esposito, A, Ferrari, A., Pelliccioni, M. and Silari, M. *A Neutron Survey Meter with Sensitivity Extended up to 400 MeV*. Radiat. Prot. Dosim. **44**(1/4), 193–197 (1992).
8. Fracas, P. *Dosimetries de Zone et Individuelle des Neutrons autour des Accélérateurs d'Ions au Moyen de Détecteurs Thermoluminescents (TLD600–700) ou de Détecteurs Solides de Traces (LR 115 avec Bore)*. Mémoire d'ingénieur (Conservatoire National des Arts et Metiers, Paris) (1988).
9. Sullivan, A. H. *A Guide to Radiation and Radioactivity Levels near High Energy Particle Accelerators* (Nuclear Technology Publishing, Ashford, Kent, England) (1992).
10. Tesch, K. *A Simple Estimation of the Lateral Shielding for Proton Accelerators in the Energy Range 50 to 1000 MeV*. Radiat. Prot. Dosim. **11**(3), 165–172 (1985).
11. Siebers, J. V., DeLuca Jr, P. M., Pearson, D. W. and Coutrakon, G. *Measurement of Neutron Dose Equivalent and Penetration in Concrete for 230 MeV Proton Bombardment of Al, Fe, and Pb Targets*. Radiat. Prot. Dosim. **44**(1/4), 247–251 (1992).
12. Alsmiller, R. G., Santoro, R. T. and Barish J. *Shielding Calculations for a 200 MeV Proton Accelerator and Comparisons with Experimental Data*. Particle Accelerators **7**, 1–7 (1975).
13. Hagan, W. K., Colborn, B. L., Armstrong, T. W. and Allen, M. *Radiation Shielding Calculations for a 70 to 250 MeV Proton Therapy Facility*. Nucl. Sci. Eng. **98**, 272–278 (1988).

NEUTRON MEASUREMENTS AROUND A HIGH ENERGY LEAD ION BEAM AT CERN

A. Aroua†, T. Buchillier†, M. Grecescu† and M. Höfert‡
†IRA, Institut de Radiophysique Appliquée
Centre Universitaire, 1015 Lausanne, Switzerland
‡CERN, Laboratoire Européen pour la Physique des Particules
1211 Genève-23, Switzerland

Abstract — Several experiments at CERN using high energy, high intensity beams of lead ions search for evidence of a quark–gluon plasma. The ions pass through an accelerating chain which supplies an energy up to 160 GeV per nucleon, 35 TeV in total. When these lead ions collide with targets or are lost due to unintentional imperfections in the transfer lines they interact with materials and produce stray radiation fields. The spectrometric and microdosimetric results are presented of the first measurements in such stray fields made at CERN, on top of a concrete shielding surrounding a lead ion beam which collided with a lead target. They are compared with the results obtained with a high energy proton beam operated in a similar configuration.

INTRODUCTION

In addition to protons of 450 GeV CERN can provide heavier ion beams. Following the construction of a special linac it recently became possible to accelerate lead ions ($^{82}Pb^+$) up to 160 GeV per nucleon or a total energy of 35 TeV. They pass through the CERN accelerating chain which, in addition to the linac, includes the Booster synchrotron, the Proton Synchrotron (PS) and the Super Proton Synchrotron (SPS). The high energy lead ions are then directed onto high Z material targets leading to collisions that produce high energy densities and high concentrations of nucleons. Ions are used by seven experiments searching for a new state of matter, the elusive quark–gluon plasma, that is thought to have existed at the beginning of the Universe during the first microsecond following the Big Bang.

As in the case of high energy protons, beam losses of lead ions, either deliberately on targets or inadvertently during acceleration or beam transfer, give rise to the formation of hadronic cascades in material and stray radiation fields around these beam loss points. It is assumed that in a first approximation the number of neutrons created in an interaction is proportional to the number of nucleons in the lost ion, i.e. many more neutrons are expected for the loss of a high energy lead ion than for a proton of the same energy per nucleon. It is of interest to study the spectral distribution of these neutrons as in the case of heavy ions their energy distribution may be different from that of neutrons generated in high energy proton interactions.

In the following, spectrometric and microdosimetric results of measurements made at CERN on top of a concrete shielding are presented. The shielding surrounded a lead ion beam colliding with a lead target. These results are compared with spectra obtained around a high energy proton beam during the H6 experiment carried out in the framework of a CEC contract[1].

MATERIAL AND METHODS

The measuring instruments used were: a multisphere neutron spectrometer with a ^{3}He thermal detector[2], ^{209}Bi and ^{232}Th fission track detectors[3,4], a low neutron sensitivity Geiger-Müller counter[5], an Andersson-Braun type rem counter[6], a HANDI[7] and a REM-500[8] type TEPCs.

The measurements were performed in CERN's West experimental hall, on top of a concrete shielding surrounding the lead target. The point of measurement was located 50 cm above the surface of the shielding. The lead beam was pulsed with a repetition time of 19.2 s.

RESULTS AND DISCUSSION

The Bonner spheres and the ^{232}Th/^{209}Bi detectors

The counting rates obtained with Bonner spheres of various diameters are plotted in Figure 1 and the form of the curve indicates coherent results. It also suggests a significant contribution of high energy neutrons in the

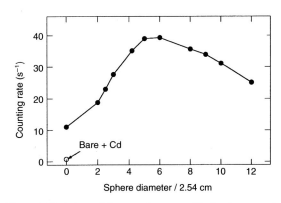

Figure 1. Variation of the counting rate with the diameter of the Bonner spheres.

stray field (high relative response of spheres of diameter greater than 8″). Indications on the hardness of the neutron spectrum are given by the position of the peak of the curve in Figure 1, by the 9″ to 3″ count ratio and by the ^{209}Bi to ^{232}Th fission ratio given in Table 1. The 9″ to 3″ ratio gives an indication about the relative contributions of thermal plus epithermal, and fission-like neutrons, whereas the ^{209}Bi to ^{232}Th ratio accounts for the high energy component of the neutron spectrum (>50 MeV). The 9″ to 3″ and the ^{209}Bi to ^{232}Th ratios obtained with the H6 spectrum are 1.0 and 10.9% respectively. This means that in the 'lead' spectrum there is a higher relative contribution of neutrons with energies above 1–2 MeV and a lower relative contribution of neutrons with energies above 50 MeV than in the H6 spectrum.

The upper plot in Figure 2 presents the neutron fluence spectrum, normalised to maximum intensity, unfolded from the data given in Figure 1 and the known response matrix of the spectrometer, using the SAND unfolding code. The 'first guess' spectrum used is a $E\Phi_E$ flat spectrum up to 100 MeV, followed by a Maxwellian tail. The spectrum is characterised by a high energy peak around 100 MeV, and another one at a few MeV (evaporation peak). It can be compared qualitatively with the spectrum obtained during the H6 experiment, where the beam was composed of 205 GeV protons and positive pions interacting with a copper target.

The 'lead' neutron spectrum is clearly different from that measured on the concrete shielding during the H6 experiment. In the H6 'concrete' spectrum the high energy peak is higher in amplitude than the 'evaporation' one, whereas, in the 'lead' neutron spectrum the situation is reversed.

Integral results computed from the neutron fluence spectrum are presented in Table 1. The ambient dose equivalent H*(10)-old according to ICRP Recommendation 21 was calculated using, below 20 MeV, conversion factors given by Wagner et al[9], while following the more recent ICRP Recommendation 60, H*(10)-60 and the personal dose equivalent for normal incidence $H_p(0°,10)$-60 were calculated using the set of factors given by Siebert and Schuhmacher[10]. These new factors based on new values of stopping powers for protons have been adopted by ISO[11]. The result is that old and new values of the ambient dose equivalent are about the same with a 4.5% increase only for H*(10)-60 compared with H*(10)-old. Above 20 MeV, the ambient dose equivalent conversion factors computed by Sannikov and Savitskaya[12] were used.

The Studsvik neutron monitor and the Geiger-Müller counter

Table 2 gives the neutron ambient dose equivalent measured with a Studsvik rem counter. For the type of spectrum measured the neutron monitor gives a value that is 23% lower than that evaluated from the results obtained with the Bonner spheres. This is in line with similar measurements in stray radiation fields caused by GeV proton losses. Table 2 also presents the photon ambient dose equivalent measured with the Geiger-Müller counter.

Table 1. Neutron integral results.

9″/3″	1.22
^{209}Bi/^{232}Th (%)	4.33
$\dot{\Phi}$ total (n.cm^{-2}.min^{-1})	1.59 10^4
$\dot{D}$*(10)(μGy.h^{-1})	33.4
$\dot{H}$*(10)-old (μSv.h^{-1})	244
$\overline{Q}$*(10)-old	7.3
$\dot{H}$*(10)-60 (μSv.h^{-1})	255
$\overline{Q}$*(10)-60	7.6
$\dot{H}_p$ (0°,10)-60 (μSv.h^{-1})	264

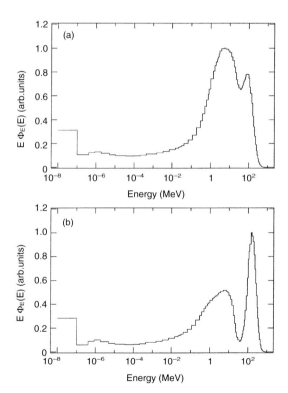

Figure 2. Neutron fluence spectrum from (a) a shielded lead ion beam colliding with a lead target, unfolded using the SAND code. A $E \Phi_E$ flat first guess spectrum up to 100 MeV, followed by a Maxwellian tail is used. The spectrum is compared with that obtained behind a concrete shield of a proton beam (b).

The TEPCs

Figure 3 presents the absorbed dose and dose equivalent distributions (y$\dot{D}$(y) and y$\dot{H}$(y)) measured with the HANDI and the REM-500 TEPCs. The REM-500 presents a low cut-off at about 5 keV.μm^{-1} and a higher one at about 256 keV.μm^{-1}. It seems to accumulate the events above 256 keV.μm^{-1} in the last few channels. The microdosimetric spectra are compared qualitatively with those obtained on top of the concrete shielding where a proton beam hits a copper target in the H6 experiment[14]. The 'lead' microdosimetric spectrum looks similar to the one measured in the H6 experiment above 10 keV.μm^{-1} and lacks a low LET contribution which in the latter was attributed to the presence of muons. The dose equivalent distribution measured in the H6 experiment shows a peak around 10 keV.μm^{-1} which is absent in the 'lead' distribution. This peak is due to protons and pions produced by high energy neutrons (100 MeV region), which were found to be relatively more abundant in the H6 experiment than in the 'lead' case (as shown in Figure 2).

Table 2. Results obtained with the Studsvik rem counter and the Geiger-Müller counter.

Studsvik	counting rate (s^{-1})	60.7 ± 0.7%
	sensitivity* (pulses per Sv)	1.17 10^9
	$\dot{H}_n$*(10) (μSv.h^{-1})	187
	H_n*(10)-Studsvik/H_n*(10)-BS	0.77
Geiger-Müller	counting rate (s^{-1})	26.3 ± 1.1%
	sensitivity** (pulses per Sv)	3.03 10^9
	$\dot{H}_\gamma$*(10) (μSv.h^{-1})	31.3

*sensitivity to ^{241}Am-Be
**sensitivity to ^{60}Co, H*(10) according to ICRU 47[13]

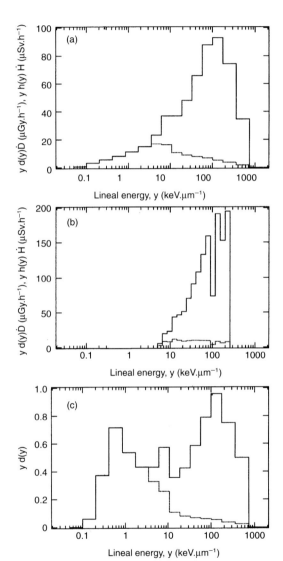

Figure 3. Microdosimetric distributions y $\dot{D}$(y) (dotted line) and y $\dot{H}$(y) (solid line) around a shielded lead ion beam colliding with a lead target plotted against the lineal energy, y, obtained with (a) the HANDI and (b) the REM500 TEPCs. The distributions are compared in shape with the y d(y) and y h(y) distributions obtained behind similar concrete shield but using a proton beam which collides with a copper target (c).

Table 3. Integral results obtained with the HANDI and REM-500 TEPCs.

	HANDI	REM-500
$\dot{D}$ total (μGy.h^{-1})	76	—
$\dot{H}$-21 total (μSv.h^{-1})	328	—
$\overline{Q\text{-}21}$ total	4.33	—
$\dot{H}$-60 total (μSv.h^{-1})	372	—
$\overline{Q\text{-}60}$ total	4.91	—
$\dot{D}$ neutron (μGy.h^{-1})	26	28
$\dot{H}$-21 neutron (μSv.h^{-1})	263	327
$\overline{Q\text{-}21}$ neutron	10.2	11.7
$\dot{H}$-60 neutron (μSv.h^{-1})	322	—
$\overline{Q\text{-}60}$ neutron	12.4	—
$\dot{D}$ gamma (μGy.h^{-1})	50	—
$\dot{H}$-21 gamma (μSv.h^{-1})	65	—
$\dot{H}$-60 gamma (μSv.h^{-1})	50	—
H_n*(10)-HANDI/H_n*(10)-BS	1.08	
H_n*(10)-REM500/H_n*(10)-BS	1.34	
H_γ*(10)-HANDI/H_γ*(10)-GM	2.08	

Table 3 gives the dosimetric results associated with these microdosimetric distributions. In terms of total absorbed dose the agreement between the two instruments is of the order of 8%, whereas the dose equivalent indicated by the REM-500 is 24% higher than that given by the HANDI. This is due to use of a higher quality factor by the REM-500 TEPC[15]. Table 3 also shows the ratio of the neutron and gamma dose equivalents indicated by the TEPCs and by the Bonner spheres system and the GM counter respectively.

CONCLUSION

This report presented first results of measurements carried out at CERN, on top of a concrete shielding, in a stray radiation field originating from interactions of a lead ion beam with a lead target. Different active and passive detectors were used to analyse the field. The spectrometric and microdosimetric results were qualitatively compared with those obtained during the H6 experiment. The neutron spectra show differences which can be explained by the physics of the different interactions.

REFERENCES

1. Aroua, A., Höfert, M., Sannikov, A. V. and Stevenson, G. R. *Reference High-Energy Radiation Fields at CERN*. Presented at the 6th Int. Symp. on Radiation Physics, Rabat, Morocco, 18–22 July 1994. Divisional Report CERN/TIS-RP/94-12/CF (European Laboratory for Particle Physics (CERN), Genève) (1994).
2. Aroua, A., Grecescu, M., Lerch, P., Prêtre, S., Valley, J.-F. and Vylet, V. *Evaluation and Test of the Response Matrix of a Multisphere Neutron Spectrometer in a Wide Energy Range*. Nucl. Instrum. Methods A**321**, 298–316 (1992).
3. Aroua, A. *High Energy Extension of a Neutron Spectrometer by Means of Thorium and Bismuth Fission Detectors*. Technical Memorandum CERN/TIS-RP/TM/93-18 (European Laboratory for Particle Physics (CERN), Genève) (1993).
4. Aroua, A. *Response Functions of Thorium and Bismuth Fission Track Detectors used with Bonner Spheres*. Technical Memorandum CERN/TIS-RP/TM/93-35 (European Laboratory for Particle Physics (CERN), Genève) (1993).
5. Wagner, E. B. and Hurst, G. S. *A Geiger–Müller γ-Ray Dosimeter with Low Neutron Sensitivity*. Health Phys. **5**, 20 (1961).
6. Aroua, A., Boschung, M., Cartier, F., Gmür, K., Grecescu, M., Lerch, P., Prêtre, S., Valley, J.-F. and Wernli, Ch. *Study of the Response of Two Neutron Monitors in Different Neutron Fields*. Radiat. Prot. Dosim. **44**(1–4), 183–187 (1992).
7. Kunz, A. W., Pihet, P., Arend, E. and Menzel, H. G. *An Easy-to-Operate Portable Pulse-Height Analysis System for Area Monitoring with TEPC in Radiation Protection*. Nucl. Instrum. Methods A**299**, 696–701 (1990).
8. Health Physics Instruments. *Model REM-500 Neutron Survey Meter, Operation and Repair Manual* (1993).
9. Wagner, S. R., Grosswendt, B., Harvey, J. R., Mill, A. J., Selbach, H.-J. and Siebert, B. R. L. *Unified Conversions Functions for the New ICRU Operational Radiation Protection Quantities*. Radiat. Prot. Dosim. **12**(2), 231–235 (1985).
10. Siebert, B. R. L. and Schuhmacher, H. *Quality Factors, Ambient and Personal Dose Equivalent for Neutrons, Based on the New ICRU Stopping Power Data for Protons and Alpha Particles*. Radiat. Prot. Dosim. **58**(3), 177–183 (1995).
11. ISO. *Reference Neutron Radiation*, ISO/TC85/SC2/WG2/SG3/WD8529/3, Draft.
12. Sannikov, A. V. and Savitskaya, E. N. *Ambient Dose and Dose Equivalent Factors for High Energy Neutrons*. Report CERN/TIS-RP/93-14 (European Laboratory for Particle Physics (CERN), Genève) (1993).
13. ICRU. *Measurement of Dose Equivalents from External Photon and Electron Radiations*. Report 47 (Bethesda, MD: International Commission on Radiation Units and Measurements) (1992).
14. Aroua, A., Höfert, M. and Sannikov, A. V. *HANDI-TEPC Results of the CERN-CEC July and September 1993 Experiments (H6J93, H6S93)*. Internal Report CERN/TIS-RP/IR/93-45 (European Laboratory for Particle Physics (CERN), Genève) (1993).
15. Aroua, A., Höfert, M. and Sannikov, A. V. *On the Use of Tissue-Equivalent Proportional Counters in High Energy Stray Radiation Fields*. Radiat. Prot. Dosim. **59**(1/4), 49–53 (1995).

SECONDARY DOSE EXPOSURES DURING 200 MeV PROTON THERAPY

P. J. Binns and J. H. Hough
National Accelerator Centre
PO Box 72, Faure 7131, South Africa

Abstract — Prior to the first scheduled treatments at a new 200 MeV proton therapy facility a study was commissioned to ascertain the nature of extraneous radiations produced by ancillary components in the beam line. Estimates of the dose and dose equivalent were obtained in the vicinity of the isocentre using a tissue-equivalent proportional counter. The character of the scattered radiation was deduced from the measured single-event distributions and varied with lateral displacement from central axis of the primary beam. A fast neutron dose component was identified at the patient position that decreased progressively with lateral displacement and radial distance from the final collimator. Extending beyond the periphery of the patient collimator a forward peaked cone of scattered high energy protons was evident. Dose equivalents determined at the patient position and attributable to scattering from the beam delivery system varied between 33×10^{-3} and 80×10^{-3} Sv per treatment gray.

INTRODUCTION

A clinical programme for treating small intracranial lesions and skull base tumours recently commenced at the proton therapy facility of the National Accelerator Centre (NAC). A unique patient positioning system using real-time stereophotogrammetry (SPG) ensures that the target volume is correctly positioned at the treatment isocentre. Initial treatments were confined to crossfire plateau irradiations with an unmodulated 200 MeV proton beam using a fixed horizontal beamline. Total dose prescriptions varied between 18 and 34 Gy and were given in as many as three fractions, the largest of which was 20 Gy. To achieve lateral uniformity in beam intensity at the isocentre for circular field sizes ranging up to 10 cm diameter the emergent proton beam, which is relatively well focused, passes through a number of passive scatterers and occluding rings. Depending on the beam entry point, the patient extends beyond the shadow of the final collimator and elastically scattered protons deflected out of the defined field will pose a potential treatment hazard.

Prior to the first scheduled treatments a study was commissioned to evaluate the effectiveness of the collimator shielding and to ascertain the nature of the extraneous radiation produced by ancillary components in the beamline. Both quantitative and qualitative information was obtained using a low pressure proportional counter constructed of A-150 tissue-equivalent plastic.

METHOD

At the NAC 200 MeV protons are produced for radiotherapy using a solid pole cyclotron that serves as an injector to the main separated sector cyclotron which operates at a frequency of 26 MHz. A schematic layout of the dedicated beam delivery system utilised for therapy has been described previously[1] and is shown in Figure 1 configured for the shoot-through conditions pertinent to this work and the first patient irradiations. The distance from the vacuum window to the isocentre is 7 m and the total mass in the path of the beam is approximately 2 g.cm^{-2} which corresponds to an energy loss of about 9 MeV for a beam with a mean incident energy of 200 MeV. Efficiency of proton transport through the beam delivery system is approximately 1% and an absorbed dose rate to the patient of about 3 Gy.min^{-1} is obtained when a beam of intensity 15 nA is extracted from the main cyclotron. This absorbed dose rate is specified on the entrance plateau of an unmodulated beam at a depth of 5 cm in water. Beam

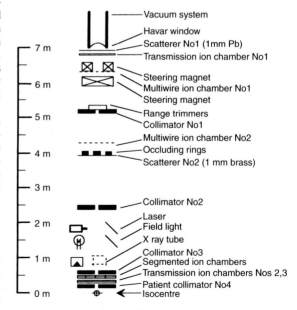

Figure 1. Layout of the 200 MeV horizontal proton beam facility indicating the approximate positions of the beamline components. The size of components is not to scale.

modification is afforded by two scatterers plus an occluding ring system. The first scatterer (No 1) is a 1 mm thick Pb plate located immediately downstream from the vacuum window. The occluding rings (50 mm thick brass) are mounted on the second scatterer (1 mm brass plate) positioned a further 2.9 m downstream. There are three antiscatter collimators in the beam, each 50 mm thick made from brass (Nos 1 and 3) and Pb (No 2). The collimators are (No 1) 20, (No 2) 50 and (No 3) 38 cm wide. The final (patient) collimator (No 4) is positioned 27.5 cm upstream of the isocentre and is also made of brass. This collimator is 23 cm wide and accommodates fixed inserts that provide treatment fields of various shapes which can be rotated around the beam axis. The inserts are made of brass (50 mm thick) and for shoot-through therapy field sizes were restricted to 50 mm diameter or less.

Beam monitoring is performed at various stages along the length of the delivery system using a variety of ionisation chambers. The absorbed dose delivered to the patient is monitored by a dual transmission ionisation chamber (5 μm thick aluminised Mylar foils, 5 mm separation) located adjacent to the final collimator and the integrated charge measured from one of these chambers served as beam monitor for each experimental run.

Measurements were performed using a spherical tissue-equivalent proportional counter (TEPC) of internal diameter 1.27 cm manufactured by Far West Technology. The counter was filled with propane-based TE gas at a pressure of 8.7 kPa to simulate a sphere of unit density tissue with a diameter of 2 μm. Detector pulses were processed using standard instrumentation[2] and pulse heights calibrated by means of an internally mounted ^{244}Cm source emitting alpha particles of 5.8 MeV that are assumed to deposit 175 keV when crossing the counter diameter. A precision pulser related the different amplifier gains to the dose calibration.

Spectral distortions due to electronic dead time and pulse pile up were minimised by regulating the proton beam intensity to restrict the registered event rate to between 2000 and 3000 s^{-1}. Depending upon the experimental conditions beam intensities were reduced to between approximately 1 and 5 nA without any retuning of the beam transport system by adjusting the widths of two pairs of slits in the transfer beamline between the injector and main cyclotrons.

Event size distributions were binned into 40 logarithmic intervals per decade of lineal energy and converted to absorbed dose distributions according to well prescribed methods[3]. The dose equivalent was similarly determined after equating LET with lineal energy and using the Q(L) relationship recommended to approximate the new radiation field weighting factors[4].

Initially the radiation produced at the patient collimator was studied with the beam aperture blocked by a solid brass plug and with the detector positioned in air at a distance of 15 cm from the end of the collimator. This approximated the collimator to skin–surface distance. Subsequently the detector was housed in a solid nylon (type 6) phantom of dimensions 10 × 10 × 10 cm^3 at a depth of 1 mm and further measurements performed with various thicknesses of nylon added to the front surface of the phantom.

The efficacy of each of the antiscatter collimators was then examined for an open circular field size of 4 cm diameter with the beam impinging upon a 5 cm thick Pb shield on the back wall of the therapy vault 3 m downstream from the isocentre. The counter was positioned 15 cm from the patient collimator at lateral displacements of 15, 30, 60 and 120 cm from central axis. At these locations the number of collimators obscuring the detector from the beam exit window was successively reduced. Further measurements were performed to investigate radiation levels at the entrance to the therapy vault and at the site of one of the charged couple device (CCD) TV cameras used to monitor patient position. The selected camera position was 1.60 m downstream from the isocentre at a radial distance of 1.90 m from the beam axis. The camera is mounted on a bracket extending 1.40 m from the back wall of the therapy vault 85 cm above the height of the beam axis whilst the vault entrance is 8 m perpendicular to the isocentre.

RESULTS AND DISCUSSION

Event size spectra measured on central axis with the beam aperture blocked are illustrated in Figures 2 and 3 for different depths inside the nylon phantom. All spectra were truncated at 0.25 keV.μm^{-1}, just above the lower threshold of pulse height measurement. The area under each curve is proportional to the evaluated absorbed dose in 10^{-2} Gy and is presented in Table 1 normalised to the number of monitor units equivalent to a prescribed dose of 1 Gy at the isocentre for a 4 cm

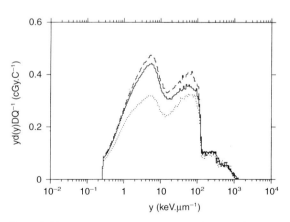

Figure 2. Single-event spectra measured on central axis 15 cm behind the plugged patient collimator (No 4 in Figure 1). Spectra are normalised per treatment gray at the isocentre and illustrate neutron dose build-up for increasing depths 0 (···), 4 (—) and 5 mm (---) inside a solid nylon phantom.

diameter field size. Throughout this study the normalisation to a prescription Gy is specified for the isocentre at a depth of 10 cm in water on the entrance plateau of an unmodulated beam. The characteristic shape of the secondary charged particle spectrum is indicative of neutron interactions in the A-150 plastic wall of the TEPC. These neutrons are produced following inelastic scattering of protons by components in the beam delivery system and the brass plug closing the beam aperture. The recoil proton component centred at approximately 5 keV.μm^{-1} is associated with high energy neutrons that are driven out of target nuclei such as in the patient collimator and brass plug after being struck by incident protons whilst that at approximately 70 keV.μm^{-1} is due to much lower energy evaporation neutrons following the decay of excited nuclei. Figure 2 shows an increasing recoil proton dose component as secondary charged particle equilibrium is attained at 5 mm depth in the phantom. This depth together with the wall thickness of the counter corresponds to a total build-up of 0.71 g.cm^{-2} which is equivalent to the range of a 26.5 MeV proton in nylon[5]. At depths greater than 5 mm the neutron dose is attenuated with the most marked reduction evident for low energy neutrons, as shown in Figure 3.

Measurements performed 15 cm off axis with and without 5 mm build-up yielded identical spectra and absorbed dose values per monitor unit, demonstrating a broad maximum for the build-up curve at this position.

Moving away from the central axis of the primary proton beam produces distinct changes in the scattered radiation that impinges upon the patient. Figure 4 shows in-air spectra obtained at lateral displacements of 15, 30 and 120 cm with the area under each curve again proportional to the evaluated absorbed dose normalised to the number of monitor units equivalent to a prescription dose of 1 Gy at the isocentre. These doses together with the dose equivalents assessed from the spectra are presented in Table 2. A marked increase in the absorbed dose is apparent 30 cm off-axis, which is immediately outside the shadow of the patient collimator. The illustrated spectrum is characterised by a prominent low LET component (0.25 $\leq$ y < 14 keV.μm^{-1}) that accounts for 87% of the measured absorbed dose. Moving further out to a lateral displacement of 120 cm where the line of sight to the beam exit window from the detector is completely unobscured by any collimator the stray absorbed dose decreases dramatically and the peak previously observed at a lineal energy of approximately 1 keV.μm^{-1} migrates to between 2 and 3 keV.μm^{-1}. The low LET component of the measured spectra is attributed mainly to events from protons elastically scattered out of the primary beam by the beam delivery system. The identified shift to higher lineal energies of the low LET component with increased lateral displacement is consistent with the notion of proton scattering[6]. At 120 cm off-axis protons reaching the detector are scattered through a greater angle and lose more energy than those closer to the beam axis which are deflected less. The neutron absorbed dose inferred from the height of the proton edge at 140 keV.μm^{-1} and the heavy recoil component both decrease steadily with lateral displacement and radial distance from the patient collimator. Although there is

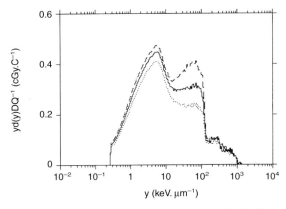

Figure 3. Single event spectra measured on central axis 15 cm behind the plugged patient collimator (No 4 in Figure 1). Spectra are normalised per treatment gray at the isocentre and illustrate neutron dose attenuation for increasing depths 5 (---), 25 (—) and 50 mm (···) inside a solid nylon phantom.

Table 1. Absorbed dose per prescription gray assessed at different depths in a nylon phantom positioned on central axis with the beam aperture blocked. The wall thickness of the TEPC was 1.27 mm.

Nylon thickness (mm)	Build-up (g.cm^{-2})	Dose (10^{-2} Gy)
0	0.14	1.68
4	0.59	2.07
5	0.71	2.22
10	1.27	2.15
15	1.83	2.09
20	2.39	1.99
25	2.96	1.99
50	5.77	1.73

Table 2. Absorbed dose and dose equivalent per prescription gray assessed at different positions off central axis for a 4 cm diameter aperture.

Lateral displacement (cm)	Dose (10^{-2} Gy)	Dose equivalent (10^{-3} Sv)
15	1.18	80
30	2.74	75
60	2.39	59
120	0.99	33

a distinct increase in the total absorbed dose when moving to 30 cm (Table 2) this is not reflected in the dose equivalent to the patient. The dose equivalents obtained behind the patient collimator are consistent with values determined using Bonner spheres in a similar position at an existing therapy facility that also utilises passive scattering[7].

Figure 5 shows spectra normalised to unit dose measured at the two sites distant to the isocentre. Adjacent to the camera position events attributed to scattered fast protons from the beam delivery system appear the main contributor to the absorbed dose whilst at the vault entrance the general radiation field is characterised by a large slow proton component due to low energy neutron interactions in the wall of the TEPC. The measured absorbed doses expressed as a percentage of the prescribed dose were 0.36 and 0.03% respectively at the camera and vault entrance. At the vault entrance the assessed dose equivalent was 3×10^{-3} Sv per treatment Gy.

CONCLUSIONS

Single-event spectra were measured at various positions inside a proton therapy vault and dose equivalents attributed to scattering from the beam delivery system were between 33×10^{-3} and 80×10^{-3} Sv per treatment gray were assessed at the patient position.

In an attempt to reduce the scattered radiation dose outside the defined field additional shielding has been introduced. This includes a 5 cm thick steel plate that extends the width of collimator No 2 by an additional 70 cm and a new brick shield 19 cm thick and 120 cm wide positioned 3.5 m upstream from the isocentre.

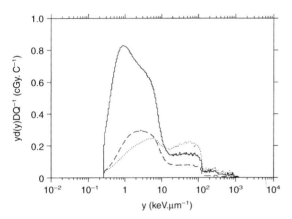

Figure 4. Single-event spectra measured in air 15 cm behind the patient collimator (No 4 in Figure 1) and at lateral displacements of 15 (···), 30 (—) and 120 cm (---) from the central axis of a circular treatment field 4 cm in diameter. Spectra are normalised per treatment gray at the isocentre.

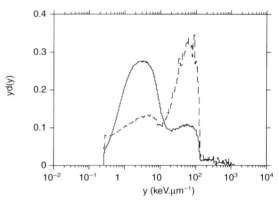

Figure 5. Single-event spectra measured adjacent to a patient positioning camera (—) and at the vault entrance (---) normalised to unit dose. The camera position was at a radial distance of 1.9 m in a forward direction and the vault entrance approximately 8 m perpendicular to the isocentre.

REFERENCES

1. Jones, D. T. L., Schreuder, A. N., Symons, J. E. and Yudelev, M. *The NAC Particle Therapy Facilities*. In: Hadrontherapy in Oncology, Eds U. Amaldi and B. Larsson (Elsevier Science BV) pp. 307–328 (1994).
2. Binns, P. J. and Hough, J. H. *Consideration of Radiation Quality in Treatment Planning with p(66)/Be(40) Neutrons*. Int. J. Radiat. Oncol. Biol. Phys. **24**, 975–981 (1992).
3. International Commission on Radiation Units and Measurements. *Microdosimetry*. Report 36 (Bethesda, MD: ICRU Publications) (1983).
4. ICRP. *1990 Recommendations of the International Commission on Radiological Protection*. ICRP Publication 60 (Oxford: Pergamon Press) (1991).
5. International Commission on Radiation Units and Measurements. *Stopping Powers and Ranges for Proton and Alpha Particles*. Report 49 (Bethesda, MD:ICRU Publications) (1993).
6. Robertson, J. B., Eaddy, J. M., Archambeau, J. O., Coutrakon, G. B., Miller, D. W., Moyers, M. F., Siebers, J. V., Slater, J. M. and Dicello, J. F. *Variation of Measured Proton Relative Biological Effectiveness as a Function of Initial Proton Energy*. In: Hadrontherapy in Oncology, Eds U. Amaldi and B. Larsson (Elsevier Science BV) pp. 706–711 (1994).
7. Koehler, A. M. Harvard Cyclotron Laboratory Cambridge, MA. Private communication (1995).

CRITICALITY ACCIDENT DOSIMETRY: AN INTERNATIONAL INTERCOMPARISON AT THE SILENE REACTOR

R. Médioni† and H. J. Delafield‡
† Institut de Protection et de Sûreté Nucléaire
Département de Protection de la santé de l'Homme et de Dosimétrie
Service de Dosimétrie, B.P.n° 6
F-92265 Fontenay-aux-Roses Cedex, France
‡ Formerly at Health Physics Services
AEA Technology
Harwell, Didcot, Oxfordshire OX11 0RA, UK

INVITED PAPER

Abstract — In criticality accident dosimetry, specialised techniques, which differ markedly from those used in routine radiological protection, are used to measure separately the gamma ray and neutron components of dose. In order to improve these techniques and to evaluate the performance of their dosimetry systems, fifteen laboratories participated in an international intercomparison at the SILENE reactor, Valduc, France. The physical techniques used in criticality dosimetry are first described, for personnel and installed dosemeters, and for the initial sorting of those exposed. The SILENE reactor is briefly described together with a review of the dosimetry and spectrometry techniques used to characterise the radiation field at a reference position and to study the uniformity of the field on an arc around the source. The experimental arrangements of the participants' dosemeters for the intercomparison irradiations (in free-air and on phantoms) are given, followed by a summary of the doses measured by the participants. From an overall analysis of this intercomparison, it is concluded that the participants' results are in good agreement with the reference measurements for both the neutron and gamma ray components of dose.

INTRODUCTION

During the processing and handling of fissile material in a facility, it is possible in exceptional circumstances that, as a result of human error or failure of a safeguards system, a critical mass or volume is reached. In such circumstances, a nuclear chain reaction occurs leading to a critical excursion with a strong emission of neutrons and gamma rays. Even with today's high level of safety technology, the risk of criticality is never absolutely excluded and in the past accidents, with differing degrees of gravity, have occurred throughout the world.

The aim of criticality accident dosimetry is to provide, for such situations, techniques and procedures to identify rapidly the exposed personnel, to assess the doses received and to give, as far as is practicable, information on the spatial distribution of the dose in the body. The ultimate goal is to improve the protection of man and the evaluation of the risks.

In such an accident, the doses received are mainly due to the initial emission of neutrons and gamma rays, with a lesser contribution of gamma ray dose coming from the residual fission products from the critical excursion and from the induced activity of the structure of the facility.

Dose assessments are necessary for guidance in the medical treatment of the exposed individual, for the evaluation of the fission yield of the accident, for public relations reports and to reassure those people who have not been seriously exposed. Moreover such dose assessments provide valuable information for studies on the effects of acute exposure in man. Specialised techniques, which differ markedly from those used in routine radiological protection, are used in criticality dosimetry to measure separately the gamma ray and neutron components of dose.

The main features of a criticality accident dosimetry system are:

(i) A quick sorting technique to identify those persons who have been exposed.
(ii) The evaluation of the maximum absorbed dose (neutron and gamma ray components) in the body. For neutrons the dose is due to heavy charged particles, to protons from the $^{14}N(n,p)^{14}C$ reaction and to gamma rays from the $^{1}H(n,\gamma)^{2}H$ reaction. These components of the dose have been calculated by Auxier *et al*[1] for 'Element 57', of a cylindrical phantom model.
(iii) The determination of the orientation of persons with respect to the source of radiation.
(iv) The measurement of the total dose which may vary in the range from 100 mGy up to 10 Gy. The dosimetry system used must be capable of giving the neutron and gamma ray components of the dose with an uncertainty of less than ±50% within 48 h and less than ±25% four days later[2].
(v) Estimation of the neutron spectrum which can improve a dose assessment. The spectrum may be derived from dosemeter measurements or from a knowledge of the critical assembly.

(vi) The estimation of the dose to any person not wearing a dosemeter.

A very comprehensive manual on dosimetry for criticality accidents was published by the IAEA[3] and more recently the subject has been reviewed by Delafield[4].

DOSIMETRY SYSTEMS

The physical dosimetry gives an evaluation of the dose received and an estimation of its distribution in the body. This is achieved directly from an interpretation of the dosemeter measurements (personnel and area), or later on from an experimental reconstruction of the incident with phantoms or by calculation with simulation codes[5].

Special systems for criticality accident dosimetry include:

(1) Area dosemeters which are installed at known locations and are easily collected following an incident. They give the characteristics of the field at each location (neutron dose and eventually the spectrum, and the gamma ray dose).
(2) Personal dosemeters which may be attached to a criticality belt worn by a worker. These provide an assessment of the neutron and gamma ray components of dose received by a person, taking into account their orientation to the incident.

Finally the results obtained would be studied together with any independent evaluation of the doses given by the biological dosimetry[6].

Sorting of exposed people

An indium foil is usually included in the personnel dosemeter. The high value of the thermal neutron cross section for the (n,γ) reaction on ^{115}In provides a high sensitivity detector. The induced ^{116}In^m activity ($T_{1/2}$ = 54 min) is measured by a portable survey meter. However the relationship between the count rate and the (n + γ) dose is subject to many factors (neutron spectrum shape, scattered neutrons, n/γ ratio and the orientation of the dosemeter relative to the incident field), so that the overall uncertainties[7] on a dose estimate based on indium monitoring may reach a factor of 10.

In the absence of any personal dosemeter, a measurement of the body sodium activation may be used for quick sorting and to estimate the dose received by an individual. Sodium is activated through the ^{23}Na(n,γ)^{24}Na reaction by neutrons thermalised in the body. The radioisotope formed ($T_{1/2}$ = 14.96 h) can be measured by portable detectors but care should be taken to avoid, or allow for, the measurement of the induced ^{38}Cl ($T_{1/2}$ = 37 min).

The capture probabilities for neutrons normally incident on a human phantom, given by Cross[8] and tabulated in the IAEA Report[3], are in good agreement with experimental values obtained for various spectra. The measured specific activity of ^{24}Na is directly related to the neutron dose. However this relationship varies with the shape of the neutron spectrum[9]. Moreover the total dose (n + γ) can only be deduced from the neutron dose if the n/γ dose ratio is known for that particular critical assembly.

Hence because of these uncertainties on any initial dose assessments based on indium or sodium, such measurements are only considered as exposure indicators. More accurate dose assessments must be based on a full interpretation of all available dosemeter measurements.

Neutron dosimetry

The neutron leakage spectrum is dependent on the type of critical assembly and its shielding[10]. Because the neutron recoil dose and kerma tissue factors are strongly dependent upon neutron energy, knowledge of the neutron spectrum is very important so that neutron dosemeters should provide some spectral information. Moreover, the spectrum is required to estimate neutron doses to individual organs and to correct gamma ray measurements for their dosemeter responses to neutrons. Bricka and Médioni[11] have given a detailed description of the detectors commonly used to measure the different components of the neutron spectrum. For thermal and intermediate energy neutrons, activation detectors based on (n,γ) reactions are used. For the more important measurement of the fast neutron component of dose, threshold detectors based on (n,p), (n,n') or (n,fission) reactions are used. In addition, silicon diodes, with an effective threshold energy of ~0.2 MeV and a negligible response to gamma rays, are now used for the measurement of fast neutron dose. The most commonly used criticality dosemeters include gold, copper, indium and sulphur.

Gamma ray dosimetry

The gamma ray dosemeters are the same as those used for monitoring occupational exposure: photographic film, TLDs (^{7}LiF, CaSO$_4$:Dy, CaF$_2$, Al$_2$O$_3$) and RPL glasses. However, care must be taken for any supralinear effects (for TLDs in particular) and special high-dose calibrations are often required. Corrections have to be applied to measurements made with these dosemeters to take into account their response to neutrons. The magnitude of these corrections will vary with the type of critical assembly and any additional shielding, being dependent upon the gamma ray to neutron dose ratio and the neutron spectrum.

INTERNATIONAL INTERCOMPARISON

Criticality accident dosimetry is required to provide a rapid assessment of the dose to personnel following an accident. Specialised techniques are needed to determine the neutron and gamma ray components of the dose. Neutron spectra models are often used for the evaluation. Determination of the gamma ray component is complicated by the response of gamma ray dosemeters to neutrons. These difficulties have been acknowledged by the IAEA who organised twenty years ago a research programme which included four international intercomparisons[12–15] using different facilities (Valduc, Oak Ridge, Vinča and Harwell). In addition, they sponsored the compilation of a compendium of spectra[10] encountered in a criticality accident and a manual[3] on dosimetry for criticality accidents.

More recently, in 1989, within the framework of EURADOS, a Working Group on criticality accident dosimetry[16] was formed to plan an international intercomparison of criticality accident dosimetry which took place at the SILENE reactor, Valduc, France, in June 1993. The aim of this working group was to establish reference dosimetry and spectrometry measurements of the leakage radiation fields available around the SILENE reactor using two different shields (lead and

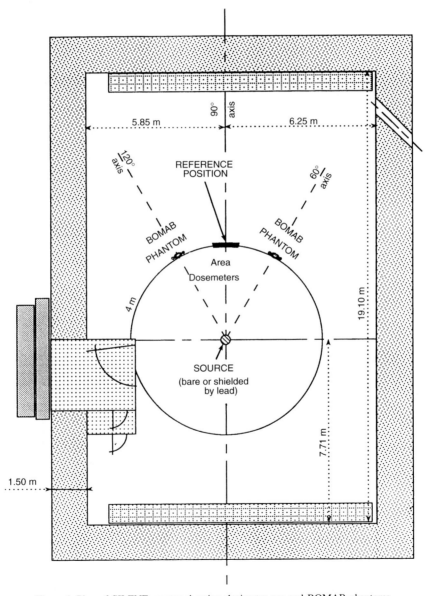

Figure 1. Plan of SILENE reactor showing dosimetry arc and BOMAB phantoms.

bare) which give different neutron spectra and a different neutron to gamma ray ratio; this characterisation was performed before the international intercomparison. This intercomparison was sponsored by the CEC in collaboration with the IAEA, supported by the US-DoE and was hosted by the Institut de Protection et de Sûreté Nucléaire (IPSN). The objectives were to provide participants with a practical training opportunity to use their criticality accident dosimetry systems under realistic conditions, to improve the techniques of measurement and the methods of interpretation in terms of the relevant dosimetric quantities and to evaluate the performance of their dosimetry systems for two different quality radiation fields (neutron spectra and neutron to gamma ray dose ratios).

The SILENE reactor

The SILENE reactor is a homogeneous experimental reactor designed and built by the IPSN at Valduc,

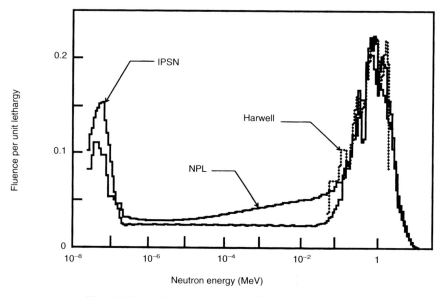

Figure 2. Neutron fluence spectrum for SILENE (lead shield).

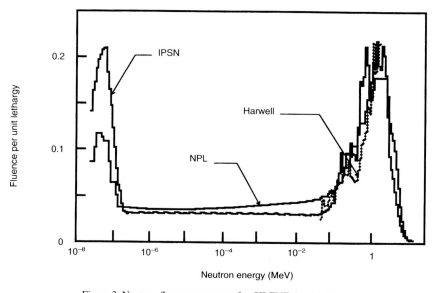

Figure 3. Neutron fluence spectrum for SILENE (unshielded).

France. The reactor and some of the experiments for which it has been used have been widely described[17-19]. The SILENE reactor (Figure 1) is located in a large concrete cell. The core is an annular vessel (360 mm diameter) containing a fissile solution.

This reactor can be operated in three different modes:

(i) steady state (used during the reference dosimetry measurements), for which a constant power, in the range from 10 mW to 10 kW can be maintained for long periods (hours);
(ii) free evolution mode, which simulates a criticality accident in a solution (this mode was used for the irradiations in the intercomparison);
(iii) pulse mode, which gives a very high peak power (1 GW) for a very short time (6 ms).

When the experiment is completed, the fissile solution is transferred into a tank in a shielded room, to allow a quick access to the reactor cell.

For the intercomparison programme, the reactor was operated both with a lead shield (thickness, 100 mm) surrounding the core and unshielded to provide leakage radiation fields of different neutron spectral quality and with a different neutron to gamma ray dose ratio.

Reference measurements

The leakage radiation fields for the reactor, both shielded by lead and bare, were characterised at a central reference position, 4 m from the reactor axis (Figure 1), at a height of 1.20 m from the floor corresponding to the mid-level of the solution in the core.

Co-ordinated measurements[20] were performed by different laboratories (IPSN/France, AEA/Harwell, NPL/Teddington and Homburg University) using various techniques to measure, in free-air, the neutron spectra and the neutron and gamma ray doses. These techniques included: a neutron spectrometry system (recoil counters associated with multispheres), ionisation chambers, TEPC, Si diodes, GM counters and TLDs (alumina).

Figures 2 and 3 show reasonable agreement, over the important energy region from 50 keV to 2 MeV (which contributes over 70% of the dose), between the neutron spectra measured by IPSN, NPL and AEA for the two configurations of the reactor. Independent measurements of the neutron and gamma ray doses were in good agreement (within ~±10%).

Following the procedures described by Médioni and Delafield[20], the Fontenay (IPSN) and NPL spectra measurements, which covered the complete energy range from thermal neutrons up to several MeV, were used to give the neutron fluences and derived doses for the energy groups required for the intercomparison, whilst the Harwell (AEA Technology) spectra measurements of intermediate energy and fast neutrons were used in a confirmatory role. Kerma values were derived using the NCRP conversion factors[21] and the Element 57 components of dose using the Auxier[1] calculations.

In a similar way, reaction rates were derived for the activation detectors. Now the neutron and gamma ray components of kerma were also measured independently by Fontenay (IPSN) using paired ionisation chambers. The best measurement of kerma, for each reactor source configuration, was then taken as the mean of the two spectrum measurements (Fontenay and NPL) and the ionisation chamber measurements. These values of kerma were adopted as the best reference values for the intercomparison. Subsequently the Fontenay and NPL spectrum measurements, and the derived Element 57 doses and detector reaction rates, were normalised to these best kerma values for both reactor fields. The reference doses, based on the combined spectrometry and ionisation chamber measurements, are given in Table 1 normalised to the fission yield of the pulses used in the intercomparison.

An extensive study of the isotropy of the radiation field around the source was performed. For the section of arc (~6 m) used for the intercomparison (for the exposure of dosemeters in air and on two BOMAB phantoms) the neutron and gamma ray doses were within ±5% of that at the reference position for both reactor configurations.

An extensive description of the techniques used, the detailed measurements obtained for the reference position and the results of the isotropy study have been published in Part 1 of the final report[20] of this intercomparison.

Intercomparison experiments

A two week programme was undertaken at Valduc (7–18 June 1993). Because of the large number of participants, and the requirement to limit the number of dosemeters on the front of the phantoms, two irradiations were performed with the reactor shielded by lead (100 mm thick) followed by a further two irradiations with the reactor bare. The programme included:

(a) exposure of dosemeters, in free-air and on BOMAB phantoms (Figure 1), including sodium activation (BOMAB phantoms were filled with a dilute solution of sodium nitrate);
(b) dosemeter and ^{24}Na activation measurements by participants;
(c) assessment and reporting of preliminary results by participants within 48 h;
(d) analysis of preliminary results, lectures and informal discussion with the presentation of papers on research and development and the dosimetry systems used by participants.

Dose quantities

Following previous international intercomparisons

the following dosimetric quantities were derived from the measurements:

For dosemeters exposed on the front of phantoms, the quantities reported were:

D_n neutron recoil + proton dose for Element 57,
$D_{n\gamma}$ neutron H(n-γ) dose for Element 57,
$D_{t\gamma}$ total gamma ray dose.

For dosemeters exposed in free-air, the principal quantities reported were:

K_n neutron kerma,
D_γ incident gamma ray dose,
and the neutron fluences above given energy limits.

In most cases, the quantities applicable to dosemeters exposed on the body, D_n and $D_{n\gamma}$ were also derived from these neutron dosemeters.

Measurement and evaluation of dosemeters

Most participants brought their own equipment for the measurement and assessment of their dosemeters. Typically, this included gamma ray spectrometers or other counting equipment and thermoluminescence readers. Immediately upon receipt of the dosemeters from the reactor, participants started making measurements to evaluate their dosemeters and so present their preliminary dose assessment to the meeting at Valduc. As many measurements as practicable were made at Valduc. In general, the longer-lived activation foils were then recounted by participants on returning to their own laboratories. Then participants sent in their final dose assessments together with a description of their accident dosimetry system[22]. These dose assessments are summarised and compared with the reference dosimetry measurements in the final report[23].

Table 1. Comparison of mean doses derived from participants' dosemeters with reference measurements and performance of criticality accident dosemeters.

Quantities	Lead shield (3×10^{17} fissions)				Bare reactor (2×10^{17} fissions)			
	Reference doses (Gy)	Mean/Reference		Performance*	Reference doses (Gy)	Mean/Reference		Performance*
		Ratio	SD (%)			Ratio	SD (%)	
Free-air:								
neutron kerma	3.36	0.99	13	0.92	2.56	1.10	10	0.92
neutron recoil dose	3.66	1.01	18	0.80	2.90	1.13	12	0.89
neutron ^{1}H(n,γ)^{2}H	1.22	0.81	25	0.44	0.84	0.87	30	0.56
incident γ ray dose	0.58	1.22	18	0.50	3.98	0.91	20	0.79
n/γ ratio	5.77	0.87	25		0.64	1.21	16	
Phantom								
neutron kerma	3.36	1.04	10	1.00	2.56	1.21	12	0.56
neutron recoil dose	3.66	1.01	12	1.00	2.90	1.13	16	0.75
neutron ^{1}H(n,γ)^{2}H	1.22	0.63	18	0.11	0.84	0.86	31	0.67
total γ ray dose	1.80	0.81	10	0.69	4.82	0.91	14	0.93
n/γ ratio	2.03	1.26	20		0.60	1.26	21	

*Performance is expressed in terms of the proportion of participants' measurements within ±25% of reference measurements. This proportion is out of unity, so that for example, all participants' measurements (proportion = 1.00) of the neutron recoil dose on phantom for the lead shielded configuration were within ±25% of the reference measurement (between 0.75 and 1.25 times this value).

Table 2. Participating laboratories to the intercomparison.

Laboratory	Country	Laboratory	Country
IRD	BRAZIL	BARC	INDIA
CRL	CANADA	ENEA	ITALY
RBI	CROATIA	IAE	POLAND
IRD	CZECH REPUBLIC	IOB	RUSSIAN FEDERATION
RNL	DENMARK	CIEMAT	SPAIN
IPSN	FRANCE (two laboratories)	AEA	UK
KfK	GERMANY	WSR	USA

Fifteen laboratories (Table 2) from fourteen countries attended this Intercomparison to test and evaluate their criticality dosimetry systems. In addition, three laboratories performed experimental studies including blood chromosome[6] and electron spin resonance[24] measurements and tests with bubble detectors[24].

Results

Figures 4 and 5 show the relative values of the dose quantities measured by the participants for the free-in-air and on-phantom irradiations for both configurations of the source. These results are normalised to the reference values. For each dose quantity, Table 1 gives the ratio of the mean value of the participants' measurements to the reference measurement. This table shows good agreement for the measurement of kerma and recoil dose (ratio of 0.99 to 1.04 for the lead shielded source and 1.10 to 1.21 for the bare reactor).

For the $^1H(n,\gamma)^2H$ component of dose, the mean values are below the reference measurements (ratio 0.63 to 0.87). However in a practical situation, the ability to derive this component of dose is of secondary importance as it is measured by a gamma ray dosemeter worn on the front of the phantom. For the gamma ray measurements, there is fair agreement between the mean values of the participants' measurements and the reference measurements for the lead shielded configuration (free-air, incident γ ray dose; ratio 1.22: phantom, total γ ray dose; ratio 0.81) and for the bare source (ratio is 0.91 for both cases).

For neutrons (kerma or recoil dose), the standard deviation on the mean (free-air, on-phantom, for lead shielded or bare source) is on average about ±13%. For

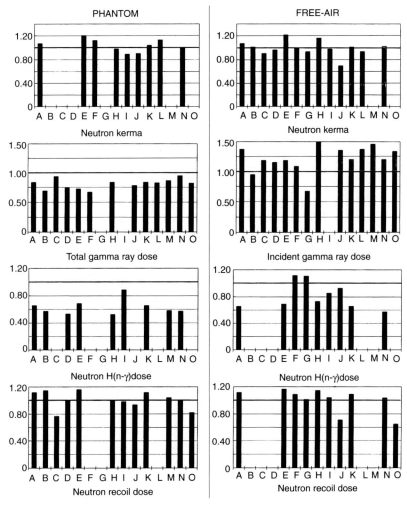

Figure 4. Participants' results normalised to reference measurements for the lead-shielded source (participants are identified by their laboratory code letter).

gamma ray quantities, the standard deviation of the mean is larger which illustrates the difficulty of measuring the gamma ray dose in a mixed radiation field for which it is necessary to take into account the response of the gamma ray dosemeter to neutrons. For the $^1\text{H}(n,\gamma)^2\text{H}$ dose, the standard deviation is on average typically about ±26%. For the incident gamma ray dose, which is measured directly by dosemeters exposed at free-air stations, the standard deviations of the mean of the results are ±18% for the lead shielded source and ±20% for the bare source. The total gamma ray dose (incident + $^1\text{H}(n,\gamma)^2\text{H}$) is measured by dosemeters exposed on the front of the phantom; the corresponding standard deviations are ±10% (lead) and ±14% (bare).

Only a few participants gave results on reaction rates for the activation detectors. The results, from five participants, given for ^{32}S were good (standard deviation of mean about ±9%, ratio to reference value about 1.15) and for ^{115}In satisfactory (standard deviation of mean about ±16%, ratio to reference value about 0.93). In reality, it is these good measurements of the principal threshold detectors which led to satisfactory evaluations of the neutron kerma and recoil doses. However the measurements of the thermal neutron and resonance (n,γ) activation detectors (^{63}Cu and ^{197}Au) used bare and cadmium covered showed a wider scatter in the results (standard deviation of mean about ±18% and ±37% respectively).

The sodium activity measurements were performed by nine laboratories for both configurations (standard

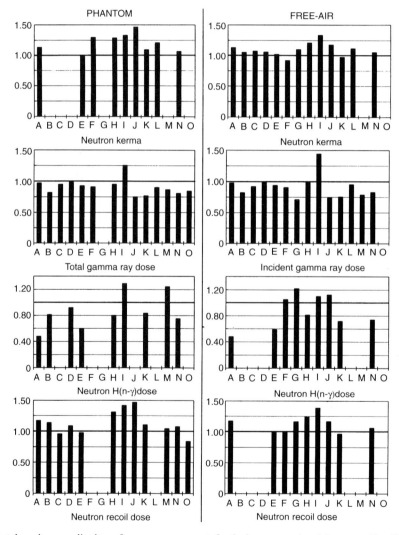

Figure 5. Participants' results normalised to reference measurements for the bare source (participants are identified by their laboratory code letter).

deviation of mean about ±10%) and results were in good agreement with reference values (ratio for bare source: 0.91, for lead shielded source: 0.87).

The performance criteria proposed by the IAEA Panel[2] recommend that the dosimetry system must be capable of giving the neutron and gamma ray components of the dose with an uncertainty of less than ±25%. Table 1 shows that only for the neutron measurements (kerma and recoil dose), made on a phantom for the lead shielded source, did all the participants satisfy the recommended criteria.

CONCLUSION

This intercomparison study provided participants with a practical training opportunity to test under realistic conditions their criticality accident dosimetry systems used for personal (on phantom) and area dosimetry (free-air). Fifteen laboratories evaluated the performance of their dosimetry systems for two different quality radiation fields having different neutron spectra and neutron to gamma ray dose ratios.

The results have shown that all participants were able to make a rapid estimate of the dose received in a simulated criticality accident, even though most of them were working well away from their home base and often with unfamiliar equipment.

For neutrons, the performance was good; typically 86% of the results were within ±25% of the reference dose values. For gamma rays, the performance was in general reasonable; typically 73% of the results were within ±25% of the reference values but only 50% were for the measurement of the incident free-air dose for the lead shielded source.

A set of three reports[20,22,23] have been published including: the reference dosimetry of the radiation field, the description of participants' systems, and the description of experiments and participants' results with analysis and conclusion.

This intercomparison has demonstrated the need and the absolute necessity to perform periodic tests of criticality accident dosimetry systems and shown the usefulness of experimental facilities like the SILENE reactor which is a perfect tool to simulate criticality accidents and to conduct experiments in this field.

ACKNOWLEDGEMENTS

This intercomparison was supported by the Commission of the European Communities under Study Contracts (DGXII: B17*0051F and BI7*0076F; DGXI: 92-PR-007 and 92-PR-008). The authors gratefully acknowledge the help received from the management and the staff of the SILENE reactor, especially Dr F. Barbry, in the provision of the irradiations.

In addition, they also wish to thank all EURADOS-WG9 members who invested much time both during the meetings and in preparing for the intercomparison; all their colleagues from France and the UK and all intercomparison participants who have contributed to this work and had a part in the success of this programme.

REFERENCES

1. Auxier, J. A., Snyder, W. S. and Jones, T. D. *Neutron Interactions and Penetration in Tissue.* In: Radiation Dosimetry, Eds F. H. Attix and W. C. Roesch (New York: Academic Press) Vol. 1, pp. 275–316 (1968).
2. IAEA. *Nuclear Accident Dosimetry Systems.* STI/PUB/241 (Vienna: IAEA) (1970).
3. IAEA. *Dosimetry for Criticality Accidents — A Manual.* Technical Reports Series No 211. STI/DOC/10/211 (Vienna: IAEA) (1982).
4. Delafield, H. J. *Nuclear Accident Dosimetry — An Overview.* Radiat. Prot. Dosim. **23**(1–4), 143–149 (1988).
5. Bottollier, J. F., Médioni, R., Plawinski, L., Bouteilloux, P. and Herault, G. *Dosimetric Reconstruction in the Case of a Radiological Accident.* Congress on Radiological Accident, 10-12 April 1995, Grenoble, France.
6. Voisin, P., Lloyd, D. and Edwards, A. *Chromosome Aberrations Scoring as Biological Dosimetry of a Criticality Accident.* Radiat. Prot. Dosim. **70**(1–4), 467–470 (This issue) (1997).
7. Swaja, R. E. and Oyan, R. *Uncertainties Associated with Using Quick-sort Techniques to Estimate Neutron Doses Following Criticality Accidents.* Health Phys. **52**(1), 65–68 (1987).
8. Cross, W. G. *Sodium Activation in the Human Body.* Radiat. Prot. Dosim. **10**(1–4), 265–276 (1985).
9. Delafield, H. J. *Nuclear Accident Dosimetry.* Radiat. Prot. Dosim. **10**(1–4), 237–249 (1985).
10. Ing, H. and Makra, S. *Compendium of Neutron Spectra in Criticality Accident Dosimetry.* IAEA Technical Reports Series No 180. STI/DOC/10/180 (Vienna: IAEA) (1978).
11. Bricka, M. and Médioni, R. *Measurement of Neutron and Gamma Radiation.* In: Dosimetry for Criticality Accidents — A Manual. Technical Reports Series No 211. STI/DOC/10/211, p. 57–82 (Vienna: IAEA) (1982).
12. Bricka, M. *First IAEA Measurement Intercomparison, Valduc (France) June 1970.* CEA Report SESRCI 71/408MB (1971).
13. Haywood, F. F. and Poston, J. W. *Second IAEA Accident Dosimetry Coordination Meeting and Intercomparison Experiments, May 3-15, 1971.* Rep. ORNL-TM-3770 (1973).
14. Mirič, I., Ubovič, Z., Mirič, P. and Velickovič, D. *Report on the Third IAEA Intercomparison Experiment at Vinča, Yugoslavia.* IAEA Report MG-140 (1977).
15. Gibson, J. A. B., Delafield, H. J. and Reading, A. H. *Nuclear Accident Dosimetry: Report on the Fourth IAEA Intercomparison Experiment at Harwell, UK., 7-18 April 1975.* Report AERE-R8520 (London: HMSO) (1976).

16. Médioni, R. *EURADOS-WG9: Criticality Accident Dosimetry. Activities during the Period 1990–1995. Contract No F13P-CT920001. Final Report.* IPSN/DPHD 95-386. (September 1995).
17. Barbry, F. and Médioni, R. *SILENE, Réacteur Source pour Irradiation.* In: Proc. Fourth Symp. on Neutron Dosimetry, Munich, 1-5 June 1981. CEC Report EUR 7448 EN. Vol. 1, pp. 443–452 (1981).
18. Barbry, F. *A Review of the SILENE Criticality Excursion Experiments.* CEA Report SRSC 93.220 (December 1993).
19. Tournier, B., Barbry, F. and Verrey, B. *SILENE, a Tool for Neutron Dosimetry.* Radiat. Prot. Dosim. **70**(1–4), 345–348 (This issue) (1997).
20. Médioni, R. and Delafield, H. J. *An International Intercomparison of Criticality Accident Dosimetry Systems at the SILENE Reactor, Valduc, Dijon, France, 7-18 June 1993. Part 1: Reactor and Reference Dosimetry of Radiation Fields.* AEA Technology and IPSN Report HPS/TR/H/1(95) (1995).
21. NCRP. *Protection against Neutron Radiation.* Appendix B.I. *Dose Distribution in a Cylindrical Phantom for Neutron Energies Up to 14 MeV.* NCRP Report No 38 (National Council on Radiation Protection and Measurements, Washington, DC, USA) (1971).
22. Delafield, H. J. and Gibson, J. A. B. (Eds) *An International Intercomparison of Criticality Accident Dosimetry Systems at the SILENE Reactor, Valduc, Dijon, France, 7–18 June 1993. Part 2: Dosimetry Systems used by Participants at the Experiment.* AEA Technology and IPSN Report HPS/TR/H/2(95) (1995).
23. Delafield, H. J. and Médioni, R. *An International Intercomparison of Criticality Accident Dosimetry Systems at the SILENE Reactor, Valduc, Dijon, France, 7-18 June 1993. Part 3: Description of the Experiment and Participants' Results.* AEA Technology and IPSN Report HPS/TR/H/3(95) (1995).
24. D'Errico, F., Alberts, W. G. and Matzke, M. *Advances in Superheated Drop (Bubble) Detectors Techniques.* Radiat. Prot. Dosim. **70**(1–4), 103–108 (This issue) (1997).

BIOLOGICAL DOSIMETRY AFTER A CRITICALITY ACCIDENT

M. Fatôme, D. Agay, S. Martin, J. C. Mestries and E. Multon
Département de Radiobiologie
Centre de recherches du service de santé des armées E. Pardé
BP 87, 38702 La Tronche, France

INVITED PAPER

Abstract — In the case of a criticality accident, there are two radiation components, i.e. neutron and gamma, and the exposure is heterogeneous. The available techniques are biophysical techniques, which can give data about the dose distribution and the neutron component dose, classical techniques (electroencephalography, lymphocytes counting, dicentric and acentric fragments counting, biochemical indicators) which can give an estimation of the mean biological dose, new biochemical parameters which are good indicators of vital prognosis and new cytogenetic techniques under study. Biological dosimetry or rather research of biological indicators of irradiation remains difficult, particularly because of the heterogeneity of exposure. A multiparametric determination seems the only solution.

INTRODUCTION

When a radiation accident by external overexposure occurs, the best possible evaluation of the abosrbed dose and the best possible assessment of the vital prognosis are matters of urgency. In particular, it is necessary to appreciate rapidly whether the victims need hospitalisation in a specialised service. Furthermore, this evaluation allows the efficient preparation of the treatment that it will be necessary to undertake during the weeks following the accident. Biological dosimetry, which is based on biological assays, is a necessary complement to physical and clinical dosimetries.

Physical dosimetry, and particularly the reconstruction of the accident, is essential and gives important information about the dose and its distribution. However, it cannot take into account individual radiosensitivity and hence cannot be a perfect indicator of the biological consequences of the exposure. For instance, a dose of 2.5 Gy can be lethal for some people whereas a dose of 5 Gy can be non-lethal for few people.

Clinical dosimetry is based on the observation of the delay, the intensity, the frequency and the duration of the early and transient neuro-vegetative symptoms, such as nausea and vomiting, fatigue, headache, diarrhea, hypotension and hyperthermia. It can give interesting data. However, it can be perturbated by different factors such as associated lesions.

It is hard to find convenient biological indicators for dosimetry. Generally, the dose–effect relationship is not good, so that the useful tests are still small in number. This problem is more acute when considering a criticality accident. In this case, there are two main difficulties:

(1) There are two radiation components, neutrons and gamma rays, with different biological efficiencies, generally higher for neutrons.
(2) In the organism the neutron component is rapidly attenuated, so that the irradiation is heterogeneous. Moreover, this heterogeneity is generally enhanced by the position of the victim relative to the reactor, as it was the case during the accident at Mole in 1965.

On the whole, the available techniques can give an estimation of the mean biological dose, but most are unable to give a good assessment of the dose heterogeneity or of the dose fractions delivered by neutron and gamma components. Some new assays can give a good assessment of the vital prognosis, which represents an important endpoint.

The available techniques can be divided into four classes:

(1) Biophysical techniques which can give data about the dose distribution in the organism and the neutron component dose.
(2) Classical techniques which can give an estimation of the mean biological dose.
(3) New biochemical parameters which are good indicators of vital prognosis after neutron–gamma irradiation.
(4) New cytogenetic techniques under study.

BIOPHYSICAL TECHNIQUES

These are at the frontier between physical and biological dosimetry. Some techniques are neutron-specific. This is the case with activation analysis.

Activation analysis

Neutron irradiation induces a transformation of stable elements into radioactive isotopes. In the organism, it induces ^{24}Na which is a gamma emitter with a half-life of 15 h and is easily detectable. It is necessary to take into account the activity of ^{38}Cl during the 3 h following irradiation. This analysis shows a linear evolution of initial activity as a function of the neutron dose in air. Furthermore, the activation that induces ^{32}P from ^{32}S in hair, nails and teeth is important because it can give an idea of the dose heterogeneity.

Electron spin resonance

The main consequence of the action of ionising radiations on organic polymers is the creation of free radicals of which the mass density is proportional to the dose. These free radicals, which have a long life in polymers, are paramagnetic and are detected by ESR. The surface under the absorption spectrum is proportional to the number of spins present in the resonance cavity and consequently to the dose[1]. The threshold is low, around 0.5 Gy. The signal decays with time as the combination of exponential kinetics. Thus ESR can allow a rapid evaluation of the dose with good accuracy. Up to now, the practical biological material is the nail.

The main drawbacks consist in the lack of prognosis indication, the importance of the detection material, and the necessity of a very specialised personnel.

A new technology under study could allow the detection of free radicals on bones *in vivo* without sampling.

CLASSICAL TECHNIQUES

Electroencephalography (EEG)

The central nervous system of the adult presents a strong functional radiosensitivity so that cerebral electrical activity can be altered after irradiation doses as low as 0.5-1 Gy. Between 1 and 2.5 Gy, the modifications are a monotonous tracing with a low amplitude, an increased number of slow waves with a relatively high amplitude and spikes or spike waves, the number of which increases with the dose[2]. Between 2.5 and 6 Gy, there are many spike waves, many slow waves and persistance of a disorganised EEG for more than several days.

After a mixed neutron–gamma irradiation, the alterations are more important and spindles of rapid waves (16-22 s^{-1}) appear, with a low amplitude for 1-1.5 s and are characteristic of neutron irradiation.

The advantages are a low dose threshold, the earliness of the response, a neutron-specific modification and the non-invasive character of the technique.

The drawbacks are the general lack of control records, the necessity of a very specialised personnel, the long duration of the recording, at least equal to 1 h, and the lack of information about the distribution of the dose.

Lymphocyte counting

Peripheral blood lymphocytes are very sensitive to ionising radiation. The decrease in their number at 24-28 h after irradiation and the decreasing curve slope can be useful indicators of the mean absorbed dose. For instance, a reduction of 50% of their number 24 h after irradiation indicates a potentially lethal exposure.

The technique has several advantages: sampling is easy, the counting can be automated, the result can be obtained with a short delay (a few hours), the sensitivity is good, the dose threshold being around 0.25 Gy, the dose–effect relationship is linear. Its main drawbacks are the rapid turnover in blood, the important inter-individual variations, the strong influence of concurrent factors, the lack of specificity for neutron exposure and the lack of information about the heterogeneity of the exposure. In the case of criticality, this technique can only give a mean and global indication.

Lymphocytes unstable chromosomic aberrations

Lymphocytes are mostly in a quiescent state so that the radiation-induced chromosome aberrations will stay latent until cell division. The method consists of counting the dicentric and acentric fragments. The principle of the technique consists of culturing the lymphocytes in the presence of phytohaemagglutinin which induces their division. After 48 h, they are blocked in their metaphase by colchicine[3]. Reference curves are determined for different radiation qualities. The dose–effect relationship is linear–quadratic for low LET radiations and linear for high LET radiations, such as neutrons. This technique can give an indication of the dose fraction given by the two components.

The dose threshold can be as low as 0.2-0.3 Gy, but it is necessary to examine several hundreds of metaphases; therefore the analysis requires a long time and specialised personnel. The automated counting of dicentrics is rarely performed, compared with the efficiency of metaphase detectors.

Another interest of dicentrics and acentric fragments consists in their specific formation by irradiation. They are good biological indicators in the case of a homogeneous irradiation. Their statistical distribution follows the Poisson law.

In the case of a heterogeneous exposure, there is a dilution of irradiated lymphocytes with the others, so that the distibution does not follow the Poisson law and the number of aberrations decreases more quickly. On the whole, this technique is not a good indicator of heterogeneity. Because of the unstable character of dicentrics and acentric fragments, it loses its efficiency within three months and can be considered as a short-term dosemeter. Nevertheless, it remains the best available technique for biological dosimetry.

Biochemical indicators

An early increase in serum amylase level has been described in humans when the parotid gland or the pancreas is irradiated[4]. The dose threshold is around 1 Gy. Until now, it was the only biochemical parameter that might be useful. The difficulties arise from the lack of a dose–effect relationship in humans, the important inter-individual variability and the great number of concurrent and confusion factors (infection, inflammation, associated lesions, etc.).

NEW BIOCHEMICAL ASSAYS WHICH ARE GOOD INDICATORS OF VITAL PROGNOSIS AFTER NEUTRON–GAMMA IRRADIATION

Recent biochemical investigations have shown that several parameters could be useful for prognostic indication. Baboons were submitted to a mixed neutron–gamma irradiation. The neutron to gamma ratio was equal to 5.5. With the reactor operating in free evolution mode, the different doses were delivered in about 2 min. During the first few hours after the plasma increase the early radio-induced inflammatory reaction mediator IL-6 can be considered as a good indicator (Figure 1). After 24 h, it is important to study the serum iron increase, which is a consequence of extravascular hyperhaemolysis and of inflammation (Figure 2). Cholesterol and apolipoproteins can give a later indication between the 3rd and the 8th days[5].

A few days before the acute phase of irradiation syndrome, a second inflammatory reaction can appear. It is characterised by a new increase in inflammation mediators (IL-1, IL-6, IL-8), acute phase proteins (CRP, fibrinogen, alpha1 antitryspin, alpha 2 macroglobulin) and lipidic peroxidation indicators (MDA). This reaction seems a good prognostic indicator[6].

There is also a biphasic alteration of haemostasis factors. The second variation which appears several days after irradiation (increase in haemostasis and fibrinolysis activation factors) could also be a prognostic indicator.

NEW CYTOGENETIC TESTS UNDER STUDY

Micronuclei

Micronuclei are chromosomal fragments that lack centromeres. They are not incorporated into the new nucleus during division and form a small satellite structure[7]. The lymphocytes must be in culture for more than two days with 24 to 32 h of incubation with cytochalasine B which blocks the separation of the cytoplasm and leads to characteristic figures of binucleated lymphoctes. Micronuclei are counted in at least 1000 cells. The counting is fast and can be completed in an hour (Table 1). The reference curves are still few and concern a reduced number of radiation qualities. They are near that found for dicentrics, and micronuclei are good indicators of radiation quality. Their distribution does not follow the Poisson law, and this dispersion increases with high LET radiation, but it is insufficient to give an exact separation of the dose in cases of heterogeneity. Micronuclei as unstable aberrations disappear rapidly after the exposure so that they are only a

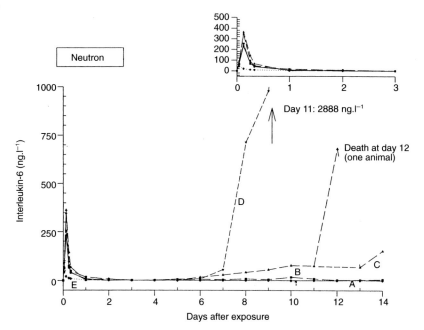

Figure 1. Plasma IL-6 concentration in baboons submitted to a mixed neutron-gamma irradiation. The neutron to gamma ratio was 5.5. The doses were delivered in about 2 mins, the reactor operating in free evolution mode. Key: Curve A (●) 2 Gy (n = 6); B (■) 4 Gy (n = 6); C (▲) 6 Gy (n = 6); D (▼) 8 Gy (n = 4) ; E (♦) control (n = 4).

short-term dosemeter. Finally, unlike dicentrics, they are not radiation-specific. They can be produced by chemical mutagens. Their spontaneous rate increases with age and is higher in women. It is also increased by alcohol and tobacco.

Fluorescent *in situ* hybridisation technique (FISH)

This has the advantage of showing stable abnormalities such as translocations. It allows the detection of specific nucleic acid sequences in tissues, cells and chromo-

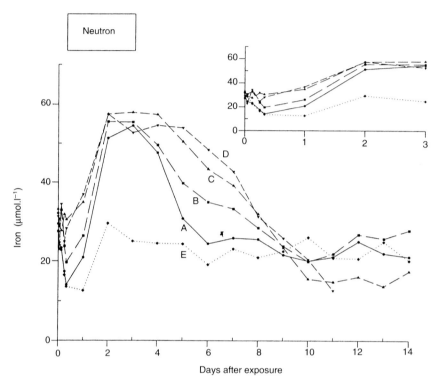

Figure 2. Serum iron concentration in baboons submitted to a mixed neutron-gamma irradiation. The neutron to gamma ratio was 5.5. The doses were delivered in about 2 mins, the reactor operating in free evolution mode. Key as Figure 1.

Table 1. Comparison of the different cytogenetic techniques.

	Detection	Preparation	Analysis	Specificity	Term	Heterogeneity indicator
Conventional cytogenetic technique	Unstable aberrations (dicentrics and acentric fragments)	48 h	(long)	Good	Short-term (< 3 months)	Imperfect
Micronuclei	Unstable abnormalities	48 h	Fast (1 h)	Not good	Short-term	Not good
FISH	Unstable and stable abnormalities	days	Relatively fast (possibility of automation)	Imperfect	Mean and long-term	Imperfect

somes. The studies concerning its application to biological dosimetry are still few. The lymphocyte is also the material used. The DNA molecule is denatured to a monochain form. Then hybridisation is practised with labelled probes specific for each chromosome. Generally a large chromosome is used. The hybridisation sites are finally detected by fluroescent antibodies[8]. Four days are necessary for preparation. To avoid confusion with dicentrics, centromere staining techniques have been developed. The counting of luminous points is relatively fast. The dose–effect curves are similar to those obtained after conventional techniques. However, many studies remain to be done for eliminating false positive and false negative responses. The observation of translocations could be useful for the detection of old exposures.

In the future, this techniques could bring more information by combination with the prematurate chromosomal condensation technique, which allows the visualisation of chromosomes[9]. After staining, it is possible to determine the number of excess fragments relative to the normal number of chromosomes. The preparation is short (2 h). However, sampling must be done early to avoid the influence of repair. Moreover, many studies remain necessary for determining the influence of dose rate, radiation quality and combined injuries.

CONCLUSION

After a criticality accident, biological indicators of irradiation are few and mostly imperfect. The main difficulty comes from the exposure heterogeneity. New biochemical and cytogenetic developments could bring further precise information about biological consequences of irradiation and particularly about vital prognosis. The new biophysical studies concerning measurement of free radicals in bone *in vivo* would certainly bring useful information. Another way could consist in the cytokinesis shown in the skin during the early inflammatory reaction. Thus a multiparametric determination seems to be a useful solution.

REFERENCES

1. Sagstuen, H., Theysen, H. and Henriksen, T. *Dosimetry by ESR Spectroscopy following a Radiation Accident.* Health Phys. **45**, 961–968 (1982).
2. Court, L., Bassant, M. H., Gourmelon, P., Gueneau, G. and Pasquier, Ch. *Impairment of Electrical Functions of CNS and Alterations in Cell Populations associated with Irradiation.* Br. J. Cancer **55**, 230–231 (1986).
3. Doloy, M. T. *Dosimétrie basée sur le Dénombrement des Anomalies Chromosomiques contenues dans les Lymphocytes Sanguins.* Radioprotection **26** (Suppl. 1), 171–184 (1991).
4. Barrett, A., Jacobs, A., Kohn, J., Raymond, J. and Powles, R. L. *Changes in Serum Amylase and its Isoenzymes after Wholebody Irradiation.* Br. Med. J. **285**, 170–171 (1982).
5. Martin, S. *Dosimétrie Biologique appliquée aux Rayonnements Gamma et Neutron-Gamma.* Rapport de synthèse finale. Commande DRET 91/1006. Juillet 1995.
6. Mestries, J. C. and Multon, E. *Le Syndrome aigu d'Irradiation: Mal des Rayons ou Maladie de l'Irradié?* Rev. Sci. Tech. Défense **3**, 5–17 (1995).
7. Fenech, M. *The Cytokinesis-block Micronucleus Technique: a Detailed Description of the Method and its Application to Genotoxicity Studies in Human Populations.* Mutat. Res. **285**, 35–44 (1993).
8. Bauchinger, M., Schmid, E., Zitzelsberger, H., Braselmann, H. and Nahrsted, U. *Radiation-induced Chromosome Aberrations analysed by Two-colour Fluorescence in situ Hybridization with Composite Whole Chromosome-specific DNA Probes and a Pancentromeric DNA Probe.* Int. J. Radiat. Biol. **64**, 179–184 (1993).
9. Evans, J. W., Chang, J. A., Giaccia, A. J., Pinkell, D. and Brown, J. M. *The Use of Fluorescence in situ Hybridization combined with Premature Chromosome Condensation for the Identification of Chromosome Damage.* Br. J. Cancer **63**, 517–521 (1991).

CHARACTERISATION OF A CLOTHING MATERIAL FOR GAMMA DOSIMETRY IN MIXED NEUTRON–GAMMA FIELDS

M. Benabdesselam†, P. Iacconi†, D. Lapraz†, A. Serbat‡, J. Dhermain‡ and J. Laugier§
†LPES-CRESA. Université de Nice-Sophia Antipolis
Parc Valrose. 06108-Nice Cedex 2, France
‡DGA/CEB. 16 bis, avenue Prieur de la Côte d'Or
94114-Arcueil Cedex, France
§DGA/CEB. Centre d'Etudes du Bouchet
BP no 3, 91710-Vert-le-Petit, France

Abstract — When an irradiation accident occurs, it is very important for medical treatment to be able to reconstruct the cartography of the dose absorbed by the irradiated person. A cotton textile coated with an α-alumina dosimetric powder is suitable to clothe persons subjected to irradiation risks. With some precautions, the dose may be measured by thermoluminescence (TL). Responses of the aluminised cotton textile relative to the absorbed dose are shown for different ratios of D_γ/D_n and relative to the photon energy.

INTRODUCTION

In the case of an accidental irradiation, it would be very helpful in setting up the medical prognosis to be able to map the surface cartography of the dose absorbed by the irradiated persons so as to determine the possible spared areas, as well as those areas most exposed. Until now, this was only possible by means of a reconstruction of the accident, but of course there are many difficulties. Several techniques have been proposed as a solution, but they present some inconveniences which limit their use.

Among these techniques are electronic spin resonance (ESR) of clothing (cotton textiles and synthetic fibres)[1] and of the dental enamel[2], and the radioyoluminescence of hair[3]. The main problem with these techniques lies in their high detection threshold (about 1 Gy). Furthermore, most of these methods do not give the position of the injured person relative to the radiation beam. The various tests carried out by thermoluminescence (TL) and thermostimulated exoelectronic emission (TSEE) on cotton textile clothing have not been efficacious[4].

A specially prepared clothing material is here described which is meant to be worn by the person subject to irradiation risks. A cotton textile impregnated with α-alumina powder has been studied within the framework of a collaboration between LPES-CRESA and ETCA-CEB. This cotton allows the manufacture of clothes to be worn by persons subject to irradiation risks (criticality accident or others). Because this material is only sensitive to γ radiation, it is possible to evaluate by TL the received γ dose as well as to determine the main irradiated areas.

This paper presents a study of the physical characteristics of the dosimetric cotton in an n–γ mixed field and its response as a function of the energy of photons. It also describes the possibility of reconstructing the way in which the injured person was irradiated.

DESCRIPTION OF THE DOSIMETRIC TEXTILE

The dosimetric textile is made of cotton (120 g.m^{-2}), mercerised and whitened, on top of which a radiosensitive layer of alpha alumina powder is added. Alpha alumina has been chosen because of its good thermoluminescent response to γ rays and its insensitivity to neutrons, its chemical inertia, and its low cost. Alumina powder was fixed on the cotton with an appropriate

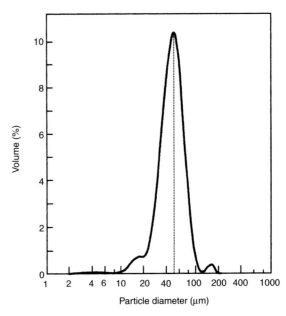

Figure 1. Granulometric distribution of α alumina 44. Measurement by laser granulometry.

binder, transparent to TL emission and resistant to washing and rubbing. A thickening substance was also introduced in order to ensure good performance. Different tests led to the use of a Desmarquest (Evreux, France) alumina powder (alumina 44) with varying content from 12 to 65 g.m^{-2}. The different constituents of the dosimetric textile are given in Table 1. The granulometric distribution of the α alumina crystals used is shown in Figure 1. The dosimetric cotton textile is manufactured at the Institut Textile de France (I.T.F.) at Mulhouse, France.

The signal induced by γ rays may be measured by TL, TSEE or PITL[5] directly on the piece of a dosimetric cotton. The most reliable detection method and the most acceptable is TL. The method has already been described elsewhere for its application in n-γ mixed fields[6]. The TL curve obtained under γ rays (Figure 2) is similar to that obtained under X rays[6]; it is characteristic of α alumina material.

THERMOLUMINESCENCE CHARACTERISTICS OF DOSIMETRIC COTTON

The use of the clothing textile in accident dosimetry requires a calibration of the thermoluminescent response of the textile. This may be carried out on a reference tissue or, after the accident, on the clothing worn by the injured person (additive method). The response to ^{60}Co γ rays has been determined up to a kerma in air of 30 Gy for a 22.1 g.m^{-2} dosimetric cotton (Figure 3, curve A); more precise measurements between 0 and 1.5 Gy are also shown for a cotton with 42 g.m^{-2} of alumina (Figure 3, curve B). With a 65 g.m^{-2} dosimetric cotton, the detection threshold is lowered to 0.1 Gy.

Response relative to photon energy

The energy response of the dosimetric textile has been obtained on a cotton with alumina content of 65 g.m^{-2} between 0.045 and 1.25 MeV (Figure 4). Irradiations were made with fluorescence X rays, and ^{60}Co and ^{137}Cs sources. The curve obtained is compared with the response of pure alumina[7] and that of a single crystal of carbon-doped alumina (TLD-500). The Al$_2$O$_3$:C contains between 100 and 5000 ppm of C[8].

For energies less than 100 keV (photoelectric effect range), hypersensitivity of the dosimetric cotton was

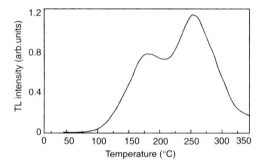

Figure 2. TL curve obtained by direct measurement on 1 cm^2 of a 65 g.m^{-2} dosimetric cotton, 30 days after γ radiation (^{60}Co, air kerma of 0.48 Gy). The low temperature peaks have disappeared, the dosimetric peak is weakly affected since the fading is about 10% in 66 days[6].

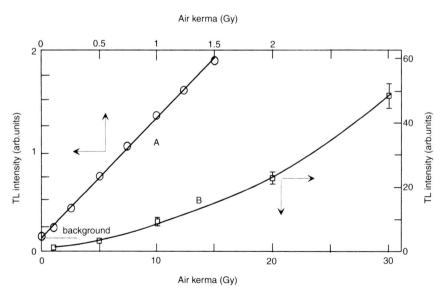

Figure 3. TL response of the dosimetric cotton plotted against kerma in air (^{60}Co). Curve A: 42 g.m^{-2} dosimetric cotton. Curve B: 22.1 g.m^{-2} dosimetric cotton.

observed, which is between that of the pure alumina and that of TLD-500. The observed differences are without doubt due to both effects of material bulk and radiation build-up.

Simulations on a mannequin in a mixed neutron–gamma fields

Two series of tests on a mannequin have been carried out in a mixed n-γ field at SILENE, Valduc, France. SILENE is an experimental reactor using a fissile uranyl nitrate solution containing highly enriched uranium as fuel. In these experiments, it operated in free evolution mode. This mode simulates an accidental criticality excursion allowed to evolve freely[9]. The average number of fissions is 3×10^{17}.

Dosimetric cotton pieces are disposed at different places on an anthropomorphic mannequin PLASTI-NAUT (tissue-equivalent plastic) as shown in Figure 5. For each series of tests, the mannequin was placed once facing the reactor (Figure 5, (a)) and once with the right shoulder towards the radiation (Figure 5, (b)). In each case, classical dosemeters of Desmarquest alpha alumina powder were used under Eppendorf covers and were directly placed next to the cotton pieces for purposes of comparison. One measurement was also made by placing a dosimetric cotton piece in free air under Eppendorf covers and a classical dosemeter in air.

At the time of the first series of tests, the mannequin was placed 4 m away from the source used without shielding and the gamma/neutron dose ratio was then equal to 1.1. The tested cotton had a content of 19.5 g.m^{-2}.

For the second series of measurements, the mannequin was placed 3 m away from the source. A reflector, contributing to diffuse γ radiation and lead shield-

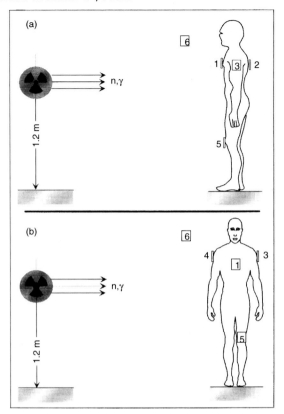

Figure 5. Positioning of dosimetric cotton pieces on the mannequin. (a) Facing the radiation. (b) Right shoulder towards the radiation.

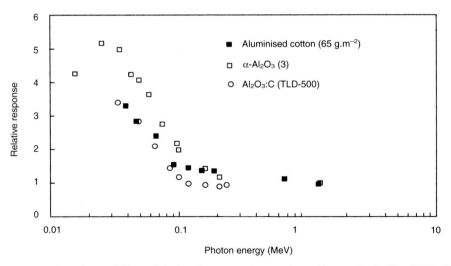

Figure 4. Photon energy dependence of 65 g.m^{-2} dosimetric cotton and comparison with pure alumina[7] and TLD-500 responses.

ing, placed in the core environment, brought down the gamma/neutron dose ratio to 0.15. The dosimetric textile then contained 42 g.m^{-2} of alumina. In these two cases, the source was at 1.20 m from the ground.

The results are reported in Tables 2 and 3. Table 2 shows that the evaluated doses by dosimetric cotton pieces and classical dosemeters agree and that these values do correspond to the orientation of the γ beam (Tables 2 and 3): for example, the measured doses behind the mannequin are lower than those measured in front of it.

DISCUSSION AND CONCLUSION

The fabrication technique of the dosimetric textile allows a range of products to be obtained in which the alumina content varies between 12 and 65 g.m^{-2} and which have textile qualities making them apt for the manufacture of dosimetric clothing.

The tests carried out show that the dosimetric cotton is particularly suited to γ accident dosimetry even in mixed n-γ fields. It can be used to dress persons subject

Table 1. Ratios of the different constituents of α-alumina 44.

Constituents	Content (%)
Binder	50
Thickening substance	5
Alumina 44	37
NH$_4$OH	1
Water	7

Table 3. Evaluated doses in case of 42 g.m^{-2} dosimetric cotton pieces.

	Positioning	Dosimetric cotton with 42 g.m^{-2}	
		Kerma in air (Gy)	σ (%)
A	Chest (1)	2.96	8.1
	Back (upper) (2)	2.08	7.7
	Left arm (3)	2.72	5.9
	Right arm (4)	2.48	6.5
	Left hand (7)	2.48	6.5
	Abdomen (8)	2.88	2.8
	Back (lower) (9)	2.00	8
	Free air (Eppendorf) (6)	2.24	2.9
B	Chest (1)	2.48	9.7
	Back (upper) (2)	1.96	6.1
	Left arm (3)	2.08	11.5
	Right arm (4)	2.41	5
	Left hand (7)	2.48	16
	Abdomen (8)	2.40	10
	Back (lower) (9)	2.62	1.5
	Free air (Eppendorf) (6)	1.84	8.7

The mannequin is 3 m away from the source used with a reflector and a shielding lead ($D_\gamma/D_n = 0.15$). A: The mannequin is facing the reactor. B: Its right shoulder is towards the reactor. Numbers refer to Figure 5. Positions 7, 8 and 9 respectively correspond to cotton pieces placed on the left hand (the arm kept horizontally), the abdomen and the lower back.

Table 2. Evaluated doses for 19.5 g.m^{-2} dosimetric cotton pieces and with TLD classical dosemeters.

	Positioning	Dosimetric cotton with 19.5 g.m^{-2}		Classical dosemeter		(b−a)/a (%)
		Kerma in air (Gy) (a)	σ (%)	Kerma in air (Gy) (b)	σ (%)	
A	Chest (1)	2.47	10.2	2.29	0.62	−7
	Back (2)	1.32	9.9	1.01	1.16	−23
	Left arm (3)	2.05	0.4	2.14	0.05	4
	Right arm (4)	2.14	0.2	2.21	1.46	3
	Knee (5)	2.47	0.2	2.15	0.44	−13
	Free air (Eppendorf) (6)	1.97	9	1.90	0.58	−4
B	Chest (1)	1.59	3.8	1.88	0.15	18
	Back (2)	2.17	8.3	2.27	0.1	5
	Left arm (3)	1.01	7.1	0.85	0.74	−16
	Right arm (4)	2.37	4.6	2.40	1.47	1
	Knee (5)	1.94	8.8	1.91	2.27	−2
	Free air (Eppendorf) (6)	1.92	9.4	1.90	0.58	−1

The mannequin is 4 m away from the source used without shielding ($D_\gamma/D_n = 1.1$). A: The mannequin is facing the reactor. B: Its right shoulder is towards the reactor. Numbers refer to Figure 5. σ columns are standard deviations on four measurements. In the last column, a comparison is made between the two methods.

to irradiation risks but may also be used as strips judiciously disposed on the person who may be irradiated. In the first case (accident), it allows the γ dose absorbed by the irradiated person to be mapped, while in the second case, the technique gives the orientation of the irradiation beam and an estimation of the dose at various points on the surface of the body. These indications are extremely important for the medical staff who will have to treat the injured because they allow an evaluation of the heterogeneity gradient of the irradiation without making a long and hazardous reconstruction of the accident.

ACKNOWLEDGEMENT

The authors thank Mr Huet (Institut Textile de Mulhouse) for the development of the dosimetric textile.

REFERENCES

1. Kamenopoulou, V., Barthe, J., Hickman, C. and Portal, G. *Accidental Gamma Irradiation Dosimetry using Clothing*. Radiat. Prot. Dosim. **17**, 185–188 (1986).
2. Davies, J. E. *Exoemission and Thermoluminescence from Human Enamel and Shark Enameloid*. Radiat. Prot. Dosim. **4**(3/4), 181–184 (1983).
3. Alekhin, A., Babenko, S. P., Kraitor, S. N. and Kurlnereva, K. K. *Use of Radioluminescence and EPR for Accident Dosimetry*. J. At. Energ. **53**(2), 91–95 (1982).
4. Barthe, J., Iacconi, P., Lapraz, D., Petel, M., Keller, P. and Portal, G. *Use of Cotton in an Accident Dosimetry*. Radiat. Prot. Dosim. **6**(1/4), 125–128 (1984).
5. Lapraz, D., Iacconi, P., Alessandri, M. F., Barthe, J. and Portal, G. *Dosimetric Characteristics of Specially Prepared Clothing Material*. Radiat. Prot. Dosim. **33**(1/4), 131–134 (1990).
6. Benabdesselam, M., Iacconi, P., Lapraz, D., Serbat, A., Laugier, J. and Dhermain, J. *Qualification of Specially Prepared Clothing Material for Gamma Dosimetry. Application to Neutron-Photon Radiation Fields*. Radiat. Prot. Dosim. **66**(1–4), 125–128 (1996).
7. Portal, G. *Etude et Développement de la Dosimétrie par Radiothermoluminescence*. Thesis, University of Toulouse (1978).
8. Akselrod, M. S., Kortov, V. S. and Gorelova, E. A. *Preparation and Properties of α-Al_2O_3:C*. Radiat. Prot. Dosim. **47**(1/4), 159–164 (1993).
9. Tournier, B., Barbry, F. and Rozain, J. P. *SILENE, a New Radiation Reference Source*. In: Proc. Topical Meeting on Physics, Safety and Applications of Pulse Reactors. Washington DC, November 1994, pp. 72–79 (1994).

CHROMOSOME ABERRATIONS SCORING FOR BIOLOGICAL DOSIMETRY IN A CRITICALITY ACCIDENT

P. Voisin†, D. Lloyd‡ and A. Edwards‡
†Institut de Protection et de Sûreté Nucléaire, Département de Protection
de la Santé de l'Homme et de Dosimétrie
Section Autonome de Radiobiologie Appliquée à la Médecine, IPSN-B.P. No 6
92265 Fontenay-aux-Roses Cedex, France
‡National Radiological Protection Board, Chilton, Didcot, Oxon, OX11 0RQ, UK

Abstract — During an international intercomparison of dosimetry systems for the simulation of a criticality accident which took place at the SILENE Reactor (Valduc, France), whole blood was exposed to reference pulses of mixed gamma and neutron irradiation. This was done with specimens supported free-in-air and attached to a phantom. Each blood sample was shared between the NRPB and IPSN laboratories, cultured and scored for chromosomal aberrations. Each laboratory made estimates of the neutron and gamma components of dose by reference to their own pre-existing *in vitro* calibration curves. It was necessary to have an estimate of the ratio of the neutron to gamma dose for this purpose. Good agreement was obtained between the results from both laboratories and with the reference doses. The conclusion is that biological dosimetry using peripheral lymphocytes gives credible estimates of personal doses in criticality accidents.

INTRODUCTION

When people are overexposed to ionising radiation, it is usually to X or gamma rays[1]. A criticality accident, which is very rare, is, however, more complex, since the radiation involved is a mixture of both fission neutrons and gamma rays. However, because the two components have very different relative biological efficiencies, their dosimetric discrimination is of importance for investigating the accident and perhaps for planning the treatment of casualties.

The induction by ionising radiations of unstable structural chromosome aberrations (dicentrics, centric rings and fragments) in peripheral blood lymphocytes is considered to be the most advanced biological dosemeter. Dose–response relationships between the dicentric yields and the physical doses can be established *in vitro*, and these vary with radiation quality (LET) and dose rate[2].

It has been suggested that by using the chromosomal method equivalent whole-body doses could be estimated for gamma rays and fission neutrons separately if the neutron/gamma ratio was known from physical dosimetry. An IAEA technical report[3] provides a guideline for such a calculation. However, there are very few studies which have tested the method[4,5].

An International Intercomparison of Criticality Accident Dosimetry Systems, hosted by the Institut de Protection et de Sûreté Nucléaire (IPSN, France) took place at the SILENE reactor at Valduc in France during June 1993[6]. This provided a good opportunity for the NRPB and IPSN cytogenetic laboratories to test the method.

MATERIALS AND METHODS

A detailed description of irradiation procedures and dosimetry systems of the criticality experiment is provided in the final report[6]. Briefly, a critical excursion was achieved by injecting a fissile solution into the core of a vessel and withdrawing a control rod. The duration of the pulse of radiation was typically 1–2 min. Two experimental configurations were used: (a) a bare source where the neutron/gamma ratio was about 1.0, and (b) the source was surrounded by a 10 cm thick lead shield so that the neutron/gamma ratio was about 5.0.

A fresh blood sample was collected by antecubital venepuncture from an apparently healthy male donor in tubes containing lithium heparin anticoagulant. Irradiation was performed at room temperature, with the tubes suspended in air and fixed on a man-phantom on a section of arc 4 m from the core. After irradiation, blood tubes were incubated, as soon as possible, in a water bath at 37°C for 2 h to allow for DNA repair[7]. Control tubes positioned outside the reactor were treated similarly.

The blood samples were divided into two aliquots and despatched to the cytogenetics laboratories. They were received within a few hours at IPSN and on the following day at NRPB. The cell cultures were therefore set up about 7 and 24 h respectively after the irradiations.

The lymphocytes were cultured using the standard protocol of each laboratory, similar to that described in the IAEA technical report[3]. Briefly, several replicate cultures from each blood sample were made by adding cells to culture medium supplemented with phytohaemagglutinin, bromodeoxyuridine, fetal calf serum and penicillin/streptomycin.

After 46 h of culture at 37°C, demecolcine was added and incubation continued for two more hours. Cells were then harvested and metaphase spreads prepared according to standard techniques. The slides were stained with fluorescence plus Giemsa so that cells in their first and second divisions could be distinguished.

Unstable chromosome aberrations (dicentrics, centric rings and fragments) were scored only in first cycle complete metaphases (with 46 centromeres). The estimation of gamma ray and fission neutron doses received by each irradiated specimen was made by reference to dose response curves for dicentrics previously published by each laboratory[4,8,9]. These curves have been produced by exposing blood *in vitro* to high dose rate (0.5 Gy.min^{-1}) ^{60}Co gamma rays and a fission spectrum of neutrons, respectively.

For the NRPB laboratory:

$$Y = 1.42 \times 10^{-2} D + 7.59 \times 10^{-2} D^2$$

$$Y = 83.5 \times 10^{-2} D$$

For the IPSN laboratory:

$$Y = 2.3 \times 10^{-2} D + 5.4 \times 10^{-2} D^2$$

$$Y = 90 \times 10^{-2} D$$

(Y = yield of dicentrics and D = dose in Gy).

RESULTS AND DISCUSSION

Although several types of aberrations were recorded, dosimetry is commonly based only on the dicentric and so only this aberration is considered further. The zero dose control cultures were consistently free from aberrations. The remaining scoring results for dicentrics from both laboratories are shown in Table 1. By examining the ratios of variance to mean these results do not show overdispersion compared with a Poisson distribution except for the first result and this is mainly caused by the observation of just one cell with 9 dicentrics. Therefore it implies irradiation homogeneity and permits use of Poisson statistics for calculating uncertainties on the dose estimates.

Since different qualities of radiation do not induce aberrations of differing appearance in the microscope the apportionment of the observed dicentrics to the neutron or gamma components of dose was carried out, assuming additivity, using an iterative procedure as described in the IAEA manual[3]. This required, for each irradiation geometry, a value for the neutron/gamma ratio to be supplied by physical dosimetrists.

In the first instance these ratio data were obtained as a mean of the values generated by the various physical dosemeters used by the other participants in the exercise. Later, formal reference values were made available by the organisers of the exercise using spectrometric and other sophisticated instruments that would not have been present at a real criticality accident[6]. Therefore two sets of biological estimates for neutron and gamma doses are shown. The first, in Table 2, uses the ratios from the physicist co-participants and, of course, is subject to errors inherent in their particular systems. In a real accident this is the type of data on which the cytogenetics method would need to rely. The later revised values are shown in Table 3. These use the reference ratios and for the purpose of the citicality exercise form a comparison that can be made between the biological method of dosimetry and the 'true' doses.

Irrespective of which neutron/gamma ratio is used the neutron dose estimates from the chromosome measurements do not vary much for the respective geometries shown in Tables 2 and 3. This is particularly so for the lead shielded configuration, and the variation is only 10–20% for the bare source measurements. The gamma measurements change slightly more, by 10–30% between Tables 2 and 3. Table 2 also shows that the dicentric yields scored by the two laboratories for each criticality configuration are generally in good agreement; only for the lead shielded phantom exposure do the standard errors on the yields fail to overlap. Overall in Table 3 the cytogenetic dose estimates are close to the reference doses; all but one (lead shield, free air, IPSN) are within statistical uncertainties at the 95% level of confidence.

Table 1. Dicentric scoring results from the two laboratories for the different irradiation conditions.

Reactor	Geometry	Lab.	Cells scored	No of dicentrics	Distribution of dicentrics									
					0	1	2	3	4	5	6	7	8	9
Lead shield	Free air	NRPB	34	100	2	6	11	4	3	3	4	0	0	1
		IPSN	50	125	1	11	11	18	7	2				
	Phantom	NRPB	40	108	1	9	10	10	2	7	1			
		IPSN	55	202	0	3	5	19	14	9	4	1		
Bare source	Free air	NRPB	28	85	0	5	9	4	6	1	1	1	1	
		IPSN	14	36	0	4	3	3	3	1				
	Phantom	NRPB	10	37	0	1	3	1	0	3	2			
		IPSN	19	53	2	1	7	4	1	2	2			

This experiment has used blood samples suspended in free air or attached to a phantom. It should be remembered that in a real accident in which blood is taken from an exposed person, one is sampling cells that were irradiated throughout the body. In this exercise it was not possible to place blood samples at sites within the phantom. However, those attached to its surface do, to some extent, simulate the presence of a body and the results obtained even in these conditions show a close correspondence with the reference values. Generally, within a body, cells would sustain a higher gamma radiation dose than those situated at the front surface. This difference however should not reduce the accuracy of the biological method provided that the neutron/gamma ratio averaged throughout the body can be derived from the physical measurements. This is feasible because the surface neutron kerma/gamma dose ratio which would be obtained from standard criticality and thermoluminescence personal dosemeters can be converted by calculation to the ratio required for the biological dosimetry.

The overall conclusion is that biological dosimetry using chromosomal aberration analysis is able to estimate doses in a criticality accident.

Table 2. Cytogenetic estimates of neutron and gamma doses (±SE) for the different irradiation conditions made by using the mean of neutron to gamma ratios measured by physical dosemeters of co-participants.

Reactor	Geometry	n/γ ratio	Lab.	Dicentric/cell ± SE	Cytogenetic doses (Gy)	
					Neutron	Gamma
Lead shield	Free air	4.76	NRPB	2.94 ± 0.29	3.5 ± 0.3	0.73 ± 0.07
			IPSN	2.50 ± 0.22	2.7 ± 0.2	0.58 ± 0.05
	Phantom	2.44	NRPB	2.70 ± 0.26	3.1 ± 0.3	1.26 ± 0.12
			IPSN	3.67 ± 0.26	3.9 ± 0.3	1.59 ± 0.11
Bare source	Free air	0.78	NRPB	3.03 ± 0.33	2.6 ± 0.3	3.3 ± 0.4
			IPSN	2.57 ± 0.43	2.3 ± 0.4	2.9 ± 0.5
	Phantom	0.71	NRPB	3.70 ± 0.61	2.9 ± 0.5	4.0 ± 0.7
			IPSN	2.79 ± 0.38	2.4 ± 0.3	3.3 ± 0.4

Table 3. Cytogenetic estimates of neutron and gamma doses (± SE) for the different irradiation conditions made by using the neutron/gamma ratios from the reference measurements.

Reactor	Geometry	n/γ ratio	Lab.	Cytogenetic doses (Gy)		Reference doses (Gy)	
				Neutron	Gamma	Neutron	Gamma
Lead shield	Free air	5.78	NRPB	3.5 ± 0.3	0.60 ± 0.06	3.42	0.59
			IPSN	2.7 ± 0.2	0.46 ± 0.04		
	Phantom	1.87	NRPB	3.0 ± 0.3	1.59 ± 0.15	3.42	1.83
			IPSN	3.8 ± 0.3	2.04 ± 0.14		
Bare source	Free air	0.64	NRPB	2.3 ± 0.3	3.7 ± 0.4	2.60	4.04
			IPSN	2.1 ± 0.4	3.3 ± 0.6		
	Phantom	0.53	NRPB	2.4 ± 0.4	4.6 ± 0.8	2.60	4.89
			IPSN	2.1 ± 0.3	3.9 ± 0.5		

REFERENCES

1. Littlefield, L. G. and Lushbaugh, C. C. *Cytogenetic Dosimetry for Radiation Accidents — "The Good, the Bad and the Ugly"*. In: The Medical Basis for Radiation Accident Preparedness. Eds R. C. Ricks and S. A. Fry pp. 461–478 (1990).
2. Bender, M. A., Awa, A. A., Brooks, A. L., Evans, H. J., Groer, P. G., Littlefield, L. G., Pereira, C., Preston, R. J. and Wachholz, B. W. *Current Status of Cytogenetic Procedures to Detect and Quantify Previous Exposures to Radiation*. Mutat. Res. **196**, 103–159 (1988).

3. IAEA. *Biological Dosimetry: Chromosomal Aberration Analysis for Dose Assessment*. Technical Report Series no 260 (Vienna: International Atomic Energy Agency) (1986).
4. Biola, M. T., Le Go, R., Vacca, G., Ducater, G., Dacher, J. and Bourguignon, M. *Efficacité Relative de Divers Rayonnements Mixtes Gamma, Neutrons pour l'Induction in vitro d'Anomalies Chromosomiques dans les Lymphocytes Humains*. In: Biological Effects of Neutron Irradiation, STI/PUB/354, pp. 221–236 (Vienna: International Atomic Energy Agency) (1974).
5. Purrott, R. J., Lloyd, D. and Dolphin, G. W. *Chromosome Aberration Dosimetry Following a Controlled Criticality Pulse to Human Peripheral Blood Lymphocytes*. In: Nuclear Accident Dosimetry — Report on the 4th IAEA Intercomparison Experiment at Harwell, UK, 7–18 April 1975. AERE-R8521, pp. 46–50 (1976).
6. Delafield, H. J. and Medioni, R. *An International Intercomparison of Criticality Accident Dosimetry Systems at the SILENE Reactor, Valduc, Dijon, France, 7–18 June 1993. Part 3: Description of the Experiment and Participants' Results*. AEA Technology Report HSP/TR/H/3(95) (1995).
7. Virsik-Peuckert, R. P. and Harder, D. *Temperature and the Formation of Radiation-induced Chromosome Aberrations. II. The Temperature Dependence of Lesion Repair and Lesion Interaction*. Int. J. Radiat. Biol. **49**, 673–681 (1986).
8. Lloyd, D., Edwards, A. A. and Prosser, J. S. *Chromosome Aberrations induced in Human Lymphocytes by in vitro Acute, X and Gamma Radiation*. Radiat. Prot. Dosim. **15**, 83–88 (1986).
9. Doloy, M. T. *Dosimétrie basée sur le Dénombrement des Anomalies Chromosomiques contenues dans les Lymphocytes Sanguins*. Radioprotection **26**(suppl. 1), 171–184 (1991).

FAST NEUTRON RADIOTHERAPY: THE UNIVERSITY OF WASHINGTON EXPERIENCE AND POTENTIAL USE OF CONCOMITANT BOOST WITH BORON NEUTRON CAPTURE

K. J. Stelzer, K. L. Lindsley, P. S. Cho, G. E. Laramore and T. W. Griffin
Department of Radiation Oncology
University of Washington Medical Center
Seattle, Washington, USA

INVITED PAPER

Abstract — Well defined, randomised clinical trials have demonstrated significant advantages for fast neutron radiotherapy over photon radiotherapy in locally advanced prostate cancer (10-year survival 46% vs 29%), salivary gland tumours (10-year local/regional control 56% vs 17%), and inoperable squamous cell carcinoma of the lung (2-year survival 16% vs 3%). Fast neutron treatment of high grade brain tumours (astrocytomas) resulted in tumour sterilisation, but also caused significant brain injury. This effect was important, however, in that conventional radiation therapy, and other conventional treatments, have not demonstrated sterilisation of high grade astrocytomas at any dose. The treatment of cancer utilising ^{10}B and its neutron capture reaction (boron neutron capture therapy, or BNCT) may have potential to be effective in brain tumours and melanoma. Recent studies at the University of Washington have demonstrated that fast neutron radiotherapy cell kill can be significantly enhanced by the boron capture reaction utilising the thermal component of our fast neutron beam. This enhancement can increase cell kill between 10 and 100 fold. The implications for concomitant BNCT/fast neutron therapy with new ^{10}B carriers will be discussed with emphasis on high grade brain tumours.

INTRODUCTION

The United States National Cancer Institute (NCI) funded the construction of several hospital-based, high energy, isocentric-capable cyclotrons for fast neutron clinical trials. The University of Washington was one of these sites of development. Scanditronix Corporation of Uppsala, Sweden, was subcontracted to design and build the University of Washington facility. This device utilises a cyclotron to deliver 50 MeV protons to a gantry-mounted beryllium target, producing a neutron beam with an average energy of 22 MeV. The clinical experience in treating malignancies with fast neutrons and the potential for concomitant BNCT/fast neutron therapy with new ^{10}B carriers will be discussed with emphasis on GBM treatment.

METHODS

The results of those clinical trials involving treatment of patients with fast neutrons at the University of Washington for prostate cancer, salivary gland cancer, non-small cell lung cancer, and glioblastoma multiforme (GBM) were reviewed.

Histopathology data after fast neutron treatment of GBM were reviewed in greater detail to provide an approximation of tumour control as a function of neutron dose. These data points were fitted to a sigmoidal function to provide an approximation of the dose–response relationship for fast neutron control of GBM. Experiments from the University of Washington hospital cyclotron testing enhancement of tumour cell kill with ^{10}B neutron capture agents present at the time of fast neutron radiation were reviewed[1,2]. The degree of tumour cell kill enhancement with BNCT after *in vitro* fast neutron radiation of a tumour cell line[2] was entered into the Poisson model as used by Porter[3,4]. In this model,

$$P = e^{-NS} \quad (1)$$

where P is the probability of tumour control, S is the probability of cell survival, and N is the number of tumour cells. If the probability of tumour control at a reference dose is known (P_1), and the tumour cell survival at a new dose relative to that at the reference dose is known (S_2/S_1), then the probability of tumour control at the new dose (P_2) can be estimated as:

$$P_2 = \exp[(S_2/S_1) \ln P_1] \quad (2)$$

Results for dose enhancement ratio in a rat brain tumour model irradiated *in vivo*[1] were used as a second estimate of the potential for enhanced control of GBM using concomitant BNCT/fast neutron therapy. The dose enhancement ratio was used to calculate the effective fast neutron radiation dose, which was then used to estimate GBM control using the sigmoidal dose–response function fitted to the histopathology data.

RESULTS

Prostate cancer

Prior to the NCI-sponsored development of hospital based cyclotrons for fast neutron radiotherapy, a randomised trial was conducted by the Radiation Therapy Oncology Group (RTOG) using physics-based generators, comparing a combination of neutrons plus photons to photons in the treatment of local/regionally advanced prostate cancer[5,6]. At 10 years, tumour control (70% versus 58%) and survival (46% versus 29%) were significantly higher in the mixed neutron/photon arm. A confirmatory study was conducted by the Neutron Therapy Collaborative Working Group (NTCWG) using hospital-based cyclotrons from 1986 to 1990. With improved beam characteristics of these generators, fast neutrons alone (13.6 Gy to the pelvis, 20.4 Gy total to the prostate) were compared to photons (50 Gy to the pelvis, 70 Gy total to the prostate) in patients with stages C, D1, and high-grade B2 prostate cancer. The 5-year results for the 172 evaluable patients showed significantly improved local/regional control (89% versus 68%) in the neutron arm[7]. Furthermore, abnormal elevation of prostate specific antigen, a serum marker for prostate cancer, was significantly higher in the photon arm (45% versus 17%). A significant survival difference has not yet been observed (68% for neutrons versus 73% for photons). However, longer follow-up is required to assess a survival difference because patients can survive several years after initial treatment failure with this slowly progressive tumour.

Salivary gland tumours

The NCI and the Medical Research Council (MRC) of Great Britain jointly sponsored a phase III randomised trial of fast neutrons versus conventional low linear energy transfer (LET) radiation (photons and/or electrons) in unresectable, locally advanced malignant salivary gland tumours. The neutron dose was 16.5 Gy to 22 Gy, based upon radiobiological intercomparison studies between the four participating facilities (hospital and physics-based)[8]. The low LET dose was 70 Gy. Thirty-two patients were entered on this trial and 25 were eligible and evaluable (12 low LET, 13 neutron). Analysis at 2 years showed significantly improved local/regional control (67% versus 17%) and a trend towards improved survival (62% versus 25%) in the neutron arm[8]. Because of these differences, the study was closed early for ethical reasons[9]. The 10-year analysis of these patients showed persistence of higher local/regional control in the neutron arm (56% versus 17%). However, the 10-year survival was no different between arms (approximately 20%), presumably due to development of distant metastases which were observed in 16 of the 25 evaluable patients.

Non-small cell lung cancer

A phase III randomised trial of fast neutrons (20.4 Gy) versus photons (66 Gy) was conducted by the NTCWG from 1986 to 1991 for patients with inoperable non-small cell lung cancer, predominantly stage III[10].

For the entire group of 193 evaluable patients there was no difference in overall survival. However, the 2-year actuarial survival was superior with neutrons in the subgroup of patients treated for squamous cell carcinoma (16% versus 3%). Presumably, this sub-set of patients with squamous cell carcinoma benefited from fast neutron therapy because of improved local/regional control in this tumour type that has less tendency to metastasise than other sub-types of non-small cell lung cancer. Toxicity was similar between the treatment arms except for an increased incidence of asymptomatic skin and soft tissue changes in the neutron arm.

Glioblastoma multiforme

A number of trials have been conducted using fast neutrons with or without a component of photon radiotherapy in the treatment of high-grade astrocytomas, the vast majority of which were GBM. Notably, the median survivals for these trials using a variety of dose schemes and treatment techniques (with or without photons) were consistently near 10 months[11–27]. Although fast neutron radiotherapy can sterilise GBM, it does so at the expense of normal brain toxicity. For example, the RTOG conducted a randomised dose searching trial for high-grade astrocytomas in which various doses of fast neutron radiation were delivered in combination with 45 Gy of photon radiation[23]. There was no difference in survival for patients receiving the lower three neutron doses (3.6–4.8 Gy) compared to those receiving the higher three neutron doses (5.2–6 Gy). Autopsy data were available for 36 patients and revealed both radiation damage and viable tumour at all dose levels. Therefore, no therapeutic window was identifiable for any of the six regimens.

The results of histological evaluation of local control as a function of fast neutron dose for high-grade astrocytoma are shown in Figure 1. Four data points are

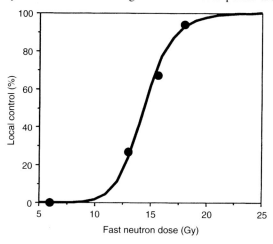

Figure 1. Local control of high-grade astrocytic brain tumours as a function of neutron dose based on histopathology data.

shown on this graph. The University of Washington experience included autopsy data in 15 of 36 reported patients[24]. Only one of these 15 patients was found to have residual tumour, and that patient received photons to a dose of 39 Gy and neutrons to a dose of 7.2 Gy. The majority of the other 14 patients who were found to have tumour sterilised received fast neutrons alone to a dose of 18 Gy. As an approximation, a local control value of 95% was assigned at 18 Gy based on those data. From the MRC data[14], 18 patients initially received neutrons to 15.6 Gy, but the dose was decreased to 13 Gy after patients were noted to be dying without evidence of viable tumour, but with fatal brain damage. At 15.6 Gy, 5 patients who lived at least 9 months had clinical evidence of dementia without focal neurological signs, without raised intracranial pressure, and without significant response to steroids. Four of these 5 patients had autopsy and either no tumour or only limited microscopic tumour foci of questionable viability were observed. Three other patients had symptoms of neutron-related dementia, but also had signs of local recurrence, and 2 of these patients received 13 Gy. Of the 4 patients treated to 13 Gy with histologic assessment, 3 had evidence of tumour sterilisation. If it is assumed that the remaining 7 patients treated to 13 Gy died with recurrent disease based on the clinical description, then the local control at this dose would be 3/11 or 27%. At 15.6 Gy, a significant proportion of patients appeared to die of neutron-induced brain injury. Therefore, the local control estimation at 15.6 Gy was based only on the 12 patients with histological evaluation, and was assigned a value of 8/12 or 67%.

These three data points were fitted to the sigmoidal dose–response function shown in Figure 1. An additional illustrative data point (not used for curve fitting) of 0% local control was placed at 6 Gy as an estimate based on the RTOG dose searching trial. In that trial, none of the 18 patients treated with up to 5.78 Gy neutron dose in addition to 45 Gy photon dose had tumour sterilisation on autopsy[24].

Concomitant boron neutron capture/fast neutron therapy

The concept of using a thermalised component of the fast neutron beam from the University of Washington cyclotron has recently been investigated in two studies. Laramore et al[2] reported that exposing tumour cells in vitro to 100–500 µg ^{10}B g^{-1} resulted in a 10–100 fold increase in cell kill upon treatment with 2–10 Gy of fast neutrons compared to fast neutrons alone. Buchholz et al[1] used an in vivo tumour model and also found a BNCT enhancement of fast neutron kill. In that study, rat glioma cells were implanted in rat flanks, and the resulting tumours were radiated with fast neutrons. Tumours were then excised and a colony forming assay was used to measure clonogenic cell survival. Half of the rats were treated with ^{10}B-enriched boronophenylalanine, and the mean ^{10}B tumour concentration was 68.4 µg.g^{-1}. The dose modifying factor for ^{10}B-enhanced fast neutron treatment was 1.32 at a surviving fraction of 0.1, which corresponded to a fast neutron dose 5 Gy.

Based on the MRC data, a reduction in morbidity from fast neutrons was clinically evident upon lowering the dose to 13 Gy. If one assumes that a total fast neutron dose of 12 Gy could be administered with reasonable safety to patients with GBM, it is possible to estimate local control with addition of concomitant BNCT from the two experiments discussed above. At that dose of fast neutrons without concomitant BNCT, the fitted function would predict a local control of 10%. From the in vitro results[2], a conservative estimate of enhanced cell kill of 10-fold was substituted into Equation 2 ($S_2/S_1 = 0.1$), to yield an estimated local control with concomitant BNCT of 79%. Using the data from the in vivo model[1], the dose modifying factor of 1.32 would be applicable with a dose per fraction of 5 Gy or less (e.g. 4 Gy times 3 fractions). The 12 Gy fast neutron dose would then become the equivalent of 15.8 Gy to the tumour, yielding an estimated local control from the fitted dose–response curve of 77%.

DISCUSSION

Fast neutron radiotherapy has been found in randomised trials to improve survival and/or local/regional control in prostate cancer, salivary gland carcinoma, and squamous cell carcinoma of the lung. Published results from non-randomised trials suggest an advantage for fast neutrons over photons in the control of unresectable sarcomas of soft tissue, bone, and cartilage[28]. Conversely, randomised and non-randomised trials have shown no advantage for fast neutrons in the treatment of squamous cell cacinoma of the head and neck[29] or for treatment of high-grade astrocytomas.

In each of these disease sites, it would be desirable to improve or create a therapeutic window between efficacious and toxic doses. For example, fast neutron radiotherapy can sterilise GBM, but at doses that are slightly higher than those that cause significant brain injury. In salivary gland tumours, local control was improved with fast neutron radiotherapy, but nearly half the patients developed local/regional recurrence at 10 years. Furthermore, there was no survival difference at 10 years due to the high incidence of distant metastatic disease. One limitation in the treatment of these salivary tumours, particularly adenoid cystic carcinoma, is the limited fast neutron dose that may be delivered prophylactically to potential sites of perineural spread along cranial nerves. That limitation in dose is due to the origin of these nerves in the brainstem, which is a structure that must be excluded from the neutron radiation field at approximately 10 Gy. These perineural regions can represent first sites of failure, with the primary mass controlled, and may also be sources of origination for distant metastases. Similar circumstances can occur with fast neutron radiation for lung cancer, in which the spinal cord dose must be limited, requiring field modification and potential underdosing at field margins of gross or microscopic tumour in the mediastinum. In all these examples, the dose-limiting nature of the central nervous system is the major obstacle to curative treatment of tumours with fast neutrons.

Tumour cell kill was experimentally enhanced with ^{10}B capture of thermal neutrons generated from a fast neutron beam. Although the degree of cell kill from the BNCT component is small in comparison to that directly from fast neutrons, the enhancement appears great enough to affect tumour control. In fact, tumour control for GBM using a fast neutron dose of 12 Gy was predicted to be 77–79% using two different models that were applicable from the experimental results. It is encouraging that the two models predicted similar values for tumour control. Assuming that newer ^{10}B carriers for BNCT are favourably distributed within tumour compared to normal brain tissue, a therapeutic window sufficient for effective treatment of GBM could exist. Obviously, there is a high degree of uncertainty for the fast neutron control estimates of high grade astrocytomas due to limitations in histopathologic data and variability in relative biological effectiveness of fast neutron beams between the different centres. Nevertheless, even if the predictions of GBM control presented here have been overestimated by a factor of three, a significant advancement in the treatment of this tumour would be achieved. Similar improvements could be attained in the treatment of other malignancies with fast neutrons due to characteristically narrow therapeutic windows.

The University of Washington Department of Radiation Oncology is cooperating with Lockheed Martin Idaho Technologies Company, Washington State University, and Ionix Corporation in developing clinical trials of new ^{10}B carrier agents. Along with a trial of BNCT in GBM using the epithermal neutron beam from the Washington State University reactor, concomitant BNCT/fast neutron therapy will be investigated for GBM, salivary gland cancer, non-small cell lung carcinoma, sarcoma, and squamous cell carcinoma of the head and neck using the fast neutron facility at the University of Washington. The hypothesis to be tested is that clinical tumour control using a fast neutron beam, amenable to some degree of field shaping and with depth–dose characteristics sufficient for tumours too deep to be treated with an epithermal beam, can be enhanced by adding a tumour-selective ^{10}B carrier. Clearly, additional information on the microdosimetry of thermalised neutrons from a fast neutron beam will be necessary for optimal study design and interpretation of results.

REFERENCES

1. Buchholz, T. A., Rasey, J. S., Laramore, G. E., Livesey, J. C., Chin, L., Risler, R., Hamlin, D. K., Spence, A. M. and Griffin, T. W. *Concomitant Boron-Neutron Capture Therapy during Fast-neutron Irradiation of a Rat Glioma.* Radiology **191**, 863–867 (1994).
2. Laramore, G. E., Wooton, P., Livesey, J. C., Wilbur, D. S., Risler, R., Phillips, M., Jacky, J., Buchholz, T. A., Griffin, T. W. and Brossard, S. *Boron Neutron Capture Therapy: a Mechanism for Achieving a Concomitant Tumor Boost in Fast Neutron Radiotherapy.* Int. J. Radiat. Oncol. Biol. Phys. **5**, 1135–1142 (1994).
3. Porter, E. H. *The Statistics of Dose/Cure Relationships for Irradiated Tumours: Part I.* Br. J. Radiol. **53**, 210–227 (1980).
4. Porter, E. H. *The Statistics of Dose/Cure Relationships for Irradiated Tumours: Part II.* Br. J. Radiol. **53**, 336–345 (1980).
5. Russell, K. J., Laramore, G. E., Krall, J. M., Thomas, F. J., Maor, M. H., Hendrickson, F. R., Krieger, J. N. and Griffin, T. W. *Eight Years Experience with Neutron Radiotherapy in the Treatment of Stages C and D Prostate Cancer: Updated Results of the RTOG 7704 Randomized Clinical Trial.* Prostate **11**, 183–193 (1987).
6. Laramore, G. E., Krall, J. M., Thomas, F. J., Russell, K. J., Maor, M. H., Hendrickson, F. R., Martz, K. L., Griffin, T. W. and Davis, L. W. *Fast Neutron Radiotherapy for Locally Advanced Prostate Cancer. Final Report of a Radiation Therapy Oncology Group Randomized Clinical Trial.* Am. J. Clin. Oncol. **16**, 164–167 (1993).
7. Russell, K. J., Caplan, R. J., Laramore, G. E., Burnison, C. M., Maor, M. H., Taylor, M. E., Zink, S., Davis, L. W. and Griffin, T. W. *Photon versus Fast Neutron External Beam Radiotherapy in the Treatment of Locally Advanced Prostate Cancer: Results of a Randomized Prospective Trial.* Int. J. Radiat. Oncol. Biol. Phys. **28**, 47–54 (1993).
8. Griffin, T. W., Pajak, T. F., Laramore, G. E., Duncan, W., Richter, M. P., Hendrickson, F. R. and Maor, M. H. *Neutron vs Photon Irradiation of Inoperable Salivary Gland Tumors — Results of an RTOG-MRC Cooperative Randomized Study.* Int. J. Radiat. Oncol. Biol. Phys. **15**, 1085–1090 (1988).
9. Laramore, G. E., Krall, J. M., Griffin, T. W., Duncan, W., Richter, M. P., Saroja, K. R., Maor, M. H. and Davis, L. W. *Neutron vs Photon Irradiation for Unresectable Salivary Gland Tumors: Final Report of an RTOG-MRC Randomized Clinical Trial.* Int. J. Radiat. Oncol. Biol. Phys. **27**, 235–240 (1993).
10. Koh, W. J., Griffin, T. W., Laramore, G. E., Stelzer, K. J. and Russell, K. J. *Fast Neutron Radiation Therapy: Results of Phase III Randomized Trials in Head and Neck, Lung, and Prostate Cancers.* Acta Oncol. **3**, 293–298 (1994).
11. Battermann, J. J. *Fast Neutron Therapy for Advanced Brain Tumors.* Int. J. Radiat. Oncol. Biol. Phys. **6**, 333–335 (1980).
12. Breteau, N., Destembert, B. and Sabattier, R. *An Interim Assessment of the Experience of Fast Neutron Boost in Glioblastomas, Rectal and Bronchus Carcinomas in Orleans.* Strahlentherapie **161**, 787–790 (1985).

13. Breteau, N., Destembert, B., Favre, A., Ph'eline, C. and Schlienger, M. *Fast Neutron Boost for the Treatment of Grade IV Astrocytomas.* Strahlenther. Onkol. **165**, 320–323 (1989).
14. Catterall, M., Bloom, H. J., Ash, D. V., Walsh, L., Richardson, A., Uttley, D., Gowing, N. F. C., Lewis, P. and Chaucer, B. *Fast Neutrons Compared with Megavoltage X-rays in the Treatment of Patients with Supratentorial Glioblastoma: a Controlled Pilot Study.* Int. J. Radiat. Oncol. Biol. Phys. **6**, 261–266 (1980).
15. Duncan, W., McLelland, J., Davey, P., Jack, W. J., Arnott, S. J., Gordon, A., Kerr, G. R. and Williams, J. R. *A Phase I Study of Mixed (Neutron and Photon) Irradiation using Two Fractions per Day in the Treatment of High-grade Astrocytomas.* Br. J. Radiol. **59**, 441–444 (1986).
16. Duncan, W., McLelland, J., Jack, W. J., Arnott, S. J., Gordon, A., Kerr, G. R. and Williams, J. R. *Report of a Randomised Pilot Study of the Treatment of Patients with Supratentorial Gliomas using Neutron Irradiation.* Br. J. Radiol. **59**, 373–377 (1986).
17. Duncan, W., McLelland, J., Jack, W. J., Arnott, S. J., Davey, P., Gordon, A., Kerr, G. R. and Williams, J. R. *The Results of a Randomised Trial of Mixed-Schedule (Neutron/Photon) Irradiation in the Treatment of Supratentorial Grade III and Grade IV Astrocytoma.* Br. J. Radiol. **59**, 379–383 (1986).
18. Griffin, T. W., Davis, R., Laramore, G., Hendrickson, F., Rodrigues-Antunez, A., Hussey, D. and Nelson, J. *Fast Neutron Radiation Therapy for Glioblastoma Multiforme. Results of an RTOG Study.* Am. J. Clin. Oncol. **6**, 661–667 (1983).
19. Herskovic, A., Ornitz, R. D., Shell, M. and Rogers, C. C. *Treatment Experience: Glioblastoma Multiforme Treated with 15 MeV Fast Neutrons.* Cancer **49**, 2463–2465 (1982).
20. Hornsey, S., Morris, C. C., Myers, R. and White, A. *Relative Biological Effectiveness for Damage to the Central Nervous System by Neutrons.* Int. J. Radiat. Oncol. Biol. Phys. **7**, 185–189 (1981).
21. Kolker, J. D., Hapern, H. J., Krishnasamy, S. et al. *"Instant-mix" Whole Brain Photon with Neutron Boost Radiotherapy for Malignant Gliomas.* Int. J. Radiat. Oncol. Biol. Phys. **19**, 414–493 (1990).
22. Kurup, P. D., Pajak, T. F., Hendrickson, F. R., Nelxon, J. S., Mansell, J., Cohen, L., Awschalom, M., Rosenberg, I. and Ten Haken, R. K. *Fast Neutrons and Misonidazole for Malignant Astrocytomas.* Int. J. Radiat. Oncol. Biol. Phys. **11**, 679–686 (1985).
23. Laramore, G. E., Diener-West, M., Griffin, T. W., Nelson, J. S., Griem, M. L., Thomas, F. J., Hendrickson, F. R., Griffin, B. R., Myrianthopoulos, L. C. and Saxton, J. *Randomized Neutron Dose Searching Study for Malignant Gliomas of the Brain: Results of an RTOG Study.* Int. J. Radiat. Oncol. Biol. Phys. **14**, 1093–1102 (1988).
24. Laramore, G. E., Griffin, T. W., Gerdes, A. J. and Parker, R. G. *Fast Neutron and Mixed (Neutron/Photon) Beam Teletherapy for Grades III and IV Astrocytomas.* Cancer **42**, 96–103 (1978).
25. Mizoe, J. E., Aoki, Y., Morita, S. and Tsunemoto, H. *Fast Neutron Therapy for Malignant Gliomas — Results from NIRS Study,* Strahlenther. Onkol. **169**, 222–227 (1993).
26. Saroja, K. R., Mansell, J., Hendrickson, F. R., Cohen, L. and Lennox, A. *Failure of Accelerated Neutron Therapy to Control High Grade Astrocytomas.* Int. J. Radiat. Oncol. Biol. Phys. **17**, 1295–1297 (1989).
27. Stephens, L. C., Hussey, D. H., Raulston, G. L., Jardine, J. H., Gray, K. N. and Almond, P. R. *Late Effects of 50 MeVd → Be Neutron and Cobalt-60 Irradiation of Rhesus Monkey Cervical Spinal Cord.* Int. J. Radiat. Oncol. Biol. Phys. **9**, 859–864 (1983).
28. Laramore, G. E., Griffith, J. T., Boespflug, M., Pelton, J. G., Griffin, T., Briffin, B. R., Russell, K. J., Koh, W., Parker, R. G. and Davis, L. W. *Fast Neutron Radiotherapy for Sarcomas of Soft Tissue, Bone, and Cartilage.* Am. J. Clin. Oncol. **12**, 320–326 (1989).
29. Maor, M. H., Errington, R. D., Caplan, R. J. et al. *Fast-neutron Therapy in Advanced Head and Neck Cancer: a Collaborative International Randomized Trial.* Int. J. Radiat. Oncol. Biol. Phys. **32**, 599–604 (1995).

ENERGY SPECTRA IN THE NAC PROTON THERAPY BEAM

F. D. Brooks†, D. T. L. Jones‡, C. C. Bowley†, J. E. Symons‡, A. Buffler† and M. S. Allie†
†Department of Physics, University of Cape Town
Rondebosch, 7700 South Africa
‡Medical Radiation Group, National Accelerator Centre
PO Box 72, Faure, 7131 South Africa

Abstract — In order to tailor a proton beam for radiation therapy several beam modification devices are used which affect both the dose distribution and the energy spectrum of the beam. Knowledge of the proton spectra is helpful for optimising the beam delivery system and for comparison with theoretical calculations. Proton elastic scattering has been used to measure spectra in the NAC 200 MeV clinical beam. A polyethylene scatterer is located at the treatment isocentre. Two scintillator detector $\Delta E/E$ telescopes are placed symmetrically about the beam axis such that the energy of both the scattered and recoil protons is half the incident beam energy. Multiparameter data acquisition and analysis are used to determine the coincident summed spectra under various irradiation conditions.

INTRODUCTION

In order to tailor a proton beam from an accelerator for use in radiotherapy several beam modification devices have to be introduced between the accelerator and the patient. These devices affect both the dose distribution and the energy spectrum of the beam. Monte Carlo and analytic calculations[1] can be undertaken to predict the effect of these devices and thus ensure beam characteristics are optimised for specific therapy requirements. The reliability of these calculations needs to be checked, however, and this can only be done, ultimately, by experimental measurement. Spectral information can also be used in treatment planning programmes and for explaining differences in clinical, radiobiological and dosimetric data obtained at various centres. No energy spectra measurements appear to have been reported for any of the high energy proton therapy facilities now in operation. At the National Accelerator Centre (NAC) a project is now under way to develop and test a system for measuring spectra in the NAC's 200 MeV proton therapy beam[2,3]. A description of this system is presented here and some preliminary measurements which have been made are discussed.

The beam line has previously been described in detail[2–4]. Briefly, a passive double scatterer and occluding ring system[5,6] is used to spread and flatten the beam. The beam is collimated at several points along the beam line,

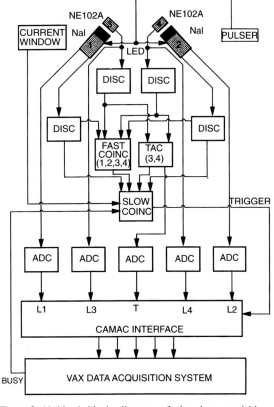

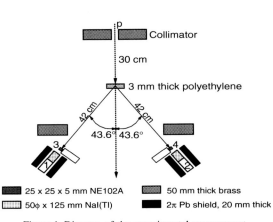

Figure 1. Diagram of the experimental arrangement.

Figure 2. Abridged block diagram of the data acquisition system.

including a final collimator, situated 31 cm upstream of the treatment isocentre, which is generally shaped specifically for each field to match the tumour shape. In addition there are multiwire and quadrant (Lawrence Berkeley National Laboratory) ionisation chambers, with feedback systems to steering magnets in order to ensure accurate beam alignment and symmetry, and a pair of parallel plate ionisation chambers for dose monitoring. The reference beam for absolute dose calibrations[7,8] has a 5 cm diameter final collimator and a range (measured to the distal 50% of the Bragg peak) in water of 24.00 ± 0.03 cm, corresponding to an energy of 191 MeV[8,9]. To ensure reproducibility of this range, proton beams with slightly higher energies than required are delivered, and the range is then trimmed by the insertion of 0.06 g.cm^{-2} thick plastic plates.

In addition to the above devices, which are always in the beam, acrylic attenuators can be manually inserted in the beam path to adjust the range of the beam to the planned treatment depth. Range modulating acrylic propellers[10] are used to provide spread-out Bragg peaks (SOBP). In the present series of measurements the effects on the spectra of the devices which are patient-specific (final collimators, attenuators and modulators) were investigated.

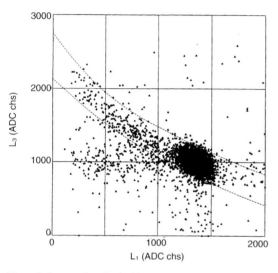

Figure 3. Scatter plot of coincident events against pulse heights L_1 and L_3 (from telescope 1–3). The dashed lines show the window used to select proton events and reject reaction tail events.

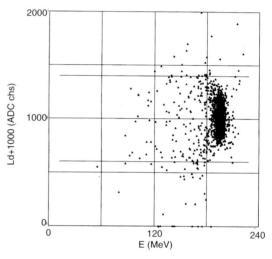

Figure 5. Scatter plot showing the same events as in Figure 4 (b) (d = 10 mm) replotted against total energy E and the pulse height difference $L_d = L_1 - L_2 + 1000$. The lines at $L_d = \pm 400$ show the window imposed to select coincidences satisfying the condition $E_1 \approx E_2$.

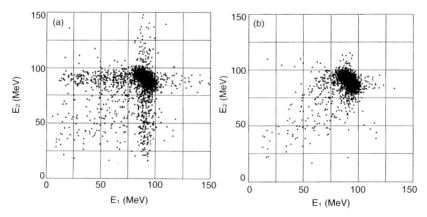

Figure 4. Scatter plots of coincident events against energies E_1 and E_2, obtained (a) without imposing; and (b) imposing, proton windows in the two telescopes. The data were taken using a final collimator of diameter d = 10 mm.

EXPERIMENTAL METHOD

A thin polyethylene radiator (30 mm diameter × 3 mm thick) was mounted at the treatment isocentre. A pair of ΔE/E detector telescopes, each consisting of a ΔE detector (25 mm × 25 mm × 5 mm thick NE102 plastic scintillator) backed by a full energy NaI(Tl) E detector (50 mm diameter × 125 mm thick), was used to detect the elastically scattered and recoil protons. A diagram of the experimental arrangement is given in Figure 1. The NaI(Tl) crystals were fitted with thin entrance windows (7 μm Havar) and enclosed in annular lead shields 20 mm thick. Additional shadow-shielding for both telescopes was also provided. The beam and telescope axes were all horizontal and intersected at the treatment isocentre. The telescope axes made equal angles of θ = 44° with the beam axis. The distance between the radiator and the entrance window of the NaI(Tl) crystals was 42 cm. Relativistic kinematics gives θ = 43.6° as the angle of maximum efficiency for detecting the scattered and recoil protons for 200 MeV incident protons, but the finite angular resolution of the detectors (3° full width at half maximum (FWHM)) allowed coincidences to be registered for the full range of energies of interest (0 < E < 205 MeV) with θ = 44°.

The energy-dependent efficiency factor φ(E) due to the relativistic effects mentioned above was determined by calculation. φ(E) was found to be a maximum for 84 MeV incident protons, declining to about 80% and 60% of maximum for 20 MeV and for 200 MeV incident protons respectively. The variation in φ(E) with θ for 200 MeV incident protons was measured to check the validity of the method used in the calculations.

A fourfold coincidence between the scintillators of the two telescopes was required to define an event. An abridged block diagram of the data acquisition system is given in Figure 2. The four detector pulse heights (L_1, L_2, L_3 and L_4) as well as the coincidence time delay (T), between the two plastic scintillators, were digitised and written to an event file using the NAC data acquisition system. The electronics also incorporated a LED pulser system to monitor the gain stability of the NaI(Tl) detectors.

DATA REDUCTION

The data analysis, which was carried out off-line, consisted of applying appropriate windows on the L (Figures 3, 4 and 5) and T parameters in order to select true p-p coincidence events. These were events for which a scattered proton detected in one telescope is in coincidence with the associated recoil proton detected in the other telescope. This excludes a large fraction of the 'reaction tail' events which occur in the scintillators. For true events the summed pulse heights of the four scintillators, all calibrated to equal gain, then gave the incident proton energy $E(= E_1 + E_2 + E_3 + E_4)$. The spectra thus obtained were then corrected for the energy dependence of the detection efficiency of the system. This efficiency, $\epsilon(E)$, was assumed to be proportional to the factor φ(E) referred to above and to the Rutherford

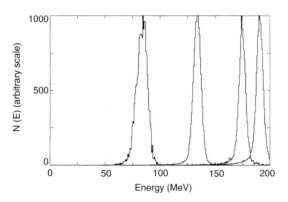

Figure 6. Proton energy spectra measured using the d = 50 mm collimator and with 0, 31, 93 and 146 mm thick acrylic degraders in the beam.

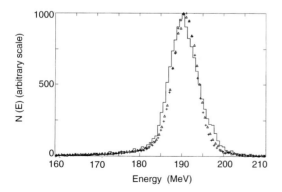

Figure 7. Proton energy spectra measured using collimator diameters d = 10 mm (histogram), 50 mm (triangles) and 100 mm (crosses).

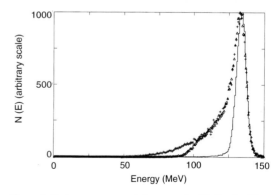

Figure 8. Proton energy spectra measured with the d = 50 mm collimator and 93 mm acrylic degrader and with modulator propellers providing 0 mm (histogram), 50 mm (triangles) and 110 mm (crosses) SOPBs.

cross section for p-p elastic scattering, which varies as E^{-2} in the laboratory frame. Thus

$$\epsilon(E) = k\,\phi(E)E^{-2}$$

was assumed, and the constant k was adjusted so as to normalise the peak intensity of each spectrum to the arbitrary value of 1000 counts per energy bin.

PROTON SPECTRA

All data presented here were obtained with the double scatterer beam spreading system in the beam. Attenuators have the expected effect on the energy spectrum as illustrated (Figure 6) by the results obtained using 0, 31, 93, and 146 mm thick acrylic attenuation in the beam. The peak energy decreases and the FWHM increases with increasing thickness.

The effect of the collimator diameter d (for d = 10, 50, and 100 mm respectively) is very small as shown in Figure 7. The comparison of different modulator wheels (providing 0, 50, and 110 mm SOBPs respectively) shows clearly that the modulator propeller spreads the high energy component of the spectrum, as desired, without introducing any significant additional low energy component. The FWHMs with the modulators in the beam are the same because the open and thinner sections of the modulator propeller are the same for both cases (Figure 8).

Moving the scatterer 270 mm upstream from the isocentre produces no significant increase in the FWHM.

DISCUSSION

This system had several advantages: firstly, it enabled measurements to be done *in situ* in the normal therapy beam at the usual beam currents used for therapy (20–80 nA) without pulse pile-up problems; secondly, it minimised the 'reaction tail' effect caused by protons undergoing nuclear reactions in the scintillators.

A feature of the data is the negligible number of low energy protons in the spectra. The preliminary measurements have also revealed limitations in the technique, some of which could be reduced or eliminated in future work. Measurements performed with the scatterer and occluding rings removed gave a FWHM of 5 MeV, which gives an indication of the resolution of the NaI(Tl) detectors. In addition, the energy thresholds of these detectors could be lower than the values of about 22 MeV used here if the ambient background (mainly from gammas and neutrons) in the experimental area could be reduced significantly. This would make it possible to extend studies of the low energy component to incident energies below about 60 MeV, the limit imposed by the present thresholds. In order to make measurements at incident proton energies less than 30 MeV (15 MeV in each telescope), it will also be necessary to improve the performance of the ΔE detectors.

It has been demonstrated that the method described here is suitable for assessing spectral characteristics of a high energy proton therapy beam. It should prove useful in providing experimental data to check theoretical calculations on which treatment planning is based.

REFERENCES

1. Gottschalk, B., Koehler, A. M., Schneider, R. J., Sisterson, J. M. and Wagner, M. S. *Multiple Coulomb Scattering of 160 MeV Protons*. Harvard Cyclotron Laboratory Report HCL 11/19/90, (Harvard Cyclotron Laboratory, Cambridge, USA) (1990).
2. Jones, D. T. L., Schreuder, A. N., Symons, J. E. and Yudelev, M. *The NAC Particle Therapy Facilities*. In: Hadrontherapy in Oncology, Eds U. Amaldi and B. Larsson. pp. 307–328 (Elsevier Science B.V.) (1994).
3. Jones, D. T. L. *NAC — the Only Proton Therapy Facility in the Southern Hemisphere*. In: Ion Beams in Tumor Therapy, Ed. U. Linz. pp. 350–359 (London: Chapman and Hall) (1995).
4. Schreuder, A. N., Jones, D. T. L., Symons, J. E., Fulcher, T. and Kiefer, A. *The NAC Proton Therapy Beam Delivery System*. In: Proc. 14th Int. Conf. on Cyclotrons and their Applications, Ed. J. Cornell, pp. 523–526 (Singapore, World Scientific) (1996).
5. Koehler, A. M., Schneider, R. J. and Sisterson, J. M. *Flattening of Proton Dose Distributions for Large Field Radiotherapy*. Med. Phys. **4**, 297–301 (1977).
6. Gottschalk, B. *Proton Nozzle Design Program NEU*. Private communication (1991).
7. Vynckier, S., Bonnett, D. E. and Jones, D. T. L. *Code of Practice for Clinical Proton Dosimetry*. Radiother. Oncol. **20**, 53–63 (1991).
8. Vynckier, S., Bonnett, D. E. and Jones, D. T. L. *Supplement to the Code of Practice for Clinical Proton Dosimetry*. Radiother. Oncol. **32**, 174–179 (1994).
9. ICRU. *Stopping Powers and Ranges for Protons and Alpha Particles*. Report 49, (Bethesda, MD: ICRU Publications) (1993).
10. Koehler, A. M., Schneider, R. J. and Sisterson, J. M. *Range Modulators for Protons and Heavy Ions*. Nucl. Instrum. Methods **131**, 437 (1975).

CALCULATED FLUENCE SPECTRA AT NEUTRON THERAPY FACILITIES

M. A. Ross†, P. M. DeLuca Jr†, D. T. L. Jones‡, A. Lennox§ and R. L. Maughan∥
†Department of Medical Physics
University of Wisconsin-Madison
Madison, WI 53706, USA
‡National Accelerator Centre, Faure, 7131 South Africa
§Midwest Institute for Neutron Therapy
Fermi National Accelerator Laboratory, Batavia, IL 60510, USA
∥Harper-Grace Hospital, Gershenson Radiation Oncology Center
Detroit, MI 48201, USA

Abstract — The Monte Carlo transport codes LAHET and MCNP were used to calculate energy fluence spectra at three neutron therapy facilities. The results compare very favourably with measured data. Kerma spectra and the ratio of ICRU muscle tissue kerma to A-150 kerma, along with the carbon to oxygen kerma ratio, were determined. Absorbed dose rate calculations are in reasonable agreement with measured values. Use of these codes to study modifications to existing therapy beams is briefly discussed.

INTRODUCTION

The production of neutrons at therapy facilities is usually obtained using low Z projectiles bombarding low Z targets. Neutron yield and penetration requirements have resulted in the Be(p,n) and Be(d,n) reactions with projectile energies between 40 and 70 MeV being most favourable[1]. Experimental information on the resulting neutron energy fluence spectra is rather limited and there are large discrepancies between the measurements that do exist. For example, recent measurements at the therapy facilities in Clatterbridge, England[2] and the National Accelerator Centre (NAC) in South Africa[3] show significant differences in the fluence spectra. These differences are surprising since the facilities are nearly identical. It is now technically feasible to calculate neutron fluence spectra directly at therapy facilities. These calculations can be used to understand the influence of various parameters on the resulting spectra. Section 2 of the present paper discusses calculations of neutron fluence spectra at NAC, Fermi National Accelerator Laboratory (FNAL), and Harper Hospital in Detroit, MI. The NAC and FNAL facilities use the Be(p,n) reaction while Harper Hospital uses the Be(d,n) reaction for production of therapy neutrons. The use of the computer codes LAHET (*Los Alamos High Energy Transport*)[4] and MCNP (*Monte Carlo N-Particle*)[5] is also discussed in Section 2. Comparisons of the calculated fluence spectra along with comparisons with measured fluence spectra are discussed in Section 3. Comparisons of kerma weighted spectra are also discussed in Section 3. Section 4 discusses conclusions derived from these calculations and some of the possibilities for future work.

FLUENCE CALCULATIONS

The Monte Carlo transport codes LAHET and MCNP were used to calculate the neutron energy fluence spectrum at the NAC, FNAL and Harper Hospital therapy facilities. The NAC facility was chosen because the calculated fluence spectra can be compared to the extensive fluence measurements of Jones et al[3]. FNAL was chosen for the large amount of clinical data generated from 25 years of operation. In addition, the NAC facility used some aspects of the FNAL facility as a model so comparisons between the two facilities are natural. Harper Hospital was chosen for comparison with a facility using the Be(d,n) reaction.

The incident proton energy at NAC and FNAL is 66 MeV, while the incident deuteron energy at Harper Hospital is 48.5 MeV. A description of the NAC facility can be found in Reference 6, of FNAL in Reference 7, and of Harper in Reference 8. Table 1 summarises information about the target and collimation assemblies at all three facilities. The important features to note are differences in the target thickness and stopping material for the incident charged particles, and the size and composition of filters and collimators.

LAHET is used to transport the 66 MeV protons and the 48.5 MeV deuterons through the Be target and also through the stopping material. Protons or deuterons below 1 MeV deposit their energy locally. LAHET is also used to transport neutrons above 20 MeV that are produced by nuclear interactions through the entire geometry of the therapy facility (i.e. target, collimators, and filters). Neutrons below 20 MeV, produced either from direct reactions or from down-scattering of higher energy neutrons have their kinematic variables stored for subsequent transport by MCNP. The MCNP code uses evaluated nuclear data from a number of sources including the ENDF/B-V and ENDL data libraries. MCNP can also be used for photon and electron transport up to 1 GeV.

The LAHET and MCNP codes use a combinatorial geometry method to define cells in a Cartesian co-ordinate system. A specified material and density are associated with each cell. The geometry of each facility is replicated as closely as possible using scale drawings. In order to make the definitions of the cells easier and to speed the calculations a few simplifications are made in defining the geometry. For example, the cooling jacket that surrounds the Be target (this was copper at NAC and aluminium at FNAL) is ignored and replaced by an equivalent volume of air. However, the composition and dimensions of the materials along the beam path are kept intact. In addition, small regions containing air are ignored as are transmission ionisation chambers or other beam monitoring devices. Finally, complicated cell boundaries not in the direct path of the beam are replaced by simple boundaries.

RESULTS AND COMPARISONS

Fluence calculations

The recent measurements by Jones et al[3] serve as a benchmark for these Monte Carlo calculations. An NE213 scintillation detector located 6 m from the target measured the neutron spectrum using time-of-flight techniques. The detector resolution was 0.7 ns at 6 m and the threshold was set to 3.5 MeV protons. Below 3.5 MeV the measured spectra were exponentially extrapolated to zero neutron energy. The results for a 100 mm by 100 mm field size with the clinical filter in place are shown in Figure 1 along with the calculated fluence spectra for the same field size. The reported detector resolution was included in the calculated spectrum. The spectra are arbitrarily normalised at 40 MeV. The agreement in shape between the measured and calculated spectra is quite good. The only difference is that the measurements indicate a slight decrease in the fluence at higher energies while the calculations predict a constant fluence. There is similar agreement between the calculated and measured fluence spectra at different field sizes.

The results of fluence calculations at all facilities for a 100 mm × 100 mm field size are shown in Figure 2. Comparing the NAC and FNAL spectra, significant differences are apparent, especially below 25 MeV. However, the spectra are nearly identical when calculations of the NAC facility are made in which the filter is replaced by an equivalent volume of air and the Be target thickness is increased to the same thickness as the FNAL target. The shapes of the Harper fluence spectra are much different from the NAC or FNAL calculations and are characterised by a peak at around

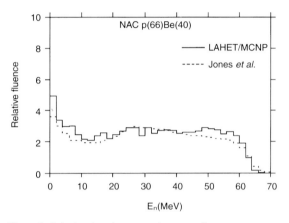

Figure 1. Calculated and measured energy fluence spectra at NAC for a 100 mm by 100 mm field size. The detector resolution from the measurements by Jones et al[3] was included in the calculated spectrum. The curves are arbitrarily normalised at 40 MeV.

Table 1. Summary of target and collimator parameters at NAC, FNAL and Harper.

	NAC	FNAL	Harper
Target	19.6 mm Be	22.1 mm Be	15.9 mm Be
Proton stopping material	Water + Cu	Au	
Filters	8 mm iron (flattening) 25 mm polyethylene (hardening)	none	18 stainless steel leaves
Pre-collimators	130 mm iron	130 mm iron	130 mm of stainless steel
Collimators	'Bookend' collimator, 5 levels of moveable jaws, 3 iron, 2 boronated polyethylene	Removable collimators, mixture of concrete, polyethylene pellets and water surrounded by Benelex*	250 mm of tungsten rods, each rod is 3.2 mm in diameter, the centres are spaced 3.6 mm apart

*Trade name, American Masonite Corporation, Chicago, IL, USA.

20 MeV with a FWHM of about 20 MeV. The calculations at Harper also indicate a large component of the fluence below 2 MeV. The shape of the spectra agrees very well with experimental data for d+Be neutron production at 50 MeV[8]. Finally, calculations at different field sizes were performed and indicate that the neutron fluence is essentially independent of field size at all facilities.

Kerma spectra and dose rate calculations

Kerma weighted spectra for ICRU muscle tissue and A-150 plastic[9] were determined from the calculated fluence spectra using neutron kerma factors from Howerton[10] for energies below 20 MeV. Above 20 MeV, kerma factors for hydrogen are taken from White et al[11], while values for carbon, nitrogen and oxygen are taken from Chadwick et al[12,13]. It has been assumed that trace elements in both ICRU muscle and A-150 plastic are oxygen. The kerma weighted spectra in ICRU muscle tissue for a 100 mm by 100 mm field at all three facilities are shown in Figure 3. The curves are normalised to the same dose (i.e. the integral of the kerma spectrum). It is obvious from Figure 3 that most of the dose at the Harper facility is delivered by neutrons between 10 and 30 MeV.

The ratios of ICRU muscle tissue kerma to A-150 kerma [$(K^{tissue}_{A150})_N$] was estimated to be 0.92, 0.93, 0.93 for NAC, FNAL and Harper, respectively. These values are in fairly good agreement with the recommended neutron dosimetry protocol[14] value of 0.95. The ratio of carbon to oxygen kerma is the prime factor in relating the dose in A-150 to that for muscle. Measurements of the carbon to oxygen kerma ratio are in progress at each of these facilities. Using the Howerton and Chadwick kerma factors and the calculated fluence spectra, the carbon to oxygen kerma ratio was determined. For a 100 mm by 100 mm field this ratio is 1.40, 1.41, and 1.40 for NAC, FNAL, and Harper, respectively.

The in-air absorbed dose rate in A-150 plastic (in units of Gy per incident particle on target) at the NAC facility was estimated and compared to the measured dose rate. The calculated dose rates are obtained using the calculated fluence spectrum for a 100 mm by 100 mm field in air at the treatment distance of 1.50 m weighted with the kerma factors for A-150 plastic. The measured dose rate was obtained using a 0.5 cm³ ionisation chamber with a 17 mm A-150 build-up cap in a 100 mm by 100 mm field. The results indicate that the calculated dose rate is about 20% lower than the measured dose rate.

Calculations of the absorbed dose rate at specified depths in a water phantom were also made and compared with the measured values at each facility. The size of the water phantom used in the calculation was chosen to be 0.6 m by 0.6 m by 0.6 m. The dose was calculated by determining the fluence spectra at the specified depth and again weighting the spectra with the kerma factors for A-150 plastic. In order to ensure a flat beam profile, the area used in determining the fluence spectra in phantom was chosen to be approximately 60% of the 100 mm by 100 mm field size. The results are summarised in Table 2 and indicate that for NAC and FNAL

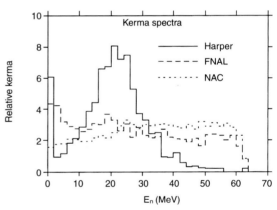

Figure 3. Kerma weighted spectra at NAC, FNAL and Harper. The curves have been normalised to the same dose (integral of the kerma spectra). The field size is 100 mm by 100 mm.

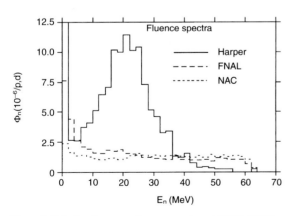

Figure 2. Calculated energy fluence spectra at NAC, FNAL, and Harper for a field size of 100 mm by 100 mm. All curves are normalised to be per incident charged particle striking the target.

Table 2. Summary of calculated and measured dose rates in A-150. Units are 10^{-17} Gy per incident particle on target.

	Depth (mm)	Calculated dose	Measured dose
NAC (in air, 1.50 m)	—	2.94	3.98
NAC SSD = 1.50 m	20	2.82	4.25
FNAL SSD = 1.80 m	100	2.23	3.28
Harper SSD = 1.829 m	9	8.77	8.99
Harper SSD = 1.829 m	50	7.10	7.73
Harper SSD = 1.829 m	100	5.20	5.75

(which use the Be(p,n) reaction) the calculated dose rates are lower than the measured rates by 30%. The results from Harper are in much better agreement with the calculated values being lower by 10% or less. It should be noted that the dose due to photons is not included in the calculations.

CONCLUSIONS AND FUTURE WORK

A considerable amount of work has been performed in calculating the fluence spectra at three neutron therapy facilities. Results using the Monte Carlo codes LAHET and MCNP show good agreement between the calculated and measured neutron spectra at NAC. The calculated fluence spectra at NAC and FNAL which both use the Be(p,n) reaction are actually quite similar. The most important difference between the two facilities seems to be the fact that NAC uses a polyethylene/iron filter while FNAL uses no filter. The fluence spectra at Harper has a distinctly different shape due to the Be(d,n) reaction. Using recent calculations of neutron kerma factors, the fluence spectra was converted to kerma spectra. Values for $(K^{tissue}_{A150})_N$ were estimated at each facility and are in reasonably good agreement with the recommended value of 0.95. A direct comparison of actual absorbed dose rates shows that the calculated values, determined from the kerma weighted fluence spectra, are reasonably close to measured values.

These calculations show that direct comparisons of different therapy facilities can be made without the need for expensive and time consuming experiments. In addition, calculations can be used to study modifications of existing therapy beams to enhance their biological effectiveness. One area of current interest is boron neutron capture therapy (BNCT). Calculations are underway to study how the existing therapy beam at FNAL could be modified to enhance the fluence of neutrons in the thermal and epithermal energy region where BNCT is most effective. The hope is to obtain a tumour dose enhancement of perhaps 10–15%.

REFERENCES

1. Amols, H. I., DiCello, J. F., Awschalom, M., Coulson, L., Johnsen, S. W. and Theus, R. B. *Physical Characterization of Neutron Beams Produced by Protons and Deutrons of Various Energies Bombarding Beryllium and Lithium Targets of Several Thicknesses*. Med. Phys. **4**, 486 (1977).
2. Crout, N. M. J., Fletcher, J. G., Green, S., Scott, M. C., and Taylor, G. C. *In situ Neutron Spectrometry to 60 MeV in a Water Phantom Exposed to a Cancer Therapy Beam*. Phys. Med. Biol. **36**, 507 (1991).
3. Jones, D. T. L., Symons, J. E., Fulcher, T. J., Brooks, F. D., Nchodu, M. R., Allie, M. S., Buffler, A. and Oliver, M. J. *Neutron Fluence and Kerma Spectra of a p(66)/Be(40) Clinical Source*. Med. Phys. **19**, 1285 (1992).
4. Prael, R. E. and Lichtenstein, H. *Users Guide to LCS: The LAHET Code System*. Los Alamos National Laboratory Report LA-UR-89-3014 (1989).
5. Briesmeister, J. F. *MCNP — A General Monte Carlo N-Particle Transport Code, Version 4A*. Los Alamos National Laboratory Report LA-12625 (1993).
6. Jones, D. T. L., Schreuder, A. N., Symons, J. E. and Yudelev, M. *The NAC Particle Therapy Facility*. In: Hadrontherapy in Oncology, Eds U. Amaldi, B. Larsson (Elsevier Science BV) p. 307 (1994).
7. Rosenberg, I. and Awschalom, M. *Characterization of a p(66)Be(49) Therapy Beam*. Med. Phys. **8**, 99 (1981).
8. Meulders, J. P., Leleux, P., Macq, P. C. and Pirart, C. *Fast Neutron Yields and Spectra from Targets of Varying Atomic Number Bombarded with Deuterons from 16 to 50 MeV*. Phys. Med. Biol. **20**, 235 (1975).
9. International Commission on Radiation Units and Measurements. *Neutron Dosimetry in Biology and Medicine*. ICRU Report 26 (1977).
10. Howerton, R. J. *Calculated Neutron Kerma Factors based on the LLNL ENDL Data File*. UCRL-50400 **27**: revised (University of California, Lawrence Livermore National Laboratory, Livermore, CA) (1986).
11. White, R. M., Broerse, J. J., DeLuca Jr, P. M., Dietze, G., Haight, R. C., Kawashima, K., Menzel, H. G., Olsson, N. and Wambersie, A. *Status of Nuclear Data for Use in Neutron Therapy*. Radiat. Prot. Dosim. **44**, 11 (1992).
12. Chadwick, M. B., Blann, M., Cox, L. J., Young, P. G. and Meigooni, A. *Calculation and Evaluation of Cross-sections and Kerma Factors for Neutrons up to 100 MeV on Carbon*. Nucl. Sci. Eng. (to be published). Also Lawrence Livermore National Laboratory Report UCRL-JC-121303 (1995).
13. Chadwick, M. B. and Young, P. G. *Calculation and Eveluation of Cross-sections and Kerma Factors for Neutrons up to 100 MeV on ^{16}O and ^{14}N*. Nucl. Sci. Eng. (submitted).
14. Miniheer, B. J., Wootton, P., Williams, J. R., Eenmaa, J. and Parnell, C. J. *Uniformity in Dosimetry Protocols for Therapeutic Applications of Fast Neutron Beams*. Med. Phys. **14**, 1020 (1987).

BIOPHYSICAL INVESTIGATIONS OF THERAPEUTIC PROTON BEAMS

R. Becker[1], J. Bienen[4], U. Carl[1], P. Cloth[2], M. Dellert[3], V. Drüke[2], W. Eyrich[3], D. Filges[2], M. Fritsch[3], J. Hauffe[3], W. Hoffmann[4], H. Kobus[2], R. Maier[2], M. Moosburger[3], P. Olko[5], H. Paganetti[1], H. P. Peterson[1], Th. Schmitz[1], K. Schwenke[1] and R. Sperl[3]
[1] Institut für Medizin, KFA-Jülich, D-52425 Jülich, Germany
[2] Institut für Kernphysik, KFA-Jülich, D-52425 Jülich, Germany
[3] Physikalisches Institut, Universität Erlangen, D-91051 Erlangen, Germany
[4] Fachbereich Physik, Bergische Universität
Gausstr 20, 42097 Wuppertal, Germany
[5] Institute of Nuclear Physics, Health Physics Laboratory
ul. Radzikowskiego 152, 31-342 Krakow, Poland

Abstract — The KFA programme to gather the necessary dosimetric and biophysical information to develop optimal dose calculation routines for conformal proton radiation therapy is introduced. First experiments in a water phantom and with a plastic phantom are presented for incident protons of 175.5 MeV. Furthermore, the first detailed biophysical investigations on the RBE dependence of monoenergetic and energetically spread out proton beams is described and the results are discussed.

INTRODUCTION

One goal in radiation therapy has always been to conform the therapeutic dose as closely as possible to the target volume and at the same time to minimise the volume of, and the dose to normal tissues, which necessarily have to be irradiated during a treatment. Conformal tumour therapy places higher requirements on the planning of treatments than do conventional irradiations. This is particularly true when protons or other light ions are used, due to the sharp edges in the dose profiles that can be produced. Wrong calculations of the maximum depth of penetration and of profiles may lead to either underdosage of parts of the target area or to unwanted overdosage of adjacent healthy tissues. Likewise, tissue inhomogeneities influence the depth dose distribution of protons more strongly than for photons or electrons[1].

In addition, the biological effectiveness of protons changes as they are slowed down in tissue. This effect is not as pronounced as for heavier ions, but should be observed in particular at the distal edge (in and behind the Bragg Peak) of the dose profiles[2]. To date, the biophysical basis for proton therapy is not satisfactory[3], since no systematic data base for different beam energies and characteristics is available.

The new synchrotron, called COSY (COoler SYnchrotron), operated at the KFA-Jülich is able to deliver proton pencil beams with energies between 100 MeV and 2.5 GeV. This covers the energy range (100 MeV to 250 MeV) of interest in radiation therapy for the treatment of deep seated tumours. A biophysical programme was started, which follows two aims: to establish a systematic biophysical basis for proton therapy and to develop realistic proton dose calculation routines for treatment planning based on Monte Carlo methods.

The paper introduces the biophysical programme carried out by the COSY-Med collaboration at KFA. First experimental results with regard to in-phantom measurements of depth–dose profiles and microdosimetric distributions are presented. Furthermore, results of biophysical simulations are discussed.

THE COSY-MED PROGRAMME

It is the aim of the programme to gather the necessary dosimetric and biophysical information for the development of optimal dose calculation routines for treatment planning[4] (Table 1). The programme includes an experimental and a calculational part, which complement each other.

Experiments are performed to determine the incident beam characteristics for data normalisation and to provide input data for the calculations. Absolute dose, microdosimetric energy deposition distributions, energy loss spectra and high resolution dose profiles are measured in a water or a plastic phantom as a function of depth. For these purposes different kinds of detectors are used (Table 1).

The initial biological experiments concentrate on the determination of cell killing as a function of position on the Bragg curve for different dose levels in the Bragg peak. The selected cells are V79 hamster cells. For the determination of the relative biological effectiveness (RBE), ^{60}Co is used as reference radiation.

Depth–dose profiles, secondary particle spectra and

Correspondence to: Dr Thomas Schmitz, Institute of Medicine, KFA-Research Centre Jülich, D-52425 Jülich, Germany

energy deposition spectra are calculated as a function of incident beam parameters and depth in matter using Monte Carlo programs. The biophysical simulations use the result of these calculations and apply two different approaches to predict the RBE of protons. One is the microdosimetric method, which uses so-called weighting functions, which are folded with microdosimetric energy deposition spectra to get information on the RBE of the radiation field[5,6]. The other one is the track structure model of Katz and co-workers[7,8].

The central part of the programme is the comparison of experiment and simulation. On the one hand, inconsistencies in the measurements can be detected. On the other hand, the precision of the calculation can be assessed. The aim is to identify which physical processes necessarily have to be included into simulations to achieve sufficient accuracy in the predictions of dose profiles and a realistic description of the secondary particle spectra. The latter are important, since they are the basis for the biophysical simulations. First results with regard to the simulation of dose profiles are presented elsewhere in this volume[9].

FIRST EXPERIMENTAL RESULTS

Experiments have been carried out in a water and a plastic phantom. The water phantom is a Plexiglas tank with a volume of 50 cm × 50 cm × 40 cm (width × depth × height). The wall thickness is 1 cm. A 3-axis scanner unit is used to position the detectors within the phantom with an accuracy of better than 1 mm. The plastic phantom is fabricated from Plexiglas and its thickness can be varied between 4 cm and 30 cm. The Si microstrip detectors and the fibre hodoscopes are either placed in front or behind the phantom. COSY delivered protons of 175.5 MeV (±1%) kinetic energy (600 MeV/c momentum). The momentum spread was about 5×10^{-3} and the beam diameters on the surfaces of the phantoms had been adjusted to about 1 cm FWHM and 0.5 cm FWHM respectively in case of the water and the plastic phantom. The extraction pulse duration was about 200 ms. About 50% of the protons were delivered within the first 50 ms, which caused serious dead time problems in the electronics of the scintillators used as beam monitors. Therefore, the measurements

Table 1. Overview of the COSY-Med programme (see text). Detector characteristics are also given.

Dosimetry	Experiments	Simulation
Beam characteristics		
Beam energy	Determined from synchrotron	Input parameters for Monte Carlo
Beam energy spread	Parameters	codes
Beam intensity	Plastic scintillators	
Beam profile	Si microstrip detectors	
	Fibre hodoscopes	
Phantom studies		
Absolute dose	TE ionisation chambers	HET, PTRAN, MC4
	TLDs	
Depth–dose profiles	TE ionisation chambers	HET, PTRAN, MC4
	TLDs	
	Si microstrip detectors	
	Fibre hodoscopes	
Energy loss spectra	Si microstrip detectors	HET, PTRAN, MC4
	Fibre hodoscopes	
Secondary particle spectra		HET, MC4
Microdosimetric distributions	Tissue-equivalent proportional counter (TEPC)	PTRAN, HET, PMIC
Biology		
RBE (radiobiological effectiveness)	Cell survival in V79 hamster cells	Track structure model, microdosimetric weighting functions
Detector characteristics		
Detector	Characteristics	
TE ionisation chambers	$0.1\ cm^3$ and $1\ cm^3$ detection volume, air filled TLD-300	
TLDs		
Tissue-equivalent proportional counter (TEPC)	Far West, 1/2″ counter, propane-based TE gas, 1 μm simulated diameter, University of Homburg	
Si microstrip detectors	Grid structure, 200 μm × 3 mm (width × height) single strip size	
Fibre hodoscopes		

cannot be normalised to the number of incident protons, which would be desirable.

Figure 1 displays a depth–dose profile measured with the small-volume (0.1 cm^3) ionisation chamber and with TLDs. The data are normalised to the monitor counts, even though this number is, as mentioned, distorted by dead time effects. It is assumed, since the extraction was very stable, that on average all monitor counts are equally affected. Nevertheless, only arbitrary units are given on the ordinate. The error bars on the ionisation chamber data reflect the standard deviation of the mean of five repetitive measurements taken at each position during one scan. Three scans taken at different times are shown.

The TLD results agree with the ionisation chamber measurements rather well (Figure 1). The higher Bragg peak value might be due to the fact that the TLDs are geometrically smaller than the ionisation chamber. When analysing the glow curve of the TLDs, the quality of the radiation field in terms of the microdosimetric parameter dose mean lineal energy, y_D[10], can be assessed from the ratio of heights of peaks 1 and 2[11]. However, the ratio has to be calibrated for the particular radiation field and this has been done by comparing the results with proportional counter measurements described below. In the future, the intention is to use the TLDs in biological stack experiments as on-line dosemeters.

Figure 2 presents microdosimetric dose distributions measured at different depth along the beam axis. The spectra are plotted against the logarithmic increment of lineal energy and are normalised to unit dose. In this presentation, equal areas of the distribution represent equal fractions of absorbed dose. The shift of the spectra to higher lineal energies reflects the increase in proton stopping as they are slowed down. The distributions measured near the entrance of the phantom at 19.2 mm and 24.2 mm are too broad for a beam of high energy protons. The reason is that the counter was 12 mm outside the beam axis in the beam penumbra at these positions. No significant contribution of high energy neutrons can be detected from the spectra. They would contribute at lineal energies above the proton edge at about 120 keV μm^{-1}. This can be expected, since the neutron production at 175 MeV incident proton energy is very low.

Finally, Figure 3 displays a two-dimensional dose profile measured with the Si microstrip detector in the horizontal plane perpendicular to the beam. The plastic phantom was used for these measurements. Clearly the beam broadening with depth can be seen, as well as the thin-down behind the Bragg peak. The drop in the energy deposition from front to the Bragg peak indicates that the incident beam was geometrically divergent.

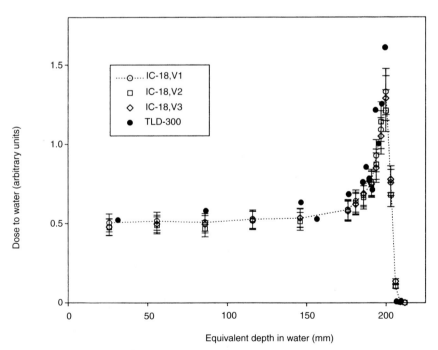

Figure 1. Bragg curve measured with a small volume (0.1 cm^3) air-filled TE ionisation chamber and with TLD-300. The three results for the ionisation chamber (IC18 V1-V3) represent three scans of the ionisation chamber through the water phantom, taken at different times. The depth is given as water equivalent depth, indicating that the thickness of any material in the beam path was recalculated to equivalent water thickness. The incident beam kinetic energy was 175.48 MeV (600 MeV.c^{-1} momentum).

BIOPHYSICAL INVESTIGATIONS

A method has been proposed[12–14] to compare the beam quality of different therapeutic installations using microdosimetric distributions together with microdosimetric weighting functions. The main assumption of this model is that a dose–biological effect relationship can be expressed as an integral convolution of two separate functions of lineal energy y[5,6]. No general assumptions about the mechanisms of radiation action are made. The model is strictly only valid for the calculation of absolute RBE values if the single event spectra are appli-

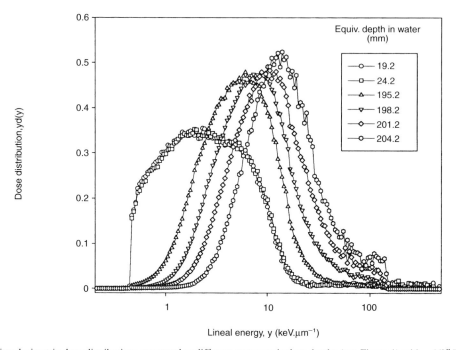

Figure 2. Microdosimetric dose distributions measured at different water equivalent depths (see Figure 1) with a 1/2″ Rossi counter, operated with propane tissue-equivalent gas at a pressure to simulate a tissue sphere of 1 μm diameter. Same beam energy as in Figure 1.

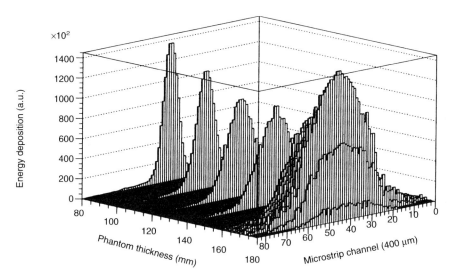

Figure 3. Depth dose profile in the horizontal beam plane. The profile was measured with a Si-microstrip detector placed behind a variable thickness plastic phantom. The beam energy (momentum) was as given in Figure 1.

cable, i.e. at 'low doses', and only linear terms contribute to the dose–response curve.

Beside this, the track structure model[7,8] is used to predict the RBE of protons. In contrast to the microdosimetric approach, this model predicts RBE values for different survival levels. It is therefore in principle better suited for calculations relating to biological effects in radiation therapy. The track structure model uses radiosensitivity parameters as biological input which are derived from the gamma ray survival curve of the respective cell line.

One great advantage of the microdosimetric approach is the possibility of determining experimentally not only the biological but also the physical side, measuring microdosimetric spectra. Using both models, a detailed investigation on the behaviour of the RBE as a function of various beam parameters has been performed.

RESULTS OF THE CALCULATIONS

Microdosimetric spectra

Energy loss curves as well as proton energy spectra at various depths in water were calculated with the Monte Carlo code PTRAN*[15]. From these spectra the corresponding yd(y) distributions can be obtained[2]. In Figure 4, a calculated yd(y) distribution in the centre of the Bragg peak for a 173 MeV proton beam with a momentum spread of $\Delta p/p = 0.002$ is compared with a measured distribution. The PTRAN input parameters were determined by fitting the calculated depth–dose curve to the ionisation chamber measurements (Figure 1). The agreement between the measured and calculated yd(y) distributions is fairly good when keeping in mind that the following simplifications were made in the calculations: Firstly, the CSDA was used to calculate the energy deposition in the 1 μm sphere. This simplification results in a small shift of the calculated yd(y) spectra to lower y due to a narrower microdosimetric f(y) spectrum ($d(y) = y/y_F f(y)$). Secondly, no delta ray components were included. The contribution of delta electrons to the target dose from protons passing outside the target (called passers or touchers) depends on proton energy and target diameter: 20 MeV protons deposit 26% of the total dose in the 1 μm spherical target by delta electrons. This dose fraction decreases to 7.1% for 5 MeV protons and to 0.3% for 1 MeV protons[16]. Thirdly, no secondary particles from nuclear interactions were included. These particles may have high RBE values. However, the contributions from secondary charged particles other than protons increased the LET_D significantly only at depths smaller than 0.9 times the CSDA range of the proton[17], which means at depths slightly smaller than the maximum of the Bragg peak. The influence of the secondary charged particles from nuclear reactions is therefore very small in the Bragg peak region and it can be expected to lead to only a small enhancement of biological effectiveness. Finally, one has to keep in mind that the proton energy distributions obtained by PTRAN are distributions as a function of depth, thus integrating radially. In the experiment a 1/2 inch counter was used. However, this effect should be very small.

The contribution of the secondary charged particles to the dose is included in our calculations. This contribution is small in the region of the Bragg peak. In the entrance region, nuclear interactions contribute about 56% to the dose for a 250 MeV beam. For a 70 MeV beam this value is only about 8%. In the Bragg peak this value is in the region of 1% for a 250 MeV beam and much lower for lower initial proton energies.

RBE for monoenergetic beams

Using the proton energy spectra, the RBE was obtained for the inactivation of V79 cells using the

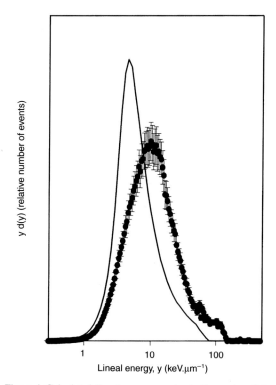

Figure 4. Calculated lineal energy spectra in the centre of the Bragg Peak for an incident beam with an energy of 173 MeV and an energy spread of $\Delta p/p = 0.002$ (solid line). Shown for comparison is a measured lineal energy spectrum at the same depth of penetration (solid circles with error bars).

---

* Additionally, the codes HET and MC4 have been used. A comparison of the results is reported in the paper of Becker *et al* in this volume.

response function of Morstin et al[6] within the microdosimetric model.

In addition, the track structure model was used for RBE calculations. As there exists a broad distribution of proton energies due to the scattering of the beam along its path, the formalism given by Katz and Sharma[7] was followed, describing the cellular response to a mixed radiation environment. The radiosensitivity parameters for the aerobic survival of Chinese hamster cells, CH2B2, were considered[18,19].

For both models it was found that the RBE increases with depth at and beyond the Bragg peak due to the presence of low energy protons. The rise decreases with increasing initial proton energy. Furthermore, the position where the RBE begins to rise is shifted towards the end of the Bragg peak increasing the initial proton energy.

Using the track structure model, the RBE (aerobic survival of Chinese hamster cells, CH2B2) is about 1 in the entrance region and gradually increases with the proton penetration depth. In the centre of the Bragg peak the values of RBE range from 1.1 for a 250 MeV beam to 1.3 for a 70 MeV beam. Distal to the Bragg peak, where only a small fraction of dose is delivered, the RBE was found to be even higher. With the microdosimetric model and for Chinese hamster V79 cells in G1/S phase, these values range from 1.3 for a 250 MeV beam to 1.7 for a 70 MeV beam. One has to keep in mind that these values are for monoenergetic beams. Defining an energy spread will lower the RBE values.

Influence of the modulation technique

For a dose conformation, the energy and the direction of the beam has to be modified during treatment to produce a so-called 'spread out Bragg peak' (SOBP). This is accomplished either by a passive or by an active method. The depth modulation in the passive procedure is based on the use of an absorbing material of variable thickness in the beam path. The active one is based on a direct variation of the beam energy during treatment. When simulating a SOBP, the modulation technique has to be taken into account. For a beam scanning device the respective dose distributions belong to energetically different proton beams. On the other hand, simulating a range modulated beam, one has to use incident proton beams each with the same energy, but penetrating different depths in a range filter (as an approximation a range shifter consisting of water was assumed). Following the expression that the dose can be described as fluence times LET, it is evident that similar doses can be achieved with different proton energy distributions. Isodose contours are isoeffect contours only if the energy spectra of the accompanying particles remain constant.

A detailed analysis of the RBE behaviour in a SOBP, considering the modulation type, with different biophysical models has been made. A SOBP was defined with the conditions of the distal and proximal 90% of the maximum dose at the positions 5.3 and 13.2 cm, respectively (Figure 5). Three different initial beam situations were studied. Firstly, a beam scanning method, considering a beam delivery system which is able to produce semi-monoenergetic beams (momentum spread $\Delta p/p = 10^{-2}$) in 1 MeV steps. Secondly, proton beams modulated using a rotational wheel (consisting of water) with energies of 137 MeV and 250 MeV ($\Delta p/p = 10^{-2}$). Here, the needed absorber thicknesses are 0 cm to 6.9 cm and 24.2 cm to 31.1 cm, respectively (step size 0.3 cm): 137 MeV is the lowest possible energy to build the considered SOBP.

The lowest initial proton beam energy offers the sharpest dose fall-offs. The beam scanning method offers no advantage compared with the passive range modulation system with a 137 MeV beam. The positions, where the dose reaches 10% of the SOBP plateau value distal to the Bragg peak, are 14.1 cm for the beam scanning method and the passive range modulation system with a 137 MeV beam. For the 250 MeV beam and the passive range modulation system this position is 14.9 cm. The reason is the absorber material which broadens the respective Bragg peaks.

The calculated RBE values and the resulting biological doses for a flat physical dose plateau, are given in Figure 5 (RBE at 10% survival for aerobic survival of Chinese hamster cells, CH2B2; relative to 250 kV$_p$ X rays; track structure model). In the distal part of the SOBP the RBE is higher the lower the maximum proton energy entering the modulator. The first reason is that the beam with the highest range has the highest weighting in the SOBP. Secondly, at the end of the SOBP only

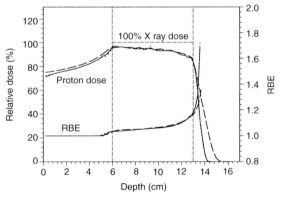

Figure 5. Depth dependence of the proton doses necessary to reach a flat biological dose distribution of 2 Gy between 6 and 13 cm (left scale). Depth dependence of the RBE values (right scale). Full curve: 88–137 MeV active modulation; dotted curve: 137 MeV passive modulation; broken curve: 250 MeV passive modulation (the plots for the active method and the 137 MeV passive method are nearly identical). The 100% X ray dose value and the target limit is sketched. Biological endpoint: aerobic survival of Chinese hamster cells, CH2B2; biophysical model: track structure[20].

this beam contributes. In the plateau of the SOBP there is an overlap of various Bragg curves with a high contribution of regions proximal to the Bragg peak where the RBE is about 1. With respect to the RBE the beam scanning method offers no advantage in comparison with the passive technique when using the lowest possible beam energy for the range modulation. As the RBE is reflected by the proton energy distirbution, differences in RBE are caused by differences in the proton energy spectra. It was found that the proton energy spectra at similar depths in the water phantom depend on the modulation technique. The higher the energy from which the beam is modulated by a range shifter or by scanning, the higher is the average proton energy at a given depth. As the RBE is nearly energy independent for energies above 20 MeV this affects not the biological effectiveness in the entrance of the SOBP but in the distal part.

SUMMARY AND OUTLOOK

A biophysical programme has been set up to gather the necessary dosimetric and biophysical information for the development of optimal dose calculation routines for therapy planning. First experiments have been performed and showed the suitability of the systems. Systematic measurements have to be performed for different initial proton beam energies between 100 MeV and 250 MeV with regard to depth–dose distributions, microdosimetric distributions and biology.

More experimental and theoretical work is needed to study the difference between the measured and the calculated microdosimetric spectra. Measurements of the proton RBE as a function of different beam modulation techniques are needed for comparison of our results concerning the SOBP modelling. Additionally, biological experiments which determine both the parameters for the microdosimetric and those for the track structure model, would offer the possibility of comparing the RBE predictions.

RBE values depend on the biological endpoint, the initial proton energy, the depth of beam penetration and the initial energy spread of the input proton beam. The RBE values increase with depth. Furthermore the depth dependence of the RBE results in a shift of the biological dose distribution to higher depths in comparison to the physical dose distribution. Even for modulated proton beams an increase of the RBE values with depth in the SOBP was found. This gradient implies that, despite a homogeneous physical dose, the distal end of a irradiated volume receives a higher biological dose than the proximal end.

A detailed analysis of the RBE behaviour in a SOBP, considering the type of beam modulation, with different biophysical models has been achieved. It was found that the beam modulation technique influences the proton energy spectra, the fluence, the LET, and the RBE, within a SOBP although the dose remains constant. Modulating the same dose distribution with a higher initial proton beam energy results in a lower average LET, a lower RBE, and a higher fluence. The RBE for the beam scanning method behaves like the respective RBE dependence of a range modulated beam with similar energies.

ACKNOWLEDGEMENTS

The authors thank Mrs M. Langen, Mrs S. Schaden, Mr D. Barthmann (Institute of Medicine, KFA Jülich) and Mr N. Paul (Institute for Nuclear Physics, KFA-Jülich) for their valuable help in setting up and running the experiments. They are also in debt to Mr Heinz Rongen (Central Electronics Department, KFA-Jülich) for his help with respect to various electronic problems that had to be solved. They thank the colleagues from the University of Saarland, Department of Biophysics and Fundamental Physics, Homburg, for giving us their 1/2″ proportional counter.

REFERENCES

1. Urie, M. *Treatment Planning for Proton Beams*. In: Ion Beams in Tumor Therapy, Ed. U. Linz (London: Chapman & Hall) pp. 279–289 (1995).
2. Paganetti, H., Olko, P., Kobus, H., Becker, R., Schmitz, Th., Waligorski, M. P. R., Filges, D. and Müller-Gärtner, H. W. *Calculation of RBE for Proton Beams using Biological Weighting Functions*. Int. J. Rad. Oncol. Biol. Phys. (in press).
3. Blakely, E. A. *Biological Beam Characterisation*. In: Ion Beams in Tumor Therapy, Ed. U. Linz (London: Chapman & Hall) pp. 63–72 (1995).
4. Bartmann, D., Drüke, V., Wyrich, W., Filges, D., Hoffmann, W., Klasen, C., Linz, U., Maier, R., Neef, R. D., Paul, N. and Schmitz, Th. *Pre-Therapeutic Physical Experiments on Phantoms with Proton Beams at Energies between 50 MeV and 250 MeV*. COSY Proposal No. 26, Forschungszentrum Jülich (1993).
5. Zaider, M. and Brenner, D. J. *On the Microdosimetric Definition of Quality Factors*. Radiat. Res. **103**, 302–316 (1985).
6. Morstin, K., Bond, V. P. and Baum, J. W. *Probabilistic Approach to Obtain Hit-size Effectiveness Functions which Relate Microdosimetry and Radiobiology*. Radiat. Res. **120**, 383–402 (1989).
7. Katz, R. and Sharma, S. C. *Response of Cells to Fast Neutrons, Stopped Pions, and Heavy Ion Beams*. Nucl. Instrum. Methds. **111**, 93–116 (1973).
8. Katz, R. and Sharma, S. C. *Heavy Particles in Therapy: An Application of Track Theory*. Phys. Med. Biol. **19**, 413–435 (1974).
9. Becker, R. et al. *Comparison of HETC and PTRAN with Phantom Measurement Data*. Radiat. Prot. Dosim. **70**(1–4), 497–500 (This issue) (1997).

10. ICRU. *Microdosimetry*. Report 36 (Bethesda, MD: ICRU Publications) (1983).
11. Hoffmann, W. and Songsiritthigul, P. *TLD-300 Dosimetry at the Chiang Mai 14 MeV Neutron Beam*. Radiat. Prot. Dosim. **44**(1–4), 301–304 (1992).
12. Menzel, H. G., Pihet, P. and Wambersie, A. *Microdosimetric Specification of Radiation Quality in Neutron Radiation Therapy*. Int. J. Radiat. Biol. **57**, 865–883 (1990).
13. Pihet, P., Menzel, H. G., Schmidt, R., Beauduin, M. and Wambersie, A. *Biological Weighting Function for RBE Specification of Neutron Therapy Beams. Intercomparison of 9 European Centres*. Radiat. Prot. Dosim. **3**, 437–442 (1990).
14. Wambersie, A. *Contribution of Microdosimetry to the Specification of Neutron Beam Quality for the Choice of the 'Clinical RBE' in Fast Neutron Therapy*. Radiat. Prot. Dosim. **52**, 453–460 (1994).
15. Berger, M. J. *Proton Monte Carlo Transport Program PTRAN*. National Institute of Standards and Technology Publication NISTIR 5113 (1993).
16. Olko, P. and Booz, J. *Energy Deposition by Protons and Alpha Particles in Spherical Sites of Nanometer to Micrometer Diameter*. Radiat. Environ. Biophys. **29**, 1–17 (1990).
17. Berger, M. J. *Penetration of Proton Beams Through Water I. Depth–dose Distribution, Spectra and LET Distribution*. National Institute of Standards and Technology Publication NISTIR 5226 (1993).
18. Roth, R. A., Sharma, S. C. and Katz, R. *Systematic Evaluation of Cellular Radiosensitivity Parameters*. Phys. Med. Biol. **21**, 491–503 (1976).
19. Katz, R., Zachariah, R., Cucinotta, F. A. and Zhang, C. *Survey of Cellular Radiosensitivity Parameters*. Radiat. Res. **140**, 356–365 (1994).
20. Paganetti, H. and Schmitz, Th. *The Influence of the Beam Modulation Technique on Dose and RBE in Proton Radiation Therapy*. Phys. Med. Biol. **41**, 1649–1663 (1996).

MICRODOSIMETRIC STUDIES ON THE ORSAY PROTON SYNCHROCYCLOTRON AT 73 AND 200 MeV

V. P. Cosgrove†, S. Delacroix‡, S. Green§, A. Mazal‡ and M. C. Scott†
†School of Physics and Space Research
University of Birmingham, Birmingham B15 2TT, UK
‡Centre de Protonthérapie d'Orsay
BP 65, 91402, Orsay Cedex, France
§Department of Medical Physics, Queen Elizabeth Medical Centre
Edgbaston, Birmingham B15 2TH, UK

Abstract — A planar microdosemeter has been developed for making measurements on charged particle radiotherapy beams. Initially used for making measurements on the 62 MeV proton therapy beam at Clatterbridge, the detector has now been used on both the 73 MeV and 200 MeV proton beams at Orsay. Measurements on this facility were complicated by the fact that it is a synchrocyclotron, with 440 Hz modulation of the primary MHz-pulsed beam. As a result, extreme care had to be exercised to prevent recording multiple events. Analysis of the microdosimetric spectra shows that for both the 73 and 200 MeV beams there is a significant increase in the dose averaged mean lineal energy, $\bar{y}_D$, towards the end of the proton range. This will be associated with a rise in RBE. The implications of this increase are discussed.

INTRODUCTION

The use of protons for tumour treatment is well established worldwide, and interest in this form of radiotherapy continues to grow. Although a number of microdosimetric studies have been made on proton beams, many of these have used microdosemeters developed for work with neutrons (see, e.g. Ref. 1). In contrast, the work reported here, together with a related programme on the Clatterbridge synchrocyclotron[2], uses a planar microdosemeter specially developed for proton beam work.

METHODS AND MATERIALS

The Orsay proton irradiation facility

The synchrocyclotron at Orsay was first used for proton therapy in 1991, when a 73 MeV beam produced by degrading the 200 MeV primary beam was used to treat optical melanomas. Since then, over 600 patients have been treated, averaging about 5 per week. More recently, the 200 MeV primary beam has been used to treat intracranial targets.

The layout of the facility is shown in Figure 1. The 73 MeV beam was produced by passing the primary beam through a 128 mm thick graphite block (density 1.855 g.cm^{-3}). Collimation and modulation of the beams was performed outside the vacuum system, and the arrangement for the 73 MeV beam is shown in Figure 2. The modulator, which is designed to produce a uniform dose over the target volume, consists of a rotating Perspex wheel with vanes of varying thickness. The total range of the 73 MeV protons is 40 mm.

The proton microdosemeter

The design of the proton microdosemeter has been shown earlier[2]. The main design criterion was to have a detector suitable for use in charged particle beams with a minimal thickness entrance window. The entrance window used was 50 μm thick aluminised Mylar. Because this bowed under differential pressure, the conducting walls defining the detection volume were made from 3 μm thick aluminised Mylar. The single central anode was 25 μm diameter, and the filling gas used was a tissue-equivalent mixture (either propane or methane

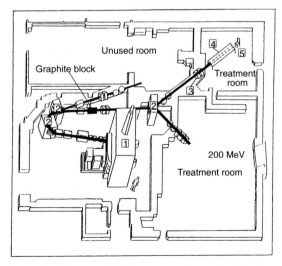

Figure 1. The layout of the Orsay proton therapy facility. Key: 1, 200 MeV synchrocyclotron; 2, beam bending magnets; 3, final beam focussing magnets; 4, 73 MeV beam line; 5, patient chair.

based) filled to a pressure equivalent to a tissue thickness of 2 μm.

Standard nuclear electronics were used, the anode normally being at 900 V. Both in order to ensure comlete charge collection and to minimise interference from radiofrequency units on the cyclotron, relatively long amplifier time constants were used (6 μs). Even so, interference limited the lower lineal energy measured to 0.5 keV.μm^{-1}. Pulse height spectra were accumulated simultaneously in high and low gain channels, the position of the proton edge being used for energy calibration: for a 2 μm tissue-equivalent thickness filling this edge was at 95.4 keV.μm^{-1}.

RESULTS

Making microdosimetric measurements on a synchrocyclotron beam poses particular problems since the machine is frequency modulated, i.e. it produces a small bunch of RF pulses (19–20 MHz) at a low frequency (440 Hz at Orsay). Consequently, each beam bunch could contain one or more protons in each RF pulse, which could not be resolved by the microdosemeter. In order to determine the optimum beam current, an NE213 scintillator was used to study the beam characteristics. This showed that it was necessary to run the machine at extremely low average beam currents in order to minimise pulse pile-up in the microdosemeter. Even at microdosemeter count rates of 350 s^{-1} (i.e. less than one count per modulated burst) significant pulse pile-up was observed, and running the beam to produce 50 counts.s^{-1} was the normal operating condition for the measurements reported here.

Measurements made in the umodulated beam used Perspex energy degraders to change the mean proton energy at the microdosemeter. Figure 3 shows the measurement positions for the 73 MeV beam together with the relative dose at each position, whilst Figure 4 shows the y d(y) spectra at these four positions. (Note, incidentally, from Figure 4 that the microdosimetric spectrum corresponding to point N is broader, and slightly shifted to higher lineal energies, than that for point M, reflecting the greater energy spread of the

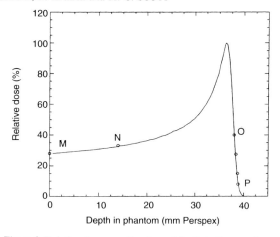

Figure 3. Relative dose as a function of depth in Perspex for a 10 mm diameter unmodulated 73 MeV Orsay proton beam: (○) microdosimetric measurement points.

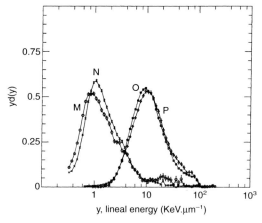

Figure 4. The y d(y) spectra for the 10 mm diameter 73 MeV proton beam at the measurement points marked in Figure 3.

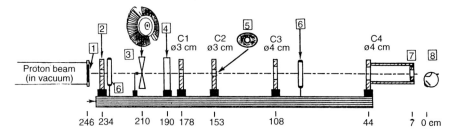

Figure 2. A schematic diagram of the Orsay 73 MeV proton therapy beam line. Key: 1, Vacuum window, made from a scintillator material. 2, 1st beam collimator, 3 cm diameter, 4.8 cm deep. 3, Perspex modulator. 4, Perspex range shifter. C1–4, Antiscatter collimators, all 4.8 cm deep. 5, Elliptical scattering foil, thickness = 0.1 mm of lead, outer cross-section = 20 × 13 mm^2, inner diameter = 5 mm. 6, Transmission monitors. 7, Patient collimator. 8, Location of patient's eye.

beam and the presence of low energy protons.) A similar series of measurements was made with the modulated beam at different depths in a Perspex phantom, and Figures 5 and 6 show the measurement positions and measured y d(y) spectra respectively. In all cases, the low lineal energy cut-off was imposed by RF noise.

Values of $\bar{y}_D$, the dose averaged lineal energy, for the modulated and unmodulated beams derived from this data are shown in Figure 7, where it can be seen that beyond 15 mm or so in Perspex $\bar{y}_D$ increases steadily in both cases: from 20 to 40 mm depth in Perspex the increase in $\bar{y}_D$ is from 4 keV.μm^{-1} to 20 keV.μm^{-1}.

The measurement depths in Perspex for the 200 MeV beam are shown in Figure 8, which also shows the depth dependence of the relative dose. The corresponding variation of $\bar{y}_D$ with depth is shown in Figure 9, where it can be seen that there is a significant drop (a factor of 2) in $\bar{y}_D$ between 0 and 200 mm depth in Perspex before the final increase at the end of the range (230 mm).

DISCUSSION

The design object for the beam modulator is, as noted earlier, to produce a uniform dose over an extended region, which Figure 5 shows has been achieved. However, the rise in $\bar{y}_D$ towards the end of the range for both the 73 and 200 MeV beams will be associated with an increase in RBE. That there is a significant increase in the RBE towards the end of the proton range for both modulated and unmodulated beams has been seen in a number of studies[3-6], and has also been observed in a radiobiological study related to the present programme, which will be published shortly.

For heavier particles, the significant variation of RBE with depth is usually compensated for by a different modulator design, which compensates for the RBE increase by diminution of the physical dose with depth, to produce a uniform RBE dose, i.e. the product of dose and RBE. In the case of protons, the increase in $\bar{y}_D$ is significant only in a narrow region towards the end of the range. To compensate for this effect, a reduction in the physical range together with a smoothed 'shoulder' in the depth–dose curve would be necessary. However, an error in the conversion model from dose to RBE dose could then lead to an under-dosage, or an over-dosage, of the target volume.

Knowledge of the RBE values for proton beams is still not well established, depending, as it does, on the biological endpoint and, in particular, on each organ tissue. Whilst the region at the end of the range may be of great importance in therapy — for example, because it may be near critical organs which need to be spared — there are big physical uncertainties involved in achiev-

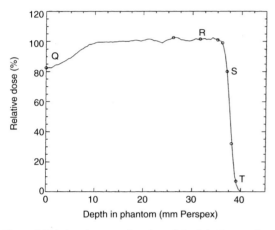

Figure 5. Relative dose as a function of depth in Perspex in a 10 mm diameter modulated 73 MeV beam: (○) microdosimetric measurement points.

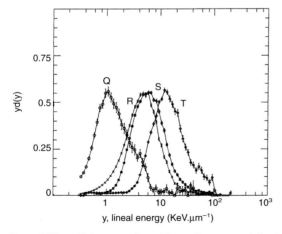

Figure 6. The y d(y) spectra for a 10 mm diameter modulated 73 MeV proton beam at the measurement points marked in Figure 5.

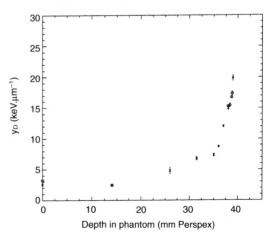

Figure 7. Variation of $\bar{y}_D$ with depth in Perspex for the modulated (X) and unmodulated (○) 73 MeV beams.

ing any improvement. These include measurement of the physical dose, uncertainties in the particle range after traversing different heterogeneous regions, and the effects of multiple scattering. All these effects will moderate the increase in RBE at the end of the proton range. However, the basic information on the change of mean lineal energy with depth obtained from microdosimetric measurements, together with related radiobiological information, should be incorporated in clinical practice when sufficient progress has been made in resolving the uncertainties which exist.

ACKNOWLEDGEMENTS

This work was funded by the Clatterbridge Cancer Research Trust, UK, the Institut Curie, Paris, and the Ligue Nationale de Lutte Contre le Cancer, France, to whom the authors express their gratitude. The contributions from colleagues at both Birmingham and Orsay is similarly warmly acknowledged.

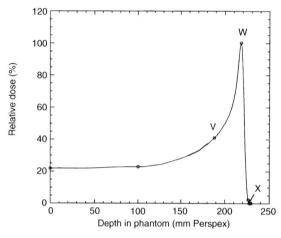

Figure 8. Relative dose as a function of depth in Perspex for the unmodulated 200 MeV beam: (○) microdosimetric measurement points.

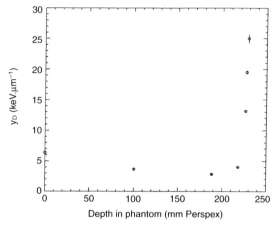

Figure 9. Variation of $\bar{y}_D$ with depth in Perspex for the unmodulated 200 MeV beam.

REFERENCES

1. Bettega, D. and Lombardi, L. *Physical and Radiobiological Parameters of Proton Beams up to 31 MeV*. Nuovo Cimento D**2**, 907–916 (1983).
2. Cosgrove, V. P., Aro, A., Green, S., Scott, M. C., Taylor, G. C., Bonnet, D. and Kacperek, A. *Studies Relating to 62 MeV Proton Cancer Therapy of the Eye*. Radiat. Prot. Dosim. **44**, 405–409 (1992).
3. Robertson, J., Williams, J., Schmidt, R., Little, J., Flynn, D. and Suit, H. *Radiobiological Studies of a High Energy Modulated Proton Beam Utilising Cultured Mammalian Cells*. Cancer **35**, 1664–1677 (1975).
4. Matsubara, S., Ohara, M., Hiraoka, J., Koike, S. and Ando, K. *Chromosome Aberration Frequencies Produced by a 70 MeV Proton Beam*. Radiat. Res. **123**, 182–191 (1990).
5. Courdi, A., Bramart, N., Herault, J. and Chauvel, P. *The Depth Dependent Radiation Response of Human Melanoma Cells Exposed to 65 MeV Protons*. Br. J. Radiol. **67**, 800–804 (1994).
6. Belli, M. *et al. Inactivation and Mutation Induction in V79 Cells by Low Energy Protons: Re-evaluation of the Results at the LNL Facility*. Int. J. Radiat. Biol. **63**, 331–337 (1993).

COMPARISON OF HETC AND PTRAN WITH PHANTOM MEASUREMENT DATA

R. Becker‡, J. Bienen||, P. Cloth†, M. Dellert§, V. Drüke†, W. Eyrich§, D. Filges†, M. Fritsch§, J. Hauffe§, U. Heinrichs‡, W. Hoffmann||, H. Kobus†, R. Maier†, M. Moosburger§, H. Paganetti‡, N. Paul†, H. P. Peterson‡, Th. Schmitz‡, K. Schwenke‡ and R. Sperl§
†IKP, Forschungszentrum Jülich GmbH (KFA), 52425 Jülich, Germany
‡IME, Forschungszentrum Jülich GmbH (KFA), 52425 Jülich, Germany
§Institute for Physics, Erlangen University, Germany
||Department of Physics, Wuppertal University, Germany

Abstract — For the treatment of several tumour types, calculated proton dose distributions predict an improvement in tumour response compared to conformal photon and electron radiotherapy. The dose can be delivered with much higher precision due to the limited proton range and the sharp distal fall-offs. For tumour treatment and dosimetric purposes, it is necessary to know exactly the proton range for a given energy in tissue or an equivalent medium such as water or acrylic. Different calculation codes, mainly based on the Bethe–Bloch formula which determines the energy loss due to Coulomb scattering, are currently used in proton dosimetry. They differ in the treatment of the nuclear collisions that also occur and in the simulated detector sizes. Sometimes, inelastic scattering is completely neglected or underestimated. Calculated dose distributions are presented for proton beams in water with a kinetic energy of around 175 MeV. Two Monte Carlo codes, PTRAN and HETC, are compared with recent experimental data from COSY, Jülich, and the discrepancies that occur are shown.

INTRODUCTION

It is predicted that the use of protons with kinetic energies between about 50 MeV and 250 MeV in cancer radiotherapy offers significant advantages, when compared to radiations conventionally employed. The reason is the characteristic depth–dose distribution of protons, which allows an improved conformation of the treatment dose to a target and, at the same time, a better sparing of healthy tissue. This has been shown for several tumour sites on the basis of comparative treatment planning calculations[1]. However, the dose calculation routines for protons in treatment planning programmes are quite simplistic and the results have to be interpreted with great care[2]. Monte Carlo calculation techniques could help to improve this situation and would enable further exploitation of the possible advantages of protons in high precision conformal radiation therapy[3].

When developing or applying a Monte Carlo code for dose calculation in treatment planning the main question is, how precise or how realistic are the achieved results. They are dependent on the nuclear data library used, the physical laws applied and the types of particle interactions considered. The best way to test the precision of dose calculations is the comparison with experiments that have been performed under known conditions and which are modelled by the calculations.

Measured depth–dose profiles have been compared with calculations performed with two different Monte Carlo codes. The measurements took place at the Low Energy Measurement Area of the Cooler Synchrotron (COSY) at KFA Jülich, Germany.

PROTON INTERACTIONS

The Monte Carlo codes use the knowledge of the different particle interactions with matter. When penetrating in matter such as tissue or water, protons interact in different ways with the atoms and molecules of the surrounding material. The main interaction occurring is electromagnetic scattering with shell electrons, the proton is decelerated and finally stopped. This process is well described by the Bethe–Bloch formula, which for this reason is the kernel of all Monte Carlo codes. The free electrons created in this process, δ electrons, are however usually not transported by the codes and their energy is locally deposited.

On their way, the protons are also laterally deflected by small angle Coulomb scattering. This leads to a broadening of the beam. The resulting penumbra is very important if the proton beam passes through inhomogeneities. Therefore, scattering must not be neglected in a detailed calculation.

Energy straggling is commonly included in the Monte Carlo programmes with a Gaussian, a Vavilov or a Landau energy loss distribution. This leads to a broadening of the Bragg peak in the energy deposition distribution. It influences the position and the shape of the distal edge of the peak.

Sometimes, especially on the high energy part of the proton path, nuclear reactions occur, in which secondary particles (mostly neutrons, but also protons, alpha particles, deuterons, tritons, etc.) are generated. Most Monte Carlo programmes do not take these reactions into account or they give only an estimation of their amount. However, there are some programmes which even transport these generated secondary particles.

It is obvious that differences in treatment of the physical interactions will lead to different results in the calculation of dose distribution and proton range. A first fit of calculated Bragg peaks to data from PSI, Switzerland[4], indicated a difference in the Monte Carlo programmes used of several per cent and made it clear that more precise experimental data are necessary for Monte Carlo code validation.

MONTE CARLO CODES USED

HETC (High Energy Transport Code), which is part of the HERMES package[5], was written for transport calculations in High Energy Physics. It is a 3-D transport programme. In a three-dimensional combinatorical geometry input, regions with different sizes and material compositions can be defined. With a variety of different detector modules, parameters like energy deposition, particle flux or energy spectra can be simulated. The detector size and the beam parameters like diameter and energy spread are variable and can be adjusted to the experimental set-up. Secondary particles like neutrons, alphas, deuterons and tritons are generated and transported. The energy of the emerging recoil nuclei is deposited locally.

PTRAN[6] is widely used in proton dosimetry. It is a 2-D transport programme and calculates the transport of proton pencil beams in water. The proton energy deposition and the proton spectra are always measured over the whole lateral field of the phantom; the detector size is not variable. Secondary particles are not produced, but the programme gives an estimation of the number of nuclear collisions which occurred on the proton path. Neither programme simulates δ electrons, but deposits their kinetic energy locally.

EXPERIMENT AND SIMULATION

Measurements took place at the Low Energy Experimental Area (NEMP) of COSY (Cooler Synchrotron) at KFA Jülich, Germany, which can deliver proton beams of kinetic energies of up to 2.5 GeV with a momentum spread of less than 0.5% ($\Delta p/p \leq 0.1\%$ for electron cooled beam). For these measurements protons with a beam momentum of p = 600 MeV/c were used, which is equal to a proton kinetic energy of 175.48 MeV.

In the water phantom, dose distributions were measured with an IC 18 ionisation chamber and with TLD-300 (see also Ref. 7 for further information).

The measured dose distributions are shown in Figure 1. The dots represent the data from the ionisation chamber measurement and the TLD data is printed as squares.

The beam and detector parameters for HETC (broken line) were close to the real parameters: $\Delta p/p = 0.5\%$, beam radius $\sigma(r) = 2.7$ mm and the size of the cylindrical detector was 1 mm in depth and a diameter d = 3.8 mm. For PTRAN (dotted line), the energy and the momentum spread were chosen accordingly. For both programs, the position of the Bragg peak was at a greater depth in the phantom as compared to the experiment. An additional HETC simulation with a reduced energy of E = 172.9 MeV was necessary in order to match the experimental peak position. All curves were normalised to the peak of the TLD measurement. The doses are given in arbitrary units since a normalisation to the number of primary protons was impossible due

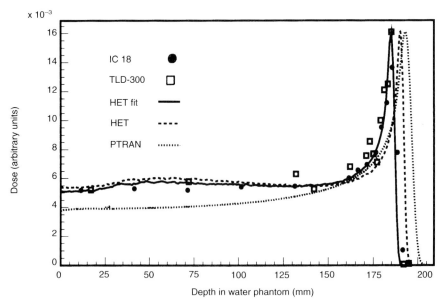

Figure 1. Measured and simulated depth–dose distributions in the water phantom.

to dead time problems of the scintillator starting counters which led to an underestimation of the number of protons per spill.

The uncertainty of the detector position was in the range of 1–2 mm.

RESULTS OF THE WATER PHANTOM MEASUREMENTS

The calculated Bragg peak position differs from the measured one; both HETC and PTRAN simulate a deeper beam penetration at the energy used in the experiment. HETC is 5 mm or 2.7% deeper, PTRAN 7 mm or 3.7%. The differences between the two programs can be explained by the use of different energy range tables. In an additional test run of both Monte Carlo codes at a proton kinetic energy of 250 MeV, the positions of the Bragg peaks showed a difference of nearly 2 cm. Because the range tables are an essential part of all therapy planning programs, great store should be set by the development of precise data.

The shift of both programs towards higher ranges compared to the experimental data must be investigated further.

The PTRAN Bragg curve has a too high peak-to-plateau ratio, if compared to the HETC and the experimental Bragg curve. This can be explained due to PTRAN's laterally integrating energy deposition detector. Whereas HETC simulates a detector with a size in the range of the beam diameter, PTRAN also counts protons which were scattered out of the region where the detector in the experiment measured. The slower rise in the plateau region of the curve of the HETC calculation and the experimental data can be explained with this multiple scattering effect. Using a detector with an infinite lateral size also leads to a broader Bragg peak; FWHM (PTRAN) = 16.5 mm, FWHM(HETC) = 10.5 mm. This fact limits the use of PTRAN to the simulation of data acquired with broad detector sytems and excludes the validation of data from small detector measurements.

OUTLOOK

Although the depth–dose distribution is an important parameter of interest in proton therapy, it is not suitable for application in biophysical calculations. Here, more detailed measurements like particle spectra, lateral beam distributions and other event-by-event taken data are required. First test measurements were therefore performed with a Silicon microstrip detector developed at the Erlangen University. The microstrip device contains two crossed linear detectors with an active area of 2×2 cm^2 and pitches of 200 μm. The energy loss of the protons was measured in the silicon behind an acrylic phantom, which is variable in thickness. The ADC spectra obtained at different phantom thicknesses are shown in Figure 2. With increasing depth, the mean energy loss of the protons shifts to a higher value while the peak flattens out. A comparison of these data with the simulation is not fully finished, but will give a more detailed look to the precision of Monte Carlo codes.

SUMMARY

The measurements showed that high precision data acquisition is necessary for Monte Carlo code validation. When using small detector devices, PTRAN is not suitable to reproduce the results. The outcome of the HETC simulation is very promising, although the shift towards higher depths needs to be further investigated at other proton beam energies. More measurements at different energies are planned at COSY in the first half of 1996. The plastic phantom data will be further analysed. The existing Monte Carlo codes will be used to validate the obtained ADC spectra. Further calculations will be extended to three dimensions, to simulate the dose distributions in the plastic and the water phantom. The development of a Monte Carlo based treatment planning program will be continued.

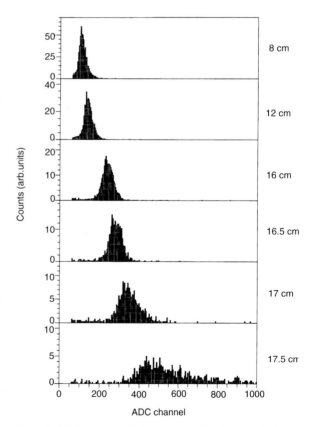

Figure 2. ADC spectra of different acrylic phantom thicknesses.

REFERENCES

1. Tatsuzaki, H., Urie, M. and Linggood, R. *Comparative Treatment Planning: Proton Vs. X-Ray Beams Against Glioblastoma Multiforme.* NCI Proton Workshop. Int. J. Radiat. Oncol. Biol. Phys. **22**, 265–273 (1992).
2. Urie, M. *Treatment Planning for Proton Beams.* In: Ion Beams in Tumor Therapy, Ed. U. Linz (London: Chapman & Hall) (1995).
3. White, R. M., Chadwick, M. B., Chandler, W. P., Hartmann-Siantar, C. L., Resler, D. A. and Weaver, K. A. (Unpublished observations).
4. Scheib, S. Private communications.
5. Cloth, P. *et al.* HERMES — *A Monte Carlo Program System for Beam Materials Interaction Studies.* Report Jülich-2203 (1988).
6. Berger, M. *Proton Monte Carlo Transport Program PTRAN.* National Institute of Standards and Technology Publication NISTIR 5113 (1993).
7. Becker, R. *et al. Biophysical Investigations of Therapeutic Proton Beams.* Radiat. Prot. Dosim. **70**(1–4), 485–492 (This issue) (1997).

INVESTIGATION OF EFFICIENCY OF THERMOLUMINESCENCE DETECTORS FOR PARTICLE THERAPY BEAMS

P. Bilski[†], M. Budzanowski[†], W. Hoffmann[‡], A. Molokanov[§], P. Olko[†] and M. P. R. Waligórski[†‖]
[†]Institute of Nuclear Physics, ul. Radzikowskiego 152, PL-31342 Kraków, Poland
[‡]University of Wuppertal, Physics Dept, Gaussstr. 20, D-42097 Wuppertal, Germany
[§]Joint Institute of Nuclear Research, Dubna, Russia
[‖]Centre of Oncology, ul. Garncarska 11, 30-115, Kraków, Poland

Abstract — Dosimetric properties of thermoluminescence detectors produced at the Institute of Nuclear Physics were investigated with respect to their possible application in the dosimetry of proton radiotherapy beams. Relative TL efficiency, η, was investigated with 5.3 MeV alpha particles for

(i) CaF_2:Tm (TLD-300 from Harshaw);
(ii) LiF:Mg,Ti (MTS-N) sintered pellets with Ti content ranging from 4.3 to 130 ppm;
(iii) LiF:Mg,Cu,P sintered pellets (MCP-N);
(iv) thin layer DA-2 ($CaSO_4$:Dy), and CaF_2:Tm plated on 0.1 mm aluminium foil. The highest values of η were found for LiF:Mg,Ti pellets with increased Ti content and TLD-300. The dose–response of all detectors was studied for 0.05, 0.5 and 5 Gy of ^{137}Cs gamma rays. LiF:Mg,Ti with increased Ti concentration and CaF_2:Tm detectors were found to be linear in this dose range, while DA-2 and standard MTS-N detectors are supralinear and MCP-N is sublinear. MTS-N detectors were used to measure the dose–depth distribution in water for a 200 MeV proton beam, and showed good agreement with the readings of an ionisation chamber.

INTRODUCTION

The application of thermoluminescence dosemeters (TLDs) in charged particle clinical beam dosimetry is difficult due to the strong dependence of their response on radiation type, energy and dose level.

At present, Harshaw CaF_2:Tm (TLD-300) and LiF:Mg,Ti (TLD-100, TLD-700) are most frequently used and recommended for dose measurements in neutron and proton radiotherapy[1–3]. The advantages of these phosphors are twofold. First, their TL efficiency for densely ionising particles is relatively high. Second, different dependence on ionisation density of certain glow peaks is observed. Glow curves of both phosphors can roughly be divided into 'main dosimetric peak' and 'high temperature peak' regions, each consisting of many overlapping peaks. TL peaks in these two regions exhibit a different dependence on ionisation density, so the ratio between 'high' and 'low' temperature regions can be used for estimating the LET of the radiation or for discriminating between high and low LET components of a mixed field[4,5]. However, any implementation of these findings into a dosimetric protocol is impractical, due, for example, to the strong dependence of response on production batch or handling procedures. Also, standard chips are too thick (typically 0.9 mm) to resolve the sharp dose gradients in the Bragg peak region.

The aim of this work was to study the dosimetric properties of some TL detectors recently developed at the Institute of Nuclear Physics (INP), Kraków, Poland, with respect to their applicability in the dosimetry of clinical proton beams, and to compare their properties with those of other commercially available INP detectors such as MTS-N (LiF:Mg,Ti), MCP-N (LiF:Mg,Cu,P), DA-2 ($CaSO_4$:Dy on Al foil) and Harshaw TLD-300 (CaF_2:Tm). Among the newly developed detectors were:

(i) thin-layer CaF_2:Tm (CFT) bonded with silicone resin to 0.1 mm Al foil, specially designed for accurate measurements of dose distributions within the Bragg peak;
(ii) a family of LiF:Mg,Ti sintered pellets with different concentrations of Ti dopant (changed in order to adjust the dose and LET response[6]).

METHODS

Definitions

Relative TL efficiency, η, is defined as the ratio of measured TL signal per unit absorbed dose of a given radiation type normalised to the TL signal per unit absorbed dose of a reference radiation, usually ^{137}Cs gamma rays. The method of determining the value of η has been described elsewhere[7]. The linearity factor, f(D), for a given (high) dose D is defined as the ratio of measured TL signal per delivered dose, normalised to this ratio at a low dose.

TL detectors, handling procedure and readout

DA-2 detectors are circular (6 mm diam.) and consist

of a thin layer of CaSO$_4$:Dy powder (grain size 10–20 μm) bonded with silicone resin to 0.1 mm thick aluminium foil. These detectors are routinely used in Polish coal mines for environmental monitoring of radon daughter products. Using the same technique, thin-layer CaF$_2$:Tm detectors (here named CFT) were prepared using powder obtained by crushing TLD-300 chips. The effective thickness of the active layer of CFT detectors, assessed by comparing their photon sensitivity with that of standard TLD-300 detectors, is 7 mg.cm^{-2}. DA-2 and CFT detectors were annealed at 300°C for 1 h so as not to destroy the silicone resin. MTS-N (LiF:Mg,Ti) are sintered pellets of diameter 4.5 mm, thickness 0.7 mm, their sensitivity, energy and dose response closely resembling those of Harshaw TLD-100. MTS-N detectors were annealed for 1 h at 400°C. In addition to standard pellets (120 ppm Mg and 13 ppm Ti) three batches of 100 detectors each were prepared with different Ti concentration (4.3, 39, and 130 ppm). MCP-N (LiF:Mg,Cu,P) are sintered pellets of the same size as MTS-N, with standard activator concentrations (0.2% Mg, 0.05% Cu and 1.25% P), annealed for 10 min at 240°C. TLD-300 (CaF$_2$:Tm), square chips from Harshaw, were annealed for 1 h at 400°C. All detectors were additionally annealed for 10 min at 100°C before readout.

Readouts were performed with a RA'94 reader, using a linear heating ramp at a rate of 5°C.s^{-1} to a maximum temperature of 400°C (330°C for detectors with silicone resin). Glow curve analysis was by TL signal integration in regions of interest, i.e. for MTS-N, main peak (sum of peaks 3 + 4 + 5) and high temperature peak (peaks 6-10); for CaF$_2$:Tm, main peak 3 and high temperature peak (peaks 4 + 5), and all of the glow curves for DA-2 and MCP-N. It is in principle possible to unfold individual glow peaks using one of the available deconvolution techniques. However, results are often uncertain and such a procedure cannot be recommended for routine applications.

Irradiations

Photon exposures were performed with a ^{137}Cs source at the calibration stand of the INP Kraków. Values of absorbed dose in tissue were 0.05, 0.5 and 5 Gy. During irradiations detectors were placed in PMMA holders to ensure charged particle equilibrium. α particle irradiations were performed using an Amersham AMR-33 α particle source, containing ^{214}Am, ^{244}Cm and ^{239}Pu isotopes which emits alpha particles of average energy 5.3 MeV. The energy of α particles was measured using a Canberra 7401 α spectrometer with a 300 mm^2, FWHM = 50 keV, surface-barrier Si detector. The proton irradiation was performed at the Joint Institute of Nuclear Research in Dubna (Russia) at the Phasotron 200 MeV medical beam. Dose in the water phantom was measured with a clinical dosemeter KD-27012 with an air-filled thimble ionisation chamber VAK-253, sensitive volume 50 mm^3 (produced by VEB RFT Messelektronik, Dresden). Dosemeter calibration was made with a ^{60}Co source in accordance with the recommendations of the 'Code of Practice for Clinical Proton Dosimetry'[8].

RESULTS

In Figure 1 glow curves of thin-layer CaF$_2$:Tm (CFT) and standard TLD-300 detectors are compared after irradiations with 50 mGy of ^{137}Cs gamma rays. After α particle irradiation, the ratio of peak 5 to peak 3, P$_{5/3}$, for thick TLD-300 is higher (P$_{5/3}$ = 4.0) than that for thin-layer detectors (P$_{5/3}$ = 2.9). In fact, stopping α particles in TLD-300 have a higher average ionisation density than α particles crossing the thin detector layer of CFT (track segment irradiation). The observed tempera-

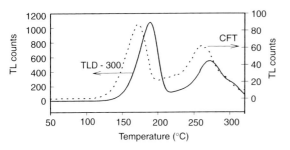

Figure 1. Glow curves of TLD-300 and thin-layer CaF$_2$:Tm detectors, for α particles and ^{137}Cs gamma rays.

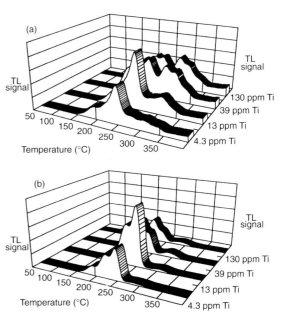

Figure 2. Glow curves of LiF:Mg,Ti with different Ti concentration for (a) α particles and (b) ^{137}Cs gamma rays.

ture shift between glow curves results from the faster heating of thin detectors.

The glow curve structure of LiF:Mg,Ti depends strongly on Ti concentration, radiation LET and dose level. In Figure 2 glow curves of LiF:Mg,Ti with different Ti content are compared after 500 mGy of ^{137}Cs γ rays and a 7×10^6 cm^{-2} fluence of α particles. With increasing Ti concentration the main dosimetric peak broadens due to the growth of peaks 3 and 4 with respect to peak 5. The amplitude of high temperature peaks increases, but these peaks could be even better separated from the main dosimetric peaks.

In Table 1 the measured values of TL efficiency η for the set of detectors available are displayed. The high TL efficiency of peak (4 + 5) in TLD-300, equal to 0.56 for 5.3 MeV α particles is consistent with the data of Hoffmann and Prediger[9], who reported η = 1 for charged particles with ionisation density y_D = 100 keV.μm^{-1} and η ≈ 0.5 for y_D = 200 keV.μm^{-1}. It is apparent that the increase of Ti concentration from 13 to 130 ppm resulted in a desirable twofold increase of main peak efficiency for α particles, which reaches η = 0.61, a value even greater than that obtained for TLD-300. Decrease of Ti content below 13 ppm has a negligible effect on η. The relative efficiency of high temperature peaks depends weakly on the amount of Ti, but is generally remarkably high.

In Table 2 linearity factors for doses 0.5 and 5 Gy, normalised to the response for 50 mGy of ^{137}Cs, are presented. No supralinearity in the response of the main peak of TLD-300 is observed up to 5 Gy. For LiF:Mg,Ti, increasing the amount of Ti improves the linearity of the response at higher doses. While for low Ti concentration (4.3 and 13 ppm) the supralinearity factor of the main dosimetric peak does not exceed 1.04 for 5 Gy, for higher Ti concentration perfectly linear response is observed over that dose region. This observation is consistent with the previous data of Rossiter et al[6], who reported an increase of supralinearity with decreasing Ti concentration. High temperature peaks exhibit strong supralinearity for all Ti contents reaching f(5 Gy) = 4 − 5.

LiF:Mg,Ti detectors with standard Ti concentration (13 ppm) were used for preliminary measurements of depth–dose distribution in a water phantom irradiated by a 200 MeV therapeutic proton beam from the Phasotron of the Joint Institute of Nuclear Research in Dubna, Russia. In Figure 3 readings of LiF:Mg,Ti detectors (main dosimetric peak), re-calculated to dose in tissue after calibration with ^{137}Cs gamma rays, are compared with results obtained with an air-filled, thimble ionisation chamber. The agreement between results is fairly good, including the Bragg peak region. On the same plot the ratio between the low and high temperature peak readings is plotted against depth in phantom, which could help in assessing the quality (LET) of the beam.

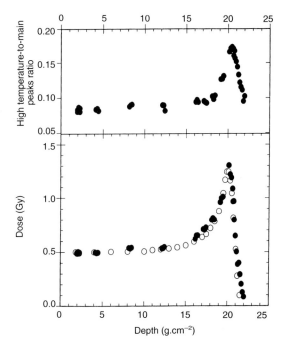

Figure 3. (a) Depth–dose curves of 200 MeV proton beam measured with LiF:Mg,Ti (●) and an ionisation chamber (○); (b) ratio of high temperature to main peak as a function of depth in phantom.

Table 1. Measured TL efficiency and high temperature-to-main peak ratio, for 5.3 MeV α particles.

Detector	Relative TL efficiency, η		Ratio: high temperature -to-main peak
	Main peak	High temperature peak	
LiF:Mg,Cu,P (MCP-N)	0.063 ± 0.007	—	—
CaF$_2$:Tm (TLD-300)	0.11 ± 0.01	0.56 ± 0.06	4.77 ± 0.45
LiF:Mg,Ti, Ti = 4.3 ppm	0.38 ± 0.04	3.8 ± 0.4	0.39 ± 0.03
LiF:Mg,Ti, Ti = 13.0 ppm	0.33 ± 0.04	4.9 ± 0.6	0.27 ± 0.05
LiF:Mg,Ti, Ti = 39.0 ppm	0.51 ± 0.07	5.4 ± 0.6	0.66 ± 0.02
LiF:Mg,Ti, Ti = 130.0 ppm	0.61 ± 0.07	6.4 ± 0.7	0.72 ± 0.06

DISCUSSION AND CONCLUSION

An ideal TL detector for dosimetry in charged particle radiotherapy beams should show a linear response over the dose range 0.1–5 Gy, relative TL efficiency close to 1 for high LET particles and the possibility of determining radiation quality from the ratio of glow curve peak areas. As these characteristics are interrelated, i.e. the degree of supralinearity and TL efficiency depend on dose and ionisation density, it may be difficult or impossible to find such a TL detector. TLD-300 features several advantages such as a lack of supralinearity up to 5 Gy and high TL efficiency for densely ionising particles. However, CaF_2:Tm cannot be used for measurements in the presence of scattered, low energy photons due to a 12 fold over-response resulting from its high effective atomic number. The possibility of producing thin-layer CaF_2:Tm detectors which can be used to measure high dose gradients of dose in the Bragg peak region of a proton beam, was shown in this work.

Lithium fluoride, with $Z_{eff} = 8.2$, is more appropriate for dose measurements in radiation fields containing a photon component; however, standard LiF:Mg,Ti detectors show a decrease of TL efficiency for high LET particles. Preliminary measurements of depth dose distribution in water produced by 200 MeV protons performed with standard LiF:Mg,Ti pellets, showed good agreement with readings of an ionisation chamber. This agreement was possible due to the large straggling of 200 MeV protons after penetrating a depth of 20 cm in water, resulting in a spreading of stopping protons within the Bragg peak. For measurements of particles with higher LET, e.g. protons of energy about 60 MeV or heavy ions, LiF:Mg,Ti detectors with higher content of Ti are recommended because they demonstrate the onset of supralinearity at higher doses and much higher TL efficiency for such radiations. More work should be devoted to optimise these detectors for the dosimetry of charged particle radiotherapy beams.

ACKNOWLEDGEMENTS

The authors would like to acknowledge the kind travel support from EURADOS (P.O., P.B.) and from the local organisers (M.B.) which enabled us to participate in this Symposium.

Table 2. Dose response per unit dose for TL dosemeters, relative to dose response after 50 mGy of ^{137}Cs gamma rays.

Detector	Relative response			
	Main dosimetric peak		High temperature peak	
	0.5 Gy	5 Gy	0.5 Gy	5 Gy
LiF:Mg,Cu,P (MCP-N)	1.00 ± 0.01	0.91 ± 0.01	—	—
$CaSO_4$:Dy (DA-2)	1.09 ± 0.03	1.23 ± 0.04	—	—
CaF_2:Tm (TLD-300)	1.02 ± 0.01	0.96 ± 0.01	1.03 ± 0.02	1.27 ± 0.04
LiF:Mg,Ti, Ti = 4.3 ppm	1.01 ± 0.005	1.03 ± 0.01	1.6 ± 0.1	4.7 ± 0.3
LiF:Mg,Ti, Ti = 13.0 ppm	1.00 ± 0.007	1.045 ± 0.007	1.7 ± 0.05	5.1 ± 0.1
LiF:Mg,Ti, Ti = 39.0 ppm	1.00 ± 0.004	1.00 ± 0.009	1.6 ± 0.02	4.2 ± 0.06
LiF:Mg,Ti, Ti = 130.0 ppm	1.00 ± 0.01	1.00 ± 0.01	1.6 ± 0.03	3.2 ± 0.08

REFERENCES

1. Fattibene, P., Calicchia, A., d'Errico, F., De Angelis, C., Egger, E. and Onori, S. *Preliminary Assessment of LiF and Alanine Detectors for the Dosimetry of Proton Therapy Beams.* Radiat. Prot. Dosim. **66**, 305–309 (1996).
2. Carlsson, C. A. and Carlsson, A. C. *Proton Dosimetry: Measurement of Depth Dose from 185 MeV Protons by Means of Thermoluminescent LiF.* Radiat. Res. **42**, 207–219 (1970).
3. Momeni, M. H., Cahill, T. A. and Horn, P. L. *Proton Irradiation of Beagle Eyes. I. Si(Li) and Thermoluminescent Dosimetry of 20, 35 and 45 MeV Protons.* Radiat. Res. **53**, 15–23 (1973).
4. Pradhan, A. S. and Rassow, J. *Radiation Induced Thermoluminescence in CaF_2:Tm Detectors.* Nucl. Instrum. Methods, **A255**, 234–237 (1987).
5. Loncol, T., Vynckier, S. and Wambersie, A. *Thermoluminescence in Protons and Fast Neutron Therapy Beams.* Radiat. Prot. Dosim. **66**, 299–304 (1996).
6. Rossiter, M. J., Rees-Evans, D. B., Ellis, S. C. and Griffiths, J. M. *Titanium as a Luminescence Centre in Thermoluminescent Lithium Fluoride.* J. Phys. D: Appl. Phys. **4**, 1245 (1971).
7. Bilski, P., Olko, P., Burgkhardt, B., Piesch, E. and Waligorski, M. P. R. *Thermoluminescence Efficiency of LiF:Mg,Cu,P (MCP-N) Detectors to Photons, Beta Electrons, Alpha-Particles and Thermal Neutrons.* Radiat. Prot. Dosim. **55**, 31–38 (1994).
8. Kovar, I. *Code of Practice for Clinical Proton Dosimetry.* JINR Communication E-16-93-310, Dubna (1993).
9. Hoffmann, W. and Prediger, B. *Heavy Particle Dosimetry with High Temperature Peaks of CaF_2:Tm and ^{7}LiF Phosphors.* Radiat. Prot. Dosim. **6**, 149–152 (1983).

MEASUREMENT OF THE HEAT DEFECT IN WATER AND A-150 PLASTIC FOR HIGH ENERGY PROTONS, DEUTERONS, AND α PARTICLES

H. J. Brede, O. Hecker and R. Hollnagel
Physikalisch-Technische Bundesanstalt (PTB)
Bundesallee 100, D-38116 Braunschweig, Germany

Abstract — An experimental set-up is described which allows the calorimetric determination of the heat defect of solids and fluids relative to that of gilded copper which is used as the reference material. The calorimeter operates in a quasi-adiabatic mode and is suitable for protons, deuterons, α particles and other heavy ions with energies above 5 MeV. Corrections have been made for secondary electron emission and also for heat losses due to the temperature gradient on the surface of the calorimeter core. Measurements over 26 months indicate a repeatability of the measured heat defect within 4×10^{-3} (68% confidence level). Measurements show an LET dependence of the heat defect for two different samples of A-150 plastic, and for distilled water.

INTRODUCTION

To date the determination of absorbed dose to tissue from fast neutrons at the PTB has been based on measurements with tissue-equivalent ionisation chambers which were calibrated in the well-known photon reference fields of the PTB. Calorimetric methods seem best suited to improve the accuracy of absorbed dose determinations and, in particular, for the practical implementation of a primary standard for absorbed dose to tissue. A calorimeter which uses the tissue substitute A-150 plastic is in operation at the Laboratoire Primaire des Rayonnements Ionisants (LPRI)[1]. PTB is developing a water calorimeter as the primary standard for absorbed dose to tissue for several kinds of radiation.

The use of an absorbed dose calorimeter requires the heat defect of the absorber to be accurately determined from an independent experiment. The heat defect, h, describes that fraction of the absorbed energy which does not result in a temperature change of the absorber. It is defined by the equation:

$$h = (W_a - W_h)/W_a \qquad (1)$$

where W_a represents the energy absorbed and W_h the energy that appears as heat.

Experiments with photons and electrons[2-4] showed that highly purified water exhibits no heat defect and that the heat defect is reproducible for well-defined water qualities. In mixed neutron–photon fields the absorbed dose to water is generated by secondary electrons and charged particles such as protons, deuterons, α particles, and recoil oxygen nuclei. The various particles produce markedly different ionisation densities along their tracks and this can be described by their linear energy transfer, LET. Knowledge of the LET dependence of the heat defect in water is therefore a prerequisite for the water calorimeter in order to reduce the main component of the uncertainty in water calorimetry and to apply a correction factor in mixed neutron–photon fields.

This calorimeter has been especially designed for the determination of the heat defect in fluids and solids for charged particles with energies above 5 MeV. In a first series of measurements, the heat defect of water and A-150 plastic relative to a reference material was determined. At present, gilded copper serves as the reference material and it is assumed that its heat defect is zero. The design of the calorimeter does, however, also allow the heat defect to be measured on an absolute scale, independent of the reference material.

EXPERIMENTAL SET-UP AND MEASUREMENTS

The measurement principle is based on the total energy loss of charged particles in a composite core of a calorimeter. Similar methods have already been used[5-7]. The material under investigation — in this case A-150 plastic or water — is contained in a cylindrical canister of gilded copper of an inner diameter of 15 mm and 5 mm in length, and which has an eccentric beam stop (see Figure 1). Two different A-150 plastic absorbers from different production lots are investigated. All samples have the same geometrical dimensions — a cylindrically shaped probe, 3.5 mm in thickness and 15 mm in diameter. They are pressed into the canister. A 7 μm thick indium-sealed molybdenum entrance window allows charged particles to enter the core.

The calorimeter (Figure 1) comprises a core, inner and outer jacket, and vacuum chamber which are thermally and electrically insulated from one another and operates in an adiabatic mode. It is directly flanged to the beam line and can be shifted laterally, i.e. perpendicular to the direction of the incident charged particle beam, for two different types of measurement. In the first case, the beam reaches the A-150 plastic or the water through the window. The temperature increase, ΔT_i (i indicates either A-150 or water) measured on the core surface is given by

$$\Delta T_i C (1 - h_i)^{-1} = Q_b U_b \qquad (2)$$

where C is the total heat capacity of the calorimeter core, h_i the heat defect in either A-150 plastic or water, Q_b the total beam charge deposited in the core and U_b the voltage which is used to accelerate the particles, i.e. the product $Q_b U_b$ is the total energy imparted in the absorber. In the other case, the beam is stopped in the gilded copper and the measured temperature increase of the core surface, ΔT_m, is given by

$$\Delta T_m C (1 - h_m)^{-1} = Q_b U_b \qquad (3)$$

where h_m represents the heat defect of gilded copper. The temperature increases (for A-150, water, and gilded copper) are each measured with the same sensor and at the same location on the core surface. Since the temperature measurements are needed for both types of experiments, differences in the temperature increase over time on the surface of the gilded copper canister must be considered. By using the Equations 1, 2 and 3, the quantity $1 - (\Delta T_i / \Delta T_m)$, which is the heat defect in water or A-150 plastic, respectively, can be determined provided that heat losses of the calorimeter core surface can be corrected for and that the heat defect h_m of gilded copper is zero.

In order to compensate for the heat transfer of the

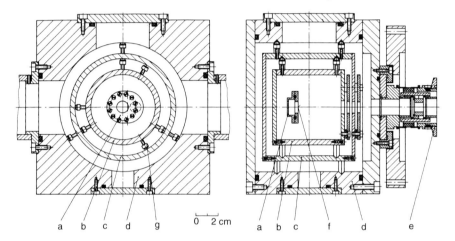

Figure 1. Core (a), inner jacket (b) and outer jacket (c), vacuum chamber (d), and flange connection to the beam line (e), entrance window (f), and screws (g) to align the core with nylon cords (not shown).

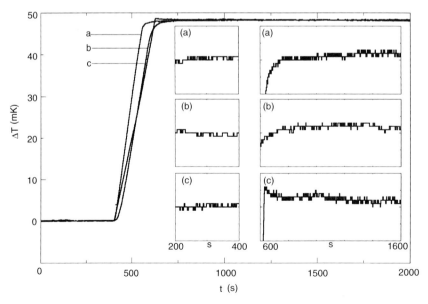

Figure 2. Time (t) plotted against temperature (ΔT) response of the calorimeter. Beam stop: A-150 plastic (a), water (b), and gilded copper (c). The inserts on the left hand side show the initial temperature drifts, the inserts on the right hand side the response after the end of irradiation. The temperature scale of all inserts is 2 mK.

vacuum chamber — at ambient temperature of 292.0 ± 0.5 K — to the outer jacket, the temperature of this jacket is stabilised to within 5 mK. The core and inner jacket act as a Faraday cup for the incident charged particles. The influence on the charge reading of the core resulting from secondary electrons which are produced on the surface of the core and on the apertures of the outer shields has been carefully studied. A negative bias voltage of 300 V between the inner jacket and the core reduces this effect to less than 1% of the core's charge.

The calorimeter design and the alignment of the beam prevent either jacket from being struck by the primary charged particles. The incident beam current is reduced to approximately 500 pA with two pairs of slits. Two quadrupole lenses and steering magnets focus the beam onto the calorimeter core. A system of apertures 11 mm in diameter for the inner jacket, 10 mm for the outer jacket, and of 9 mm in diameter in front of the calorimeter localises the beam, reduces scattering and provides diagnostic information. The pressure in the beam line and the calorimeter is maintained at less than 10^{-2} Pa.

The temperature measurements of the core and the jackets are made with cylindrically shaped thermistors (NTC) 0.35 mm in diameter and 0.5 mm in length (type 6331 from Philips) and are connected to thin copper wires 0.06 mm in diameter and 30 cm in length. The NTC comprises one leg of a DC Wheatstone Bridge followed by a DC amplifier with about 10^3 times voltage amplification. The amplified DC voltages from the inner jacket and core, as well as the temperature of the outer jacket, vacuum chamber and ambient room, are recorded with analogue-to-digital converters with 12 bit resolution. In addition, the charge collected by the apertures and the calorimeter core are recorded during the entire measurement. All data are transferred to the main computer in real time.

Table 1. Heat defect in A-150 plastic irradiated with protons (p), deuterons (d), and α particles (α) for different samples from LPRI and PTB. LET is the linear energy transfer. S (secondary electrons), R (reactions), T (temperature gradient), and C (coulomb scattering) are correction factors. The uncertainties (last digits, in parentheses) correspond to a 68% confidence level.

Beam	Energy (MeV)	LET range (keV.μm^{-1})	Correction factors				Heat defect (10^{-2})	Remark
			S-1 (10^{-3})	R-1 (10^{-4})	T-1 (10^{-4})	C-1 (10^{-5})		
p	19.1	3–100	3.2 (8)	−16 (8)	3 (2)	5.3 (8)	1.1 (2)	PTB
p	17.2	4–100	3.2 (8)	−12 (4)	3 (2)	6.0 (8)	3.2 (1)	LPRI
p	10.1	5–100	3.0 (8)	−2.5 (8)	3 (2)	8.0 (8)	2.3 (1)	PTB
p	4.65	8–100	3.0 (8)	−0.3 (2)	3 (2)	9.7 (8)	2.8 (3)	PTB
d	12.8	6–96	6 (1)	−12 (4)	3 (2)	3.9 (8)	3.6 (3)	PTB
d	6.35	13–96	6 (1)	−2.0 (6)	3 (2)	7.5 (9)	4.0 (2)	PTB
d	4.87	14–96	6 (1)	−1.0 (5)	3 (2)	8.0 (9)	4.8 (1)	LPRI
d	4.75	14–96	6 (1)	−1.0 (5)	3 (2)	8.0 (9)	3.3 (4)	PTB
α	25.5	27–260	10 (3)	−13 (5)	3 (2)	1.8 (6)	5.9 (3)	LPRI
α	24.4	27–260	10 (3)	−13 (5)	3 (2)	1.8 (6)	4.2 (3)	PTB
α	8.80	60–260	10 (3)	−0.3 (1)	3 (2)	4.5 (8)	7.9 (4)	LPRI
α	8.60	60–260	10 (3)	−0.3 (1)	3 (2)	4.5 (8)	4.5 (3)	PTB

Table 2. Heat defect in the distilled water used at PTB with protons (p), deuterons (d), and α particles (α). LET is the linear energy transfer. S (secondary electrons), R (reactions), T (temperature gradient), and C (coulomb scattering) are correction factors. The uncertainties (last digits, in parentheses) correspond to a 68% confidence level.

Beam	Energy (MeV)	LET range (keV.μm^{-1})	Correction factors				Heat defect (10^{-2})
			S-1 (10^{-3})	R-1 (10^{-4})	T-1 (10^{-4})	C-1 (10^{-5})	
p	19.00	3–100	2.3 (4)	−16 (6)	3 (2)	5.3 (8)	0.4 (3)
p	4.65	8–100	1.8 (5)	−0.2 (3)	3 (2)	9.7 (9)	2.0 (3)
d	12.90	6–96	2.5 (4)	−12 (4)	3 (2)	3.9 (8)	1.8 (2)
d	3.27	14–96	1.8 (5)	−1.0 (5)	3 (2)	8.5 (9)	1.9 (4)
α	24.40	25–250	7.5 (8)	−13 (5)	3 (2)	1.8 (8)	2.6 (2)
α	5.40	70–250	3.0 (6)	−0.2 (1)	3 (2)	5.3 (8)	4.9 (7)

The calorimetric measurement starts with an analysis of the initial drift for about 400 s. After an irradiation time of 200 s, the system equilibrates during a time of approximately 500 s. During the next 500 s, the final temperature drift is recorded (see Figure 2). The temperature increase is determined from the linear region of the initial and final drifts and is normalised to the beam charge collected from the core. The measurements are carried out alternately with gilded copper and water or A-150 plastic as the beam stop. The quantity $(\Delta T_i/\Delta T_m)$ is determined from sequential measurements in order to reduce systematic errors.

About 150 measurements for A-150 plastic over the course of 10 months and about 200 measurements for water over the course of 26 months were performed with beams of different particles and of various energies. The core was loaded with distilled water or A-150 plastic, respectively, at the beginning of the measurements and its content was not replaced during the whole period. The beam power was about 4 mW for all irradiations. The repeatability of the mean value from different measurement campaigns that were carried out at different times was within 0.4% (68% confidence level).

DISCUSSION OF THE RESULT

The LET dependence of the heat defect was studied for two different samples of A-150 plastic and for distilled water. The results are summarised in Tables 1 and 2 and indicate that A-150 plastic and water have positive heat defects relative to that of gilded copper. The heat defect increases with increasing LET for A-150 plastic as well as for the water used. There is a significant difference in the heat defect of A-150 plastic between LPRI's and PTB's samples which can only be explained by the different production lots of the samples under investigation.

Accumulated doses of up to approximately 50 kGy were delivered to the different lots. Within the uncertainty of this method no dependence of the heat defect values on the accumulated dose in the absorber was observed.

Experiments are in progress in order to verify the assumption that gilded copper has a negligible heat defect. In these experiments, the heat defect will be measured on an absolute scale by comparing the dissipation of heat in the calorimeter from various beams and from ohmic heating elements.

REFERENCES

1. Caumes, J. and Simoen, J. P. *TE-calorimeter as a Primary Standard for Neutron Absorbed Dose Calibrations*. J. Eur. Radiothér. **5**, 235–239 (1984).
2. Ross, C. K., Klassen, N. V., Shortt, K. R. and Smith, G. D. *Water Calorimetry with Emphasis on the Heat Defect*. NRC-Publ. 29637 (Ottawa: NRC) pp. 69–76 (1988).
3. Roos, M., Grosswendt, B. and Hohlfeld, K. *An Experimental Method for Determining the Heat Defect of Water Using Total Absorption of High-Energy Electrons*. Metrologia **29**, 59–65 (1992).
4. Selbach, H. J., Hohlfeld, K. and Kramer, H. M. *An Experimental Method for Measuring the Heat Defect of Water Using Total Absorption of Soft X-Rays*. Metrologia **29**, 341–347 (1992).
5. Fleming, D. M. and Glass, W. A. *Endothermic Processes in Tissue-Equivalent Plastic*. Radiat. Res. **37**, 316–322 (1969).
6. McDonald, J. C. and Goodman, L. J. *Measurements of the Thermal Defect for A-150 Plastic*. Phys. Med. Biol. **27**, 229–233 (1982).
7. Schulz, R. J., Venkataramanan, N. and Saiful Hug, M. *The Thermal Defect of A-150 Plastic and Graphite for Low-Energy Protons*. Phys. Med. Biol. **35**, 1563–1574 (1990).

3D TREATMENT PLANNING FOR 14 MeV NEUTRONS

R. Schmidt, T. Frenzel, A. Krüll, L. Lüdemann and T. Matzen
Abteilung Strahlentherapie
Universitäts-Krankenhaus Eppnedorf
Universität Hamburg, Germany

Abstract — Three-dimensional treatment planning is now widely used in radiotherapy with high energy photons. Special tools, for example beam's-eye view (BEV) and dose–volume histograms (DVH) are developed to simulate the shape and position of the beams and to evaluate the dose distribution in the target and the organs at risk. 3D visualisation is used to assess the dose distribution with respect to the target and the organs at risk. Standard blocks made from steel are used at our department to reduce the dose to critical organs. The ring model, originally developed for photon treatment, could be modified to be used with neutrons, too. This implies that the neutron dose is separated into its primary and secondary components, arising from concentric rings around the point of interest. A comparison between measurements in a water phantom and corresponding calculations demonstrates the attainable accuracy. For the DT neutron generator used for patients' treatment at our institution a commercially available treatment planning system was modified to enable the coplanar calculations of three-dimensional treatment plans. Furthermore, a visualisation system developed at our hospital is used to render the 3D dose matrix, together with anatomical data and the regions of interest. The mode of shaded surfaces and transparency simulates a three-dimensional impression of the rendered structures. A realistic case of a neutron treatment is used to clarify the different steps of the 3D treatment planning procedure and its implementation into clinical use.

INTRODUCTION

Three-dimensional treatment planning is now widely used with high energy photons. Based on a series of coplanar CT slices the dose distribution can be made conformal to the three-dimensional shape of the target. Special tools, as the beam's eye view BEV and dose–volume histograms (DVH) are used to position and shape the beams and to assess the resulting treatment plan.

In principle the procedures developed for the photon treatment planning can also be used for neutrons, if some particular properties of the neutrons are taken into account. In this paper the modifications necessary for the use of 14 MeV DT neutrons are described. The usefulness of these investigations will be demonstrated with a realistic 3D treatment plan, where some of the tools mentioned above are applied.

METHODS AND MATERIAL

Measurements and calculations are performed for a 14 MeV DT neutron generator, in clinical use for 20 years. As neutron fields are always contaminated with photons, the dose distributions have to be measured and calculated separately for both types of radiation. The heterogeneity of the irradiated area has to be taken into account by appropriate use of attenuation and kerma corrections[1]. The use of blocks and absorbers in neutron treatment require the use of algorithms that calculate the effects on the primary and scattered radiation. At our institute standardised steel blocks are used to modify the rectangular treatment fields. For the determination of the primary and scatter contribution to the dose, the ring model, commonly used for photons, was applied[2]. This model divides the area around the point of interest into rings. The relative contribution of the scattered neutrons additional to the primary dose is then calculated (Figure 1). The depth dose distributions of 32 rectangular neutron fields, ranging from 4 cm × 4 cm to 15 cm × 25 cm, were used to analyse the contribution of primary and scattered neutrons to the dose. The corresponding coefficients for the model are then incorporated into the commercial treatment planning system, MEVAPLAN.

RESULTS

By comparison of the measured and calculated dose distribution of a field with a lateral block the applica-

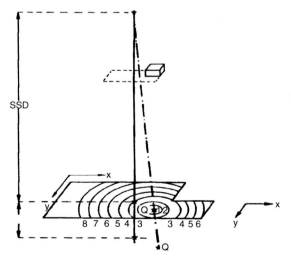

Figure 1. Sketch of the ring method taken from Reference 1.

bility of the ring model can be shown (Figure 2). The left part of the dose distributions are affected by the steel block of 12 cm thickness. Obviously the calculations result in a steeper dose gradient compared to the measurements. For neutrons the scatter effects are much more important than for photons so that the photon algorithm leads to the deviation shown. For further application account has to be taken of the simulated dose gradient at the border of a block, which is steeper than in reality. However, the calculated dose in the open field and in the shaded region is generally consistent with the measured value.

As an example, a treatment simulation of a patient with a salivary gland tumour is given. The tumour and the organs at risk, i.e. the eyes, myelon, brain stem and brain, are marked on the 32 coplanar CT slices. For the positioning of the beams and blocks the beam's eye

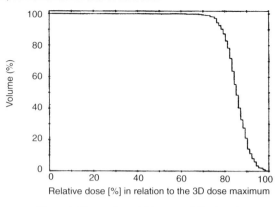

Figure 4. Dose-volume histogram of the target.

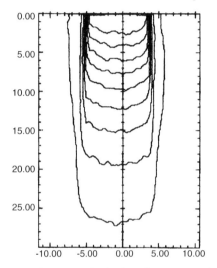
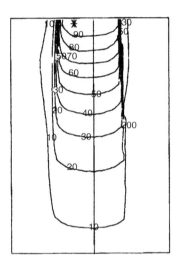

Figure 2. Comparison of the dose distribution of an asymmetric neutron field, generated by use of a lateral steel block (left, measurement; right, calculation).

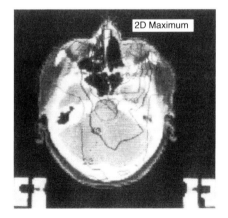
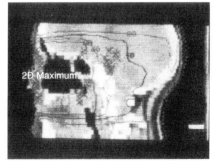

Figure 3. Treatment plan of a patient with a salivary gland tumour in transversal (left) and sagittal (right) planes.

view is used. The dose distribution is then calculated, based on the anatomical information of the CT. In Figure 3 four individually weighted fields are used to generate the displayed isodoses. From the sagittal reconstruction it can be concluded that the target is well covered by the 90% isodose. A dose homogeneity of 25% for the target was obtained. This is demonstrated by the corresponding DVH (Figure 4). As some regions of the left eye were inside the target, it partially obtained the same dose as the target, whereas the dose to the right eye could be substantially reduced by the use of the blocks. A visualisation system (VOXELMAN) was used to assess the physical dose distribution in relation to the organs. In Figure 5 the 80% isodose surface can be seen. The target is mainly covered by this dose; only at the surface of the nose can a coverage not be totally achieved because of the build-up effect. On the other hand the dose sparing at the right eye can clearly be seen.

DISCUSSION

Though the algorithms used for irregularly shaped fields for photons are not perfectly suitable for neutrons, a sufficient accuracy for 3D treatment planning can be obtained. The common tools are very helpful for the individual beam arrangement. The main objective for the use of three-dimensional treatment plans is the application of irregularly shaped fields for the reduction of the dose to organs at risk. With new machines with high energy neutrons and integrated multileaf collimators, conformation therapy will be possible. For high energy neutrons the photon algorithms can be more easily adapted as the relative scatter contribution is less than for DT neutrons.

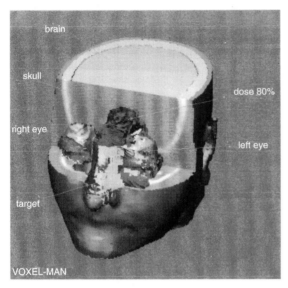

Figure 5. 3D visualisation of the treatment plan of Figure 3.

REFERENCES

1. Schmidt, R. and Thom, E. *Neutron Dose Planning with the MEVAPLAN System.* Strahlenther. Onkol. **166**, 301–305 (1990).
2. Jacobs, F. W. and van Kleffens, H. J. *The Ring Method: a Semi-empirical Method to Calculate Dose Distributions of Irregularly Shaped Photon Beams.* Radioth. Oncol. **7**, 363–369 (1986).

EVALUATION OF THE DOSE ALGORITHM BASED ON THE SCATTER MODEL APPLIED FOR NEUTRON THERAPY TREATMENT PLANNING

M. Yudelev, R. L. Maughan, T. He and D. P. Ragan
Gershenson Radiation Oncology Center, Karmanos Cancer Institute
Harper Hospital and Wayne State University
3990 John R Street, Detroit, Michigan 48201, USA

Abstract — The algorithm based on the scatter model was successfully applied in total dose (neutron plus gamma) calculations for Harper Hospital d(48.5) + Be neutron beam. The central axis and the lateral beam profiles as well as the peak scatter factors were measured for a number of rectangular and irregularly shaped fields. These data were used to calculate the tissue- and scatter-to-air ratios (TAR and SAR) as well as to determine the beam parameters pertinent to the treatment planning algorithm and to evaluate its accuracy in clinical cases. The agreement between the measured data and the data derived from the treatment planning program is usually within 2% for the central axis points and about 5% for the lateral beam profiles.

INTRODUCTION

The similar nature of neutron beam transport mechanisms (i.e. indirect ionisation and exponential attenuation) makes it possible to apply the algorithms used in photon beam calculations in fast neutron therapy treatment planning. The physical characteristics of the Harper Hospital neutron therapy beam, produced in a unique superconducting cyclotron by d(48.5) + Be reaction[1], resemble those of 4 MV photons[2]. In spite of these similarities the distinctive properties of the neutron beam as well as the use of a novel multi-rod collimator[3] set limitations on the use of photon algorithms in neutron beam calculations. The treatment planning systems, Theraplan V05 (Theratronics International Ltd, PO Box 13140, Kanata, Ontario, Canada K2K 2B7) and VRSplan[4] employed at Harper Hospital for teletherapy dose calculations, utilise Cunningham's scatter model[5].

METHODS AND MATERIALS

Scatter model

The model calculates the dose to an arbitrary point in a beam of indirectly ionising radiation as a sum of primary and scatter components. It requires a knowledge of the tissue-air ratios for the zero field area (TAR_0), the scatter-air ratios (SAR) and a function defined as the 'Profile Function' (PF) which accounts for the effect of the collimator, wedge filters and other beam modifiers. Unlike the photon beam, in the neutron calculations the algorithm was modified so that the PF is described by the same set of parameters regardless of the field shape. In addition, a parameter describing the output as a function of the field size (OF) is also required. The dose D_P to an arbitrary point at depth d in a homogeneous tissue-equivalent phantom at distance f from the source can be calculated from the following equation[5]:

$$D_P = D_0 \times OF \times \left(\frac{f + d_m}{f + d}\right)^2 \times (TAR_0 \times PF_P + SAR_P) \quad (1)$$

For the central axis points PF_P is 1 and the relationship between TAR and the dose at P (central axis depth dose, CADD) normalised to the dose at the depth of dose maximum (d_m) becomes:

$$TAR_P = CADD_P \times \left(\frac{f + d}{f + d_m}\right)^2 \times PSF \quad (2)$$

where TAR_P is the sum of TAR_0 and SAR_P and PSF is the peak scatter factor for the field defined at d_m.

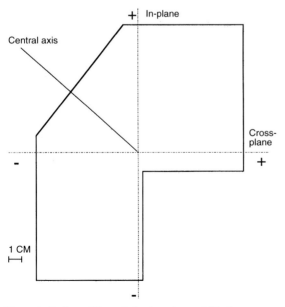

Figure 1. Outline of irregularly shaped test field. Scans were done along beam depth, cross-plane and in-plane axes.

Measurements

The TAR_0 and the SAR_P as function of depth and radial distance to the field edge were derived from the CADD and PSF measurements in a set of 13 rectangular fields ranging from $2\,cm \times 2\,cm$ to $25\,cm \times 25\,cm$. The parameters of the profile function were optimised to match the dose profiles measured at 1.2, 5, 10 and 20 cm depths for opened and wedged rectangular fields.

To determine the PSF the ionisation chamber measurements were done first at the depth of the dose maximum in the water phantom and then at the same distance from the target in air. The charge accumulated by $0.3\,cm^3$, 5 mm wall tissue-equivalent (TE) plastic ionisation chamber (Type IC30/A150-2.5, Wellhöfer Dosimitrie, Germany) was measured with a high quality electrometer (Type 35617 EBS, Keithley Instruments, Cleveland, Ohio).

The profiles along beam depth, cross-plane and in-plane axes were measured in a water phantom (Type NWP, Nucletron, Inc., The Netherlands) at the source to surface distance (SSD) of 182.9 cm by means of the same ionisation chamber. In all measurements the chamber was flushed with methane-based TE gas.

The model was evaluated by defining the CADDs, TARs, and lateral dose distribution in a $5 \times 25\,cm^2$ rectangular field and a test field of irregular shape, the outline of which is shown in Figure 1, and comparing these with the parameters obtained from the treatment planning program, as described below.

RESULTS

To define the PSFs the ionisation chamber readings in the phantom and in air were extrapolated to a zero size field. The ratios were obtained and normalised to the value at zero field (Figure 2). These PSFs were then used to derive the TARs from the central axis per-cent depth dose curves measured for the rectangular fields. The values of TAR_0 were obtained by extrapolation at

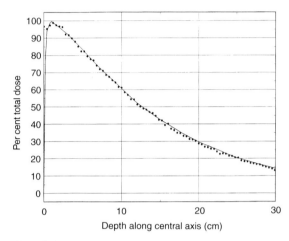

Figure 3. Central axis depth doses computed (solid line) and measured (●) for a $5 \times 25\,cm^2$ field at the source to surface distance of 182.9 cm.

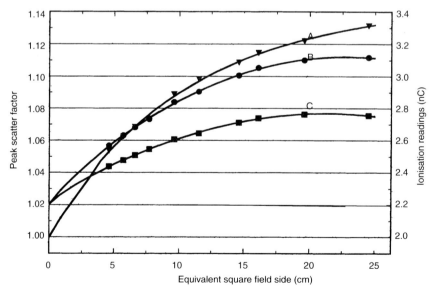

Figure 2. Peak scatter factors (curve A, ▼) and ionisation readings (curve B, ●) in water and (curve C, ■) in air. Ionisation readings shown on right hand axis are extrapolated to a unique value at a zero size field.

each depth. The SAR as a function of the radius of equivalent circular field together with TAR_0 as given in Table 1 are used in the dose computations.

The CADD curves measured and computed for the 5×25 cm^2 rectangular field are shown in Figure 3. The TARs measured and computed along the central axis of the irregularly shaped test field are presented in Figure 4. The agreement of the measured and computed values is within 2%.

The beam profiles measured at 5 cm depth in 5×25 cm^2 agree in general with that derived by the treatment planning program (Figure 5). The computed values are underestimated in the region just outside the penumbra and overestimated at points located more than

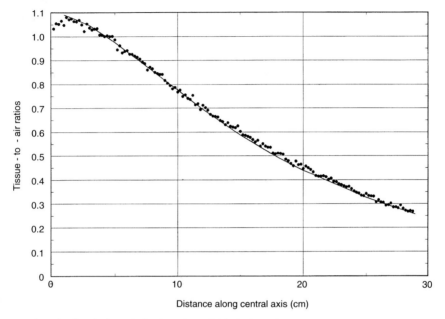

Figure 4. Tissue-to-air ratios for the irregularly shaped test field of Figure 1 computed (solid line) and measured (●) along the central axis.

Table 1. TAR_0 and SAR used in Harper Hospital d(48.5) + Be neutron therapy treatment planning.

Depth (cm)	TAR_0	Radius of equivalent circular field (cm)							
		1.1	2.2	3.4	5.6	7.9	10.1	12.4	14.6
0.0	0.408	0.016	0.031	0.046	0.074	0.098	0.120	0.143	0.165
0.3	0.935	0.017	0.031	0.046	0.070	0.086	0.098	0.108	0.114
0.9	1.012	0.017	0.034	0.050	0.076	0.093	0.106	0.117	0.123
2.0	0.954	0.026	0.051	0.075	0.112	0.135	0.149	0.160	0.167
3.0	0.894	0.035	0.069	0.101	0.152	0.178	0.193	0.205	0.211
5.0	0.752	0.052	0.105	0.152	0.223	0.258	0.279	0.292	0.298
7.0	0.629	0.060	0.119	0.175	0.260	0.310	0.335	0.348	0.354
10.0	0.491	0.059	0.115	0.171	0.264	0.329	0.364	0.387	0.401
12.0	0.385	0.060	0.124	0.184	0.285	0.353	0.393	0.417	0.428
15.0	0.307	0.051	0.104	0.155	0.245	0.315	0.364	0.399	0.416
18.0	0.238	0.044	0.089	0.135	0.215	0.283	0.334	0.367	0.390
20.0	0.192	0.042	0.086	0.129	0.203	0.266	0.315	0.356	0.386
25.0	0.110	0.038	0.076	0.113	0.177	0.232	0.274	0.306	0.321
30.0	0.072	0.026	0.052	0.078	0.129	0.172	0.209	0.240	0.265
35.0	0.037	0.020	0.042	0.063	0.102	0.135	0.165	0.192	0.218
40.0	0.031	0.013	0.027	0.041	0.067	0.092	0.114	0.136	0.156

16 cm away from the central axis for the 25 cm dimension of the beam. The lateral profiles along the in- and cross-plane axes measured and computed at 0.9, 5, 10 and 20 cm depths in the irregularly shaped test field are plotted in Figure 6. The maximum discrepancy between measured and predicted values is 5%.

DISCUSSION AND CONCLUSIONS

The results demonstrate that the scatter model used in photon beam calculations can be successfully applied in neutron therapy treatment planning if separation of primary and secondary components is available. Yudelev et al[6] showed that differences between measured and derived tissue-to-maximum ratios for a p(66) + Be(40) neutron beam can be as high as 7%. Since the penetration characteristics of the neutron beam analysed in the present work resemble those of 4 MV X rays the 'dose to air' is more easily achievable which results in better agreement of the model with the measurements.

Higher discrepancies (about 5%) were observed between measured and computed values in the beam profiles adjacent to the penumbra of the irregularly shaped field than in the case of the rectangular field. This anomaly may be explained by the boundary conditions at which the measurements and calculations were performed (i.e. close proximity to the beam edge and finite ionisation chamber size). In the model the primary beam profile is described by an empirical function symmetrical about beam geometrical edge[5]. For neutron beams characterised by a high scatter component[3] and 1.5% transmission through collimator off-axis[4] such a function results in an underestimation of the dose in the umbra by as much as 5%, since the beam is not symmetrical about the geometric edge.

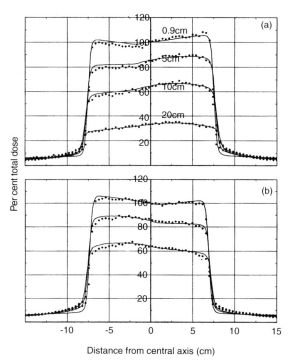

Figure 6. Dose profiles along (a) the in-plane axis and (b) the cross-plane axis of the irregularly shaped test field computed (solid line) and measured (●) at 0.9, 5, 10 and 20 cm depths and source to surface distance of 182.9 cm.

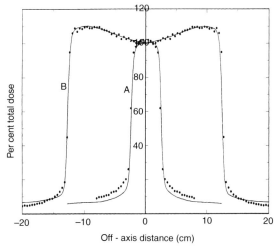

Figure 5. Dose profiles computed (solid line) and measured (●) at 5 cm depth along short (curve A) and long (curve B) axes in a 5×25 cm² field at the source to surface distance of 182.9 cm.

REFERENCES

1. Maughan, R. L., Powers, W. E. and Blosser, H. G. *A Superconducting Cyclotron for Neutron Radiation Therapy*. Med. Phys. **21**(6), 779–785 (1994).
2. Maughan, R. L. and Yudelev, M. *Physical Characteristics of a Clinical d(48.5) + Be Neutron Therapy Beam Produced by a Superconducting Cyclotron*. Med. Phys. **22**(9), 1459–1465 (1995).
3. Maughan, R. L., Blosser, G. F., Blosser, E. B., Yudelev, M., Forman, J. D., Blosser, H. G. and Powers, W. E. *A Multi-rod Collimator for Neutron Therapy*. Int. J. Radiat. Oncol. Biol. Phys. **34**(2), 411–420 (1996).
4. He, T., Ragan, D. P. and Mesina, C. F. *Clinical Experience in 3D CT Simulation and Treatment Planning*. Med. Phys. **21**(6), 928 (1994).
5. Jones, H. E. and Cunningham, J. R. *Physics of Radiology*. 4th edn (Springfield, IL: Charles C. Thomas) (1983).
6. Yudelev, M., Schreuder, A. N. and Jones, D. T. L. *Tissue-Maximum Ratios for a p(66) + Be(40) Neutron Therapy Beam*. Radiat. Prot. Dosim. **44**(1), 417–420 (1992).

SPECIFICATION OF ABSORBED DOSE AND RADIATION QUALITY IN HEAVY PARTICLE THERAPY (A REVIEW)

A. Wambersie† and H. G. Menzel‡
†Université Catholique de Louvain (UCL)
Cliniques Universitaires St-Luc 1200 Brussels, Belgium
‡European Commission
DGXII F-6 (Radiation Protection Research Action)
200, rue de la Loi, B-1049 Brussels, Belgium

INVITED PAPER

Abstract — The introduction of heavy particles (hadrons) into radiation therapy aims at improving the physical selectivity of the irradiation (e.g. proton beams), or the radiobiological differential effect (e.g. fast neutrons), or both (e.g. heavy ion beams). Each of these new therapy modalities requires several types of information before prescribing doses to patients, as well as for recording and reporting the treatments: (i) absorbed dose measured in a homogeneous phantom in reference conditions; (ii) dose distribution computed at the level of the target volume(s) and the normal tissues at risk; (iii) radiation quality from which an evaluation on the RBE could be predicted; and (iv) RBE measured on biological systems or derived from clinical observation. The ICRU has published recommendations for fast neutrons and a similar report is in preparation for proton beams. These recommendations are now universally applied. The single beam isodoses and thus the dose distributions are similar in neutron and photon therapy. Similar algorithms can then be used for treatment planning and the same rules can be followed for dose specification for prescribing and reporting a treatment. In hadron therapy, the RBE of the different beams raises specific problems. For fast neutrons, the RBE varies within wide limits (about 2 to 5) depending on the neutron energy spectrum, dose, and biological system. For protons, the RBE values range between smaller limits (about 1.0 to 1.2). A clinical benefit is thus not expected from RBE differences. However, the proton RBE problem cannot be ignored since dose differences of about 5% can be detected clinically in some cases. The situation is most complex with heavy ions since the RBE variations, as a function of particle type and energy, dose and biological system, are at least as large as for fast neutrons. In addition, the RBE varies with depth. Radiation quality thus has to be taken into account when prescribing and reporting a treatment. This can be done in different ways: (a) description of the method of beam production; (b) computed LET spectra and/or measured microdosimetric spectra at the points clinically relevant; (c) RBE determination. The most relevant data are those obtained for late tolerance of normal tissues at 2 Gy per fraction ('reference RBE'). The 'clinical RBE' selected by the radiation oncologist when prescribing the treatment will be close to the reference RBE, but other factors (such as heterogeneity in dose distribution) may influence the selection of the clinical RBE. Combination of microdosimetric data and experimental RBE values improves the confidence in both sets of data.

INTRODUCTION

The introduction of heavy particles (hadrons) into radiation therapy aims at improving the physical selectivity of the irradiation (e.g. proton beams), or the radiobiological differential effect (e.g. fast neutrons), or both (e.g. heavy ion beams).

The rationale for proton beam therapy is straightforward: increasing the dose to the target (tumour) volume for the same absorbed dose to the surrounding normal tissues. Clinical experience as well as radiobiological data indicate that a small increase in dose can improve significantly the local control rate and the complication rate, at least in some patient series.

The rationale for fast neutron therapy (high LET radiation) is based on radiobiological arguments, briefly[1]:

(i) a reduction in the oxygen enhancement ratio (OER) with high LET radiations, i.e. a selective efficiency against hypoxic cancer cells;
(ii) a greater selective efficiency against cells in the most radioresistant phases of the mitotic cycle;
(iii) smaller role of the sub-lethal lesions, i.e. less importance of the dose per fraction.

These factors can bring an advantage or a disadvantage, depending on the characteristics of the cancer cell population and of the normal tissues at risk. This raises the important problem of patient selection[2].

Heavy charged particles are high LET radiations and thus combine the advantage of the high physical selectivity of the protons with the selective radiobiological advantage of the neutrons for selected patient series.

The radiobiological rationale on which fast neutron therapy (and heavy charged particle therapy) was based initially is still valid today and has not been refuted by more recent experimental results after more than thirty years.

To be applied in optimal conditions, each of these new therapy modalities requires several types of information before the prescription of the treatment, as well as for recording and reporting the treatment:

(i) absorbed dose measured in a homogeneous phantom in reference conditions;

(ii) dose distribution computed at the level of the target volume(s) and the normal tissues at risk;
(iii) radiation quality from which an evaluation on the RBE could be predicted;
(iv) RBE measured on biological systems and/or derived from clinical observation.

PRESENT SITUATION IN HEAVY PARTICLE THERAPY AND SHORT REVIEW OF THE CLINICAL DATA

Since 1970, neutron therapy programmes have been initiated, first in Hammersmith, and then in more than twenty centres worldwide[3]. More than 20,000 patients have been treated so far with fast neutrons (alone or in combination with other radiation therapy techniques). For some patient series, the follow-up extends over 25 years. Although these figures are impressive, there is, surprisingly, up to now no general agreement on the value of fast neutron therapy and its place among the other radiation therapy techniques. In some centres, the neutron therapy programmes had to be terminated, mainly for technical reasons. In contrast, in the centres where the programmes could be continued, encouraging results were reported each time the neutrons were applied in adequate technical conditions[1].

The value of fast neutrons for the treatment of locally advanced salivary gland tumours is recognised. According to NCI, fast neutron therapy should be the treatment of choice for advanced stage salivary gland tumours, and surgery should be limited to cases where there is a high likelihood of achieving a negative surgical margin and where the risk of facial nerve damage is high.

For locally extended prostatic adenocarcinoma, two prospective randomised trials performed in the USA have shown the superiority of fast neutrons over photons. In the first trial, mixed beams (combination of neutrons and photons) were more efficient than photons alone as far as local control and survival is concerned.

In the second trial comparing neutrons (alone) and photons, a statistically significant advantage for neutrons was observed for local control (11% as against 32% loco-regional failure at 5 years). In contrast, no benefit has been observed so far for survival. However, an improvement in survival can be expected in the coming years with neutrons because of a significantly lower proportion of patients with elevated PSA level in the neutron treated group (17% rather than 45%, at 5 years, for neutrons and photons respectively).

Definitive conclusions concerning the place of fast neutrons in the management of patients with prostatic adenocarcinoma are of special interest for the future of neutron therapy because of a high potential recruitment.

In addition, randomized trials and pilot studies have shown an advantage for neutron therapy compared to conventional photon therapy for several selected patient series: e.g. with inoperable or recurrent soft tissue sarcomas (slowing growing, well differentiated), inoperable tumours of the maxillary sinus, head and neck tumours (locally extended, with fixed adenopathies), non-small cell bronchus carcinomas, melanomas (palliation), inoperable or recurrent rectum carcinomas (pain relief), and radioresistant histologies.

Altogether these patients represent about 10–15% of the patients referred to the radiation therapy departments today. It is worthwhile pointing out that the tumours for which good clinical results were obtained with neutrons are those for which a clinical benefit could have been predicted from radiobiological data (e.g. slowly growing tumours[4]).

High LET radiation therapy can also be applied using heavy ions. After termination of the therapy programme at the Berkeley cyclotron (USA), heavy ion beam therapy is used today only at NIRS in Chiba (Japan). In Europe, several programmes are under development: the most advanced one is at the GSI in Darmstadt (Germany). In addition, the TERA project (CERN/Geneva and Italy) and the Austron project in Austria are actively being developed by experienced and enthusiastic teams.

According to the Berkeley experience, the clinical results obtained with heavy ions are comparable to those obtained with neutrons: benefits were observed for the same localisations (e.g. salivary glands, prostate) and poor results were observed with both types of radiations for other localisations (e.g. brain, oesophagus). It can thus be hypothesised that the similarity in the clinical results reflects a specific effect of high LET radiations.

Proton beam therapy is developing quite rapidly worldwide: more than 15,000 patients have been treated as of the end of December 1995. Uveal melanoma is recognised today to be an excellent clinical indication for protons and this technique is applied in more than 10 centres worldwide. A proton energy of 55–60 MeV is generally sufficient. As a consequence, cyclotrons designed for fast neutron therapy can be used for proton therapy of uveal melanomas. In Europe, more than 1700 patients with uveal melanoma were treated at PSI-Villigen by the end of 1994. Around the same time, in France, 400 patients were treated at Orsay (close to Paris) and 470 at the Centre Antoine-Lacassagne in Nice.

Besides treating uveal melanoma, protons were used successfully at MGH in Boston for the treatment of some tumours of the base of the skull, such as chordomas and chondrosarcomas. A recent re-evaluation of the MGH results indicates that proton beam therapy is especially efficient for chondrosarcomas, but to a lesser extent for chordomas (local recurrence-free survival at 5 years: 95% and 62% respectively). However, accurate histological differential diagnosis between these two types of tumours may be difficult. With conventional photon therapy, the reported local control rates do not exceed 30%.

For many years, the development of proton beam therapy was delayed by the lack of availability of suitable accelerators. In 1990, the first hospital-based proton therapy centre was established at Loma Linda (California, USA). Four treatment rooms (three with isocentric gantry) are fully available for patient treatment with 250 MeV protons. At MGH in Boston, a 250 MeV cyclotron, designed by the IBA company in Belgium, is now being installed and will be operational in about one year. Both these facilities will certainly allow us to extend the indications for protons; they are opening a new era in the field of proton beam therapy.

DOSE DETERMINATION AND DOSE SPECIFICATION FOR REPORTING

As far as neutron dosimetry is concerned, the International Commission on Radiation Units and Measurements (ICRU) published, in 1989, Report 45 on *Clinical Neutron Dosimetry — Part I: Determination of Absorbed Dose in a Patient Treated by External Beams of Fast Neutrons*[5]. This report was the result of an agreement between the US and European neutron therapy groups and is applied today worldwide[6,7].

For proton beam therapy, ICRU Report *Clinical Proton Dosimetry — Part I: Beam Production, Beam Delivery and Measurement of Absorbed Dose* is now in press[8]. It contains recommendations on how to measure absorbed dose in a homogeneous phantom in reference conditions. They can be summarised as follows:

(a) A standard A-150 tissue-equivalent or graphite thimble chamber having a standard ^{60}Co calibration factor is the recommended reference dosemeter for clinical proton dosimetry.
(b) The recommended numerical values of the required quantities are the following:
  (i) the mass electronic stopping powers in the various materials, relative to the csda range in water, should be taken from in ICRU Report 49[9];
  (ii) for $W_{p,air}/e$ above about 1 MeV, a value of 34.8 + 0.7 J.C^{-1} is recommended.
(c) A water, graphite or A-150 calorimeter should be used, when possible, to confirm the proton calibration factor of the reference chamber.
(d) A fluence-based dosimetry technique, such as a Faraday cup, can be useful to verify independently the calibration of the reference chamber.
(e) The proton absorbed dose should be specific in water.
(f) Measurements should be made in a phantom at a point where the dose is fairly uniform (centre of spread out Bragg peak, SOBP).
(g) Direct intercomparisons with ionisation chambers, or indirectly through the use of travelling calorimeters, Faraday cups, or even mailed integrating dosemeters are recommended. When possible, dosimetric intercomparisons should be combined with microdosimetric and radiobiological intercomparisons.

This protocol, which has been widely discussed at several meetings, is now widely accepted by the proton therapy community. It can easily extended and adapted to heavy charged particles.

ICRU Report 46, *Photon, Electron, Proton and Neutron Interaction Data for Body Tissues* provides a vast amount of basic interaction data[10]. Finally, an ICRU Reporting Committee has recently been appointed to collect, analyse and compare the measured and computed nuclear data needed in heavy particle therapy.

As far as dose specifications for reporting is concerned, the recommendations proposed by the ICRU for photons[11] can be followed and adapted to neutrons with only limited alteration. As a matter of fact, the isodose curves for single neutron and photon beams are very similar and thus also the dose distributions resulting from their combination. In particular, the recommendations concerning the selection of the ICRU Reference Point can be followed rather easily. However, attentions should be paid to the atomic composition of the various body tissues, and one should avoid specifying the dose at a point close to the border between tissues of different atomic composition. The atomic composition of the tissues is indeed of greater importance for neutrons than for photons. ICRU Report 46 mentioned above[10] also contains a vast amount of information on the composition of various body tissues.

The problem of dose specification for reporting is certainly more complex for proton beams (and heavy ion beams). First, an ICRU reference point must be selected as clinically relevant, following the same criteria as for photons. It should be easy to define in a clear and unambiguous way, in a region where the dose could be measured accurately and where there is no steep dose gradient. The ICRU reference point should thus logically be located in the central part of the target volume and (when possible) on the beam axis. However, it has to be checked that there is no significant dose fluctuation at the level of the selected ICRU reference point, due for example to inadequate beam modulation.

In addition to the dose at the ICRU reference point, the dose variation at the level of the target volume(s) and of the normal tissues at risk should be reported. In particular, dose–volume histograms should be available in centres involved in proton beam therapy.

Comparison of the dose–volume histograms for the target volume(s) and the normal tissues at risk computed for proton and conventional photon irradiation is one way to evaluate the potential gain expected with protons. However the best beam arrangements should be compared for both radiation qualities (optimisation procedure). From the comparison of the dose–volume histograms, different mean dose values and weighted mean dose values could be derived; their clinical relevance is still to be evaluated.

SPECIFICATION OF RADIATION QUALITY

Importance of the problem

Specification of radiation quality is an important issue in heavy particle therapy since the biological effectiveness of the beams is related to radiation quality. In other words, the RBEs of the neutron, proton and heavy ion beams are higher than unity and depend significantly on particle type and energy and radiation field composition. Radiation quality must be specified in an accurate and unambiguous way.

The introduction of neutron therapy has led to a large number of radiation quality investigations using biological and physical approaches. Radiation quality specifications in radiation therapy must meet specific criteria. The main criterion is derived from the accuracy requirement for absorbed dose delivery. Any weighting factor to be evaluated from biological and physical experiments and to be applied to the absorbed dose must fulfil comparable accuracy requirements[12]. The method for evaluating the weighting factor needs to be valid only within the range of neutron energies used in therapy. The weighting factor would have to be used in support of establishing reference RBEs with increased reliability and to enable RBE variations of clinical relevance at a given facility to be taken into account.

As will be discussed later, radiobiological experiments in neutron therapy beams have been used to determine RBE values and RBE ratios. Such experiments were aimed at providing the basis for the reference RBE approach. These radiobiological investigations are absolutely necessary for the determination of the reference RBE and are required in any attempt to find a suitable radiation quality specification. However, their usefulness is limited by the inherently large uncertainties and unavoidable experimental difficulties. An alternative approach is to identify physical beam parameters that can be related to RBE[13,14].

Physical beam parameters

An evaluation of the radiation quality can be derived from the description of the beam production technique.

For fast neutrons, this includes the neutron-producing reaction (p+Be or d+Be), the energy of the incident particles, the target composition and geometry, and other geometrical factors (e.g. collimation system, additional polyethylene filter, etc.). An indication of the neutron energy can be derived from these parameters. The neutron energy spectrum can also be measured by spectroscopic methods. As in current practice, single parameters appear more useful than complete spectral information. The mean neutron energy or related parameters, such as the half-value thickness (HVT), have also been used (Figure 1)[15].

In the beginning of fast neutron therapy, it was common practice to describe the neutron beam in terms of the 'neutron component' and 'gamma component'. Also at that time (and for many applications), an RBE value of 3 was assigned to the neutron component and an RBE value of 1 to the gamma component. The development of the microdosimetric programmes has shown that this approach is indeed an oversimplification and that several types of secondary charged particles must be taken into account (see for example Figure 2[16]).

Contribution of microdosimetry to radiation quality specification

The neutron energy, in terms of spectrum or average value, is primary information. Neutrons deposit energy through secondary charged particles whose distributions critically depend on the shape of the neutron spectrum and on the types and thresholds of nuclear reactions with the different tissues. The biological properties of neutron beams are therefore more directly correlated with the secondary radiation spectra and only indirectly with the neutron energy. However, the spectra of secondary charged particles are not easily accessible information, especially for high energy neutron beams. Furthermore, there exists no biophysical model that describes adequately biological observations relevant for therapy.

The measurement of microdosimetric spectra offers, at least for the restricted field of neutron therapy, a sufficiently accurate description of the secondary radiation components and provides comprehensive and detailed information on radiation quality.

Microdosimetric spectra have also been obtained in therapeutic proton beams. As can be seen in Figure 3, for the 85 MeV beam of Louvain-la-Neuve, the y spectra are progressively shifted towards the high y values when measurements are performed at the level of the initial plateau, at the beginning, middle, and end of the spread out Bragg peak. However, the largest part of the various spectra correspond to a range of y where the biological weighting function (see below) remains close to unity, thus only a small RBE increase (if any) can be expected[17].

THE RBE PROBLEM IN HEAVY PARTICLE THERAPY

The RBE problem is of great relevance in heavy particle therapy, but it is raised in different terms and is of a different magnitude for neutrons, protons and heavy ions. However it is important, especially for interpreting and comparing the data, to try to approach this problem in the same way for all radiation qualities.

For fast neutrons, the RBE values are high, ranging from about 2 to 5 (relative to gamma rays). They increase with decreasing dose, they are usually higher for late effects than for early effects, and they decrease significantly with increasing neutron energy[2,15,18]. In the energy range used in therapy, RBE variations up to 40% are observed for some biological systems and endpoints[15,19]. In contrast, in a given neutron beam, the RBE does not vary as a function of depth (Figure 2) (except in special conditions, e.g. when neutrons are produced by incident protons and no additional polyethylene filtration is added[20]).

For proton beams, the observed RBEs are lower, ranging from about 1.0 to 1.2. This range is clinically significant but only slightly larger than the experimental uncertainties on RBE determinations (Figure 4)[21]. The proton energy decreases with depth, and there is thus a possible increase in RBE.

For heavy ions, the RBE variations are large, even larger than for fast neutrons. The RBE varies with dose and biological system, as for neutrons; in addition it varies greatly with the type of particles, their initial energy, and the depth in the irradiated medium.

The RBE concept

The RBE is a clear, unambiguous, and well-defined radiobiological concept. An RBE value is the result of an experiment and is thus associated with an experimental uncertainty. The biological system, type and level of effect, the dose, and the experimental conditions in which a given RBE value has been obtained must be specified[22–24].

Strictly speaking, the commonly used jargon that a certain radiation, e.g. fast neutrons from a given nuclear reaction, have a certain RBE is fundamentally incorrect and misleading. RBE values cited in this way are usually not the result of a given experiment but are judgements based on experimental RBEs and sometimes clinical experience. Although this practice is common and reflects some convenience of procedure for a complex issue, it must be stressed very clearly that the use of such comparative values for different types of radiations should not be confused with RBE values[25]. When comparing two radiation qualities there is no single but a large number of RBE values, i.e. one RBE value for every set of system, effect, and experimental condition (with the corresponding confidence intervals).

The reference RBE

Since the beginning of fast neutron therapy, and even

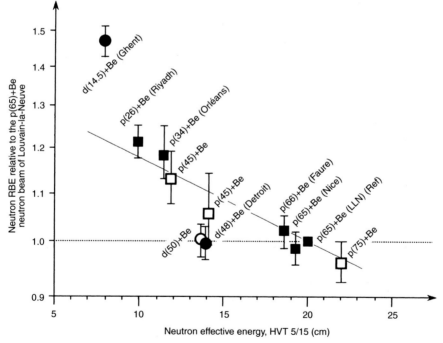

Figure 1. Variation of RBE on different neutron beams as a function of energy. The p(65) + Be neutron beam of Louvain-la-Neuve is taken as the reference (RBE = 1). The closed squares and circles correspond to neutron beams produced at six different visited facilities. The gray squares and circles correspond to neutron beams produced at the variable-energy cyclotron of Louvain-la-Neuve. On the abscissa, the effective energy of the neutron beams is expressed by the half-value-thickness (HVT 5/15) measured in defined conditions (see Ref. 15). Intestinal crypt regeneration in mice after a single fraction irradiation is taken as the biological system for RBE determination. The error bars indicate the 95% confidence intervals. A straight line is fitted through the point corresponding to neutron beams produced by the (p + Be) reaction (squares). For comparison, the neutron beams produced by the (d + Be) reaction are represented by the circles. From Gueulette et al[15].

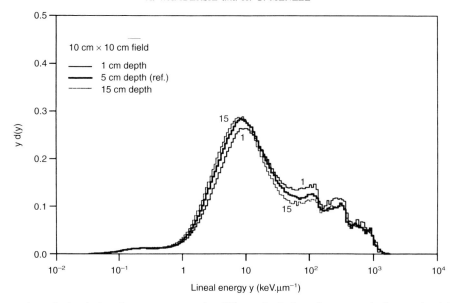

Figure 2. Comparison of microdosimetric spectra measured at different depths in a phantom on the beam axis of the p(65) + Be neutron therapy beam of Louvain-la-Neuve (the highest energy used in neutron therapy). The curves represent the distributions of individual energy deposition events in a simulated volume of tissue 2 μm in diameter. The parameter y (lineal energy) represents the energy deposited by a single charged particle traversing the sphere, divided by the mean cord length. Four peaks can be identified in the p(65) + Be neutron spectra: the first is at 8 keV.μm^{-1} and corresponds to high energy protons, the second is at 100 keV.μm^{-1} and corresponds to low energy protons, the third is at 300 keV.μm^{-1} and is due to α particles and the last at 700 KeV.μm^{-1} and is due to recoil nuclei. The (rather flat) peak corresponding to the gamma component of the beam is at 0.3 keV.μm^{-1}. The difference between the spectra measured at different depths is relatively small and no significant RBE variation has been detected. From Pihet et al[16].

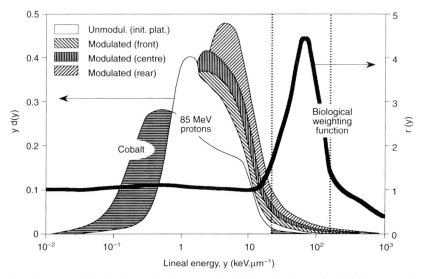

Figure 3. Microdosimetric spectra (distribution in dose), measured at different depths in the 85 MeV proton beam produced at the cyclotron of Louvain-la-Neuve (left ordinate). There is a progressive shift of the spectra to the right as a function of depth. The spectrum for ^{60}Co gamma rays is shown for comparison. The right ordinate represents the biological weighting function (see text) for intestinal crypt regeneration in mice after single dose irradiation. Only a small part of the proton spectra overlaps the rising part of the biological weighting function: only a discrete (if any) variation in RBE is thus to be expected. From Gueulette et al[17].

before, a large number of determinations of RBE values for a variety of systems and endpoints have been made[1,15,19]. In some centres, comprehensive sets of RBE measurements were performed, e.g. the full RBE/dose relationships for the tolerance of different normal tissues (including late tolerance) and different types of tumours[26,27].

However, most of the RBE determinations performed during the radiobiological pre-therapeutic experiments, as well as during the radiobiological intercomparisons, used biological systems chosen because they were suitable for RBE determination: they were well codified, reliable, easy to transport, providing rapidly reproducible results (e.g. mammalian cells *in vitro*, *Vicia faba*, intestinal crypt cell systems, etc.)[15,28]. A general and formal agreement does not exist as yet on the choice of the radiobiological system for radiobiological intercomparisons.

Due to the wide variation of RBE with the biological system and criterion, dose, and experimental conditions, it is necessary to select reference conditions for RBE specification when transferring or exchanging clinical information. These reference conditions should be as indicative as possible for the clinical situations.

The following conditions appear to be relevant:

(i) dose level: 2 Gy (photon equivalent) per fraction;
(ii) biological system: a system and endpoint 'representative' of the RBE for average or overall late tolerance of normal tissues.

The RBE defined in such reference conditions could be called reference RBE ($_{ref}RBE_{n/\gamma}$)[25,29].

The reference RBE is a radiobiological approach and, in principle, there should be only one reference RBE value for a given neutron beam. This implies the assumption that a single RBE value can be defined for an 'overall' or 'average' late tolerance for the normal tissues in patients. The fact that the alpha/beta ratios for late tolerance of different normal tissues are similar supports the validity of this assumption.

The clinical RBE

In contrast to the reference RBE, the 'clinical RBE' is a clinical and operational concept. It is the RBE value that the radiation oncologist has to select and adopt when prescribing the treatment. It is based on the available radiobiological data (experimental and clinical). However, selection of the clinical RBE is a medical decision (since it directly influences the dose to the patient). It depends on the judgement and experience of the radiation oncologist and thus implies his medical responsibility.

The term 'clinical RBE' has been used especially in the US (e.g. RTOG neutron therapy studies) and it was understood as a ratio of the absorbed doses which would be given with photons and neutrons respectively at a given neutron therapy facility (and for a given tumour localisation). This definition is reasonable in neutron therapy to the extent that the isodoses for single neutron and photon beams are comparable, at least for the high energy neutron beams (i.e. produced by protons of at least 50 MeV).

When a new type of radiation beam is applied in therapy for which clinical observation does not exist yet, the reference RBE should logically be taken as a first

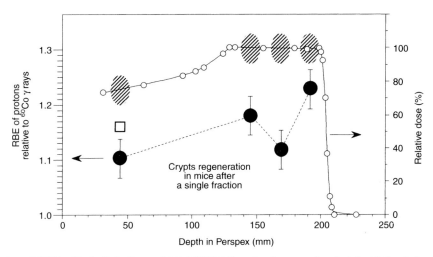

Figure 4. Variation of RBE with depth in the modulated 200 MeV proton beam produced at the National Accelerator Centre (NAC), Faure (left ordinate, closed circles). The width of the spread out Bragg peak is 7 cm (from 13 to 20 cm in depth). The error bars indicate the 95% confidence intervals. The open square corresponds to the RBE for the unmodulated beam. The depth–dose curve in the modulated beam is superimposed for comparison (right ordinate, small open circles). The hatched areas indicate the different positions where the mice were irradiated. The width of the shaded areas corresponds to the size of the mice abdomens. From Gueulette et al[21].

indication for selecting the clinical RBE. Later on, to the extent that clinical experience is being built up (in particular with dose escalation phase I studies), the clinical RBE can be progressively adjusted. For example, today in a new neutron therapy centre, the clinical RBE would be derived nearly exclusively from the accumulated clinical observation (with, of course, the weighting for differences in neutron energy spectra).

Some adjustment required to the clinical RBE, relative to the reference RBE, could be due to the influence of the dose distribution at the level of the different normal tissues (importance of dose heterogeneities, the volume effect, and the tissue heterogeneities). It could also be due to the fact that the RBE for late tolerance of the tissues involved in a particular treatment is not identical to the reference RBE derived from a series of radiobiological experiments.

When the clinical RBE is derived only from the reference RBE, it implies that the different radiation therapy protocols are compared on the basis of the late normal tissue tolerance only. This is actually often the case when testing new therapeutic modalities. Tumour control could also be taken into account, and the prescription could be based on the proportion of uncomplicated cures (optimisation would then be based on the comparison of the dose–effect curves for tumour control and normal tissue complications).

If the physical selectivity of the neutron beam used in a particular centre is significantly inferior to that achieved with photons [e.g. 16 MeV cyclotron or (d + T) generator], a dose reduction factor must be introduced when prescribing the target dose. This factor should be clearly separated from the clinical RBE. It is, however, recognised that it is not always possible to separate completely the influence of the dose distribution (dose heterogeneity and volume effect) from the clinical RBE.

The dose distributions for proton and heavy ion beams are totally different from those obtained with photons (and in general superior). Here again, the dose prescription should take into account both the differences in dose distribution and the clinical RBE.

In conclusion the clinical RBE is the quantity the radiation oncologist has to select when prescribing the irradiation (e.g. in terms of monitor units). Although it is a dose ratio for the two compared radiation qualities, it is not in a strict sense a RBE: it is in principle the reference RBE weighted by an empirical account of the clinical experience (when available) and (to some extent) the physical selectivity. It is a medical decision, which determines the dose to the patient and thus implies the responsibility of the radiation oncologist.

When reporting the treatment technique, the measured RBE values, the best estimate of the reference RBE and the selected clinical RBE value have to be indicated.

Contribution of microdosimetry for RBE prediction using an empirical biological weighting function

In order to solve the problem of specifying radiation quality in heavy particle therapy, it would be useful to identify a parameter with a relative variation similar to that of RBE. It would then be necessary to prove that such a parameter could be determined with sufficiently low uncertainty[30,31].

The approach briefly described here was developed by Pihet[16,32] for fast neutron therapy and is based on the principle of using a weighting biological function applied to the dose distribution in lineal energy for a given radiation field in order to determine a single parameter that estimates its quality. This procedure is well known in microdosimetry and has been applied in an analogous way, for example, in radiation protection[33] for determining the mean quality factor for a given radiation field:

$$Q = \int q(y) \; d(y) \; dy$$

Applying this principle to the field of neutron therapy, the problem of specifying the radiation quality for a neutron beam with a given energy compared with that of another neutron beam chosen as a reference may be solved by optimising a weighting function $r(y)$ so that the integral R:

$$R = \int r(y) \; d(y) \; dy$$

reproduced the RBE ratio between the two neutron beams. This approach only assumes a correlation between the RBE of a given neutron beam and the shape of its microdosimetric dose distribution. It does not require further assumption regarding the biophysical meaning of the energy actually deposited in the site.

The specification of radiation quality for neutron therapy beams requires that the parameter R is determined with an uncertainty of less than 5%. The crucial problem therefore remains as to how accurately the weighting function $r(y)$ can be optimised in order to fulfil this requirement. During the 1980s, biological intercomparisons of neutron therapy beams in the energy range between d(14) + Be and p(65) + Be became available. At the same time, a microdosimetric intercomparison programme carried out under the auspices of the EORTC (European Organization for Research and Treatment of Cancer) enabled the measurement of the microdosimetric characteristics for 14 different neutron beams including those used in the biological experiments.

By using the microdosimetric spectra and the RBE ratios determined for the same neutron beams under identical irradiation conditions as input data, the weighting function $r(y)$ could be optimised numerically by an iterative procedure. This unfolding method has been applied several times in microdosimetry to evaluate empirically biological weighting functions. Assuming an initial guess function (the RBE against LET relation of Barendsen was used[34–36], the parameters of the function r(y) are optimised by successive iterations in order to match the calculated parameter type R and the experimental RBE ratio for each neutron beam. The main limitations of this approach are the energy range, the biological endpoint, and the dose level for the RBE values used as input data.

This type of calculation has actually been performed using the data for nine neutron beams of different energies ranging from [d(4) + Be] to [p(65) + Be][16]. The shapes of the optimised weighting functions found by using two different series of RBE ratios are similar to that of the RBE against LET curves. The integral of the weighted dose distribution, R, gives an estimate of the RBE of the beam. In the case of (p(65) + Be) neutrons, this integral is equal to 1, as this neutron beam was taken as reference radiation quality. The weighted dose distribution enable the identification of the secondary radiation components that are mainly responsible for the differences in RBE.

The numerical procedure used enables the determination of the uncertainty of the weighing function and the parameter R according to the uncertainties of RBE input data. Assuming that the ideal weighing function r(y) has been found, the ratio of experimental RBE and calculated R parameters would be expected to be equal to 1. The statistical analysis of the curve RBE/R against R therefore indicates the accuracy achievable in estimating the RBE ratio for a given neutron beam from the microdosimetric parameter R. With the RBE values and microdosimetric data available at present, the overall uncertainty of R is found to be <3% (±1SD), including the uncertainty of input RBE values and uncertainties of microdosimetric measurements. The empirical procedure proposed for the determination of the biological weighting function therefore gives an adequate solution of the problem of accounting for differences in radiation quality between different neutron therapy facilities (Figure 5)[16].

The calculation of the weighing function may be improved if further progress in biological data including data for different endpoints is achieved. In that respect, systematic determination of RBE values for late tolerance of normal tissues performed in similar experimental conditions for a large number of neutron and other heavy particle beams would be of greatest interest[37].

CONCLUSION

When new treatment modalities (in particular new types of radiations) are introduced in radiation therapy, it is important to be able to transfer, exchange, and report the information about these treatments and their results as quickly as possible and in the most objective way.

Exchange of the medical information (e.g. type of patient recruitment, complete description of the initial

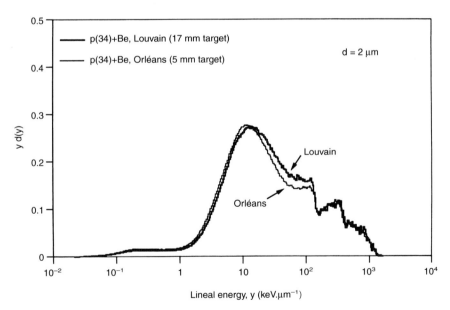

Figure 5. Comparison of the microdosimetric distributions obtained for two neutron beams of the same nominal energy p(34) + Be at the neutron therapy facility of Orléans (5 mm thick Be target) and at Louvain-la-Neuve (17 mm thick Be target). The major difference in the spectra occurs between 20 and 100 keV.µm^{-1} which is mainly due to the contribution of protons with relatively low energy. This difference is physically explained by the different target thickness used in the two centres. Biologically, using intestinal crypt regeneration in mice as endpoint, a RBE difference of 6% has been observed between the two neutron beams, which is clinically significant. From Pihet et al[16].

clinical status, response to treatment, complications, follow-up, etc.) should be made using universally agreed-upon terminology, definitions, classifications, etc.

From a more technical point of view, a consensus must be aimed at about dosimetry and dose specification for reporting. During the past decade, the ICRU has been actively involved in reaching this consensus and improving uniformity. ICRU Report 45[5] contains a protocol for clinical neutron dosimetry. A similar report is in press with a protocol for clinical proton dosimetry. Actually, several direct dosimetric intercomparisons for neutron and proton beams have shown that an agreement on absorbed dose better than 1% could be achieved when applying these protocols, and in particular the recommended numerical values of the required quantities.

Concerning dose specification for reporting, the approach of ICRU Report 50[11] recommended for photons can be easily adapted for fast neutrons. For protons, the ICRU reference point (as described in Report 50) should be adopted. In addition, accurate methods to describe the dose variation at the level of the target volume(s) and the normal tissues at risk are in principle available in all centres involved in proton beam therapy.

Specification of radiation quality, and the related problem of RBE, in the field of heavy particle therapy is complex. Microdosimetry provides an objective description of the radiation quality at the point of interest under the actual irradiation conditions. For neutrons, microdosimetry spectra describe the spectra of the secondary particles through which the neutrons directly produce their biological effect.

RBE determinations are absolutely needed for applying the new types of beams in optimal conditions. The data most clinically relevant are those obtained for late normal tissue tolerance at a standard fractionation scheme of 2 Gy per fraction.

Besides the classical concept of RBE introduced jointly by the ICRP and the ICRU in 1963, the concepts of 'reference RBE' and 'clinical RBE' are proposed for application in heavy particle therapy. When reporting the technical conditions of the treatment, the results of the experimental RBE determination(s), the best estimate of the reference RBE, and the selected clinical RBE must be clearly mentioned. Correlation of the experimental RBE values and of the microdosimetric description of the beams improves the confidence in both sets of data.

REFERENCES

1. Wambersie, A. *Neutron Therapy from Radiobiological Expectation to Clinical Reality*. Radiat. Prot. Dosim. **44**, 379–395 (1992).
2. Tubiana, M., Dutreix, J. and Wambersie, A. *Introduction to Radiobiology* (London: Taylor & Francis) (1990).
3. Catterall, M. and Bewley, D. K. *Fast Neutrons in the Treatment of Cancer* (London: Academic Press) (1979).
4. Batterman, J. J., Breur, K., Hart, G. A. M., van Peperzeel, H. A. *Observations on Pulmonary Metastasis in Patients after Single Doses and Multiple Fractions of Neutrons and Cobalt-60 Gamma Rays*. Eur. J. Cancer **17**, 539–548 (1981).
5. International Commission on Radiation Units and Measurements (ICRU). *Clinical Neutron Dosimetry. Part I: Determination of Absorbed Dose in a Patient Treated by External Beams of Fast Neutrons*. ICRU Report 45 (Bethesda, MD: ICRU Publications) (1989).
6. American Association of Physicists in Medicine (AAPM). *Protocol for neutron beam dosimetry*. Report No 7 (New York: American Institute of Physics) (1980).
7. Broerse, J. J., Mijnheer, B. J. and Williams, J. R. *European Protocol for Neutron Dosimetry for External Beam Therapy*. Br. J. Radiol. **54**, 882 (1981).
8. International Commission on Radiation Units and Measurements (ICRU). *Clinical Proton Dosimetry. Part I: Beam Production, Beam Delivery and Measurement of Absorbed Dose*. (Bethesda, MD: ICRU Publications) (in press).
9. International Commission on Radiation Units and Measurements (ICRU). *Stopping Powers and Ranges for Protons and Alpha Particles*. Report 49 (Bethesda, MD: ICRU Publications) (1993).
10. International Commission on Radiation Units and Measurements (ICRU). *Photon, Electron, Proton and Neutron Interaction Data for Body Tissues*. ICRU Report 46 (Bethesda, MD: ICRU Publications) (1992).
11. International Commission on Radiation Units and Measurements (ICRU). *Prescribing, Recording and Reporting Photon Beam Therapy*. ICRU Report 50 (Bethesda, MD: ICRU Publications) (1993).
12. Mijnheer, B. J., Battermann, J. J. and Wambersie, A. *What Degree of Accuracy is Required and can be Achieved in Photon and Neutron Therapy?* Radiother. Oncol. **8**, 237–252 (1987).
13. Kellerer, A. M. and Rossi, H.H. *The Theory of Dual Radiation Action*. Curr. Topics Radiat. Res. Q. **8**, 85–157 (1972).
14. Menzel, H. G., Pihet, P. and Wambersie, A. *Microdosimetric Specification of Radiation Quality in Neutron Radiation Therapy*. Int. J. Radiat. Biol. **57**, 865–883 (1990).
15. Gueulette, J., et al. *RBE Variations between Fast Neutron Beams as a Function of Energy. Intercomparison Involving 7 Neutron Therapy Facilities*. Bull. Cancer/Radiothér. **83** (Suppl 1), 55s–63s (1996).
16. Pihet, P., Menzel, H. G., Schmidt, R., Beauduin, M. and Wambersie, A. *Biological Weighting Function for RBE Specification of Neutron Therapy Beams. Intercomparison of 9 European Centres*. Radiat. Prot. Dosim. **31**, 437–442 (1990).

17. Gueulette, J., Grégoire, V., Octave-Prignot, M. and Wambersie, A. *Measurements of Radiobiological Effectiveness in the 85 MeV Proton Beam produced at the Cyclotron CYCLONE of Louvain-la-Neuve, Belgium.* Radiat. Res. **145**, 70–74 (1996).
18. Stone, R. S. *Neutron Therapy and Specific Ionization.* Am. J. Roentgenol. **59**, 771–785 (1948).
19. Gregoire, V., Beauduin, M., Gueulette, J., De Coster, B. M., Octave-Prignot, M., Vynckier, S. and Wambersie, A. *Radiobiological Intercomparison of p(45) + Be and p(65) + Be Neutron Beams for Lung Tolerance in Mice after Single and Fractionated Irradiation,* Radiat. Res. **133**, 27–32 (1993).
20. Hornsey, S., Myers, R., Parnell, C. J., Bonnett, D. E., Blake, S. W. and Bewley, D. K. *Changes in Relative Biological Effectiveness with Depth of the Clatterbridge Neutron Therapy Beam* Br. J. Radiol. **61**, 1058–62 (1988).
21. Gueulette, J., Böhm, L., De Coster, B. M. Vynckier, S., Octave-Prignot, M., Schreuder, A. N., Symons, J. E., Jones, D. T. L., Wambersie, A. and Scalliet, P. *RBE Variation as a Function of Depth in the 200 MeV Proton Beam produced at the National Accelerator Centre of Faure (South Africa).* Radiother. Oncol. (submitted).
22. International Commission on Radiation Units and Measurements (ICRU). *Quantitative Concepts and Dosimetry in Radiobiology.* ICRU Report 30 (Bethesda, MD: ICRU Publications) (1979).
23. International Commission on Radiation Protection (ICRP). *Report of the RBE Subcommittee to the International Commission on Radiation Protection and the International Commission on Radiation Units and Measurements.* Health Phys. **9**, 357–386 (1963).
24. International Commission of Radiological Protection (ICRP). *1990 Recommendations of the International Commission on Radiation Protection.* ICRP Publication 60. Ann. ICRP **21**(1–3) (Oxford: Pergamon Press) (1991).
25. Wambersie, A. and Menzel, H. G. *RBE in Fast Neutron Therapy and in Boron Neutron Capture Therapy. A Useful Concept or a Misuse?* Strahlentherapie **169**, 57–64 (1993).
26. Hornsey, S. *Experimental Central Nervous System Injury from Fast Neutrons.* In: Radiation Injury to the Nervous System. Eds P. H. Gutin et al (New York: Raven Press) pp. 137–48 (1991).
27. Van Der Kogel, A. J. *Late Effects of Radiation on the Spinal Cord: Dose–Effect Relationships and Pathogenesis.* Thesis, University of Amsterdam, Publication of the Radiobiological Institute TNO, Rijswijk, The Netherlands (1979).
28. Hall, E. J., Astor, M. and Brenner, D. J. *Biological Intercomparisons of Neutron Beams used for Radiotherapy Generated by p-Be in Hospital-based Cyclotrons.* Br. J. Radiol. **65**, 66–71 (1992).
29. Wambersie, A. and Batterman, J. J. *Practical Problems Related to RBE in Neutrontherapy.* In: Progress in Radio-oncology III. Eds K. H. Kärcher et al (Vienna: ICRO) pp. 155–162 (1987).
30. International Commission on Radiation Units and Measurements (ICRU). *Microdosimetry.* ICRU Report 36 (Bethesda, MD: ICRU Publications) (1983).
31. Pihet, P., Gueulette, J., Menzel, H. G., Grillmaier, R. E. and Wambersie, A. *Use of Microdosimetric Data of Clinical Relevance in Neutron Therapy Planning.* Radiat. Prot. Dosim. **23**, 471–474 (1988).
32. Pihet, P. *Etude Microdosimétrique de Faisceaux de Neutrons de Haute Énergie. Applications Dosimétriques et Radiobiologiques.* Thesis, Université Catholique de Louvain, Louvain-la-Neuve, Belgium (1989).
33. International Commission on Radiation Units and Measurements (ICRU). *Determination of Dose Equivalents Resulting from External Radiation Sources.* ICRU Report 39 (Bethesda, MD: ICRU Publications) (1985).
34. Barendsen, G. W. *Impairment of the Proliferative Capacity of Human Cells in Culture by Alpha-particles of Different Linear Energy Transfer.* Int. J. Radiat. Biol. **8**, 453–466 (1964).
35. Barendsen, G. W. *Sublethal Damage and DNA Double-strand Breaks have Similar RBE-LET Relationships; Evidence and Implications.* Int. J. Radiat. Biol. **63**, 325–330 (1993).
36. Barendsen, G. W. *The Relationships between RBE and LET for Different Types of Lethal Damage in Mammalian Cells: Biophysical and Molecular Mechanisms.* Radiat. Res. **139**, 257–270 (1994).
37. Wambersie, A., Pihet, P. and Menzel, H. *The Role of Microdosimetry in Radiotherapy.* Radiat. Prot. Dosim. **31**, 421–432 (1990).

A UNIFIED SYSTEM OF RADIATION BIO-EFFECTIVENESS AND ITS CONSEQUENCES IN PRACTICAL APPLICATIONS

D. E. Watt
School of Physics and Astronomy
University of St Andrews
St Andrews, Fife, KY16 9SS, UK

Abstract — A model has been formulated as a function of the quality parameter, 'mean free path for linear primary ionisation', denoted here by lambda, for prediction of the risk of inactivation of irradiated mammalian cells. The model is intended to be a unified one applicable to any radiation type. In effect, it determines the yield of double-strand breaks per charged particle fluence in the medium of interest. It includes provision for direct and indirect action with a repair capability that is dependent on the duration of the irradiation, rather than the dose rate, and the known variation in radiosensitivity during progress of the cell cycle. The general validity of the model, derived from probe experiments with heavy charged particles, is tested against the results of Hofer *et al* in their experiments on synchronised CHO cells, labelled with the Auger electron emitter ^{125}I. Good agreement is obtained for an allocated mean repair time of 2.3 h. Analysis of the predictions of the model leads to new interpretations on the damaging role of heavy particles in therapy and protection, viz: that delta rays around fast heavy ion tracks contribute minimally, possibly negligibly, to the damage; that protons, and neutrons because of the short ranges of the recoils at their most damaging energy, can never reach the saturation level found for the heavier ions at the same lambda and that there is no difficulty in extrapolation of effects in cellular targets for high and low LET radiations to low doses. Maximum RBEs, known to be LET dependent, are found to have a common maximum near 2 nm when the radiation quality is defined as 'the mean free path for linear primary ionisation'. On the basis of this work there seem to be realistic prospects of defining a new unified fluence-based system of dosimetry which, with conceptually new instrumentation designed to have a radiation response simulating that of the nanometre sites, possibly a segment of the DNA, will be more rigorous and meaningful than the currently adopted system. Examples of application of the new quality concept leads to proposals for improved heavy particle therapy and of risk evaluation in radiation protection.

INTRODUCTION: DAMAGE MECHANISM

Either of the two physical quantities, $L_{100,D}$, the dose restricted LET[1], or lambda (λ), the mean free path for linear primary ionisation of the equilibrium charged particles[2], have been shown to have special properties which enable a more generalised and accurate description of radiation quality to be made — judged on the criterion of unified correlation for various biological endpoints and varied radiation types[2–4]. Lambda, being the more fundamental quantity, is applied here to the interpretation of damage mechanisms and to the development of a generalised model of biological effectiveness. Note that the quantity lambda, the linear primary ionisation, which excludes the delta rays should not be confused with the volume quantity, the specific primary ionisation used by Lea[5] and by Harder[1], which includes the ionisation from secondary electrons. Tables of the commonly used quality parameters for any radiation type can be found in Reference 24.

The unified system expresses the cross section for induction of the effect, for a specified biological endpoint, as a function of lambda. Use of effect cross sections for interpretation of damage mechanisms has two distinct advantages: cross sections provide a means of using the tracks as probes to explore for target structure within the cell nucleus and they can be expressed as functions of the various possible contenders for physical quality parameters to determine which best unifies the results for all radiation types. Furthermore, the data for different biological targets can also be correlated in a generalised way by taking ratios with respect to the recorded saturation cross sections at $\lambda < \sim 1.8$ nm. The results, obtained from the experimental data for eukaryotic and prokaryotic cells, provide a clear indication of the quantities to be used and the mechanisms involved for quantitative expression of the radiation action.

The most important findings are as follows.

(1) The fundamental mechanism initiating damge appears to be due to the spatial correlation of two random interactions along single charged particle tracks (determined by λ) with the strands in the intranuclear DNA to produce double strand breaks. This is a template effect. Similarly spaced interactions at other orientations are predicted to have negligible effect which, if correct, negates the relevance of energy imparted.

(2) As the radiation quality is defined by the zeroth moment of energy transfer, it follows that absorbed dose and LET which involve the first moment of energy transfer, even at the sub-microscopic level, cannot be physically relevant factors for the description of radiation effects in mammalian cells.

The work leads to interesting conclusions some of which are already well established in radiobiology but others are conceptually new. Arguments in justification of the conclusions reached and listed below are given in the references indicated.

(1) Radiation quality is best expressed in terms of the 'mean free path for linear primary ionization'[3,4].

(2) Double strand breaks (dsbs) in the intracellular DNA of mammalian cells are the dominantly important lesions[3] for the endpoints of eukaryotic cell inactivation, chromosome aberrations, mutations and oncogenic transformations[6].

(3) DNA segments within the cell nucleus constitute the multiple sensitive sites. About 15 to 20 sites are at risk upon a mean chord traversal by a charged particle track in the nucleus[3]. The probability of damaging all sites at risk is an order of magnitude larger for fast heavy charged particles than for electrons — because of the interplay of particle range, which is often less than the nuclear diameter for the most damaging electrons, and the value of the quality parameter lambda for damage (averaged over the spectrum of electrons generated)[4].

(4) From simple consideration of the role of secondary electrons and the short diffusion length of water radicals (~3 nm) estimates can be made of the size of radiosensitive sites and the probability that these will be activated[7,8]. This leads to the conclusion that action is due to single tracks in the equilibrium spectrum, at doses up to a few tens of gray[4].

(5) If damage is due solely to single track action then a mechanism must be proposed to account for the observation of sigmoidal dose–survival curves. Also, as single track action is incompatible with dose rate effects, which have been widely reported in the literature[9], another explanation must be found. Evidence is presented below to demonstrate that the solution may lie in the treatment of the repair characteristics of the cell. The conclusion is that for external irradiations with X and γ rays or electrons, the duration of the irradiation and not the dose rate appears to be a key factor[3]. For internal irradiations, with for example Auger emitters such as ^{125}I, the localised simultaneous emission of several electrons (average ~7) means that there is possibly a small contribution (<10%) from dual track effects (see Figures 1(a) and (b)).

(6) The magnitude of the saturation ('overkill') effect cross section for normal mammalian cells is determined by the projected cross-sectional area of the DNA at risk multiplied by the segment overlap factor (target multiplicity) along the mean chord traversal, i.e. ~50 μm² for fast heavy particles and ~3.5 μm² for equilibrium electron spectra (see 3 above).

(7) For fast heavy charged particles, delta rays are found to have almost negligible effect — damage probability ≤10^{-4} per electron[2,3]. However, for slow heavy particles (Z > 6), in the saturation region (γ < 2 nm), the yield of delta rays produced in the cell environment becomes so large that the effect cross section increases by up to a factor of three times the saturation level for heavy particle irradiation of mammalian cells. The effect is greater for smaller targets such as yeast cells and bacillus subtilis[10–12] and is attributed to action in neighbouring cells by the combined effects of the very high delta ray yield and their ranges which cause optimal damage if they stop in the neighbouring cell nucleus. Nevertheless the damage efficiency of an individual δ ray remains at ≤10^{-4}. The reason proposed for the low effectiveness per delta ray is that electrons are only efficient at inducing DNA dsbs at the very end of their range when their mean free path for linear primary ionisation is near 2 nm (~100 eV). Then their range is only a few nm, they have to stop in the DNA and they have to produce two ionisations which correlate exactly with the double strands in the DNA. The combined probabilities of achieving these requirements is clearly very small[2,3].

(8) Extrapolation of observed cellular effects to low doses is different for electrons and heavy particles and is determined automatically in the application of Equation 3 by the projected cross-sectional area of the DNA, the target multiplicity at risk (n_0) — modified by the particle range if less than the mean chord length, and the value of lambda. The initial slopes of the survival curves, are different for electrons compared to heavy particles due to the combined influences of the repair function and the degree to which the respective radiations can sustain two nanometre-spaced events throughout the cell nucleus to cause a significant degree of damage. This is achievable with heavy charged particles (Z > 1) but not with electrons. Alpha particles can just achieve optimum damage but protons cannot because their range is less than the cell nuclear diameter when λ ~ 2 nm. Consequently the effect cross section for neutrons, which damage mainly via the proton recoils, must always be less than that of other heavy charged particles at the same lambda[4]. Use of RBE in the dose-based system conveys the opposite message, because it is equivalent to the fundamental effect cross section weighted by the LET ratio, as shown in Figure 2(a) and (b).

(9) RBEs plotted conventionally as a function of LET are shown to lie on independent curves for each radiation type and, therefore, it is invalid to draw a single curve through values of RBE obtained with different radiation types[3] (and different LETs — see Figure 2(a)). In the past it has been demonstrated that RBE reaches a maximum value for any fast heavy charged particle type when z^2/β^2 ~ 2000 to 3000[10]. The parameter z^2/β^2 is not valid for slow heavy particles. However it can now be replaced by the general rule that for particles trav-

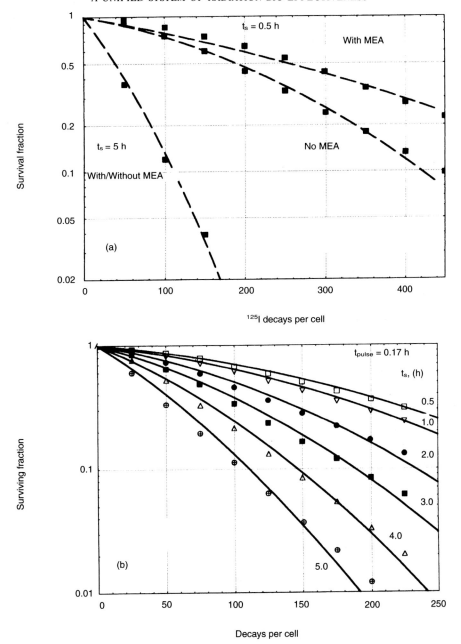

Figure 1. (a) Survival fractions, calculated (– – –) from the relation $F = \exp[-\sigma_B \cdot \phi_{eq}]$ and Equation 3, are compared with the measured results obtained by Hofer and Bao[19] (■) for irradiations with the Auger electron emitter ^{125}I at 0.5 h and 5 h into the S phase of the cell cycle of synchronised cells. The 0.5 h data were obtained with and without the presence of the radioprotector cysteamine (MEA). (b) A comparison is made between the calculated (——) survival fractions and the observations of Hofer et al[18] (■) for the survival of cells irradiated at the various times, t_S, indicated into the S phase of the cell cycle. The allocated mean repair time, t_r, is 2.3 h and a small amount of dual track action ranging from 10% ($t_c = 0.5$ h) to 3% ($t_c = 5$ H) was included (see text).

ersing the cell nucleus the maximum RBE for any specified endpoint in mammalian cells will always occur when the net 'mean free path for linear primary ionisation', λ, is equal to 2 nm (Figure 2(b)).

For λ < 2 nm, saturation damage occurs and so the significant excess damage caused by the δ ray contribution, although large per unit fluence, is small in terms of the increasing dose and therefore does

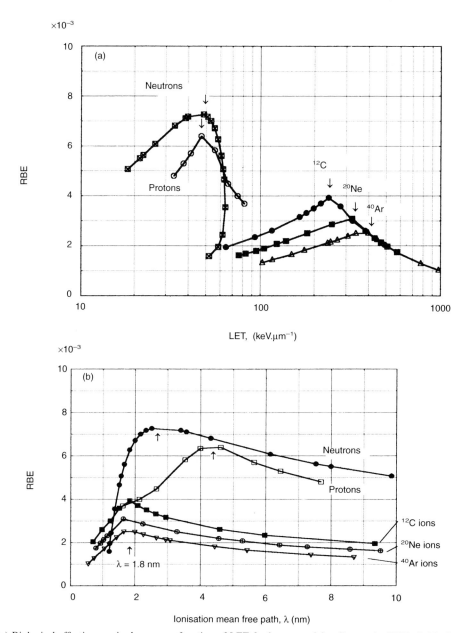

Figure 2. (a) Biological effectiveness is shown as a function of LET for heavy particles. To get the RBE, divide the ordinate by 1.56×10^{-3} or 4.65×10^{-4} for reference radiation of 250 kV X rays or ^{60}Co γ rays respectively. Note that the maxima RBE, calculated from a fitted curve to the experimental cross sections are strongly dependent on LET and radiation type. The source of data is listed in reference 25. (b) RBEs from Figure 2(a) are expressed as a function of λ, the mean free path for primary ionisation. Note that the position of the maxima RBEs is at a common value of λ ~ 2 nm independently of radiation type. The slight displacement for protons and neutrons is due to the limited range (less than the cell nuclear diameter) of the charged particles (protons) — see text.

not affect the position of the maximum RBE as a function of λ. For example, the observed excess damage[11,12], amounting to about 300% in the inactivation cross section, due to delta rays produced by slow accelerated argon ions in mammalian cells, at lambda values of ~0.2 nm, is not revealed using RBE thereby emphasising the inappropriateness of absorbed dose as a dosimetry parameter.

(10) As the effect cross section is an absolute measure of the biological effect per unit radiation fluence, for the irradiation conditions used, it should be possible to model the absolute biological effectiveness in terms of the relevant equilibrium charged particle fluence weighted appropriately for the probabilities of direct and indirect actions expressed in terms of appropriate quality parameters. A model for cell inactivation is attempted below in which the direct action is expressed as a function of lambda. The contribution to damage by indirect action is determined by treating the particle tracks as a line source of radicals and utilising the track-restricted LET to determine the yield of OH radicals from the known chemical G value[13]. The diffusion length is used to estimate the radical interaction cross section for production of sub-lesions. In this way the bio-effectiveness can be expressed in terms of basic physico-chemical quantities, obtainable from independent sources, and biological parameters such as the probability for production of sub-lesions, taken to be single strand breaks in the DNA, the change in radiosensitivity through the cell cycle and the mean repair rate. As some of the biological parameters cannot yet be determined independently, empirical means have been used. Consequently validation tests of the model have been attempted for various radiation types applied in a range of different situations.

Whatever the outcome of the modelling, the empirical identification of quality parameters that give a better correlation of radiation effects for any radiation type, as assessed by using the experimental effect cross section, means that better control should be achievable over the effectiveness of radiation fields in therapy and protection. One consequence is that neutrons are demonstrated to be less damaging to mammalian cells than are the heavier particles, on an equal fluence basis. RBEs, quality factors, or radiation weighting factors, in their present form become redundant. Also, the curve of effect cross section as a function of lambda, constitutes a suitable response function for new instrumentation to measure biological effectiveness[14,15].

CALCULATION OF BIOLOGICAL EFFECTIVENESS

Repair of damage, the $U(t_i)$ function

The duration of the irradiation, the nature of the cell population, whether it is synchronised or asynchronised, and the link to the position in the cell cycle are important factors in determining the shape of the cell survival curve from the present model. The function $U(Z,t_i)$ represents the probability that induced double stranded breaks in the DNA will not be repaired by completion of the cell cycle. The temporal expression for repair is similar in form to those used by Lea[5], Harder[16], and Kiefer[17] but introduces a specified damage fixation time, allowance for variation in radiosensitivity in the cell cycle and synchronisation when relevant. Further study is being pursued to investigate the validity of the proposed function and to extract values for the mean repair times for typical irradiation conditions.

(a) *For synchronised cells*

$$U(Z,t_i) = \frac{1}{t_i} \int_{t_c}^{t_c+t_i} Z(F,t) \exp[-(t_m - t)/t_r] dt \quad (4)$$

(b) *For asynchronised cells*

$$U(Z,t_i) = \frac{1}{t_i} \frac{1}{t_m} \int_0^{t_i} \int_0^{t_m} Z(F,t_j) \exp[-(t_j - t)/t_r] dt_j dt \quad (5)$$

$Z(F,t)$ is an approximate function which takes into account the change in radiosensitivity during the cell cycle. t_i, t_c and t_m are respectively the duration of the irradiation, the time into the cell cycle and the time at which damage is considered fixed (mitosis).

MODEL FOR CELL INACTIVATION BY IONISING RADIATION: PROBABILITY OF DOUBLE-STRAND BREAKS IN INTRACELLULAR DNA

The cumulative probability for the inactivation of mammalian cells is written as:

$$\frac{\sigma_B}{\sigma_S} = \left[\epsilon_{dsb}(\lambda) + \left(\frac{\sigma_{ssb}}{\sigma_S}\right)^2 + 2\epsilon_{ssb}(\lambda)\left(\frac{\sigma_{ssb}}{\sigma_S}\right)\right] U(Z,t_i)$$

$$\sigma_S = \left[\sigma_{g,DNA} \, n_0 \, \frac{\overline{R}}{d}\right] \quad (3)$$

where

n_0 = number of dsb segments at risk per track traversal (~15 for fast ions);

d = mean chord length through the cell nucleus;

R = the mean projected range of the relevant tracks; if $R > d$,

R/d = 1 which partially allows for the reduced number of DNA segments at risk for 'stopper and insider' tracks;

$\sigma_{g,DNA}$ = the projected area of the intranuclear DNA — varies with cell type;

σ_S = the saturation cross section = $\sigma_{g,DNA} n_0 R/d$; if $R > d$, $R/d = 1$;

$\epsilon_{dsb}(\lambda)$ = the efficiency of dsb production by direct action;

$\epsilon_{dsb}(\lambda)$ = $e^{-\lambda_0/\lambda}(1 - e^{-1.0/\lambda})^2$; $\lambda_0 = 1.8$ nm and λ = mfp for linear primary ionisation;

$\epsilon_{ssb}(\lambda)$ = $1 - e^{-1.0/\lambda}$ = efficiency for ssb production by direct action in a single DNA strand;

σ_{ssb} = $a_1 e^{-a_2}(1 - e^{-a_3 L})$ where a_1 is a radical interaction cross section determined by the diffusion length, a_2 is a scavenging efficiency for radicals and $a_3 L_{T,100}$ is the mean number of reactions with DNA strands[8];

$U(Z, t_i)$ = probability that dsbs remain unrepaired,

t_i = duration of irradiation,

t_r = mean repair time.

APPLICATION OF THE MODEL TO INCORPORATED RADIONUCLIDES

As a validation test of the model, the cell survival experiments of Hofer et al[18,19] are of special interest because they are unique in the manner by which the exposure times are controlled to specific points in the time cycle of synchronised cells. The experiments were performed using ^{125}I-iodo-deoxyuridine-labelled Chinese hamster ovary cells. The irradiation is conducted at a constant dose rate. By accumulating the decays at liquid nitrogen temperatures, the initial radiation damage is effectively constrained to a known point in the S phase of the cell cycle. Electron-emitting radionuclides, especialy those decaying by Auger electron emission, are known to be excessively damaging when incorporated into the nuclear component of mammalian cells. The subject is of considerable topical interest because of the importance in nuclear medicine as well as the general implications for radiation protection. Conventional dosimetry is widely recognised as being inadequate to quantify the observed effects[20]. The claim is made that Auger electron emitters such as ^{125}I exhibit the damage properties of high LET particles. Although the model given here is evolved from fundamental principles comprising well-established physical quantities there are two key parameters which are not known with any certainty — the mean time taken for the self-repair of DNA double strand breaks and the proportion of damage that may be attributable to radical diffusion from two separate tracks. The model predicts that for doses less than a few tens of gray, the action will be by single charged particle tracks and should give rise to pure exponential survival curves in the Hofer-type experiments. However in the special circumstances of an Auger electron emitter of highly localised electrons, ~7 per decay, there is quite likely to be a small number of coincident events due to multiple tracks. The reasonably good agreement obtained between the calculated and experimental values is seen in Figures 1(a) and (b). A mean repair time of 2.3 h is required to obtain consistency with the observed data. Additionally a small portion of damage of between 3 and 10% is attributed to simultaneous dual track events. The general trends of survival with irradiation time and cycle time are found to be well described as is the effect of adding the radioprotector (Figure 1(a)). Despite the good agreement, it should be noted that Hofer et al express doubts about dsbs in DNA being the actual cause of cell death.

HEAVY PARTICLE THERAPY

Much discussion has ensued on the optimum type of accelerated ion to be used in heavy particle therapy[21,22]. From the mechanisms proposed here, the maximum RBE will always occur when the mean free path for linear primary ionisation of the charged particle is uniquely equal to ~2 nm in the cell nucleus. (The corresponding LETs are multi-valued viz: p, 75 keV.μm^{-1}; α, 125 keV.μm^{-1}; ^{12}C, 217 keV.μm^{-1}; ^{20}Ne, 254 keV.μm^{-1}; ^{40}Ar, 340 keV.μm^{-1}. Consequently the biophysical therapeutic advantage is expected to be the same for any ion type having the requisite mean free path of 2 nm. There is an exception to this rule for protons (and proton recoils from neutrons) which, because of their short ranges (less than the mean chord distance through the cell nucleus) have maxima RBE at ~47 keV.μm^{-1} (λ = 4.6 nm) and 52 keV.μm^{-1} (λ = 2.3 nm) respectively (see Figures 2(a) and (b)). Thus other factors to be considered are the ranges over which the 2 nm spacing can be sustained in the cell nucleus, and the possible effects on surrounding healthy tissue of the delta ray penumbra emitted by slow ions in the region of saturated damage (λ < 2 nm). The delta ray dose contribution does not alter the position of the maximum RBE. In terms of effect cross sections, protons and neutrons are somewhat less damaging, because of the range limitation of protons at λ ~ 2 nm, than are accelerated helium ions and other heavy ions with Z > 2. The latter ions with the same λ will have identical effects. Consequently in radiation therapy there seems to be no justification for building large machines to accelerate ions heavier than helium or carbon as apparently no biophysical advantage is to be gained. Beam penetration to deep-seated tumours can be achieved more readily and economically with the lighter ions.

In boron neutron capture therapy, it is interesting to note that the ^{10}B(n,α)^{7}Li reaction products are at the position of optimum saturation damage, free from delta ray effects. Consequently, for intracellular action, damage should be 100% efficient making fractionation of treatment unnecessary, if judged solely on the biophysical arguments.

PROPOSED SYSTEM OF RISK CONTROL FOR RADIOLOGICAL PROTECTION

On the basis of the proposed unified system of 'dosimetry' an improved system of 'dose' limitation based on fluence can be constructed for better risk control. When compared with the current regulatory system, points of anomaly emerge, e.g. the inappropriateness of using same ICRP radiation weighting factor of 20 for fast neutrons, heavy accelerated ions and natural alpha particles. For a fluence-based system, alpha particles have a maximum effect cross section three times smaller than those of the most damaging heavy particles. The maximum effect cross sections for neutrons are smaller than those for alpha particles because the most damaging proton recoils from neutrons have ranges less than the cell nucleus. The significant differences in effectiveness, as a function of photon energy for X and gamma rays and fast electrons, per unit equilibrium fluence is appropriately quantified (Figure 3a). In the case of neutron irradiations, the evaluation leads to a simple smooth effect curve which harmonises with the histogram of radiation weighting factors recommended by ICRP making the latter's step function unnecessary[23], (Figure 3b).

The ICRP risk coefficients are currently as follows:

(1) For radiation workers, the risk corresponding to a dose limit of 20 mSv per year is determined from the cancer risk coefficient (R_{ICRP}) of 4×10^{-2} Sv^{-1} (called the 'nominal probability coefficient' by ICRP) and is equal to $20 \times 10^{-3} \times 4 \times 10^{-2} = 8 \times 10^{-4}$ per year.

(2) For the general population, the risk corresponds to a dose limit of 1 mSv per year. The cancer risk coefficient (R_{ICRP}) is 5×10^{-2} Sv^{-1} and the probability is $1 \times 10^{-3} \times 5 \times 10^{-2} = 5 \times 10^{-5}$ per year. Risk factors, R_f per unit fluence, proposed in the new system, are related to the current ICRP risk factors by:

$$R_{f,\gamma} = R_{ICRP} \frac{K_{f,v_c}}{\sigma_{B,v_c}} Q_{v_c} \sigma_{B,\gamma} = 0.4 \times \sigma_{B,\gamma}$$

$$R_{f,n} = R_{ICRP} \frac{K_{f,n_c}}{\sigma_{B,n_c}} Q_{f,n_c} \sigma_{B,n} = 2.11 \times \sigma_{B,n}$$

$$R_{Tot} = R_{f,\gamma} \Phi_\gamma + R_{f,n} \Phi_n \qquad (4)$$

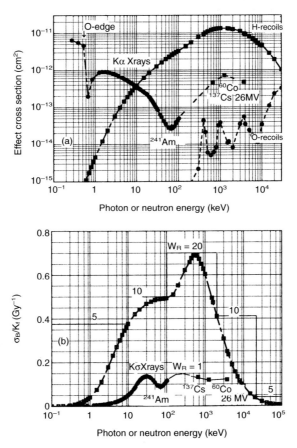

Figure 3. (a) Effect cross sections (absolute biological effectiveness) in water, determined from the model, are plotted as a function of photon or neutron energy. This fluence-dependent system reveals the widely varying changes in effect with radiation energy. (b) The results for (a) can be converted to the dose-based system of ICRP by dividing by the kerma factors. The radiosensitivities obtained are seen to fall nicely within the ICRP 60 recommendations without the need for a step function.

where Ks are kerma factors, Qs are quality factors for the reference radiations, subscript 'c'. R_{Tot}, the net risk for a mixed field of photons and neutrons, is obtained by substitution into Equation 4.

ACKNOWLEDGEMENT

This work was partially supported by the Commission of the European Communities, Radiation Protection Programme under contract F13P-CT92-0041.

REFERENCES

1. Bartels, E. R. and Harder, D. *Restricted LET Represents Particle Track Structure in Nanometre Targets.* In: Proc. 10th Int. Congr. of Radiation Research, August 27–Sept. 1, Wurzburg, Germany (1995).
2. Watt, D. E. *Identification of Biophysical Mechanisms of Damage by Ionising Radiation.* Radiat. Prot. Dosim. **13**(1–4), 285–294 (1985).
3. Watt, D. E., Al-Afran, I. A. M., Chen, C. Z. and Thomas, G. E. *On Absolute Biological Effectiveness and Unified Dosimetry.* J. Radio. Prot. **9**(1), 33–49 (1989).
4. Simmons, J. A. and Watt, D. E. *Radiation Protection — a Radical Reappraisal* (Submitted to Medical Physics Publishing, USA) (1997).
5. Lea, D. E. *Actions of Radiations in Living Cells.* 2nd edn (Cambridge University Press) (1955).
6. Alkharam, A. S. and Watt, D. E. *Risk Scaling Factors from Inactivation to Chromosome Aberrations, Mutations and Oncogenic Transformations in Mammalian Cells.* Radiat. Prot. Dosim. **70**(1–4), 537–540 (This issue) (1997).
7. Chaterjee, A. and Holley, W. R. *Energy Deposition Mechanisms and Biochemical Aspects of DNA Strand Breaks by Ionizing Radiation.* Int. J. Quantum Chem. **39**, 709 (1991).
8. Watt, D. E. and Hill, S. J. A. *An Empirical Model for the Induction of Double-strand Breaks in DNA by the 'Indirect' Action of Ionising Radiation.* Radiat. Prot. Dosim. **52**(1/4), 17–20 (1994).
9. Hall, E. J. *Radiobiology for the Radiobiologist.* (London: Harper and Row) (1978).
10. Watt, D. E., Alkharam, A. S., Child, M. B. and Salikin, M. S. *Dose as a Damage Specifier in Radiobiology for Radiation Protection.* Radiat. Res. **139**(2), 249–251 (1994).
11. Kiefer, J., Rase, S., Schneider, E., Stratten, H., Kraft, G. and Liesem, H. *Heavy Ion Effects on Yeast Cells: Induction of Canavine Resistant Mutants.* Int. J. Radiat. Biol. **42**, 591–600 (1982).
12. Kraft, G., Blakely, E. A., Huber, L., Kraft-Weyrather, W., Miltenburger, H. G., Muller, W., Shuber, M., Tobias, C. A. and Wulf, H. Adv. Space Res. **4**, 219–226 (1984).
13. Watt, D. E., Kadiri, L. A. and Glodic, S. *Observed Cellular Effects Lead to a Track 'Core' Model of Radiation Action.* In: Biophysical Modelling of Radiation Effects. Eds K. H. Chadwick, G. Moschini and M. N. Varma. (Adam Hilger, London) pp. 201–209 (1992).
14. Watt, D. E. *An Approach Towards a Unified Theory of Damage to Mammalian Cells by Ionising Radiation for Absolute Dosimetry.* Radiat. Prot. Dosim. **27**(2), 73–84 (1989).
15. Breskin, A., Chechik, R., Colautti, P., Conte, V., De Nardo, L., Pansky, A., Shchemelinin, S., Talpo, G. and Tornielli, G. *A Single-electron Counter for Nanodosimetry.* Radiat. Prot. Dosim. **61**(1–3), 199–204 (1995).
16. Harder, D. and Virsik-Peuckert, P. *Kinetics of Cell Survival as Predicted by the Repair/Interaction Model.* Br. J. Cancer **49** (Suppl. VI), 243–247 (1984).
17. Kiefer, J. *A Repair Fixation Model Based on Classical Enzyme Kinetics.* In: Quantitative Mathematical Models in Radiation Biology (Schloss Rauisch-Holzhausen, Germany: Springer-Verlag) pp. 171–179 (1987).
18. Hofer, K. G., Van Loon, N., Schneiderman, M. H. and Charlton, D. E. *The Paradoxical Nature of DNA Damage and Cell Death Induced by ^{125}I Decay.* Radiat. Res. **130**, 121–124 (1992).
19. Hofer, K. G. and Bao, Shi-Ping. *Low-LET and High-LET Radiation Action of ^{125}I Decays in DNA: Effect of Cysteamine on Micronucleus Formation and Cell Killing.* Radiat. Res. **141**, 183–192 (1995).
20. Humm, J. L., Howell, R. W. and Rao, D. V. *Dosimetry of Auger-electron Emitting Radionuclides.* Report No 3 of AAPM Nuclear Medicine Task Group No 6. Med. Phys. (in press).
21. *Medical Satellite Meeting of the Second European Particle Accelerator Conference, Nice. June 14–16.* Eds P. Marin and P. Mandrillon (Editions Frontieres B.P. 33. Gif-sur-Yvette, Cedex, France) (1990).
22. *Fourth Workshop on Heavy Particles in Biology and Medicine.* Sept. 23–25, GSI, Darmstadt, Germany. Ed. G. Kraft (1991).
23. ICRP. *1990 Recommendations of the International Commission on Radiological Protection.* Publication 60. Ann. ICRP **21**(1–3) (Oxford: Pergamon Press) (1991).
24. Watt, D. E. *Quantities for Dosimetry of Ionizing Radiations in Liquid Water.* (Basingstoke: Taylor and Francis) (1996).
25. Watt, D. E. and Alkharam, A. S. *Charged Particle Track Structure Parameters for Application in Radiation Biology and Radiation Chemistry.* Int. J. Quantum Chem: Quantum Biol. Symp. **21**, 195–207 (1994).

RISK SCALING FACTORS FROM INACTIVATION TO CHROMOSOME ABERRATIONS, MUTATIONS AND ONCOGENIC TRANSFORMATIONS IN MAMMALIAN CELLS

A. S. Alkharam and D. E. Watt
School of Physics and Astronomy
University of St. Andrews
St Andrews, Fife, KY16 9SS, UK

Abstract — Analyses of bio-effect mechanisms of damage to mammalian cells in terms of the quality parameter 'mean free path for primary ionisation', for heavy charged particles, strongly suggests that there is a common mechanism for the biological endpoints of chromosome aberrations, mutations and oncogenic transformation. The lethal lesions are identified as unrepaired double-strand breaks in the intracellular DNA. As data for the various endpoints studied can be represented in a unified scheme, for any radiation type, it follows that radiation risk factors can be determined on the basis of simple ratios to the inactivation cross sections. There are intrinsic physical reasons why neutrons can never reach the saturation level of heavier particles for equal fluences. The probabilities of risk with respect to inactivation, for chromosome dicentrics, mutation of the HPRT gene and of oncogenic transformation are respectively 0.24, 5.8×10^{-5}, and 4.1×10^{-3}.

INTRODUCTION: DAMAGE MECHANISMS AND RISK FACTORS

Analysis of published results on the inactivation of mammalian cells exposed to ionising radiations has led to the identification of good quality parameters which have the capability of unifying data for any radiation type. These are the dose-restricted LET, $L_{100,D}$[1], and 'the mean free path for primary ionisation', λ_i[2]. Another quality parameter which has been used is z^2/β^2[3] but surprisingly there seems to have been little use of the track-restricted LET, $L_{100,T}$[4]. In Figure 1 a comparison is made of the trend with neutron energy of these quantities. λ_i, the zeroth moment of energy transfer, is the most fundamental of these quantities and is best for interpretive purposes. It is also more appropriate dimensionally. Therefore, it is proposed that radiation damage can be more accurately and more rigorously specified in terms of a fluence-based system using λ_i for the equilibrium charged particles generated by the radiation field[5]. However, for assessment of the deleterious effects of ionising radiations, as is required in radiological protection, it is necessary to be able to determine the risk of oncogenic transformation. Some years ago, preliminary work indicated that other important biological endpoints such as chromosome aberrations, mutations and oncogenic transformations all arose from a common mechanism of damage, viz. double-strand breaks in the intranuclear DNA[2]. It may be that certain specific double-strand breaks are involved[6] but the important point is that if the basic damage mechanism is the same, then it should be possible to determine scaling factors to enable the probability of occurrence of the other endpoints to be deduced from the results for inactivation. Yang et al in 1989[7] and Frankenberg in 1994[8] have demonstrated that unrepaired or misrepaired double-strand breaks in the DNA can lead to chromosome aberrations, gene mutations or oncogenic transformations. Here a graphical comparison is made between the quality parameters, λ and $L_{100,D}$ but the main objective is to conduct a more detailed study to test the feasibility of scaling risk for the various endpoints for application in a fluence-based system of radiological protection.

METHOD OF APPROACH

Cross sections for induction of the biological endpoints of interest (σ_B in cm^2) were extracted from published survival data using the relation:

$$\sigma_B = 1.6 \times 10^{-9} \frac{L_T}{\rho D} \qquad (1)$$

where L_T is the track average LET (in keV.μm^{-1}) for the equilibrium spectrum of charged particles involved and D is the dose (in Gy). Using Equation 1 cross sections were determined for the initial slopes of the survival curves, as this avoids any complications with cell

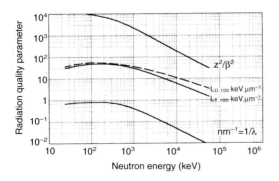

Figure 1. Radiation quality parameters fluence-weighted for the proton recoil equilibrium from neutrons in water.

recovery. In those cases where the survival curve is sigmoid, the initial slopes were determined by fitting a linear-quadratic expression for the yield of lesions and the corresponding cross sections were plotted as functions of the quality parameters, the mean free path for primary ionisation, λ_i and the dose-restricted LET, $L_{100,D}$. Results, for V79 chinese hamster cells, were obtained for inactivation, chromosome dicentrics, and mutations of the HPRT gene. Results are given for oncogenic transformation in C3H10T$_{1/2}$ mouse embryo cells.

INACTIVATION OF MAMMALIAN CELLS

Cross sections for inactivation of V79 cells[4] are shown as a function of λ_i and $L_{100,D}$ in Figures 2(a) and 2(b) respectively, for a wide range of heavy charged particle and neutron irradiations. Visual inspection confirms that both quality parameters give good grouping of the data within the spread of the physical and biological errors and it is not possible to distinguish clearly between their respective merits. More detailed examination reveals that the $\sigma-\lambda_i$ curve has a clear point of inflection at $\lambda_i = 1.4 \pm 0.5$ nm which is attributed to the mean chord spacing of the strands in the DNA segment and identifies the double-strand break as the critical lesion for inactivation for all the radiation types tested. Although neutrons are indirectly ionising radiations, their damage mechanism is seen to be identical to that of all the other heavy particles. Neutrons, however, can never achieve the saturation cross section at $\lambda \leq 2$ nm because of the limited range of protons at optimum λ. Similarly proton beams can never quite achieve optimum damage. This conclusion reflects the target multiplicity within the cell nucleus. The saturation cross section for inactivation of V79 cells is 38 ± 4 μm^2, obtained from the point of inflection and is independent of radiation type. For values of $\lambda > 2$ nm, note that the gradient of the curve is -1.22 ± 0.07, which will be used later to justify scaling of data.

CHROMOSOME DICENTRICS[9–26]

Cross sections for production of chromosome dicentrics in human lymphocytes are shown in Figure 3(a) plotted as a function of λ and in Figure 3(b) against the reciprocal $L_{100,D}$. Again there is no immediately obvious advantage in which quality parameter to use, especially as the spread of data is greater than for inactivation. However, in Figure 3(a), the greater power of λ_i is clearly evident in the interpretation of the neutron data. For neutrons with $\lambda_i < \sim 3$ nm, the decrease in effect cross section is due to short-range proton recoils. The ^{20}Ne point at $\lambda_i \sim 0.75$ nm has an apparently anomalously low cross section because its range is very much less than the mean chord through the cell nucleus, reflecting sub-nuclear target multiplicity. For chromosome aberrations, the saturation cross section is 9.1 ± 0.8 μm^2 and the slope for non-saturating particles is -1.18.

HPRT MUTATIONS[27–33]

Figure 4 shows the results for HPRT mutations in V79 chinese hamster cells. Here the spread of results is very large but the same general features are observed as for the other biological endpoints. The saturation cross section is 1.56×10^{-3} μm^2 and the slope of non-saturating radiations is -1.14.

ONCOGENIC TRANSFORMATIONS[34–39]

Figure 5 shows data for radiation-induced oncogenic transformation of mouse embryo C3H10T$_{1/2}$ cells. De-

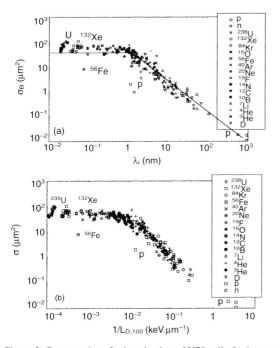

Figure 2. Cross sections for inactivation of V79 cells for heavy particles plotted as a function of (a) λ_i, and (b) reciprocal dose-restricted LET.

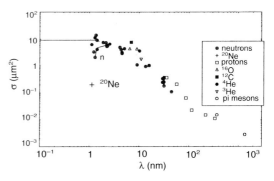

Figure 3. Cross sections for chromosome aberrations in V79 cells for heavy particles plotted as a function of λ_i.

spite the large spread of data the general trends are consistent with those observed above for the other endpoints indicating that the basic damage mechanisms are identical. The saturation cross section is 0.022 μm^2 and the slope of cross section as a function of λ_i is -1.12.

CONCLUSIONS

Examination of details at the molecular and sub-molecular level reveals that the radiation-induced biological endpoints of inactivation, chromosome aberrations, gene mutations and oncogenic transformation are initiated by the same basic damage mechanism, viz. the production of double-strand breaks in the intracellular DNA due to a correlation between the primary ionisations and the spacing of the strands in the DNA. As the shape of the $\sigma-\lambda$ curves is closely similar for all the endpoints studied, the ratio of the saturation effect cross sections to those for inactivation can be used as a scaling factor for estimation of risk of occurrence. The importance here is that if a good model can be derived for predicting the inactivation of mammalian cells, the scaling factors can be incorporated to yield the probability of cancer induction — a factor which is of major importance in radiation protection.

As the slopes of the curves shown in Figures 2 to 5, for the various biological endpoints are the same within the experimental errors, it seems to be justifiable to take ratios with respect to inactivation to determine the scaling factors for risk estimation. The scaling factors are found to be 0.24, 5.8×10^{-4} and 4.1×10^{-5} respectively for induction of chromosome dicentrics, oncogenic transformation and gene mutations.

ACKNOWLEDGEMENT

Partial support for this research was received from the Commission of the European Communities, DG XII, under contract: F13P-CT92-0041.

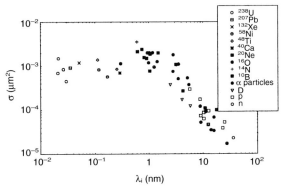

Figure 4. Cross sections for induction of mutations in the HPRT gene by heavy charged particles, as a function of λ_i.

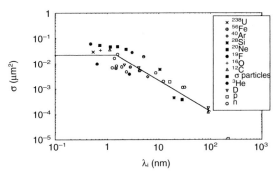

Figure 5. Cross sections for oncogenic transformation of C3H10T$_{1/2}$ mouse embryo cells by heavy particles, plotted as a function of λ_i.

REFERENCES

1. Harder, D., Blohm, R. and Kessler, M. *Restricted LET Remains a Good Parameter of Radiation Quality*. Radiat. Prot. Dosim. **23**, 1–4 (1988).
2. Watt, D. E. *On Absolute Biological Effectiveness and Unified Dosimetry*. J. Radiol. Prot. **9**, 33–49 (1989).
3. Katz, R., Sharma, S. C. and Homayoonfar, M. *The Structure of Particle Tracks*. In: Topics in Radiation Dosimetry. Radiation Dosimetry. Suppl. 1, pp. 317–383 (1972).
4. Watt, D. E. and Alkharam, A. S. *Charged-Particle Track Structure Parameters for Application in Radiation Biology and Radiation-Chemistry*. Int. J. Quantum Chem. **21**, 195–207 (1994).
5. Watt, D. E. *A Unified System of Radiation Bio-effectiveness and its Consequences in Practical Applications*. Radiat. Prot. Dosim. **70**(1–4), 529–536 (This issue) (1997).
6. Ward, J. E. *The Yield of DNA Double-strand Breaks Produced Intracellularly by Ionizing Radiation: a Review*. Int. J. Radiat. Biol. **57**(6), 1141–1150 (1990).
7. Yang, T. C., Craise, L. M., Mei, M. and Tobias, C. A. *Neoplastic Cell-Transformation by High-LET Radiation — Molecular Mechanisms*. Life Sci. Space Res. XXIII **10**, 131–140 (1989).
8. Frankenberg, D. *Repair of DNA Double-Strand Breaks and its Effect on RBE*. Adv. Space Res. **14**, 235–248 (1994).
9. Barjaktarovic, N. and Savage, R. K. *RBE for d(42 MeV)-Be Neutrons Based on Chromosome-type Aberrations Induced in Human Lymphocytes and Scored in Cells at First Division*. Int. J. Radiat. Biol. **37**, 667–675 (1980).
10. Bauchinger, M., Schmid, E., Rimpl, G. and Kuhn, H. *Chromosome Aberrations in Human Lymphocytes after Irradiation with 15.0 MeV Neutrons in vitro. I. Dose-response Relation and RBE*. Mutat. Res. **27**, 103–1–9 (1975).
11. Bocian, E., Pszona, S. and Ziemaba-Zak, B. *Dose-response Curve for Chromosome Aberrations in Human Lymphocytes Irradiated with 7.4 MeV Protons in vitro*. Stud. Biophys. **39**, 167–176 (1973).

12. Bettega, D., Dubini, S., Conti, A. M. F., Pelucchi, T. and Lombardi, L. T. *Chromosome Aberrations Induced by Protons up to 31 MeV in Cultured Human Cells*. Radiat. Environ. Biophys. **19**, 91–100 (1981).

13. Edwards, A. A., Lloyd, D. C., Prosser, J. S., Finnon, P. and Moquet, J. E. *Chromosome Aberrations Induced in Human Lymphocytes by 8.7 MeV Protons and 23.5 MeV Helium-3 Ions*. Int. J. Radiat. Biol. **50**, 137–145 (1986).

14. Edwards, A. A., Finnon, P., Moque, J. E., Lloyd, D. C., Darroudi, F. and Natarajan, A. T. *The Effectiveness of High Energy Neon Ions in Producing Chromosomal Aberrations in Human Lymphocytes*. Radiat. Prot. Dosim. **52**, 299–303 (1994).

15. Edwards, A. A. *Progress Report, F13P-CT920064i. The Induction of Chromosomal Changes in Human and Rodent Cells by Accelerated Charged Particles: Early and Late Effects*. (Didcot, UK: NRPB). (1994).

16. Fabry, L., Leonard, A. and Wamberssie, A. *Induction of Chromosome Aberrations in Go Human Lymphocytes by Low Doses of Ionizing Radiations of Different Quality*. Radiat. Res. **103**, 122–134 (1985).

17. Lloyd, D. C., Purrott, R. J., Dolphin, G. W. and Edwards, A. A. *Chromosome Aberrations Induced in Humans by Neutron Irradiation*. Int. J. Radiat. Biol. **29**, 169–182 (1976).

18. Lloyd, D. C., Edwards, A. A., Prosser, J. S., Bolton, D. and Sherwin, A. G. *Chromosome Aberrations Induced in Human Lymphocytes by d-T Neutrons*. Radiat. Res. **98**, 561–573 (1984).

19. Matsubara, S., Ohara, H., Hiraoka, T., Koike, S., Ando, K., Yamaguchi, H., Kuwarbara, Y., Hoshina, M. and Suzuki, S. *Chromosome Aberration Frequencies Produced by a 70-MeV Proton Beam*. Radiat. Res. **123**, 182–191 (1990).

20. Purrott, R. J., Edwards, A. A., Lloyd, D. C. and Stather, J. W. *The Induction of Chromosome Aberrations in Human Lymphocytes by in vitro Irradiation with α-particles from Plutonium-239*. Int. J. Radiat. Biol. **38**, 277–284 (1980).

21. Rimpl, G. R., Schmid, E., Braselmann, H. and Bauchunger, M. *Chromosome Aberrations Induced in Human Lymphocytes by 16.5 MeV Protons*. Int. J. Radiat. Biol. **58**, 999–1007 (1990).

22. Sasaki, M. *Radiation-induced Chromosome Aberrations in Lymphocytes: Possible Biological Dosimeter in Man*. In: Proc. Int. Symp. on Biological Aspects of Radiation Protection, Kyoto, Oct. 1969. pp. 81–90 (1971).

23. Sevankaev, A. V., Zherbin, E. A., Lunchnik, N. V., Obatturov, G. M., Kozlov, V. M., Tjatte, E. G. and Kapchigashev, S. P. *Cytogenetic Effects Produced by Neutrons in Lymphocytes of Human Peripheral Blood in vitro*. Genetica **15**, 1046–1060 (1979).

24. Takatsuji, T. and Takekoshi, H. *Induction of Chromosome Aberrations by 4.9 MeV Protons in Human Lymphocytes*. Int. J. Radiat. Biol. **44**, 553–562 (1983).

25. Todorov, S. L. *Radiation-induced Chromosome Aberrations in Human Peripheral Lymphocytes. Exposure to X-rays or Protons*. Strahlentherapie **149**, 197–204 (1975).

26. Vulpis, N. and Tognacci, L. *Chromosome Aberration as a Dosimetric Technique for Fission Neutrons over the Dose-range 0.2–50 rad*. Int. J. Radiat. Biol. **33**, 301–306 (1978).

27. Belli, M. *Progress Report, Contract F13P-CT920053. Molecular and Cellular Effectiveness of Charged Particles (Light and Heavy Ions) and Neutrons* (1994).

28. Belli, M., Goodhead, D. T., Ianzini, F., Simone, G. and Tabocchini, M. A. *Direct Comparison of Biological Effectiveness of Proton and Alpha-particles of the Same LET. II. Mutation Induction at the HPRT Locus in V79 Cells*. Int. J. Radiat. Biol. **61**, 625–629 (1992).

29. Cox, R., Thacker, J., Goodhead, D. T. and Munson, R. J. *Mutation and Inactivation of Mammalian Cells by Various Ionising Radiations*. Nature **267**, 425–427 (1977).

30. Hei, T. K., Chen, D. J., Brenner, D. J. and Hall, E. J. *Mutation Induction by Charged Particles of Defined Linear Energy Transfer*. Carcinogenesis **9**, 1233–1236 (1988).

31. Kranert, T., Stoll, U., Schneider, E. and Kiefer, J. *Mutation Induction in Mammalian Cells by Very Heavy Ions*. Adv. Space Res. **12**(2), 111–118 (1992).

32. Thacker, J., Stretch, A. and Stephens, M. A. *Mutation and Inactivation of Cultured Mammalian Cells Exposed to Beams of Accelerated Ions. II. Chinese Hamster V79 Cells*. Int. J. Radiat. Biol. **36**, 137–148 (1979).

33. Stoll, U., Schmidt, A., Schneider, E. and Kiefer, J. *Killing and Mutation of Chinese Hamster V79 Cells Exposed to Accelerated Oxygen and Neon Ions*. Radiat. Res. **142**, 288–294 (1995).

34. Bettega, D., Calzolari, P., Chiorda, G. N. and Tallone-Lombardi, L. *Transformation of C3H10T1/2 Cells with 4.3 MeV α Particles at Low Doses: Effects of Single and Fractionated Doses*. Radiat. Res. **131**, 66–71 (1992).

35. Hei, T. K., Komatsu, K., Hall, E. J. and Zaider, M. *Oncogenic Transformation by Charged Particles of Defined LET*. Carcinogenesis **9**, 747–750 (1988).

36. Hieber, L., Ponsel, G., Roos, H., Fromke, E. and Kellerer, A. M. *Absence of a Dose-rate Effect in the Transformation if C3H10T1/2 Cells by α-particles*. Int. J. Radiat. Biol. **52**, 859–869 (1987).

37. Miller, R. C., Geard, C. R., Brenner, D. J., Komatsu, K., Marino, S. A. and Hall, E. J. *Neutron-energy-dependent Oncogenic Transformation of C3H10T1/2 Mouse Cells*. Radiat. Res. **117**, 114–127 (1989).

38. Miller, R. C., Marino, S. A., Brenner, D. J., Martin, S. G., Richard, M., Randers-Pehrson, G. and Hall, E. J. *The Biological Effectiveness of Radon-Progeny Alpha Particles. II. Oncogenic Transformation as a Function of Linear Energy Transfer*. Radiat. Res. **142**, 54–60 (1995).

39. Yang, T. C., Craise, L., Mei, M. -T. and Tobias, C. *Neoplastic Cell Transformation by Heavy Charged Particles*. Radiat. Res. **104**, S177–S187 (1985).

NANODOSIMETRIC RESULTS AND RADIOTHERAPY BEAMS: A CLINICAL APPLICATION?

L. Lindborg and J.-E. Grindborg
Swedish Radiation Protection Institute
S-171 16 Stockholm, Sweden

Abstract — Clinical RBE values observed in radiation therapy with neutrons, X rays and ^{60}Co γ rays have been compared with ratios of the values for the dose mean lineal energy, $\bar{y}_D$, measured in similar beams. Following from the linear-quadratic (LQ) dose effect relation used for fractionated radiotherapy, it is estimated that the ratios of the α values of survival curves for early reacting tissues determined for fractionated irradiations, are approximately equal to the ratios of the $\bar{y}_D$ values, if this quantity is determined for a sphere diameter of only a few nanometres.

INTRODUCTION

A knowledge of the absorbed dose alone is not sufficient to describe accurately the effectiveness of ionising radiation on a biological system. A quantity characterising the individual tracks of the ionising radiation is usually also needed. In the field of radiation protection, a radiation weighting factor or a quality factor related to LET (or the lineal energy) is used to weight the absorbed dose to arrive at a quantity aimed at being an indication of stochastic effects[1]. In the field of radiation therapy the relative biological effectiveness (RBE) is used. It is mainly based on clinical observations. To take account of differences in radiation quality in treatment dose specification, RBE values need to be determined very accurately[2]. Attempts to relate differences in RBE between different radiation beams with microdosimetric single event distributions of the lineal energy have been made[3,4]. By comparing RBE values obtained in beams used for radiotherapy with measured single event distributions in a 2 μm tissue-equivalent sphere, a biological weighting function was derived. When the function was applied to any of the beams, it was possible to predict the RBE within a few per cent. Such careful investigations have shown that microdosimetric energy deposition distributions contain information relevant to the description of biological effectiveness variation between radiation beams in a phenomenological way. They possibly also contain more basic information related to the relative response of a biological system at a given level of cell survival. However this information may have to be related to a smaller scale energy deposition representation.

There is good evidence that severe damage to the DNA molecule may lead to cell death. Pairs of double-strand breaks (DSB) have been hypothesised by Barendsen[5] as the cause of direct lethal damage. Locally multiply damaged sites (LMDS) involving DNA have been mentioned by Brenner and Ward[6] and dual double-strand breaks on the nucleosome have been suggested by Tilikidis and Brahme[7] as the cause of cell death. The former concluded that LMDS are probably caused by at least two to five ionisations in sites of diameters of 1 nm to 4 nm. The DNA is a long molecule shaped as a double helix with a distance of about 2 nm in between. While for low-LET radiations indirect action of the radiation is supposed to dominate, direct action of the radiation on the DNA is supposed to be dominant for neutrons[8]. The indirect action also means that radicals coming from a distance 2 nm away from the DNA molecule may interact and damage it. There is thus reason to believe that distributions of the energy deposition for objects in the nanometre range would be at least as relevant as distributions in the micrometre range for analysing the RBE for cell killing.

Measured single event distributions for objects in the nanometre range are rare[9,10]. Measurements with the variance–covariance method[11], in which the dose average mean lineal energy, $\bar{y}_D$, of such distributions is determined, have been reported for some radiation qualities[12–14]. A linear relation between $\bar{y}_D$ measured in the nanometre range and the yield of dicentric chromosome aberrations, which are known to be related to cell killing, has been discussed by Zaider and Rossi[16]. The intention with this report is to see if $\bar{y}_D$ values in the nanometre range correlate with clinical RBE values and can thus characterise the radiation quality in a meaningful way. The primary quantity measured has been the number of ions collected on an electrode inside a gas cavity. These numbers have been converted to lineal energy in analogy with other studies.

THE LINEAR QUADRATIC APPROXIMATION IN RADIOTHERAPY

A linear-quadratic approximation is often used in radiotherapy to estimate the biological effect of a given absorbed dose[8,17]. The cell fraction, S, that survives a single absorbed dose, D, is represented by a relation:

$$S = \exp(-\alpha D - \beta D^2) \qquad (1)$$

The two coefficients α and β are adjusted to fit the cell survival curve. If the single dose D is replaced by a number, n, of equally large dose fractions, d, so that

$D = nd$, and if instead the logarithm of cell killing, E, is used to define the biological effect, then

$$E = nd\alpha[1 + d/(\alpha/\beta)] \quad (2)$$

The term 'linearly dependent on the absorbed dose' is generally assumed to estimate the lethal lesions due to single particles, while the second term — dependent on the square of the dose per fraction — estimates the lethal lesions created by interacting sub-lesions, which have appeared along the tracks of independent particles.

The relative biological effectiveness, RBE, is the ratio between absorbed doses giving the same fraction of cell survival. For a X ray beam (index X) and a γ ray beam (index γ) the RBE is

$$RBE \equiv D_\gamma/D_X = \alpha_X[1 + d_X/(\alpha/\beta)_X]/\alpha_\gamma[1 + d_\gamma/(\alpha/\beta)_\gamma] \quad (3)$$

For low-LET radiations, and early-reacting tissues and tumours, the ratio of α/β is typically 10 Gy (as determined from multifraction experiments). A common dose fraction, d, in conventional therapy is 2 Gy. The first term within the brackets of Equation 2 will then dominate[18] and the second term will affect E in very much the same way in all low-LET beams. If $d_X = d_\gamma$, Equation 3 may be approximated by

$$RBE \equiv D_\gamma/D_X \approx \alpha_X/\alpha_\gamma \quad (4)$$

A cell survival curve for a neutron beam (index n) has hardly any shoulder and the ratio $(\alpha/\beta)_n$ for neutrons becomes much larger than that for a low-LET beam. In this case Equation 3 may be approximated by

$$RBE \equiv D_\gamma/D_n \approx \alpha_n/\alpha_\gamma[1 + d_\gamma/(\alpha/\beta)_\gamma] \quad (5)$$

Reported RBE values are 1.25 for X rays (usually with 250 kV accelerating potential) as compared to cobalt gamma rays, and 3.2 for 14 MeV neutrons as compared to cobalt gamma rays[19]. With typical clinical values for a low-LET beam and for early reacting tissues and tumours ($d = 2$ Gy and $(\alpha/\beta) = 10$ Gy), Equations 4 and 5 give $(\alpha_X/\alpha_\gamma) = 1.25$ and $(\alpha_n/\alpha_\gamma) = 3.8$ respectively.

DOSE MEAN LINEAL ENERGY, $\bar{y}_D$ AND THE INTEGRAL PROXIMITY FUNCTION, $T(x)$

Values of $\bar{y}_D$ for a wide range of object sizes are available for a few radiation qualities. Uncollimated

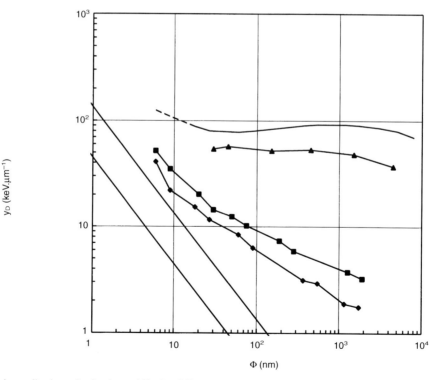

Figure 1. Experimentally determined values of $\bar{y}_D$ for different simulated tissue-equivalent sphere diameters, Φ. For 15 MeV neutrons the line fitted to the data points by the authors[20] has been used. The data points have been connected by straight lines between consecutive points for the other three radiation qualities. The two straight lines to the left correspond to 30 eV (1 ion pair) and 90 eV (3 ion pairs) divided by the mean chord lengths for the different spheres. Measurement results are shown for a 15 MeV neutron beam[20] (—); 5.7 MeV neutrons[14] (▲); 100 kV X rays[12] (■), and ^{60}Co γ[12] (♦).

neutron beams with energies of 5.7 MeV and 15 MeV were investigated for sphere diameters down to about 20 nm[13,14]. The measurements were made with wall-less spherical chambers working either in the proportional or ionisation chamber mode. In a later communication Goldhagen and Randers-Pehrson[20] reported new $\bar{y}_D$ values for the 15 MeV neutron beam that were significantly smaller below 300 nm and these included an extrapolation down to 8 nm. The more recent results are considered to be the more accurate[21] and are used in this report. Collimated low-LET beams were investigated by Grindborg et al[12]. Measurements were made with walled spherical ionisation chambers in an X ray beam (100 kV and HVL 0.141 mm Cu), and a ^{60}Co γ ray beam. The uncertainties of all results are about 10%, but for the low-LET beams a larger uncertainty was reported at 6 nm and 9 nm.

In Figure 1, $\bar{y}_D$, as a function of the simulated tissue-equivalent (unit density) sphere diameter, Φ, is shown for the four radiation qualities. Also shown are lines corresponding to one and three ionisations (30 eV per ion pair created) occurring in spheres with different diameters. Both low-LET beams show a quite strong increase in $\bar{y}_D$ with decreasing sphere diameter, while the increase for the two neutron beams is less pronounced. A general remark is that the differences seen between the radiation qualities in the micrometre range, are different from the differences seen in the nanometre range. This is so for all radiation qualities and, in particular, when values for the low-LET beams are compared with values for the neutron beams.

Since $\bar{y}_D$ was known for a wide range of sphere diameters, it was possible to calculate the integral proximity function[22,23] for any of the four radiation beams. That function, $T(x)$, defines the mean energy imparted within a radius, x, from randomly chosen ionisations. The functions for the 15 MeV neutron beam and the ^{60}Co γ ray beam are shown in Figure 2. The figure also includes the mean energy imparted, $\bar{\epsilon}$, as a function of the sphere radius for the absorbed doses 2 Gy, 22 Gy and 60 Gy, and $\bar{\epsilon}$ was calculated from $\bar{\epsilon} = 4\pi\rho D x^3/3$, where ρ is the density of tissue. The figure shows that in the 15 MeV neutron beam, a dose of 22 Gy imparts the same amount of energy as the single track within a sphere with the radius 500 nm. For the ^{60}Co γ ray beam and the absorbed dose 60 Gy, the corresponding radius is

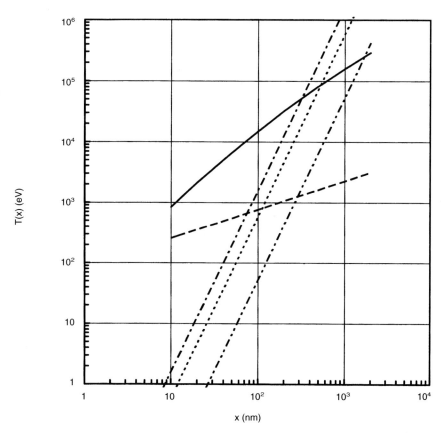

Figure 2. The integral proximity function, $T(x)$, for various radii, x, and 15 MeV neutrons (—) and ^{60}Co γ (– – – –). Also shown is the mean energy imparted within various radii at 2 Gy (–··–··–), 22 Gy (·····) and 60 Gy (–·–·–).

70 nm. For a dose of 2 Gy the corresponding radii become 2000 nm and 300 nm for the neutron and gamma ray beams, respectively. This illustrates that for a target of a few tens of nanometres, it is unlikely that more than one particle track (one event) will have deposited energy in the target, even after a dose as high as the total doses used in radiation therapy.

DISCUSSION

In radiation therapy the absorbed dose is split into a number of smaller doses (typically 2 Gy), often given 5 days a week, until a total dose of about 60 Gy or about 22 Gy are delivered in a ^{60}Co γ ray beam and a 14 MeV neutron beam, respectively. This procedure allows the repair mechanisms of the cells to eliminate less severe lesions between the dose fractions. Whereas such sub-lethal lesions may interact with each other at larger doses to cause cell death, directly lethal lesions are mainly due to energy deposition by single tracks at doses for which the inter-track term, i.e. the second term of Equation 2, is negligible. The effect of the single track per unit dose, the first term of Equation 2, is equal to α. If the assumption is made that the effect of a single particle is proportional to the lineal energy of the particle for a relevant sphere diameter, and if it is further assumed that the mean, $\bar{y}_D$, is a sufficient measure of its distribution, then α and $\bar{y}_D$ become proportional to each other, and, for instance, the ratio α_n/α_γ should be equal to the ratio $\bar{y}_{D,n}/\bar{y}_{D,\gamma}$. By testing such an equality for different radiation qualities, the size of a biological target adequate to represent cell death as a function of energy deposition may be derived. Although, for different radiation qualities, direct and indirect effects are expected to contribute to various degrees to the observed effect, the equality is expected to occur for spheres of very similar diameters.

The ratio of the $\bar{y}_D$ values for X rays to cobalt gamma rays and the corresponding ratio for the 15 MeV neutron to cobalt gamma rays are shown in Figure 3 for different sphere diameters, Φ. The ratio for the low-LET beams decreases with decreasing sphere diameter. The two experimental values at 9 nm and 6 nm are most likely too large since it was impossible to correct the values for the X rays at those diameters for the covariance of the signals in a proper way. At 20 nm and 30 nm the ratio between the experimental points is 1.3 and is expected to be equal to or smaller than 1.25 at 6 nm. The $\bar{y}_D$ ratio in this range (6 nm to 30 nm) is in reasonable agreement with the α ratio 1.25.

For the 15 MeV neutron beam there was no experimental $\bar{y}_D$ value below 20 nm, but an extrapolation was reported down to 8 nm. For 15 MeV neutrons and 2 nm a calculated $\bar{y}_D$ value of 133 keV.μm^{-1} has been reported$^{(24)}$ and gives some support for the extrapolation. A $\bar{y}_D$ ratio equal to 3.8 occurs at about 7 nm where $\bar{y}_D$ for the neutrons is estimated to be 119 keV.μm^{-1}. The uncertainty of the $\bar{y}_D$ values for the cobalt beam increased to ±30% at 6 nm and 9 nm. If values ±30% different from the best estimates are chosen here, the $\bar{y}_D$ ratio becomes close to the α ratio for sphere diameters in the range 5 nm to 9 nm. The uncertainty of the clinical RBE values is also relevant but will not be discussed here.

Thus, for both the low-LET beams and the 15 MeV neutron beam, a sphere diameter seems to exist at which the $\bar{y}_D$ ratio equals the α ratio. From this analysis it is estimated that its size is between 6 nm and 9 nm, but the uncertainty is quite large.

A lesion that most likely leads to cell death is a cluster of double-strand breaks (DSBs) or pairs of DSBs in close proximity$^{(5-7)}$. The target sizes mentioned vary from 2 nm up to about 8 nm, if both the ranges for indirect action in low-LET beams and the influence of the mean chord length are considered. This range is in agreement with the finding reported above.

In an analysis of dicentric chromosome aberrations in human lymphocytes, Virsik et al$^{(15)}$ concluded that the number of exchange-type aberrations or lethal lesions per unit dose, varies in proportion to the track core LET below 70 keV.μm^{-1}. The lateral restriction of this concept was chosen to be only a few nanometres. It was suggested that this quantity is equivalent to the dose average lineal energy determined in the nanometre range. There is good agreement between their prediction of target size and the dependence of $\bar{y}_D$ and the results above.

The microdosimetric results used in this analysis have important limitations and uncertainties and the interpretation has to be verified by measurements in the

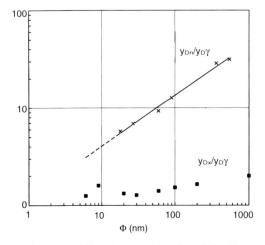

Figure 3. Ratios of $\bar{y}_D$ values as a function of the diameter. When possible the ratio was calculated from the experimental data points shown in Figure 1. However, small interpolations have been made to get results at the same sphere diameter. The line in this figure is a linear fit to the calculated ratios and is used for the extrapolation to smaller diameters. The extrapolation region is dashed.

particular beams used for determination of the clinical RBE. This is most important for the neutron beam, but the X ray beam, for which the $\bar{y}_D$ was taken, was of a lower energy than that for which the RBE was taken. Another important uncertainty is introduced by the extrapolation of the $\bar{y}_D$ between 20 nm and about 6 nm. The low-LET results had not been corrected for the wall effect.

A proportionality between $\bar{y}_D$ and the initial slope, α, of the cell survival curve may be of practical interest. Not only is $\bar{y}_D$ well suited to be measured at high dose rates, the derivation of a relative α value from physical measurements may qualify $\bar{y}_D$, measured in the nanometre range, as a quantity for specifying the radiation quality.

CONCLUSIONS

From clinical RBE values, ratios of α values were calculated using the linear quadratic dose effect relation for fractionated radiotherapy. An approximate numerical agreement between ratios of $\bar{y}_D$ values for spheres of a few nanometres and the ratios of the α values for early-reacting tissues and tumours was found. The observation is interpreted as further evidence of a biological target for cell killing of a few single nanometres. An extension of this investigation to other beams for radiation therapy seems interesting, and will expose whether the reasonable agreement found here between relative $\bar{y}_D$ values and relative α values for fractionated irradiation schemes is more general. If so, it may make $\bar{y}_D$ a quantity suitable for characterising clinical radiation beams as used in radiation therapy.

ACKNOWLEDGEMENTS

The authors are grateful for several clarifying discussions with Prof. Anders Brahme at the Department of Medical Radiation Physics at the Stockholm University, Stockholm.

REFERENCES

1. ICRP. *1990 Recommendations of the International Commission on Radiological Protection*. Publication 60, Ann. ICRP. **21**(1–3) (Oxford: Pergamon Press) (1990).
2. Brahme, A. *Dosimetric Precision Requirements in Radiation Therapy*. Acta Radiol. Oncol. **23**, 379–391 (1984).
3. Tilikidis, A., Lind, B., Näfstadius, P. and Brahme, A. *An Estimation of the Relative Biological Effectiveness of 50 MV Bremsstrahlung Beams by Microdosimetric Techniques*. Phys. Med. Biol. **41**, 55–69 (1996).
4. Menzel, H. G., Wambersie, A. and Pihet, P. *Microdosimetric Specification of Radiation Quality in Neutron Radiation Therapy*. Int. J. Biol. **57**(4), 865–883 (1990).
5. Barendsen, G. W. *The Relationships between RBE and LET for Different Types of Lethal Damage in Mammalian Cells: Biophysical and Molecular Mechanisms*. Radiat. Res. **139**, 257–270 (1994).
6. Brenner, D. and Ward, J. F. *Constraints on Energy Deposition and Target Size of Multiply Damaged Sites Associated with DNA Double-Strand Breaks*. Int. J. Radiat. Biol. **61**(6), 737–748 (1992).
7. Tilikidis, A. and Brahme, A. *Microdosimetric Description of Beam Quality and Biological Effectiveness in Radiation Therapy*. Acta Oncol. **33**(4), 457–469 (1994).
8. Hall, E. J. *Radiobiology for the Radiologist*. Fourth Edition (Philadelphia: Lippincott) (1994).
9. Anachkova, E., Kellerer, A. M. and Roos, H. *Neutron Energy Deposition Spectra at Simulated Diameters down to 50 nm*. Radiat. Prot. Dosim. **70**(1–4), 207–210 (This issue) (1997).
10. Kliauga, P. *Nanodosimetry of Heavy Ions Using a Miniature Cylindrical Counter of Wall-Less Design*. Radiat. Prot. Dosim. **52**(1–4), 317–321 (1994).
11. Kellerer, A. M. and Rossi, H. H. *On the Determination of Microdosimetric Parameters in Time Varying Radiation Fields. The Variance-Covariance Method*. Radiat. Res. **97**, 237–245 (1984).
12. Grindborg, J.-E., Samuelson, G. and Lindborg, L. *Variance-Covariance Measurements in Photon Beams for Simulated Nanometer Objects*. Radiat. Prot. Dosim. **61**(1–3), 119–124 (1995).
13. Goldhagen, P., Randers-Pehrson, G., Marino, S. and Kliauga, P. *Variance-Covariance Measurements of $\bar{y}_D$ for 15 MeV Neutrons in a Wide Range of Site Sizes*. Radiat. Prot. Dosim. **31** (1–4), 167–170 (1990).
14. Lindborg, L., Marino, S., Kliauga, P. and Rossi, H. H. *Microdosimetric Measurements and the Variance-Covariance Method, Some Experimental Experience*. Radiat. Environ. Biophys. **28**, 251–263 (1989).
15. Virsik, R. P., Blohm, R., Hermann, K.-P., Modler, H. and Harder, D. *Chromosome Aberrations and the Mechanism of the "Primary Lesion Interaction"*. In: Proc. Eighth Symp. on Microdosimetry, 1982, Commission of the European Communities, Radiation Protection, Report EUR 8395 EN (1983).
16. Zaider, M. and Rossi, H. H. *On the Application of Microdosimetry to Radiobiology*. Radiat. Res. **113**, 15–24 (1988).
17. Fowler, J. F. *What Next in Fractionated Radiotherapy*. Br. J. Cancer **49**, Suppl. VI, 285–300 (1984).
18. Barendsen, G. W. *Mechanisms of Cell Reproductive Death and Shapes of Radiation Dose-Survival Curves of Mammalian Cells*. Int. J. Radiat. Biol. **57**(4), 885–896 (1990).
19. Battermann, J. J., Hart, G. A. M. and Breuer, K. *Dose-Effect Relations for Tumour Control and Complication Rate after Fast Neutron Therapy for Pelvic Tumours*. Br. J. Radiol. **54**, 899–904 (1981).

20. Goldhagen, P. and Randers-Pehrson, G. *Variance-Covariance Measurements: A Practical Method for Microdosimetry in Submicroscopic Volumes.* In: Radiation Research: A Twentieth-Century Perspective. Volume II: Congress Proceedings (New York: Academic Press) (1992).
21. Goldhagen, P. Private communication (1995).
22. ICRU. International Commission on Radiation Units and Measurements. *Microdosimetry.* Report 36 (Bethesda, Maryland 20814, USA: ICRU Publications) (1983).
23. Zaider, M., Brenner, D. J., Hanson, K. and Minerbo, G. N. *An Algorithm for Determining the Proximity Distribution from Dose-Average Lineal Energies.* Radiat. Res. **91**, 95–103 (1982).
24. Coyne, J. J. and Caswell, R. S. *Neutron Energy Deposition on the Nanometer Scale.* Radiat. Prot. Dosim. **44**(1–4), 49–52 (1992).

BNCT: STATUS AND DOSIMETRY REQUIREMENTS

R. Gahbauer[1], N. Gupta[1], T. Blue[2], J. Goodman[3], J. Grecula[1], A. H. Soloway[4] and A. Wambersie[5]
[1] Division of Radiation Oncology, [2] Nuclear Engineering Program
[3] Division of Neurosurgery, [4] Department of Medicinal Chemistry
The Ohio State University, Columbus, OH 43210, USA
[5] Universite Catholique Louvain, Cliniques Universitaires St Luc, Brussels, Belgium

INVITED PAPER

Abstract — BNCT is a binary cancer treatment modality, consisting of the delivery of a suitable boron compound to tumour cells followed by irradiation of the tumour by thermal neutrons. Originally proposed by Locher in 1936, the first clinical trials at Brookhaven and at MIT in the 1950s were unsuccessful because of the non-selectivity of the boron compound used. New classes of boron carriers have since been developed and neutron sources have been optimised. Since 1968 more than 100 patients have been treated in Japan. Clinical studies have again started in the USA (1994 at MIT and BNL) and are expected soon to begin in Europe. Basic principles of this treatment modality and general requirements for boron compounds and reactor or accelerator based neutron sources are reviewed. Complexities involved in macro- and microdosimetry and thus the biological evaluation of boron compounds are discussed.

INTRODUCTION

With high doses of radiation, tumour control can be achieved very effectively. In clinical practice the amount of radiation, however, is limited by the tolerance of normal tissues harbouring the tumour. Ingenious diagnostic, geometrical, physical and biological methods have been developed to provide sparing of normal tissues and preferential or selective effects on tumour cells relative to normal tissues, as illustrated by the dose response modifications for normal tissue and tumour in Figure 1. In the treatment of malignant brain tumours the use of conventional, low LET radiation has been shown to increase median survival 2.5 fold[1], however permanent tumour control is very rarely achieved even at very high doses[2]. Fast neutron trials likewise failed to show a therapeutic window, i.e. tumour control within the tolerance limits of normal tissues. However, in most reports from these trials it was suggested that a better tumoricidal effect was observed than seen with equivalent high doses of low LET radiation. Laramore *et al* concluded in 1988[3] that clinical trials could only resume if it was possible to deliver High LET radiation to tumour cells with greater selectivity, such as mono-

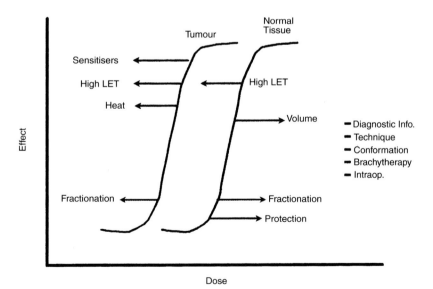

Figure 1. Sparing of normal tissues, and preferential or selective effects on tumour cells relative to normal tissues obtained by different treatment techniques.

clonal antibodies, etc. Conceptually boron neutron capture therapy is an ideal method to deliver selectively high LET radiation to tumour cells. Beyond its intrinsic cellular specificity, the capture reaction can be restricted to a limited target volume, since it requires activation by thermal neutrons — a binary mode of activation.

PRINCIPLES

BNCT is based on the nuclear fission that occurs when non-radioactive ^{10}B (in nature 20% of elemental boron) absorbs thermal neutrons (0.025 eV), as shown in Figure 2. Alpha particles and lithium nuclei emanate from this reaction with a range of 4–7 μm, sharing between them 2.3–2.8 MeV. The main effect is due to the alpha particle. The gamma radiation produced, contributes very little to the local or normal tissue effect. The rationale for BNCT is based on the fact that the capture cross section values of normal tissue elements for thermal neutrons are two orders of magnitude less than of ^{10}B. The thermal neutron capture cross sections for a few capture agents of interest and for components of normal tissue are listed in Tables 1 and 2. Based on the percentage of the elements in tissue, their capture cross section and the types of radiation emitted in this reaction, only hydrogen (gammas) and nitrogen (protons) contribute significantly to the dose delivered to normal tissues and tumour by the capture reactions ^{1}H(n,γ)^{2}H and ^{14}N(n,p)^{14}C. The energy loss of epithermal neutrons in tissue by scattering with hydrogen generates energetic recoil protons which deposit significant radiation dose to tissues. Gamma radiation emanating from the reactor or accelerator also has to be considered.

To meet the criteria of selectivity and effectiveness, there must be a significant differential boron concentration in tumour rather than normal tissues and approximately 20–40 μg per gram of B in tumour (10^9 ^{10}B atoms per cell). There are a number of nuclides with even higher cross sections than boron of which some have been evaluated as possible capture agents, e.g. gadolinium[4,5]. The main interest, however, is focused on boron, since

(1) the range of the high LET particle is limited to approximately 1 cell diameter, resulting in a very local (selective) energy deposition, and
(2) boron compounds rival carbon in their extensive covalent chemistry and stability, allowing for the synthesis of diverse chemical entities, from low to macromolecular species.

STATUS AND DEVELOPMENTS

Compounds

If one were to conceive the ideal compound as boron carrier to the tumour cell, it would very selectively target at least the surface of all tumour cells or even better the tumour DNA for an additional order of magnitude increase in effect[6]. If a compound of that selectivity was available it should penetrate the blood–brain barrier to reach even microscopic extensions of tumour. For less selective compounds the blood–brain barrier provides some selectivity between tumour (with BBB breakdown) and normal brain. Today's compounds (BSH, BPA) were developed ~30 years ago, yet their metabolism and the precise biochemical basis of tumour accretion is not completely understood.

One trend in modern compound development aims to design low molecular weight agents related to natural

Table 1. Thermal neutron capture cross sections for some capture agents.

Nuclide	Neutron capture cross section values	Nuclide	Neutron capture cross section values
^{6}Li	942	^{157}Gd	225,000
^{10}B	3,838	^{164}Dy	1,800
^{113}Cd	19,800	^{168}Yb	3,500
^{149}Sm	42,000	^{184}Os	3,000
^{151}Eu	5,800	^{196}Hg	3,000
^{155}Gd	61,000	^{199}Hg	2,000

Table 2. Composition of tissue and thermal neutron capture cross sections of each component.

Nuclide	Weight in tissue (%)	Neutron capture cross section	Nuclide	Weight in tissue (%)	Neutron capture cross section
H	10.0	0.332	P	1.16	0.18
C	18.0	0.0034	S	0.2	0.53
N	3.0	1.82	Cl	0.16	32.68
O	65.0	1.8×10^{-4}	K	0.2	2.1
Na	0.11	0.43	Ca	2.01	0.4
Mg	0.04	0.053	Fe	0.01	2.57

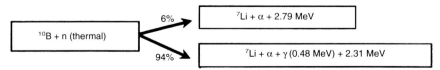

Figure 2. The ^{10}B(n,α)^{7}Li reaction, which is the basis of boron neutron capture therapy.

biochemical constituents, based on the rationale that cycling cells (tumour cells vs non-cycling normal brain cells) have a greater requirement for 'building blocks'. Among the boron compounds that have been developed are phosolipid ethers that may selectively accumulate in plasma membranes[7], amino acids that may be incorporated into tumour proteins[8], and nucleosides that can be converted into nucleotides to be incorporated into DNA or RNA[9].

Other approaches have involved the development of porphyrins[10], DNA intercalators[11], radiation sensitisers[12] and polyamines[13]. Progress has been made with macromolecular species with the synthesis of boronated liposomes[14], low density lipoproteins[15], antisense oligonucleotides[16], monoclonal antibodies and their fragments, bispecific antibodies[17] and tumour receptor binders such as epidermal growth factor[18]. Disruption of the blood–brain barrier is being investigated for those compounds otherwise unable to reach microscopic extensions of tumour[19].

BNCT neutron sources

It has been recommended that for BNCT to be successful a thermal neutron fluence of about 5×10^{12} n.cm^{-2} be delivered to a tumour[20] with ^{10}B concentration of 30 μg.g^{-1}. Early efforts in BNCT research[21] used thermal neutron beams (neutron energies <1 eV) for treating patients with deep seated tumours in clinical trials. Due to the short mean free path of thermal neutrons, the flux levels for deep seated tumours were not sufficient to provide the necessary fluence levels for successful treatment without too great damage to normal tissue; and the clinical trials were a failure. It was then decided that epithermal neutrons (1 eV < neutron energies <10 keV) were necessary for the treatment of deep seated tumours. Epithermal neutrons thermalise at a depth of about 2.5 cm. They provide a maximum thermal neutron fluence at the tumour site with a minimum of damage to normal tissue. Superficial tumours like melanoma, however, are best treated with a thermal neutron beam.

Reactor neutron sources

Initially, only reactors were thought to be capable of delivering the necessary fluence of neutrons in suitable lengths of time. The first clinical trials of BNCT were carried out at the nuclear reactors at the Brookhaven National Laboratory (BNL) and the Massachusetts Institute of Technology (MIT) — both utilising thermal neutron beams. Subsequently, in Japan, BNCT treatments have been carried out at the Hitachi Training Reactor (HTR), the Japan Atomic Energy Research Institute Research Reactor (JRR), the Musachi Institute of Technology Reactor, and the Kyoto University Reactor (KUR). All these reactors provided thermal neutron beams for BNCT treatments. Of these, only the JRR and the KUR are currently available for BNCT treatments. The KUR staff has plans to remodel their filter design to provide an epithermal beam along with the currently available thermal neutron beam[22]. The participants in the European Collaboration for BNCT have modified the high flux reactor (HFR) at Petten in the Netherlands to obtain an epithermal beam[23] for their animal studies, leading to possible human clinical trials in the near future.

In the United States the two reactors that are currently being used for BNCT clinical trials with epithermal beams are the Brookhaven Medical Research Reactor (BMRR) and the Massachusetts Institute of Technology Reactor (MITR). Prompted by the advent of clinical trials other research reactors like Georgia Institute of Technology Research Reactor (GITRR), and the Power Burst Reactor Facility (PBF) at Idaho Falls, Idaho are also being evaluated for BNCT. In addition, the staff of a number of research reactors all over the US are considering upgrading their facilities for possible BNCT use. The possible use of fission plates to improve the epithermal neutron flux at a relatively low operating power is being actively considered by a number of research reactors[24,25].

Accelerator-based neutron sources

Since it is highly unlikely that reactors can be sited in populated areas as a part of major medical centres, alternative sources of thermal and epithermal neutrons for BNCT are being investigated. The most appealing of the alternatives which has been suggested is the use of low energy proton accelerators with low z targets[26]. Accelerators are currently in use in radiation therapy, and have very good acceptance in the medical community. With the advancement of technology, current accelerator designs are very compact compared to accelerators built a few decades ago. Among different types of proton accelerators proposed are radio-frequency quadrupole (RFQ) linacs[27,28], tandem electrostatic accelerators[29,30], electro-static quadrupole (ESQ) accelerators[31], and an RF focused drift tube linac[32]. The main components of an accelerator-based neutron source for BNCT consists of a high current proton linac, a target to produce neutrons by a suitable (p,n) reaction (with its associated heat removal system), and a moderator assembly to filter the energetic neutrons born at the target to energies suitable for therapy.

Reactors are capable of producing higher flux levels at the patient than accelerator based sources of neutrons. Hence, for a single treatment session with the latter for BNCT, the treatment time would be longer. However, with current thoughts on fractionated treatments to allow for sub-lethal damage repair from the low LET dose components to the normal tissue, thus increasing its tolerance[33], this disadvantage of the accelerator-based neutron source is less important. For example, it has been estimated that it may take 36 min to treat a patient in a single treatment session with an accelerator based neutron source[34]. If the treatment is fractionated over four treatment sessions, then the treatment time per session is a very reasonable time of 9 min.

Other alternative neutron sources

Besides the accelerator based neutron source, other alternative neutron sources proposed for BNCT are spallation neutron sources[35], a ^{252}Cf neutron source[36], a photoneutron source[37,38] (which uses an electron accelerator), and an inertial electrostatic confinement (IEC) source[39]. In addition, BNCT has also been suggested as a means of boosting the dose to the tumour in external beam fast neutron therapy[40] and fast neutron brachytherapy[41]. Among the proposed non-reactor neutron sources, accelerator-based neutron sources have probably received the most attention.

Macro- and microdosimetry and treatment planning

The various dose components in BNCT represent a mix of high and low LET radiations. The mixed radiation field presents a unique problem in treatment planning in BNCT. The RBEs of the different components are not only different, they also change with depth in tissue (resulting from energy changes), as does the overall mix of high and low LET radiations[42]. The RBEs of the individual components cannot be determined directly, but are usually derived based on some simplifying assumptions from *in vitro* or *in vivo* experimental data, or data generated for various endpoints in small or large animal models. Depending on these simplifying assumptions, the RBEs determined by different methods may vary considerably. Also, the quality of each epithermal neutron beam to be used for BNCT is different, resulting in perceived differences in the RBEs of their components, as well as the necessity to re-evaluate each beam, to generate the RBE data.

Another complicating factor in the planning and evaluation of BNCT experiments and treatments are the unknowns in localisation of the ^{10}B compound, and its dosimetric implications. The knowledge of the localisation of the ^{10}B compound in the normal brain and in the tumour are necessary for treatment planning and predicting or modelling possible tumour response. A compound taken up intracellularly (by the cell cytoplasm, or better still, by the cell nucleus) will have a much higher biological effectiveness than a compound which remains in the extracellular spaces. Further, some compounds, like BSH, remain primarily in the intraluminar spaces, and provide an added 'geometric protection'[43,44] to the capillary endothelium.

From the macrodosimetric perspective the unknowns in boron localisation and the geometric protection have been combined into the so called 'compound factor'[45]. The compound factor may be determined by small and large animal experiments with and without boron by comparing isoeffects. The task of evaluating every compound by performing *in vitro*, *in vivo*, small and large animal studies in order to determine the compound factors is insurmountable. Hence, microdosimetric techniques need to be developed to determine microdosimetric parameters for each compound and as a method of evaluating different compounds. Also, each compound shows variations in uptake from patient to patient. There is a need for having microdosimetric tools to evaluate the sub-factors of the compound factor.

Analytical and stochastic models for calculating the microdosimetric parameters based on various scenarios of boron subcellular distributions have been developed over the past decade — the most popular of these methods being Monte Carlo simulations[6,43,44,46–49]. These models have been validated by microdosimetric measurement methods using tissue-equivalent proportional counters (with boron-lined walls) to mimic various target sizes[50]. High resolution autoradiographic methods have been recently developed to provide an additional tool to measure subcellular boron uptakes[51], and will further validate these calculational methods of calculating the microdosimetric parameters. The above methods of determining the microdosimetric parameters will make the task of evaluating boron compounds less formidable, and also provide the necessary parameters (compound factors) to perform accurate treatment planning for BNCT.

Accurate methods of measuring the thermal neutron and fast neutron fluence distributions, and the measurement of the neutron and gamma dose distributions in tissue-equivalent phantoms has been well developed and documented[52,53]. With the development of microdosimetric techniques for determining the microdosimetric parameters for each boron compound, the boron dose may then be calculated based upon measured gross boron uptakes by different standard methods which have been described very well in the literature. These include prompt gamma spectroscopy[54], alpha track autoradiography[55], and atomic emission spectroscopy[56]. Monte Carlo based treatment planning techniques have been developed and are in use for treatment planning in BNCT[57–59]. These methods have been validated with measurements and with other calculation methods[60]. Monte Carlo based treatment planning methods suffer from the limitation that they are time consuming. Some faster calculation methods are currently under development, in order to implement faster treatment planning for routine BNCT use[61,62].

PROBLEMS SPECIFIC TO BNCT

In the main attraction of BNCT, namely to be able to deliver quite selectively high LET radiation to tumour, there still remains the biggest unknown and challenge. Since the range of the alpha particle is limited to one cell diameter, cells not loaded with sufficient amounts of boron will only receive radiation doses unavoidably received by normal tissues. Cells are most resistant to conventional radiation if they are non-cycling, metabolically inactive, i.e. hypoxic: these same cells may prove the most difficult to load with sufficient amounts of boron. A variety of strategies may have to be employed to overcome this problem, e.g. prolonged infusions, fractionated infusion and treatment, compound combinations with different mechanisms of uptake, blood–brain barrier disruptions, etc.

In order to achieve adequate tumour control it is almost certainly necessary to use doses of a magnitude that will bring the dose to normal tissues close to tolerance levels (adventitious radiations: fast neutrons, protons from nitrogen captures, gammas, boron from normal tissue uptake). The determination of a tolerance dose in BNCT is extraordinarily difficult, due to the complex mix of high and low LET radiations involved, the constituents of which change rapidly with depth in tissue at different rates[42]. At the same time the possibility that a boron compound may be taken up more avidly by critical regions of the brain favouring a certain metabolic pathway, must be considered for every compound.

CLINICAL APPLICATIONS

Clinical trials at Brookhaven National Laboratory and at MIT failed because of limitations of the neutron beams available and because boron compounds had no selectivity for tumour and inadequate clearance from blood. Utilising BSH as a new compound, Hatanaka reported some encouraging results from Japan, thus stimulating renewed interest in the technique worldwide. Epithermal beams were developed[63] to allow for the treatment of deeper seated tumours and to avoid the necessity of skull reflection in the treatment room, thus also permitting fractionated therapy in the future.

At present clinical trials have resumed for brain tumours at BNL and for melanomas at MIT and are planned to begin in Europe in the near future. To date, 12 patients have been treated in BNL. No early morbidities have been seen so far from either the boron compound or radiation at BNL or MIT[64,65].

CONCLUSION

The developments in epithermal neutron beam technology have reached a high level of sophistication, and both reactor and non-reactor (accelerator-based) epithermal neutron sources are or can be made available with the availability of funding. Methods for characterising by measurement and verifying epithermal beam performance have also been developed. Macrodosimetry tools for routine treatment planning of patients have been developed and are in use in the US clinical trials. However, faster calculation methods need to be developed for routine treatment planning in BNCT, if it becomes an approved clinical modality. The need for microdosimetry in BNCT is even more acute than its need in conventional radiotherapy, because the localisation of the ^{10}B compound affects the biological effect by orders of magnitude. Also the radiation field is highly variable, both with respect to the LET spectrum and the dose rates with depth. Finally, the compounds in clinical use today are aged, but ongoing clinical trials may help define the needs for future compounds. Newer compounds with higher tumour specificity and persistence need to be developed and tested.

REFERENCES

1. Walker, M. D., Alexander, E. and Hunt, W. *Evaluation of BCNU and or Radiation Therapy in the Treatment of Anaplastic Gliomas (a Cooperative Clinical Trial)*. J. Neurosurg. **49**, 333–343 (1978).
2. Salazar, O. M., Rubin, P. and Feldstez, R. L. *High Dose Radiation Therapy with Treatment of Localized Gliomas, Final Report*. Int. J. Radiat. Oncol. Biol. Phys. **5**, 1733–1740 (1979).
3. Laramore, G. E., Diener-West, M., Griffin, T. W., Nelson, J. S., Griem, M. L., Tomas, F. G., Hendrickson, F. R., Griffin, B. R., Myranthopoulas, L. C. and Saxton, J. *Randomized Neutron Dose Searching Study for Malignant Gliomas of the Brain: Results of an RTOG Study*. Int. J. Radiat. Oncol. Biol. Phys. **14**, 1093–1102 (1988).
4. Shih, J. A. and Brugger, R. M. *Gadolinium as a Neutron Capture Therapy Agent*. Med. Phys. **19**(3), 733–744 (1992).
5. Matsumoto, T. *Evaluation of Depth-dose Distributions for Gadolinium Neutron Capture Therapy*. In: Advances in Neutron Capture Therapy. Eds A. H. Soloway *et al* (New York: Plenum Press) pp. 235–240 (1993).
6. Gabel, D., Foster, S. and Fairchild, R. *The Monte Carlo Simulation of the Biological Effect of the $^{10}B(n,a)^7Li$ Reaction in Cells and Tissue and its Implication for Boron Neutron Capture Therapy*. Radiat. Res. **111**, 14–25 (1987).
7. Lemmen, P., Werner, B. and Streiches, B. *Ether Lipids as Potential Boron Carriers for Boron Neutron Capture Therapy: Synthesis of rac-1-(9-0-carbororanyl)nonyl-2-methyl-glycero-3-phosphocholine (B-Et-11-Ome)*. In: Advances in Neutron Capture Therapy, Eds A. H. Soloway *et al* (New York: Plenum Press) pp. 297–300 (1993).
8. Wyzlic, I. M., Tjarks, W., Soloway, A. H., Anisuzzaman, A. K. M., Rong, Fj.-G. and Barth, R. F. *Strategies for the Design and Synthesis of Boronated Nucleic Acid and Protein Components as Potential Delivery Agents for Neutron Capture Therapy*. Int. J. Radiat. Oncol. Biol. Phys. **28**, 1203–1213 (1994).
9. Rong, F.-G., Soloway, A. H., *Synthesis of 5-tethered Carborane-containing Pyrimidine Nucleosides as Potential Agents for DNA Incorporation*. Nucleosides & Nucleotides, **13**, 2021–2034 (1994).
10. Kahl, S. B. and Koo, M.-S. *Synthesis and Properties of Tetrakis-carborane-carboxylate esters of 2,4-bis-(amb-dihydroxyethyl)*

deuteroporphyrin IX, In: Progress in Neutron Capture Therapy for Cancer. Eds B. J. Allen *et al* (New York: Plenum Press) pp. 223–226 (1992).

11. Tjarks, W., Malmquist, J., Gedda, L., Sjoberg, S. and Carlsson, J. *Synthesis and Initial Biological Evaluation of Carborane-containing Phenanthrifinum Derivatives*. In: Proc. Sixth Int. Symp on Neutron Capture Therapy for Cancer. Kobe, 31 Oct–4 Nov 1994, p. 112.

12. Scobie, M. and Threadgill, M. D. *Synthesis of Carborane-containing Nitroimidazole Compounds via Mild 1.3-dipolar Cycloaddition*. J. Chem. Soc. Chem. Commun. **13**, 939–941 (1992).

13. Hariharan, J. R., Wyzlic, I. M. and Soloway, A. H. *Synthesis of Novel Boron-containing Polyamines Agents for DNA Targeting in Neutron Capture Therapy*. Polyhedron **14**, 823–825 (1995).

14. Feakes, D. A., Shelly, K. J., Hawthorne, M. F., Schmidt, P. G., Elstad, C. A., Meadows, G. G. and Bauer, W. F. *Liposomal Delivery of Boron to Murine Tumors for Boron Neutron Capture Therapy*. In: Advances in Neutron Capture Therapy. Eds A. H. Soloway *et al* (New York: Plenum Press) pp. 395–398 (1993).

15. Kahl, S. B., Pate, D. W. and Waunschel, L. A. *Low Density Lipoprotein Reconstitutions with Alkyl and Aryl Carboranes*. In: Advances in Neutron Capture Therapy. Eds A. H. Soloway *et al* (New York: Plenum Press) pp. 399–402 (1993).

16. Spielvogel, B. F., Soad, A., Powell, W., Tomasz, J., Porter, K. and Shaw, B. R. *Chemical and Enzymatic Incorporation of Boron into DNA*. In: Advances in Neutron Capture Therapy. Eds A. H. Soloway *et al* (New York: Plenum Press) pp. 389–393 (1993).

17. Liu, L., Barth, R. F., Adams, D. M., Soloway, A. H. and Reisfeld, R. A. *Bispecific Antibodies as Targeting Agents for Boron Neutron Capture Therapy of Brain Tumors*. J. Hematother. **4**, 477–483 (1995).

18. Capala, J., Barth, R. F., Bendayan, M., Lauzon, M., Adams, D. M., Soloway A. H., Fenstermaker, R. A. and Carlsson, J. *Boronated Epidermal Growth Factors as a Potential Targeting Agent for Boron Neutron Capture Therapy of Brain Tumors*. Bioconjugate Chem. (in press).

19. Yang, W., Barth, R. F., Carpenter, D. E., Moeschberger, M. L. and Goodman, J. H. *Enhanced Delivery of Boronophenylalanine by means of Intracarotid Injection and Blood Brain Barrier Disruption for Neutron Capture Therapy*. J. Neuro-oncol. (submitted).

20. Barth, R. F., Soloway, A. H. and Fairchild, R. G. *Boron Neutron Capture Therapy of Cancer*. Cancer Res. **50**, 1061–1070 (1990).

21. Sweet, W. *Practical Problems of the Past in the Use of Boron-Slow Neutron Capture Therapy in the Treatment of Glioblastoma Multiforme*. In: Proc. First Int. Symp. on Neutron Capture Therapy. Eds G. Brownell and R. G. Fairchild. BNL-51730, p. 376 (1983).

22. Kobayashi, T., Sakurai, Y. and Kanda, K. *Study on Remodelling the Heavy Water Facility of the Kyoto University Reactor for Neutron Capture Therapy from the Concept of Neutron Energy Spectrum Control*. In: Advances in Neutron Capture Therapy. Eds A. H. Soloway *et al* (New York: Plenum) pp. 29–32 (1993).

23. Moss, R. L. *Progress Towards Boron Neutron Capture Therapy at the High Flux Reactor Petten*. In: Neutron Beam Design, Development and Performance for Neutron Capture Therapy. Eds O. K. Harling *et al*. Basic Life Sciences **54** (New York: Plenum Press) pp. 169–183 (1985).

24. Liu, H. B. and Brugger, R. M. *A Study of the concept of a Fission Plate as a Source for an Epithermal Neutron Beam*. In: Proc. Sixth Int. Symp. on Neutron Capture Therapy, 31 Oct.–5 Nov. 1994 (in press).

25. Harling, O. K., Kiger, S. and Redmond II, E. L. *High Intensity Fission Converter Based Epithermal Neutron Beam for Boron Neutron Capture Therapy*. In: Proc. Sixth Int. Symp. on Neutron Capture Therapy, 10 Oct.–5 Nov. 1994 (in press).

26. Blue, J. W., Roberts, W. K., Blue, T. E., Gahbauer, R. A. and Vincent, J. S. *A Study of Low Energy Proton Accelerators for Neutron Capture Therapy*. In Neutron Capture Therapy Proc. Second International Symposium on Neutron Capture Therapy. Ed. H. Hatanaka. (Tokyo) pp. 147–158 (1985).

27. Wangler, T. P., Stovall, J. E., Bhatia, T. S., Wang, C. K., Blue, T. E. and Gahbauer, R. A. *Conceptual Design of an RFQ Accelerator-based Neutron Source for Boron Neutron Capture Therapy*. Los Alamos National Laboratory LA-UR-89-912. Particle Accelerator Conference, Chicago, IL, 20–23 March 1989.

28. Hamm, R. W. and Shubaly, M. R. *A Pre-clinical Radio-frequency Quadrupole Linac for Boron Neutron Capture Therapy Measurements*. In: Proc. First Int. Workshop on Accelerator-Based Neutron Sources for Boron Neutron Capture Therapy, Jackson, WY, 11–14 Sept. 1994. INEL Report Conference-940976, pp. 55–66 (1995).

29. Shefer, R. E., Klinkowstein R. E., Yanch, J. C. and Brownell, G. L. *An Epithermal Neutron Source for BNCT using a Tandem Cascade Accelerator*. In: Progress in Neutron Capture Therapy for Cancer. Eds B. N. Allen *et al* (New York: Plenum Press) pp. 119–122 (1992).

30. Shefer, R. E., Klinkowstein R. E., Yanch, J. C. and Howard, W. B. *Tandem Electrostatic Accelerators for BNCT*. In: Proc. First Int. Workshop on Accelerator-Based Neutron Sources for Boron Neutron Capture Therapy, Jackson, WY, 11–14 Sept. 1994. INEL Report Conference-940976, pp. 89–98 (1995).

31. Kwan, J. W., Anderson, O. A, Reginato, L. L., Vella, M. C. and Yu, S. S. *Electrostatic Quadrupole DC Accelerators for BNCT Applications*. In: Proc. First Int. Workshop on Accelerator-Based Neutron Sources for Boron Neutron Capture Therapy, Jackson, WY, 11–14 Sept. 1994. INEL Report Conference-940976, pp. 111–120 (1995).

32. Swenson, D. A. *A New Linac Structure for the BNCT Application*. In: Proc. First Int. Workshop on Accelerator-Based

Neutron Sources for Boron Neutron Capture Therapy, Jackson, WY, 11–14 Sept. 1994. INEL Report Conference-940976, pp. 121–128 (1995).

33. Gahbauer, R. G., Goodman, J. and Blue, T. *Some Thoughts on Tolerance, Dose, and Fractionation in Boron Neutron Capture Therapy*. In: Clinical Aspects of Neutron Capture Therapy — Proceedings of a Workshop on Boron Neutron Capture Therapy, Feb. 1988, Brookhaven National Laboratory. Eds Ralph G. Fairchild *et al*. Basic Life Sciences **50**, 81–85 (New York: Plenum Press (1989).

34. Blue, T. E., Woollard, J. E., Gupta, N. and Gahbauer, R. A. *Beam Design and Evaluation for BNCT*. In: Proc. First Int. Workshop on Accelerator-Based Neutron Sources for Boron Neutron Capture Therapy, Jackson, WY, 11–14 Sept. 1994. INEL Report Conference-940976, pp. 197–210 (1995).

35. Grusell, E., Condé, H., Larsson, B., Rönnqvist, T., Sornsuntisook, O., Crawford, J., Reist, H., Dahl, B., Sjöstrand, N. G. and Russel, G. *The Possible Use of a Spallation Neutron Source for Neutron Capture Therapy with Epithermal Neutrons*. In: Neutron Beam Design, Development, and Performance for Neutron Capture Therapy, Eds O. K. Harling *et al*. Basic Life Sciences **54** (New York: Plenum Press) pp. 249–258 (1985).

36. Kim, J., Yanch, J. C. and Wilson, M. J. *Californium-based Epithermal Neutron Beams for Neutron Capture Therapy*. In: Advances in Neutron Capture Therapy, Proc. Fifth Int. Symp. on Neutron Capture Therapy, Columbus, OH, Sept. 1992. Eds A. H. Soloway *et al* (New York: Plenum Publishing) pp. 131–134 (1993).

37. Nigg, D. W., Mitchell, H. E., Harker, Y. D., Yoon, W. Y., Jones, J. L. and Harmon, J. F. *Epithermal Photoneutron Source Studies for BNCT*. In: Proc. First Int. Workshop on Accelerator-Based Neutron Sources for Boron Neutron Capture Theapy, Jackson, WY, 11–14 Sept. 1994. INEL Report Conference-940976, pp. 373–386 (1995).

38. Jones, J. L. and Yoon, W. Y. *Feasibility Study of the Application of a Linear Electron Accelerator to BNCT*. In: Proc. 12th Int. Conf. on The Application of Accelerators in Research and Industry, University of North Texas, Denton, TX, November 1992.

39. Miley, G. H., Javedani, J. B., Gu, Y. B., Satsangi, A. J., Heck, P. F., Tzonev, I. D., Williams, M. J., Del Medico, S. G., Nebel, R. A., Barnes, D. C., Turner, L., Nadler, J. H. and Sved, J. *The IEC: a Novel Source for Boron Neutron Capture Therapy*. In: Proc. First Int. Workshop on Accelerator-Based Neutron Sources for Boron Neutron Capture Therapy, Jackson, WY, 11–14 Sept. 1994. INEL Report Conference-940976, pp. 111–120 (1995).

40. Laramore, G. E., Wootton, P., Livesey, J. C., Wilbur, D. S., Risler, R., Phillips, M., Jacky, J., Buchholz, T. A., Griffin, T. W., Bossard, S., *Boron Neutron Capture Therapy: a Mechanism for Achieving a Concomitant Tumor Boost in Fast Neutron Radiotherapy*. Int J. Radiat. Oncol. Biol. Phys. **28**(5), pp. 1135–1142 (1994).

41. Beach, J. L., Schroy, C. B., Ashtari, M., Harris, M. R. and Maruyama, Y. *Boron Neutron Capture Enhancement of Cf-252 Brachytherapy*. Int. J. Radiat. Oncol. Biol. Phys. **18**, 1421–1427 (1990).

42. Gupta, N., Gahbauer, R. A., Blue T. E. and Wambersie, A. *Dose Prescription in Boron Neutron Capture Therapy*. Int. J. Radiat. Oncol. Biol. Phys. **28**(5), pp. 1157–1166 (1994).

43. Deutsch, O. L. and Murray, B. W. *Monte Carlo Dosimetry Calculation for Boron Neutron Capture Therapy in the Treatment of Brain Tumors*. Nucl. Technol. **26**, 320 (1976).

44. Rydin, R. A., Deutsch, O. L. and Murray, B. W. *The Effect of Geometry on Capillary Wall Dose for Boron Neutron Capture Therapy*. Phys. Med. Biol. **21**, 134–138 (1976).

45. Gahbauer, R. A., Fairchild, R. G., Goodman, J. H. and Blue, T. E. *Can Relative Biological Effectiveness be Used for Treatment Planning in Boron Neutron Capture Therapy?* In: Tumor Response Monitoring and Treatment Planning. Ed. A. Breit (Berlin: Springer Verlag) (1992).

46. Blue, T. E., Gupta, N. and Wollard, J. E. *A Calculation of the Energy Dependence of the RBE of Neutrons*. Phys. Med. Biol. **38**, 1693–1712 (1993).

47. Wheeler, F. J. *Microdosimetric Calculations for Boron-drug Compound Factors for Boron Neutron Capture Therapy*. Trans. Am. Nucl. Soc. **65**, 147 (1992).

48. Kobayashi, T. and Kanda, K. *Analytical Dose Calculation in Cell Nucleus from Spherical ^{10}B Containing Region and a Concept of RBE for Neutron Capture Therapy*. In: Advances in Neutron Capture Therapy. Eds A. H. Soloway *et al* (New York: Plenum) pp. 217–220 (1993).

49. Kalend, A. M., Bloomer, W. D. and Epperly, M. W. *Dosimetric Consequences of $^{10}B(n,\alpha)^7Li$ Reaction Occuring at the Cellular Membrane*. Int. J. Radiat. Oncol. Biol. Phys. **31**(1), 171–178 (1995).

50. Wu, C. S., Amols, H. I., Kliauga, P., Reinstein, L. E. and Saraf, S. *Microdosimetry for Boron Neutron Capture Therapy*. Radiat. Res. **130**(3), pp. 355–359 (1992).

51. Solares, G. R. and Zamenhof, R. G. *A Novel Approach to the Microdosimetry of Neutron Capture Therapy, Part I: High-Resolution Quantitative Autoradiography Applied to Microdosimetry in Neutron Capture Therapy*. Radiat. Res. (submitted).

52. Raaijmakers, C. P. J., Konijnenberg, M. W., Verhagen, H. W. and Mijnheer, B. J. *Determination of Dose Components in Phantom Irradiated with an Epithermal Neutron Beam for Boron Neutron Capture Therapy*. Med. Phys. **22**(3), 321–329 (1995).

53. Harling, O. K., Moulin, D., Chabeuf, J.-M. and Solares, G. R. *On-Line Beam Monitoring for Boron Neutron Capture Therapy at the MIT Research Reactor*. In: Proc. Sixth Int. Symp. on Neutron Capture Therapy, 31 Oct–5 Nov 1994 (in press).

54. Kobayashi, T. and Kanda, K. *Microanalysis System of ppm-order ^{10}B Concentrations in Tissue for Neutron Capture Therapy by Prompt Gamma-ray Spectrometry*. Nucl. Instrum. Methods **204**, 525 (1983).

55. Wollard, J. E., Blue, T. E., Curran, J. F., Dobelbower, M. C. and Busby, H. R. *An Alpha Autoradiographic Technique for Spatial Quantification of ^{10}B Concentrations in Tissue.* Nucl. Sci. Eng. **110**, 96–103 (1992).

56. Barth, R. F., Adams, D. M., Soloway, A. H., Mechetner, E. B., Alam, F. and Anisuzzaman, A. K. M. *Determination of Boron in Tissues and Cells using Direct-current Plasma Atomic Emission Spectroscopy.* Anal. Chem. **63**, 890–893 (1991).

57. Wheeler, F. J. and Nigg, D. W. *Three Dimensional Radiation Dose Distribution Analysis for Boron Neutron Capture Therapy.* Nucl. Sci. Eng. **110**, 16–31 (1992).

58. Zamenhoff, R. G., Clement, S. D., Harling, O. K., Brenner, J. F., Wazer, D. E., Madoc-Jones, H. and Yanch, J. C. *Monte Carlo Based Dosimetry and Treatment Planning for Neutron Capture Therapy of Brain Tumors.* In: Proc. Int. Workshop on Neutron Beam Design, Development, and Performance for Neutron Capture Therapy. Eds O. K. Harling *et al* (Boston: Plenum) Basic Life Sci. **54**, 283 (1989).

59. Reinstein, L. E., Ramsay, E. B., Gajewski, J., Rammamoorthy, S. and Meek, A. G. *SBNCT-PLAN: A 3-dimensional Treatment Planning System for Boron Neutron Capture Therapy.* In: Advances in OH, Sept 1992. Eds A. H. Soloway *et al* (New York: Plenum Publishing) pp. 171–175 (1993).

60. Nigg, D. W., Randolph, P. D. and Wheeler, F. J. *Demonstration of Three Dimensional Deterministic Radiation Transport Theory Dose Distribution Analysis for Boron Neutron Capture Therapy.* Med. Phys. **18**, 43–53 (1991).

61. Niemkiewicz, J., Blue, T. E. and Gupta, N. *A Calculation of Neutron Flux Distributions in BNCT Using Removal-Diffusion Theory.* Trans. Am. Nucl. Soc. **70**, 13–15 (1994).

62. Niemkiewicz, J., Gupta, N. and Blue, T. E. *Point Kernel $^{1}H(n,\gamma)^{2}H$ Dose Calculations in BNCT.* Trans. Am. Nucl. Soc. **65**, 155–157 (1992).

63. Fairchild, R. G., Saraf, S. K., Kalef-Ezra, J. and Laster, B. H. *Comparison of Measured Parameters from a 24-Key and a Broad Spectrum Epithermal Neutron Beam for Neutron Capture Therapy: an Identification of Consequential Parameters.* Med. Phys. **17**, 1045–1052 (1990).

64. Joel, D. personal communications.

65. Madoc-Jones, H., Zamenhof, R., Solares, G., Harling, O., Yam, C.-S., Riley, K., Wazer, D., Rogers, G. and Atkins, M. *A Phase-I Dose-Escalation Trial of Boron Neutron Capture Therapy For Subjects with Metastatic Subcutaneous Melanoma of the Extremities* (submitted).

MICRODOSIMETRIC ANALYSIS OF ABSORBED DOSE IN BORON NEUTRON CAPTURE THERAPY

C. Kota and R. L. Maughan
Gershenson Radiation Oncology Center
Harper Hospital and Wayne State University
3990 John R., Detroit, MI 48201, USA

Abstract — The biological effectiveness of absorbed dose in BNCT depends on relative contributions from the gamma, neutron and the thermal neutron capture dose components, and on the resulting spectrum of secondary charged particles. This spectrum is a variable, likely to change within the patient, with subcellular boron location, and between different beams. In the present work, a dual proportional counter microdosimetric technique to measure and analyse the absorbed dose and its single event spectrum is described. Single event spectra of the gamma, neutron and ^{10}B dose components have been separately calculated from experimental measurements. Spectra for a 6 μm diameter volume (simulating a cell nucleus), with a uniform distribution of ^{10}B and with ^{10}B localised external to it have been calculated. For the same macroscopic ^{10}B dose, the dose to the nucleus with the ^{10}B localised outside the nucleus is reduced to half of that when it is uniformly distributed. The reduced dose is of a lower average lineal energy compared to the uniform dose; its relative effectiveness depends on the radiosensitivity of the irradiated tissue, being larger for sensitive tissues and smaller for radioresistant tissues.

INTRODUCTION

The aim of boron neutron capture therapy (BNCT) is to deliver a local radiation dose to a tumour volume preferentially loaded with ^{10}B, by irradiating it with thermal neutrons. One approach to provide a thermal neutron fluence at the tumour location involves irradiating the patient with external beams of epithermal neutrons produced from reactors or accelerators[1]. In this technique, a non-negligible dose is delivered to the normal tissues and tumour by fast neutrons and gamma rays accompanying the epithermal neutron beam. The boron neutron capture reaction can also be used to obtain preferential tumour dose enhancements in fast neutron therapy and in ^{252}Cf brachytherapy, which could be clinically significant (boron neutron capture enhanced fast neutron therapy — BNCEFNT)[2].

The radiation fields employed in all of the above scenarios consist of gamma rays and neutrons of a range of energies. The absorbed dose in tissue is deposited by: secondary electrons produced by gamma rays; protons produced by neutron interactions; alpha particles and lithium ions produced by the ^{10}B neutron capture reaction. These secondary charged particles of different types have a range of energies and, consequently, different biological effectiveness. In addition, the relative contributions from the different dose components, which determine the secondary charged particle spectrum (microdose spectrum), depend on the spatial location within the patient, subcellular location of boron, etc. As a result, the relative biological effectiveness (RBE) of the absorbed dose is a complex function of its microdose spectrum. From various radiobiological studies published in the literature, it is seen that the total effective dose for a given effect level is often calculated using the following isoeffect equation[3]:

Effective dose =
$$D_g \times RBE_g + D_n \times RBE_n + D_b \times RBE_b \quad (1)$$

where D_g, D_n and D_b are the dose contributions from the gamma, neutron and boron components; RBE_g, RBE_n and RBE_b are the RBEs of the gamma, neutron and the boron dose components when applied alone to produce the same effect. In general, it is difficult to estimate the RBEs for pure neutron and boron dose components. Even when these RBEs are known in a given situation, they must be used with caution in other situations because of their dependence on the neutron spectrum, and on the subcellular boron location. The quantification of the effectiveness of the absorbed dose in boron neutron capture therapy therfore presents a formidable challenge to the physicist.

The present work describes a dual counter (microdosimetric) technique, which is an extension of proportional counter microdosimetry, to the field of BNCT. The dual counter technique can be used to measure the absorbed dose and microdose spectrum of the different dose components in various simulated microscopic volumes. The microdose spectra can be used to quantify changes in radiation quality of the absorbed dose in different situations. In this study, the dual counter technique has been used to study the effect of subcellular localisation of ^{10}B on the dose to the nucleus and its radiation quality.

METHODS AND MATERIALS

The dual counter microdosimetric technique is based on the measurement of single event spectra using two Rossi type, tissue-equivalent (TE) proportional counters at a given point in the radiation field of interest. The tissue-equivalent plastic (TEP) A-150 wall of one of the counters is uniformly loaded with 50 μg.g^{-1} of ^{10}B, to simulate a tissue containing a uniform distribution of 50 μg.g^{-1} of ^{10}B. When placed in the radiation field of interest, the counter with no boron in its wall (henceforth referred to as the TE counter) measures the microdose spectrum of secondary charged particles generated by gamma interactions, nitrogen capture reaction and neutron interactions in the A-150 wall. The counter with boron in its wall (henceforth referred to as the ^{10}B counter), in addition to the spectrum measured by the TE counter, measures the microdose spectrum of secondary charged particles produced by the ^{10}B thermal neutron capture reactions in its wall.

In the present work, two identical 1/2″ LET counters with 8 mm thick A-150 TEP walls and an internal diameter of 1.27 cm (manufactured by Far West Technologies, Goleta, CA) were used to make the microdosimetric measurements. The experimental details pertinent to the calibration of the counters and the data acquisition procedures have been described earlier[4]. The two counters were simultaneously filled with a propane-based TE gas to pressures of 2.2 kPa and 26.4 kPa, to simulate unit density tissue volumes of 0.5 μm and 6 μm diameters, and were operated at 600 V and 750 V respectively. The counters were calibrated in terms of lineal energy by positioning the proton edge in the measured spectra at ~150 keV.μm^{-1} for the 0.5 μm diameter volume, and at ~105 keV.μm^{-1} for the 6.0 μm diameter volume. The 0.5 μm diameter volume is assumed to adequately approximate a Bragg–Gray cavity for secondary charged particle equilibrium for all components of the dose. On the other hand, while the simulated volume of 6.0 μm diameter is assumed to represent a Bragg–Gray cavity for the gamma and neutron components of the dose, for the boron reaction dose component, it is assumed to represent a cell nucleus with the ^{10}B localised external to its volume.

Single event spectra of the absorbed dose were measured for the two simulated volumes, at a distance of 10 cm from a ^{252}Cf source in a large water phantom and were assumed to be representative of the spectra in an epithermal beam. This assumption was verified by comparison with a spectrum measured in an epithermal neutron beam for the 0.5 μm diameter volume. The ^{10}B dose was obtained by subtracting the dose measured by the TE counter from the dose measured by the ^{10}B counter. The single event spectra corresponding to the gamma dose, the TE counter and the ^{10}B counter were recalculated for equal y intervals, using linear interpolation between the y and yd(y) values. Single event spectra corresponding to the gamma, neutron and boron dose components were obtained by subtracting the appropriate spectra. The single event spectrum of each dose component was then renormalised to unit dose. The complete single event spectrum corresponding to specific fractional contributions from the different dose components was obtained by summing the weighted component spectra. These spectra were then convoluted with different weighting functions to estimate the effectiveness of the absorbed dose.

RESULTS AND DISCUSSIONS

Figure 1 shows the single event spectra measured by the two counters for a simulated volume of 0.5 μm, in ^{252}Cf radiation and in an epithermal neutron beam. The spectra of the individual dose components in the two radiation fields are shown in Figure 2. It is seen that while the spectra of the gamma and boron dose in the two radiation fields are quite similar, the spectrum of the neutron dose is quite different. This is because the ^{252}Cf radiation contains a larger fraction of fast neutrons than the epithermal beam, which results in a larger recoil proton component of lower lineal energy. To study the effect of ^{10}B localised only in the cytoplasm,

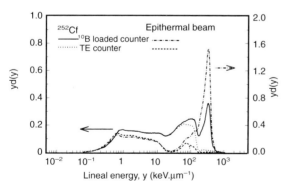

Figure 1. Microdosimetric single event spectra measured with the TE and the ^{10}B counters for a 0.5 μm volume, normalised to a dose of 10 mGy measured by the TE counter. The solid and the dotted lines represent the spectra measured in ^{252}Cf radiation, while the dashed and the dot-dashed lines represent spectra measured in an epithermal neutron beam.

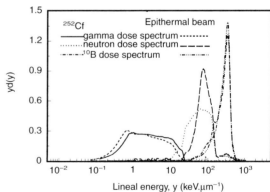

Figure 2. Microdosimetric single event spectra of the different dose components, calculated from the spectra shown in Figure 1. The single event spectrum from each dose component is normalised to unit dose.

a single event spectrum of the absorbed dose was measured for a simulated volume of 6.0 μm. In this case, the dose measured for the neutron and the gamma components is representative of the true dose. However, the dose measured for the ^{10}B component corresponds to the dose in a 6.0 μm diameter volume devoid of ^{10}B, but surrounded by a uniform distribution of 50 μg.g^{-1} of ^{10}B. The ^{10}B dose in this volume was measured to be 0.4 times that measured for a uniform ^{10}B distribution.

Recently, Gabel et al[5] calculated the effective dose to a nucleus (7.6 μm in diameter) with the ^{10}B localised in the cytoplasm of the cell (13.0 μm in diameter), to be ~0.43. To relate the present measurements to these calculations, it was assumed that the reduction in the absorbed dose in a volume devoid of boron is the same for a volume 6 μm or 7.6 μm in diameter. This assumption was based on the argument that, with an average range of ~7.5 μm, the ^{10}B reaction products originating in the cytoplasm (2.7 μm or 3.5 μm thick shell) would result in approximately the same dose in a nuclear volume 6 μm or 7.6 μm in diameter. For an average cellular concentration of 50 μg.g^{-1} of ^{10}B, the concentration in the cytoplasm with the nuclear volume devoid of ^{10}B is ~62.5 μg.g^{-1}. After correcting for this change in the ^{10}B concentration, the nuclear dose measured in this work is calculated to be 0.5 times that measured with the ^{10}B uniformly distributed. This is in fair agreement with the effective dose calculated by Gabel et al.

To estimate the effectiveness of the total dose from its single event spectrum, a representative situation was considered, with fractional contributions from the gamma, neutron and boron dose components equal to 0.17, 0.17 and 0.66 respectively. A complete spectrum was reconstructed from the ^{252}Cf radiation components and was assumed to be representative of epithermal neutron beam spectra. Figure 3 shows the reconstructed spectra for a 6 μm volume corresponding to boron present uniformly and localised to the cytoplasm, for the same macroscopic absorbed dose. From this figure it is seen that the decreased nuclear dose with the boron localised to the cytoplasm is also of lower average lineal energy.

To assess the importance of this change in the lineal energy of the dose to the nucleus, the two spectra were convoluted with three different weighting functions, shown in Figure 4. These weighting functions represent the saturation in the biological effectiveness of the absorbed dose, arising from the excessive ionisation densities at higher lineal energies. The first two functions are analytical saturation correction functions with saturation parameters equal to 100 keV.μm^{-1} and 160 keV.μm^{-1}. These saturation parameters were scaled for the 6 μm volume from values of 124 keV.μm^{-1} and 200 keV.μm^{-1} used by Kellerer and Rossi[6] for a 2 μm volume. This scaling was based on the change in the position of the proton edge between these two site diameters. The third function used was similarly scaled from the RBE(y) against y function derived by Menzel et al for early effects in mice[7].

The ratio of the macroscopic absorbed dose in the nucleus, with boron localised to the cytoplasm to boron distributed uniformly is 0.67. The effective nuclear dose is the ratio of the y* or RBE values, calculated with the boron localised to the cytoplasm to boron distributed uniformly. This effective nuclear dose was calculated using the three weighting functions described above to be 0.7, 0.62 and 0.74 respectively. The first and the third functions represent a smaller saturation parameter compared to the second function, implying a greater radiosensitivity. Therefore, these results indicate that the effective nuclear dose depends not only on the decrease in the physical dose, but also on the intrinsic sensitivity of the cell to radiation.

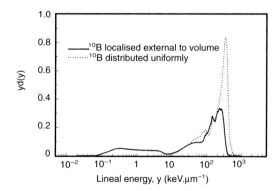

Figure 3. Reconstructed single event spectra for a 6 μm volume, corresponding to ^{10}B present uniformly throughout the volume and to ^{10}B localised exterior to the volume. Fractional gamma dose = 0.17, fractional neutron dose = 0.17, fractional ^{10}B dose = 0.66.

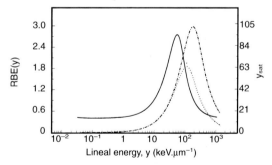

Figure 4. Biological weighting functions to account for the variation in the RBE with lineal energy. The solid line represents the function derived by Menzel et al, modified to the 6 μm volume. The dotted line and the dashed line represent the correction functions used to obtain the saturation corrected dose mean lineal energy y*, with saturation parameters of 100 keV.μm^{-1} and 160 keV.μm^{-1} respectively.

CONCLUSIONS

The dual counter microdosimetric technique has been used to measure the single event spectra of the absorbed dose corresponding to the gamma, neutron and boron components for a 0.5 μm volume and a 6 μm volume. The measured spectra have been manipulated using a spectral subtraction technique to obtain the single event spectra of the absorbed dose in a 6 μm volume (representing a cell nucleus) with and without any boron in it. For the same macroscopic absorbed dose, the dose to the nucleus is reduced by half if the boron is localised to the cytoplasm. The reduction in the effective dose, however, is likely to depend on the intrinsic radiosensitivity of the cell. The dual counter microdosimetric technique appears to be a valuable tool for quantifying the quality of the absorbed dose in BNCT.

ACKNOWLEDGEMENTS

The authors would like to thank Dr A. T. Porter for his support and encouragement for this work.

REFERENCES

1. Allen, B. J., Moore, D. E. and Harrington, B. V. *Progress in Neutron Capture Therapy for Cancer* (New York: Plenum) (1992).
2. Kota, C. *Microdosimetric Considerations in the Use of the Boron Neutron Capture Reaction in Radiation Therapy*. Ph.D. Thesis, Wayne State University (1996).
3. Lam, G. K. Y. *An Isoeffect Approach to the Study of Combined Effects of Mixed Radiations — The Nonparametric Analysis of in vivo Data*. Radiat. Res. **119**, 424–431 (1989).
4. Kota, C. and Maughan, R. L. *A Dosimetry System for Boron Neutron Capture Therapy Based on the Dual Counter Microdosimetric Technique*. Bull. Cancer/Radiother. **83**(suppl. 1), 1735–1755 (1996).
5. Gabel, D., Foster, S. and Fairchild, R. G. *The Monte Carlo Simulation of the Biological Effect of the $^{10}B(n, \alpha)^{7}Li$ Reaction in Cells and Tissue and its Implication for Boron Neutron Capture Therapy*. Radiat. Res. **111**, 14–25 (1987).
6. Kellerer, A. M. and Rossi, H. H. *The Theory of Dual Radiation Action*. Curr. Topics Radiat. Res. **8**, 85–158 (1972).
7. Menzel, H. G., Pihet, P. and Wambersie, A. *Microdosimetric Specification of Radiation Quality in Neutron Radiation Therapy*. Int. J. Radiat. Biol. **57**, 865–883 (1990).

CHARACTERISATION OF AN ACCELERATOR-BASED NEUTRON SOURCE FOR BNCT OF EXPLANTED LIVERS

S. Agosteo[1], P. Colautti[2], M. G. Corrado[3], F. d'Errico[4], M. Matzke[5], S. Monti[6], M. Silari[7] and R. Tinti[6]
[1]Dipartimento di Ingegneria Nucleare, Politecnico di Milano, via Ponzio 34/3, I-20133 Milano, Italy
[2]INFN, Laboratori Nazionali di Legnaro, via Romea 4, I-35020 Legnaro, Padova, Italy
[3]Università degli Studi di Milano, Dipartimento di Fisica, via Celoria 16, I-20133 Milano, Italy
[4]Dipartimento di Costruzioni Meccaniche e Nucleari, Università degli Studi di Pisa, via Diotisalvi 2, I-56126 Pisa, Italy
[5]Physikalisch-Technische Bundesanstalt, Bundesallee 100, D-38116 Braunschweig, Germany
[6]ENEA–ERG–FISS–FIRE, via Martiri di Monte Sole 4, I-40129 Bologna, Italy
[7]Consiglio Nazionale delle Ricerche, ITBA, via Ampère 56, I-20131 Milano, Italy

Abstract — An accelerator-based thermal neutron source for BNCT of the explanted liver was designed using the MCNP code. Neutrons are generated via (d,n) reactions by 7 MeV deuterons bombarding a beryllium target. The therapy constraints were approached by simulating an irradiation cavity placed inside a graphite reflector parallelepiped containing a heavy-water moderator in turn enclosing the beryllium target. The experimental verification was performed at the Laboratori Nazionali di Legnaro (Italy). The thermal and epithermal neutron flux was measured at various positions in the irradiation cavity by means of activation techniques employing bare and cadmium covered indium foils. Further measurements were performed with BF_3 detectors. The fast neutron component of the dose equivalent and the energy spectrum above 100 keV were assessed by means of a recently developed technique employing variable threshold superheated drop detectors. The prompt gamma ray dose was measured with ^{7}LiF TLDs.

INTRODUCTION

Liver explant, tumour removal and subsequent reimplant of the organ is a surgical technique developed at the Policlinico S. Matteo in Pavia[1] Italy, up to now only performed on a few patients. In the case of multiple tumours when surgery is impossible, the idea of treating the explanted organ with boron neutron capture therapy (BNCT) was proposed.

As is well known, reactors provide intense neutron fluxes and are the only neutron sources presently utilised for BNCT. The possibility of employing accelerator-based neutron sources is also currently discussed[2]. The accelerator-based neutron source for BNCT of the explanted liver, described in the present work, is intended for the Centro Nazionale di Adroterapia Oncologica (CNAO, National Centre of Oncological Hadrontherapy), a hospital-based facility for proton and light ion therapy proposed in Italy and whose feasibility study has recently been completed[3–7]. The beam current delivered by the 11 MeV proton linac or the 3 MeV.u^{-1} ion linac (the two injectors of the main accelerator) is almost entirely available for neutron generation via (p,n) and (d,n) reactions.

A liver radiation treatment requires a thermal neutron fluence of 5×10^{12} cm^{-2} and should be imparted within three hours, thus needing a thermal neutron flux of at least 5×10^8 cm^{-2}.s^{-1}. The flux spatial distribution over the irradiated organ should be uniform in order to minimise neutron attenuation with depth inside the liver, as is pointed out in Ref. 8. It is worth mentioning that the use of broad thermal neutron beams (15–22 cm in diameter) extends the treatment depth to about 6 cm[9].

As neutrons do not cross other structures before impinging on the liver, a strong epithermal (from a few eV to 10 keV) component is not necessary in the present case. The neutron yields required for BNCT are generally achieved using accelerated particles of a few MeV. Therefore, fast neutrons are inevitably present and should be strongly attenuated before reaching the organ. For this reason, a moderator is obviously needed. The prompt gamma rays produced in the moderator and in the other structures of the irradiation facility should be kept to a minimum. The cumulative absorbed dose due to fast neutrons and prompt gamma rays should be kept below 1 Gy for a treatment dose of 20 Gy (due to the $^{10}B(n,\alpha)^7Li$ reaction products).

A 7 MeV deuteron beam impinging on a beryllium target was considered here. Although in the present design the ion injector can deliver only up to 6 MeV deuterons, 7 MeV deuterons are discussed here, as complete information about the double differential neutron yield is available in the literature. A brief description of the moderator structure designed by Monte Carlo simulations with the MCNP[10] code is given. The main part of the paper discusses the experimental results obtained at the Van de Graaff accelerator of the Laboratori Nazionali di Legnaro (LNL), Italy.

MODERATOR DESIGN

The moderator geometry and materials were chosen in order to meet the thermal fluence uniformity and fast neutron attenuation requirements mentioned in the Introduction. The moderator optimisation was performed through Monte Carlo simulations using the

MCNP code. The double differential distribution of neutrons with energies higher than 0.8 MeV, produced by 7 MeV deuterons bombarding a thick beryllium target, measured in Ref. 11, was employed as the simulation source. The fraction below 0.8 MeV[12] was assumed equal to the value at 0° for all angles. The neutron yield above 0.05 MeV was estimated[11,12] to be 0.85×10^{10} s^{-1}·μA^{-1}.

A comprehensive description of the optimisation of the moderator is given in Ref. 8. The geometry adopted consists of a $120 \times 120 \times 100$ cm^3 graphite reflector parallelepiped containing a $55 \times 30 \times 40$ cm^3 heavy-water moderator parallelepiped, in turn enclosing the beryllium target. The deuteron beam impinges perpendicularly on the target, which is at a 75° angle with respect to the base of the heavy-water tank. The required neutron fluence is obtained in a hemispherical irradiation cavity, with an 18 cm diameter, located over the heavy-water tank and behind the beryllium target. A cross-sectional view of the moderator structure resulting from the simulation optimisation is shown in Figure 1. The heavy-water tank was not considered at that stage.

EXPERIMENTAL

The experimental verification of the accelerator-based source was performed at the CN Van De Graaff accelerator of the LNL. The deuteron beam energy was 7 MeV. The 1 mm thick beryllium target completely stopped the beam. The deuteron current impinging on the target was measured using an air-cooled Faraday cup with an electron suppressor.

Some changes were introduced with respect to the simulation design to facilitate the assembling of the experimental structure. The irradiation cavity was made in a cubic form ($18 \times 18 \times 18$ cm^3) and the target was perpendicular to the base of the heavy-water tank. The Faraday cup was inserted in a pipe placed inside the heavy-water tank. This tank ($55 \times 30 \times 40$ cm^3) was made with 5 mm thick plates of stainless steel to hinder corrosion. The reflector ($120 \times 120 \times 100$ cm^3) was made of nuclear graphite (density 1.7 g.cm^{-3}).

The experimental structure was further studied with simulations. While the modifications introduced in the target inclination and in the irradiation cavity shape have negligible effects, a reduction of the thermal fluence and an increase of the prompt gamma ray dose in the irradiation cavity were caused by neutron absorption in the stainless steel heavy-water tank. A further simulation was performed by considering a container made of a corrosion resistant aluminium alloy (Al-1100) with the same geometry and thickness. The neutron absorption effect was found to be lower. Table 1 reports the average neutron flux in the energy groups of interest and the average prompt gamma ray dose in the irradiation cavity for the different situations described above. The prompt gamma ray dose was estimated with the conversion factors of ICRP 21[13].

The deuteron current was estimated with a charge integrator connected to the Faraday cup described above. A BF$_3$ proportional counter was placed at a fixed position close to the reflector to check the charge integrator stability during every measurement. The result of a single measurement was accepted when the ratio of the BF$_3$ counts to the measured current was consistent with those of the whole set of irradiations. In the follow-

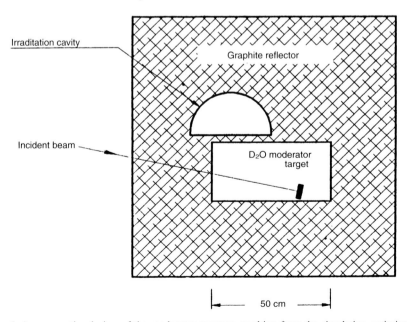

Figure 1. A cross-sectional view of the moderator structure resulting from the simulation optimisation.

ing the various measurement techniques adopted for the source characterisation are described separately for clarity.

Activation techniques

The thermal and epithermal components of the neutron flux were measured with bare and cadmium covered indium foils (diameter 1 cm, thickness from 90.4 mg.cm^{-2} to 96.13 mg.cm^{-2}) at various positions in the irradiation cavity. Various nine-foils sets placed on a plane perpendicular to the floor of the irradiation cavity were irradiated separately to minimise flux perturbation. Five different planes were selected, therefore the spatial distribution of the neutron flux was investigated at 45 different positions. The gamma activity of the irradiated foils was measured with a 2″ × 2″ NaI scintillator. The nominal deuteron current impinging on the beryllium target was 100 nA.

The uncertainties due to the counting statistics and those due to measurement (i.e. normalisation uncertainties) were estimated separately[14]. In particular, a uniform distribution was given to every source of normalisation uncertainty and the probability distribution of their sum was assessed with the Monte Carlo method. The number 'i' of Monte Carlo histories necessary to provide reliable results was estimated by comparing the theoretical variance of every uniform distribution with that calculated after the ith history. This method was adopted as the number of the measured positions in the cavity and the long irradiation times did not permit repeated measurements. The normalisation uncertainty sources considered were:

(i) *NaI scintillator peak efficiency*: estimated with MCNP simulations. The random number interval is limited by the value calculated with the indium foil strictly contiguous to the aluminium cover of the scintillator and by that with the foil at a distance of 1 mm.

(ii) *Thermal flux depression factor*: estimated from the values of Ref. 15. The random number interval is limited by the values estimated for the thinner and the thicker foils used in the measurements.

(iii) *Epithermal flux depression factor*: estimated with Monte Carlo simulations performed with both the MCNP and FLUKA[16] codes. In particular, as only neutron dosimetry cross sections were available for indium in the adopted MCNP libraries, the infinitely thin target reaction rate was calculated with MCNP and those related to the thicker and the thinner targets used in the measurements were estimated with FLUKA.

(iv) *Average deuteron current striking the beryllium target*: an uncertainty of 10% was assigned to the value measured by the charge integrator in each irradiation. This results from a comparison between charge values and counts of BF$_3$ used as a monitor.

(v) *Foil weight*: an uncertainty of 1 LSB (least significant bit) was given, as a digital scale was used.

(vi) $\sigma_{eff} = \int_{0.4\,eV}^{\infty} \sigma(E)\Phi(E)dE / \int_{0.4\,eV}^{\infty} \Phi(E)dE$.

This effective cross-section was estimated with MCNP simulations as the ratio of the infinitely thin target reaction rate to the flux in the irradiation cavity between 0.4 eV and the maximum neutron energy.

(vii) *Cadmium correction factor* F_{Cd}: estimated from Ref. 17. The random number interval is limited by the values estimated for the thinner and the thicker foils used in the measurements.

(viii) *A further uncertainty* of 10% was conservatively assigned to the recorded counts to account for other sources of bias (such as foil positioning, deuteron energy, irradiation and waiting time, etc.) which, even when affected by low individual uncertainties, can jointly influence the results.

The thermal and epithermal fluxes per unit deuteron current averaged over all positions were 1.710 × 10^6 cm^{-2}.s^{-1}.μA^{-1} and 2.964 × 10^5 cm^{-2}.s^{-1}.μA^{-1}, respectively. The counting uncertainties referring to the

Table 1. Average neutron flux density and prompt gamma ray dose per unit deuteron current in the irradiation cavity resulting from the simulation of the designed and experimental structures.

	Neutron flux density (cm^{-2}.s^{-1}.μA^{-1})			Prompt gamma ray dose (Gy.s^{-1}.μA^{-1})
	Thermal	0.4 eV–10 keV	>10 keV	
Design (without tank)	6.8 × 10^6 ± 3 × 10^4	4.0 × 10^5 ± 8 × 10^3	7.3 × 10^4 ± 4 × 10^3	5.3 × 10^{-7} ± 9 × 10^{-9}
Experimental structure with stainless steel tank	1.6 × 10^6 ± 5 × 10^3	3.1 × 10^5 ± 1 × 10^3	6.3 × 10^4 ± 6 × 10^2	4.5 × 10^{-6} ± 2 × 10^{-8}
Experimental structure with aluminium tank	4.0 × 10^6 ± 8 × 10^3	3.5 × 10^5 ± 1 × 10^4	6.1 × 10^4 ± 3 × 10^3	2.3 × 10^{-6} ± 1 × 10^{-8}

considered positions were between 0.38% and 0.5% for the thermal flux and between 1.17% and 1.75% for the epithermal flux. The normalisation uncertainties were about 7% in all cases.

The flux spatial distribution was numerically estimated with MCNP. In particular, the geometry and the materials of the experimental facility were carefully simulated. The neutron thermal and epithermal fluences were scored within cubic cells (1 cm side) located inside the irradiation cavity and centred in the same positions as the indium foils. Figure 2 shows the measured and calculated thermal and epithermal fluxes at the positions of interest.

BF_3 proportional counters

The thermal neutron flux inside the irradiation cavity was also measured with 20th Century Electronics™ BF_3 counters, model 5EB40/13. Their thermal neutron sensitivity was determined with a moderated Am-Be neutron source. As the active length of these detectors is 5 cm, the measured flux fits the indium foil results averaged over the positions covered by the counter. Repeated measurements could be performed in the same positions since the irradiation times were shorter in this case. However, a Monte Carlo method analogous to that described above was employed. The considered normalisation uncertainties affecting the measurement results were the detector sensitivity, the counting time, the deuteron current and the detector positioning.

In order to minimise pulse pile-up, a nominal deuteron beam current of 1 nA was used. Table 2 reports the measured fluxes, compared with the values resulting from the activation technique averaged over the positions covered by the BF_3 counter. Both normalisation uncertainties resulting from five repeated measurements and calculated with the Monte Carlo technique are reported. The repeated measurement uncertainty obviously includes those originating from counting statistics and biases.

Superheated drop detectors

A recently developed neutron spectrometry technique called BINS (bubble interactive neutron spectrometer)[18,19] was also employed for the characterisation of the fast neutron flux distribution within the irradiation cavity. Specific advantages offered by the system in our case were a small probe size (4 cm³), allowing for measurements within the cavity, a broader dynamic range and higher sensitivity than threshold activation detectors, and a complete gamma discrimination.

BINS relies on the use of two different superheated emulsions (octafluorocyclobutane and dichlorotetrafluoroethane) whose detection thresholds are selected by controlling the operating temperature. The response matrix of the system, measured at the PTB Braunschweig with monoenergetic neutrons, is virtually orthogonal (i.e. the response functions are linearly independent) which allows for effective few-channel unfolding procedures. With their sharp detection thresholds, the two emulsions scan the 0.01–1 MeV and the

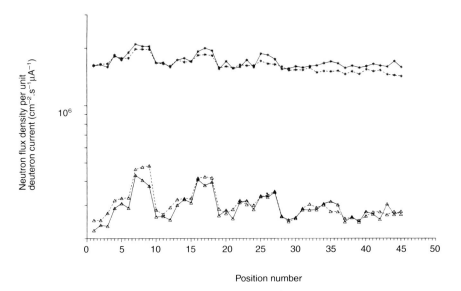

Figure 2. Calculated and measured (in foils) thermal and epithermal fluxes in the irradiation cavity. The connecting lines are only eyeguides. Upper lines (∗) thermal flux; full line, experimental; dashed line MCNP. Lower lines (△) 0.4 eV–10 keV flux: full line, experimental; dashed line, MCNP.

1–10 MeV ranges, respectively. While any threshold may be virtually generated within that interval, the current spectrometer prototype operates on a seven-isolethargic-bin basis and unfolded results cover the 0.1–12 MeV range.

Measurements were carried out sequentially: the two detectors were first irradiated at 25°C and their temperature then successively increased in 5°C steps, by means of a time proportioning controller. At each step, counting statistics of at least a thousand bubbles were accumulated by detecting acoustically the neutron induced vaporisation events. Results were then analysed by means of two unfolding codes, MSITER and UNFANA, and are illustrated in Figure 3.

MSITER requires pre-information on the spectrum, i.e. the fluence values and their covariance matrix[20]. Starting from these data, the code proceeds by minimising the weighted absolute differences between measured and calculated counting rates. In our case, the MCNP calculated spectrum was used as pre-information for the unfolding algorithm. An uncorrelated uncertainty of 10% on the scaling factor of each response function was assumed and added quadratically to the 10% uncertainties on each measured data point. Uncorrelated relative uncertainties on the MCNP a priori spectrum were assumed as 10% between 0.1 and 7 MeV, and 50% above 7 MeV. Under these assumptions, input and output spectra virtually coincided. The value of the χ^2 per degree of freedom was lower than 1, indicating complete consistency between MCNP-calculated and BINS-measured data. The other code employed, UNFANA, is an analytical development of the Bayesian-statistical Monte Carlo unfolding methods[21]. The probability distribution from which Monte Carlo unfolding methods sample possible spectra, is replaced here by a suitable exponential distribution from which the code proceeds analytically. The method reduces the computing time required by Monte Carlo methods while extracting the maximum information, with the associated uncertainties, from the measured data. This unfolding generated a smoother spectral distribution which better reflects the intrinsic resolution of BINS; being free of binding pre-information. Results were quite satisfactory: spectra unfolded with the two methods differ from

Table 2. Thermal flux density per unit deuteron current in the irradiation cavity measured with BF_3 counters.

Detector position	BF_3 thermal flux density ($cm^{-2}.s^{-1}.\mu A^{-1}$)	Counting uncertainty (%)	Normalisation uncertainty (%)	Repeated measurement uncertainty (%)	Average In thermal flux density ($cm^{-2}.s^{-1}.\mu A^{-1}$)
1	1.795×10^6	2.8×10^{-1}	3.19	3.0×10^{-1}	1.825×10^6
2	1.644×10^6	3.0×10^{-1}	3.19	2.0×10^{-1}	1.631×10^6
3	1.654×10^6	2.0×10^{-1}	3.19	5.0×10^{-1}	1.613×10^6
4	1.784×10^6	2.8×10^{-1}	3.19	3.0×10^{-1}	1.822×10^6

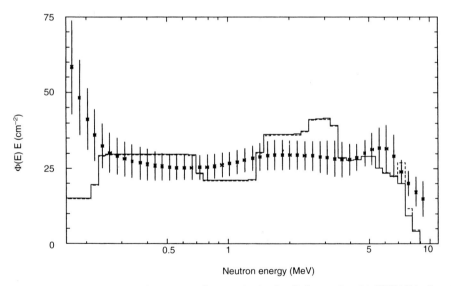

Figure 3. Calculated and measured (BINS) fast neutron fluences in the irradiation cavity. (∗) UNFANA, (– – –) MSITER, (———) MCNP.

each other by less than the associated uncertainties, which confirmed the overall structure of the energy distribution. Fluence integrals over broader energy intervals were also calculated from both codes (Table 3) and were affected by much smaller uncertainties thanks to negative correlations between adjacent groups.

Prompt gamma ray dose: TLDs

The gamma ray dose in the irradiation cavity was assessed by ^{7}LiF TLDs, model GR207, provided by ENEA[22]. These dosemeters are characterised by a low thermal neutron sensitivity (0.57 pGy.cm^2). The LiF detector is combined with Al and Cu filters in order to establish the electronic equilibrium. Table 4 reports the resulting air kerma at five positions inside the irradiation cavity. These five positions correspond to the five In-foil irradiation planes (see above). The uncertainties were calculated by placing different dosemeters at the same position during different irradiations. The results reported in Table 4 without uncertainty refer to a single measurement. It should be noted that the values of Table 4 are larger than the MCNP estimate (4.5×10^{-6} Gy.s^{-1}.µA^{-1}), by a factor ranging from 1.45 to 1.80. This discrepancy could be explained by the thermal neutron activation of the Cu filter that was observed in the NaI spectrometry of the irradiated detector. Moreover, MCNP only treats prompt gamma ray production and transport and does not consider the de-excitation radiation of the activated materials (mainly the heavy-water tank in our case).

DISCUSSION AND CONCLUSIONS

A satisfactory agreement between measured and simulated data is illustrated in Figures 2 and 3. The numerical thermal flux estimates are lower than the experimental values from position 30 to 45, i.e. in the part of the irradiation cavity furthest away from the beryllium target. This may derive from the assumption made in the simulations for the neutron yield below 0.05 MeV. The thermal flux uniformity appears quite satisfactory; it should be pointed out that the experimental structure was assembled using nuclear grade graphite blocks manufactured to fit in a reactor core. Therefore some narrow channels, allowing neutron streaming, are inevitably present in the reflector. The larger discrepancies from uniformity are observed at positions 7–9 and 16–18, at the bottom of the irradiation cavity. A smoother but analogous behaviour characterises the numerical data which do not take into account the above mentioned streaming channels. This warrants shielding the irradiation cavity vertically by a few centimetres.

When considering the thermal flux averaged over all positions, we conclude that the thermal fluence required for the irradiation (i.e 5×10^{12} cm^{-2}) can be achieved in the present situation (i.e. with the stainless steel tank) with a beam current of about 270 µA, for an irradiation lasting three hours. The prompt gamma ray dose in the entire treatment would be unacceptably large, i.e. 13.1 Gy (see Table 1). A situation closer to the therapy requirements could be obtained with the adoption of an aluminium tank. From the simulation results of Table 1 follows that the required current for a three hour irradiation would be about 120 µA and the prompt gamma ray dose about 3.0 Gy.

The doses to normal liver and to the tumour were calculated with MCNP, referring to the configuration with an Al tank, in order to outline some clinical remarks. A sphere with the same volume and composition of a liver[23] was placed in the irradiation cavity. The ^{10}B content was assumed to be 10 µg.g^{-1} and 50 µg.g^{-1} in the normal liver and in a spherical tumour (radius 1 cm) placed at the liver centre, respectively. The doses referring to a three hour treatment with a deuteron current of 120 µA are listed in Table 5. It should be noted that the ^{10}B dose to the tumour is lower than the expected value of 20 Gy, because the fluence attenu-

Table 3. Fluence integrals of the high energy part of the spectrum assessed with MCNP simulations and BINS-measurements (unfolded with MSITER and UNFANA).

Energy interval (MeV)	MCNP	MSITER*	UNFANA*
0.2–20.0	1.0	1.0	1.0
0.5–20.0	0.759	0.761 (6)	0.749 (6)
1.0–20.0	0.592	0.596 (6)	0.584 (7)
2.0–20.0	0.405	0.410 (6)	0.400 (8)
3.0–20.0	0.256	0.263 (6)	0.289 (9)
4.0–20.0	0.162	0.170 (7)	0.214 (11)
5.0–20.0	0.102	0.110 (8)	0.153 (13)
6.0–20.0	0.057	0.065 (11)	0.099 (14)
8.0–20.0	0.005	0.006 (33)	0.029 (31)

*In parentheses: relative uncertainty (%).

Table 4. Air kerma rate of the prompt gamma rays in the irradiation cavity measured with TLDs. Position 1 is in the same plane perpendicular to the cavity bottom, close to the cavity side towards the Be target, position 5 is close the opposite side, positions 2, 3 (cavity centre) and 4 are intermediate.

Position	Air kerma rate per unit deuteron current (Gy.s^{-1}.µA^{-1})
1	$7.5 \times 10^{-6} \pm 9 \times 10^{-7}$
2	8.1×10^{-6}
3	$7.8 \times 10^{-6} \pm 3 \times 10^{-7}$
4	6.8×10^{-6}
5	$6.5 \times 10^{-6} \pm 6 \times 10^{-7}$
Averaged over all positions	$7.4 \times 10^{-6} \pm 7 \times 10^{-8}$

ation inside the liver. Moreover, the liver tolerance of 10 Gy for a single fraction treatment is exceeded. The dose to normal liver could be reduced by increasing the tumour to normal tissue ^{10}B content ratio and/or the prompt gamma ray production in the strucutral materials of the moderating structure.

In conclusion, our findings led to the adoption of an aluminium tank and the vertical shift of the irradiation cavity, while other materials with a lower prompt gamma ray production are evaluated for the heavy-water tank. Moreover, the study of filters attenuating the prompt gamma ray fluence in the irradiation cavity will be the matter for future research. Microdosimetric measurements of the dose to the normal liver and to the tumour are planned to be performed inside the irradiation cavity.

ACKNOWLEDGEMENTS

This work was supported by Istituto Nazionale di Fisica Nucleare (INFN, Italy) and in part by National Institutes of Health (NIH, USA). The authors thank Mr Giovanni D'Angelo (Dipartimento di Ingegneria Nucleare, Milano, Italy) and Mr Aldo Del Gratta (Dipartimento di Costruzioni Meccaniche e Nucleari, Pisa, Italy) for their contribution to performing the measurements.

Table 5. Neutron and prompt gamma ray doses to the tumour and to normal tissue in a liver phantom inside the irradiation cavity, referring to an irradiation of three hours and a deuteron current of 120 µA, with the Al tank configuration.

	Dose to normal liver[a] (Gy)	Dose to the tumour[a] (Gy)
^{10}B(n,α)^{7}Li	3.5 (0.2)	14.0 (3.3)
^{14}N(n,p)^{14}C	0.9 (0.2)	0.02 (0.2)
Fast neutrons	0.5 (2.5)	0.01 (2.5)
Prompt gamma rays[b]	3.1 (0.6)	0.06 (0.6)
Prompt gamma rays[c]	3.0 (0.5)	0.06 (0.6)
Total	11.0 (0.3)	14.1 (3.3)

[a] In parentheses: relative uncertainty (%).
[b] Produced in the liver. The 480 keV gamma rays produced by the ^{10}B(n,α)^{7}Li reacton are included.
[c] Produced in the moderating structure (see Table 1).

REFERENCES

1. Vischi, S., Spada, M., Alessiani, M., Fossati, G. S., Maestri, M., Guagliano, A., Dionigi, P, and Zonta, A. Proceedings XXI Congresso Società Italiana di Chirurgia d'Urgenza, Posa, Italy, October 1992 (in Italian).
2. Proceedings of the First International Workshop on Accelerator-Based Sources for Boron Neutron Capture Therapy, Jackson, Wyoming, September 1994, Idaho National Engineering Laboratory CONF-940976 (1994).
3. Amaldi, U. and Silari, M. (Eds) *The TERA Project and the Centre for Oncological Hadrontherapy*, 2nd edn, April 1995 (INFN-LNF, Frascati, Roma, Italy) (1995).
4. Campi, D. and Silari, M. (Eds) *The National Centre for Oncological Hadrontherapy — Updates and Revisions* (INFN-LNF, Frascati, Roma, Italy) (1995).
5. Amaldi, U., Arduini, G., Badano, L., Cambria, R., Campi, D., Gerardi, F., Gramatica, F., Leone, R., Manfredi, G., Nonis, M., Pullia, M., Rossi, S., Sangaletti, L., Silari, M. and Tosi. G. *A Hospital-based Hadrontherapy Complex* In: Proc. Fourth European Particle Accelerator Conference, London (UK), 27 June–1 July 1994, (World Scientific) pp. 49–51 (1994).
6. Amaldi, U., Arduini, G., Badano, L., Cambria, R., Campi, D., Gerardi, F., Gramatica, F., Leone, R., Manfredi, G., Nonis, M., Pullis, M., Rossi, S., Sangaletti, L., Silari, M. and Tosi. G. *The Italian Project for a Hadrontherapy Centre*. Nucl. Instrum. Methods **A360**, 297–301 (1995).
7. Arduini, G., Leone, R., Martin, R. L., Rossi, S. and Silari, M. *An H⁻/Light Ion Synchrotron for Radiation Therapy*. Nucl. Instrum. Methods A (in press).
8. Agosteo, S., Bodei, G., Leone, R. and Silari, M. *Monte Carlo Study of a Thermal Neutron source generated by 11 MeV Protons and 7 MeV Deuterons*. In: Proc. First Int. Workshop on Accelerator-Based Sources for Boron Neutron Capture Therapy, Jackson, Wyoming, September 1994. Idaho National Engineering Laboratroy CONF-940976, pp. 255–268 (1994).
9. Zamenhof, R. G. *The Design of Neutron Beams for Neutron Capture Therapy*. Nucl. Sci Appl. **4**, 303–316 (1991).
10. Briesmeister, J. F. (Ed.) *MCNP — A General Monte Carlo N-Particle Transport Code, Version 4A*. Los Alamos National Laboratory LA-12625-M (1993).
11. Meadows, J. W. *The ^{9}Be(d,n) Thick-Target Neutron Spectra for Deuteron Energies between 2.6 and 7.0 MeV* Nucl. Instrum. Methods **A324**, 239–246 (1993).
12. Smith, D. L., Meadows, J. W. and Guenther, P. T. *Neutron Emission from the ^{9}Be(d,n)^{10}B Thick-Target Reaction for 7 MeV Deuterons*. Nucl. Instrum. Methods **A241**, 507–510 (1985).
13. ICRP Committee 3 Task Group. *Data for Protection against Ionizing Radiation from External Sources: Supplement to ICRP Publication 15*. ICRP 21 (Oxford: Pergamon Press) (1971).
14. Cohen, E. R. *Uncertainty and Error in Physical Measurements*. In: Proc. Int. School of Physics "Enrico Fermi", Course CX. Eds L. Crovini and T. J. Quinn. (North Holland) pp. 11–31 (1992).
15. Sangiust, V. and Terrani, M. *Misure di Flussi di Neutroni Termici ed Epitermici con il Metodo dell'Attivazione* CESNEF FN001, Politecnico di Milano (in Italian) (1969).

16. Fassò, A. Ferrari, A., Ranft, J. and Sala, P. R. *FLUKA: Performances and Applications in the Intermediate Energy Range* In: Proc. Specialists' Meeting on Shielding Aspects of Accelerators, Targets and Irradiation Facilities, Arlington, Texas, 28–29 April 1994. NEA/OECD, pp. 287–304 (1995).
17. Price, W. J. *Nuclear Radiation Detection* (New York: McGraw-Hill) (1964).
18. d'Errico, F., Alberts, W. G., Curzio, G., Guldbakke, S., Kluge, H. and Matzke, M. *Active Neutron Spectrometry with Superheated Drop (Bubble) Detectors*. Radiat. Prot. Dosim. **61**(1/3), 159–162 (1995).
19. d'Errico, F., Alberts, W. G. and Matzke, M. *Advances in Superheated Drop (Bubble) Detector Techniques*. Radiat. Prot. Dosim. **70**(1–4), 103–108 (This issue) (1996).
20. Matzke, M. *Unfolding of Pulse Height Spectra: The HEPRO Program System* (Physikalisch-Technische Bundesanstalt, Braunschweig) Report PTB-N-19 (1994).
21. Weise, K. *Mathematical Foundation of an Analytical Approach to Bayesian-Statistical Monte Carlo Spectrum Unfolding.* (Physikalisch-Technische Bundensanstalt, Braunschweig) Report PTB-N-24 (1995).
22. Monteventi, F. ENEA AMB/IRP/DOSE Bologne (Italy), private communication (1995).
23. ICRP Committee 2 Task Group. *Reference Man: Anatomical, Physiological and Metabolic Characteristics*. ICRP 23 (Oxford: Pergamon) (1975).

REACTOR BASED EPITHERMAL NEUTRON BEAM ENHANCEMENT AT ŘEŽ

M. Marek†, J. Burian†, J. Rataj†, J. Polák† and F. Spurný‡
†Reactor Physics Dept. 803
Nuclear Research Institute Řež
Cz 250 68 Řež, Czech Republic
‡Nuclear Physics Institute of Czech Academy of Sciences
Dept of Radiation Dosimetry
Na Truhlarce 39-64, 180 86 Praha 8, Czech Republic

Abstract — The characterisation is described of the epithermal neutron beam at the LVR-15 reactor. The epithermal neutron fluence rate at the reactor power of 10 MW was found to be 1.07×10^{12} $m^{-2}.s^{-1}$. The fast neutron dose rate was 0.5 $Gy.h^{-1}$ and the incident gamma dose rate 2.25 $Gy.h^{-1}$. At this beam whole-body human phantom measurements were designed wtih the aim of determining both the neutron and gamma dose over the whole body during the irradiation. The thermal neutron fluence rate inside the phantom along its backbone was also measured. A new therapy room for BNCT purposes has been designed to enable irradiation studies without any effect on other users of the LVR-15 reactor. The ground plot of the room has dimensions 3.5×2.5 m and the room height is 2 m. The beam filter has been optimised to increase the epithermal fluence rate and decrease the fast neutron dose rate. Influence of the collimator geometry, the Be reflector outside the reactor core, the material and thickness of the filter, and the U converter outside the Be reflector were studied. The calculations were performed with the codes DOT and MCNP. The so far optimal variant shows that it is possible to get an epithermal fluence rate of at least 5.92×10^{12} $m^{-2}.s^{-1}$ with the fast neutron dose rate of 2.92 $Gy.h^{-1}$.

INTRODUCTION

At the reactor LVR-15 at NRI Řež a BCNT facility has been designed and the existing epithermal neutron beam has parameters as follows: the epithermal neutron fluence rate 1.07×10^{12} $m^{-2}.s^{-1}$, the fast neutron dose rate 0.5 $Gy.h^{-1}$ and the incident gamma dose rate 2.25 $Gy.h^{-1}$. The beam was used for a detailed study of the neutron and gamma dose distribution on a whole human body phantom. But the parameters of the facility needed be improved. A therapy room has been designed and a detailed calculational optimisation of the beam has begun with the aim of increasing the epithermal fluence rate with the fast neutron dose rate as low as possible.

BEAM CHARACTERISATION

The filter for the BNCT epithermal neutron beam consists of cylindrical blocks of 100 cm in diameter. The total filter length is 265 cm and it begins 75 cm from the edge of the reactor core. In the space between the core and the filter there is a row of Be reflector cassettes, two rows of displacements and a shutter filled with water or air (in operation). The active part of the filter consisted of 55 cm of aluminium, 15 cm of sulphur and 1 cm of titanium. The outer diameter of the beam of 11.5 cm is adjusted by the final shutter.

The spectral beam characteristics were determined with a Bonner sphere detector which consists of a cylindrical Li(Eu) thermal neutron detector and a standard set of polyethylene spheres (diameters: 50, 76, 101, 127 and 203 mm)[1]. The SAND-II code with the Diesten-feld's response functions was used for unfolding. As an input a spectrum calculated by the DOT code with the SAILOR-DLC76 cross section library was used. With respect to the good detector efficiency the measurement had to be performed at the reactor power level of 1 kW and then recalculated to a power of 10 MW. Incident gamma dose and fast neutron dose in the free beam were determined by a dual ionisation chamber[2] at the reactor power of 8 MW. The beam parameters are presented in Table 1.

The fast neutron fluence rate above approximately 1.5 MeV was also determined by a ^{232}Th fission chamber. The detector calibrated at the ^{252}Cf neutron source has

Table 1. Characteristics of the BNCT epithermal neutron beam at the LVR-15 (power of 10 MW).

Total neutron fluence rate ($m^{-2}.s^{-1}$)	1.47×10^{12}
Thermal neutron fluence rate (<0.414 eV) ($m^{-2}.s^{-1}$)	2.43×10^{11}
Epithermal neutron fluence rate (0.414 eV to 30 keV) ($m^{-2}.s^{-1}$)	1.07×10^{12}
Fast neutron fluence rate (>30 keV) ($m^{-2}.s^{-1}$)	1.57×10^{11}
Ratio of fast to epithermal fluence	0.15
Ratio of fast to total fluence	0.11
'Average' neutron energy (keV)	13.3
Fast neutron dose rate ($Gy.h^{-1}$)	0.5
Fast neutron dose per epithermal neutron ($Gy.m^2$)	1.3×10^{-16}
Incident gamma dose rate ($Gy.h^{-1}$)	2.25
Incident gamma dose per epithermal neutron ($Gy.m^2$)	7.3×10^{-16}

an efficiency of 5×10^{-7} cm^2. The value derived from Bonner spectrometer measurement (5.2×10^{10} m^{-2}.s^{-1}) and determined by the fission chamber (5.8×10^{10} m^{-2}.s^{-1}) are in a good agreement.

The axial distribution of the epithermal neutron fluence rate was measured with the 50 mm Bonner sphere (its 90% activity limits in the beam spectrum is from 0.1 eV to 16 keV). The ratio of the detector response at the distances of 3.5 cm from the edge of the filter and 23.5 cm respectively is 2.5.

The radial epithermal neutron beam profile was measured at two different distances from the edge of the beam filter: 3.5 cm and 18.5 cm. At the closer position the decrease from the beam centre to its radius (5.75 cm) is 0.8. At a distance of 18.5 cm the profile is more flat and the same decrease of 0.8 is found at more than twice the radius.

PHANTOM MEASUREMENTS

The measurements were made with the help of a human whole-body phantom consisting of a human skeleton and Plexiglas instead of skin. The phantom was filled with water to simulate tissue. Along the back bone and through the head a measuring channel is situated. Measurements were performed for two horizontal configurations which differed in the angle between the body axis and the beam axis. In the first case the angle between axes was 45° and in the second one the axes were equivalent.

The thermal neutron distribution was measured in the channel with a Si semiconductor detector with a natural Li radiator[3]. Natural Li was chosen to decrease the detector efficiency. The thermal neutron distribution is presented in Figure 1. The maximum of thermal neutron fluence rate is 3.15×10^{12} m^{-2}.s^{-1} in the depth of 2.5–3 cm (at reactor power 10 MW).

The gamma and fast neutron dose distribution on the phantom was measured in positions as follows (detectors placed on the phantom surface):

(1) on ear (No 1 for the angle 45°, No 7 for the angle 0°)
(2) under chin on the neck (Nos 2 and 8)
(3) on the right side of chest (Nos 3 and 9)
(4) on the left side of chest (Nos 4 and 10)
(5) on the abdomen (Nos 5 and 11)
(6) on the gonads (Nos 6 and 12)

The gamma dose was determined with two independent sets of TLDs in Cd[2] and both were evaluated using a standard personal dosimetry process. Results are presented in Table 2. They are in a good agreement, differences are caused by slightly different locations of dosemeters on the phantom.

To determine the neutron dose a set of personal dosemeters based on ^{238}U and ^{232}Th in contact with a film and solid state track dosemeters (SSTD) were used. The personal dosemeters were evaluated using a standard process as the dose equivalent in the element 59 of the Snyder phantom considering the change of neutron spectrum in the different measured positions[4]. The SSTDs based on CR-39 in contact with the natural boron metal radiator[2] were evaluated as the dose equivalent from ^{252}Cf neutrons. Results are presented in Table 2. The differences in dose equivalent are probably also caused by slightly different location of dosemeters on the phantom.

In consideration of the fact that the maximal dose rate in a tumour during treatment (for the described filtered beam) is supposed to be 15 Gy.h^{-1}[1] the measured dose rates along the phantom are low.

THERAPY ROOM

A new therapy room that has inner dimesions of 3.5×2.5 m^2 and a height of 2 m has been designed. The room enables experiments and measurements to be made without any effect on other users of the reactor with respect to neutron and gamma background in the reactor hall. The room walls and roof are of heavy concrete blocks 30 cm thick. Their inner surface as well as the floor are covered by 8 cm layer of borated (3%) polyethylene to decrease the thermal neutron background in the room and to improve neutron shielding abilities of the walls. The entrance to the room is through a 60 cm width labyrinth.

To improve the quality of the BNCT facility a new end-shutter has been designed. It consists of layers of borated polyethylene and lead. Its total thickness is 60 cm and when it is open the outgoing beam is reduced to 14 cm in diameter.

BEAM ENHANCEMENT

The beam intensity described in the Table 1 is too low for the therapy applications and needs to be increased. A schematic picture of the whole facility is in Figure 2. There are four ways for the epithermal neutron fluence rate, ϕ_e optimisation with as low as possible fast neutron dose rate $\dot{D}_f$:

(i) optimise geometry of the collimator between the filter and the end-shutter,
(ii) optimise the filter materials and thickness,
(iii) replace the Be reflector outside the core by a row of cassettes with air,
(iv) add a row of fuel elements behind the Be reflector.

To study all the possibilities computer codes DOT and MCNP were used. The DOT calculations were performed with the SAILOR-DLC 76 coupled cross section library condensed to 47 neutron energy groups and 21 gamma energy groups. The code was used for the optimisation of the collimator geometry and converter calculations. With the Monte Carlo MCNP3A and MCNPDAT4 library the filter optimisation and Be reflector influence were calculated.

An optimum variant for the collimator between the filter and the end-shutter was found to be a cone geometry with 5 cm of Al and 8 cm of graphite wall. This shape enable optimal collecting of neutrons going out of the filter and the Al + graphite walls are a good neutron reflector and additional moderator.

The filter composition was optimised with respect to the highest ϕ_e and the lowest ratio of the fast neutron dose rate and the epithermal neutron fluence rate $\dot{D}_f/\phi_e$. The materials used were as follows: graphite, Al, Al_2O_3, S and 1 cm of Ti. Some of the results of the MCNP calculations are presented in Figure 3. The variants without any graphite layer are undermodulated; they have the highest ϕ_e but too high ratio $\dot{D}_f/\phi_e$. After adding a 10 cm graphite layer as the first ϕ_e decreases but faster than the ratio $\dot{D}_f/\phi_e$. The optimum seems to be 5 cm of graphite — ϕ_e is higher then for both other groups and the ratio $\dot{D}_f/\phi_e$ lies between. For 5 cm graphite the thickness of Al can be reduced to 40 cm. The resulted variant consists of 5 cm graphite, 40 cm Al and 15 cm S: $\phi_e = 5.92 \times 10^{12}$ m^{-2}.s^{-1} and $\dot{D}_f/\phi_e = 1.37 \times 10^{-16}$ Gy.m^2.

Calculations of the influence of the beryllium reflector outside the reactor core for this variant shows that the ϕ_e decreases 1.76 times but $\dot{D}_f$ is 2.01 times less.

The additional rows of fission material out of the reactor core behind the Be reflector were calculated by the DOT code. Results for the variant 55 cm Al and 15 cm S show that the epithermal neutron fluence rate increased 1.99 times.

Table 2. Gamma and neutron dose measurement on the human whole body phantom. 'Angle' describes the angle between the human body axis and the beam axis.

Position	Gamma dose rate (mGy.h^{-1})		Neutron dose rate (mSv.h^{-1})	
	TLD set No 1	TLD set No 2	Personal dosemeter	SSTD
	Angle 45°	Angle 45°	Angle 45°	Angle 45°
1. Ears	371	363	167.5	
2. Neck	251	250	>180	
3. R. chest	56.8	74.5	58.2	
4. L. chest	81.5	137	59	
5. Abdomen	53.8	41.7	56	
6. Gonads	21.7	20.5	15	
	Angle 0°	Angle 0°	Angle 0°	Angle 0°
7. Ears	444	542	>180	542
8. Neck	154	167	89.8	15
9. R. chest	60.7	84.3	59	27
10. L. chest	73.3	106.7	60	76
11. Abdomen	37.2	46.7	64	5
12. Gonads	23.8	20.6	21	0.8

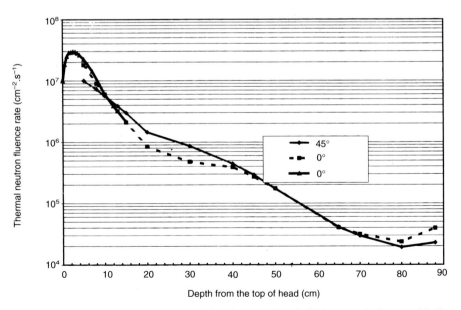

Figure 1. Thermal neutron fluence rate distribution inside the phantom. 0° and 45° between the beam and body axes. 10 MW reactor power.

CONCLUSIONS

The total epithermal neutron fluence rate in the free beam of 1.07×10^{12} m^{-2}.s^{-1} was too low for appropriate treatment application (more than 3 times less than on HFR in Petten). But the results of calculation show that the optimal variant with additional moderator (5 cm of graphite) and the filter consisting of 40 cm Al + 15 cm S can give a therapeutic beam with the epithermal neutron fluence rate of $\phi_e = 5.92 \times 10^{12}$ m^{-2}.s^{-1} and ratio $\dot{D}_f/\phi_e = 1.37 \times 10^{-16}$ Gy.m^2. The influence of the U converter in connection with Be reflector is comparable with the case when the Be reflector is replaced by an air displacement.

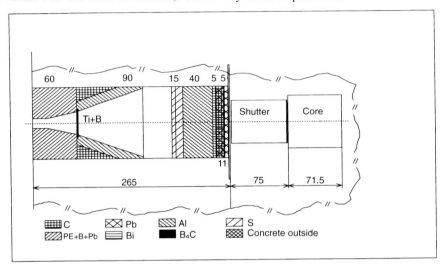

Figure 2. The epithermal neutron beam filter. Dimensions in cm.

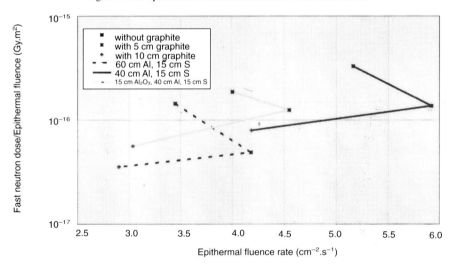

Figure 3. The beam filter optimisation. MCNP calculations for different filter materials.

REFERENCES

1. Marek, M., Burian, J., Prouza, Z., Rataj, J. and Spurný, F. *Neutron Capture Therapy Beam on the LVR-15 Reactor.* In: Radiat. Prot. Dosim. **44**(1/4), 453–456 (1992).
2. Spurný, F., Votočková, I. and Turek, K. *Determination of Dosimetry Characteristic for Epithermal BNCT Beam.* (Institute of Radiation Dosimetry) UDZ-346/92 (1992).
3. Marek, M. *Small Si-Li Detectors for Thermal Neutron On-line Measurements.* Safety Nucl. Energ. **3**(41), 7/8, 165–169 (1995).
4. Singer, J. and Trousil, J. *Radioizotopy* **15**, 375 (1974).

FRICKE-INFUSED AGAROSE GEL PHANTOMS FOR NMR DOSIMETRY IN BORON NEUTRON CAPTURE THERAPY AND PROTON THERAPY

G. Gambarini†, C. Birattari†, D. Monti‡, M. L. Fumagalli†, A. Vai†, P. Salvadori§, L. Facchielli∥ and A. E. Sichirollo∥
†Dipartimento di Fisica dell'Università di Milano and I.N.F.N., V. Celoria 16, 20133 Milano, Italy
‡Dip. di Chimica Organica e Industriale dell'Università, V. Venezian 21, 20133 Milano, Italy
§Istituto di Fisiologia Clinica del C.N.R., V. Savi 8, 56100 Pisa, Italy
∥Istituto Nazionale per lo Studio e la Cura dei Tumori, V. Venezian 1, 20133 Milano, Italy

Abstract — In recent radiotherapy techniques, such as boron neutron capture therapy (BNCT) and proton therapy, the high non-uniformity of the spatial distribution of absorbed dose makes mandatory 3-D dose determinations in order to carry out good treatment planning. The investigated dosimetric technique is based on a chemical dosemeter (ferrous sulphate solution) incorporated in a gel (agarose). Ionising radiation causes a variation in certain parameters of the system such as the relaxation rates of hydrogen nuclei, measurable by nuclear magnetic resonance imaging (NMR) or the gel optical absorption in the visible spectrum, measurable by spectrophotometry. Gel containing ^{10}B in the amount typically accumulated in tumours for BNCT was analysed. The isodose curves were obtained from NMR analysis of a phantom of borated gel after irradiation in the thermal column of a nuclear reactor. The results show that the gel is a promising dosimetric system. A method for depth–dose profiling in tissue irradiated by a proton beam is also suggested. In the profile the Bragg peak position has been determined within 0.1 mm and the widening in the peak ramp proves to be about 1 mm.

INTRODUCTION

The aim of radiotherapy is the achievement of a significant energy deposition in tumours, with low energy release in surrounding healthy tissues. This selective energy deposition may be achieved by boron neutron capture therapy (BNCT) and by hadrontherapy, in particular proton therapy.

BNCT takes advantage of the possibility of accumulating boron nuclei in cancerous tissue, and of the high cross section (3840 b) of the reaction $^{10}B(n,\alpha)^{7}Li$ with thermal neutrons. The short range of the emitted particles, α and ^{7}Li, allows for a localised energy deposition. Proton therapy takes advantage of the fact that charged particles, and in particular protons, release a high dose at the end of their range in a medium, at a depth that is a function of both the energy of the charged particles and the composition of the medium itself.

In both cases, information on the spatial distribution of absorbed dose or on depth–dose profiles is mandatory to plan the treatment. The study and the setting up has been undertaken of a system that allows for the three-dimensional determination of absorbed dose. The system consists of a chemical dosemeter infused in a tissue-equivalent gel, with which it is possible to make phantoms. The phantoms act as dosemeters: after exposure they may be conveniently analysed, and the spatial distribution of absorbed dose is detectable.

THE DOSEMETER

The system investigated consists of a chemical dosemeter infused into a gel. The chosen gelling agent, whose role is that of maintaining the spatial localisation of the signal, is the polysaccharide Agarose SeaPlaque (from the Fluka Chemical Corporation), which has shown interesting properties[1]. The chemical dosemeter is a ferrous sulphate solution, which is the active constituent of the standard Fricke dosemeter. In such a solution, as is known, ionising radiation produces a conversion of ferrous ions, Fe^{2+}, to ferric ions, Fe^{3+}, and the number of oxidised ions is linearly correlated to the absorbed dose. The ferric ion yield, and consequently the interval of linearity, are dependent on the dosemeter composition. For a ferrous sulphate solution incorporated in a gelatine, the G value (i.e. Fe^{2+} oxidised per 100 eV absorbed energy) is higher than that of the standard Fricke dosemeter.

In a $FeSO_4$-doped gel, the absorbed dose after exposure to ionising radiation can be determined by measuring certain parameters of the system, which undergoes a modification owing to the conversion of ferrous to ferric ions. The conventional method for the Fricke dosemeter reading is founded on the variation in the optical absorption in the visible spectrum, measurable by means of spectrophotometric instrumentation. Indeed, spectrophotometry provides a valid methodology because it is practical and highly reliable, and it is convenient in most circumstances. However, if a 3-D determination of the absorbed dose is required, more attractive parameters are the spin relaxation times of hydrogen nuclei in the aqueous solution. Ferrous and ferric ions are both paramagnetic, but they cause a different reduction in the spin relaxation rates of hydrogen. A linear correlation between absorbed dose and either transverse or longitudinal relaxation rate has been found[2-7]. Hence, after a phantom made with FeSO$_4$-infused gel is exposed to ionising radiation, one can determine the three-dimensional distribution of the absorbed dose by analysing the phantom via a nuclear magnetic resonance (NMR) imaging system.

In NMR analysis, a good result is achieved[8] if the dosemeter response R is defined as the difference between the relaxation rate (1/T) measured in the irradiated sample and that measured, at the same time, in an unirradiated sample from the same gel preparation:

$$R = (1/T)_{irr} - (1/T)_{blank}$$

The highest sensitivity has been obtained with the following gel composition[8]:

Ferrous sulphate solution   1 mM Fe(NH$_4$)·6H$_2$O, 50 mM H$_2$SO$_4$ in the amount of 50% of the final weight,

Agarose SeaPlaque   [C$_{12}$H$_{14}$O$_5$(OH)$_4$] in the amount of 1% of the final weight,

Highly purified water   H$_2$O in the amount of 49% of the final weight.

The sensitivity of this gel is better than that of the standard Fricke dosemeter: in the γ ray field of a ^{137}Cs biological irradiator, delivering a dose rate of 0.14 Gy.s^{-1}, the dose–response curve, obtained by measuring the transverse relaxation rates (1/T$_2$) with a Somatom Siemens imaging analyser operating at 1.5 T, 63 MHz, shows good linearity up to ≈40 Gy, with a slope equal to 0.2 s^{-1}.Gy^{-1}.

FRICKE-GEL DOSEMETER IN BNCT

In the design of a phantom to be utilised in the determination of absorbed dose in a neutron field, particular attention has to be paid to tissue equivalence. In fact, in thermal neutron fields the absorbed energy derives from reaction products, and consequently in a tissue substitute both the cross sections for thermal neutrons and the secondary particle spectra have to be similar to those of the tissue to be simulated. The only way to achieve this result is to produce a tissue substitute with isotopic composition identical to that of the specific tissue of interest, at least as regards the isotopes that give the main contributions to the absorbed dose.

In tissue, thermal and epithermal neutrons release a very low energy, the main contributions coming from nuclear reactions with nitrogen and hydrogen:

^{14}N(n,p)^{14}C

^{1}H(n,γ)^{2}H

In small volumes (less than 1 cm^3) the dominant contribution to the absorbed dose from thermal neutrons is due to protons (approximately 590 keV) emitted in the reaction with nitrogen. So, the percentage of N is of fundamental importance. Photons emitted in the reaction with hydrogen, having an energy of 2.2 MeV, interact at a distance from the reaction site, and they appreciably contribute to the absorbed dose in relatively large biological objects (ICRU 26)[9]. Oxygen and carbon, which are the other main components of tissue, give a negligible contribution to the absorbed dose in the thermal and epithermal region of neutron energy, and they play a very similar role[10]. For these elements only the total percentage (C + O) may be considered. (In bone the contribution of calcium is also to be taken into consideration.)

In this work brain tissue has been simulated, because brain is the actual target of BNCT. The gel previously described has been augmented with the organic compound carbonyldiamide (urea [CH$_4$N$_2$O] from the Fluka Chem. Corp.) in the amount of 4% of the final weight. The elemental composition of brain, from ICRU 44[11], and of the gel are reported in Table 1. The tissue equivalence of this gel for thermal neutrons is very good. The tissue equivalence for γ rays is also good. In fact, the effective atomic number (photoelectric effect) proves to be 7.7 for brain and 7.8 for the gel.

The response of the gel to γ radiation has been investigated by analysing various dosemeters irradiated with ^{137}Cs gammas. The slope of the dose–response curve is found to be equal to 0.065 s^{-1}.Gy^{-1}. This value is lower than that obtained with the dosemeters optimised for γ rays, but it is still higher than that of the standard Fricke dosemeter. In contrast, the linearity of the response is good up to 50 Gy, that is, for a wider range than that of the previous gel, as one might expect considering that lower ion conversion allows wider linearity.

Some dosemeters have also been prepared by augmenting gel with 40 µg.g^{-1} of ^{10}B, which is the quantity typically accumulated in tumours. Both the dosemeters (i.e. with and without ^{10}B) showed the same sensitivity to gamma radiation.

The dosemeter response was then analysed after exposure in the thermal column of a TRIGA MARK II reactor (L.E.N.A., Pavia, Italy) up to various fluences, as determined in a previous work[12] utilising activation techniques. As expected, when exposed to thermal neutrons, even in high fluences, the gel without boron absorbs a low dose. The absorbed dose per unit fluence for ICRP tissue exposed to thermal neutrons[13] is equal to 2.01×10^{-13} Gy.cm^2. The additional absorbed dose due to 40 µg.g^{-1} of ^{10}B has been evaluated by means of the relation:

Table 1. Elemental composition of brain tissue and of the gel dosemeter.

	Percentage by mass				
	H	N	C	O	Other
Brain	10.7	2.2	14.3	71.2	1.4
Gel	10.9	2.2	1.4	85.4	0.1

$$D = 1.602 \times 10^{10} \sigma F N E\phi \text{ (Gy)}$$

where

D is the absorbed dose in Gy,
$\sigma = 3.837 \times 10^{-21}$ cm^2 is the reaction cross section,
$F = 4 \times 10^{-5}$ is the ^{10}B fraction by weight,
$N = 6 \times 10^{22}$ is the number of atoms per gram of ^{10}B,
$E = 2.28$ MeV is the energy released per event, and
Φ is the thermal neutron fluence in n.cm^{-2}.

For the absorbed dose per unit fluence:

$$D = 3.364 \times 10^{-12} \text{ (Gy.cm}^2) \tag{1}$$

Dosemeters made with both borated and not borated gels have been analysed, contained in small cylindrical Teflon vials, with 0.5 mm thick walls, 22 mm internal diameter, and 45 mm height. The dosemeters were exposed to various thermal neutron fluences in the thermal column of the reactor[14]. From differential measurements with borated and regular gels exposed in the same thermal column position and in the same fluence, it has been found that the response of borated gel per unit fluence is $R_B/\phi = 0.108 \times 10^{-12}$ s^{-1}.cm^2 and that of the regular gel is $R/\phi = 0.015 \times 10^{-12}$ s^{-1}.cm^2. So

$$0.093 \times 10^{-12} \text{ s}^{-1}.\text{cm}^2 \tag{2}$$

(i.e. 86% of the total) is attributable to the absorbed dose due to ^{10}B. The remaining 14% of the response, includes contributions from thermal neutrons in tissue, fast neutrons, reactor background and activation of the materials of the containers and holders. By comparing the experimental value of Equation 2 with the calculated dose (Equation 1) due to ^{10}B, the sensitivity of the borated dosemeter to secondary particles produced by thermal neutrons reacting with ^{10}B has been found to be equal to 0.028 (s^{-1}.Gy^{-1}). This value is lower than that obtained for γ rays, reported above, but it is sufficient to obtain satisfactory dose evaluations.

The results reported above, which confirm the promising features of this NMR gel dosemeter, have encouraged us to pursue further investigations, by analysing broader samples. Therefore, the 3-D response in larger phantoms was studied, to define the modality for spatial dose distribution or isodose curve determination.

Cylindrical Teflon containers were prepared, having 8 cm internal diameter, 15 cm height, and 2.3 mm thickness. Teflon was chosen in order to have low activation of the container material and thus reduce this spurious effect on the response of the dosemeter.

Borated and non-borated phantoms were inspected with the NMR imaging analyser before and after irradiation to assess the feasibility of drawing isodose curves. In Figure 1 the NMR image of a borated gel phantom after exposure in the thermal column of the reactor is shown; the thermal neutron fluence at the entrance in the phantom was about 10^{13} n.cm^{-2}. The NMR analysis has been made four hours after exposure, to avoid any effects due to ion diffusion.

The results obtained, showing the feasibility of 3-D dose measurements in phantoms, are promising, and they encourage improvement of the technique with the aim of increasing the sensitivity of the gel dosemeter, so as to reduce the contribution of spurious effects, deriving, for example, from walls, from temperature differences during gelling and during irradiation or from mechanical shocks. Moreover, ion diffusion has to be reduced, to avoid a prompt NMR analysis after irradiation.

FRICKE-GEL DOSEMETER IN PROTON THERAPY

Protons, and in general heavy charged particles, offer advantages in radiotherapy. In fact, the low scattering and straggling events and the high ionisation near the end of particle tracks (Bragg peak) enable high doses to be delivered in tissue also at depths of many centimetres while maintaining a low entrance dose.

The Bragg peak of monoenergetic proton beams is relatively sharp and, in order to obtain a uniform dose in a tumour, a convenient spread in depth is obtained by modulating the energy of the beam. Measurements of unmodulated Bragg peaks for various energies are therefore necessary to define the parameters of the beam correctly, and measurements of the depth–dose profiles resulting from the modulated beam are necessary for correct planning of the therapy.

The feasibility of utilising the Fricke-infused agarose gel for depth–dose profile determination in tissue exposed to a proton beam has been investigated.

A shortcoming of ferrous sulphate gels is ion diffusion; this causes a progressive loss of localisation of the information on the absorbed dose. This fact causes some limitations on the suitability of FeSO$_4$-infused phantom dosemeters for three-dimensional determination, because a prompt analysis after irradiation is necessary if good spatial resolution is required.

In consideration of such an ion diffusion, and of the consequent necessity of performing a prompt NMR analysis after sample exposure, and furthermore in consideration of the fact that sometimes a swift NMR analysis is not possible, the feasibility has been investigated of a technique for determining the depth–dose profile by analysing the response of gel contained in thin capillaries pierced through tissue, at increasing distances from the surface. The capillary material brings a negligible, or at least calculable contribution to the results of the measurements. Thus, the possibility has been verified of analysing small samples of the Fricke-infused gel. In particular, the feasibility has been tested of evaluating the relaxation rates when the gel dosemeter is contained in thin glass capillaries of 1.1 mm internal diameter, using the gel without urea. The capillaries were exposed in the ^{137}Cs irradiator, up to different doses. The samples were examined in a research NMR analyser (Bruker AC-300) operating at 7.05 T and

300 MHz. In this high field, the water–proton relaxation in the dosemeter gives a very narrow signal, and therefore longitudinal relaxation times T_1 are more measurable than transverse relaxation times T_2. The longitudinal relaxation rates $(1/T_1)$ were therefore determined here, utilising an inversion recovery procedure. The NMR spin-lattice relaxation rates depend on field frequency, and in Fricke solutions the NMR sensitivity is lower for higher frequencies[15]. The response to γ rays has proved to be 0.006 $s^{-1}.Gy^{-1}$.

Two Plexiglas phantoms, with a rank of capillaries placed at increasing distances from the front surface, were exposed to a proton beam from the cyclotron Cypris 325 (CGR-MeV). The cyclotron provides a (1–50 μA) proton beam with 16 ± 0.6 MeV fixed energy. The extracted beam, from a titanium window 25 μm thick, passes through a flattening system composed of a nylon stopper (circular tablet 3 mm thick) and an Al scatterer (sheet 0.057 mm thick). The beam energy results reduced to 13.5 MeV, and the transversal profile of the beam, measured by an array of thermoluminescence dosemeters (TLD-700), is satisfactorily flat in the region of interest. After exposure, the capillaries were inspected in the NMR analyser. The results are shown in Figure 2, where each point is obtained as a mean of two values. The specific energy released in Plexiglas

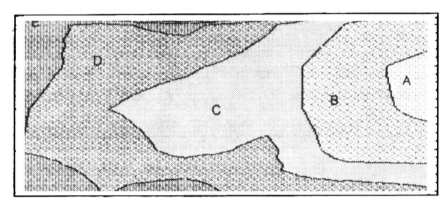

Figure 1. Imaging of a cylindrical borated gel phantom exposed in the thermal column of the reactor. The resulting dose values are: A, 26–28 Gy; B, 24–26 Gy; C, 22–24 Gy; D, 20–22 Gy; E, 18–20 Gy.

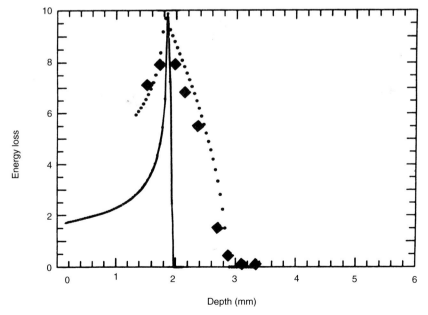

Figure 2. Depth-dose profile obtained from 13.5 MeV protons in Plexiglas. (♦) Experimental points, (———) Computer Bragg curve, (. . .) computed loss energy in the dosemeter.

by 13.5 MeV protons was evaluated with the computer program TRIM-91 (IBM). A computer program has also been written, in order to calculate the energy loss in the experimental conditions, that is, in the gel contained in glass capillaries placed in the Plexiglas phantom. In the program the stopping powers from ICRU 49[16] have been utilised. In Figure 2 the computed curves are shown: (a) (solid line) specific energy released in Plexiglas by 13.5 MeV protons as a function of depth, and (b) (dotted line) energy released in a gel dosemeter enclosed in glass containers with square section, 1 mm internal side, 0.2 mm thickness, located at increasing depths in Plexiglas. These results confirm the potentiality of the technique. In fact, the position of the maximum of the Bragg peak is obtained within 0.1 mm in respect of the calculated value, and the widening in the ramps is within 1 mm, equal to the dimension of the dosemeter, and so the goodness of the results will be presumably the same, for whichever proton energy.

Moreover, the dosemeter gel has also been tested by means of spectrophotometric analysis, in order to verify the consistence and correctness of the results obtained from the NMR analysis. The results of this work, presented to the IRRMA '96 congress[17], will be successively published.

CONCLUSIONS

The results described above regarding 3-D dose determination for BNCT and depth–dose profiling in proton therapy are very promising and they encourage continuation of experiments with the aim of removing the difficulties (such as ion diffusion) and improving the dosemeter.

REFERENCES

1. Schulz, R. J., deGuzman, A. F., Nguyen, D. B. and Gore, J. C. *Dose-response Curves for Fricke-infused Gels as obtained by Nuclear Magnetic Resonance*. Phys. Med. Biol. **35**, 1611–1622 (1990).
2. Gore, J. C., Kang, Y. S. and Schulz, R. J. *Measurement of Radiation Dose Distributions by Nuclear Magnetic Resonance (NMR) Imaging*. Phys. Med. Biol. **29**, 1189–1197 (1984).
3. Gore, J. C., Kang, Y. S. and Schulz, R. J. *The Measurement of Radiation Dose Distributions by Nuclear Magnetic Resonance (NMR) Imaging*. Magn. Res. Imaging **2**, 244 (1984).
4. Appleby, A., Christman, A. E. and Leghouz, A. *Imaging of Spatial Radiation Dose Distribution in Agarose Gels using Magnetic Resonance*. Med. Phys. **14**, 382–384 (1987).
5. Olsson, L. E., Petersson, S., Ahlgren, L. and Mattsson, S. *Ferrous Sulphate Gels for Determination of Absorbed Dose Distributions using MRI Technique: Basic Studies*. Phys. Med. Biol. **34**, 43–52 (1989).
6. Olsson, L. E., Fransson, A., Ericsson, A. and Mattsson, S. *MR Imaging of Absorbed Dose Distributions for Radiotherapy using Ferrous Sulphate Gels*. Phys. Med. Biol. **35**, 1623–1631 (1990).
7. Hazle, J. D., Hefner, L., Nyerick, C. E., Wilson, L. and Boyer, A. L. *Dose-response Characteristics of a Ferrous-sulphate-doped Gelatin System for Determining Radiation Absorbed Dose Distributions by Magnetic Resonance Imaging (Fe MRI)*. Phys. Med. Biol. **36**, 1117–1125 (1991).
8. Gambarini, G., Arrigoni, S., Cantone, M. C., Molho, N., Facchielli, L. and Sichirollo, A. E. *Dose-response Curve Slope Improvement and Result Reproducibility of Ferrous-sulphate-doped Gels Analysed by NMR Imaging*. Phys. Med. Biol. **39**, 703–717 (1994).
9. ICRU. *Neutron Dosimetry for Biology and Medicine*. Report 26 (Washington, DC: ICRU Publications) (1977).
10. Nutton, D. H. and Harris, S. J. *Tissue Equivalence in Neutron Dosimetry*. Phys. Med. Biol. **25**, 1173–1180 (1980).
11. ICRU. *Tissue Substitutes in Radiation Dosimetry and Measurement*. (Bethesda, MD: ICRU Publications) (1989).
12. Frigerio, F. *Realizzazione di un Campo Neutronico per la Cura dei Tumori Epatici Mediante TCN*. Physics graduation thesis, Università di Pavia, Italy (1991/1992).
13. Caswell, R. S., Coyne, J. J. and Randolph, M. L. *Kerma Factors for Neutron Energies below 30 MeV*. Radiat. Res. **83**, 217–254 (1980).
14. Gambarini, G., Arrigoni, S., Bonardi, M., Cantone, M. C., deBartolo, D., Desiati, S., Facchielli, L. and Sichirollo, A. E. *A System for 3-D Absorbed Dose Measurements with Tissue Equivalence for Thermal Neutrons*. Nucl. Instrum. Methods A**353**, 706–710 (1994).
15. Podgorsak, M. B. and Schreiner, L. J. *Nuclear Magnetic Relaxation Characterization of Irradiated Fricke Solution*. Med. Phys. **19**, 87–95 (1992).
16. ICRU. *Stopping Powers and Ranges for Protons and Alpha Particles*. (Bethesda, MD: ICRU Publications) (1993).
17. Gambarini, G., Monti, D., Fumagalli, M. L., Birattari, C. and Salvadori, P. *Phantom-dosimeters Examined by NMR Analysis: a Promising Technique for 3-D Determination of Absorbed Dose*. In: Proc. 3rd Topical Meeting on Industrial Radiation and Radioisotope Measurements and Applications, Raleigh, North Carolina, 6–9 October 1996.

Proceedings of the Eighth Symposium on Neutron Dosimetry,
Paris, November 13–17 1995

AUTHOR INDEX

Agay, D.	(455)	Bowley, C.C.	(477)	Dollo, R.	(181)	Großwendt, B.	(219)
Agosteo, S.	(559)	Brede, H.J.	(13)	Domingo, C.	(87)	Gualdrini, G.F.	(109)
Alberts, W.G.	(17)		(17)	Dörschel, B.	(117)	Guldbakke, S.	(109)
	(97)		(505)	Drake, P.	(235)	Gupta, N.	(547)
	(103)	Breskin, A.	(219)	Drüke, V.	(485)		
	(117)	Bronić, I.K.	(33)		(497)	Habrand, J.L.	(429)
	(127)	Brooks, F.D.	(477)	d'Errico, F.	(79)	Hager, L.G.	(121)
	(323)	Buchillier, T.	(437)		(103)	Haight, R.C.	(1)
Alekseev, A.G.	(341)	Budzanowski, M.	(501)		(109)	Harvey, J.R.	(127)
Alevra, A.V.	(251)	Buffler, A.	(477)		(113)		(149)
	(295)	Burgkhardt, B.	(143)		(559)		(265)
	(327)		(361)			Hauffe, J.	(485)
Alkharam, A.S.	(537)	Burian, J.	(567)	Edwards, A.	(467)		(497)
Allie, M.S.	(477)			Egger, E.	(109)	He, T.	(513)
Anachkova, E.	(207)	Carl, U.	(485)	Eisenhauer, C.M.	(323)	Hecker, O.	(505)
Angelone, M.	(169)	Chadwick, M.B.	(1)	Engelmann, H.J.	(251)	Heilmann, C.	(413)
Apfel, R.E.	(109)	Chang, S.Y.	(349)	Esposito, A.	(169)	Heinrich, W.	(395)
	(113)	Chartier, J.-L.	(xiii)	Eyrich, W.	(485)		(405)
Aroua, A.	(279)		(305)		(497)	Heinrichs, U.	(497)
	(285)		(313)			Hietel, B.	(337)
	(437)		(323)	Facchielli, L.	(571)	Hirning, C.R.	(67)
		Chechik, R.	(219)	Fatôme, M.	(455)	Höfert, M.	(437)
Baek, W.Y.	(37)	Chen, J.	(55)	Fernandez-Moreno, F.	(79)	Hoffmann, W.	(485)
Baixeras, C.	(87)	Cherevatenko, E.P.	(215)	Fernández, F.	(87)		(497)
Barbry, F.	(345)	Cho, P.S.	(471)	Ferrand, R.	(429)		(501)
Barelaud, B.	(79)	Clapier, F.	(429)	Fiechtner, A.	(157)	Hoflack, C.	(73)
Barlow, K.	(265)	Clech, A.	(181)	Fieg, G.	(361)	Hollnagel, R.A.	(387)
Barthe, J.	(59)	Clifford, T.	(273)	Filges, D.	(485)		(505)
	(73)	Cloth, P.	(485)		(497)	Hough, J.H.	(247)
	(79)		(497)	Fracas, P.	(429)		(441)
Bartlett, D.T.	(121)	Colautti, P.	(559)	French, A.P.	(139)		
	(127)	Corcalciuc, V.	(21)		(149)	Iacconi, P.	(461)
	(161)	Corrado, M.G.	(559)		(265)	Ing, H.	(273)
	(235)	Cosgrove, V.P.	(493)	Frenzel, T.	(509)	Itié, C.	(331)
	(323)	Cousins, T.	(273)	Fritsch, M.	(485)		
	(395)	Cross, W.G.	(419)		(497)	Jackson, M.	(139)
Batistoni, P.	(169)	Curzio, G.	(79)	Fukuda, A.	(47)		(149)
Beaujean, R.	(413)		(109)	Fumagalli, M.L.	(571)		(265)
Becker, R.	(485)					Jahr, R.	(xiii)
	(497)	Dangendorf, V.	(219)	Gahbauer, R.	(547)	Jakes, J.	(133)
Bednář, J.	(153)	Deboodt, P.	(187)	Gall, K.	(429)		(337)
Benabdesselam, M.	(461)	Decossas, J.L.	(79)	Gambarini, G.	(175)		(405)
Benck, S.	(21)	DeLuca, P.M. Jr.	(1)		(571)	Jansen, J.T.M.	(27)
Bienen, J.	(485)		(13)	Gilvin, P.J.	(161)	Jansky, B.	(305)
	(497)		(17)	Gmür, K.	(157)	Jaunich, G.	(299)
Bilski, P.	(501)		(481)		(395)	Jetzke, S.	(327)
Binns, P.J.	(247)	Delacroix, D.	(429)	Golnik, N.	(211)	Jones, D.T.L.	(477)
	(441)	Delacroix, S.	(429)		(215)		(481)
Birattari, C.	(571)		(493)	Goodman, J.	(547)	Józefowicz, K.	(143)
Blue, T.	(547)	Delafield, H.J.	(445)	Grecescu, M.	(285)		
Bordy, J.M.	(59)	Dellert, M.	(485)		(285)	Kase, K.R.	(49)
	(73)		(497)		(437)	Kellerer, A.M.	(55)
	(79)	Devita, A.	(181)	Grecula, J.	(547)		(207)
Bottollier-Depois, J.-F.	(203)	Dhermain, J.	(273)	Green, S.	(493)	Khamis, M.	(251)
	(353)		(461)	Griffin, T.W.	(471)	Kharlampiev, S.A.	(341)
	(395)	Dietz, E.	(97)	Grillmaier, R.E.	(395)	Khoshnoodi, M.	(357)
	(429)		(109)	Grindborg, J.-E.	(541)	Kim, J.L.	(349)
Boutaine, J.L.	(193)	Dietze, G.	(395)	Großwendt, B.	(37)	Kim, J.S.	(349)

AUTHOR INDEX

Klein, H.	(225)	Mazal, A.	(203)	Renouf, M.C.	(149)	Stewart, R.D.	(165)
	(313)		(429)	Roesler, S.	(405)	Strauch, K.	(413)
Klett, A.	(361)		(493)	Rokni, S.H.	(425)	Sychev, B.S.	(215)
Kluge, H.	(261)	McAulay, I.R.	(395)	Roos, H.	(207)	Symons, J.E.	(477)
	(305)	McDonald, J.C.	(165)	Rosenstock, W.	(299)	Szabo, J.L.	(193)
	(327)		(323)	Ross, M.A.	(481)	Szornel, K.	(197)
Knauf, K.	(251)	McLean, T.	(273)				
	(327)	Médioni, R.	(445)	Salvadori, P.	(571)	Takeda, N.	(365)
Köble, T.	(299)	Menzel, H.-G.	(xiii)	Sannikov, A.V.	(291)	Tanner, J.E.	(165)
Kobus, H.	(485)		(395)		(383)	Tanner, R.J.	(121)
	(497)		(517)		(405)		(127)
Kopp, J.	(413)	Mestries, J.C.	(455)	Saupsonidis, O.	(79)		(161)
Kota, C.	(555)	Meulders, J.-P.	(21)	Savitskaya, E.N.	(383)	Taylor, G.C.	(313)
Kralik, M.	(279)	Michaud, S.	(429)	Savvidis, E.	(79)	Thai, L.-X.	(425)
Krüll, A.	(509)	Mijnheer, B.J.	(27)		(83)	Thomas, D.J.	(255)
Kruziniski, G.	(299)	Minsart, G.	(187)	Scacco, A.	(175)		(313)
Kudo, K.	(365)	Moiseev, T.	(93)	Schmidt, R.	(509)		(323)
Kurkdjian, J.	(353)	Molokanov, A.	(501)	Schmitz, Th.	(485)	Thomas, R.H.	(371)
		Monti, D.	(571)		(497)	Tinti, R.	(559)
Lahaye, T.	(59)	Monti, S.	(559)	Schnuer, K.	(395)	Titt, U.	(219)
	(73)	Moosburger, M.	(485)	Schraube, G.	(337)	Tommasino, L.	(395)
	(79)		(497)	Schraube, H.	(127)		(419)
Landre, F.	(73)	Mori, C.	(365)		(133)	Tournier, B.	(345)
Lapraz, D.	(461)	Multon, E.	(455)		(279)	Toyokawa, H.	(365)
Laramore, G.E.	(471)				(291)	Turek, K.	(153)
Laugier, J.	(461)	Naismith, O.F.	(241)		(305)		
Lavelle, J.	(121)		(255)		(323)	Uritani, A.	(365)
	(161)		(313)		(337)		
Leicher, M.	(413)	Nath, R.	(109)		(379)	Vai, A.	(571)
Lennox, A.	(481)	Nauraye, C.	(429)		(391)	Valley, J.-F.	(285)
Lequin, S.	(73)	Nelson, W.R.	(49)		(395)	Vanhavere, F.	(187)
Leuthold, G.	(379)	Newhauser, W.D.	(13)		(405)	Vareille, J.C.	(79)
	(391)		(17)	Schrewe, U.J.	(13)	Vermeersch, F.	(187)
Lim, T.	(395)	Niehues, N.	(251)		(17)	Verrey, B.	(345)
Lindborg, L.	(395)	Novotny, T.	(279)		(295)	Vilgis, M.	(143)
	(541)	Nunes, J.	(197)		(395)	Voigt, D.	(133)
Lindsley, K.L.	(471)			Schuhmacher, H.	(219)	Voisin, P.	(467)
Liu, J.C.	(49)	Olko, P.	(485)	Schwartz, R.B.	(323)	Votočková, I.	(353)
	(425)		(501)	Schwenke, K.	(485)	Vylet, V.	(425)
Lloyd, D.	(467)				(497)		
Louis, M.	(429)	Paganetti, H.	(485)	Scott, M.C.	(493)	Waker, A.J.	(67)
Lüdemann, L.	(509)		(497)	Serbat, A.	(461)		(197)
Luguera, E.	(79)	Paul, D.	(331)	Serov, A.Y.	(215)	Waligórski, M.P.R.	(501)
	(87)		(353)	Shaw, P.V.	(161)	Wambersie, A.	(517)
Luszik-Bhadra, M.	(97)	Paul, N.	(497)	Shvidkij, S.V.	(215)		(547)
	(261)	Pelcot, G.	(331)	Sichirollo, A.E.	(175)	Watt, D.E.	(529)
		Pelliccioni, M.	(169)		(571)		(537)
Maier, R.	(485)	Pescayre, G.	(181)	Siebert, B.R.L.	(97)	Webb, W.	(273)
	(497)	Peterson, H.P.	(485)		(109)	Weeks, A.R.	(149)
Mao, X.S.	(49)		(497)		(117)	Weise, K.	(327)
Marek, M.	(567)	Piesch, E.K.A.	(127)		(241)	Weitzenegger, E.	(337)
Mares, V.	(279)		(143)		(255)		(405)
	(291)	Pillon, M.	(169)		(361)	Wernli, C.	(157)
	(337)	Plawinski, L.	(203)		(371)		(165)
	(379)	Plewnia, A.	(361)	Silari, M.	(559)	Wiegel, B.	(279)
	(391)	Polák, J.	(567)	Simpson, B.R.S.	(247)		(305)
Martin, J.D.	(113)	Posny, F.	(313)	Slypen, I.	(21)	Willems, G.	(37)
Martin, S.	(455)	Prêtre, S.	(285)	Sohrabi, M.	(357)	Wittstock, J.	(251)
Martini, M.	(175)	Pszona, S.	(269)	Soloway, A.H.	(547)		(327)
Martone, M.	(169)			Sperl, R.	(485)		
Matzen, T.	(509)	Raaijmakers, C.P.J.	(27)		(497)	Yoon, S.C.	(349)
Matzke, M.	(13)	Rado, V.	(169)	Spurný, F.	(153)	Yudelev, M.	(513)
	(103)	Raffaglio, C.	(175)		(203)		
	(261)	Ragan, D.P.	(513)		(353)	Zamani, M.	(83)
	(327)	Rannou, A.	(xiii)		(395)	Zamani-Valassiadou, M.	(79)
	(559)	Rataj, J.	(567)		(409)	Zielczyński, M.	(215)
Maughan, R.L.	(481)	Reitz, G.	(395)		(567)	Zoetelief, J.	(27)
	(513)		(413)	Steele, J.D.	(161)		
	(555)			Stelzer, K.J.	(471)		

Proceedings of the Eighth Symposium on Neutron Dosimetry,
Paris, November 13–17 1995

LIST OF PARTICIPANTS

Agosteo, S.
Politecnico di Milano
Dipartimento di Ingegneria Nucleare
Via Ponzio 34/3
MILANO
I-20133
ITALY

Alberts, W.G.
Physikalisch Technische Bundesanstalt
Bundesallee 100
Postfach 3345
BRAUNSCHWEIG
D-38116   GERMANY

Alevra, A.V.
Physikalisch Technische Bundesanstalt
Bundesallee 100
PO Box 3345
BRAUNSCHWEIG
D-38116   GERMANY

Ambroise, J.
DPHD-SDOS-LEMDI
Institut de Protection et de Surete Nucleaire
BP 6, FONTENAY-AUX-ROSES CEDEX
F-92265   FRANCE

Angelone, M.
Associazione EURATOM-ENEA sulla
Fusione
Centro Ricerche Frascati
Via E. Fermi 27
CP 65, FRASCATI (ROMA)
I-00044   ITALY

Apfel, R.E.
Apfel Enterprises Incorporated
Yale University   25 Science Park   Suite 312
PO Box 2159
NEW HAVEN
CT 06520-8286   USA

Aroua, A.
Centre University
Dept. de l'Interieur de la Sante Publique
LAUSANNE
CH-1015   SWITZERLAND

Audoin, G.
Institut de Protection et de Surete Nucleaire
DPHD/SDOS/LRDE
BP 6, FONTENAY-AUX-ROSES CEDEX
F-92265   FRANCE

Barbry, F.
Commissariat A L'Energie Atomique
Centre D'Etudes de Valduc
BP 21, IS-SUR-TILLE
F-21120   FRANCE

Barelaud, B.
Universite de Limoges   LEPOFI
Faculte des Sciences
123 Rue Albert Thomas
LIMOGES CEDEX
F-87060   FRANCE

Barlow, K.
Defence Radiological Protection Service
Institute of Naval Medicine
A lverstoke, GOSPORT
Hampshire   PO12 2DL   UK

Barthe, J.R.
French Atomic Energy Commission
CEA.DTA.DAMRI CE SACLAY
BP 52, GIF-SUR-YVETTE CEDEX
F-91193   FRANCE

Bartlett, D.T.
National Radiological Protection Board
Chilton
DIDCOT
Oxon   OX11 ORQ   UK

Bednar, J.
Academy of Sciences of the Czech Republic
Nuclear Physics Inst.
Na Truhlarce 39/64
PRAGUE 8
CS-180 86   CZECH REPUBLIC

Benabdesselam, M.
Universite de Nice-Sophia Antipolis
LPES-CRESA
Parc Valrose
NICE CEDEX 2
F-06108   FRANCE

Berard, P.
Commissariat A L'Energie Atomique
Inst. de Protection et de Surete Nucl.
DPS/SHI
BP 38, PIERRELATTE CEDEX
F-26701   FRANCE

Bernard, D.M.
Bicron NE
Bath Road
Beenham, READING
Berkshire   RG7 5PR   UK

Biau, A.
OPRI
31 Rue de l'Ecluse
LE VESINET
F-78110   FRANCE

Bilski, P.
Institute of Nuclear Physics
ul. Radzikowskiego 152
KRAKOW
PL-31-342   POLAND

Binns, P.J.
National Accelerator Centre
PO Box 72, FAURE
7131   SOUTH AFRICA

Blanc, D.
Universite Paul Sabatier de Toulouse
Place de l'Eglise
LANTA
F-31570   FRANCE

Bordy, J-M.
Commissariat A L'Energie Atomique
DPT/SIDR/CEN
Av. Division Lecler
BP 6, FONTENAY-AUX-ROSES CEDEX
F-92265   FRANCE

Boschung, M.
Paul Scherrer Institute
Radiation Hygiene Division
VILLIGEN PSI
CH-5232   SWITZERLAND

Bottollier-Depois, J-F.
Institut de Protection et de Surete Nucleaire
Human Health Protection & Dosimetry
BP No. 6, FONTENAY-AUX-ROSES
CEDEX
F-92265   FRANCE

Brambilla, A.
CEA/DEIN/SPE
CEA Saclay
GIF-SUR-YVETTE CEDEX
F-91191   FRANCE

Brault, J-P.
COGEMA
2 rue Paul Dautier
BP 4, VELIZY CEDEX
F-78141   FRANCE

Brede, H.J.
Physikalisch Technische Bundesanstalt
Bundesallee 100
BRAUNSCHWEIG
D-38116   GERMANY

Brossier, J.
DPHD/SDOS/LRDE
Institut de Protection et de Surete Nucleaire
BP 6, FONTENAY-AUX-ROSES CEDEX
F-92265   FRANCE

LIST OF PARTICIPANTS

Budzanowski, M.
Institute of Nuclear Physics
Ul. Radzikowskiego 152
KRAKOW
PL-31-342   POLAND

Burgkhardt, B.
Nuclear Research Centre Karlsruhe
Dosimetry & Safety Division HS/D
PO Box 3640, KARLSRUHE 1
D-76021   GERMANY

Casson, W.H.
Los Alamos National Laboratory
Health Physics Measurements Group ESH-4
LOS ALAMOS
NM 87545   USA

Castelo, J.
Hospitalet de l'Infant
Apartado de Correos 27
TARRAGONA
E-43890   SPAIN

Chartier, J-L.
Commissariat a l'Energie Atomique
Centre d'Etudes Nucleaires
IPSN/DPHD/SDOS
60-68 Av. du gal Leclerc
BP 6, FONTENAY-AUX-ROSES CEDEX
F-92265   FRANCE

Cherubini, R.
Istituto Nazionale di Fisica Nucleare
Laboratori Nazionali di Legnaro
Via Romea 4
LEGNARO (PADOVA)
I-35020   ITALY

Clech, A.
CEA/COGEMA Etablessement de Marcoule
BP 170
BAGNOLS-SUR-CEZE CEDEX
F-30206   FRANCE

Colautti, P.
Istituto Nazionale di Fisica Nucleare
Laboratorio Nazionale di Legnaro
Via Romea 4
LEGNARO (PADOVA)
I-35020   ITALY

Comte,
MGP Instruments
BP 1, LAMANON
F-13113   FRANCE

Constantinescu, B.
Institute of Atomic Physics
Cyclotron Laboratory
PO Box MG-6
BUCHAREST
ROMANIA

Conti, A.
CISAM
S. PIERO A GRADO (PISA)
I-56010   ITALY

Corato, C.
ANPA-AMM
Via V. Brancati 48
ROMA
I-00144   ITALY

Court, L.
Electricite de France
Service de Radioprotection
3 Rue de Messine
PARIS CEDEX 08
F-75384   FRANCE

Crossman, J.S.P.
University of Saint Andrews   Rad. Biophysics
Bute Buildings
North Haugh, ST. ANDREWS
Fife   Scotland   KY16 9SS   UK

Cutarella, D.
Commissariat a l'Energie Atomique
Inst. de Prot. et de Surete Nucleaire.
BP 6, FONTENAY-AUX-ROSES CEDEX
F-92265   FRANCE

D'Errico, F.
Universita degli Studi di Pisa
Via Diotisalvi 2
PISA
I-56126   ITALY

Dangendorf, V.
Physikalisch Technische Bundesanstalt
Bundesallee 100
BRAUNSCHWEIG
D-38116   GERMANY

Darmon, G.
Eurisys Mesures
ZA de l'Observatoire
4 Avenue des Frenes
Montigny-le-Bretonneux
ST. QUENTIN EN YVELINES CEDEX
F-78067   FRANCE

De Lillie, L.
UWC RPC (Physics)
Private Bag X17, BELLVILLE
7535   SOUTH AFRICA

De Luca, P.M.
University of Wisconsin
1300 University Avenue Room 1530 MSC
MADISON
WI 53706-1532   USA

De Padua, L.
CEA-Cadarache
SPR/SMSE
BP 1, ST. PAUL-LEZ-DURANCE CEDEX
F-13108   FRANCE

Deboodt, P.
SCK-CEN
Belgian Nuclear Research Centre
Boeretang 200
MOL
B-2400   BELGIUM

Decossas, J-L.
Universite de Limoges
123 Rue Albert Thomas
LIMOGES CEDEX
F-87060   FRANCE

Denis, J-M.
Universite Catholique de Louvain
Institut de Physique Nucleaire
Chemin du Cyclotron 2
LOUVAIN-LA-NEUVE
B-1348   BELGIUM

Dietze, G.
Physikalisch Technische Bundesanstalt
Division 6
Bundesallee 100
Postfach 33 45, BRAUNSCHWEIG
D-38116   GERMANY

Dollo, R.
Electricite de France
DEPT-EPN Etat Major
13-27 Esplanade Charles de Gaulle
Immeuble PB 26
PARIS LA DEFENSE 10
F-92060   FRANCE

Draaisma, F.
IRI afd. Cursorisch Onderwys
Mekelweg 15
DELFT
NL-2629 JB   NETHERLANDS

Drake, P.
Ringhals Nuclear Power Plant
Ringhalsverket
Vattenfall, VAROBACKA
S-430 22   SWEDEN

Espagnan, M.
Etablissement COGEMA de Marcoule
Head of Dosimetry Laboratory
BP 170, BAGNOLS/CEZE CEDEX
F-30206   FRANCE

Fatome, M.
Ctr. de Rec. du Serv. de Sante des Armees
24 Avenue des Maquis du Gresivaudan
BP 87, LA TRONCHE CEDEX
F-38702   FRANCE

Faure, M-L.
COGEMA
Service Medical du Travail
PO Box 16, PIERRELATTE CEDEX
F-26701   FRANCE

Fehrenbacher, G.
GSF-Institut fur Strahlenschutz
Ingolstadter Landstr. 1
Postfach 1129, OBERSCHLEISSHEIM
D-85758   GERMANY

Fernandez-Moreno, F.
Universidad Autonoma de Barcelona
CERDENYOLA DEL VALLES
(BARCELONA)
E-08193   SPAIN

LIST OF PARTICIPANTS

Festag, J.G.
Gesellschaft fur Schwerionenforschung mbH
Darmstadt Abt. Sicherheit und Strahlenschutz
Planckstrasse 1
Postfach 11 05 52, DARMSTADT
D-64291   GERMANY

Fiechtner, A.
Paul Scherrer Institute
Division for Radiation Hygiene
VILLIGEN PSI
CH-5232   SWITZERLAND

Fieg, G.
Forschungszentrum Karlsruhe
PO Box 3640, KARLSRUHE
D-76021   GERMANY

Frontier, J-P.
Commissariat a l'Energie Atomique/INSTN
CEA Saclay
GIF-SUR-YVETTE CEDEX
F-91191   FRANCE

Fukuda, A.
Electrotechnical Laboratory
1-1-4 Umezono
Tsukuba, IBARAKI
305   JAPAN

Gahbauer, R.A.
Ohio State University Hospitals
Division of Radiation Oncology
300 West 10th Avenue
COLUMBUS
OH 43210-1228   USA

Gaillard-Lecanu, E.
DPHD/SDOS/LRDE
Institut de Protection et de Surete Nucleaire
BP 6, FONTENAY-AUX-ROSES CEDEX
F-92265   FRANCE

Gambarini, G.
INFN
Via Celoria 20
MILANO
I-20133   ITALY

Gerdung, S.
Universitat des Saarlandes
Strahlenbiophysik
Am Markt. 5   Zeile 4
DUDWEILER
D-66125   GERMANY

Goldhagen, P.
Environmental Measurements Laboratory
376 Hudson Street
NEW YORK
NY 10014-3621   USA

Golnik, N.
Institute of Atomic Energy (A-1)
Radiation Protection Department
OTWOCK-SWIERK
PL-05-400   POLAND

Gourmelon, P.
IPSN/DPHD
PO Box 6, FONTENAY-AUX-ROSES
CEDEX
F-92265   FRANCE

Grecescu, M.
Centre Universitaire
Institut de Radiophysique Appliquee
LAUSANNE
CH-1015   SWITZERLAND

Grosswendt, B.
Physikalisch Technische Bundesanstalt
Bundesallee 100
BRAUNSCHWEIG
D-38116   GERMANY

Gualdrini, G.F.
Ente per le Nuove Tecnologie
L'Energia e l'Ambiente (ENEA)
Via dei Colli 16
BOLOGNA
I-40136   ITALY

Guibbaud, Y.
Electricite de France
EPN Generation & Transmission
6 Rue Ampere
BP 114, SAINT-DENIS CEDEX
F-93202   FRANCE

Haan, S.
Commissariat a l'Energie Atomique
DEIN-SPE
GIF-SUR-YVETTE CEDEX
F-91191   FRANCE

Hansen, W.
Technische Universitat Dresden
Im FZ Rossendorf (FWSM)
Postfach 510119, DRESDEN
D-01314   GERMANY

Harvey, J.R.
Dosimetry Consultant
9 Torchacre Rise
DURSLEY
Gloucestershire GL11 4LW   UK

Hecker, O.
Physikalisch Technische Bundesanstalt
Bundesallee 100
BRAUNSCHWEIG
D-38116   GERMANY

Heinmiller, B.
Atomic Energy of Canada Limited Research
Chalk River Laboratories
CHALK RIVER (Ontario)
KOJ 1JO   CANADA

Herbaut, Y.
Commissariat a l'Energie Atomique
Centre d'Etudes Nucleaires de Grenoble
SPR/LMR
17 Rue des Martyrs
GRENOBLE CEDEX 9
F-38054   FRANCE

Herry, C.
ZA de l'Observatoire
4 Avenue des Frenes
Montigny-le-Bretonneux
ST. QUENTIN EN YVELINES CEDEX
F-78067   FRANCE

Herve, J-Y.
Commissariat a l'Energie Atomique
Centre d'etudes de Valduc
BP No. 14, IS-SUR-TILLE
F-21120   FRANCE

Hirning, C.R.
Ontario Hydro
1549 Victoria Street East
WHITBY (Ontario)
L1N 9E3   CANADA

Hoffmann, W.
Bergische Universitat Wuppertal F10.05
Gausstrabe 20
WUPPERTAL
D-42097   GERMANY

Hoflack, C.
IPSN SDOS
BP No. 6, FONTENAY-AUX-ROSES
CEDEX
F-92265   FRANCE

Hollnagel, R.A.
Physikalisch Technische Bundesanstalt
Bundesallee 100
BRAUNSCHWEIG
D-38116   GERMANY

Iacconi, P.
Universite de Nice-Sophia Antipolis
Laboratoire de Physique Experimentale
CRESA
Parc Valrose
NICE CEDEX 2
F-06108   FRANCE

Ing, H.
Chalk River Bubble Technology Industries
Highway # 17 E
Box 100
CHALK RIVER (Ontario)
KOJ 1JO   CANADA

Itie, C.
Commissariat a l'Energie Atomique
Inst. de Prot. et de Surete Nucleaire   DPT
SIDR
BP 6, FONTENAY-AUX-ROSES CEDEX
F-92265   FRANCE

Izak Biran, T.
Israel Atomic Energy Commission
YAVNE
81800   ISRAEL

Jackson, M.
Defence Radiological Protection Service
Institute of Naval Medicine
Alverstoke GOSPORT
Hampshire PO12 2DL   UK

LIST OF PARTICIPANTS

Jahr, R.
PTB Physikalisch Technische Bundesanstalt
Bundesallee 100
BRAUNSCHWEIG
D-38116   Germany

Jakes, J.
GSF
Institut fur Strahlenschutz
Ingolstadter Landstr. 1
OBERSCHLEISSHEIM
D-85758   GERMANY

Jansky, B.
Nuclear Research Institute Rez
REZ
Nr. Prague
CZ-250 68   CZECH REPUBLIC

Jozefowicz, K.
Institute of Atomic Energy
Radiation Protection Department
OTWOCK-SWIERK
PL-05-400   POLAND

Jung, M.
Universite Louis Pasteur
Centre de Recherches Nucleaires - Phase
23 rue de Loess
BP 20, STRASBOURG CEDEX 2
F-67037   FRANCE

Kharlampiev, S.A.
Institute for High Energy Physics
PROTVINO
Moscow Region
142284   RUSSIA

Khosnoodi, M.
Atomic Energy Organisation of Iran
North Cargar Street
PO Box 14155
TEHRAN
4494   IRAN

Klein, H.
Physikalisch Technische Bundesanstalt
Postfach 3345, BRAUNSCHWEIG
D-38116   GERMANY

Klett, A.
Laboratorium Prof. Berthold GmbH
Zahlrohrentwicklung
Calmbacher Str. 22
PO Box 100163, BAD WILDBAD
D-75323   GERMANY

Kluge, H.
Physikalisch Technische Bundesanstalt
Bundesallee 100
Postfach 3345, BRAUNSCHWEIG
D-38116   GERMANY

Knauf, K.
Physikalisch Technische Bundesanstalt
Bundesallee 100
Postfach 3345, BRAUNSCHWEIG
D-38116   GERMANY

Koble, T.
Fraunhofer-Institut fur Naturwissenschaftlich-
Technische Trendanalysen (INT)
PO Box 1491, EUSKIRCHEN
D-53864   GERMANY

Kobus, H.
Forschungszentrum Julich GmbH
Postfach 1913, JULICH
D-52425   GERMANY

Kockerols, P.
Belgonucleaire
Europalaan 20
DESSEL
B-2480   BELGIUM

Kota, C.
Harper Hospital & Wayne State University
3990 John R. Street
DETROIT
MI 48201   USA

Krajcar-Bronic, I.
Ruder Boskovic Institute
Bijenicka 54
PO Box 1016, ZAGREB
10001   CROATIA

Kralik, M.
Czech Metrological Institute
Radiova 1
PRAHA 10
CZ-102 00   CZECH REPUBLIC

Kustarjov, V.N.
Institute for High Energy Physics
Serpukhov
MOSCOW
142284   RUSSIA

Lachet, B.
DPHD-EC
BP 6, FONTENAY-AUX-ROSES CEDEX
FRANCE

Lahaye, T.
Commissariat a l'Energie Atomique
IPSN/DPHD/SDOS
BP 6, FONTENAY-AUX-ROSES CEDEX
F-92265   FRANCE

Le Teurnier, D.
Aries
44 Bis Boulevard Felix Faure
CHATILLON
F-92320   FRANCE

Lebaron-Jacobs, L.
DPHD-SARAM
Institut de Protection et de Surete Nucleaire
BP 6, FONTENAY-AUX-ROSES CEDEX
F-92265   FRANCE

Lee, Y-K.
GIST
119-121 Grande Rue
SEVRES
F-92310   FRANCE

Legee, F.
Commisariat a l'Energie Atomique
60-68 Avenue du General Leclerc
BP 6, FONTENAY-AUX-ROSES CEDEX
F-92290   FRANCE

Leuthold, G.
GSF-Inst. Strahlenschutz
Ingolstadter Landstrasse 1
Postfach 1129, OBERSCHLEISSHEIM
D-85758   GERMANY

Levrard, J.
COGEMA Etablissement de Marcoule
BP 170, BAGNOLS-SUR-CEZE CEDEX
F-30206   FRANCE

Lindborg, L.
Swedish Radiation Protection Institute
Box 60204, STOCKHOLM
S-171 16   SWEDEN

Liu, J.C.
Stanford Linear Accelerator Center
2575 Sandhill Road
PO Box 4349 MS 48, STANFORD
CA 94309   USA

Loncol, Th.
Cliniques Universitaires St-Luc
Avenue Hippocrate 10
BRUSSELS
B-1200   BELGIUM

Luszik-Bhadra, M.
PTB Physikalisch Technische Bundesanstalt
Bundesallee 100
BRAUNSCHWEIG
D-38116   GERMANY

Marek, M.
Nuclear Research Institute Rez
REZ
Nr Prague
CZ-250 68   CZECH REPUBLIC

Mares, V.
GSF
Postfach 1129, OBERSCHLEISSHEIM
D-85758   GERMANY

Martin, L.
DPHD/SDOS/LRDE
Institut de Protection et de Surete Nucleaire
BP 6, FONTENAY-AUX-ROSES CEDEX
F-92265   FRANCE

Masson, B.
EG&G Instruments BERTHOLD
ZI Petite Montagne sud
1 Rue du Gevaudan CE 1734
EVRY CEDEX
F-91047   FRANCE

Matzke, M.
Physikalisch Technische Bundesanstalt
Bundesallee 100
Postfach 3345, BRAUNSCHWEIG
D-38116   GERMANY

LIST OF PARTICIPANTS

Maughan, R.L.
Harper Hospital & Wayne State University
Gershenson Radiation Oncology Center
3990 John R Street
DETROIT
MI 48201   USA

Mazal, A.
Centre de Protontherapie d'Orsay
BP 65, ORSAY CEDEX
F-91402   FRANCE

McDonald, J.C.
Pacific Northwest National Laboratory
902 Battelle Boulevard
PO Box 999 MS P7-03, RICHLAND
WA 99352   USA

Medioni, R.
Commissariat a l'Energie Atomique
DPR-STEP-STID-CEA
BP 6, FONTENAY-AUX-ROSES CEDEX
F-92265   FRANCE

Menzel, H.G.
European Commission
XII/F/6, MO75 4/7
Rue Montoyer 75, BRUSSELS
B-1049   BELGIUM

Moiseev, T.
Institute for Physics & Nuclear Engineering
PO Box MG-6 Sect. 5, BUCHAREST
RO-76900   ROMANIA

Muller, H.
DPHD-SDOS-LRDE-CAD
BP 1, ST. PAUL LEZ DURANCE CEDEX
F-13108   FRANCE

Nemecek, S.
Division of ICN Biomedicals Incorporated
3300 Hyland Avenue
COSTA MESA
CA 92626   USA

Newhauser, W.D.
Physikalisch Technische Bundesanstalt
Bundesallee 100
BRAUNSCHWEIG
D-38116   GERMANY

Nguyen, A.
Commissariat a l'Energie Atomique Saclay
DEIN-SPE
GIF-SUR-YVETTE CEDEX
F-91191   FRANCE

Niehues, N.
DBE
Woltorfer Strasse 74
Postfach 1169, PEINE
D-31224   GERMANY

Novotny, T.
Czech Metrological Institute
Radiova 1
PRAGUE 10
CZ-102 00   CZECH REPUBLIC

Olko, P.
Institute of Nuclear Physics
ul Radzikoswkiego 152
KRAKOW
PL-31-342   POLAND

Paganetti, H.
Forschungszentrum Julich GmbH
Postfach 1913, JULICH
D-52425   GERMANY

Paillard, P.
CEA, DAM-BIII
BP 12, BRUYERES-LE-CHATEL
F-91680   FRANCE

Parize, J-M.
Electricite de France
12-14 Avenue Dutrievoz
VILLEURBANNE
F-69628   FRANCE

Paul, D.
IPSN/DPHD/SDOS CE Cadarache
BP 1, ST. PAUL-LEZ-DURANCE CEDEX
F-13108   FRANCE

Pauli, E.
Commissariat a l'Energie Atomique
IPSN/DPHD/Dir
BP 6, FONTENAY-AUX-ROSES CEDEX
F-92265   FRANCE

Pelcot, G.
DPHD/SDOS/LRDE/CAD
BP 1, ST. PAUL-LEZ-DURANCE CEDEX
F-13108   FRANCE

Perrin, M.L.
Electricite de France
DER
1 avenue du General de Gaulle
CLAMART CEDEX
F-92265   FRANCE

Pihet, P.
Commissariat á l'Energie Atomique
DPHD/S.DOS
BP 6, FONTENAY-AUX-ROSES CEDEX
F-92265   FRANCE

Plawinski, L.
Institute for Protection and Nuclear Safety
Division
BP 6, FONTENAY-AUX-ROSES CEDEX
F-92265   FRANCE

Portal, G.
12 Rue Fustel de Conlanges
MASSY
F-91300   FRANCE

Praca, C.
CEA
DAM B-III
BP 12, BRUYERES-LE-CHATEL
F-91680   FRANCE

Pszona, S.
Soltan Institute of Nuclear Studies
OTWOCK-SWIERK
PL-05-400   POLAND

Rabatin, K.
ICN Pharmaceuticals Incorporated
3300 Hyland Avenue
COSTA MESA
CA 92626   USA

Rannou, A.
Commissariat A L'Energie Atomique
CEA/DPHD/SDOS
BP 6, FONTENAY-AUX-ROSES CEDEX
F-92265   FRANCE

Reitz, G.
DLR Institut fur Luft-und Raumfahrtmedizin
Linder Hohe, KOLN
D-51147   GERMANY

Renouff, M.G.
NE Technology
Bath Road, Beenham, READING
Berkshire RG7 5PR   UK

Rimpler, A.
Bundesamt fur Strahlenschutz (BfS)
Waldowallee 177
BERLIN
D-10312   GERMANY

Roos, H.
Technical University Munchen
Schillerstrasse 42
MUNICH
D-80336   GERMANY

Rosenstock, W.
Fraunhofer-Institut INT
Appelsgarten 2
Postfach 1491, EUSKIRCHEN
D-53864   GERMANY

Ross, A.
University of Wisconsin
1300 University Avenue
MADISON
WI 53706   USA

Rottner, B.
Eurisys Mesures
4 Avenue des Frenes
Montigny-le-Bretonneux
ST. QUENTIN EN YVELINES CEDEX
F-78067   FRANCE

Rousseau, F.
DGCCRF
59 Boulevard Vincent Auriol
PARIS CEDEX 13
F-75703   FRANCE

Sabattier, R.
Centre Hospitalier Regional d'Orleans
BP 6709,
ORLEANS-LA-SOURCE CEDEX 2
F-45067   FRANCE

LIST OF PARTICIPANTS

Sabbatini, V.
CISAM
S. PIERO A GRADO (PISA)
I-56010   ITALY

Saez-Vergara, J.C.
CIEMAT   Radiation Protection Division
Avda. Complutense 22
MADRID
E-28040   SPAIN

Sannikov, A.V.
Institute for High Energy Physics
Druzhba Street 14 121
PROTVINO
Moscow Region
142284   RUSSIA

Savitskaya, E.N.
Institute for High Energy Physics
PROTVINO, Moscow Region
142284   RUSSIA

Schmidt, R.
Universitats-Krankenhaus
Hamburg-Eppendorf
Martinistrasse 52
HAMBURG
D-20251   GERMANY

Schmitz, Th.
Forschungzentrum Julich GmbH
Postfach 1913
JULICH
D-52425   GERMANY

Schraube, H.D.E.
GSF-Forschungszentrum Neuherberg
Ingolstadter Landstrasse 1
OBERSCHLEISSHEIM
D-85758   GERMANY

Schrewe, U.J.
Physikalisch Technische Bundesanstalt
Postfach 3345, BRAUNSCHWEIG
D-38023   GERMANY

Schuhmacher, H.
Physikalisch Technische Bundesanstalt
Box 3345
BRAUNSCHWEIG
D-38116   GERMANY

Serbat, A.
ETCA DE
16 Bis Avenue Prieur de la Cote D'Or
ARCUEIL CEDEX
F-94114   FRANCE

Serviere, H.
DPHD/SDOS/LRDE
BP 6, FONTENAY-AUX-ROSES CEDEX
F-92265   FRANCE

Shamai, Y.
Israel Atomic Energy Commission
YAVNE
81800   ISRAEL

Shvidkij, S.V.
Joint Institute for Nuclear Research
DUBNA, Moscow Region
141980   RUSSIA

Siebert, B.R.L.
Physikalisch Technische Bundesanstalt
Postfach 3345, BRAUNSCHWEIG
D-38116   GERMANY

Sigala, M.
Bureau National de Metrologie
22 Rue Monge, PARIS
F-75005   FRANCE

Slypen, I.
Catholique Universite de Louvain
Chemin du cyclotron 2
LOUVAIN-LA-NEUVE
B-1348   BELGIUM

Stelzer, K.J.
University of Washington Medical Center
1959 Northeast Pacific Street
Box 356043, SEATTLE
WA 98195-6043   USA

Symons, J.E.
National Accelerator Centre
PO Box 72, FAURE
7131   SOUTH AFRICA

Szabo, J-L.
CEA/DTA-DAMRI
BP 52, GIF-SUR-YVETTE CEDEX
F-91193   FRANCE

Tanner, J.E.
Pacific Northwest National Laboratory
70 Battelle Boulevard
PO Box 999, RICHLAND
WA 99352   USA

Tanner, R.J.
National Radiological Protection Board
Chilton, DIDCOT
Oxon, OX11 ORQ   UK

Taylor, G.C.
National Physical Laboratory
Queens Road,
TEDDINGTON
Middlesex TW11 OLW   UK

Taymaz, A.
Istanbul University
Vezneciler Campus, ISTANBUL
TR-34459   TURKEY

Temple, C.E.
Health & Safety Executive
St. Peters House Balliol Road
BOOTLE, Merseyside
L20 3LZ   UK

Thevenin, J-C.
DPHD/SDOS
BP 6, FONTENAY-AUX-ROSES
F-92265   FRANCE

Thomas, D.J.
National Physical Laboratory   CIRA
Building 47 Queens Road
TEDDINGTON
Middlesex, TW11 OLW   UK

Tillie, J-L.
Laboratoires de Repression des Fraudes
369 Rue Jules-Guesdes
VILLENEUVE D'ASCQ
F-59237   FRANCE

Titt, U.
Physikalisch Technische Bundesanstalt
Bundesallee 100
BRAUNSCHWEIG
D-38116   GERMANY

Tommasino, L.
A.N.P.A. AMM
Via Vitaliano Brancati 48
ROME (RM)
I-00144   ITALY

Tournier, B.
CEA Valduc
BP 14, IS-SUR-TILLE
F-21120   FRANCE

Toyokawa, H.
Nagoya University
Furo-cho, Chikusa-ku
NAGOYA
464-01   JAPAN

Truffert, H.
COGEMA
BEAUMONT LA HAGUE CEDEX
F-50444   FRANCE

Umiatowski, K.
Commissariat a l'Energie Atomique Saclay
DEIN-SPE
GIF-SUR-YVETTE CEDEX
F-91191   FRANCE

Van Cauteren, J.
AV-Controlatom
Avenue du Roi 157
BRUXELLES
B-1060   BELGIUM

Vanhavere, F.
SCK-CEN
Belgian Nuclear Research Centre
Boeretang 200, MOL
B-2400   BELGIUM

Vareille, J-C.
Universite de Limoges   LEPOFI
123 Rue Albert Thomas
LIMOGES CEDEX
F-87060   FRANCE

Vaz, P.
OECD Nuclear Energy Agency
12 Boulevard des Iles
ISSY-LES-MOULINEAUX
F-92130   FRANCE

LIST OF PARTICIPANTS

Vermeersch, F.
CEN-SCK
Boeretang 200, MOL
B-2400 BELGIUM

Vesseron, P.
Centre D'Etudes Nucleaires
BP 6, FONTENAY-AUX-ROSES CEDEX
F-92265 FRANCE

Voigt, J.
GSF-Forschungszentrum Neuherberg
Ingolstadter Landstr. 1
OBERSCHLEISSHEIM
D-85758 GERMANY

Voisin, P.
DPHD/SARAM
BP 6, FONTENAY-AUX-ROSES CEDEX
F-92265 FRANCE

Vylet, V.
Stanford Linear Accelerator Center
PO Box 4349 MS 48, STANFORD
CA 94309 USA

Wahl, W.
GSF-Institut fur Strahlenschutz
Ingolstadter Landstr. 1
Postfach 1129, OBERSCHLEISSHEIM
D-85758 GERMANY

Waker, A.J.
Atomic Energy of Canada Limited - Research
Chalk River Laboratories
CHALK RIVER (Ontario)
KOJ 1JO CANADA

Wambersie, A.
Cath. Univ. de Louvain Cliniques Univ.
Avenue Hippocrate 10/4752
BRUSSELS
B-1200 BELGIUM

Watt, D.E.
Univ. of St Andrews
Westburn Lane, North Haugh
ST. ANDREWS
Fife Scotland KY16 9SS

Weeks, A.R.
Magnox Electric
Berkeley Technology Centre
BERKELEY
Gloucestershire GL13 9PB UK

Weinstein, M.
Israel Atomic Energy Commission
PO Box 9001, BEER SHEVA
84190 ISRAEL

Wernli, Ch.
Paul Scherrer Institute
VILLIGEN PSI
CH-5232 SWITZERLAND

White, R.M.
Lawrence Livermore National Laboratory
PO Box 808, LIVERMORE
CA 94550 USA

Wiegel, B.
Physikalisch Technische Bundesanstalt
Bundesallee 100, BRAUNSCHWEIG
D-38116 GERMANY

Yoon, S-C.
Korea Atomic Energy Research Institute
PO Box 105, Daedukdanji
TAEJUN
305-353 REPUBLIC OF KOREA

Yudelev, M.
Harper Hospital & Wayne State University
3990 John R. Street
DETROIT
MI 48201 USA

Zafiropoulos, D.
INFN
Lab. Nationali di Legnaro
Via Romea 4
LEGNARO
I-35020 ITALY

Zamani-Valasiadou, M.
Nuclear Physics Laboratory
Aristotle University of Thessaloniki
THESSALONIKI
GR-54006 GREECE

Zanini, A.
Universite di Torino
INFN
Via Pietro Giuria 1
TORINO
I-10125 ITALY

Zoetelief, J.
TNO Medical Biological Laboratory
151 Lange Kleiweg
PO Box 5815
RIJSWIJK
NL-2280 HV NETHERLANDS

ADVERTISEMENT

THE PHYSICS OF RADIATION PROTECTION
B. Dörschel, V. Schuricht and J. Steuer

ISBN 1 870965 42 6

Contents:
Radiation sources and radiation fields,
Interaction of radiation with matter,
Radiation effects and radiation damage,
Basic concept of radiation protection,
Radiation exposure of man,
Radiation protection measuring techniques,
Physical fundamentals for limiting radiation exposure,
Calculation of radiation fields,
Calculation of radiation exposure of man,
Calculations in the design and assessment of measuring methods for radiation protection,
Calculations in the design of activities for the reduction of radiation exposure,
Fundamentals of radiation measuring techniques,
Sources of ionising radiation and their radiation fields,
Fundamentals of radiation protection measuring techniques,
Practical radiation protection measuring techniques.

Scope:
Dealing with problems of radiation protection is necessary because mankind will not be able to manage without application of ionising radiation, since the exploitation of nuclear energy is the only alternative in a number of countries in the foreseeable future to cover the energy demand. Therefore radiation and environmental protection are of great importance. Furthermore problems of radiation risk play an important role in public discussions. The book is aimed at both practising specialists and scientists wishing to learn about the fundamental science of radiation protection. It is neither a book in favour of nor against nuclear energy or the application of ionising radiation in any other specific field. The description of a few applications is incidental to the demonstration of the application of physics to radiation protection. The first part of the book, 'Physical Fundamentals of Radiation Protection', presents a concise description of radiation sources and radiation fields, interaction of radiation with matter, radiation effects and radiation damage, basic concept of radiation protection, radiation exposure of man, radiation protection measuring techniques and physical fundamentals for limiting radiation exposure. The second part, 'Calculational Exercises for Radiation Protection' is intended to supplement the first part by carrying out relevant calculations, amending and adding special aspects and to give guidance in solving practical problems. The book is written for scientists as well as for students and staff working in nuclear facilities, hospitals and institutions responsible for radiation and environmental protection.

Readership: Radiation protection scientists and practitioners, regulators and advanced students.

Publication date: February 12 1996 Hardback (320 pp),

Price: £65.00 (UK), US$ 123.50 (outside UK) (including postage and packing)

Orders: This book may be ordered from any scientific or technical bookshop or direct from:

Nuclear Technology Publishing
P.O. Box No 7
Ashford, Kent TN23 1YW
England

Telephone: (44)(0)(1233) 641683 Facsimile: (44)(0)(1233) 610021

Form DL April 1996

ADVERTISEMENT

THERMOLUMINESCENCE DOSIMETRY MATERIALS: PROPERTIES AND USES

ISBN 1 870965 19 1

S.W.S. McKeever, M. Moscovitch and P.D. Townsend

Contents:

Introduction
Thermoluminescence and thermoluminescence dosimetry
Fluorides
Oxides
Sulphates and borates

Scope

This book selects a range of the most popular thermoluminescence dosemeter (TLD) materials in use today and provides a critical account of their thermoluminescence (TL) and dosimetric properties. The information provided includes in-depth discussions of TL mechanisms, including an account of luminescence properties, and relevant information regarding dosimetric characteristics. The book is intended for those involved in TLD materials research, and for technicians and workers involved in the practical application of these materials in TL dosimetry. The advent of modern spectroscopic methods for measuring TL emission spectra (the so-called "3D" presentation) seemed to the authors to be an invitation to compile such spectra for all the major TLD materials. Further consideration led to an expansion of the initial idea to include a compilation of dosimetric properties. One intention is to provide a synopsis of the TL and dosimetric properties of the most widely used TLD materials currently available and and to form a link between the solid state defect properties of these materials and their actual dosimetric properties. A second intention is to provide a solid framework from which future studies could be launched. Too often in the past materials research into TLD materials has been haphazard, to say the least. By illustrating the links between solid state physics and the radiation dosimetry properties of these materials the book points to the future and to the pressing need for enhanced research on TLD materials.

Readership: Scientists, dosimetry practitioners, researchers and students in solid state dosimetry.

Publication date: December 19 1995 (Hardback, 214 pp)

Price: £50.00 (UK), US$95.00 outside UK (including postage and packing)

Orders: This book may be ordered from any scientific or technical bookshop or direct from:

Nuclear Technology Publishing
P.O. Box No 7
Ashford, Kent TN23 1YW
England.

Telephone: (44)(0)(1233) 641683   Facsimile: (44)(0)(1233) 610021

Nuclear Technology Publishing
21 Years 1974 – 1995

1997 Journal Prices

APPLIED HEALTH PHYSICS ABSTRACTS AND NOTES (1997 Volume 23)
ISSN 0305-7615 (commenced publication 1974)

An international abstracts journal in applied health physics covering radiation protection, radiation dosimetry, measurement techniques, radiation effects and radiation accidents. The journal is published quarterly and includes approximately 2,500 abstracts per year.

1997:   1 year subscription cost: £175.00 (UK), US$360.00 (outside UK)

****

RADIATION PROTECTION DOSIMETRY   (1997 Volumes 69, 70, 71, 72, 73 and 74)
ISSN 0144-8420 (commenced publication 1981)

An international journal covering all aspects of personal and environmental dosimetry and monitoring for ionising and non-ionising radiations, including biological aspects, physical concepts, external dosimetry and monitoring, internal dosimetry and monitoring, environmental and workplace monitoring and dosimetry related to patient protection. There are **six** volumes of four issues per volume for 1997.

1997:   1 year subscription cost: £620.00 (UK), US$1330.00 (outside UK)
1997/8: 2 year subscription cost: £1240.00 (UK), US$2660.00 (outside UK)
1997/9: 3 year subscription cost: £1860.00 (UK), US$3990.00 (outside UK)

Radiation Protection Dosimetry is rated 1st out of 650 energy related journals published in the UK, and 30th out of 4992 journals published throughout the world, according to listings in the recently published Oak Ridge Energy Technology Data Exchange Journal Productivity Listings (1993-1995). It is 1st in the world out of all radiation protection related journals.

****

INTERNATIONAL JOURNAL OF RADIOACTIVE MATERIALS TRANSPORT   (1997 Volume 8)
ISSN 0957-476X (commenced publication 1990)

An international journal covering all aspects of transport of radioactive materials including regulations, package design, safety assessments, package testing, transport experiences and accidents. The journal is published quarterly and includes an abstracts section and regular news features from the International Atomic Energy Agency.

1997:   1 year subscription cost   £108.00 (UK), US$220.00 (outside UK)
1997/8: 2 year subscription cost: £216.00 (UK), US$440.00 (outside UK)
1997/9: 3 year subscription cost: £324.00 (UK), US$660.00 (outside UK)

****

Free sample copies of any journal and a complete catalogue of all of our radiation protection publications are available, on request. All back issues are available.

POSTAGE:- Journal subscription copies outside the UK are sent by PRINTFLOW AIR. First class AIRMAIL, if requested, is charged at US$30.00 extra, per volume.

Please send your orders to: **Nuclear Technology Publishing**
P.O. Box No 7, Ashford
Kent TN23 1YW,   England

Telephone: 44 (0) 1233 641683 Fax: 44 (0) 1233 610021
E-mail:subscriptions@ntp.eunet.co.uk

PRICES.97(October 1996)